Formulas for Perimeter *P*, Circumference *C*, and Area

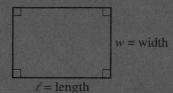

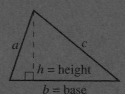

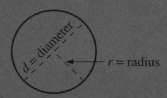

Square

$P = 4s$
$A = s^2$

Rectangle

$P = 2\ell + 2w$
$A = \ell w$

Triangle

$P = a + b + c$
$A = \frac{1}{2}bh$

Circle

$C = \pi d$ or $C = 2\pi r$
$A = \pi r^2$

Formulas for Volume *V*

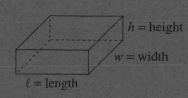

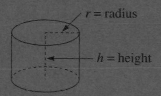

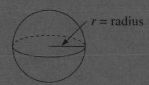

Rectangular Solid

$V = \ell wh$

Cylinder

$V = \pi r^2 h$

Sphere

$V = \frac{4}{3}\pi r^3$

Pythagorean Theorem

In a right triangle with legs of length *a* and *b* and hypotenuse of length *c*,

$$c^2 = a^2 + b^2.$$

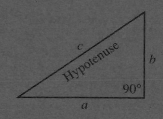

Beginning Algebra

Beginning Algebra

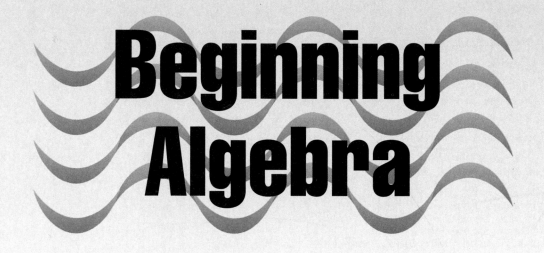

Dennis T. Christy
Robert Rosenfeld

Nassau Community College

WCB **Wm. C. Brown Publishers**

Dubuque, Iowa • Melbourne, Australia • Oxford, England

Book Team

Editor *Paula-Christy Heighton*
Developmental Editor *Theresa Grutz*
Photo Editor *Diane S. Saeugling*
Publishing Services Coordinator/Design *Barbara J. Hodgson*

Wm. C. Brown Publishers
A Division of Wm. C. Brown Communications, Inc.

Vice President and General Manager *Beverly Kolz*
Vice President, Publisher *Earl McPeek*
Vice President, Director of Sales and Marketing *Virginia S. Moffat*
Marketing Manager *Julie Joyce Keck*
Advertising Manager *Janelle Keeffer*
Director of Production *Colleen A. Yonda*
Publishing Services Manager *Karen J. Slaght*
Permissions/Records Manager *Connie Allendorf*

Wm. C. Brown Communications, Inc.

President and Chief Executive Officer *G. Franklin Lewis*
Corporate Senior Vice President, President of WCB Manufacturing *Roger Meyer*
Corporate Senior Vice President and Chief Financial Officer *Robert Chesterman*

Copyedited by Carol I. Beal

Cover photo by Jeff Rotman Photography

Cover/interior design by Schneck-DePippo Graphics

Illustrations by Schneck-DePippo Graphics

Printed in the United States of America by Wm. C. Brown Communications, Inc.,
2460 Kerper Boulevard, Dubuque, IA 52001

10 9 8 7 6 5 4 3 2 1

Chapter 1
p. 2: © W. H. Alburty/London Stock Exchange/Unicorn Stock Photos; p. 12:
© Bob Daemmrich Photos/Stock Boston; p. 25: © John A. Schakel, Jr./Unicorn
Stock Photos; p. 32: © R. J. Mathews/Unicorn Stock Photos; p. 43: © 1989 Gary
Gladstone/The Image Bank

Chapter 2
p. 78: © George Holton/Photo Researchers, Inc.; pp. 89, 95: © Tom McCarthy/
Unicorn Stock Photos; p. 104: © Bohdan Hrynewych/Stock Boston; p. 112:
W. Strode/Superstock

Chapter 3
p. 139: © US Library of Congress/Science Photo Library/Photo Researchers, Inc.;
p. 144: © Martha McBride/Unicorn Stock Photos

Chapter 4
p. 202: © Aneal Vohra/Unicorn Stock Photos

Chapter 5
p. 219: © Frank Siteman 1990/The Picture Cube; p. 236: © Rivera Collection/
Superstock, Inc.

Chapter 6
p. 268: © Joseph L. Fontenot/Unicorn Stock Photos; p. 275: © 1989 Michael
Salas/The Image Bank; p. 302: © F. Rickard-Artdia/Agence Vandystadt/Photo
Researchers, Inc.; p. 308: © Betts Anderson/Unicorn Stock Photos

Chapter 7
p. 320: © Tom Hollyman, 1986/Photo Researchers, Inc.; p. 327: © Spencer Grant/
The Picture Cube; p. 348: Comstock, Inc./Comstock, Inc.

Chapter 8
p. 395: © Rich Baker-10905/Unicorn Stock Photos

Chapter 9
p. 425: © Fred D. Jordan, 1989/Unicorn Stock Photos; p. 432: © Aneal Vohra/
Unicorn Stock Photos

Appendix
p. 448: © S. Barrow/Superstock, Inc.

To Margaret and Leda, Thank You

Contents

Preface

Audience

This book is intended for students who need a concrete approach to mathematics. It is written for college students who have never studied algebra or who need a review course in beginning algebra. The thorough pedagogical features of the text and the associated ancillary package ensure that the student has a wealth of helpful material.

Approach

This book is written with the belief that current textbooks must provide a dynamic approach to problem solving that allows students to monitor their progress and that effectively integrates graphics and calculator use. Our approach in these four areas is explained next.

Problem-Solving Approach

Our experience is that students who take beginning algebra learn best by "doing." Examples and exercises are crucial since it is usually in these areas that the students' main interactions with the material take place. The problem-solving approach contains brief, precisely formulated paragraphs, followed by many detailed examples. A relevant word problem introduces *every* section of the text, and word problems or other motivational problems are included in *every* section exercise set.

Graphics

A major component of a problem-solving approach with intuitive concept development is a strong emphasis on graphics. Students are given, and are encouraged to draw for themselves, visual representations of the concepts they are analyzing and the problems they are solving. Effective use of color enhances the many images in concept developments and in exercise sets.

Interactive Approach

Because students learn best by doing, progress check exercises are associated with each example problem in the text. By doing these exercises, students obtain immediate feedback on their understanding of the concept being discussed.

Calculator Use

The text encourages the use of calculators and discusses how they can be used effectively. It is assumed in the discussions that students have scientific calculators that use the algebraic operating system (AOS). Calculator illustrations show primarily the keystrokes required on a Texas Instruments TI-30-SLR+.

Features

- Problem-solving approach with intuitive concept developments
- Extensive and varied word problems and application problems
- "Think About It" exercises to develop critical-thinking skills
- Section introductions that include an interest-getting applied problem that is solved as an example in the text
- Discussions about basic concepts in geometry
- Effective use of color with an emphasis on graphics
- Over 6,300 exercises and 430 examples
- "Progress Check" exercises that allow for instant self-evaluation
- Problem sets of graduated difficulty that are closely matched to the example problems
- "Remember This" exercises that end each section and are crafted to provide smooth transition to the next section as well as spiral review of previous material
- Abundant chapter review exercises

- A chapter test and a cumulative review test for each chapter
- Boxes with labels for important definitions and rules
- "Note" and "Caution" remarks that provide helpful insights and point out potential student errors
- Unique chapter summaries that highlight specific objectives and key terms and concepts at the end of each chapter
- Instructions on calculator use
- Rigorous accuracy checking to avoid errors in the text
- Complete instructional package

Pedagogy

Section Introductions

In the spirit of problem solving, each section opens with a problem that should quickly involve students and teachers in a discussion of an important section concept. These problems are later worked out as an example in each section.

Keyed Section Objectives

Specific objectives of each section are listed at the beginning of the section, and the portion of the exposition that deals with each objective is signaled by numbered symbols such as $\boxed{1}$.

Systematic Review

Students benefit greatly from a systematic review of previously learned concepts. At the end of each chapter there are a detailed chapter summary that includes a checklist of objectives illustrated by example problems, abundant chapter review exercises, a chapter test, and a cumulative review test. In addition, each section exercise set is concluded with a short set of "Remember This" exercises that review previous concepts, with particular emphasis on skills that will be needed in the next section.

"Think About It" Exercises

Each exercise set is followed immediately by a set of "Think About It" exercises. Although some of these problems are challenging, this section is not intended as a set of "mind bogglers." Instead, the goal is to help develop critical-thinking skills by asking students to create their own examples, express concepts in their own words, extend ideas covered in the section, and analyze topics slightly out of the mainstream. These exercises are an excellent source of nontemplate problems and problems that can be assigned for group work.

For the Instructor

The *Annotated Instructor's Edition* provides a convenient source for answers to exercise problems. Each answer is placed near the problem and appears in red. This feature eliminates the need to search through a separate answer key and helps instructors forecast effective problem assignments. Suggestions and comments based on our experiences in developmental mathematics are also provided to complement your teaching techniques, and relevant historical asides that can enliven the course material are often given. We mention three sources helpful for this historical material. *A History of Mathematics* by Victor J. Katz from HarperCollins, which emphasizes the influence of the most important textbooks of the past; *A History of Mathematical Notations* by Florian Cajori from Open Court, an older work that traces the origins and development of familiar mathematical symbols; and *The History of Mathematics: An Introduction,* 2e, by David M. Burton from Wm. C. Brown Publishers, a good discoursive general introduction.

The *Instructor's Resource Manual* includes a guide to supplements that accompany *Beginning Algebra,* reproducible tests, and transparency masters of key concepts and procedures.

The *Instructor's Solutions Manual* contains solutions to every problem in the text, including solutions to the "Think About It" problems. These solutions are intended for the use of the instructor only.

WCB Computerized Testing Program provides you with an easy-to-use computerized testing and grade management program. No programming experience is required to generate tests randomly, by objective, by section, or by selecting specific test items. In addition, test items can be edited and new test items can be added. Also included with the *WCB Computerized Testing Program* is an on-line testing option which allows students to take tests on the computer. Tests can then be graded and the scores forwarded to the grade-keeping portion of the program.

The *Test Item File* is a printed version of the computerized testing program that allows you to examine all of the prepared test items and choose test items based on chapter, section, or objective. The objectives are taken directly from *Beginning Algebra.*

For the Student

The *Student's Solutions Manual and Study Guide* provides a summary of the objectives, vocabulary, rules and formulas, and key concepts for each section. Detailed solutions are given for every-other odd-numbered section exercise. Additional practice is available for each section, and each chapter ends with a sample practice test. The *Student's Solutions Manual and Study Guide* is available for student purchase through the bookstore.

Videotapes and Software that are text specific have been developed to reinforce the skills and concepts presented in *Beginning Algebra*. Contact your Wm. C. Brown Publishers representative for detailed descriptions.

Acknowledgments

A project of this magnitude is a team effort that develops over many years with the input of many talented people. We are indebted to all who contributed. In particular, we wish to thank the reviewers of this text, who are listed separately; Carol Hay, Linda J. Murphy, and Nancy K. Nickerson, who checked for accuracy in our final manuscript; Carole D. Carney, who checked for accuracy in the typeset text; Mary Trerice, who helped formulate exercise answers; Amy Driscoll, who did a prompt and accurate job of typing the manuscript; Carol Beal, who skillfully copyedited the manuscript; Dwala Canon, Eugenia M. Collins, Theresa Grutz, Barbara Hodgson, Earl McPeek, Linda Meehan, and Diane Saeugling, Wm. C. Brown Publishers; Deborah Schneck, Schneck-DePippo Graphics. To our parents, a special thank you. In each case they have always supported our efforts and taught us to persevere and overcome obstacles. Finally, but most important, we thank our wives, Margaret and Leda, and dedicate this book to them. They have given that special help and understanding only they could provide.

Dennis Christy
Robert Rosenfeld

Reviewers

Donald Bardwell
Nicholls State University

Carole Bauer
Triton College

Marybeth Beno
South Suburban College

Russell R. Blankenfeld
Rochester Community College

Dr. Ben Bockstege
Broward Community College

William L. Drezdzon
Oakton College

Judith Gebhart
Sinclair Community College

Elizabeth Harold
University of New Orleans

Vicky Lymbery
Stephen F. Austin State University

Vince McGarry
Austin Community College

J. Larry Martin
Missouri Southern State College

Debbie Singleton
Lexington Community College

Gerry Vidrine

Calculator Use TO THE STUDENT

A scientific hand-held calculator is now standard equipment for beginning algebra and beyond. These ten-dollar wonders provide you with the benefits of electronic computation that is fast, accurate, and easy to learn. Most important, efficient calculator use helps you focus on important mathematical ideas. To understand and apply mathematical concepts is our fundamental aim, and calculators are marvelous aids in attaining this goal.

A scientific calculator (the type you need) contains at least the following special features: algebraic keys x^2, $\sqrt{x}$, $1/x$, y^x or x^y, $\sqrt[x]{y}$; parentheses keys (,); a scientific notation key EE or EXP; and one memory that can store and recall.

In this book we also assume a scientific calculator using the algebraic operating system (AOS). Texas Instruments, Sharp, and Casio produce scientific calculators using this system. With AOS you can key in the problem exactly as it appears, and the calculator is programmed to use the order of operations discussed in Section 1.2. For example, since multiplication is done before addition, 2 $\boxed{+}$ 3 $\boxed{\times}$ 4 $\boxed{=}$ 14. If your calculator displays 20 when you key in this sequence, it is operating on left-to-right logic. You must then be careful to key in the problem so the correct order of operations is followed. Calculator illustrations in this text show primarily the keystrokes required on a Texas Instruments TI-30-SLR+. In any case, you should read the owner's manual that comes with your calculator to familiarize yourself with its specific keys and limitations.

One other introductory note—a calculator *computes,* that's all. You do the important part—you *think.* You analyze the problem, decide on the significant relationships, and determine if the solution makes sense in the real world. It's nice not to get bogged down in certain calculations and tables, but critical thinking has always been the main goal.

Beginning Algebra

1 From Numbers to Algebra

One share of stock in Apple Computer opens a week with a share price of $39\frac{7}{8}$ and closes the week with a share price of $41\frac{1}{4}$ dollars. For this week, find the dollar gain for an investor who owns 400 shares of this stock. (See Example 10 of Section 1.1.)

THE TRANSITION from arithmetic to algebra requires a firm grasp of number concepts. In this chapter we review some important procedures and vocabulary from arithmetic, and we discuss in detail the basic properties and operations associated with real numbers. In the process we will begin to study numerical relations in a more general way by using symbols, such as *x,* that may be replaced by numbers. By using such symbols, we create a generalized version of arithmetic, which we call algebra, that is a powerful tool in analyzing concepts in a wide variety of fields.

1.1   Whole Numbers and Fractions

1 Graph a set of whole numbers.

2 Identify and use the symbols $=, <, >, \neq, \leq, \geq$.

3 Write a number as a product of its prime factors.

4 Express a fraction in lowest terms.

5 Multiply, divide, add, and subtract fractions.

1   To discuss number concepts, it is useful to have names for some special collections of numbers. Our most basic need is for the numbers that are used in counting: 1, 2, 3, 4, and so on. We write three dots ". . . ," called **ellipses,** to mean "and so on," and we call this set* of numbers the natural numbers.

$$\text{Natural numbers} = \{1,2,3,4, \ . \ . \ .\}$$

By attaching 0 to the list of natural numbers, we obtain the set of whole numbers.

$$\text{Whole numbers} = \{0,1,2,3,4, \ . \ . \ .\}$$

A good way to describe sets of numbers is to picture them on a number line. For example, Figure 1.1 shows the picture, or graph, of the set of numbers $\{1,3,5,7\}$. To construct this graph, draw a horizontal line; then place a mark to represent 0 at a convenient point on the line. Choose a unit length, and mark off and label consecutive whole numbers moving to the right. Now to graph $\{1,3,5,7\}$, just place dots at the points corresponding to 1, 3, 5, and 7. We call a dot corresponding to a number the **graph** of the number.

Figure 1.1

EXAMPLE 1   Graph the set of even whole numbers from 0 to 8.

Solution   Even numbers are exactly divisible by 2, so the set of even whole numbers from 0 to 8 is $\{0,2,4,6,8\}$. To graph this set of numbers, place dots on the number line at 0, 2, 4, 6, and 8, as shown in Figure 1.2.

Figure 1.2

Note   An important objective of this chapter is to develop numbers so that we can fill in the number line. Then each point on the line will correspond to a number, and each number will correspond to a point on the number line.

PROGRESS CHECK 1   Graph the set of odd whole numbers from 1 to 9. Odd whole numbers are whole numbers that are not exactly divisible by 2.

2   An important property of any set of numbers that can be associated with points on the number line is that the numbers can be put in numerical order. If we compare two numbers, a and b, then either a is less than b, a is greater than b, or a equals b. These order relations are symbolized as follows.

*A **set** is simply a collection of objects, and we may describe a set by listing the objects or members of the collection within braces.

Progress Check Answer

1.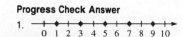

Statement	Read	Comment
$a = b$	a equals b	a and b represent the same number
$a < b$	a is less than b	the symbol $<$ points to a, the smaller number
$a > b$	a is greater than b	the symbol $>$ points to b, the smaller number

To illustrate, because 2 is smaller than 5, we may write

$$2 < 5, \quad \text{read} \quad 2 \text{ is less than 5,}$$
$$\text{or} \quad 5 > 2, \quad \text{read} \quad 5 \text{ is greater than 2.}$$

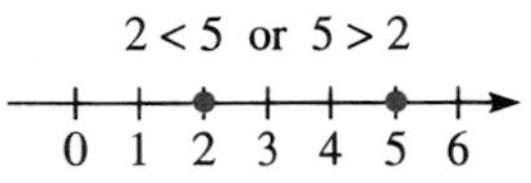

2 < 5 or 5 > 2

+--+--+●--+--+●--+--+
0 1 2 3 4 5 6

Figure 1.3

Relations of "less than" and "greater than" can be seen easily on the number line, as shown in Figure 1.3. The graph of the larger number is to the right of the graph of the smaller number. Note that $2 < 5$ may be written as $5 > 2$, and vice versa.

EXAMPLE 2 Insert the proper symbol ($<$, $>$, $=$) to indicate the correct order.

a. 9 _____ 4   **b.** 4 _____ 9
c. 2 + 3 _____ 7 − 2   **d.** 7 − 5 _____ 5 − 5

Solution

a. Because 9 is greater than 4, the correct order is given by $9 > 4$.
b. Because 4 is less than 9, write $4 < 9$.
c. 2 + 3 and 7 − 2 both represent 5, so $2 + 3 = 7 - 2$.
d. $7 - 5 > 5 - 5$ because 7 − 5 is 2, 5 − 5 is 0, and $2 > 0$.

PROGRESS CHECK 2 Insert the proper symbol ($<$, $>$, $=$) to indicate the correct order.

a. 0 _____ 5   **b.** 5 _____ 0
c. 9 − 3 _____ 2 + 3   **d.** 3 + 7 _____ 7 + 3   ⌐

The symbols $\neq$, $\leq$, and $\geq$ are also commonly used to indicate an inequality relation.

$a \neq b$ means a is not equal to b.

$a \leq b$ means a is less than or equal to b.

$a \geq b$ means a is greater than or equal to b.

In the following example note that statements like $a \leq b$ are true if either the "less than" part is true or the "equal" part is true. It is never possible for both the "less than" part and the "equal" part to be true simultaneously.

EXAMPLE 3 Classify each statement as true or false.

a. $2 \leq 5$ **b.** $2 \geq 2$
c. $5 \leq 2$ **d.** $2 + 3 \neq 6$

Solution

a. $2 \leq 5$ is read "2 is less than or equal to 5." This statement is true because "2 is less than 5" is true.
b. $2 \geq 2$ is read "2 is greater than or equal to 2." Because "2 is equal to 2" is true, $2 \geq 2$ is true.
c. $5 \leq 2$ is false because $5 < 2$ is false and $5 = 2$ is false.

d. $2 + 3 \neq 6$ is read "2 + 3 does not equal 6." Because $2 + 3$ represents 5, the statement is true.

PROGRESS CHECK 3   Classify each statement as true or false.

a. $5 \geq 2$ **b.** $0 \geq 1$

c. $5 \leq 5$ **d.** $4 + 6 \neq 6 + 4$ ⌐

3   Another important set of numbers is the prime numbers. A **prime number** is a natural number greater than 1 that is exactly divisible only by itself and 1. The first 10 prime numbers are

$$2, 3, 5, 7, 11, 13, 17, 19, 23, 29, \ldots ,$$

and the list continues indefinitely. It is often useful to write a number as a product of prime factors. To illustrate what this means, consider the terminology associated with the multiplication problem below that uses a raised dot to indicate multiplication.

$$\overbrace{2 \cdot 3 \cdot 5}^{\text{product}}$$
$$\underset{\text{factors}}{\uparrow \ \uparrow \ \uparrow}$$

We call $2 \cdot 3 \cdot 5$ a **product,** and 2, 3, and 5 **factors** in the product. Because $2 \cdot 3 \cdot 5$ equals 30, we express 30 as a product of prime factors by writing $2 \cdot 3 \cdot 5$, and we say this product expresses 30 in **prime-factored form.** A procedure for expressing a number as a product of its prime factors is shown in the next example.

EXAMPLE 4   Express 350 as a product of its prime factors.

Solution   Perform successive divisions using prime numbers until the result of the division is a prime number. Step 1 is at the bottom of the calculations.

$$
\begin{array}{r}
7 \\
5\overline{\smash{)}35} \\
5\overline{\smash{)}175} \\
2\overline{\smash{)}350}
\end{array}
$$

Step 4. The result (7) is a prime number.
Step 3. Divide the result (35) by 5.
Step 2. Divide the result (175) by 5.
Step 1. Divide 350 by 2.

Thus, $350 = 2 \cdot 5 \cdot 5 \cdot 7$. Check the answer by doing the multiplication to obtain 350.

PROGRESS CHECK 4   Express 114 as a product of its prime factors.   ⌐

4   Prime numbers have important applications in work with fractions. Recall from arithmetic that fractions such as

$$\frac{1}{2}, \qquad \frac{6}{18}, \qquad \frac{15}{11}, \qquad \text{and} \qquad \frac{0}{1}$$

are numerals written in the form a/b, where a is called the **numerator** and b is called the **denominator.** The fraction a/b is equivalent to the division $a \div b$, so the denominator of a fraction cannot equal 0 because division by 0 is not defined. (We show why in Section 1.8.) Prime numbers may be used to formulate a general procedure for simplifying fractions. For example, to simplify $\frac{6}{9}$, we note that $6 = 2 \cdot 3$ while $9 = 3 \cdot 3$. Thus,

$$\frac{6}{9} = \frac{2 \cdot \overset{1}{\cancel{3}}}{3 \cdot \underset{1}{\cancel{3}}}.$$

Because 3 is a common factor of both 6 and 9, we may divide out this common factor

Progress Check Answers

3. (a) T   (b) F   (c) T   (d) F

4. $2 \cdot 3 \cdot 19$

by dividing both the numerator and the denominator by 3 to obtain

$$\frac{6}{9} = \frac{2 \cdot 3}{3 \cdot 3} = \frac{2}{3}.$$

This example illustrates a general principle in fractions that is used often.

Fundamental Principle of Fractions

If b and k are not zero, then

$$\frac{a \cdot k}{b \cdot k} = \frac{a}{b}.$$

When we simplify fractions by the fundamental principle, it is important to recognize that we may divide out only *nonzero factors* of the numerator and the denominator. In the next example the directions ask for a fraction in lowest terms. A fraction is in **lowest terms** when there is no natural number besides 1 that is a factor of both the numerator and the denominator. If the numerator and denominator of a fraction are equal (and are not both zero), then the fraction is equal to 1.

EXAMPLE 5 Express each fraction in lowest terms.

a. $\frac{12}{28}$
b. $\frac{15}{56}$

Solution

a. Express 12 as $2 \cdot 2 \cdot 3$, and 28 as $2 \cdot 2 \cdot 7$. Then apply the fundamental principle.

$$\frac{12}{28} = \frac{2 \cdot 2 \cdot 3}{2 \cdot 2 \cdot 7} \quad \text{Write 12 and 28 in prime-factored form.}$$

$$= \frac{3}{7} \quad \text{Divide out the common factor } 2 \cdot 2 \text{ according to the fundamental principle.}$$

In lowest terms, $\frac{12}{28}$ is expressed as $\frac{3}{7}$.

b. Express 15 and 56 in prime-factored form.

$$\frac{15}{56} = \frac{3 \cdot 5}{2 \cdot 2 \cdot 2 \cdot 7}$$

Because 15 and 56 share no common prime factors, $\frac{15}{56}$ is in lowest terms.

Note To check that two fractions are equal, use the principle

$$\frac{a}{b} = \frac{c}{d} \quad \text{provided} \quad ad = bc \quad (b, d \neq 0).$$

For example, we know $\frac{12}{28} = \frac{3}{7}$ is a true statement because $12 \cdot 7 = 28 \cdot 3$ is a true statement.

PROGRESS CHECK 5 Express each fraction in lowest terms.

a. $\frac{42}{63}$
b. $\frac{45}{76}$

Example 5 shows that we may use the following procedure to express a fraction in lowest terms.

> **To Express a Fraction in Lowest Terms**
>
> 1. Express the numerator and the denominator as a product of prime factors.
> 2. Divide out nonzero factors that are common to the numerator and the denominator according to the fundamental principle.

5   We now review how to perform the basic operations with fractions.

Multiplication

The product of two or more fractions is the product of their numerators divided by the product of their denominators. In symbols, the principle is

$$\frac{a}{b} \cdot \frac{c}{d} = \frac{a \cdot c}{b \cdot d} \qquad b, d \neq 0.$$

To express products in lowest terms, use the procedure above if the simplification is not obvious.

EXAMPLE 6   Multiply and express the answer in lowest terms.

a. $\frac{3}{8} \cdot \frac{5}{7}$ b. $5 \cdot \frac{7}{10}$ c. $\frac{4}{7} \cdot \frac{7}{4}$

Solution

a. $\dfrac{3}{8} \cdot \dfrac{5}{7} = \dfrac{3 \cdot 5}{8 \cdot 7}$ Multiply fractions.

$= \dfrac{15}{56}$ Simplify.

b. $5 \cdot \dfrac{7}{10} = \dfrac{5}{1} \cdot \dfrac{7}{10}$ Write 5 as $\dfrac{5}{1}$.

$= \dfrac{5 \cdot 7}{1 \cdot 10}$ Multiply fractions.

$= \dfrac{5 \cdot 7}{1 \cdot 5 \cdot 2}$ Write 10 in prime-factored form.

$= \dfrac{7}{2}$ Divide out the common factor 5 according to the fundamental principle.

c. $\dfrac{4}{7} \cdot \dfrac{7}{4} = \dfrac{4 \cdot 7}{7 \cdot 4}$ Multiply fractions.

$= 1$ Divide out common factors according to the fundamental principle.

PROGRESS CHECK 6   Multiply and express the answer in lowest terms.

a. $\frac{5}{8} \cdot \frac{7}{4}$ b. $\frac{3}{8} \cdot \frac{8}{3}$ c. $3 \cdot \frac{5}{6}$

Division

Division of factors involves the concept of a reciprocal. Two numbers are called **reciprocals** of each other if the product of the two numbers is 1. Thus Example 6c shows

Progress Check Answers

6. (a) $\frac{35}{32}$ (b) 1   (c) $\frac{5}{2}$

that $\frac{4}{7}$ and $\frac{7}{4}$ are reciprocals of each other. A division problem like $\frac{2}{3} \div \frac{4}{7}$ may be done as follows.

Write division in fraction form.		Fundamental principle.		Product of reciprocals is 1.
$\dfrac{2}{3} \div \dfrac{4}{7} = \dfrac{\frac{2}{3}}{\frac{4}{7}}$	$=$	$\dfrac{\frac{2}{3} \cdot \frac{7}{4}}{\frac{4}{7} \cdot \frac{7}{4}}$	$=$	$\dfrac{\frac{2}{3} \cdot \frac{7}{4}}{1} = \dfrac{2}{3} \cdot \dfrac{7}{4}$

The end result shows that to divide two fractions, we should invert the fraction by which we are dividing to find its reciprocal, and then multiply. In symbols, the principle is

$$\frac{a}{b} \div \frac{c}{d} = \frac{a}{b} \cdot \frac{d}{c} \qquad b, c, d \neq 0.$$

EXAMPLE 7   Divide and express the answer in lowest terms.

a. $\frac{2}{3} \div \frac{5}{9}$ **b.** $\frac{5}{11} \div 7$

Solution   Convert to multiplication and simplify.

a. $\dfrac{2}{3} \div \dfrac{5}{9} = \dfrac{2}{3} \cdot \dfrac{9}{5}$ Multiply $\dfrac{2}{3}$ by the reciprocal of $\dfrac{5}{9}$.

$\qquad\quad = \dfrac{2 \cdot 9}{3 \cdot 5}$ Multiply fractions.

$\qquad\quad = \dfrac{2 \cdot 3 \cdot 3}{3 \cdot 5}$ Write 9 in prime-factored form.

$\qquad\quad = \dfrac{6}{5}$ Divide out the common factor 3 according to the fundamental principle.

b. $\dfrac{5}{11} \div 7 = \dfrac{5}{11} \div \dfrac{7}{1}$ Write 7 as $\dfrac{7}{1}$.

$\qquad\quad = \dfrac{5}{11} \cdot \dfrac{1}{7}$ Multiply $\dfrac{5}{11}$ by the reciprocal of $\dfrac{7}{1}$.

$\qquad\quad = \dfrac{5 \cdot 1}{11 \cdot 7}$ Multiply fractions.

$\qquad\quad = \dfrac{5}{77}$ Simplify.

PROGRESS CHECK 7   Divide and express the answer in lowest terms.

a. $\frac{3}{4} \div \frac{7}{8}$ **b.** $\frac{9}{11} \div 18$

Addition and Subtraction

The sum (or difference) of two or more fractions that have the same denominator is given by the sum (or difference) of the two numerators divided by the common denominator. Symbolically,

$$\frac{a}{b} + \frac{c}{b} = \frac{a + c}{b} \qquad \text{and} \qquad \frac{a}{b} - \frac{c}{b} = \frac{a - c}{b} \qquad b \neq 0.$$

EXAMPLE 8   Perform the indicated operations.

a. $\frac{4}{11} + \frac{5}{11}$ **b.** $\frac{10}{13} - \frac{4}{13}$

Solution

a. $\dfrac{4}{11} + \dfrac{5}{11} = \dfrac{4+5}{11} = \dfrac{9}{11}$ **b.** $\dfrac{10}{13} - \dfrac{4}{13} = \dfrac{10-4}{13} = \dfrac{6}{13}$

PROGRESS CHECK 8   Perform the indicated operations.

a. $\frac{2}{9} + \frac{5}{9}$ **b.** $\frac{11}{4} - \frac{3}{4}$ ⌐

When fractions have different denominators, we can rewrite them using the fundamental principle so that the fractions have the same denominator. In such cases, computation is simpler if we use the smallest possible common denominator, called the **least common denominator,** or **LCD.** One way of finding the LCD follows.

To Find the LCD

1. Express each denominator as a product of prime factors.
2. The LCD is the product of all the different prime factors, with each prime number appearing the most number of times it appears in any one factorization.

This procedure is illustrated in Example 9.

EXAMPLE 9   Add $\frac{17}{18} + \frac{7}{30}$ and express the answer in lowest terms.

Solution   To find the LCD, first express 18 and 30 as products of prime factors.

$$18 = 2 \cdot 3 \cdot 3$$
$$30 = 2 \cdot 3 \cdot 5$$

The LCD will contain the factors 2, 3, and 5. In both factorizations the greatest number of times that 2 appears is once, that 3 appears is twice, and that 5 appears is once. Thus,

$$\text{LCD} = 2 \cdot 3 \cdot 3 \cdot 5 = 90.$$

Because $90 = 18 \cdot 5$ and $90 = 30 \cdot 3$, we may add the fractions as follows.

$$\begin{aligned}\dfrac{17}{18} + \dfrac{7}{30} &= \dfrac{17 \cdot 5}{18 \cdot 5} + \dfrac{7 \cdot 3}{30 \cdot 3} \quad \text{Fundamental principle.} \\[2mm] &= \dfrac{85}{90} + \dfrac{21}{90} \\[2mm] &= \dfrac{85 + 21}{90} \quad \text{Add the fractions.} \\[2mm] &= \dfrac{106}{90}\end{aligned}$$

Now simplify $\frac{106}{90}$.

$$\dfrac{106}{90} = \dfrac{2 \cdot 53}{2 \cdot 45} = \dfrac{53}{45}$$

In lowest terms the sum of fractions is $\frac{53}{45}$.

Note   In many cases inspection of the multiples of the numbers in the denominators will quickly produce the LCD. For the denominators in this example:

Multiples of 18:   18, 36, 54, 72, 90, . . .
Multiples of 30:   30, 60, 90, . . .

Thus, the smallest number that 18 and 30 divide exactly is 90.

PROGRESS CHECK 9 Add $\frac{3}{14}+\frac{9}{10}$ and express the answer in lowest terms. ⌐

In Example 9 the answer $\frac{53}{45}$ is called an **improper fraction** because the numerator is greater than the denominator. In many applied problems such fractions are written as **mixed numbers.** For instance, $\frac{53}{45}=1\frac{8}{45}$. The mixed number $1\frac{8}{45}$ means $1+\frac{8}{45}$ and is read "one and eight-forty-fifths." The opening problem for the chapter illustrates an application of mixed numbers.

EXAMPLE 10 Solve the problem in the chapter introduction on page 2.

Solution The dollar gain for one share is given by

$$41\frac{1}{4}-39\frac{7}{8}.$$

To subtract the mixed numbers, first rewrite the numbers as improper fractions.

$$41\frac{1}{4}=41+\frac{1}{4}=\frac{164}{4}+\frac{1}{4}=\frac{165}{4}$$

$$39\frac{7}{8}=39+\frac{7}{8}=\frac{312}{8}+\frac{7}{8}=\frac{319}{8}$$

Now subtract using 8 as the LCD.

$$\frac{165}{4}-\frac{319}{8}=\frac{165\cdot 2}{4\cdot 2}-\frac{319}{8}=\frac{330-319}{8}=\frac{11}{8}$$

The gain for one share is $\frac{11}{8}$ dollars, so the gain for 400 shares is the product of 400 and $\frac{11}{8}$.

$$400\cdot\frac{11}{8}=\frac{400}{1}\cdot\frac{11}{8}=\frac{400\cdot 11}{1\cdot 8}=550$$

For the week, the investor's gain is $550.

PROGRESS CHECK 10 One share of stock in AT&T opens a week with a share price of $38\frac{3}{4}$ and closes the week with a share price of $40\frac{3}{8}$. For this week, find the dollar gain for a mutual fund that owns 8,000 shares of this stock. ⌐

Progress Check Answers
9. $\frac{39}{35}$

10. $13,000

EXERCISES 1.1

In Exercises 1–12, graph the set of numbers described.

1. Even whole numbers from 2 to 10

2. Odd whole numbers from 3 to 11

3. Whole numbers less than 4

4. Whole numbers less than or equal to 5

5. Whole numbers which are more than 7 but less than 12

6. Whole numbers from 6 up to and including 10

7. Prime numbers between 20 and 30

8. Prime numbers less than 10

9. The prime factors of 10

10. The prime factors of 60

11. The set of all even prime numbers

12. All whole numbers which are factors of 24

In Exercises 13–24, insert the proper symbol ($<$, $>$, $=$) to indicate the correct order.

13. $10 \underline{\hspace{1cm}} 5$

14. $3 \underline{\hspace{1cm}} 10$

15. $4 + 3 \underline{\hspace{1cm}} 4 + 5$

16. $8 + 8 \underline{\hspace{1cm}} 4 + 16$

17. $13 - 5 \underline{\hspace{1cm}} 13 - 6$

18. $15 - 4 \underline{\hspace{1cm}} 15 - 3$

19. $\frac{1}{2} \cdot 4 \underline{\hspace{1cm}} \frac{1}{2} \cdot 6$

20. $\frac{1}{4} \cdot 4 \underline{\hspace{1cm}} \frac{1}{4} \cdot 5$

21. $\frac{3}{2} \cdot \frac{2}{3} \underline{\hspace{1cm}} \frac{4}{3} \cdot \frac{3}{4}$

22. $\frac{6}{5} \cdot \frac{10}{6} \underline{\hspace{1cm}} \frac{7}{6} \cdot \frac{12}{7}$

23. $12 \div 3 \underline{\hspace{1cm}} 12 \div 4$

24. $12 \div 15 \underline{\hspace{1cm}} 12 \div 18$

In Exercises 25–36, classify each statement as true or false.

25. $3 \le 6$

26. $5 \ge 1$

27. $4 \ge 1$

28. $1 \ge 3$

29. $3 - 1 \le 3 - 2$

30. $5 - 3 \le 5 - 2$

31. $2 \cdot 5 \le 2 \cdot 6$

32. $3 \cdot 8 \ge 3 \cdot 7$

33. $\frac{1}{3} \cdot 3 \ge \frac{1}{4} \cdot 4$

34. $\frac{1}{5} \cdot 5 \le 7 \div 7$

35. $3 - \frac{1}{2} \ne 4 - \frac{1}{2}$

36. $\frac{1}{2} + \frac{1}{3} \ne \frac{2}{5}$

In Exercises 37–48, express the given number as a product of its prime factors. Check your answer by finding the product of the prime factors.

37. 30

38. 42

39. 90

40. 60

41. 16

42. 81

43. 720

44. 1680

45. 1625

46. 675

47. 2310

48. 1365

In Exercises 49–60, express each fraction in lowest terms by writing the numerator and denominator in prime-factored form and applying the fundamental principle of fractions. Check that the two fractions are equal by using the principle $a/b = c/d$ provided $ad = bc$, as explained in Example 5.

49. $\frac{18}{27}$

50. $\frac{24}{60}$

51. $\frac{30}{105}$

52. $\frac{90}{315}$

53. $\frac{24}{54}$

54. $\frac{36}{300}$

55. $\frac{42}{125}$

56. $\frac{70}{27}$

57. $\frac{720}{675}$

58. $\frac{90}{42}$

59. $\frac{66}{88}$

60. $\frac{404}{1212}$

In Exercises 61–72, find the product of the given fractions and express it in lowest terms.

61. $\frac{4}{9} \cdot \frac{2}{5}$

62. $\frac{2}{3} \cdot \frac{2}{5}$

63. $3 \cdot \frac{1}{4}$

64. $4 \cdot \frac{1}{5}$

65. $\frac{9}{5} \cdot \frac{10}{18}$

66. $\frac{12}{5} \cdot \frac{15}{36}$

67. $3 \cdot \frac{5}{21}$

68. $5 \cdot \frac{3}{20}$

69. $\frac{1}{6} \cdot 15$

70. $\frac{1}{12} \cdot 18$

71. $1\frac{3}{4} \cdot 3\frac{5}{7}$

72. $2\frac{1}{3} \cdot 3\frac{2}{7}$

In Exercises 73–78, find the reciprocal of the given number and check that the product of the number and its reciprocal is 1.

73. $\frac{3}{5}$

74. $\frac{9}{8}$

75. 2

76. 3

77. $\frac{1}{4}$

78. $\frac{1}{5}$

In Exercises 79–90, do the division and express the quotient in lowest terms. Answers may be left as improper fractions.

79. $\frac{3}{4} \div \frac{6}{10}$

80. $\frac{4}{5} \div \frac{1}{10}$

81. $\frac{3}{2} \div \frac{3}{2}$

82. $\frac{5}{2} \div \frac{2}{5}$

83. $\frac{1}{5} \div 2$

84. $\frac{2}{3} \div 3$

85. $12 \div \frac{1}{2}$

86. $18 \div \frac{1}{3}$

87. $3\frac{1}{3} \div 3\frac{2}{3}$

88. $4\frac{1}{2} \div 4\frac{3}{4}$

89. $10\frac{1}{2} \div 3\frac{1}{2}$

90. $5\frac{3}{4} \div 2\frac{1}{4}$

In Exercises 91–108, perform the indicated operations and express the answer in lowest terms. Answers may be left as improper fractions.

91. $\frac{5}{7} + \frac{9}{7}$

92. $\frac{11}{17} + \frac{1}{17}$

93. $\frac{9}{7} - \frac{2}{7}$

94. $\frac{11}{17} - \frac{1}{17}$

95. $\frac{1}{2} + \frac{1}{3}$

96. $\frac{1}{3} + \frac{1}{4}$

97. $\frac{1}{2} - \frac{1}{3}$

98. $\frac{1}{3} - \frac{1}{4}$

99. $\frac{5}{12} + \frac{5}{8}$

100. $\frac{7}{10} + \frac{7}{15}$

101. $\frac{4}{6} + \frac{6}{4}$

102. $\frac{12}{10} + \frac{10}{12}$

103. $\frac{3}{2} - \frac{2}{3}$

104. $\frac{4}{3} - \frac{3}{4}$

105. $1\frac{1}{2} + 1\frac{1}{3}$

106. $2\frac{1}{3} + 3\frac{1}{4}$

107. $2\frac{3}{4} - 1\frac{1}{2}$

108. $3\frac{7}{8} - 2\frac{5}{16}$

109. What is $\frac{1}{2}$ of $\frac{3}{4}$?

110. What is $\frac{1}{4}$ of $\frac{4}{7}$?

111. What is $\frac{3}{4}$ divided by 2?

112. What is $\frac{3}{4}$ divided by $\frac{1}{2}$?

113. **a.** Find the sum of $\frac{3}{4}$ and $\frac{4}{3}$.

　　b. Find the product of $\frac{3}{4}$ and $\frac{4}{3}$.

114. **a.** Find the sum of $\frac{2}{3}$ and $\frac{3}{2}$.

　　b. Find the product of $\frac{2}{3}$ and $\frac{3}{2}$.

115. In prime-factored form the denominator of one fraction is $2 \cdot 3 \cdot 5$, and the denominator of another is $3 \cdot 5 \cdot 7$. Find the LCD.

116. In prime-factored form the denominator of one fraction is $2 \cdot 2 \cdot 2 \cdot 3$, and the denominator of another is $2 \cdot 2 \cdot 3 \cdot 3$. Find the LCD.

117. If 100,000 people each lose $5\frac{1}{4}$ lb, how much weight is lost altogether?

118. If 50,000 people each walk $33\frac{1}{3}$ mi in a walkathon, how many miles were walked altogether?

119. The price of a share of stock goes up from $12\frac{3}{8}$ dollars to $14\frac{1}{8}$ dollars. If you own 350 shares, how much did their value increase?

120. The price of a share of stock falls from $7\frac{1}{8}$ to $6\frac{5}{8}$ dollars. If you own 600 shares, how much did their value fall?

121. An empty bottle weighs $1\frac{5}{8}$ oz. A manufacturer fills it with $3\frac{1}{2}$ oz of liquid, and seals it with a $\frac{1}{4}$-oz lid. What is the total weight for 500 of these filled bottles?

122. Three items which each weigh $4\frac{1}{2}$ lb are put in a container which weighs $\frac{1}{4}$ lb. Then 72 of these containers are put in a crate which weighs 50 lb. What is the weight of the full crate?

THINK ABOUT IT

1. What is meant by an improper fraction? What is meant by a proper fraction?

2. When you multiply a natural number by an improper fraction, the answer is larger than the original number. Why? What happens when you multiply by a proper fraction?

3. True or false? The larger a number is, the more prime factors it has. (*Hint:* Compare 30 and 94.)

4. What is the smallest whole number that has four different prime factors?

5. Show that $\frac{1313}{2424}$ reduces to $\frac{13}{24}$ and that $\frac{2323}{4141}$ reduces to $\frac{23}{41}$. Explain why, if you repeat any two-digit number in the numerator and any two-digit number in the denominator, the fraction will reduce like these did.

REMEMBER THIS

1. Compute $1.05 \times 1.05 \times 1.05$.

2. Compute $15{,}000 \times 2.6533$.

3. Compute $\frac{3}{4} \times \frac{3}{4} \times \frac{3}{4}$.

4. Compute $3 \times 4 \times 5$; compute $5 \times 3 \times 4$; compute $4 \times 5 \times 3$. Do they all have the same answer? In a multiplication expression does it matter in what order the numbers are written?

5. Compute $3 + 4 + 5$; compute $5 + 3 + 4$; compute $4 + 5 + 3$. Do they all have the same answer? In an addition expression does it matter in what order the numbers are written?

6. What number is 3 more than 12?

7. What number is 3 times as big as 12?

8. What number is 3 less than 12?

9. Which is larger, the product of 2 and 3 or the sum of 2 and 3?

10. Which is larger, the product of 1 and 2 or the sum of 1 and 2?

1.2 Exponents, Order of Operations, and Calculator Computation

If we assume a 5 percent annual rate of inflation in the price of a new car, then a new car that costs $15,000 today will cost $15{,}000(1 + 0.05)^{20}$ dollars in 20 years. Use a scientific calculator to evaluate this expression and predict the price in 20 years to the nearest dollar. (See Example 6.)

OBJECTIVES

1 Apply the definition of a natural number exponent.

2 Evaluate an expression by following the order of operations.

3 Evaluate an expression on a scientific calculator.

4 Determine reasonable answers to computations by estimation.

1   In Section 1.1 we encountered products that contained repeated factors. For instance, in prime-factored form we express 16 as $2 \cdot 2 \cdot 2 \cdot 2$. An efficient way of writing $2 \cdot 2 \cdot 2 \cdot 2$ is 2^4. We call 2^4 an **exponential expression** with **base** 2 and **exponent** 4. The number 2^4, or 16, is called the fourth **power** of 2, and in general we define a to a natural number power as follows.

Natural Number Exponent

If n is a natural number, then

$$a^n = \underbrace{a \cdot a \cdot a \cdots a.}_{n \text{ factors}}$$

We read a^2, or the second power of a, as "a squared," and we read a^3, or the third power of a, as "a cubed." For the first power of a, the exponent is understood to be 1, so note that $a^1 = a$.

EXAMPLE 1   Evaluate each expression.

a. 3^5 **b.** $3^2 \cdot 2^3$ **c.** $\left(\frac{3}{4}\right)^4$

Solution

a. The definition of the fifth power of 3 gives

$$3^5 = \underbrace{3 \cdot 3 \cdot 3 \cdot 3 \cdot 3.}_{5 \text{ factors of } 3}$$

Through repeated multiplication, we determine that $3^5 = 243$.

b. The expression $3^2 \cdot 2^3$ is read "3 squared times 2 cubed."

$$3^2 \cdot 2^3 = 3 \cdot 3 \cdot 2 \cdot 2 \cdot 2 = 72$$

c. Multiply fractions as shown in Section 1.1 to compute $\frac{3}{4}$ to the fourth power.

$$\left(\frac{3}{4}\right)^4 = \frac{3}{4} \cdot \frac{3}{4} \cdot \frac{3}{4} \cdot \frac{3}{4} = \frac{81}{256}$$

PROGRESS CHECK 1   Evaluate each expression.

a. 10^3 **b.** $2^5 \cdot 4^2$ **c.** $\left(\frac{2}{3}\right)^2$ ⌐

2   Unless we establish an agreed-upon priority for performing operations, different answers may be possible when evaluating an expression with more than one operation. For instance, $2 + 3 \cdot 4$ might equal 20 (add 2 and 3, then multiply by 4), or $2 + 3 \cdot 4$ might equal 14 (multiply 3 and 4, then add 2). The agreed-upon evaluation of $2 + 3 \cdot 4$ is 14, because multiplication is assigned a higher priority in the order of operations than addition. If we wish to override this order and add first, then we group the addition using parentheses and write the problem as

$$(2 + 3) \cdot 4,$$

with the understanding that we agree to perform all operations within grouping symbols first. Brackets [ ], braces { }, and fraction bars are also used as grouping symbols.

A complete statement of the agreed-upon order of operations follows.

Order of Operations

1. Perform all operations within grouping symbols first. If there is more than one symbol of grouping, simplify the innermost symbol of grouping first, and simplify the numerator and denominator of a fraction separately.
2. Evaluate powers of a number.
3. Multiply or divide working from left to right.
4. Add or subtract working from left to right.

Progress Check Answers

1. (a) 1,000   (b) 512   (c) $\frac{4}{9}$

When a product involves a grouping symbol, it is standard notation to omit the multiplication dot. For instance, $2 \cdot (3 + 4)$ is usually written $2(3 + 4)$, while $2 \cdot 3$ may be expressed using parentheses as $2(3)$ or $(2)(3)$.

EXAMPLE 2 Evaluate each expression.

a. $2(3 + 4)$ b. $15 - 3 \cdot 4$
c. $7 + 2 \cdot 6^3$ d. $(5 + 2)^2 + (4 - 1)^2$

Solution Follow the order of operations given above.

a. $2(3 + 4) = 2(7)$ Operate within parentheses.
   $\qquad\quad = 14$ Multiply.
b. $15 - 3 \cdot 4 = 15 - 12$ Multiply.
   $\qquad\qquad = 3$ Subtract.
c. $7 + 2 \cdot 6^3 = 7 + 2 \cdot 216$ Evaluate a power.
   $\qquad\qquad = 7 + 432$ Multiply.
   $\qquad\qquad = 439$ Add.
d. $(5 + 2)^2 + (4 - 1)^2 = 7^2 + 3^2$ Operate within parentheses.
   $\qquad\qquad\qquad = 49 + 9$ Evaluate powers.
   $\qquad\qquad\qquad = 58$ Add.

PROGRESS CHECK 2 Evaluate each expression.

a. $4 + 7 \cdot 5$ b. $(4 + 7)5$
c. $3 \cdot 2^4 - 1$ d. $(5 - 2)^2 + (3 + 3)^2$

When a problem involves two symbols of grouping, one inside the other, then parentheses are used for the inner grouping and brackets are used for the outer grouping. The next example also illustrates that a fraction bar groups and separates the numerator and the denominator of a fraction.

EXAMPLE 3 Evaluate each expression.

a. $5 + 3[8 - 2(6 - 2)]$ b. $\dfrac{(9 - 1)^2}{3^2 - 5}$

Solution

a. First simplify inside the parentheses, which are the innermost symbol of grouping; then simplify inside the bracket.

$$5 + 3[8 - 2(6 - 2)] = 5 + 3[8 - 2(4)]$$ Operate within parentheses.
$$= 5 + 3[8 - 8]$$ Multiply within brackets.
$$= 5 + 3[0]$$ Subtract within brackets.
$$= 5 + 0$$ Multiply.
$$= 5$$ Add.

b. The numerator and the denominator are simplified independently. Then simplify the resulting fraction.

$$\frac{(9 - 1)^2}{3^2 - 5} = \frac{8^2}{9 - 5}$$ Operate within parentheses in the numerator; evaluate a power in the denominator.

$$= \frac{64}{4}$$ Evaluate a power in the numerator; subtract in the denominator.

$$= 16$$ Simplify.

PROGRESS CHECK 3 Evaluate each expression.

a. $9 + 2[1 + 3(5 - 2)]$ b. $\dfrac{5 + 7^2}{(7 - 4)^3}$

3   Calculator computation

A scientific calculator is ideal for evaluating expressions, but you need to understand the algebraic rules that have been programmed into your calculator. A calculator which uses the algebraic system (AOS) calculates in the following order (which is the same as the order you would use even without a calculator).

1. Keys that operate on the single number in the display are done immediately. Such keys are called function keys, and common function keys used in this course are square $\boxed{x^2}$, square root $\boxed{\sqrt{}}$, reciprocal $\boxed{1/x}$, and sign change $\boxed{+/-}$.
2. Powers $\boxed{y^x}$ or $\boxed{x^y}$ and roots $\boxed{\sqrt[x]{y}}$ or $\boxed{y^{1/x}}$ are calculated after the function keys.
3. Multiplication $\boxed{\times}$ and division $\boxed{\div}$ have the next priority.
4. Addition $\boxed{+}$ and subtraction $\boxed{-}$ come last.

The equals key $\boxed{=}$ completes all operations. Parentheses keys $\boxed{(}$, $\boxed{)}$ can be used to change these built-in priorities when required. Pressing close parentheses $\boxed{)}$ automatically completes any operations done after the previous open parentheses $\boxed{(}$. Consider carefully the logic involved in the following examples, and keep in mind that in many cases other keystroke sequences are possible.

EXAMPLE 4   Evaluate $2(3 + 4)$ on a scientific calculator.

Solution $2(3 + 4) = 14$ as shown in Example 2a. There are two common approaches for evaluating this expression on a calculator.

Method 1   Input the problem by following the rules for order of operations. First, add 3 and 4; then multiply by 2.

$$3 \boxed{+} 4 \boxed{=} \boxed{\times} 2 \boxed{=} \boxed{14}$$

Method 2   Input the problem as it appears, including the parentheses, and let the calculator compute according to the agreed-upon order of operations. Note that the multiplication symbol is omitted in the problem statement but must be keyed in for the calculator sequence.

$$2 \boxed{\times} \boxed{(} 3 \boxed{+} 4 \boxed{)} \boxed{=} \boxed{14}$$

PROGRESS CHECK 4   Evaluate $8(7 - 4)$ on a scientific calculator.   ⌟

EXAMPLE 5   Evaluate $\dfrac{6^2 - 4}{3^2 - 5}$ on a scientific calculator.

Solution   To evaluate a fraction, divide the numerator by the denominator. Use the square key $\boxed{x^2}$ to evaluate the second power of a number, and consider carefully both of the following methods.

Method 1   Group the numerator and denominator with parentheses, because the given problem may be expressed as

$$(6^2 - 4) \div (3^2 - 5).$$

Now key in the problem as it appears.

$$\boxed{(} 6 \boxed{x^2} \boxed{-} 4 \boxed{)} \boxed{\div} \boxed{(} 3 \boxed{x^2} \boxed{-} 5 \boxed{)} \boxed{=} \boxed{8}$$

Method 2   Compute the denominator and store it. Then compute the numerator and divide by the stored number.

$$3 \boxed{x^2} \boxed{-} 5 \boxed{=} \boxed{STO} 6 \boxed{x^2} \boxed{-} 4 \boxed{=} \boxed{\div} \boxed{RCL} \boxed{=} \boxed{8}$$

Progress Check Answer

4. 24

On some calculators the store key looks like $\boxed{M_{in}}$ or $\boxed{X \to M}$, while the recall key is labeled $\boxed{MR}$ or $\boxed{RM}$.

PROGRESS CHECK 5   Evaluate $\dfrac{5^2 + 2}{4^2 - 7}$ on a scientific calculator.   ⌐

EXAMPLE 6   Solve the problem in the section introduction on page 12.

Solution   To evaluate $15{,}000(1 + 0.05)^{20}$, we utilize the power key $\boxed{y^x}$. Once again, we may input the problem by following the order of operations (method 1), or we may key in the problem as it appears (method 2) and let the priorities built into the calculator produce the desired result.

Method 1   1 $\boxed{+}$ .05 $\boxed{=}$ $\boxed{y^x}$ 20 $\boxed{\times}$ 15,000 $\boxed{=}$ $\boxed{39799.466}$

Method 2   15,000 $\boxed{\times}$ $\boxed{(}$ 1 $\boxed{+}$ .05 $\boxed{)}$ $\boxed{y^x}$ 20 $\boxed{=}$ $\boxed{39799.466}$

To the nearest dollar, the predicted price in 20 years is $39,799.

PROGRESS CHECK 6   Evaluate $12{,}000(1 + 0.06)^{15}$ on a scientific calculator. This number predicts the price in 15 years of an item that costs $12,000 today and increases in price by 6 percent each year.   ⌐

4   Estimation

It is particularly important to check to see that calculator results are reasonable, because it is easy to accidentally hit the wrong calculator buttons. Little input slips often produce large output errors that are easy to catch if we make a rough estimate of the expected result. One technique for estimating a result is to round off the numbers in a given problem so that the computation can be done mentally.

EXAMPLE 7   One share of stock in Apple Computer opens a week with a share price of $39\frac{7}{8}$ dollars and closes the week with a share price of $42\frac{1}{8}$ dollars. Select the choice that best estimates the dollar gain for the week of an investor who owns 290 shares of the stock.

  a. $6,000   b. $60,000   c. $600   d. $60

Solution   Round off $39\frac{7}{8}$ to 40, $42\frac{1}{8}$ to 42, and 290 to 300. Then mentally determine that the investor's gain is about

$$300(42 - 40) = 300(2) = 600.$$

Thus, $600, or choice **c**, estimates the solution. For the sake of comparison, the exact answer is

$$290(42\tfrac{1}{8} - 39\tfrac{7}{8}) = 652.50.$$

Note   An expression like $15{,}000(1 + 0.05)^{20}$ from Example 6 is difficult to estimate at this point. However, from the context of the problem you should be able to decide that an answer like 1 (which results from a common input error) is not a reasonable result.

PROGRESS CHECK 7   One share of stock in IBM opens a week with a share price of $105\frac{3}{4}$ dollars and closes the week with a share price of $109\frac{1}{8}$ dollars. Select the choice that best estimates the dollar gain for the week of an investor who owns 920 shares of this stock.

  a. $2,700   b. $27,000   c. $270   d. $270,000   ⌐

Progress Check Answers

5. 3

6. $28,759

7. $2,700

EXERCISES 1.2

In Exercises 1–42, evaluate the given expression.

1. 2^4
2. 2^5
3. 1^{400}
4. 1^{37}
5. 0^2
6. 0^3
7. $4^3 \cdot 3^4$
8. $5^2 \cdot 2^5$
9. $3^2 \cdot 3^4$
10. $2^4 \cdot 2^4$
11. $(\frac{2}{3})^3$
12. $(\frac{3}{4})^3$
13. $3(4 + 5)$
14. $4(5 + 6)$
15. $12 - 2 \cdot 5$
16. $14 - 3 \cdot 4$
17. $6 + 3 \cdot 4^2$
18. $4 + 5 \cdot 2^3$
19. $3 + 4^2 \cdot 2$
20. $5 + 3^2 \cdot 4$
21. $(3 + 4)^2 \cdot 2$
22. $(5 + 3)^2 \cdot 4$
23. $(3 + 4)^2 - (5 - 2)^2$
24. $(8 - 3)^2 + (7 - 4)^2$
25. $5 + 4^2 \cdot 2 + 3^2$
26. $3 + 3^3 \cdot 3 - 3^3$
27. $(4 + 4)^2 - (3 + 3)^2 - (2 + 2)^2$
28. $(5 + 5)^2 - (4 + 4)^2 - (3 + 3)^2$
29. $\frac{1}{2}(\frac{3}{2} + \frac{1}{2}) + 3^2$
30. $\frac{1}{3}(\frac{9}{2} + \frac{9}{2}) + 2^3$
31. $4 + 2[7 - 3(5 - 3)]$
32. $1 + 3[8 + 3(8 - 3)]$
33. $[4(7 - 4) - 10]^2$
34. $[5(8 - 3) - 20]^2$
35. $3[5 - 4(3 - 2)]^2$
36. $4[6 - 3(7 - 5)]^2$
37. $\dfrac{(8 - 2)^2}{2^3 - 5}$
38. $\dfrac{(7 - 3)^3}{2 \cdot 2^3}$
39. $\dfrac{2^4 - 10}{(5 - 2)^2}$
40. $\dfrac{10 - 2^3}{(10 - 2)^3}$
41. $\dfrac{3 + 4[10 - 2(4 - 1)]}{5 + 2^2}$
42. $\dfrac{12 - 8[10 - 3(5 - 2)]}{3 + 5^2}$

In Exercises 43–72, evaluate the given expression using a scientific calculator. Round off to two decimal places.

43. $8(6 + 3)$
44. $2(3 + 4)$
45. $10(15 - 9)$
46. $9(14 - 10)$
47. $2.1(5.3 + 1.1)$
48. $3.2(4.1 + 6.9)$
49. $0.6(5.2 - 4.8)$
50. $0.7(11.5 - 8.6)$
51. $1 + 2(3 + 4)$
52. $2 + 3(4 + 5)$
53. $2.1 - 3.2(4.5 - 4.4)$
54. $4.5 - 1.2(7.8 - 6.6)$
55. $\dfrac{5^2 + 5}{4^2 - 10}$
56. $\dfrac{6^2 - 6}{14 - 3^2}$
57. $\dfrac{(3.1 + 5.9)^2}{2.5^2 + 1}$
58. $\dfrac{(3.84 - 1.16)^2}{2.4^2 - 2}$
59. 2^{10}
60. 3^{12}
61. $4^5 - 2^{10}$
62. $9^6 - 3^{12}$
63. $3(1 + 2)^6$
64. $4(5 - 2)^7$
65. $(3^4 - 5)^3$
66. $(2^3 - 4)^3$
67. 1.04^{20}
68. 1.15^{30}
69. $12,000(1.03)^7$
70. $13,000(1.12)^5$
71. $11,000(1 + 0.05)^{15}$
72. $12,500(1 + 0.08)^{10}$

In Exercises 73–76, round answers to the nearest dollar.

73. In 1994 tuition at a college was $9,000. If the inflation rate is 6 percent per year, how much will tuition cost 20 years later in 2014? This is solved by evaluating $9,000(1 + 0.06)^{20}$.

74. An investment increases in value 3 percent per year. Its initial value is $30,000. What will it be worth after 15 years? Evaluate $30,000(1 + 0.03)^{15}$.

75. A restaurant meal cost 5 cents in the year 1776. If annual inflation is 3 percent, how much will a meal like it cost in 1996? Evaluate $0.05(1 + 0.03)^{220}$.

76. A car cost about $300 in 1930. If annual inflation is 5 percent, what would an equivalent car cost in 2000? Evaluate $300(1 + 0.05)^{70}$.

In Exercises 77–84, just think about the problem and then select the choice that best estimates the answer. Check your choice by doing the exact calculation with a scientific calculator.

77. One share of stock in a company opens the week with a share price of $25\frac{7}{8}$ dollars and closes at $23\frac{1}{8}$ dollars. If a person owns 316 shares, what is the investor's approximate dollar loss for the week?
   a. $9   b. $90   c. $900   d. $9,000   e. $90,000

78. The price of a share of stock in a company falls one week from $13\frac{1}{8}$ to $10\frac{7}{8}$ dollars. What is the approximate dollar loss for someone who owns 96 shares?
   a. $2   b. $20   c. $200   d. $2,000   e. $20,000

79. On an income tax form a person who uses an automobile for business purposes can deduct 19.5 cents per mile. On one trip the odometer started at 12,033.7 and ended at 12,740.1. What is the approximate allowable deduction for this trip?
   a. $1.40   b. $14   c. $140   d. $1,400   e. $14,000

80. On an income tax form a businessperson can deduct 10.3 cents per mile for any miles above 15,000 for which an automobile was used. During one trip the record of business miles went from 13,445.8 to 15,897.2. What is the approximate allowable deduction for this trip?
   a. $9   b. $90   c. $900   d. $9,000   e. $90,000

81. At a supermarket sliced turkey sells for $2.99 a half pound and sliced ham is $2.49 per half pound. In a major purchase for a big party, you order 11 lb of turkey and 9 lb of ham. Which of these answers gives a good estimate of the total cost?
   a. $10   b. $50   c. $100   d. $500   e. $1,000

82. A market sells walnuts for 89 cents a quarter pound and cashews for $1.05 per quarter pound. If you buy $3\frac{3}{4}$ lb of each, which of these answers gives a good estimate of the total cost?
   a. $10   b. $30   c. $100   d. $300   e. $1,000

83. In a rural community the property tax rate is $2.12 per each $100 of assessed value. What is the approximate tax for a house assessed at $70,000?
   a. $1.40   b. $14   c. $140   d. $1,400   e. $14,000

84. In a small town the property tax rate is $2.89 per each $100 of assessed value. Estimate the property tax on a house assessed at $98,800.
   a. $3   b. $30   c. $300   d. $3,000   e. $30,000

THINK ABOUT IT

1. Suppose n stands for any natural number. Why does 1^n equal 1? Why does 0^n equal 0?
2. Would $(\frac{1}{2})^{30}$ be a very small number or a very large number? Why?
3. An eccentric employer offers you work for exactly 30 days. Each day your pay doubles, but you must start at 1 penny for the first day. On the 30th day your pay will be 2^{29} pennies. About how much is this daily wage in dollars?
4. More than 300 years ago the French mathematician Fermat investigated numbers of the form $2^m + 1$. When m itself has the form 2^n, where n is a whole number, then the number is called a Fermat number. Fermat guessed that all such numbers would be prime numbers. Find the Fermat numbers for $n = 0$, 1, 2, and 3, and confirm that they are prime numbers. By the way, Fermat was wrong; some Fermat numbers are not prime. In 1990 mathematicians found prime factors for the Fermat number with $n = 9$. This number is 155 digits long and has three prime factors, the largest of which is a 99-digit number.
5. Which is worth more after 20 years, $10,000 invested at 5 percent annual interest, or $5,000 invested at 10 percent interest? You must compare $10,000(1 + 0.05)^{20}$ and $5,000(1 + 0.10)^{20}$.

REMEMBER THIS

1. True or false? $3 \le 3$.
2. True or false? Zero is an even number.
3. Graph the set of whole numbers from 3 to 8.
4. Find (a) the product and (b) the sum of $\frac{8}{3}$ and $\frac{3}{8}$.
5. Find 5 divided by $\frac{1}{2}$.
6. Express 18 as the product of prime factors.
7. Express 24 as the product of prime factors.
8. Express $\frac{24}{720}$ in lowest terms.
9. What number is 3 more than half of 7?
10. What number is 3 times as large as half of 7?

1.3   Algebraic Expressions and Geometric Formulas

A family buys a backyard aboveground swimming pool and selects a circular model that is 24 ft in diameter. If each cubic foot of volume holds about 7.5 gal of water, then to the nearest gallon, find how many gallons of water are needed to fill the pool to a depth of 4 ft. Use 3.14 for π. (See Example 8.)

OBJECTIVES

1 Translate between verbal expressions and algebraic expressions.

2 Evaluate algebraic expressions.

3 Apply perimeter and area formulas.

4 Apply volume formulas.

1 In algebra two types of symbols are used to represent numbers: variables and constants. A **variable** is a symbol that may be replaced by different numbers in a particular problem, while a **constant** is a symbol that represents the same number throughout a particular problem. An expression that combines variables and constants using the operations of arithmetic is called an **algebraic expression**. Consider carefully the following chart, which shows various ways to write and read algebraic expressions that involve the basic operations of addition and subtraction.

Operation	Verbal Expression	Algebraic Expression
Addition	The sum of a number x and 2	$x + 2$, or $2 + x$
	A number y plus 5	$y + 5$, or $5 + y$
	A number w increased by 1	$w + 1$, or $1 + w$
	8 more than a number b	$b + 8$, or $8 + b$
	Add 3 and a number d	$d + 3$, or $3 + d$
Subtraction	The difference of a number x and 2	$x - 2$
	A number y minus 5	$y - 5$
	A number w decreased by 1	$w - 1$
	8 less than a number b	$b - 8$
	Subtract 10 from a number d	$d - 10$
	Subtract a number d from 10	$10 - d$

Note that the order in which we write numbers in an addition statement does not affect their sum. Thus, the sum of numbers a and b may be written as $a + b$ or $b + a$. However, subtracting 10 from a number d is not the same as subtracting a number d from 10, so $d - 10$ has a different meaning from $10 - d$.

EXAMPLE 1   Translate each statement to an algebraic expression.

a. The price, p dollars, of a stereo increased by 10 dollars.
b. 5 less than the average a.

Solution   Translate according to the chart given above.

a. p dollars increased by 10 dollars is expressed as $p + 10$ or $10 + p$ dollars.
b. 5 less than a is expressed as $a - 5$. Remember that $5 - a$ is not correct here.

PROGRESS CHECK 1   Translate each statement to an algebraic expression.

a. The price, p dollars, of a stereo decreased by 5 dollars.
b. 10 more than the average a. ⌐

When multiplying numbers a and b, it is common notation to omit the multiplication dot and just write ab for the product of a and b. Therefore, each of the following expressions symbolizes "a times b."

$$ab \qquad a \cdot b \qquad a(b) \qquad (a)b \qquad (a)(b)$$

A common word associated with the operation of multiplication is *of*. For instance, a phrase like "$\frac{1}{2}$ of a number n" may be symbolized $\frac{1}{2}n$, where the word *of* translates to multiplication.

As mentioned in Section 1.1, to symbolize the division of a by b, we may write

$$a \div b, \qquad a/b, \qquad \text{or} \qquad \frac{a}{b}.$$

A division statement is called a **quotient.**

EXAMPLE 2   Translate each statement to an algebraic expression.

a. 8 more than the product of ℓ and w.
b. 15 percent of the amount of sales x.
c. Divide the sum of a, b, and c by 3.

Solution

a. ℓw is the most common way to symbolize the product of ℓ and w. We may represent 8 more than the product by $\ell w + 8$.

Progress Check Answers
1. (a) $p - 5$ (b) $a + 10$

b. In decimal form 15 percent is written 0.15. Now translate as follows:

$$\underbrace{15\text{ percent}}_{0.15}\text{ of }\underbrace{\text{the amount of sales}}_{x}.$$

We may omit the multiplication dot and represent the product by $0.15x$. (*Note:* For a review of decimals and percentages, consult the Appendix.)

c. The sum of a, b, and c is represented $a + b + c$. To represent the division of the sum by 3, write

$$\frac{a + b + c}{3} \quad \text{or} \quad (a + b + c) \div 3.$$

PROGRESS CHECK 2   Translate each expression to an algebraic expression.

a. 28 percent of the taxable income i.
b. Divide 8 less than the number x by 2.
c. The product of w and 4 more than w.

EXAMPLE 3   Translate each algebraic expression to a verbal expression.

a. $2x + 3y$ **b.** $a - 4b$ **c.** $2(x + 5)$

Solution   Understanding the order of operations is essential to interpreting the statements.

a. The sum $2x + 3y$ may be read as "the sum of 2 times x and 3 times y."
b. The difference $a - 4b$ means "subtract 4 times b from a."
c. The product $2(x + 5)$ may be stated as "2 times the sum of x and 5" or "twice the sum of x and 5."

Note   As you learn the language of algebra, it is usually much easier to understand an algebraic expression than the corresponding verbal expression.

PROGRESS CHECK 3   Translate each algebraic expression to a verbal expression.

a. $3a - 2b$ **b.** $x + 5y$ **c.** $2(\ell + w)$

2   If we are given numerical values for the symbols, we can evaluate an algebraic expression by substituting the given values and performing the indicated operations.

EXAMPLE 4   Evaluate each expression given that $x = 3$ and $y = 4$.

a. $2x - y$ **b.** $x^2 + 5y^2$ **c.** $\dfrac{y + 6}{x - 1}$

Solution   Substitute 3 for x and 4 for y; then simplify.

a. $2x - y = 2 \cdot 3 - 4$ Replace x by 3 and y by 4.
       $= 6 - 4$ Multiply.
       $= 2$ Subtract.

b. $x^2 + 5y^2 = 3^2 + 5 \cdot 4^2$ Replace x by 3 and y by 4.
       $= 9 + 5 \cdot 16$ Evaluate powers.
       $= 9 + 80$ Multiply.
       $= 89$ Add.

c. $\dfrac{y + 6}{x - 1} = \dfrac{4 + 6}{3 - 1}$ Replace x by 3 and y by 4.
       $= \dfrac{10}{2}$ Simplify the numerator and denominator independently.
       $= 5$ Divide.

Progress Check Answers

2. (a) $0.28i$ (b) $\frac{x-8}{2}$ (c) $w(w + 4)$

3. (a) Subtract 2 times b from 3 times a.   (b) Add x and 5 times y.   (c) 2 times the sum of ℓ and w.

PROGRESS CHECK 4   Evaluate each expression given that $x = 2$ and $y = 5$.

a. $y - 2x$ **b.** $3x^2 + y^2$ **c.** $\dfrac{6x}{8 - y}$ ⌐

3   An application of evaluating an algebraic expression often occurs when we apply formulas. A **formula** is an equality statement that expresses the relationship between two or more variables. For this course, a useful category of formulas is found in geometry, and the following discussion provides basic perimeter, area, and volume relations.

Perimeter and Area

Squares, rectangles, triangles, and circles are fundamental two-dimensional figures. The **perimeter** of these figures is a measure of the distance around the figure, while the **area** is a measure of the space inside the figure. For a circle, the perimeter is called the **circumference** of the circle. Perimeter is a measure of length and is measured in units like feet or meters. Area is measured in square units, such as square feet or square meters (which are usually abbreviated in algebra as ft² or m²). You need to memorize the perimeter and area formulas in Figure 1.4. In this figure right angles, which measure 90°, are symbolized with boxes.

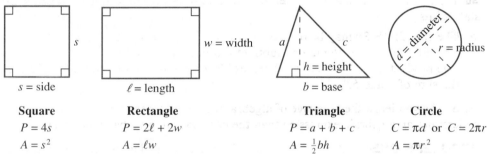
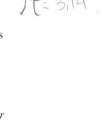

Square	Rectangle	Triangle	Circle
$P = 4s$	$P = 2\ell + 2w$	$P = a + b + c$	$C = \pi d$ or $C = 2\pi r$
$A = s^2$	$A = \ell w$	$A = \frac{1}{2}bh$	$A = \pi r^2$

Figure 1.4 Formulas for perimeter *P*, circumference *C*, and area *A*.

EXAMPLE 5   Find the perimeter and the area of a rectangular driveway that is 42 ft long and 12 ft wide. Will a 5-gal pail of driveway sealer be sufficient to cover the driveway if the directions on the pail read: "Recommended cover rate: Weathered or unsealed surfaces—from 300 to 400 ft² per 5-gal pail"?

Solution   The problem states that the driveway is a *rectangular* driveway. Therefore, sketch Figure 1.5 and apply the formulas for the perimeter and the area of a rectangle.

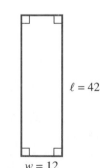

$$P = 2\ell + 2w \qquad \text{Given formula.}$$
$$= 2(42) + 2(12) \qquad \text{Replace } \ell \text{ by 42 and } w \text{ by 12.}$$
$$= 108 \qquad \text{Simplify.}$$
$$A = \ell w \qquad \text{Given formula.}$$
$$= 42 \cdot 12 \qquad \text{Replace } \ell \text{ by 42 and } w \text{ by 12.}$$
$$= 504 \qquad \text{Simplify.}$$

Figure 1.5 $w = 12$

The perimeter is 108 ft and the area is 504 ft². Because the area of the driveway exceeds 400 ft², a 5-gal pail of sealer is not sufficient to cover the driveway at the recommended cover rate.

PROGRESS CHECK 5   Find the perimeter and the area of a rectangular driveway that is 27 ft long and 11 ft wide. Will a 5-gal pail of driveway sealer be sufficient to cover the driveway using the recommended cover rate given in Example 5?   ⌐

Progress Check Answers

4. (a) 1   (b) 37   (c) 4

5. $P = 76$ ft; $A = 297$ ft²; the 5-gal pail should be sufficient.

The next example shows how to use the formulas given for a circle and includes a discussion of the number pi (π).

EXAMPLE 6   Find, to the nearest whole number, the circumference and the area of a circle with a radius of 9 meters (abbreviated m). Use 3.14 for π.

Solution   Recall that the **radius** of a circle is the distance from the center to any point on the circle, and sketch the given information as in Figure 1.6. Now substitute 9 for r in the formulas for the circumference and the area of a circle and use 3.14 for π.

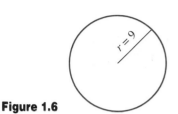

Figure 1.6

$C = 2\pi r$	Given formula.
$= 2(3.14)(9)$	Replace π by 3.14 and r by 9.
$= 56.52$	Simplify.
$A = \pi r^2$	Given formula.
$= 3.14(9^2)$	Replace π by 3.14 and r by 9.
$= 254.34$	Simplify.

To the nearest whole number, the circumference is 57 m and the area is 254 m².

Note   The number pi, symbolized π, is a nonterminating decimal. To eight decimal places,

$$\pi = 3.14159265. \ . \ . \ .$$

We usually use 3.14 as an approximation for π, but if your calculator has a $\boxed{\pi}$ key, it is easier and more accurate to use this key. For instance, you may compute the circumference in this example as follows:

$$2 \ \boxed{\times} \ \boxed{\pi} \ \boxed{\times} \ 9 \ \boxed{=} \ 56.548668.$$

The circumference is 57 m if the output is rounded off to the nearest whole number.

PROGRESS CHECK 6   Find, to the nearest whole number, the circumference and the area of a circle with a radius of 6 m. Use 3.14 for π.   ⌐

4   Volume

Volume is a measure of the space inside a three-dimensional figure. Volume is measured in cubic units such as cubic feet, which means feet · feet · feet and is abbreviated ft³. Figure 1.7 gives the formulas for finding the volume of some common three-dimensional figures.

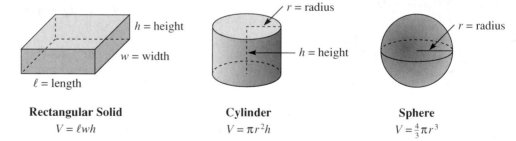

Rectangular Solid   Cylinder   Sphere
$V = \ell wh$ $V = \pi r^2 h$ $V = \frac{4}{3}\pi r^3$

Figure 1.7   Formulas for volume V.

EXAMPLE 7   Find the volume of a rectangular packaging carton that is 18 in. long, 12 in. wide, and 8 in. high.

Solution   Sketch Figure 1.8 and apply the formula for the volume of a rectangular solid.

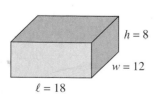

Figure 1.8

$V = \ell wh$	Given formula.
$= 18 \cdot 12 \cdot 8$	Replace ℓ by 18, w by 12, and h by 8.
$= 1,728$	Simplify.

The volume of the carton is 1,728 in.³.

Progress Check Answer
6. $C = 38$ m, $A = 113$ m²

PROGRESS CHECK 7   Find the volume of a cardboard box (a rectangular solid) that is 17 in. long, 10 in. wide, and 9 in. high.   ⌐

EXAMPLE 8   Solve the problem in the section introduction on page 18.

Solution   Sketch Figure 1.9 and apply the formula for the volume of a cylinder. Because the radius is one-half of the diameter, the radius of the pool is 12 ft.

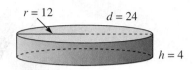

$$V = \pi r^2 h \qquad \text{Given formula.}$$
$$= 3.14(12^2)(4) \qquad \text{Replace } \pi \text{ by 3.14, } r \text{ by 12, and } h \text{ by 4.}$$
$$= 1{,}808.64 \qquad \text{Simplify.}$$

Figure 1.9

Each cubic foot of volume (which is 1,808.64 ft³) holds about 7.5 gal of water, so multiply to find how much water fills the pool.

$$1{,}808.64 \times 7.5 = 13{,}564.8$$

To the nearest gallon, the pool holds 13,565 gal of water when filled to a depth of 4 ft. (*Note:* If you use a more accurate value for π by using the $\boxed{\pi}$ key on a calculator, the answer to the nearest gallon is 13,572 gal.)

PROGRESS CHECK 8   If the family switches to a pool model that is 18 ft in diameter, then to the nearest gallon, find how many gallons of water are needed to fill the pool to a depth of 4 ft. Use 3.14 for π.   ⌐

Progress Check Answers
7. $V = 1{,}530$ in.³

8. $V = 1{,}017.36$ ft³ $= 7{,}630$ gal

EXERCISES 1.3

In Exercises 1–36, translate the given statement to an algebraic expression.

1. The sum of a number x and 3.
2. The sum of the price p and the tax t.
3. A salary s increased by $1,000.
4. A concert ticket price t increased by a $2 fee.
5. The number that is 5 more than a.
6. The number that is 6 more than b.
7. The number that is 7 less than c.
8. The number that is 8 less than d.
9. A price p minus a $5 discount.
10. A weight w minus a 10-lb weight loss.
11. A forest area A decreased by 200,000 acres.
12. A population of deer decreased by 8,000.
13. The product of c and d.
14. The product of 0.10 and x.

15. The quotient of x and y. (Divide x by y.)

16. The quotient of $3x$ and $4y$. (Divide $3x$ by $4y$.)

17. Three times as much as the number x.
18. Four times as much as the quantity $7y$.
19. One-half of the number z.
20. Two-thirds of the quantity $15x$.
21. Twelve percent of a rental fee r.
22. Six percent of a sales price s.
23. 10 more than the product of x and y.
24. 5 less than the product of a and b.
25. A price (x) increased by 8 percent of itself.
26. The cost of a restaurant meal (m) increased by 6 percent of itself.

27. The value of a car (v) decreased by 10 percent.
28. The cost of a shirt (c) marked down 30 percent.
29. Twice the sum of x and y.
30. Three times the sum of $4w$ and 5.
31. The sum of twice a and three times b.
32. The product of twice a and three times b.
33. The product of a number n and 2 more than n.
34. The product of a number m and 3 less than m.

35. Divide 5 more than x by 10.

36. Divide 10 more than x by 10 less than x.

In Exercises 37–48, translate each algebraic expression into a verbal expression. There may be several ways to do each one.

37. $2x + y$
38. $3y + 5x$
39. $a - 2b$
40. $2c - 4d$
41. $3(x + 1)$
42. $2(x - 1)$
43. $3x + 4$
44. $2y + 5$
45. $\dfrac{a + b}{2}$

46. $\dfrac{a - b}{2}$

47. $\dfrac{a}{2} + \dfrac{b}{3}$

48. $\dfrac{1}{3}x - \dfrac{1}{5}x$

In Exercises 49–66, evaluate the given expression using $x = 4$ and $y = 1$.

49. $2x + y$

50. $x + 3y$

51. $2(x + y)$

52. $(x + 3)y$

53. $\dfrac{x - 1}{y + 1}$

54. $\dfrac{2x + 1}{3x - 1}$

55. $\dfrac{x}{2} + 3y$

56. $\dfrac{x}{4} + 2y$

57. $\dfrac{15y - 3x}{2}$

58. $\dfrac{(15y - 3)x}{2}$

59. $x^2 + y^2$

60. $2x^2 + y^2$

61. $2(x^2 + 3y^2)$

62. $(2x^2 + 3)y^2$

63. $(x - y)^2$

64. $(2x - y)^2$

65. $\dfrac{4(y - 1)^{43}}{511.23}$

66. $\dfrac{47y^{99}(x - 4)}{4.7x}$

In Exercises 67–78, draw a sketch; then find the perimeter and the area of the shape named. Be sure to state the units for the answer.

67. Square; side = 3 m
68. Square; side = 1 ft
69. Square; side = 5 mi
70. Square; side = 4 yards (yd)
71. Square; side = 0.3 cm
72. Square; side = 0.25 in.
73. Rectangle; length = 5 in., width = 4 in.
74. Rectangle; length = 12 cm, width = 3 cm
75. Rectangle; length = 8 cm, width = 5 cm
76. Rectangle; length = 3 in., width = 2 in.
77. Rectangle; length = 3.5 ft, width = 2.5 ft
78. Rectangle; length = 0.7 yd, width = 0.2 yd

In Exercises 79–84, find the perimeter and area of the given triangle.

79.

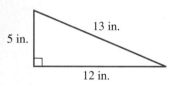

80.

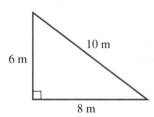

81.

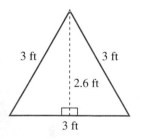

82.

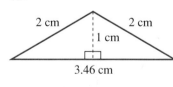

83.

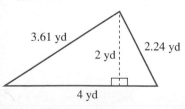

84.

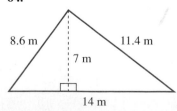

In Exercises 85–90, draw and label a sketch, and then find the circumference and the area of the given circles. Use 3.14 for π. If your calculator has a more exact value for π, use it but realize that your answer may differ from the answers given in the text. State units clearly. Round answers to two decimal places.

85. Radius = 4 in.
86. Radius = 5 in.
87. Diameter = 12 cm
88. Diameter = 2 cm
89. Radius = 10.4 ft
90. Radius = 30.1 ft

In Exercises 91–102, draw a sketch, and then compute the volume of the figure described. Follow the instructions for π given before Exercise 85. Round answers to two decimal places.

91. Rectangular solid; length = 8 in., width = 4 in., height = 2 in.
92. Rectangular solid; length = 10 cm, width = 8 cm, height = 4 cm
93. Rectangular solid; length = 12 ft, width = 12 ft, height = 6 ft
94. Rectangular solid; length = 1 ft, width = 1 ft, height = 10 in.
95. Cylinder; radius = 1 m, height = 10 m
96. Cylinder; radius = 2 m, height = 10 m
97. Cylinder; radius = 0.5 in., height = 5.0 in.
98. Cylinder; radius = 0.5 in., height = 10.0 in.
99. Sphere; radius = 1 ft
100. Sphere; radius = 2 ft
101. Sphere; radius = 6 cm
102. Sphere; radius = 3 cm
103. The wall of an industrial building is to be painted with a special paint which covers from 200 to 300 ft² per gallon can. The wall is rectangular with no windows. It is 12 ft high and 35 ft long. Compute the area of the wall. Can it be painted with one can of paint?
104. The floor of an old one-room schoolhouse is 26 ft by 24 ft. It is going to be sanded and sealed with polyurethane. The sealer covers 210 to 300 ft²/gal. How many gallons are needed to cover the floor with one coat of sealer?
105. A manufacturer of baking flour ships the flour to health food stores in rectangular boxes. The dimensions of the boxes are length = 1.5 ft, width = 1 ft, and height = 0.75 ft. Find the volume of the box. If the flour weighs 70 pounds per cubic foot (lb/ft³), how much does a shipment of 24 boxes weigh?

106. A block of lead in the shape of a cube 1 in. on each side weighs about 0.41 lb. Find the volume of a block of lead which is 10 in. on each side. What is the weight of the larger cube? Could you pick it up?
107. At a certain excavation site each cubic foot of earth weighs about 120 lb. This dirt is placed in rectangular containers to be moved. Find the volume of such a container which is 6 ft deep, 20 ft long, and 10 ft wide. Find the weight of this volume of earth. These containers are available in different materials and construction. Some will support 100,000 lb, some 200,000 lb, and some 300,000 lb. Which of these are strong enough for this job?

108. Each cubic foot of seawater weighs about 64 lb. For a project in salt removal this water is to be transported in large spherical containers which are 10 ft in radius. Find the volume of such a container. If the container weighs 10,000 lb when it is empty, how much does it weigh when it is full?

THINK ABOUT IT

1. a. Find several rectangles that have perimeter 18.
   b. Find several rectangles that have area 18.
   c. Find one rectangle that has perimeter and area both equal to 18.
2. You have a roll of 100 ft of chicken wire fencing which you will use to enclose a rectangular garden. Does it matter, as far as the enclosed area is concerned, how much you use for the length and how much for the width? That is, for a given perimeter of 100 ft, will different dimensions yield different areas? If so, what dimensions appear to give the maximum area?
3. The Greek scholar Eratosthenes who lived about 200 B.C. estimated the circumference of the earth (from North to South Poles) to be 250,000 "stadia." Divide his figure by π to estimate the earth's diameter in stadia. Scholars think that one stadium is about 600 ft (the length of a Greek sports stadium). Use this information, and the fact that there are 5,280 ft/mi, to give his estimate of the diameter of the earth in miles. The current estimate of this distance is about 7,900 mi.
4. If you double the dimensions of a solid cube, what happens to the volume? What happens to its weight? Does it double? Make up a specific example to check your intuition.
5. Why is it wrong to translate a 10 percent price increase as $x + 0.10$?

REMEMBER THIS

1. Evaluate $150(1 + 0.08)^{13}$.
2. Graph all the prime numbers between 25 and 35.

3. Estimate the cost (excluding tax) of 102 copies of a book which sells for $19.95.
4. True or false? $5 \geq 5$.

5. Compute $3 + 4 \cdot 5$.
6. Compute $\frac{1}{2} + \frac{1}{3} - \frac{1}{4}$.
7. Which is larger, 2^8 or 8^2?
8. True or false? $(3 \cdot 4)^2 = 3^2 \cdot 4^2$.
9. Write an expression for the product of 3 more than x and 5 less than twice y.
10. Express 60 as the product of prime factors.

1.4 Real Numbers and Their Graphs

The velocity of a projectile is a signed number that indicates both the speed and the direction of the projectile. Because up is usually denoted as a positive direction, a projectile traveling at a speed of 100 feet/second (ft/second) has a velocity of $+100$ ft/second when it is rising and -100 ft/second when it is falling. Speed and velocity are related by the absolute value formula

$$\text{speed} = |\text{velocity}|.$$

Find the speed of a projectile with the following velocities:
a. -30 m/second   **b.** 128 ft/second
(See Example 6.)

OBJECTIVES

1 Graph real numbers.

2 Order real numbers.

3 Find the negative or opposite of a real number.

4 Find the absolute value of a real number.

5 Identify integers, rational numbers, irrational numbers, and real numbers.

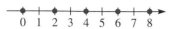

Figure 1.10

The main goal of this section is to fill in the number line so that each point on the line corresponds to a number and each number corresponds to a point on the line. In the process, we give names to different sets of numbers. Depending on the use of the number, the distinction between the types of numbers may be important.

1 Section 1.1 showed how to graph a set of whole numbers. For instance, in that section we graphed the set of numbers {0,2,4,6,8} as shown in Figure 1.10. Besides the whole numbers which are used for counting, we need positive and negative numbers to designate direction or to indicate whether a result is above or below some reference point. To graph negative numbers, we extend the number line in Figure 1.10 to the left of 0, as shown in Figure 1.11.

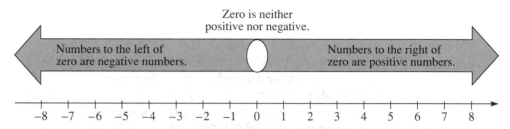

Figure 1.11

This number line is called the **real number line,** or simply the number line, and may be used to define a real number. Note that we usually omit the sign on positive numbers, so $+1 = 1$, $+2 = 2$, and so on.

Real Number

A real number is a number that can be represented by a point on the number line.

The first real numbers that we will graph are called **integers.** The set of integers is given by

$$\{ \ldots , -5, -4, -3, -2, -1, 0, 1, 2, 3, 4, 5, \ldots \}.$$

$$\underbrace{}_{\text{negative integers}} \qquad \underbrace{}_{\text{positive integers}}$$

Integers have both a sign that indicates if the number is to the left or right of 0 and a magnitude that indicates the distance between the number and 0. Zero itself is neither positive nor negative.

EXAMPLE 1 Graph the integers -4, 0, 3, and -2 on the number line.

Solution To graph the given integers, place dots on the real number line at -4, 0, 3, and -2, as in Figure 1.12.

Figure 1.12

PROGRESS CHECK 1 Graph the integers 4, -3, 0, and -1 on the number line. ⌐

The gaps on the real number line between the integers are filled in using rational numbers and irrational numbers. A **rational number** is a number that can be written as a fraction with an integer in the numerator and a nonzero integer in the denominator. Examples of rational numbers are $\frac{2}{3}$, -1.3 (which may be written as $-\frac{13}{10}$) and $2\frac{1}{2}$ (or $\frac{5}{2}$). Also, all integers are rational numbers because each integer can be written with a denominator of 1. For instance, $2 = \frac{2}{1}$, $-5 = \frac{-5}{1}$, and $0 = \frac{0}{1}$. Real numbers that

Progress Check Answer

1.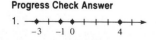

are not rational numbers are called **irrational numbers.** Real numbers like the square root of 2, written $\sqrt{2}$, and pi (π) are irrational numbers because they cannot be written as a quotient of two integers. Example 2 shows how we graph rational numbers and irrational numbers.

EXAMPLE 2   Graph the following numbers on the number line.

a. $\frac{2}{3}$ **b.** -1.3 **c.** $\sqrt{2}$

Solution   See Figure 1.13.

Figure 1.13

a. To graph the rational number $\frac{2}{3}$, divide the segment from 0 to 1 into 3 equal parts and then place a dot on the second slash mark to the right of 0.
b. To graph the rational number -1.3, divide the segment from -1 to -2 into 10 equal parts and then place a dot on the third slash mark to the left of -1.
c. We approximate the graphs of irrational numbers like $\sqrt{2}$. First, approximate $\sqrt{2}$ on a calculator.

$$2 \boxed{\sqrt{}} \boxed{1.4142136}$$

Because $\sqrt{2}$ is slightly more than 1.4, divide the segment from 1 to 2 into 10 equal parts and then place a dot slightly to the right of the fourth slash mark.

Note   In practice it is usually sufficient to give a rough estimate of the graph of the types of numbers in this example.

PROGRESS CHECK 2   Graph the following numbers on the number line.

a. $\frac{4}{5}$ **b.** -2.4 **c.** $\sqrt{3}$

2   Real numbers can be put in numerical order by using the number line in the same way that whole numbers were ordered in Section 1.1.

Order on the Number Line

For any two real numbers, the graph of the larger number is to the right of the graph of the smaller number on the number line.

EXAMPLE 3   Insert the proper symbol ($<$, $>$) to indicate the correct order.

a. -3 ____ 2   **b.** $-\frac{1}{2}$ ____ 0   **c.** -2 ____ -3

Solution   The given numbers are graphed in Figure 1.14.

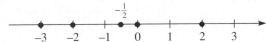

Figure 1.14

a. -3 is less than 2, written $-3 < 2$, because -3 is to the left of 2 on the number line. In general, any negative number is less than any positive number.
b. $-\frac{1}{2} < 0$, because $-\frac{1}{2}$ is graphed to the left of 0. In general, any negative number is less than zero.
c. -2 is greater than -3, written $-2 > -3$, because -2 is to the right of -3 on the number line.

Progress Check Answer

2.
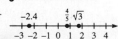

PROGRESS CHECK 3 Insert the proper symbol ($<$, $>$) to indicate the correct order.

a. -1 _____ $\frac{1}{2}$ **b.** 0 _____ -2 **c.** -4 _____ -3 ⌐

3 Two numbers on the real number line that are the same distance from 0, but on opposite sides of 0, are called **negatives** or **opposites** of each other. For instance, Figure 1.15 illustrates that the negative (or opposite) of 2 is -2, and the negative (or opposite) of -2 is 2.

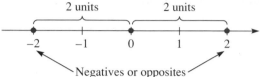

Figure 1.15

To symbolize the negative of a number, place the symbol $-$ in front of the number. Thus, the negative of a is written $-a$, and the negative of $-a$ is written $-(-a)$. In Figure 1.15 we see that the negative (or opposite) of -2 is 2, so

$$-(-2) = 2.$$

The example illustrates the double-negative rule.

Double-Negative Rule

If a is a real number, then

$$-(-a) = a.$$

The negative of 0, written -0, is 0. Zero is the only number that is its own negative.

EXAMPLE 4 Find the negative, or opposite, $-a$, of each number.

a. $a = -3$ **b.** $a = 2.5$ **c.** $a = 0$

Solution

a. By the double-negative rule, if $a = -3$, then

$$-a = -(-3) = 3.$$

In words, the negative or opposite of -3 is 3.

b. If $a = 2.5$,

$$-a = -(2.5) = -2.5.$$

c. Because zero is its own negative, if $a = 0$,

$$-a = -(0) = 0.$$

Note Our examples to this point show that we may find the negative or opposite of a real number by simply changing the sign of the number.

PROGRESS CHECK 4 Find the negative, or opposite, of each number a.

a. $a = -1.7$ **b.** $a = 0$ **c.** $a = 9$ ⌐

4 Sometimes the numerical size of a number is more important than its sign, and in such cases, the concept of absolute value is useful. The **absolute value** of a real number a, denoted $|a|$, is the distance between a and 0 on the number line. For instance $|3| = 3$, and $|-3| = 3$, as shown in Figure 1.16.

Progress Check Answers

3. (a) $-1 < \frac{1}{2}$ (b) $0 > -2$ (c) $-4 < -3$

4. (a) 1.7   (b) 0   (c) -9

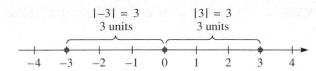

Figure 1.16

Because absolute value measures distance on the number line, the absolute value of a number is never negative. The geometric interpretation of absolute value may be translated to the following algebraic rule.

Absolute Value

For any real number a,

$$|a| = a, \text{ if } a \text{ is a positive number or zero,}$$
$$|a| = -a, \text{ if } a \text{ is a negative number.}$$

To illustrate this definition, note from Figure 1.16 that $|-3| = 3$. We arrive at this answer using the algebraic rule as follows: Because -3 is a negative number, use the formula $|a| = -a$; substituting gives $|-3| = -(-3) = 3$.

EXAMPLE 5   Evaluate each expression.

a. $|-2|$ b. $\left|\frac{1}{2}\right|$ c. $-|-2|$

Solution

a. From a geometric viewpoint $|-2| = 2$, as shown in Figure 1.17. Algebraically, -2 is a negative number, so use

$$|a| = -a \quad \text{to obtain} \quad |-2| = -(-2) = 2.$$

b. Figure 1.17 shows $\left|\frac{1}{2}\right| = \frac{1}{2}$. Algebraically, $\frac{1}{2}$ is a positive number, so

$$|a| = a \quad \text{yields} \quad \left|\frac{1}{2}\right| = \frac{1}{2} .$$

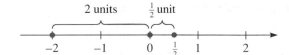

Figure 1.17

c. Treat absolute value bars as a grouping symbol and remove absolute value bars first.

$$-|-2| = -(2) = -2$$

Do not confuse $-|-2| = -2$ with $-(-2)$, which equals 2 by the double-negative rule.

PROGRESS CHECK 5   Evaluate each expression.

a. $|-4|$ b. $-|-4|$ c. $-(-4)$

EXAMPLE 6   Solve the problem in the section introduction on page 25.

Solution   Use the given formulas and simplify the absolute value.

a. Speed $= |\text{velocity}| = |-30 \text{ m/second}| = 30 \text{ m/second}$
b. Speed $= |\text{velocity}| = |128 \text{ ft/second}| = 128 \text{ ft/second}$

Progress Check Answers

5. (a) 4   (b) -4 (c) 4

PROGRESS CHECK 6 Use speed = |velocity| to find the speed of a projectile with the following velocities.

a. −64 ft/second   **b.** 0 m/second

5 The definitions given in this section for integers, rational numbers, irrational numbers, and real numbers may be used to classify numbers according to type. Figure 1.18 summarizes these definitions and shows the relationships among the various sets of numbers. In set theory when every element in set A is an element in set B, we call A a **subset** of B. Note that all of the sets we have discussed are subsets of the real numbers.

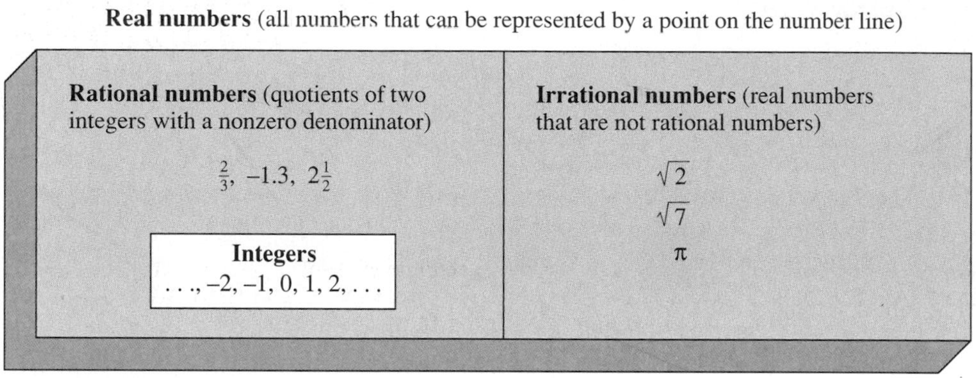

Real numbers (all numbers that can be represented by a point on the number line)

Rational numbers (quotients of two integers with a nonzero denominator)	**Irrational numbers** (real numbers that are not rational numbers)
$\frac{2}{3}$, −1.3, $2\frac{1}{2}$	$\sqrt{2}$
	$\sqrt{7}$
Integers ..., −2, −1, 0, 1, 2, ...	π

Figure 1.18

EXAMPLE 7 From the set $\{-7, 1.6, \sqrt{5}, 0, -\frac{3}{4}, \pi, 2\frac{1}{2}, \sqrt{9}\}$, list all numbers that are (a) integers, (b) rational numbers, (c) irrational numbers, and (d) real numbers.

Solution

a. −7, 0, $\sqrt{9}$ (which equals 3) are the integers in the set.
b. All integers are rational numbers, so −7, 0, and $\sqrt{9}$ are all rational numbers. The other rational numbers in the set are 1.6 (or $\frac{16}{10}$), $-\frac{3}{4}$, and $2\frac{1}{2}$ (or $\frac{5}{2}$).
c. The irrational numbers in the set are $\sqrt{5}$ and π. These numbers cannot be written as a quotient of two integers.
d. All the numbers in the set are real numbers. We will consider numbers like $\sqrt{-1}$, which are not real numbers, in Section 9.4.

Progress Check Answers

6. (a) 64 ft/sec   (b) 0 m/second

7. (a) −2, $\sqrt{4}$, 0   (b) $\frac{15}{8}$, −2, 0, 0.7, $\sqrt{4}$, $\frac{11}{35}$
(c) $\sqrt{3}$, $-\pi$ (d) All

PROGRESS CHECK 7 From $\{\frac{15}{8}, -2, 0.7, \sqrt{3}, \sqrt{4}, 0, -\pi, \frac{11}{35}\}$, list all numbers that are (a) integers, (b) rational numbers, (c) irrational numbers, and (d) real numbers.

EXERCISES 1.4

In Exercises 1–12, graph the given set of numbers. For irrational numbers place the dot approximately in the correct position.

1. $\{-3, 0, -1\}$

2. $\{-5, -2, 1\}$

3. $\{-\frac{1}{2}, \frac{3}{4}\}$

4. $\{-\frac{2}{5}, \frac{4}{3}\}$

5. Even integers from −2 to 2, inclusive

6. Odd integers from −3 to 3, inclusive

7. $\{-0.3, 0, 0.6\}$

8. $\{-0.6, -0.1, 1.2\}$

9. $\{\pi, 3, 4\}$

10. $\{-\pi, -3\}$

11. $\{\sqrt{8}, \sqrt{9}, \sqrt{10}\}$

12. $\{\sqrt{12}, \sqrt{16}, \sqrt{20}\}$

In Exercises 13–24, insert the proper symbol ($<$,$>$,$=$) to indicate the correct order.

13. -4 _____ 0

14. 0 _____ -1

15. $\frac{1}{2}$ _____ $\frac{3}{12}$

16. $\frac{1}{2}$ _____ $-\frac{2}{3}$

17. π _____ 3.14

18. $\sqrt{2}$ _____ 1.414

19. -4 _____ $-\sqrt{16}$

20. 3.2 _____ $\sqrt{10.24}$

21. 1.1 _____ 1.012

22. -0.23 _____ -0.24

23. 3^2 _____ 2^3

24. 2^4 _____ 4^2

In Exercises 25–30, find the value of $-a$, the negative of the given number.

25. $a = 3$

26. $a = -5$

27. $a = -1.1$

28. $a = 0.05$

29. $a = \frac{1}{3}$

30. $a = -\frac{3}{4}$

In Exercises 31–36, evaluate the given expression.

31. $|-3|$

32. $-|3|$

33. $|\frac{2}{3}|$

34. $-|-\frac{2}{3}|$

35. $|-4 + 5|$

36. $|-4| + |5|$

In Exercises 37–42, use the fact that speed is the absolute value of velocity to help answer the questions.

37. Find the speed of a diving submarine when its velocity is -20 ft/second.

38. Find the speed of a rocket at takeoff if its velocity is 100 ft/second.

39. Which object is moving faster, one with a velocity of -10 ft/second or one with a velocity of -15 ft/second?

40. True or false? If two projectiles have the same speed, then they must also have the same velocity.

41. The speed of a projectile is 35 m/second. Find two possible values for its velocity.

42. The speed of a projectile is 250 mi/hour. Find two possible values for its velocity.

In Exercises 43–52, list all numbers in the given set that are

a. integers

b. rational numbers

c. irrational numbers

d. real numbers

43. $\{-1,0,1,2\}$

44. $\{1,3,5,7,9\}$

45. $\{\frac{1}{2},\frac{2}{2},\frac{3}{2}\}$

46. $\{-\frac{5}{4},-\frac{7}{4},-\frac{9}{4}\}$

47. $\{1.2,2.3,3.0\}$

48. $\{-4.23,1.97,2.00\}$

49. $\{\sqrt{2},-\sqrt{3},\sqrt{4}\}$

50. $\{-\sqrt{9}, -\sqrt{10}, -\sqrt{11}\}$

51. $\{-3.4,-\sqrt{2},0,\pi\}$

52. $\{-\frac{4}{3},-0,\frac{8}{2}\}$

53. Use the graph to answer these questions.

a. What was the "profit" in 1993?

b. In what year did the company suffer the biggest loss?

c. In what year did the company make its biggest profit?

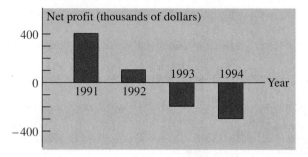

54. Use the graph to answer these questions.

a. What was the temperature on the coldest day?

b. On which of the days from Monday to Friday was it warmer than on the previous day?

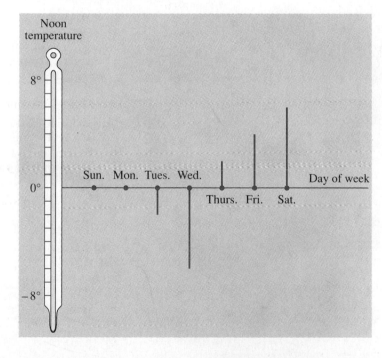

THINK ABOUT IT

1. If x represents a negative number, what does $-x$ represent?

2. Write an expression for the negative of $-x$.

3. Find values for a and b so that $|a + b| \neq |a| + |b|$.

4. Which one of these is correct?

a. When a ball is thrown up in the air, its velocity at the beginning is equal to its velocity just as it lands on the way down.

b. When a ball is thrown up in the air, its speed at the beginning is equal to its speed just as it lands on the way down.

5. Find an example where the difference of two irrational numbers is a rational number.

REMEMBER THIS

1. Give an algebraic expression for a price (p) minus a 10 percent discount.
2. Translate this algebraic expression into a verbal expression: $\dfrac{3a + 4}{5}$.
3. Evaluate $x - y^2$ when $x = 5$ and $y = 2$.
4. Find the perimeter and the area of a rectangle with length 8 in. and width 7 in.
5. Both legs of a right triangle are 5 cm in length. Find the area.

6. Find the area of a circle with diameter equal to 1 m.
7. Calculate $1.02 - 0.4^3$.
8. $1,000 invested at 7.25 percent interest for 10 years will be worth $1,000(1 + 0.0725)^{10}$ dollars. Compute this amount.
9. Insert the proper symbol ($<, >, =$) to indicate the correct order: $\frac{1}{2} \div 2$ _____ $2 \div \frac{1}{2}$.
10. Graph the prime factors of 27.

1.5   Addition of Real Numbers

On an income tax return an individual lists $1,676 in capital gains and $2,977 in capital losses. Find the net capital gain or loss for the year. (See Example 7.)

OBJECTIVES

1 Add two real numbers using the number line.

2 Add real numbers using absolute values.

3 Find the additive identity and the additive inverse for a real number a.

4 Solve applied problems by adding real numbers.

1 Because there is a point on the number line for every real number, we can use the number line to visualize the addition of real numbers. A real number will be represented by an arrow. The length of the arrow corresponds to the absolute value of the number. Positive numbers have arrows pointing to the right. Negative numbers have arrows pointing to the left. The addition $a + b$ is diagramed as "arrow for a followed by arrow for b," as shown in Example 1.

EXAMPLE 1   Find each sum by using the number line.

  a. $2 + 4$ **b.** $2 + (-4)$ **c.** $-2 + 4$ **d.** $-2 + (-4)$

Solution

  a. Consider Figure 1.19. The addition $2 + 4$ is shown on the number line by drawing an arrow with a length of 2 units starting at zero and pointing to the

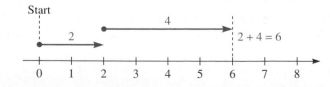

Figure 1.19

right. From the tip of the arrow at 2 we then draw an arrow pointing to the right with a length of 4 units. Because the tip of the arrow is at 6, we conclude that $2 + 4 = 6$.

b. On the number line in Figure 1.20, we start as in part **a** by drawing an arrow with a length of 2 units starting at zero and pointing to the right. To diagram $2 + (-4)$, from the tip of the arrow at 2, draw an arrow to the left with a length of 4 units. Because the tip of this arrow is at -2, we conclude that $2 + (-4) = -2$.

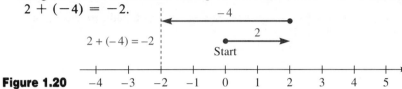

Figure 1.20

c. $-2 + 4 = 2$, as shown in Figure 1.21.

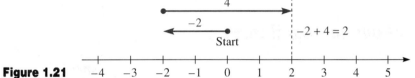

Figure 1.21

d. $-2 + (-4) = -6$, as shown in Figure 1.22.

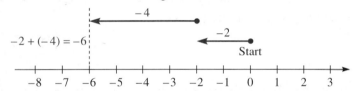

Figure 1.22

PROGRESS CHECK 1   Find each sum by using the number line.

a. $-3 + (-1)$ **b.** $3 + (-1)$ **c.** $-3 + 1$ **d.** $3 + 1$ ⌐

2   Although number line diagrams are useful for visualizing the addition of real numbers, they do not provide an efficient computational procedure. In practice, the concept of absolute value is used to add real numbers mentally. The two procedures that follow are suggested by our number line procedure.

Adding Real Numbers

Like Signs To add two real numbers with the same sign, add their absolute values and attach the common sign.
Unlike Signs To add two real numbers with different signs, subtract the smaller absolute value from the larger and attach the sign of the number with the larger absolute value.

EXAMPLE 2   Find each sum.

a. $-3 + (-5)$ **b.** $3.7 + 4.1$ **c.** $7 + (-4)$ **d.** $-\frac{5}{7} + \frac{2}{7}$

Solution   Perform these computations mentally.

a. $-3 + (-5) = -8$ by the like-signs procedure, because $|-3| = 3$, $|-5| = 5$, $3 + 5 = 8$, and the sign that -3 and -5 have in common is negative.

Progress Check Answers

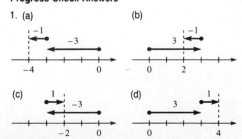

b. $3.7 + 4.1 = 7.8$ by the like-signs procedure, since $|3.7| = 3.7$ and $|4.1| = 4.1$, $3.7 + 4.1 = 7.8$, and the common sign is positive. Because the positive numbers match the numbers in ordinary arithmetic, in practice, treat the sum of two positive numbers as simply an arithmetic problem.

c. $7 + (-4) = 3$ by the unlike-signs procedure, because $|7| = 7$, $|-4| = 4$, $7 - 4 = 3$, and the positive number 7 has the larger absolute value.

d. $-\frac{5}{7} + \frac{2}{7} = -\frac{3}{7}$ by the unlike-signs procedure, because $\left|-\frac{5}{7}\right| = \frac{5}{7}$, $\left|\frac{2}{7}\right| = \frac{2}{7}$,

$$\frac{5}{7} - \frac{2}{7} = \frac{5 - 2}{7} = \frac{3}{7},$$

and the negative number $-\frac{5}{7}$ has the larger absolute value.

PROGRESS CHECK 2   Find each sum.

a. $-7 + (-6)$ **b.** $-9 + 2$ **c.** $6.5 + 2.6$ **d.** $\frac{3}{5} + \left(-\frac{2}{5}\right)$ ⌐

EXAMPLE 3   True or false? $-5 + [1 + (-3)] = [-5 + 1] + (-3)$.

Solution   First, simplify the expression on the left of the equal sign.

$$-5 + [1 + (-3)] = -5 + [-2] \quad \text{Operate within brackets.}$$
$$= -7 \quad \text{Add.}$$

Then simplify the expression on the right of the equal sign.

$$[-5 + 1] + (-3) = [-4] + (-3) \quad \text{Operate within brackets.}$$
$$= -7 \quad \text{Add.}$$

Both expressions equal -7, so the statement is true.

PROGRESS CHECK 3   True or false? $-4 + [2 + (-7)] = [-4 + 2] + (-7)$.   ⌐

    Example 3 shows that we obtain the same result if we change the grouping of the numbers in an addition problem. The order of the numbers also does not affect the sum. We can use these properties when adding more than two numbers because they allow us to find the sum by adding the positive numbers, adding the negative numbers, and then adding the results.

EXAMPLE 4   Find the sum: $11 + (-16) + (-9) + 8$.

Solution   The sum is -6, as shown below.

$$
\begin{aligned}
\text{Add the positives:} \quad & 11 + 8 && = 19 \\
\text{Add the negatives:} \quad & -16 + (-9) && = -25 \\
\text{Add the results:} \quad & 19 + (-25) && = -6
\end{aligned}
$$

PROGRESS CHECK 4   Find the sum: $-10 + 7 + (-3) + 14$.   ⌐

3   Two special addition problems involve adding 0 and adding numbers that are opposites or negative of each other. For any real number a,

$$\boxed{a + 0 = 0 + a = a,}$$

and 0 is called the additive identity. A sum, like $4 + (-4)$, involving a number and its opposite is always 0. In symbols, this important relationship is stated as follows.

Additive Inverse

For any real number a,

$$a + (-a) = -a + a = 0.$$

The opposite or negative of a number is called its **additive inverse** for this reason.

EXAMPLE 5   Find each sum.

a. $0 + (-6)$ **b.** $-8 + 8$ **c.** $-3 + [5 + (-5)]$

Solution   The additions involve the additive identity, additive inverses, or both.

a. $0 + (-6) = -6$
b. $-8 + 8 = 0$
c. $-3 + [5 + (-5)] = -3 + [0] = -3$

PROGRESS CHECK 5   Find each sum.

a. $10 + (-10)$ **b.** $-1 + 0$ **c.** $(-7 + 7) + (-4)$ ⌐

4   To solve applied problems using signed numbers, translate the common phrases below using the direction given.

   Positive direction to the right, up, rise, above, forward, gain

   Negative direction to the left, down, fall, below, backward, loss

Interpret like phrases in a similar way.

EXAMPLE 6   The temperature rose 2 degrees in 1 hour and then dropped 6 degrees in the next hour. Find the net change in temperature for the 2-hour period.

Solution   A 2-degree rise may be represented by the positive number 2, while a 6-degree drop may be represented by the number -6. The sum of 2 and -6 gives the net change in temperature.

$$2 + (-6) = -4$$

For the 2-hour period, the temperature dropped 4 degrees.

PROGRESS CHECK 6   The Dow-Jones Industrial Average fell 11 points on Monday and then rose 18 points on Tuesday. Find the net change for the two-day period.   ⌐

EXAMPLE 7   Solve the problem in the section introduction on page 32.

Solution   Represent the capital gain by the positive number 1,676 and the capital loss by the number $-2,977$. The net capital gain or loss is

$$1,676 + (-2,977) = -1,301.$$

For the year, the individual has a capital loss of $1,301.

PROGRESS CHECK 7   Find the net change in weight for an individual who loses 27 lb while dieting and then gains 46 lb when going off the diet.   ⌐

Progress Check Answers

5. (a) 0   (b) −1   (c) −4

6. +7 points

7. 19-lb gain

EXERCISES 1.5

In Exercises 1–12, find the sum by sketching arrows on the number line.

1. $4 + 2$

2. $3 + 3$

3. $5 + 0$

4. $0 + 3$

5. $1 + (-2)$

6. $4 + (-3)$

7. $2 + (-2)$

8. $-3 + 3$

9. $-1 + 5$

10. $-4 + 1$

11. $-3 + (-4)$

12. $-1 + (-2)$

In Exercises 13–42, find the sum.

13. $34 + (-27)$

14. $49 + (-22)$

15. $23 + (-30)$

16. $18 + (-19)$

17. $-35 + (-300)$

18. $-13 + (-2)$

19. $-4.7 + 1.9$

20. $-12.3 + 34.5$

21. $-4.56 + (-6.54)$

22. $-10.01 + (-1.1)$

23. $\frac{5}{3} + (-\frac{1}{3})$

24. $\frac{4}{9} + (-\frac{3}{9})$

25. $-\frac{2}{3} + \frac{1}{2}$

26. $-\frac{5}{6} + \frac{11}{9}$

27. $2 + (-3) + 2$

28. $-5 + 3 + (-2)$

29. $-1 + (-3 + 5)$

30. $-7 + (-3 + 10)$

31. $-2.3 + (2.3) + 5$

32. $1.9 + (-3.9) + (-0.1)$

33. $-3.2 + [5.5 + (-3.3)]$

34. $7.5 + [-1.5 + (4.5)]$

35. $-5 + 7 + (-9) + 11$

36. $6 + (-2) + 8 + (-10)$

37. $1.1 + (-2.2) + 3.3 + (-4.4)$

38. $-1.2 + 2.3 + (-3.4) + 4.5$

39. $-2 + \{-3 + [5 + (-7)]\}$

40. $[12 + (-5) + (-18)] + [20 + (-15)]$

41. $[-\frac{3}{5} + (\frac{1}{7} + \frac{3}{5})] + (-\frac{1}{7})$

42. $\frac{1}{2} + (-\frac{4}{9}) + \frac{6}{18} + (-\frac{1}{3}) + \frac{8}{18}$

43. What number should be added to each of these to make a sum equal to zero?

    a. 4   **b.** -7 **c.** $-\frac{3}{5}$ **d.** 1.078   **e.** 0

44. What is the additive inverse of each of these numbers?

    a. 3   **b.** -4.5 **c.** 0   **d.** $-\frac{7}{4}$ **e.** 0.03

In Exercises 45–48, fill in the blanks to make a true statement.

45. $-5 +$ _____ $= 0$

46. $184 +$ _____ $= 0$

47. _____ $+ 0 = 13.2$

48. _____ $+ 0 = -\frac{1}{3}$

In Exercises 49–54, fill in the blank with words to make a true statement.

49. Negative four plus negative two equals _____ .

50. Three plus negative nine equals _____ .

51. Negative one plus five equals _____ .

52. Three plus the additive inverse of three equals _____ .

53. Negative five plus _____ equals zero.

54. The additive identity plus four equals _____ .

In Exercises 55–60, translate the algebraic expression into an English sentence.

55. $-4 + (-5) = -9$

56. $-4 + 5 = 1$

57. $4 + (-5) = -1$

58. $-(4 + 5) = -9$

59. $-(-4) + (-5) = -1$

60. $4 + -(-5) = 9$

In Exercises 61–66, evaluate both sides of the equality and decide if the statement is true or false.

61. $|-2| + |-3| = |-2 + (-3)|$

62. $|-2| + |3| = |-2 + 3|$

63. $|2| + |-3| = |2 + (-3)|$

64. $-|-2| + 3 = -|2| + 3$

65. $2 + (-|3|) = 2 + |-3|$

66. $-|2 + (-3)| = |-2 + 3|$

In Exercises 67–78, answer the given question by using an addition calculation.

67. On an income tax return an individual lists $847 in capital gains and $912 in capital losses. Find the net capital gain or loss.

68. An investor records losses of $1,048 and $2,057 for the first two months of the year and profits of $369 and $1,186 for the next two months. Find the net profit or loss.

69. The midday temperatures in a city for five consecutive days were -3, -2, -5, 4, and 8 degrees. Find the average midday temperature by adding these numbers and dividing by 5.

70. An individual loses 35 lb on a diet and then gains back 27 lb. Find the net change in weight.

71. An undersea explorer is at a depth of 300 ft (represented by -300), before rising 110 ft. What is the depth at that point?

72. Under laboratory conditions nitrogen is a liquid when it is colder than about -200 degrees Celsius, and it is a gas when it is warmer than that. At the beginning of an experiment some nitrogen is at -275 degrees Celsius. During the experiment the temperature rises 92 degrees. What is the temperature at the end? In which state, gas or liquid, is the nitrogen at the conclusion of the experiment?

73. On four successive downs in a football game a team loses 14 yd, then gains 3, then loses 6, then gains 8. What is the net yardage gained or lost?

74. In playing five hands of a card game, a player's point scores are as follows: 36, 55, -20, -9, -12. What is the net score after these five hands?

75. At the start of the month a bank balance is $586. The activity during the month is as follows: a deposit of $150; withdrawals of $50, $75, and $125; a deposit of $88; a withdrawal of $100. What is the final balance?

76. On four consecutive days the Dow-Jones Industrial Average fell 20 points, fell 19 points, fell 7 points, and then rose 40 points. What was the net change for this period?

77. A child is playing on a long moving sidewalk in an airport. The sidewalk moves in one direction at 3 mi/hour (call this the negative direction), and the child runs along it in the opposite direction at 3 mi/hour. What is the net velocity of the child? In which direction will the child be moving relative to the building?

78. A motorboat is going against the flow of current in a river. The river is flowing downstream at 10 mi/hour. The boat is going upstream at 13 mi/hour. What is the net velocity of the boat? Will it be making any progress?

79. Calculate the entries for this addition table.

+	−1	−2	−3	−4
1				
2				
3				

80. Calculate the entries for this addition table.

+	−2	−1	0	1	2
−2					
−1					
0					
1					
2					

81. Refer to the table and determine the net profit or loss for the 12 weeks shown. Profit/loss entries are in thousands of dollars. Values in parentheses are losses.

Week	1	2	3	4	5	6	7	8	9	10	11	12
Profit	1.1	2.1	3.4	3.5	2.9	1.2	(0.2)	(1.8)	(2)	(0.3)	1.2	2.0

82. Use the graph to answer these questions.
  a. In which year was the difference in performance of the two products greatest?
  b. In which year was the difference in performance of the two products least?
  c. Find the net profit each year in combined sales for both products.

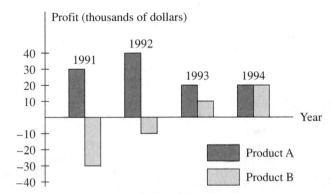

THINK ABOUT IT

1. Why is the word *identity* used for a number which leaves another one unchanged?

2. Put the numbers which follow in six different orders, and each time, add them from left to right. Show that you get the same answer each time. The numbers are -2, 5, -7. This supports the fact that order does not matter in addition.

3. If the absolute value of x is greater than the absolute value of y, then the sign of $x + y$ is
  a. the same as the sign of x **b.** the same as the sign of y

4. Find the sum: $1 + (-2) + 3 + (-4) + \cdots + 99 + (-100)$.

5. Find the sum: $-1 + 3 + (-5) + 7 + \cdots + (-97) + 99$.

REMEMBER THIS

1. Graph the set $\{\sqrt{2}, -\sqrt{5}\}$.

2. Evaluate $|-3 + 2| + |-3| + |2|$.

3. Which of these are not integers? $\{-3, -\frac{1}{2}, 0, \frac{1}{2}\}$.

4. Write an expression for the area of this triangle.

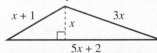

5. Evaluate $x + y^2$ when $x = -2$ and $y = 3$.

6. True or false? 15 is a prime factor of 30.

7. Write an expression for 10 more than half of x.

8. What number is called the additive identity? Why is it called that?

9. What number added to -2 equals 5?

10. Fill in the blank: $-3 +$ _____ $= -10$.

1.6 Subtraction of Real Numbers

At a popular ski resort, the following midday temperatures in degrees Celsius were recorded during a week in January:

$$2, -3, -1, 4, 0, -5, 1.$$

In statistics the difference between the extreme values gives a measure of variability in the data, called the range of the data. That is,

range = largest number − smallest number.

For the given set of midday temperatures, find the range. (See Example 5.)

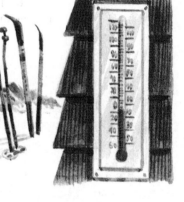

OBJECTIVES

1 Subtract real numbers.

2 Evaluate algebraic expressions involving addition and subtraction.

3 Distinguish between the $\boxed{-}$ and $\boxed{+/-}$ keys on a calculator.

4 Solve applied problems by subtracting real numbers.

1 In algebra (and other areas of mathematics), subtraction is defined as the addition of the opposite. To illustrate, we know from arithmetic that $5 - 2 = 3$. Algebraically speaking, we may convert this subtraction problem to the following addition problem and then apply the procedures from the previous section.

$$5 - 2 = 5 + (-2) = 3$$

To subtract 2, we may add the opposite of 2; and in general, we will often switch between subtraction and addition by using the following definition.

Definition of Subtraction

If a and b are any real numbers, then

$$a - b = a + (-b).$$

In words, "a minus b" is the same as "a plus the opposite of b."

EXAMPLE 1   Subtract by converting to addition.

a. $4 - 9$ **b.** $9 - 4$ **c.** $-3 - (-8)$ **d.** $-\frac{1}{2} - \frac{3}{4}$

Solution

a. To subtract 9 from 4, add the opposite of 9 to 4.

$$4 - 9 = 4 + \overbrace{(-9)}^{\text{opposite of 9}} = -5$$

$$\underset{\text{change} - \text{to} +}{}$$

b. To subtract 4 from 9, add the opposite of 4 to 9.

$$9 - 4 = 9 + (-4) = 5$$

c. To compute $-3 - (-8)$, find the sum of -3 and the opposite of -8. By the double-negative rule, the opposite of -8 is 8.

$$-3 - (-8) = -3 + 8 = 5$$

d. Use $a - b = a + (-b)$ and the fraction principles from Section 1.1.

$$-\tfrac{1}{2} - \tfrac{3}{4} = -\tfrac{1}{2} + (-\tfrac{3}{4}) = -\tfrac{2}{4} + (-\tfrac{3}{4}) = -\tfrac{5}{4}$$

PROGRESS CHECK 1   Subtract by converting to addition.

a. $5 - 13$ **b.** $13 - 5$ **c.** $-7 - 2$ **d.** $-\frac{3}{8} - (-\frac{3}{4})$ ⌐

Parts **a** and **b** in Example 1 illustrate that $a - b$ and $b - a$ give opposite answers. Therefore, be careful to write numbers in the correct order in a subtraction problem, especially when translating from a verbal expression to a numerical expression.

EXAMPLE 2   Translate to a numerical expression and simplify.

a. From the sum of 7 and -15, subtract -3.
b. Subtract the sum of -4.3 and -6.1 from 1.5.

Solution   With respect to the minus sign, the number we subtract "from" goes on the left and the number to subtract goes on the right. In both parts use a grouping symbol to enclose the indicated sum, and simplify within this grouping symbol first.

a. $[7 + (-15)] - (-3) = [-8] - [-3]$ Operate within brackets.
$\qquad\qquad\qquad\quad = [-8] + 3$ $a - b = a + (-b)$.
$\qquad\qquad\qquad\quad = -5$ Add.

b. $1.5 - [-4.3 + (-6.1)] = 1.5 - [-10.4]$ Operate within brackets.
$\qquad\qquad\qquad\qquad = 1.5 + 10.4$ $a - b = a + (-b)$.
$\qquad\qquad\qquad\qquad = 11.9$ Add.

PROGRESS CHECK 2   Translate to a numerical expression and simplify.

a. Subtract the sum of 5 and -11 from 8.
b. From the sum of -2.6 and -7.5, subtract -4.8.   ⌐

2 Algebraic expressions that involve addition and subtraction can now be evaluated for real number replacements of the variables.

EXAMPLE 3   Evaluate each expression given that $x = -3$ and $y = 4$.

a. $x + y$ **b.** $-x - y$ **c.** $-(x - y)$

Solution   Substitute -3 for x and 4 for y and simplify.

a. $x + y = -3 + 4$ Replace x by -3 and y by 4.
$\qquad\quad = 1$ Add.

Progress Check Answers
1. (a) -8 (b) 8   (c) -9 (d) $\frac{3}{8}$

2. (a) $8 - [5 + (-11)] = 14$
(b) $[-2.6 + (-7.5)] - (-4.8) = -5.3$

b. $-x - y = -(-3) - 4$ Replace x by -3 and y by 4.

$\quad\quad\quad\;\; = 3 - 4$ Double-negative rule.

$\quad\quad\quad\;\; = 3 + (-4)$ $a - b = a + (-b)$.

$\quad\quad\quad\;\; = -1$ Add.

c. $-(x - y) = -(-3 - 4)$ Replace x by -3 and y by 4.

$\quad\quad\quad\;\; = -(-7)$ Operate within parentheses, using $a - b = a + (-b)$.

$\quad\quad\quad\;\; = 7$ Double-negative rule.

PROGRESS CHECK 3   Evaluate each expression given that $x = -4$ and $y = 2$.

a. $y + x$ **b.** $-x - y$ **c.** $-(x - y)$ ⌟

3   The expression $-(-3) - 4$ contained in Example 3b is read as follows:

$$\underbrace{\text{the opposite of}}_{-} \underbrace{\text{negative 3}}_{(-3)} \underbrace{\text{minus}}_{-} \underset{\downarrow}{4}.$$

Note that the symbol $-$ is associated with the opposite of a number, a number that is negative, and the operation of subtraction. On a calculator the sign change key $\boxed{+/-}$ is used to enter a negative number and to compute the opposite of a number, while the key $\boxed{-}$ is used for subtraction.

EXAMPLE 4   Evaluate each expression on a scientific calculator.

a. $-(-3) - 4$ **b.** $-(-3 - 4)$

Solution

a. Enter -3, change the sign, and then subtract 4.

$$3 \;\boxed{+/-}\; \boxed{+/-}\; \boxed{-}\; 4 \;\boxed{=}\; \boxed{\quad\quad -1}$$

However, by the double-negative rule, note that it is unnecessary to press the sign change key $\boxed{+/-}$ twice in succession.

b. Enter -3, subtract 4, and then change the sign.

$$3 \;\boxed{+/-}\; \boxed{-}\; 4 \;\boxed{=}\; \boxed{+/-}\; \boxed{\quad\quad 7}$$

PROGRESS CHECK 4   Evaluate each expression on a scientific calculator.

a. $-12 - (-17)$ **b.** $-(-9 - 25)$ ⌟

4   We now use our method for subtracting real numbers to solve an applied problem.

EXAMPLE 5   Solve the problem in the section introduction on page 38.

Solution   In the given data, 4 is the largest number and -5 is the smallest number. Substitute these numbers in the given formula and simplify.

$$\begin{aligned} \text{Range} &= \text{largest number} - \text{smallest number} \\ &= 4 - (-5) \\ &= 4 + 5 \\ &= 9 \end{aligned}$$

The range is 9 degrees.

PROGRESS CHECK 5   Find the range for the following midnight temperatures in degrees Celsius during a week in February: $-6, -10, -2, -7, -13, -5, -3$. ⌟

Progress Check Answers

3. (a) -2 (b) 2   (c) 6

4. (a) 5   (b) 34

5. 11 degrees Celsius

EXERCISES 1.6

In Exercises 1–12 we have given either an addition expression or a subtraction expression. First, state which type of expression is given; then by using the principle that $a - b = a + (-b)$, rewrite it as an equivalent expression of the other type.

1. $7 - 4$ **2.** $5 - 18$
3. $7 + (-2)$ **4.** $5 + (-4)$
5. $-3 + (-2)$ **6.** $-6 + (-1)$
7. $-8 - (-f)$ **8.** $-9 - (-g)$
9. $1 - a$ **10.** $2 - x$
11. $d + 2$ **12.** $c + (-3)$

In Exercises 13–30, first rewrite the subtraction as an addition; then calculate the answer.

13. $8 - 3$ **14.** $13 - 4$
15. $0 - 8$ **16.** $4 - 13$
17. $-5 - 2$ **18.** $-6 - 4$
19. $-9 - (-1)$ **20.** $-10 - (-17)$
21. $4.3 - 5.6$ **22.** $5.9 - 15.2$
23. $\frac{1}{2} - (-\frac{1}{2})$ **24.** $\frac{4}{5} - (-\frac{4}{5})$
25. $0 - (-\frac{1}{5})$ **26.** $-\frac{3}{5} - (-\frac{3}{5})$
27. $0 - \frac{4}{5}$ **28.** $-\frac{1}{3} - \frac{2}{3}$
29. $-\frac{3}{8} - (-\frac{5}{4})$ **30.** $-\frac{8}{5} - (-\frac{8}{6})$

In Exercises 31–42, translate the given phrase into a numerical expression and simplify.

31. -7 minus 2.
32. -5 minus -3.
33. From the sum of 8 and -3, subtract -4.
34. From the sum of -4 and -3, subtract 12.

35. From the sum of -5 and 0, subtract the sum of -6 and 7.

36. From the sum of -12 and 12, subtract the sum of -6 and -1.

37. Subtract the sum of -1 and -1 from -1.

38. Subtract the sum of -6 and 2 from 0.
39. -1 minus the sum of -9 and -12.
40. -8 minus the sum of 4 and -4.
41. Subtract the sum of 5 and negative 6 from the opposite of 2.

42. Subtract the opposite of 3 from the additive inverse of negative 1.

In Exercises 43–48, translate the algebraic expression into an English phrase. (Answers may vary.)

43. $6 - (-1)$
44. $-2 - (-3)$
45. $-(-a) - b$
46. $c - (-b)$
47. $(-2 + x) - 5$
48. $-4 - [7 + (-y)]$

In Exercises 49–54, evaluate each expression given that $x = -2$ and $y = 3$.

49. $x - y$ **50.** $y - x$
51. $x + (-y)$ **52.** $-x + y$
53. $-(x - y)$ **54.** $-x - y$

In Exercises 55–60, evaluate each expression given that $x = 2$, $y = -3$, and $z = -1$. Remember to follow the order of operations as given in Section 1.2.

55. $x - y - z$ **56.** $x - (y - z)$
57. $x - y + z$ **58.** $(x - y) + z$
59. $-(x + y) + z$ **60.** $-x - (-y) - (-z)$

In Exercises 61–66, evaluate each expression on a scientific calculator.

61. $-3.1 - 4.56$ **62.** $-0.125 - (-0.76)$
63. $6.34 - (-2.34)$ **64.** $-9.895 - (-12.222)$
65. $\frac{100.25}{50} - (-4.995)$ **66.** $-3.4 - \frac{73.4}{16}$

In Exercises 67–78, write an appropriate subtraction expression and do the subtraction to solve the problem.

67. Find the range (as defined in the section-opening example) for this set of numbers, which represent temperatures in degrees Fahrenheit at the top of a mountain in New Hampshire at 7 A.M. on six consecutive days in early spring: $\{37,45,-20,0,-4,19\}$.

68. Find the range (as defined in the section-opening example) for this set of numbers, which represent temperature settings in degrees Celsius for five different versions of a chemical engineering experiment: $\{-2.3,-1.1,0.8,1.9,2.3\}$.

69. How many years are there between these years?
   a. 1996 and 1865
   b. 1776 and 200 B.C. [You can represent 200 B.C. as -200. Note that after you subtract $1776 - (-200)$, you must subtract 1, because there was no year 0.]
   c. 412 B.C. and 623 B.C.

70. A great early Greek contributor to the development of astronomy was Hipparchus, who began working about 160 B.C. He is credited with inventing trigonometry and giving a good estimate of the distance from the earth to the moon. We know about his work because it was described in much detail by another great astronomer, Ptolemy, who worked about 150 A.D. How many years passed between these two dates? (See the note in Exercise 69.)

71. If the temperature of liquid nitrogen falls from -197 to -209 degrees Celsius, how many degrees did it fall?

72. One winter day in Vermont the temperature dropped from 15 above zero (Fahrenheit) to 17 below zero in a period of a few hours. How much of a drop is that?

73. A chemical reaction is called endothermic if it takes in heat from the surrounding air. Such an experiment caused the temperature of the surrounding air to drop from 30.56 to -5.12 degrees Celsius. How many degrees did the temperature fall?

74. A chemical reaction is called exothermic if it gives off heat to the surrounding air. Such an experiment caused the temperature of the surrounding air to rise from -70.56 to -7.00 degrees Celsius. How many degrees did the temperature rise?

75. In the United States the settlement with the highest elevation is Climax, Colorado, at 11,560 ft. The lowest settlement is Calipatria, California, at -185 ft (this means it is below sea level). Find the difference in elevation between these two settlements.

76. In one coastal area of the United States the elevations of two nearby locations are 40 ft above sea level and 24 ft below sea level. What is the difference in elevation between these two places?

77. To find the distance between two points on a number line, you take the absolute value of the difference between their values. Find the distance between 1.45 and -4.77.

78. To find the distance between two points on a number line, you take the absolute value of the difference between their values. Find the distance between -42.85 and -21.45.

79. Complete this subtraction table for $a - b$.

a \ b	-3	0	3
-2			
0			
2			

80. Complete this subtraction table for $a - b$.

a \ b	-4	-3	-2
-3			
-2			
-1			

Exercises 81 and 82 relate to a part of mathematics called numerical analysis. In numerical analysis patterns are sometimes analyzed by repeated subtraction, where each row of numbers consists of differences between adjacent numbers in the previous row. This process is repeated until all zeros are recorded. The number of rows it takes to reach all zeros and the patterns within the rows give clues to the formula which was behind the first row.

81. The numbers in the first row are derived by substituting consecutive integers into the formula $5 - x - x^2$. Complete the calculations until you get a row of all zeros.

Row 1: 5		3		-1	-7	-15	-25
Row 2:	2		4	6		
Row 3:		-2				
Row 4:						

82. The numbers in the first row are derived by substituting consecutive integers into the formula $100 - x^3$. Complete as many rows as you need to get all zeros.

Row 1: 100	99	92	73	36	-25	-116
Row 2:	1	7				
Row 3:						
Row 4:						
Row 5:						

THINK ABOUT IT

1. a. Let x and y stand for two different positive numbers. Explain why you can tell the sign of $x + y$ but not $x - y$.
   b. Let c and d stand for two different negative numbers. Explain why you can tell the sign of $c + d$ but not $c - d$.

2. Let p stand for a positive number. Let n stand for a negative number. Can you tell the sign of any of these: $p + n$, $p - n$, $n - p$?

3. When you subtract a positive number from a quantity, the answer is smaller than the number you started with. What happens when you subtract a negative number from a given quantity?

4. Simplify: $-3 - \{-4 - [-5 - (-6 - 7)]\}$.

5. Try each of these sequences of keys on your calculator. See if you can explain what each does. (Different models of machines may produce different results.)
   a. 5 $\boxed{+/-}$ $\boxed{+/-}$ $\boxed{+/-}$ $\boxed{+/-}$ 2 $\boxed{=}$
   b. 5 $\boxed{-}$ $\boxed{-}$ $\boxed{-}$ $\boxed{-}$ 2 $\boxed{=}$

REMEMBER THIS

1. Compute $-4 + (-5)$.

2. What number is the additive identity?

3. Graph the negatives of these numbers: $\{-3, 0, 5\}$.

4. Show that 3.4 is a rational number by writing it as a fraction with integers in the numerator and denominator.

5. Translate into an algebraic expression: 4 less than the product of g and h.

6. Evaluate $a + b^2$ when $a = 1$ and $b = 2$.

7. Use a calculator to evaluate $\dfrac{1.2^2 - 3}{(4.5 - 2.1)^2}$.

8. Which is larger, 4^5 or 5^4?

9. Express 120 as the product of prime factors.

10. Which of these is true?
   a. The sum of any number and its reciprocal is 1.
   b. The product of any number and its reciprocal is 1.

1.7   Multiplication of Real Numbers

Assuming certain depreciation rules, the value V of a $7,000 desktop publishing system after t years is given by the formula

$$V = -1,400t + 7,000,$$

where t is a nonnegative integer less than or equal to 5. Find the value of this equipment after three years. (See Example 7.)

O B J E C T I V E S

1   Multiply two real numbers.

2   Multiply more than two real numbers.

3   Evaluate expressions containing signed numbers by following the order of operations.

4   Solve applied problems involving multiplication of real numbers.

---

1   An important theme in the development of math concepts asks these two questions:

> What do we already know? How can we build on this without having to undo previously established properties?

In terms of multiplication, there are three rules from arithmetic that extend to multiplication of real numbers. Remember that the result of a multiplication is called a product and the numbers that are multiplied are called factors.

Property	Comment
$a \cdot 1 = a$	Multiplication by 1 does not change any real number. For this reason, 1 is called the multiplicative identity.
$a \cdot 0 = 0$	The product of any real number and 0 is 0.
$ab = ba$	The order of the factors in a multiplication problem does not affect their product.

Another important carryover from arithmetic is the definition of multiplication as repeated addition. For example,

$$3(2) = \underbrace{2 + 2 + 2}_{3 \text{ twos}} = 6.$$

By considering multiplication by a positive integer as repeated addition, a product like $3(-2)$ yields

$$3(-2) = \underbrace{(-2) + (-2) + (-2)}_{3 \text{ negative twos}} = -6.$$

Because we do not want the order of the factors to affect the product, we also have

$$(-2) \cdot 3 = -6.$$

These examples suggest the following rule.

Different-Sign Multiplication

To multiply two real numbers with different signs, multiply their absolute values and make the sign of the product negative.

EXAMPLE 1   Multiply.

a. $7(-3)$ **b.** $(-9)(6)$ **c.** $(0)(-4.6)$ **d.** $(-12)(1)$

Solution

a. $7(-3) = -21$ *Think:* $7 \cdot 3 = 21$ and the sign of the product is negative.
b. $(-9)(6) = -54$ *Think:* $9 \cdot 6 = 54$ and the sign of the product is negative.
c. $(0)(-4.6) = 0$ *Think:* The product of any real number and 0 is 0.
d. $(-12)(1) = -12$ *Think:* Multiplication by 1 does not change any real number.

PROGRESS CHECK 1

a. $5(-12)$ **b.** $(-7)(0)$ **c.** $(1)(-8)$ **d.** $(-4.3)(2)$ ⌐

It is natural to define the product of any two positive numbers as positive. Thus, $2 \cdot 3 = 6$. To determine the sign of the product of two negative numbers, consider how we can build on our current properties by observing the pattern in the following products.

Factor		Product
decreases	$3 \cdot (-3) = -9$	increases
by 1	$2 \cdot (-3) = -6$	by 3
	$1 \cdot (-3) = -3$	
	$0 \cdot (-3) = 0$	

The continuation of this pattern for three more products is shown below.

Factor		Product
decreases	$-1 \cdot (-3) = 3$	increases
by 1	$-2 \cdot (-3) = 6$	by 3
	$-3 \cdot (-3) = 9$	

Thus, the observed pattern leads to results like $-2 \cdot (-3) = 6$ and suggests the following rule.

Same-Sign Multiplication

To multiply two real numbers with the same sign, multiply their absolute values and make the sign of the product positive.

To illustrate in practical terms that the product of two negative numbers is positive, consider the problem in Example 2.

EXAMPLE 2   The following chart analyzes the asset value of a company that declares bankruptcy in 1994 (assumed to be the present) because it continually lost $3 million a year until its asset value was $0.

Year	Number of years from present	·	Change each year (in millions)	=	Asset value (in millions)		
1997 Factor decreases	3		(−3)	=	−9	Predicted future asset value	Product increases by 3
1996 by 1	2		(−3)	=	−6		
1995	1		(−3)	=	−3		
1994 (present)	0		(−3)	=	0	Declares bankruptcy	
1993	−1		(−3)	=	3	Past asset value	
1992	−2		(−3)	=	6		
1991	−3		(−3)	=	9		

Continue this analysis and determine the asset value of the company in 1990. Also, predict the future asset value in the year 2000.

Solution Assuming that 1994 is the present, then 1990 is 4 years in the past, represented by -4, and 2000 is 6 years in the future, represented by $+6$. Then we have the following asset values.

$$\text{Asset value in 1990:} \quad -4(-3) = 12$$
$$\text{Asset value in 2000:} \quad 6(-3) = -18$$

In 1990 the company had $12 million in assets, and if previous trends continue, by the year 2000 the company will be $18 million in debt.

PROGRESS CHECK 2   Assume that 1994 is the present and that the current asset value of a small company is $0 because it has continuously lost $10,000 a year for the past 10 years. What was the asset value of the company in 1987? If previous trends continue, predict the asset value of the company in 1999.

EXAMPLE 3 Multiply.

a. $(-6)(-5)$ **b.** $(-1)(-7.4)$ **c.** $(-\frac{1}{2})(-\frac{3}{4})$

Solution

a. $(-6)(-5) = 30$ *Think:* $6 \cdot 5 = 30$ and the sign of the product is positive.
b. $(-1)(-7.4) = 7.4$ *Think:* $1 \cdot 7.4 = 7.4$ and the sign of the product is positive.
c. $(-\frac{1}{2})(-\frac{3}{4}) = \frac{3}{8}$ *Think:* $\frac{1}{2} \cdot \frac{3}{4} = \frac{3}{8}$ and the sign of the product is positive.

PROGRESS CHECK 3 Multiply.

a. $(-9)(-7)$ **b.** $(-4)(-1.3)$ **c.** $(-\frac{5}{7})(-\frac{2}{3})$

2   When a product contains more than two factors, it is helpful to use the following sign rule.

Sign Rule

A product of nonzero factors is positive if the number of negative factors is even. The product is negative if the number of negative factors is odd.

Progress Check Answers

2. 1987: $70,000; 1999: $−$50,000

3. (a) 63   (b) 5.2   (c) $\frac{10}{21}$

When all the factors are positive, the product is obviously positive. This situation is covered by the sign rule since the number of negative factors in this case is zero (an even number). Note also that the sign rule applies to a product of *nonzero* factors. If one or more factors are 0, then the product is 0.

EXAMPLE 4 Multiply.

a. $(-5)(3)(-4)$ **b.** $(-2)^3$

Solution Consider carefully both methods shown. In practice, method 2 is generally used.

a. *Method 1:*
$$(-5)(3)(-4) = (-15)(-4)$$
$$= 60$$

Multiply -5 and 3.
Then multiply -15 and -4.

Method 2:
$$(-5)(3)(-4) = 60$$

Think: $5 \cdot 3 \cdot 4 = 60$ and the sign of the product is positive because there are an even number of negative factors.

b. *Method 1:*
$$(-2)^3 = (-2)(-2)(-2)$$
$$= (4)(-2)$$
$$= (-8)$$

Natural number exponent definition.
Multiply -2 and -2.
Then multiply 4 and -2.

Method 2:
$$(-2)^3 = -8$$

Think: $2^3 = 8$ and the sign of the product is negative because there are an odd number of negative factors.

Note Use the power key $\boxed{y^x}$ on your calculator to compute $(-2)^3$, as shown below.

$$2 \boxed{+/-} \quad \boxed{y^x} \quad 3 \boxed{=}$$

Many calculators will display the correct answer, -8. However, some calculators restrict usage of the power key to a positive base and display an error message for the above sequence. In this case you must compute 2^3 and then change the sign to negative yourself according to the sign rule (because there are an odd number of negative factors).

PROGRESS CHECK 4 Multiply.

a. $7(-2)(-6)$ **b.** $(-3)^2$

3 Now that we have covered multiplication and evaluating powers for signed numbers, be careful to follow the order of operations given in Section 1.2 when evaluating expressions.

EXAMPLE 5 Evaluate each expression.

a. $3(-5) - (-4)(7)$ **b.** $(-6 + 2)^2 + (1 - 4)^2$

Solution

a. $3(-5) - (-4)(7) = -15 - (-28)$
$$= -15 + 28$$
$$= 13$$

Multiply.
$a - b = a + (-b)$.
Add.

b. $(-6 + 2)^2 + (1 - 4)^2 = (-4)^2 + (-3)^2$
$$= 16 + 9$$
$$= 25$$

Operate within parentheses.
Evaluate powers.
Add.

PROGRESS CHECK 5 Evaluate each expression.

a. $7(-2) - (-1)(-7)$ **b.** $(-9 + 4)^2 + (-3 + 3)^2$

EXAMPLE 6   Evaluate each expression given that $x = 4$ and $y = -3$.

  a. $2x + 5y$ **b.** $y(1 - y)$ **c.** $-16x^2 + 256$

Solution

  a. $2x + 5y = 2(4) + 5(-3)$ Replace x by 4 and y by -3.
  $\qquad\qquad = 8 + (-15)$ Multiply.
  $\qquad\qquad = -7$ Add.

  b. $y(1 - y) = -3[1 - (-3)]$ Replace y by -3.
  $\qquad\qquad = -3[4]$ Operate within brackets.
  $\qquad\qquad = -12$ Multiply.

  c. $-16x^2 + 256 = -16(4)^2 + 256$ Replace x by 4.
  $\qquad\qquad = -16(16) + 256$ Evaluate a power.
  $\qquad\qquad = -256 + 256$ Multiply.
  $\qquad\qquad = 0$ Add.

PROGRESS CHECK 6   Evaluate each expression given that $x = 4$ and $y = -3$.

  a. $4y - 2x$ **b.** $x(1 - x)$ **c.** $-9y^2 + 100$ ⌐

4 Applied problems involving multiplication of real numbers are illustrated by Example 2 from this section and by the section-opening problem.

EXAMPLE 7   Solve the problem in the section introduction on page 43.

Solution   Substitute 3 for t in the given formula and simplify.

$$V = -1{,}400t + 7{,}000 \qquad \text{Given formula.}$$
$$= -1{,}400(3) + 7{,}000 \qquad \text{Replace } t \text{ by 3.}$$
$$= -4{,}200 + 7{,}000 \qquad \text{Multiply.}$$
$$= 2{,}800 \qquad \text{Add.}$$

After three years the value of the equipment is $2,800 for depreciation purposes.

PROGRESS CHECK 7   Find the value of the equipment considered in Example 7 after four years.   ⌐

EXERCISES 1.7

In Exercises 1–24, compute the given product.

  1. $4(-3)$ **2.** $5(-4)$ **3.** $-6(16)$
  4. $-2(2)$ **5.** $0(\frac{3}{4})$ **6.** $-344(0)$
  7. $-4(-5)$ **8.** $-9(-9)$ **9.** $-1(-5)$
  10. $-6.7(-1)$ **11.** $\frac{2}{3}(-\frac{6}{7})$ **12.** $-\frac{8}{3}(-\frac{3}{8})$
  13. $(-2)(3)(-4)$ **14.** $4(-5)(6)$
  15. $-3(-\frac{4}{5})(-\frac{10}{3})$ **16.** $(\frac{5}{9})(-\frac{3}{5})(-\frac{2}{3})$
  17. $-[-2(7)(3)]$ **18.** $-[-9(3)(-2)]$
  19. $(-1)^5$ **20.** $(-1)^6$
  21. -2^6 **22.** $(-2)^6$
  23. $-[(3)(-3)(3)(-3)]$ **24.** $-[(-3)(4)(-5)(6)]$

In Exercises 25–48, evaluate the given expression.

  25. $4(5 - 9)$ **26.** $-3(-8 + 1)$
  27. $5 + 6(7 - 8)$ **28.** $-2 + 4(-3 + 3)$
  29. $7 - 2(1 - 5)$ **30.** $8 - 3(2 - 5)$
  31. $(5 - 6)(7 - 8)$ **32.** $(-2 - 4)(-3 + 3)$
  33. $-[3 + (-4)][2 + (-5)]$

  34. $-(-5 + 6)(-5 - 9)$
  35. $-8(3) + (-2)(5)$
  36. $12(-4) + 8(-4)$
  37. $-6(-2) - (-9)(10)$
  38. $-5(-5) - (-5)(-5)$
  39. $-5(6) - (-5)(2) + (-5)(2)$
  40. $7(-1) - 3(-1) - 4(-1)$
  41. $(-5 + 8)^2 + (-4 + 4)^3$
  42. $(-3 + 2)^2 + (-1 - 1)^2$
  43. $[5 - (-5)]^3 - [4 - (-4)]^2$
  44. $(4 - 9)^3 + (9 - 4)^3$
  45. $-[-3 + (-3)][-7 + 10]^2$
  46. $-[-2 + (-2)][-7 - 1]^2$
  47. $[(-1)^2 + (-2)^2](3 - 4)^2$
  48. $5[(2 - 1)^2 - (1 - 2)^2]$

In Exercises 49–66, evaluate each expression given that $x = 3$ and $y = -2$.

  49. $3x + 4y$ **50.** $2x + 3y$

51. $3 + xy$

52. $3 - xy$

53. $-1 + 2(x + y)$

54. $3 + 4(x + y)$

55. $x - y^2$

56. $(x - y)^2$

57. $x + y(x + y^2)$

58. $x - y(x - y^2)$

59. $(x - y)(x + y)$

60. $(x + y)^2$

61. $3y^2 - 5y + 10$

62. $-3y^2 + 5y - 10$

63. $x^2 + xy + y^2$

64. $x^2 - xy - y^2$

65. $x^3 - y^3$

66. $-y^3 + x^3$

67. If a represents a negative number, what is the sign of a^{99}?

68. If a represents a negative number, what is the sign of a^{100}?

69. Which one of these expressions equals -16: -2^4 or $(-2)^4$?

70. Which one of these expressions is larger: -3^4 or -3^5?

71. True or false? The opposite of 3^5 equals the fifth power of negative 3.

72. True or false? The opposite of 3^4 equals the fourth power of negative 3.

In Exercises 73–78, find a replacement for x that will make the given expression equal to zero.

73. $7x$

74. $1.46x$

75. $5(x - 4)$

76. $-7(x - 5)$

77. $-12(x + 2)$

78. $11(x + 8)$

In Exercises 79–84, translate the verbal expression into an algebraic expression; then simplify.

79. 5 times 3 more than -4

80. 6 times 8 less than 5

81. The product of 5 more than 5 and 6 less than 5

82. 7 times the product of 10 and its additive inverse

83. The negative of the sum of the square and the cube of -3

84. The negative of the square of the product of -3 and 4

85. Complete this multiplication table.

$\cdot$	0	1	2	3
0				
-1				
-2				
-3				

86. Complete this multiplication table.

$\cdot$	0	-1	-2	-3
0				
-1				
-2				
-3				

In Exercises 87–92, compute the product $a(-b)$ and the difference $a - b$ for the given values of a and b.

87. $a = 1, b = 2$

88. $a = 1, b = -2$

89. $a = -3, b = 4$

90. $a = -3, b = -4$

91. $a = 0, b = 5$

92. $a = 0, b = -5$

93. Find the number in this list, $\{0,2,4\}$, that makes the expression $x^2 - 6x + 8$ the largest.

94. Find the number in this list, $\{0,3,-3\}$, that makes the expression $x^2 - 6x$ the smallest.

In Exercises 95–100, write down an expression for the perimeter and the area. Then evaluate that expression for the given value of x.

95. $x = -1$

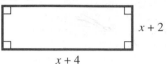

$x + 2$

$x + 4$

96. $x = -1$

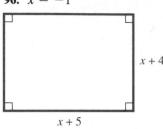

$x + 4$

$x + 5$

97. $x = -2$

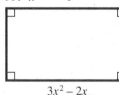

$x^2 - 2$

98. $x = -3$

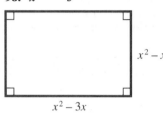

$x^2 - 3$

$x^2 - x$

$x^2 - 3x$

99. $x = -1$

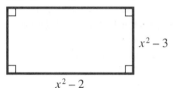

$3x^2 - 2x$

100. $x = -2$

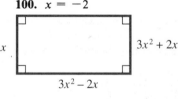

$2x^2 - x$

$3x^2 + 2x$

$3x^2 - 2x$

In Exercises 101–110, solve the given problem.

101. Assume that 1994 is the present year and that the current asset value V of a small company is $0 because it has continually lost $20,000 a year for the past eight years. What was the asset value of the company in 1989? If this trend continues, what will be the asset value in 1997? You can use the formula

$$V = -20{,}000t,$$

where positive numbers for t represent years in the future and negative numbers for t represent years in the past.

102. Assume that 1994 is the present year, that the current asset value V of a small company is $38,000, and that it has continually lost $7,500 a year for the past five years. What was the asset value of the company in 1991? If this trend continues, what will be the asset value in 2000? You can use the formula

$$V = -7{,}500t + 38{,}000,$$

where positive numbers for t represent years in the future and negative numbers for t represent years in the past.

103. In some problems involving rocket flight, the sign of the velocity indicates whether the rocket is going up or down. Here is a formula for the velocity in one such problem: velocity $= -32t + 300$, where t is time elapsed in seconds and velocity is given in feet per second. Evaluate the formula when $t = 0, 9$, and 10 seconds, and decide in each case if the rocket is on its way up or on its way down. Time equal to zero is the moment of blast off.

104. In some problems involving rocket flight, the sign of the velocity indicates whether the rocket is going up or down. Here is a formula for the velocity in one such problem: velocity $= -32t + 160$, where t is time elapsed in seconds and velocity is given in feet per second. Evaluate the formula when $t = 4, 5$, and 6 seconds, and decide in each case if the rocket is on its way up or on its way down. When the velocity equals zero, the rocket is not moving up or down but is at its highest point.

105. Assuming certain depreciation rules, the value V of an $18,000 automobile after t years is given by the formula
$$V = -3,600t + 18,000,$$
where t is an integer less than or equal to 5. Find the value of the automobile after three years.

106. The value V of a piece of manufacturing equipment depreciates to its salvage value according to the formula
$$V = -35,000t + 450,000,$$
where t is an integer less than or equal to 10. Find the value of the equipment after eight years. In how many years will it be worth less than half its original value?

107. In some experiments time equal to zero represents one particular moment in the experiment, whereas negative and positive time indicate time before and after that, respectively. In one such experiment the formula for the predicted value on a meter is given by $(t + 9)(t - 6)t$. Find the predicted value on the meter 7 seconds before zero, at time 0, and 7 seconds after zero.

108. In some experiments time equal to zero represents one particular moment in the experiment, whereas negative and positive time indicate time before and after that, respectively. In one such experiment the formula for the position of a particle on a marked track is given by $t(t + 0.3)(t - 1.5)$. Find the position of the particle 0.1 second before zero.

109. Some economists have a formula for predicting the behavior of a stock market index I near the time of a crash. Time equal to zero represents the day of the crash. The formula is
$$I = 1,000 + t(t - 10)(t + 20),$$
where t represents days before $(-)$ or after $(+)$ the crash. Use the formula to predict the value of the index five days before and five days after the crash.

110. In statistics, observations X are often converted to scaled scores which are useful for making certain comparisons. The formula for recovering the observed scores from the scaled scores is
$$X = \mu + z\sigma.$$
Here z is the scaled score, μ is the mean of all the observed scores, and σ is a measure of the variability among the observed scores. In one such study $\mu = 408$ and $\sigma = 18$. Use this information to recover the observations which correspond to z scores of $-2.5, -1.5$, and 0.

THINK ABOUT IT

1. There is a famous vaudeville joke where a salesman says, "I lose a little on each item, but I make it up in the volume." If his loss on each item is 30 cents and he sells 600 items, use a multiplication expression to show that this may be a funny joke, but it's not funny business. What is the net loss or gain?
2. How many different ways can you arrange and group the three numbers 2, -3, and 5 to get a product of -30?
3. Why is $(a - b)^2 = (b - a)^2$ no matter what numbers you use for a and b?
4. a. How can 3 times a number be equal to the original number?
   b. How can 3 times a number be less than the original number?

5. The numbers in this square are arranged according to a trick. Pick any three numbers in such a way that no two are in the same row and column, and compute their product. You will always get the same answer. Can you explain how this works? Can you make up another similar square?

4	-8	12
5	-10	15
-6	12	-18

REMEMBER THIS

1. Rewrite $x - a$ as an addition expression.
2. Rewrite $x + (-4)$ as a subtraction expression.
3. From the sum of 6 and -8, subtract -7.
4. Evaluate $x - (y - z)$ when $x = -1, y = -2$, and $z = -3$.

5. Find the area of the shaded region.

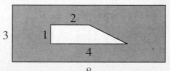

6. Find the additive inverse and the multiplicative inverse of 5.

7. Why is $\frac{3}{5}$ called a "rational" number?

8. Write an expression for the absolute value of the product of x and y.

9. For which of these pairs does the sum equal the product?
   a. 2,2   **b.** $\frac{3}{2}$, 3   **c.** $-3, \frac{3}{4}$ **d.** $-2, \frac{2}{3}$

10. Find the perimeter and the area for this right triangle.

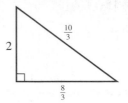

1.8   Division of Real Numbers

An investor's monthly statement from a brokerage company computes the account's percent change using the formula

$$\frac{\text{percent}}{\text{change}} = \frac{\text{closing net value} - \text{opening net value}}{\text{opening net value}} \cdot 100.$$

If the net value of an account opens the month at \$44,965.11 and closes at \$44,431.50, find the percent change to the nearest hundredth of a percent. (See Example 7.)

OBJECTIVES

1 Divide with nonzero real numbers.

2 Evaluate quotients involving 0.

3 Evaluate quotients by following the order of operations.

4 Find the reciprocal (or multiplicative inverse) of a nonzero number.

5 Divide using the reciprocal definition of division.

6 Solve applied problems involving division of real numbers.

1   In arithmetic, where we work just with positive numbers, the operation of division is defined in terms of multiplication. We will carry over these definitions and extend them to the division of all real numbers.
   Recall that

$$\frac{8}{2} = 4 \qquad \text{because} \qquad 8 = 2 \cdot 4,$$

$$\frac{5.5}{1.1} = 5 \qquad \text{because} \qquad 5.5 = 1.1(5).$$

Extending the definition that

$$\frac{a}{b} = c \qquad \text{means} \qquad a = b \cdot c$$

to quotients of real numbers yields results like

$$\frac{-8}{-2} = 4 \qquad \text{because} \qquad -8 = -2(4),$$

$$\frac{-8}{2} = -4 \qquad \text{because} \qquad -8 = 2(-4),$$

$$\frac{8}{-2} = -4 \qquad \text{because} \qquad 8 = -2(-4).$$

These examples suggest that if we take our procedure for multiplying real numbers and replace "multiply" by "divide," then we obtain a procedure for dividing real numbers.

Quotients of Nonzero Real Numbers

Same-Sign Division   To divide with two real numbers which have the same sign, divide their absolute values and make the sign of the quotient positive.
Different-Sign Division   To divide with two real numbers which have different signs, divide their absolute values and make the sign of the quotient negative.

EXAMPLE 1   Divide.

a. $\dfrac{-12}{-4}$
b. $\dfrac{80}{-16}$
c. $-18 \div 6$
d. $\dfrac{-10.5}{2.1}$

Solution

a. $\dfrac{-12}{-4} = 3$ *Think:* $\dfrac{12}{4} = 3$ and the sign of the quotient is positive.

b. $\dfrac{80}{-16} = -5$ *Think:* $\dfrac{80}{16} = 5$ and the sign of the quotient is negative.

c. $-18 \div 6 = -3$ *Think:* $18 \div 6 = 3$ and the sign of the quotient is negative.

d. $\dfrac{-10.5}{2.1} = -5$ *Think:* $\dfrac{10.5}{2.1} = 5$ and the sign of the quotient is negative.

PROGRESS CHECK 1   Divide.

a. $\dfrac{-90}{10}$
b. $(-9) \div (-3)$
c. $\dfrac{-5}{5}$
d. $\dfrac{12.8}{-3.2}$ ⌐

2   Evaluating quotients involving the number 0 deserves special attention. Consider the cases of 0/4, 4/0, and 0/0.

$$\frac{0}{4} = 0 \text{ because } 0 = 4 \cdot 0.$$

$$\frac{4}{0} \text{ is undefined because no number times 0 is 4.}$$

$$\frac{0}{0} \text{ is undefined because any number times 0 is 0.}$$

These examples illustrate the two outcomes we will use to evaluate a quotient involving 0.

> **Division Involving 0**
>
> **1.** 0 divided by any nonzero number is 0.
> **2.** Division by 0 is undefined.

EXAMPLE 2 Evaluate each quotient.

a. $\dfrac{0}{-4}$ 　　　　　　 b. $7 \div 0$ 　　　　　　 c. $\dfrac{0}{-5 + 5}$

Solution

a. $\dfrac{0}{-4} = 0$ 　　　　　　　　 *Think:* 0 divided by any nonzero number is 0.

b. $7 \div 0$ is undefined. 　　　　　　 *Think:* Division by 0 is undefined.

c. $\dfrac{0}{-5 + 5} = \dfrac{0}{0}$ is undefined. 　　 *Think:* Division by 0 is undefined.

PROGRESS CHECK 2 Evaluate each quotient.

a. $\dfrac{-3}{0}$ 　　　　　　 b. $\dfrac{0}{0}$ 　　　　　　 c. $\dfrac{0}{-3 - 3}$

3 When working a division problem in fraction form, notice that expressions like $-\dfrac{6}{2}$, $\dfrac{-6}{2}$, and $\dfrac{6}{-2}$ are all equal to -3. And, in general, if a and b are any real numbers with $b \neq 0$, then

$$-\frac{a}{b} = \frac{a}{-b} = \frac{-a}{b}.$$

Although all of these forms represent the same number, the form $\dfrac{a}{-b}$ is rarely used to express quotients, as shown in the following examples.

EXAMPLE 3 Evaluate each expression.

a. $\dfrac{3 - (-5)}{-1 - 2}$ 　　　　　　 b. $\dfrac{4(2 - 7)}{10 - 4^2}$

Solution The numerator and the denominator are simplified independently. Then, simplify the resulting fraction.

a. $\dfrac{3 - (-5)}{-1 - 2} = \dfrac{3 + 5}{-1 + (-2)}$ 　　 $a - b = a + (-b)$.

$= \dfrac{8}{-3}$ 　　　　　　　 Add.

$= -\dfrac{8}{3}$ 　　　　　　　 $\dfrac{a}{-b} = -\dfrac{a}{b}$.

b. $\dfrac{4(2 - 7)}{10 - 4^2} = \dfrac{4(-5)}{10 - 16}$ 　　 Operate within parentheses in the numerator; evaluate a power in the denominator.

$= \dfrac{-20}{-6}$ 　　　　　 Multiply in the numerator; subtract in the denominator.

$= \dfrac{10}{3}$ 　　　　　　 $\dfrac{20}{6} = \dfrac{10}{3}$ and the sign of the quotient is positive.

Progress Check Answers

2. (a) Undefined 　 (b) Undefined 　 (c) 0

PROGRESS CHECK 3 Evaluate each expression.

a. $\dfrac{-1 - (-2)}{4 - 7}$

b. $\dfrac{4^2 - 5^2}{3(2 - 6)}$

EXAMPLE 4 Evaluate each expression given that $x = -5$ and $y = 2$.

a. $\dfrac{x - y}{y - x}$

b. $\dfrac{3}{x + 5}$

c. $\dfrac{x^2 + y^2}{x + y}$

Solution

a. $\dfrac{x - y}{y - x} = \dfrac{-5 - 2}{2 - (-5)}$ Replace x by -5 and y by 2.

$= \dfrac{-7}{7}$ Subtract.

$= -1$ Divide.

b. $\dfrac{3}{x + 5} = \dfrac{3}{-5 + 5}$ Replace x by -5.

$= \dfrac{3}{0}$ Add.

Because any division by 0 is undefined, the given expression is undefined when $x = -5$.

c. $\dfrac{x^2 + y^2}{x + y} = \dfrac{(-5)^2 + 2^2}{-5 + 2}$ Replace x by -5 and y by 2.

$= \dfrac{25 + 4}{-3}$ Evaluate powers in the numerator; add in the denominator.

$= \dfrac{29}{-3}$ Add in the numerator.

$= -\dfrac{29}{3}$ $\dfrac{a}{-b} = -\dfrac{a}{b}$.

PROGRESS CHECK 4 Evaluate each expression given that $x = -5$ and $y = 2$.

a. $\dfrac{2 - y}{y - 2}$

b. $\dfrac{x + y}{y + x}$

c. $\dfrac{x^2 - y^2}{x - y}$

4   As in Section 1.1, division of fractions involves the concept of reciprocals. Two numbers are called **reciprocals** of each other if the product of the two numbers is 1. For example, 5 and $\frac{1}{5}$ are reciprocals (of each other) and $-\frac{2}{3}$ and $-\frac{3}{2}$ are reciprocals, because

$$5 \cdot \tfrac{1}{5} = 1 \quad \text{and} \quad (-\tfrac{2}{3})(-\tfrac{3}{2}) = 1.$$

In general form, the reciprocal of a nonzero number a is $1/a$ and the reciprocal of the nonzero fraction a/b is b/a. The number 0 has no reciprocal, because no number multiplied by 0 is 1. The reciprocal of a is also called the **multiplicative inverse** of a.

EXAMPLE 5 Find the reciprocal (or multiplicative inverse) of each number.

a. -4

b. $\frac{1}{8}$

c. $\frac{7}{9}$

Solution   Note that a number and its reciprocal have the same sign.

a. The reciprocal of -4 is $\frac{1}{-4}$ or $-\frac{1}{4}$, because $(-4)(-\frac{1}{4}) = 1$.
b. The reciprocal of $\frac{1}{8}$ is 8, because $\frac{1}{8} \cdot 8 = 1$.
c. The reciprocal of $\frac{7}{9}$ is $\frac{9}{7}$, because $\frac{7}{9} \cdot \frac{9}{7} = 1$.

Progress Check Answers

3. (a) $-\frac{1}{3}$ (b) $\frac{3}{4}$

4. (a) Undefined   (b) 1   (c) -3

PROGRESS CHECK 5 Find the reciprocal (or multiplicative inverse) of each number.

a. $\frac{1}{7}$ **b.** -6 **c.** $\frac{5}{8}$ ⌐

5 We may now define division in terms of multiplication by using the concept of a reciprocal.

Definition of Division

If a and b are any real numbers with $b \neq 0$, then

$$a \div b = a \cdot \frac{1}{b}.$$

In words, to divide a by b, multiply a by the reciprocal of b. From the definition, you can see that the sign in division is determined in the same manner as in multiplication.

EXAMPLE 6 Divide using the reciprocal form of division.

a. $(-9) \div (-3)$ **b.** $\left(-\frac{2}{5}\right) \div 7$ **c.** $\left(-\frac{7}{12}\right) \div \left(-\frac{5}{9}\right)$

Solution

a. $(-9) \div (-3) = -9 \cdot \left(-\frac{1}{3}\right)$ Product of -9 and the reciprocal of -3.

$\qquad\qquad\qquad = 3$ Multiply.

b. $-\frac{2}{5} \div 7 = -\frac{2}{5} \cdot \frac{1}{7}$ Product of $-\frac{2}{5}$ and the reciprocal of 7.

$\qquad\qquad = -\frac{2}{35}$ Multiply.

c. $\left(-\frac{7}{12}\right) \div \left(-\frac{5}{9}\right) = \left(-\frac{7}{12}\right) \cdot \left(-\frac{9}{5}\right)$ Product of $-\frac{7}{12}$ and the reciprocal of $-\frac{5}{9}$.

$\qquad\qquad\qquad\qquad = \frac{63}{60}$ Multiply.

$\qquad\qquad\qquad\qquad = \frac{21}{20}$ Simplify.

PROGRESS CHECK 6 Divide using the reciprocal definition of division.

a. $(-8) \div 4$ **b.** $\left(\frac{5}{6}\right) \div (-8)$ **c.** $\left(-\frac{5}{9}\right) \div \left(-\frac{7}{12}\right)$

6 To solve the application problem that opens the section, you will want to use a calculator.

EXAMPLE 7 Solve the problem in the section introduction on page 50.

Solution Substitute the opening and closing net values in the given formula and simplify.

$$\text{Percent change} = \frac{44{,}431.50 - 44{,}965.11}{44{,}965.11} \cdot 100$$

$$= \frac{-533.61}{44{,}965.11} \cdot 100 \qquad \text{Subtract by calculator.}$$

$$= (-0.0118672)(100) \qquad \text{Divide by calculator.}$$

$$= -1.18672 \qquad \text{Multiply.}$$

Progress Check Answers

5. (a) 7   (b) $-\frac{1}{6}$ (c) $\frac{8}{5}$

6. (a) -2 (b) $-\frac{5}{48}$ (c) $\frac{20}{21}$

To the nearest hundredth of a percent, the change in the account is -1.19 percent.

PROGRESS CHECK 7   The net value of an account opens the month at $12,017.58 and closes at $11,459.08. Use the formula in Example 7 to find the percent change for the month to the nearest hundredth of a percent.

Progress Check Answer

7. -4.65 percent

EXERCISES 1.8

In Exercises 1–42, do the indicated division. Check your result by doing the appropriate multiplication.

1. $\dfrac{-24}{6}$

2. $\dfrac{-81}{81}$

3. $\dfrac{119}{-1}$

4. $\dfrac{391}{-23}$

5. $\dfrac{-48}{-12}$

6. $\dfrac{-60}{-5}$

7. $\dfrac{-14.4}{-0.12}$

8. $\dfrac{-3}{-0.12}$

9. $\dfrac{-3(-4)}{2}$

10. $\dfrac{-15(14)}{6}$

11. $0 \div 4.692$

12. $0 \div (-8)$

13. $\dfrac{-17}{0}$

14. $\dfrac{1}{0}$

15. $\dfrac{3}{2} \div \dfrac{-2}{3}$

16. $\dfrac{4}{5} \div \dfrac{-4}{5}$

17. $\dfrac{4-7}{3}$

18. $\dfrac{8-20}{-4}$

19. $\dfrac{3-(-1)}{-2}$

20. $\dfrac{5-(-7)}{8}$

21. $\dfrac{-1-9}{-2}$

22. $\dfrac{-25-11}{12}$

23. $[-3-(-4)] \div 2$

24. $[0-(-12)] \div (-6)$

25. $\dfrac{-12-12}{-3-3}$

26. $\dfrac{0-4}{4-0}$

27. $\dfrac{1-(-2)}{3-(-3)}$

28. $\dfrac{5-(-4)}{4-(-5)}$

29. $\dfrac{-5(2-3)}{4-(-1)}$

30. $\dfrac{-8(5-7)}{3-(-5)}$

31. $\dfrac{3-5^2}{20-3^2}$

32. $\dfrac{-5-3^2}{2(16-3^2)}$

33. $\dfrac{(-2)(-3)+(2)(-3)}{1+3.9(-5.5)}$

34. $\dfrac{6(-4)-12(-2)}{3+4(5)}$

35. $\dfrac{-1+2(3)}{1+2(-3)}$

36. $\dfrac{-7-6(7)}{(-5+6)7}$

37. $\dfrac{(1+2+3)^2}{(1+2-3)^2}$

38. $\dfrac{3^2+4^2+5^2}{3^2+4^2-5^2}$

39. $\dfrac{-(-4-8)^2}{-1+10}$

40. $\dfrac{-(2-5)^2}{(-2+5)^2}$

41. $\dfrac{24-(-6-2)}{[24-(-6)]-2}$

42. $\dfrac{-60-[30-(-2)]}{[-60-30]-(-2)}$

In Exercises 43–48, evaluate each expression given that $x = -2$ and $y = -3$.

43. $\dfrac{x-2y}{2y-x}$

44. $\dfrac{x+2y}{2x+y}$

45. $\dfrac{2(x-y)}{2x-y}$

46. $\dfrac{-2(x-y)}{-2x-y}$

47. $\dfrac{x^2+y^2}{x^2-y^2}$

48. $\dfrac{2x^2-3y^2}{(2x-3y)^2}$

In Exercises 49–60, find the reciprocal of the given number. Then check the answer by showing that the product of the two reciprocal values equals 1.

49. -5

50. -1

51. $-\frac{2}{7}$

52. $-\frac{8}{3}$

53. $\frac{1}{3}$

54. $\frac{1}{10}$

55. $-\frac{5}{4}$

56. $-\frac{4}{9}$

57. 1.6

58. -0.5

59. -1.25

60. 0.25

In Exercises 61–66, do each problem two ways: first as division, second as multiplication by the reciprocal. Show that both answers are the same.

61. $10 \div (-2)$

62. $16 \div (-8)$

63. $-12 \div 3$

64. $0 \div 5$

65. $\dfrac{15}{4} \div \dfrac{3}{2}$

66. $\dfrac{25}{12} \div \dfrac{5}{4}$

In Exercises 67–72, rewrite any division expressions as multiplication expressions and rewrite any multiplication expressions as division expressions.

67. $a \div b$

68. $-c \div d$

69. $\dfrac{x}{4f}$

70. $\dfrac{-v}{x}$

71. $5 \cdot \dfrac{1}{x}$

72. $-x \cdot \dfrac{1}{ab}$

In Exercises 73–78, fill in the blanks to make a true statement.

73. $\dfrac{15}{-3} = \dfrac{}{3}$

74. $\dfrac{-24}{8} = \dfrac{24}{}$

75. $-\dfrac{10}{7} = \dfrac{-10}{}$

76. $-\dfrac{1}{3} = \dfrac{}{-3}$

77. $\dfrac{-42}{-7} = -\dfrac{-42}{}$

78. $\dfrac{18}{6} = \dfrac{-18}{}$

In Exercises 79–82 you can solve the word problems by using the formula

$$\text{percent change} = \frac{\text{new value} - \text{old value}}{\text{old value}} \cdot 100.$$

In each problem, round the percent change to the nearest hundredth of a percent.

79. Over the period of one year the population of a town increases from 11,985 to 12,250. Show the percent change for the population. The town clerk assumes that this percent change is a good estimate of the population change for the year to come. On this basis what is the projected population for the town after one more year?

80. Over the period of one year the value of an investment decreases from $87,000 to $75,000. Find the percent change for the investment. At this rate of change what would be the loss on an investment of $100,000?

81. Over the period of one year the registered deer kill in a state decreases from 4,200 to 3,700. Express this as a percent change. The wildlife officer assumes that this percent change is a good estimate of the percent change in the total deer population as well. If the old population size was about 10,000 deer, what is a good estimate of the new population size?

82. A company ranks TV shows on a special rating scale. Over the period of one month the rating for a certain TV show has dropped from 45.8 to 42.3. Express this drop as a percent change and round to the nearest hundredth of a percent. If the show used to have about 30 million viewers, what is the current estimate of the number of viewers?

THINK ABOUT IT

1. Use $a = 10$ and $b = 2$ to illustrate the property that

$$\frac{a}{-b} = \frac{-a}{b} = -\frac{a}{b}.$$

2. Why doesn't zero have a reciprocal?

3. Find all numbers that are equal to their own reciprocals.

4. In the problem $\frac{35}{7} = 5$ which number is called the quotient? Which is the divisor?

5. John said $\frac{0}{0}$ is 1 because any number divided by itself is 1. Jill said $\frac{0}{0}$ is 0 because zero divided by any number is zero. Jean said $\frac{0}{0}$ is 17 because $17 \cdot 0 = 0$. Who is right?

REMEMBER THIS

1. Evaluate $-3(4)(-5)$.

2. Evaluate $2 - 3(4 - 5)$.

3. Evaluate $(-1)^{1,002}$.

4. Evaluate $-1(15 - 8)$.

5. If $x = 4$, what is the value of $7(x + 8)(x - 9)(x - 4)$?

6. Translate into an algebraic expression: d times 2 more than $-x$.

7. Find the area and perimeter of a square with side 0.5 cm.

8. Find the shaded area.

3 units   ← Semicircle

4 units

9. Graph the set $\{2, -1, \frac{1}{2}, -\frac{1}{4}, \frac{1}{8}, -\frac{1}{16}\}$.

10. Express 30 as the product of prime factors.

1.9 Properties of Real Numbers

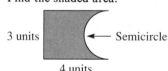

To find the perimeter of the rectangular enclosure shown, a landscaper may choose between the formulas

$$P = 2\ell + 2w \quad \text{or} \quad P = 2(\ell + w)$$

because $2\ell + 2w = 2(\ell + w)$. Which property of real numbers is illustrated by this equation? (See Example 8.)

1   Identify and use commutative and associative properties.

2   Identify and use identity and inverse properties.

3   Identify and use the distributive property.

4   Use $-a = -1 \cdot a$ and the distributive property to remove parentheses.

Unless it is stated otherwise, you may assume in algebra that a symbol like x may be replaced by any real number. Consequently, the rules that govern real numbers determine our methods of computation in algebra. In this section we discuss basic properties of real numbers with respect to addition and multiplication. Keep in mind that all rules in algebra must be in agreement with these properties.

1   In Section 1.5 it was pointed out that the order of the numbers in an addition problem does not affect their sum. Similarly, it was stated in Section 1.7 that we may reorder the factors in a product. These properties are called commutative properties.

Commutative Properties

For any real numbers a and b

$$a + b = b + a$$ Commutative property of addition

$$ab = ba.$$ Commutative property of multiplication

Both commutative properties are illustrated in Example 1.

EXAMPLE 1   If $a = -8$ and $b = 3$, show that

a. $a + b = b + a$ **b.** $ab = ba$

Solution

a. Replace a by -8 and b by 3, and find each sum separately.

$$a + b = -8 + 3 = -5$$
$$b + a = 3 + (-8) = -5$$

Thus, $a + b = b + a$ for the given values.

b. Replace a by -8 and b by 3, and find each product separately.

$$ab = -8(3) = -24$$
$$ba = 3(-8) = -24$$

Thus, $ab = ba$ if $a = -8$ and $b = 3$.

PROGRESS CHECK 1   If $a = -4$ and $b = -3$, show that

a. $a + b = b + a$ **b.** $ab = ba$

Another important pair of properties say that we obtain the same result if we change the grouping of numbers in an addition problem and in a multiplication problem. These properties are called associative properties.

Progress Check Answers

1. (a) Both equal -7.   (b) Both equal 12.

> ## Associative Properties
>
> For any real numbers a, b, and c
>
> $$(a + b) + c = a + (b + c) \qquad \text{Associative property of addition}$$
> $$(ab)c = a(bc). \qquad \text{Associative property of multiplication}$$

Example 2 illustrates these two properties.

EXAMPLE 2   If $a = -5$, $b = 2$, and $c = 7$, show that

a. $(a + b) + c = a + (b + c)$ **b.** $(ab)c = a(bc)$

Solution   According to the order of operations, perform all operations within parentheses first.

a. Find each sum separately using the given values.

$$(a + b) + c = (-5 + 2) + 7 = -3 + 7 = 4$$
$$a + (b + c) = -5 + (2 + 7) = -5 + 9 = 4$$

Thus, for the specified numbers, $(a + b) + c = a + (b + c)$.

b. Find each product separately using the given values.

$$(ab)c = (-5 \cdot 2) \cdot 7 = (-10) \cdot 7 = -70$$
$$a(bc) = -5 \cdot (2 \cdot 7) = -5 \cdot (14) = -70$$

Thus, $(ab)c = a(bc)$ if $a = -5$, $b = 2$, and $c = 7$.

PROGRESS CHECK 2   If $a = 4$, $b = -8$, and $c = 6$, show that

a. $(a + b) + c = a + (b + c)$ **b.** $(ab)c = a(bc)$

The next example will help you distinguish between commutative properties and associative properties. Remember that in sums and products, commutative properties permit a change in *written order,* while associative properties permit a change in *grouping.*

EXAMPLE 3   Label each statement as an example of one of the commutative properties or one of the associative properties.

a. $5(xy) = 5(yx)$ **b.** $(x + 2) + 6 = x + (2 + 6)$

Solution

a. This statement illustrates the commutative property of multiplication because the factors x and y are written in different order on the two sides of the equal sign.

b. This statement illustrates the associative property of addition. On the left side of the equal sign parentheses group x and 2, but on the right side parentheses group 2 and 6.

PROGRESS CHECK 3   Label each statement as an example of one of the commutative properties or one of the associative properties.

a. $2(ax) = (2a)x$ **b.** $3 + (x + 7) = 3 + (7 + x)$

2   Identity properties and inverse properties are also basic properties of the real numbers with respect to addition and multiplication. Once again, the statement of these properties is a formalization of previous work in this chapter.

Progress Check Answers

2. (a) Both equal 2.   (b) Both equal -192.

3. (a) Associative, multiplication   (b) Commutative, addition

Identity and Inverse Properties

	Addition	**Multiplication**
Identity Properties	There is a unique real number 0 such that for every real number a $$a + 0 = 0 + a = a.$$	There is a unique real number 1 such that for every real number a $$a \cdot 1 = 1 \cdot a = a.$$
Inverse Properties	For every real number a, there is a unique real number, denoted by $-a$, such that $$a + (-a) = (-a) + a = 0.$$	For every real number a (except zero), there is a unique real number, denoted by $1/a$, such that $$a \cdot \frac{1}{a} = \frac{1}{a} \cdot a = 1.$$

With respect to these properties, 0 is called the **additive identity,** 1 is called the **multiplicative identity,** $-a$ is called the **additive inverse** (or opposite) of a, and $1/a$ is called the **multiplicative inverse** (or reciprocal) of a.

EXAMPLE 4 Use one of the identity or inverse properties to simplify each statement, and indicate the property that was used.

a. $0 + (-3)$ **b.** $-4(-\frac{1}{4})$ **c.** $-9 \cdot 1$ **d.** $8 + (-8)$

Solution

a. By the addition identity property, $0 + (-3) = -3$.
b. By the multiplication inverse property, $-4(-\frac{1}{4}) = 1$.
c. By the multiplication identity property, $-9 \cdot 1 = -9$.
d. By the addition inverse property, $8 + (-8) = 0$.

PROGRESS CHECK 4 Use one of the identity or inverse properties to simplify each statement, and indicate the property that was used.

a. $1 \cdot (-5)$ **b.** $-5 + 5$ **c.** $-5 + 0$ **d.** $\frac{1}{5} \cdot 5$ ⌟

3 Up to this point the properties have involved only addition or only multiplication. The crucial property that involves both addition and multiplication is the distributive property. To illustrate this property, consider Figure 1.23 and note that we may express the area of the large rectangle as a product or as a sum.

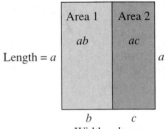

Figure 1.23

$$\text{Area as a product} = \text{length} \cdot \text{width} = a(b + c)$$
$$\text{Area as a sum} = \text{area 1} + \text{area 2} = ab + ac$$

Because both expressions represent the same number, we have

$$a(b + c) = ab + ac.$$

This equation, which relates a product and a sum, expresses the distributive property.

Distributive Property

For any real numbers a, b, and c

$$a(b + c) = ab + ac.$$

Progress Check Answers
4. (a) -5; multiplication identity   (b) 0; addition inverse   (c) -5; addition identity   (d) 1; multiplication inverse

Example 5 illustrates the distributive property in arithmetic terms.

EXAMPLE 5   If $a = 2$, $b = 3$, and $c = 4$, show that $a(b + c) = ab + ac$.

Solution   Evaluate the expression on the left of the equal sign and then on the right.

$$a(b + c) = 2(3 + 4) = 2(7) = 14$$
$$ab + ac = 2(3) + 2(4) = 6 + 8 = 14$$

Thus, $a(b + c) = ab + ac$ for the given values of a, b, and c.

PROGRESS CHECK 5   If $a = 5$, $b = 2$, and $c = 4$, show that $a(b + c) = ab + ac$. ⌐

Using properties and definitions from this chapter, the distributive property may take alternative forms. For instance, by the commutative property

$$a(b + c) = ab + ac \qquad \text{may be expressed as} \qquad (b + c)a = ba + ca,$$

and using the definition of subtraction in terms of addition, we have

$$a(b - c) = ab - ac \qquad \text{and} \qquad (b - c)a = ba - ca.$$

The distributive property also can be extended to forms like

$$a(b + c + d + \cdots + n) = ab + ac + ad + \cdots + an.$$

An important application of the distributive property is to multiply and remove parentheses, as shown in the next example.

EXAMPLE 6   Multiply using the distributive property.

a. $3(x + 4)$ **b.** $(y - 3)6$ **c.** $-7(5x + 2)$ **d.** $-2(x - y + 5)$

Solution   In parts **c** and **d** we can use the subtraction definition $a - b = a + (-b)$.

a. $3(x + 4) = 3 \cdot x + 3 \cdot 4 = 3x + 12$
b. $(y - 3)6 = y \cdot 6 - 3 \cdot 6 = 6y - 18$
c. $-7(5x + 2) = -7(5x) + (-7)(2) = -35x - 14$
d. $-2(x - y + 5) = -2 \cdot x + (-2)(-y) + (-2)(5)$
$$= -2x + 2y - 10$$

PROGRESS CHECK 6   Multiply using the distributive property.

a. $5(x - 9)$ **b.** $(x + 10)8$
c. $-3(4x + 1)$ **d.** $-6(y + x - 1)$ ⌐

All steps shown in Example 6 are reversible. So by considering the distributive property in a form like

$$ab + ac = a(b + c),$$

we may express certain sums (or differences) as products.

EXAMPLE 7   Use the distributive property to rewrite each expression as a product.

a. $5 \cdot 3 + 5 \cdot 4$ **b.** $cx + cy$ **c.** $7x - 2x$

Solution

a. $5 \cdot 3 + 5 \cdot 4 = 5(3 + 4)$
b. $cx + cy = c(x + y)$
c. $7x - 2x = (7 - 2)x$

PROGRESS CHECK 7   Use the distributive property to rewrite each expression as a product.

a. $4 \cdot 5 + 4 \cdot 7$ **b.** $9x - 9y$ **c.** $3y + 4y$ ⌐

EXAMPLE 8   Solve the problem in the section introduction on page 56.

Solution   By the distributive property,

$$2\ell + 2w = 2(\ell + w).$$

Thus, these alternative forms of a perimeter formula show an application of the distributive property.

PROGRESS CHECK 8   Alternative formulas for the area of a trapezoid are $A = \frac{1}{2}h(a + b)$ and $A = \frac{1}{2}(a + b)h$. Is $\frac{1}{2} \cdot h(a + b) = \frac{1}{2} \cdot (a + b)h$ an illustration of the distributive property?   ⌐

4   We may find the opposite of a number by simply changing the sign of the number. Multiplying any number by -1 also only changes the sign of that number. Therefore, for any real number a,

$$-a = -1 \cdot a.$$

This property and the distributive property enable us to remove parentheses, when parentheses are preceded by a $-$ sign.

EXAMPLE 9   Rewrite each statement without parentheses.

a. $-(2x + 5)$ **b.** $-(3 - x)$

Solution

a. $-(2x + 5) = -1 \cdot (2x + 5)$ $-a = -1 \cdot a.$
$\qquad\qquad\quad = -1(2x) + (-1)(5)$ Distributive property.
$\qquad\qquad\quad = -2x - 5$ Simplify.

b. $-(3 - x) = -1 \cdot (3 - x)$ $-a = -1 \cdot a.$
$\qquad\qquad\quad = -1(3) + (-1)(-x)$ Distributive property.
$\qquad\qquad\quad = -3 + x$ Simplify.

Note then that the effect of removing parentheses which follow a negative sign is to change the sign of every term inside the parentheses.

PROGRESS CHECK 9   Rewrite each statement without parentheses.

a. $-(4x - 7)$ **b.** $-(-3 + x)$ ⌐

Progress Check Answers
8. No; commutative property of multiplication
9. (a) $-4x + 7$ (b) $3 - x$

EXERCISES 1.9

In Exercises 1–6, fill in the blank to satisfy a commutative property. State whether you are using the commutative property for addition or for multiplication.

1. $(-a)b = b$ _____
2. $x(-r) = (-r)$ _____
3. $ab + c = c +$ _____
4. $2x + (-2) = -2 +$ _____
5. _____ $(x + y) = (x + y)7$
6. $-4(3x + 5y) = (3x + 5y)$ _____

In Exercises 7–12, use a commutative property to write an expression equivalent to the given one. State whether you are using the commutative property for addition or for multiplication.

7. $v + w$ **8.** $7 + t$
9. $-a + b$ **10.** $x + (-y)$
11. $c(-d)$ **12.** $x(-2)$

In Exercises 13–18, fill in the blank to satisfy an associative property. State whether you are using the associative property for addition or for multiplication.

13. $d(gk) = ($ _____ $)k$
14. $(7x)y = 7($ _____ $)$
15. $7 + (d + y) = (7 + d) +$ _____
16. $($ _____ $) + h = -3 + (2x + h)$
17. $-2s + [4f + (-3d)] = [$ _____ $] + (-3d)$
18. $(-2s + 21f) + (-8d) = -2s + [$ _____ $]$

In Exercises 19–24, use an associative property to write an expression equivalent to the given one. State whether you are using the associative property for addition or for multiplication.

19. $3(xy)$
20. $-4(ax)$
21. $-2u + (7x + s)$

22. $(-10x + 8u) + rt$

23. $3[a(-5h)]$

24. $[4x + (-3)] + g$

In Exercises 25–30, use the given values of x and y to show that

a. $x + y = y + x$ **b.** $xy = yx$

25. $x = -1, y = 2$

26. $x = 10, y = -3$

27. $x = -1.5, y = -2.4$

28. $x = -4, y = 4$

29. $x = \frac{1}{2}, y = -1$

30. $x = -\frac{3}{4}, y = \frac{1}{2}$

In Exercises 31–36, use the given values of $x, y,$ and z to show that

a. $(x + y) + z = x + (y + z)$ **b.** $(xy)z = x(yz)$

31. $x = 3, y = 5, z = 2$

32. $x = 10, y = 2, z = 4$

33. $x = -3, y = -2, z = 2$

34. $x = -1, y = 1, z = 9$

35. $x = \frac{3}{2}, y = -\frac{2}{3}, z = \frac{2}{3}$

36. $x = -\frac{1}{3}, y = \frac{1}{3}, z = -\frac{2}{3}$

In Exercises 37–42, label the statement as an example of one of the commutative or one of the associative properties.

37. $a + (b + c) = a + (c + b)$

38. $a(bc) = a(cb)$

39. $3(dg) = (3d)g$

40. $7 + (a + c) = (7 + a) + c$

41. $-1(3x) = (-1 \cdot 3)x$

42. $2[s(fg)] = (2s)(fg)$

In Exercises 43–54, use identity or inverse properties to simplify the given statement and indicate any property that was used. Assume that n does not represent zero.

43. $0 + (-8)$

44. $-45 + 0$

45. $1 \cdot (-2)$

46. $-6 \cdot 1$

47. $3\left(\frac{1}{3}\right)$

48. $-5\left(-\frac{1}{5}\right)$

49. $-3.1 + 3.1$

50. $\frac{3}{5} + \left(-\frac{3}{5}\right)$

51. $n + (-n)$

52. $n\left(\frac{1}{n}\right)$

53. $\left[-n\left(-\frac{1}{n}\right)\right]b$

54. $(-n + n) - a$

In Exercises 55–60, fill in any blanks to illustrate the distributive property.

55. $3(x + y) = $ _____ $x + $ _____ y

56. $4x(a + b) = $ _____ $a + $ _____ b

57. $5t(c - d) = 5tc$ _____ $5td$

58. $2w(x - y) = $ _____ x _____ _____ y

59. ($ $ _____ $+$ _____ $)v = sv + 3tv$

60. _____ $(q - r) = wq - wr$

In Exercises 61–66, use the given values for $a, b,$ and c to show that $a(b + c) = ab + ac$.

61. $a = 3, b = 4, c = 5$

62. $a = 2, b = 4, c = 6$

63. $a = -1, b = 2, c = 3$

64. $a = -2, b = 1, c = 7$

65. $a = -5, b = 4, c = -4$

66. $a = -10, b = -10, c = 0$

In Exercises 67–90, multiply using the distributive property.

67. $5(a + x)$ **68.** $s(t + w)$

69. $(x + y)7$ **70.** $(x + z)2$

71. $2(3s + t)$ **72.** $5(a + 4c)$

73. $(4x + 2y)6$ **74.** $(5a + 6b)12$

75. $-3(x + 2y)$

76. $-4(4c + 3d)$

77. $(8x + y)(-2)$

78. $(15m + 10n)(-1)$

79. $\frac{1}{5}(10x - 5y)$

80. $\frac{1}{3}(3x + 21)$

81. $10(2.3x - 0.2)$

82. $5(1.2d - 0.4)$

83. $-1(3x + 4)$

84. $-1(-5x - 7y)$

85. $a(3 + t + 5d)$

86. $b(3t - 4r + 5)$

87. $-2(5 - x - xy)$

88. $-3(-a + 2b - cd)$

89. $\frac{2}{3}(3x - 6 - 15y)$

90. $-\frac{5}{4}(8 - 4x - y)$

In Exercises 91–102, use the distributive property to rewrite each expression as a product.

91. $3 \cdot 9 + 3 \cdot 11$ **92.** $23 \cdot 13 - 23 \cdot 3$

93. $3x - 3y$ **94.** $5n + 5m$

95. $7x + 8x$ **96.** $22x - 6x$

97. $2x - 6x$ **98.** $x - 6x$

99. $3d - 3d$ **100.** $-7y + 7y$

101. $2x + 5x - x$ **102.** $y - 2y - 3y$

In Exercises 103–114, rewrite each expression without parentheses.

103. $-(3x + 6)$ **104.** $-(4 + 7y)$

105. $-(x - 3)$ **106.** $-(5 - r)$

107. $-(-4 + 2x)$ **108.** $-(-7y + 1)$

109. $-(2x + 3y - 4z)$

110. $-(x - y - z)$

111. $-(-2x - 5y)$

112. $-(-1 - 2x)$

113. $-(-3 + 2x - x^2)$

114. $-(4y^2 - 2y + 3)$

In Exercises 115–120 we have given a well-known formula and then rewritten it using one or more of the properties described in this section.

a. Name the properties.

b. Evaluate both versions using the given values of the variables. See if one version seems easier to use than the other.

115. Perimeter (P) of a rectangle: $P = 2l + 2w$; $P = 2(l + w)$; $l = 5.7, w = 4.3$

116. Area (A) of a trapezoid: $A = \frac{1}{2}(a + b)h$; $A = \frac{1}{2}h(a + b)$;
$a = 9, b = 6, h = 4$
117. Circumference (C) of a circle (use $\pi = 3.14$): $C = (2\pi)r$;
$C = 2(\pi r)$; $r = 10$
118. Height (y) of a point on a line: $y = mx + b$; $y = b + mx$;
$m = 2, x = 3, b = -5$

119. Sales tax (T) on buying two items: $T = 0.08p_1 + 0.08p_2$;
$T = 0.08(p_1 + p_2)$; $p_1 = \$121, p_2 = \$79$
120. Conversion of Fahrenheit (F) to Celsius (C) temperatures:
$C = \frac{5}{9}(F - 32)$; $C = \frac{5}{9}F - \frac{5}{9} \cdot 32$; $F = 41$

THINK ABOUT IT

1. Many people use a trick to mentally compute a 15 percent tip in a restaurant. The trick is really the distributive property, and it depends on the fact that finding 10 percent of a number mentally is easy. Since 15 is $10 + 5$, and 5 is half of 10, they can use the equation $0.15x = 0.10x + \frac{1}{2} \cdot 0.10x$. For example, 10 percent of \$12 is \$1.20 and half of that is \$0.60, so a 15 percent tip on \$12 is $\$1.20 + \$0.60$, or \$1.80. Use this trick to find 15 percent of the given amounts.
   a. \$200   b. \$160   c. \$36
2. On a strange planet a civilization has a weird arithmetic. One of the operations is $*$. The identity for this operation is @. One of their "numbers" is &. What is the answer to $\& * @$?
3. Why is *distributive* a good name for the property it describes?
4. True or false? You get the additive inverse of a number when you multiply the number by -1. Explain.
5. A number is called an identity for an operation if it does not change the value of other numbers when it is written on the left and also when it is written on the right. For example, 0 is the identity for addition because $0 + n = n$ and also $n + 0 = n$. Explain why 0 is not called the identity for subtraction.

REMEMBER THIS

1. Evaluate $\dfrac{-3(4)}{6}$.
2. Evaluate $\dfrac{-5 + 6 - 1}{-5 + 5 - 2}$.
3. Evaluate $-\dfrac{3}{2} - \left(-\dfrac{3}{2}\right)$.
4. Evaluate $\dfrac{x - 3}{3 - x}$ when $x = 1, 2, 3,$ and 4.
5. Show that $-\dfrac{12}{3}$ is equal to $\dfrac{-12}{3}$ and to $\dfrac{12}{-3}$.
6. Find the percent change for an investment that falls from \$2,500 to \$2,000.
7. Find the reciprocal and also the negative of $-\frac{5}{8}$.
8. Insert the proper symbol ($<, >, =$) to indicate the correct order. -4 _____ -10
9. Express 12 as the product of prime factors.
10. Write an algebraic expression for 15 percent of the sum of x and y.

Chapter 1 SUMMARY

OBJECTIVES CHECKLIST   Specific chapter objectives are summarized below along with numbered example problems from the text that should clarify the objectives. If you do not understand any objectives or do not know how to do the selected problems, then restudy the material.

1.1 **Can you:**
1. **Graph a set of whole numbers?**
   Graph the set of even whole numbers from 0 to 8.   [Example 1]
2. **Identify and use the symbols $=, <, >, \neq, \leq, \geq$?**
   True or false? $2 \geq 2$.   [Example 3b]
3. **Write a number as a product of its prime factors?**
   Express 350 as a product of its prime factors.   [Example 4]
4. **Express a fraction in lowest terms?**
   Express $\frac{12}{28}$ in lowest terms.   [Example 5a]
5. **Multiply, divide, add, and subtract fractions?**
   A share of stock goes from $39\frac{7}{8}$ to $41\frac{1}{4}$ dollars. What is the increase in value of 400 shares?   [Example 10]

1.2 Can you:

1. **Apply the definition of a natural number exponent?**
   Evaluate $3^2 \cdot 2^3$.   [Example 1b]

2. **Evaluate an expression by following the order of operations?**
   Evaluate $7 + 2 \cdot 6^3$.   [Example 2c]

3. **Evaluate an expression on a scientific calculator?**
   Evaluate $\dfrac{6^2 - 4}{3^2 - 5}$ on a scientific calculator.   [Example 5]

4. **Determine reasonable answers to computations by estimation?**
   A share of stock opens with a share price of $39\frac{7}{8}$ dollars and closes at $42\frac{1}{8}$. Select the choice that best estimates the dollar gain for an investor who owns 290 shares.
   a. $6,000   **b.** $60,000   **c.** $600   **d.** $60   [Example 7]

1.3 Can you:

1. **Translate between verbal expressions and algebraic expressions?**
   Translate "15 percent of the amount of sales x" to an algebraic statement.   [Example 2b]

2. **Evaluate algebraic expressions?**
   Evaluate $x^2 + 5y^2$ given that $x = 3$ and $y = 4$.   [Example 4b]

3. **Apply perimeter and area formulas?**
   Find the perimeter and the area of a rectangular driveway that is 42 ft long and 12 ft wide. Will a 5-gal pail of driveway sealer be sufficient to cover the driveway if the directions on the pail read: "Recommended cover rate: Weathered or unsealed surfaces—from 300 to 400 ft² per 5-gal pail"?   [Example 5]

4. **Apply volume formulas?**
   A family buys a backyard aboveground swimming pool and selects a circular model that is 24 ft in diameter. If each cubic foot of volume holds about 7.5 gal of water, then to the nearest gallon, find how many gallons of water are needed to fill the pool to a depth of 4 ft. [Example 8]

1.4 Can you:

1. **Graph real numbers?**
   Graph $\frac{2}{3}$, -1.3, and $\sqrt{2}$ on the number line.   [Example 2]

2. **Order real numbers?**
   Insert the proper symbol $(<, >)$ to indicate the correct order: -2 _____ -3. [Example 3c]

3. **Find the negative or opposite of a real number?**
   Find the negative of -3.   [Example 4a]

4. **Find the absolute value of a real number?**
   Evaluate $-|-2|$.   [Example 5c]

5. **Identify integers, rational numbers, irrational numbers, and real numbers?**
   From the set $\{-7, 1.6, \sqrt{5}, 0, -\frac{3}{4}, \pi, 2\frac{1}{2}, \sqrt{9}\}$, list all numbers that are integers. [Example 7a]

1.5 Can you:

1. **Add two real numbers using the number line?**
   Find the sum $-2 + 4$ by using the number line.   [Example 1c]

2. **Add real numbers using absolute values?**
   Find the sum $-3 + (-5)$.   [Example 2a]

3. **Find the additive identity and the additive inverse for a real number a?**
   Find the sum $-3 + [5 + (-5)]$.   [Example 5c]

4. **Solve applied problems by adding real numbers?**
The temperature rose 2 degrees in 1 hour and then dropped 6 degrees in the next hour. Find the net change in temperature for the 2-hour period. [Example 6]

1.6 **Can you:**
1. **Subtract real numbers?**
Subtract $-3 - (-8)$ by converting to addition. [Example 1c]

2. **Evaluate algebraic expressions involving addition and subtraction?**
Evaluate $-x - y$ given that $x = -3$ and $y = 4$. [Example 3b]

3. **Distinguish between the** $\boxed{-}$ **and** $\boxed{+/-}$ **keys on a calculator?**
Evaluate $-(-3 - 4)$ on a scientific calculator. [Example 4b]

4. **Solve applied problems by subtracting real numbers?**
Find the range for the given set of temperatures: $\{2, -3, -1, 4, 0, -5, 1\}$. Recall that range = largest number − smallest number. [Example 5]

1.7 **Can you:**
1. **Multiply two real numbers?**
Multiply $(-6)(-5)$. [Example 3a]

2. **Multiply more than two real numbers?**
Multiply $(-2)^3$. [Example 4b]

3. **Evaluate expressions containing signed numbers by following the order of operations?**
Evaluate $3(-5) - (-4)(7)$. [Example 5a]

4. **Solve applied problems involving multiplication of real numbers?**
Assuming certain depreciation rules, the value V of a $7,000 desktop publishing system after t years is given by the formula

$$V = -1,400t + 7,000,$$

where t is a nonnegative integer less than or equal to 5. Find the value of this equipment after three years. [Example 7]

1.8 **Can you:**
1. **Divide with nonzero real numbers?**
Divide $\dfrac{80}{-16}$. [Example 1b]

2. **Evaluate quotients involving 0?**
Evaluate $\dfrac{0}{-4}$. [Example 2a]

3. **Evaluate quotients by following the order of operations?**
Evaluate $\dfrac{x^2 + y^2}{x + y}$ given that $x = -5$ and $y = 2$. [Example 4c]

4. **Find the reciprocal (or multiplicative inverse) of a nonzero number?**
Find the reciprocal of -4. [Example 5a]

5. **Divide using the reciprocal definition of division?**
Divide $-\dfrac{2}{5}$ by 7 using the reciprocal form of division. [Example 6b]

6. Solve applied problems involving division of real numbers?

An investor's monthly statement from a brokerage company computes the account's percent change using the formula

$$\text{percent change} = \frac{\text{closing net value} - \text{opening net value}}{\text{opening net value}} \cdot 100.$$

If the net value of an account opens the month at \$44,965.11 and closes at \$44,431.50, find the percent change to the nearest hundredth of a percent.   [Example 7]

1.9 Can you:

1. Identify and use commutative and associative properties?

If $a = -5$, $b = 2$, and $c = 7$, show that $(ab)c = a(bc)$.   [Example 2b]

2. Identify and use identity and inverse properties?

Use one of the identity or inverse properties to simplify the expression $-4(-\frac{1}{4})$, and indicate the property that was used.   [Example 4b]

3. Identify and use the distributive property?

Multiply using the distributive property: $-2(x - y + 5)$.   [Example 6d]

4. Use $-a = -1 \cdot a$ and the distributive property to remove parentheses?

Rewrite $-(3 - x)$ without parentheses.   [Example 9b]

KEY TERMS

Absolute value (1.4)
Additive identity (1.9)
Additive inverse (1.5, 1.9)
Algebraic expression (1.3)
Area (1.3)
Base (1.2)
Circumference (1.3)
Constant (1.3)
Denominator (1.1)
Exponent (1.2)
Exponential expression (1.2)
Factor (1.1)
Formula (1.3)

Improper fraction (1.1)
Integer (1.4)
Irrational number (1.4)
Least common denominator (1.1)
Lowest terms (1.1)
Mixed number (1.1)
Multiplicative identity (1.9)
Multiplicative inverse (1.8, 1.9)
Negative (1.4)
Numerator (1.1)
Opposite (1.4)
Perimeter (1.3)

Power (1.2)
Prime-factored form (1.1)
Prime number (1.1)
Product (1.1)
Quotient (1.3)
Radius (1.3)
Rational number (1.4)
Real number line (1.4)
Reciprocal (1.1, 1.8)
Set (1.1)
Subset (1.4)
Variable (1.3)

KEY CONCEPTS AND PROCEDURES

Section	Key Concepts or Procedures to Review
1.1	■ Notation for the order of real numbers

Statement	Read	Comment
$a = b$	a equals b	a and b represent the same number
$a < b$	a is less than b	the symbol $<$ points to a, the smaller number
$a > b$	a is greater than b	the symbol $>$ points to b, the smaller number

■ Fundamental principle of fractions: If b and k are not zero, then

$$\frac{a \cdot k}{b \cdot k} = \frac{a}{b}$$

Section	Key Concepts or Procedures to Review

■ To express a fraction in lowest terms

1. Express the numerator and the denominator as a product of prime factors.

2. Divide out nonzero factors that are common to the numerator and the denominator according to the fundamental principle.

■ Multiplication of fractions: The product of two or more fractions is the product of their numerators divided by the product of their denominators. In symbols, the principle is

$$\frac{a}{b} \cdot \frac{c}{d} = \frac{a \cdot c}{b \cdot d} \qquad b, d \neq 0$$

■ Division of fractions
$$\frac{a}{b} \div \frac{c}{d} = \frac{a}{b} \cdot \frac{d}{c} \qquad b, c, d \neq 0$$

■ Addition and subtraction: The sum (or difference) of two or more fractions that have the same denominator is given by the sum (or difference) of the two numerators divided by the common denominator. Symbolically,

$$\frac{a}{b} + \frac{c}{b} = \frac{a + c}{b} \qquad \frac{a}{b} - \frac{c}{b} = \frac{a - c}{b} \qquad b \neq 0$$

When fractions have different denominators, we can rewrite them using the fundamental principle so that the fractions have the same denominator.

■ To find the LCD

1. Express each denominator as a product of prime factors.

2. The LCD is the product of all the different prime factors, with each prime number appearing the most number of times it appears in any one factorization.

1.2

■ Natural number exponent: If n is a natural number, then

$$a^n = \underbrace{a \cdot a \cdot a \cdots a}_{n \text{ factors}}$$

■ Order of operations

1. Perform all operations within grouping symbols first. If there is more than one symbol of grouping, simplify the innermost symbol of grouping first, and simplify the numerator and denominator of a fraction separately.

2. Evaluate powers of a number.

3. Multiply or divide working from left to right.

4. Add or subtract working from left to right.

1.3

■ Common verbal expressions and their corresponding algebraic expressions

Operation	Verbal Expression	Algebraic Expression
Addition	The sum of a number x and 2	$x + 2$ or $2 + x$
	A number y plus 5	$y + 5$ or $5 + y$
	A number w increased by 1	$w + 1$ or $1 + w$
	8 more than a number b	$b + 8$ or $8 + b$
	Add 3 and a number d	$d + 3$ or $3 + d$
Subtraction	The difference of a number x and 2	$x - 2$
	A number y minus 5	$y - 5$
	A number w decreased by 1	$w - 1$
	8 less than a number b	$b - 8$
	Subtract 10 from a number d	$d - 10$
	Subtract a number d from 10	$10 - d$

Section	Key Concepts or Procedures to Review

■ Perimeter and area formulas

Square	Rectangle	Triangle	Circle
$P = 4s$	$P = 2\ell + 2w$	$P = a + b + c$	$C = \pi d$ or $2\pi r$
$A = s^2$	$A = \ell w$	$A = \frac{1}{2}bh$	$A = \pi r^2$

■ Volume formulas

Rectangular Solid	Cylinder	Sphere
$V = \ell wh$	$V = \pi r^2 h$	$V = \frac{4}{3}\pi r^3$

1.4

■ Real number: A real number is a number that can be represented by a point on the number line.

■ Order on the number line: For any two real numbers, the graph of the larger number is to the right of the graph of the smaller number on the number line.

■ Double-negative rule: If a is a real number, then

$$-(-a) = a$$

■ Absolute value: For any real number a,

$$|a| = a, \text{ if } a \text{ is a positive number or zero,}$$
$$|a| = -a, \text{ if } a \text{ is a negative number.}$$

■ Relationship of number sets

Real numbers (all numbers that can be represented by a point on the number line)

Rational numbers (quotients of two integers with a nonzero denominator)	**Irrational numbers** (real numbers that are not rational numbers)
$\frac{2}{3}$, -1.3, $2\frac{1}{2}$	$\sqrt{2}$
	$\sqrt{7}$
Integers $\ldots, -2, -1, 0, 1, 2, \ldots$	π

1.5

■ Adding real numbers
Like signs: To add two real numbers with the same sign, add their absolute values and attach the common sign.
Unlike signs: To add two real numbers with different signs, subtract the smaller absolute value from the larger and attach the sign of the number with the larger absolute value.

■ Additive inverse: For any real number a,

$$a + (-a) = -a + a = 0$$

The opposite or negative of a number is called its additive inverse for this reason.

1.6

■ Definition of subtraction: If a and b are any real numbers, then

$$a - b = a + (-b)$$

In words, "a minus b" is the same as "a plus the opposite of b."

1.7

■ Special properties of multiplication

Property	Comment
$a \cdot 1 = a$	Multiplication by 1 does not change any real number. For this reason, 1 is called the multiplicative identity.
$a \cdot 0 = 0$	The product of any real number and 0 is 0.

Section	Key Concepts or Procedures to Review

$ab = ba$ The order of the factors in a multiplication problem does not affect their product.

■ Different-sign multiplication: To multiply two real numbers with different signs, multiply their absolute values and make the sign of the product negative.

■ Same-sign multiplication: To multiply two real numbers with the same sign, multiply their absolute values and make the sign of the product positive.

■ Sign rule: A product of nonzero factors is positive if the number of negative factors is even. The product is negative if the number of negative factors is odd.

1.8

■ Quotients of nonzero real numbers
Same-sign division: To divide with two real numbers which have the same sign, divide their absolute values and make the sign of the quotient positive.
Different-sign division: To divide with two real numbers which have different signs, divide their absolute values and make the sign of the quotient negative.

■ Division involving 0
  1. 0 divided by any nonzero number is 0.
  2. Division by 0 is undefined.

■ Definition of division: If a and b are any real numbers with $b \neq 0$, then

$$a \div b = a \cdot \frac{1}{b}$$

In words, to divide a by b, multiply a by the reciprocal of b.

1.9

■ Commutative properties: For any real numbers a and b

$$a + b = b + a$$ Commutative property of addition
$$ab = ba$$ Commutative property of multiplication

■ Associative properties: For any real numbers a, b, and c

$$(a + b) + c = a + (b + c)$$ Associative property of addition
$$(ab)c = a(bc)$$ Associative property of multiplication

■ Identity properties

Addition	**Multiplication**
There is a unique real number 0 such that for every real number a $a + 0 = 0 + a = a$	There is a unique real number 1 such that for every real number a $a \cdot 1 = 1 \cdot a = a$

■ Inverse properties

Addition	**Multiplication**
For every real number a, there is a unique real number, denoted by $-a$, such that $a + (-a) = (-a) + a = 0$	For every real number a (except zero), there is a unique real number, denoted by $1/a$, such that $a \cdot \dfrac{1}{a} = \dfrac{1}{a} \cdot a = 1$

With respect to these properties, 0 is called the additive identity, 1 is called the multiplicative identity, $-a$ is called the additive inverse (or opposite) of a, and $1/a$ is called the multiplicative inverse (or reciprocal) of a.

■ Distributive property: For any real numbers a, b, and c

$$a(b + c) = ab + ac$$

■ The opposite of a number is found by multiplying it by negative 1:

$$-a = -1 \cdot a$$

CHAPTER 1 REVIEW EXERCISES

1.1

1. Graph the set of prime numbers less than 15.

2. Insert the proper symbol ($<$, $>$, $=$) to indicate the correct order: $16 - 6$ _____ $16 - 8$.

3. True or false? $5 \geq 5$.

4. Express 1,440 as a product of its prime factors.

5. Write $\frac{48}{120}$ in lowest terms; then check that the two fractions are equal.

6. Find the product: $5 \cdot \frac{4}{30}$.

7. Divide 6 by $\frac{1}{6}$.

8. Compute $\frac{5}{4} - \frac{4}{5}$.

9. A pair of rods consists of one which is $1\frac{1}{2}$ in. long and one next to it which is $\frac{1}{10}$ in. long. If 35 pairs are placed end to end, what is the total length covered?

1.2

10. Evaluate $4^3 \cdot 3^2$.

11. Evaluate $1 + 2[12 - 2(11 - 8)]$.

12. Use a scientific calculator to evaluate $125(1 + 0.04)^{30}$.

13. A market sells cooked roast beef for $1.79 per quarter pound and cooked turkey breast for $1.69 per quarter pound. Which of these is a good estimate of the total cost of an order which contains 2 lb of each?
   a. $3   b. $10   c. $30   d. $100   e. $300

1.3

14. Translate to an algebraic expression: A price x increased by 15 percent of itself.

15. Translate to a verbal expression: $3x - y$.

16. Evaluate $\dfrac{x - 1}{y - 2}$ if $x = 7$ and $y = 4$.

17. Draw a sketch, and then find the perimeter and area for a rectangle with length 1 ft and width 6 in. Be careful to state the units for each answer.

18. Find the area and circumference of a circle with diameter 6 cm.

19. A solid sphere 1 ft in radius is made of material which weighs 10 pounds per cubic foot (lb/ft³). What is the weight of the sphere?

1.4

20. Graph the set $\{-1.2, \sqrt{2}, \pi\}$. For irrational numbers place the dot approximately in the correct position.

21. Insert the proper symbol ($<$, $>$, $=$) to indicate the correct order: -1 _____ -4.

22. If $a = -4.5$ what is the value of $-a$?

23. Evaluate $-|-4| - |4|$.

24. List all the numbers in the set $\{-2, 0, \frac{1}{2}, \sqrt{5}\}$ that are
   (a) integers and (b) rational numbers.

25. Use the graph to answer these questions.
   a. What was the temperature on the coldest day?

   b. For which of the days shown is it clear that it was colder than the previous day?

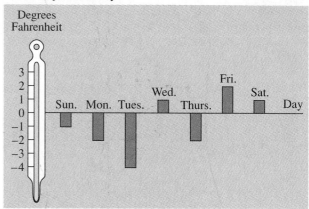

1.5

26. Find the sum by sketching arrows on the number line: $-3 + 2$.

27. Find the sum: $-2 + (-2 + 4) + (-4)$.

28. What number should be added to $-\frac{1}{3}$ to make a sum of zero?

29. Fill in the blank with word(s) to make a true statement: Two _____ negative two equals zero.

30. Translate into an English statement: $-2 + (-5) = -7$.

31. True or false? $|-3| + |2| = |-3 + 2|$.

32. Answer the question by writing an addition expression. On Monday at noon the temperature was -10 degrees. Then on four consecutive days the noon temperature rose 5 degrees, fell 3 degrees, fell 3 degrees, and rose 1 degree. What was the noon temperature on Friday?

1.6

33. Rewrite the given subtraction expression as an addition expression: $-4 - (-x)$.

34. Rewrite the given expression as an addition expression; then calculate the answer: $-4 - (-4)$.

35. Translate into a numerical expression and simplify: Subtract the sum of 3 and -4 from -1.

36. Translate into English: $x - (-y)$.

37. Given $x = -2$ and $y = 3$, evaluate $-y - (-x)$.

38. Given $x = -2$, $y = 3$, and $z = 4$, evaluate $x - (y - z)$.

39. In 431 B.C. the Temple of Apollo was built in Rome; in 1547 Michelangelo was working on St. Peter's Church in that same city. Use negative numbers for B.C. dates and express the elapsed time between these dates by writing a subtraction expression. Then compute the answer. (*Hint:* See Exercise 69 in Section 1.6.)

1.7

40. Compute: $(-3)(4)(-5)$.
41. Evaluate: $6 - 3(-5 + 2)$.
42. Given $x = 3$ and $y = -2$, evaluate $y^3 + 3y^2x + 3yx^2 + x^3$.
43. If n represents a negative number, what is the sign of $2n^5$?

44. Find a replacement for x that will make $-5(x + 2)$ equal to 0.

45. Translate into an algebraic expression: The sum of the squares of x and its additive inverse.
46. Which is larger, $5(-3)$ or $5 - 3$?
47. If $x + 2$ stands for the length of a rectangle, and $x - 2$ stands for the width, then find the perimeter when $x = 4$.

1.8

48. Divide: $\frac{3}{2} \div \left(-\frac{2}{3}\right)$.

49. Given $x = -2$ and $y = -3$, evaluate $\dfrac{x^2 - y^2}{x^2 + y^2}$.

50. What is the reciprocal of -2?

51. Rewrite the expression $c \div (-s)$ as an equivalent multiplication expression.

52. Fill in the blank to make this a true statement: $\dfrac{-36}{9} = \dfrac{8}{2}$.

53. What percent change is a drop in mortgage interest rates from 9 percent to 8.5 percent? (The formula for percent change is given before Exercise 79 in Section 1.8.)

1.9

54. Choose the correct words: When you rewrite $x(y + z)$ as $(y + z)x$, you are using the (commutative, associative) property of (addition, multiplication).
55. Fill in the blank to satisfy the associative property of multiplication: $(5x)y = $ _____ .
56. Given $x = -2$ and $y = 3$, show that $3(xy)$ is equal to $(3y)x$, which is found by applying both the commutative and associative properties of multiplication.
57. What property is illustrated by the expression $a(b + c) = ab + ac$?
58. Use the distributive property to rewrite $4x - 7x$ as a product.

59. Rewrite $-(2x + 3)$ without parentheses.

CHAPTER 1 TEST

1. True or false? $5 - 2 \le 7 - 5$.
2. Express numerator and denominator as products of prime factors, and then express $\frac{54}{120}$ in lowest terms.

3. Graph all the prime factors of 30.

4. Rewrite $\frac{3}{4} \div 2$ as a multiplication expression and evaluate it.

5. A mail order company finds a new way to wrap a product which reduces the weight of each package from $7\frac{1}{4}$ oz to $6\frac{3}{8}$ oz. What is the total reduction in weight for a shipment of 144 packages?

6. Evaluate $3 + 5^2[8 - 4(5 - 3)]$.
7. An investment of $10,500 increases 6 percent per year. Evaluate $10,500(1 + 0.06)^{15}$ to find its value after 15 years.
8. Translate into an algebraic expression: The product of a number m and 5 more than m.

9. Given that $x = 5$ and $y = 3$, evaluate $\dfrac{x^2 - y}{y^2 - x}$.

10. Compute the volume of a cylinder with radius and height each 3 cm.
11. Given that $a = -4.5$, find the value of $-a$.
12. Evaluate $|-3| + |3|$.
13. Assign the correct labels (integer, rational, irrational, real) to each number in the set $\{-3, \frac{1}{2}, \sqrt{3}\}$. A number may get more than one label.

14. Compute: $-6 - 8 + (-1)$.
15. Subtract the sum of -5 and 7 from 0.
16. Find the difference in temperature between 32 and -10 degrees Fahrenheit.
17. Given $x = 3$ and $y = -2$, evaluate $3x - 2y(x + y^2)$.
18. From the sketch, write expressions for the perimeter and the area. Then evaluate them for $x = 1$.

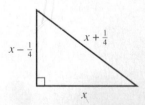

19. Use percent change $= \dfrac{\text{new value} - \text{old value}}{\text{old value}} \cdot 100$ to find the percent change for an investment which falls in value from $2,000 to $1,000 in one year. Would you get the same answer for any investment which lost half its value?
20. Rewrite the expression $f(h - w)$ by using the distributive property. Show that both expressions yield the same value when $f = 2$, $h = 3$, and $w = -5$.

2 Equations and Inequalities

An architect wants to design a track with a rectangular inner section and circular turns, as shown in the diagram. If the length of the inner section is triple the width, and the width is given by *x* ft, then write an expression in simplest form for the distance around the inner portion of the track in terms of *x*. Use 3.14 for π. (See Example 5 of Section 2.1.)

■■■■■■

AN ESSENTIAL problem-solving skill is the ability to set up and solve equations and inequalities. In this chapter we introduce some basic properties of equality and use them in a systematic way to solve linear equations, literal equations or formulas, and word problems. The chapter concludes with a discussion of inequalities, because the procedures for solving linear inequalities are similar to the procedures for solving linear equations.

2.1 Combining Like Terms

1 Identify the terms and the numerical coefficients of the terms in an algebraic expression.

2 Identify like terms and unlike terms.

3 Simplify algebraic expressions by combining like terms.

4 Simplify algebraic expressions by removing parentheses and combining like terms.

1 A central theme of this chapter is to develop techniques for solving equations and inequalities. To achieve this objective, we often need to combine algebraic expressions that are added or subtracted. Those parts of an algebraic expression separated by plus *or minus (−)* (+) signs are called **terms** of the expression. The following example will help you recognize terms.

EXAMPLE 1 Identify the term(s) in the following expressions.

a. $3x + 5x$ **b.** $x^2 - 3x + 4$ **c.** $7xy$ **d.** -7

Solution

a. $3x + 5x$ is an algebraic expression with two terms, $3x$ and $5x$.

b. $x^2 - 3x + 4$ may be written as $x^2 + (-3x) + 4$ and is an algebraic expression with three terms, x^2, $-3x$, and 4.

c. Because $7xy$ is a product (instead of a sum or difference), it is an algebraic expression with one term, $7xy$.

d. Single numbers are always single terms. Thus, the single term in this expression is -7.

PROGRESS CHECK 1 Identify the term(s) in the following expressions.

a. $3y^2 + 5y + 4$ **b.** 0   **c.** $5 - 4x$ **d.** $-5x/3$ ⌐

Example 1 illustrates that a **term** is either a number or the product or quotient of constants and variables. If a term is a product or quotient of constants and variables, the constant factor is called the **numerical coefficient** of the term. If a term consists only of a number with no variable, it is called a constant term and its numerical coefficient is the number itself. For instance, the numerical coefficient of the term 5 is just the number 5.

EXAMPLE 2 Specify the numerical coefficients of the following terms.

a. $7xy$ **b.** $x/3$ **c.** y **d.** $-x$ **e.** -3

Solution

a. The constant factor or numerical coefficient of $7xy$ is 7.

b. Because $x/3$ may be written as $\frac{1}{3}x$, the numerical coefficient of $x/3$ is $\frac{1}{3}$.

c. The numerical coefficient of y is 1, since $y = 1 \cdot y$.

d. The numerical coefficient of $-x$ is -1, since $-x = -1 \cdot x$.

e. The numerical coefficient of the term -3 is the number -3.

PROGRESS CHECK 2 Specify the numerical coefficient of the following terms.

a. $-5x^2$ **b.** $-t$ **c.** -6 **d.** xy **e.** $x/7$ ⌐

Progress Check Answers

1. (a) $3y^2$, $5y$, 4 (b) 0 (c) 5, $-4x$ (d) $-5x/3$

2. (a) -5 (b) -1 (c) -6 (d) 1 (e) $\frac{1}{7}$

2   To combine terms, we need to identify like terms and unlike terms. First, consider some examples.

Examples of Like Terms	Examples of Unlike Terms
$7x, 2x$	$7x, 2y$ (variables are different)
$7, 5$	$5x, 5$ (only one term has a variable)
$-3x^2, 4x^2$	$-3x^2, 4x$ (exponents are different)
$5xy^2, 8xy^2$	$5xy^2, 8x^2y$ (exponents are different)

Terms that have exactly the same variables with the same exponents are **like terms.** Also, constant terms like 7 and 5 are like terms. In contrast, terms that differ in either the variables or the exponents are called **unlike terms.**

EXAMPLE 3   Identify any like terms in the following expressions.

a. $9x + 4x - 3$ **b.** $5 + 7x - 3x + 4$ **c.** $6x + 2y - 3$

Solution

a. $9x$ and $4x$ are like terms.
b. $7x$ and $-3x$ are like terms, and the constant terms 4 and 5 are like terms.
c. There are no like terms in this expression.

PROGRESS CHECK 3   Identify any like terms in the following expressions.

a. $y - 11y - 11$ **b.** $1 - 11y$ **c.** $-5x + 1 + 5x + 10$ ⌐

3   To combine algebraic expressions, we identify like terms and use the distributive property in the forms

$$ac + bc = (a + b)c \quad \text{and} \quad ac - bc = (a - b)c.$$

For example, $6x + 4x$ may be simplified as follows.

$$6x + 4x = (6 + 4)x = 10x$$

In similar fashion, we may combine the terms of $5y - 8y$.

$$5y - 8y = (5 - 8)y = -3y$$

Thus, the distributive property indicates that we combine like terms by combining their numerical coefficients, and that the only kinds of terms we can combine are like terms.

EXAMPLE 4   Simplify the following expressions.

a. $5x + 11x$ **b.** $y - 11y$ **c.** $5 + 2x - 2x$

Solution   The direction *simplify* here means to find an equivalent expression that has fewer terms by combining like terms.

a. $5x + 11x = (5 + 11)x$ Distributive property.
$ = 16x$ Simplify.
b. $y - 11y = 1y - 11y$ $y = 1 \cdot y.$
$ = (1 - 11)y$ Distributive property.
$ = -10y$ Simplify.
c. $5 + 2x - 2x = 5 + (2 - 2)x$ Distributive property.
$ = 5 + 0 \cdot x$ Simplify.
$ = 5$ Simplify.

PROGRESS CHECK 4   Simplify the following expressions.

a. $-5x - 4x$ **b.** $15y + y$ **c.** $7x - 7x - 4$ ⌐

Recall from Section 1.9 that the distributive property extends to more than two terms. We need this extension to solve the problem that opens this chapter.

EXAMPLE 5   Solve the problem in the chapter introduction on page 72.

Solution   Each circular turn is a semicircle. From $C = \pi d$, the formula for the circumference of a circle, the distance around a semicircle is given by $\pi d/2$. Because the diameter is x and the length of the rectangular inner section is $3x$, we find the requested perimeter as follows.

Perimeter =	length of inner section	plus	turn distance	plus	length of inner section	plus	turn distance
=	$3x$	+	$\dfrac{\pi x}{2}$	+	$3x$	+	$\dfrac{\pi x}{2}$

$$= 3x + \frac{3.14x}{2} + 3x + \frac{3.14x}{2} \qquad \text{Use 3.14 for } \pi.$$
$$= 3x + 1.57x + 3x + 1.57x \qquad \text{Simplify.}$$
$$= (3 + 1.57 + 3 + 1.57)x \qquad \text{Distributive property.}$$
$$= 9.14x \qquad \text{Simplify.}$$

Thus, in terms of x, the distance around the inner portion of the track is $9.14x$.

PROGRESS CHECK 5   Find the perimeter in terms of x of the tract of land in Figure 2.1.

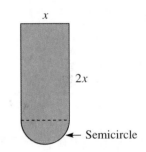

Figure 2.1

x

$2x$

← Semicircle

4   To simplify algebraic expressions, we sometimes need to remove the parentheses which group together certain terms. Parentheses are removed by applying the distributive property, as discussed in Section 1.9.

EXAMPLE 6   Simplify the following expressions.

  a. $10x + 2(90 + x)$ **b.** $5(x + 3) + 2(1 - x)$

Solution

  a.
$$10x + 2(90 + x) = 10x + 2(90) + 2(x) \qquad \text{Distributive property.}$$
$$= 10x + 180 + 2x \qquad \text{Multiply.}$$
$$= 12x + 180 \qquad \text{Combine like terms.}$$
  b.
$$5(x + 3) + 2(1 - x) = 5(x) + 5(3) + 2(1) - 2(x) \qquad \text{Distributive property.}$$
$$= 5x + 15 + 2 - 2x \qquad \text{Multiply.}$$
$$= 3x + 17 \qquad \text{Combine like terms.}$$

Note   Remember that we may reorder and regroup terms in any useful way because of the commutative and associative properties of addition, and that we usually do this *mentally*. A more formal solution to part **a** that shows all the steps follows.

$$10x + 2(90 + x) = 10x + (2 \cdot 90 + 2x) \qquad \text{Distributive property.}$$
$$= 10x + (2x + 2 \cdot 90) \qquad \text{Commutative property.}$$
$$= (10x + 2x) + 2 \cdot 90 \qquad \text{Associative property.}$$
$$= (10 + 2)x + 2 \cdot 90 \qquad \text{Distributive property.}$$
$$= 12x + 180 \qquad \text{Simplify.}$$

This level of detail is not often written in practice and is therefore not shown in the sample problems.

PROGRESS CHECK 6   Simplify the following expressions.

  a. $7y + 3(25 + 2y)$ **b.** $4(2 + x) + 3(x - 1)$

Progress Check Answers

5. $6.57x$

6. (a) $13y + 75$ (b) $7x + 5$

In the next example, note that we frequently switch between addition and subtraction by using the subtraction definition, $a - b = a + (-b)$.

EXAMPLE 7 Simplify the following expressions.

a. $4(x + 1) - 6(x + 2)$ **b.** $-(3 - x) + 2(x + 1)$

Solution

a. $4(x + 1) - 6(x + 2)$

$= 4(x + 1) + (-6)(x + 2)$ Use $a - b = a + (-b)$.

$= 4(x) + 4(1) + (-6)x + (-6)2$ Distributive property.

$= 4x + 4 + (-6x) + (-12)$ Multiply.

$= -2x - 8$ Combine like terms.

b. $-(3 - x) + 2(x + 1)$

$= -1(3 - x) + 2(x + 1)$ Use $-(3 - x) = -1(3 - x)$.

$= -1(3) + (-1)(-x) + 2(x) + 2(1)$ Distributive property.

$= -3 + x + 2x + 2$ Multiply.

$= 3x - 1$ Combine like terms.

Progress Check Answers

7. (a) $-6x - 2$ (b) $7x + 17$

PROGRESS CHECK 7 Simplify the following expressions.

a. $2(x + 3) - 8(x + 1)$ **b.** $-(7 - x) + 3(2x + 8)$ ⌐

EXERCISES 2.1

In Exercises 1–12, identify the term(s) in the given expression.

1. $2x + 4x$ **2.** $-x + 3x^2$

3. $x^2 + 5x - 2$ **4.** $3 - 4y - 5y^2$

5. $7xy + 8x - 9y$ **6.** $x^2 - 2xy + y^2$

7. $-4xy$ **8.** $29y/3$

9. $-\frac{4}{7}$ **10.** 1

11. x^2y^2 **12.** $x^2 - y^2$

In Exercises 13–24, specify the numerical coefficient of the given terms.

13. $3x$ **14.** $7y$ **15.** x^2

16. y^3 **17.** $-y$ **18.** $-xy$

19. 1 **20.** -2 **21.** $-\frac{2}{5}x$

22. $\frac{4}{3}x$ **23.** $\frac{x}{2}$ **24.** $-\frac{x}{2}$

In Exercises 25–36, identify any sets of like terms.

25. $8x + 3x - 3$ **26.** $x - 5 + 2x$

27. $5 + 7x + 2$ **28.** $4 - 2y + 8.6$

29. $\frac{x}{2} - \frac{3x}{4} + x$ **30.** $\frac{y}{3} - \frac{y}{5} + 2y$

31. $1 + 2x + 3y$ **32.** $3x + 3y - 3z$

33. $2x^2 + 3x^2 + 4$ **34.** $2x^3 + 3x - 5x^3$

35. $1 + 2x + 3x^2$ **36.** $y + 2y^2 + 3y^3$

In Exercises 37–60, simplify the given expressions by combining like terms.

37. $6x + 12x$ **38.** $2x + 8x$

39. $7x - 2x$ **40.** $8y - 5y$

41. $\frac{1}{2}x + \frac{1}{2}x$ **42.** $\frac{1}{3}y + \frac{2}{3}y$

43. $0.2x + 0.8x$ **44.** $0.3y - 1.3y$

45. $x + 8x$ **46.** $3y + y$

47. $2x - x$ **48.** $-x + 3x$

49. $-y - 12y$ **50.** $-x - 15x$

51. $5 + 3x - 3x$ **52.** $6 + 4x - 4x$

53. $2x - 4x - 5$ **54.** $3y + 4y + 5x$

55. $-3x + 3x + 7$ **56.** $-5y + 2 + 5y$

57. $-4x + 5 - 6x + y$

58. $-4 + 5x - 6y + x$

59. $x - 2y + 3x - 4y$

60. $4y - 3x + 2x - y$

In Exercises 61–72, find an algebraic expression for the perimeter of the given figures. Use 3.14 for π, and simplify as much as possible.

61.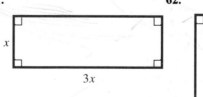

62.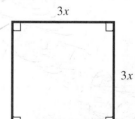

63.

64.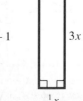

65.

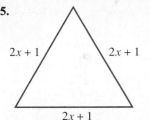

$2x + 1$ $2x + 1$

$2x + 1$

66.

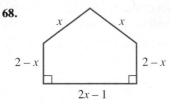

$5x$

$3x$

$4x$

67.

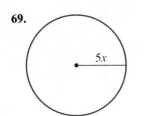

$x - 2$ $x + 1$

$x + 2$

68.

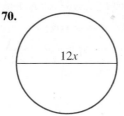

x x

$2 - x$ $2 - x$

$2x - 1$

69.

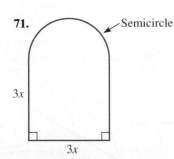

$5x$

70.

$12x$

71.   Semicircle

$3x$

$3x$

72.   Semicircle

$6x$ $10x$

$8x$

In Exercises 73–96, simplify the given expressions. First, remove the parentheses when necessary.

73. $6x + 3(14 + x)$
74. $x + 2(x + 2)$
75. $5x + (3x - 4)$
76. $3x + (5 - x)$
77. $2 + 3(x + 4)$
78. $3 + 4(5 - y)$
79. $x + \frac{1}{2}(4x - 6)$
80. $3x + \frac{2}{3}(6x + \frac{3}{2})$
81. $3(x - 1) + 2(1 - x)$
82. $4(2x - 1) + 4(1 - 2x)$
83. $0.2(0.4x + 10) + 0.3(30 - 0.3x)$
84. $0.08(1,000 + x) + 0.92(1,000 + x)$
85. $4(x + 2) - 6(x + 3)$
86. $2(x + 3) - 3(x + 6)$
87. $\frac{1}{2}(2x - 4) - \frac{1}{3}(3x + 9)$
88. $\frac{4}{3}(3 - 3x) - (x + 1)$
89. $-(3x - 4) - 2(3x - 4)$
90. $-(3x + 4) + (3x + 4)$
91. $2(5x - 4) - 2(4 - 5x)$
92. $3(2x - 5) - 3(5 - 2x)$
93. $x - 3(x + 1)$
94. $5x - 5(x + 2)$
95. $6 - 2(3 - x)$
96. $4 - 3(x + 1)$

In Exercises 97–108, simplify the given expressions.

97. $5x^2 - 1 + x^2 + 4$
98. $4y^2 - 2 - y^2 + 3$
99. $x - x^2 - 3x^2 + x$
100. $x^2 - (2 - x^2)$
101. $y^2 - (3y^2 + 2)$
102. $x^2 + 3(x + x^2)$
103. $2y^2 - 4(y + y^2)$
104. $3x^2 - 2(x - x^2)$
105. $4(x - x^2) + 4(x^2 - x)$
106. $3(y^2 - y) - (y^2 - y)$
107. $2(y^2 - 2y + 1) - 3(y^2 + 2y - 3)$
108. $3(x^2 - 4x + 5) - 2(5x^2 + x - 4)$
109. An architect designs a rectangular hallway to be 4 times longer than it is wide. Find a simple formula for the perimeter of the hallway. Use x for the width of the rectangle.
110. A certain type of rectangle is called a golden rectangle. In a golden rectangle the length is (approximately) 1.618 times the width. Write a simple formula for the perimeter of a golden rectangle whose width is x.

THINK ABOUT IT

1. Show, by using the distributive property, that $3x - 3x = 0$.
2. Show, by using the distributive property, that $3 - 3x$ does not simplify to 0.
3. Simplify $(4 - 5x)(3x - 3x)$.
4. This expression can be simplified very easily if you use the distributive property cleverly. Try it.

$$1.4(-5x^3 + 2x - 6) - 0.4(-5x^3 + 2x - 6)$$

5. A mind reader at a show says, "Think of any number and write it down. Add 50. Double your answer. Subtract 100, and tell me your answer." Then the mind reader announces your number to the audience.
   a. Let x stand for the original number, and give the algebraic expression corresponding to each step.
   b. Simplify the algebraic expression corresponding to the last step to show why the trick works. How do you figure out the original number from the player's answer?

REMEMBER THIS

1. What is the reciprocal of -4?
2. Which is larger, $2 - 3$ or $2(-3)$?
3. Write an expression for 15 percent of x.
4. Write an expression to represent 2 less than 5 times a number.
5. What is the formula for the area of a circle?
6. What number added to 7 gives a sum of 0?
7. What number multiplied by -5 gives a product of 1?
8. Write an expression for the product of x and y^2.
9. What addition expression is equivalent to $4 - 9$?
10. Evaluate $3(x + 4)$ when $x = -4$.

2.2 The Addition and Subtraction Properties of Equality

A meteorologist reports that because of the greenhouse effect, the average summer temperature at the South Pole will go up 5 degrees in the next decade to reach −2 degrees Fahrenheit. What is the average summer temperature now? (See Example 9.)

OBJECTIVES

1 Classify an equation as a conditional equation, an identity, or a false equation.

2 Determine if a number is a solution of an equation.

3 Solve linear equations by applying the addition or subtraction properties of equality.

4 Use the addition or subtraction properties to show that certain equations are identities or false equations.

5 Solve applied problems by solving linear equations.

If you can write and solve equations, you can answer the section-opening problem in a straightforward manner. An essential way to use mathematics for solving real-life problems is to express the relationships in the situation by an equation and then use the solution to the equation to answer the original problem. We begin with some definitions that answer the question "What is an equation?"

1 An equation is a statement which says that two expressions are equal. For example,

$$x + 1 = 3 \qquad \text{and} \qquad x + 1 = x \qquad \text{and} \qquad x + 1 = 1 + x$$

are equations. In these examples the expression $x + 1$ is called the left-hand side of the equation, while the expression on the right side of the equality sign is called the right-hand side of the equation. Note that an equation must always contain an equality sign, $=$. An equation may be always true or always false, or its truth may depend on the value substituted for the variable. These three truth possibilities lead to the following types of equations.

Types of Equation

1. A **conditional equation** is an equation in which some replacements for the variable make the statement true, while others make it false.
2. An **identity** is an equation that is a true statement for all admissible values of the variable.
3. A **false equation** is an equation that is never true.

Each of these cases is discussed in Example 1.

EXAMPLE 1   Classify each equation as a conditional equation, an identity, or a false equation.

a. $x + 1 = 3$ b. $x + 1 = x$ c. $x + 1 = 1 + x$

Solution

a. The equation $x + 1 = 3$ is a conditional equation, because replacing x by 2 makes the statement true while other replacements make it false.

b. The equation $x + 1 = x$ is a false equation, because the left-hand side of the equation is one greater than the right-hand side for all values of x. It may be surprising, but false statements can be useful in telling us when a problem has no solution.

c. The equation $x + 1 = 1 + x$ is an identity, because of the commutative property of addition. That is, the statement is true when x is replaced by *any* real number, because the order in which we write numbers in an addition problem does not affect their sum.

PROGRESS CHECK 1   Classify each equation as a conditional equation, an identity, or a false equation.

a. $2x + 3x = 5x$ **b.** $x = 5$ **c.** $3 + 2 = 6$ ⌐

2   The concept of solving equations is usually associated with conditional equations (like $x + 1 = 3$). To solve an equation means to find the set of all values for the variable that make the equation a true statement. This set of numbers is called the **solution set** of the equation. Each number in the solution set is called a solution or a root of the equation. Example 2 shows how to determine if a number is a solution of an equation.

EXAMPLE 2   Determine if the indicated number is a solution of the given equation.

a. $2x + 5 = 11$; 3   **b.** $5 - x = 4$; -1

Solution

a. Replace x by 3 and see if the resulting statement is true.

$$2x + 5 = 11 \qquad \text{Given equation.}$$
$$2(3) + 5 \overset{?}{=} 11 \qquad \text{Replace } x \text{ by 3.}$$
$$6 + 5 \overset{?}{=} 11 \qquad \text{Simplify.}$$
$$11 \overset{\checkmark}{=} 11 \qquad \text{Simplify.}$$

Since $11 = 11$ is a true statement, 3 is a solution of $2x + 5 = 11$.

b. Replace x by -1 and see if the resulting equation is true.

$$5 - x = 4 \qquad \text{Given equation.}$$
$$5 - (-1) \overset{?}{=} 4 \qquad \text{Replace } x \text{ by } -1.$$
$$5 + 1 \overset{?}{=} 4 \qquad \text{Simplify.}$$
$$6 \neq 4$$

Since $6 = 4$ is a false statement, -1 is not a solution of $5 - x = 4$.

PROGRESS CHECK 2   Determine if -4 is a solution of $3x - 2 = -14$.   ⌐

3   We can describe equations by whether they can be placed in certain forms. One of the simpler forms, and one of the simpler equations to solve, is a **linear equation.** By definition, a first-degree or linear equation* in one variable is an equation that can be written in the form

$$ax + b = 0,$$

where a and b are real numbers, with $a \neq 0$. For example, the equation $x + 8 = 2$ can also be written as $1x + 6 = 0$, so $x + 8 = 2$ is a linear equation.

*The equation is called first degree because the highest power of the variable is 1. It is called linear because (as shown in Section 6.2) it is associated with a graph that is a straight line.

The general method for solving a linear equation is to start with the given equation and then replace it by simpler and simpler equations that have the same solution set. If two equations have the same solution set, then they are called **equivalent equations.** For a linear equation the simplest equivalent equation we can write has the form

$$x = \text{number.}$$

One very useful way to find a simpler equivalent equation is to add the same number to both sides of an equation. This principle is expressed formally as the addition property of equality.

Addition Property of Equality

If A, B, and C are algebraic expressions, then the equations

$$A = B \quad \text{and} \quad A + C = B + C$$

are equivalent.

This property ensures that when you add the same quantity to both sides, the new equation has the same solution set as the old one.

EXAMPLE 3   Solve the equation $x - 3 = 8$.

Solution
The goal is to isolate x. Because 3 is subtracted from x, we will add 3 to both sides of the equation.

$$
\begin{array}{ll}
x - 3 = 8 & \text{Given equation.} \\
x - 3 + 3 = 8 + 3 & \text{Add 3 to both sides.} \\
x = 11 & \text{Simplify.}
\end{array}
$$

Thus 11 is the solution to the equation.
We can check the solution by replacing x by 11 in the original equation.

$$
\begin{array}{ll}
x - 3 = 8 & \text{Original equation.} \\
11 - 3 \overset{?}{=} 8 & \text{Replace } x \text{ by 11.} \\
8 \overset{\checkmark}{=} 8 & \text{Simplify.}
\end{array}
$$

Since $8 = 8$ is a true statement, the solution checks and the solution set is $\{11\}$.

PROGRESS CHECK 3   Solve the equation $x - 5 = -8$.

EXAMPLE 4   Solve the equation $-\frac{7}{4} = x - \frac{1}{5}$.

Solution   We isolate x on the right-hand side by adding $\frac{1}{5}$ to both sides.

$$
\begin{array}{ll}
-\frac{7}{4} = x - \frac{1}{5} & \text{Given equation.} \\
-\frac{7}{4} + \frac{1}{5} = x - \frac{1}{5} + \frac{1}{5} & \text{Add } \frac{1}{5} \text{ to both sides.} \\
-\frac{35}{20} + \frac{4}{20} = x - \frac{1}{5} + \frac{1}{5} & \text{Write in terms of the LCD.} \\
-\frac{31}{20} = x & \text{Simplify.}
\end{array}
$$

Now check that $-\frac{31}{20}$ is the solution to the equation.

Progress Check Answer
3. $\{-3\}$

$$-\tfrac{7}{4} = x - \tfrac{1}{5} \qquad \text{Original equation.}$$

$$-\tfrac{7}{4} \stackrel{?}{=} -\tfrac{31}{20} - \tfrac{1}{5} \qquad \text{Replace } x \text{ by } -\tfrac{31}{20}.$$

$$-\tfrac{7}{4} \stackrel{?}{=} -\tfrac{31}{20} - \tfrac{4}{20} \qquad \text{Write in terms of the LCD.}$$

$$-\tfrac{7}{4} \stackrel{?}{=} -\tfrac{35}{20} \qquad \text{Simplify.}$$

$$-\tfrac{7}{4} \stackrel{\checkmark}{=} -\tfrac{7}{4} \qquad \text{Simplify.}$$

The solution checks and the solution set is $\{-\tfrac{31}{20}\}$.

Note   The symmetric property of equality states that

$$\text{if } a = b, \text{ then } b = a.$$

Thus, at any point we are allowed to interchange the left and right sides of the equation. In this case you might find it helpful to rewrite the original equation at the start of your solution as $x - \tfrac{1}{5} = -\tfrac{7}{4}$.

PROGRESS CHECK 4   Solve the equation $-\tfrac{5}{7} = x - \tfrac{2}{5}$.   ⌐

   A second important way to find a simpler equivalent equation is to subtract the same number from both sides of an equation. This principle follows from the addition property of equality, since subtraction is defined in terms of addition, as discussed in Chapter 1.

Subtraction Property of Equality

If A, B, and C are algebraic expressions, then the equations

$$A = B \qquad \text{and} \qquad A - C = B - C$$

are equivalent.

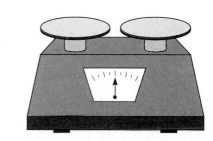

Figure 2.2

   Both the addition and subtraction properties of equality can be visualized in term of a balanced scale, as shown in Figure 2.2. To keep the scale in balance, whatever i added or subtracted on one side must also be done on the other side.

EXAMPLE 5   Solve the equation $y + 8 = -1$.

Solution   Because 8 is added to y, we will subtract 8 from both sides of the equation.

$$y + 8 = -1 \qquad \text{Given equation.}$$
$$y + 8 - 8 = -1 - 8 \qquad \text{Subtract 8 from both sides.}$$
$$y = -9 \qquad \text{Simplify.}$$

To check this solution, replace y by -9 in the original equation and simplify.

$$y + 8 = -1 \qquad \text{Original equation.}$$
$$-9 + 8 \stackrel{?}{=} -1 \qquad \text{Replace } y \text{ by } -9.$$
$$-1 \stackrel{\checkmark}{=} -1 \qquad \text{Simplify.}$$

Thus, the solution set is $\{-9\}$. Note that you also can solve the original equation by adding -8 to both sides of the equation.

PROGRESS CHECK 5   Solve the equation $p + 13 = -7$.   ⌐

4   The next two examples show how the addition and subtraction properties of equality can be used to show that certain equations are identities or false equations.

Progress Check Answers

4. $\{-\tfrac{11}{35}\}$

5. $\{-20\}$

EXAMPLE 6   Show that the equation $x + 1 = 1 + x$ is an identity.

Solution   Because 1 is added to x, we will subtract 1 from both sides of the equation.

$$x + 1 = 1 + x \qquad \text{Given equation.}$$
$$x + 1 - 1 = 1 + x - 1 \qquad \text{Subtract 1 from both sides.}$$
$$x = x \qquad \text{Simplify.}$$
$$0 = 0 \qquad \text{Subtract } x \text{ from both sides.}$$

The resulting equation $0 = 0$ is always true. Since the original equation is equivalent to an equation that is always true, $x + 1 = 1 + x$ is an identity.

Note   As soon as we reached $x = x$, it was obvious that we had reached an identity. When the two sides match precisely, subtraction at the next step will always result in $0 = 0$.

PROGRESS CHECK 6   Show that the equation $2x + 3x = 5x$ is an identity.   ⌐

EXAMPLE 7   Show that $x + 1 = x$ is a false equation.

Solution   If we proceed in the usual way, we have the following steps.

$$x + 1 = x \qquad \text{Given equation.}$$
$$x + 1 - x = x - x \qquad \text{Subtract } x \text{ from both sides.}$$
$$1 = 0 \qquad \text{Simplify.}$$

Since $1 = 0$ is never true, the original equation has no solution. When the solution set contains no elements, we say that the solution set is $\emptyset$, called the empty set.

PROGRESS CHECK 7   Show that $x - 4 = x + 5$ is a false equation.   ⌐

5   Now that we can solve some equations, we can use our methods to solve applied problems.

EXAMPLE 8   The sum of the angle measures in a triangle is 180°. Find the measure of angle B in Figure 2.3.

Solution   If we let A, B, and C represent the angle measures, then we are given that

$$A + B + C = 180.$$

Now replace A by 51 and C by 36, and solve for B.

$$51 + B + 36 = 180$$
$$B + 87 = 180 \qquad \text{Simplify.}$$
$$B + 87 - 87 = 180 - 87 \qquad \text{Subtract 87 from both sides.}$$
$$B = 93 \qquad \text{Simplify.}$$

Since the sum of 51, 93, and 36 is 180, the solution checks. The measure of angle B is 93°.

PROGRESS CHECK 8   In triangle ABC the measures of angles A and B are 29° and 107°, respectively. What is the measure of angle C?   ⌐

EXAMPLE 9   Solve the problem in the section introduction on page 78.

Solution   If x represents the average summer temperature now, then

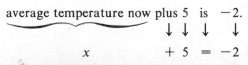

To solve the problem, we solve the equation $x + 5 = -2$.

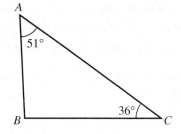

Figure 2.3

Progress Check Answers
6. $5x = 5x$ is an identity.
7. $-4 = 5$ is never true.
8. 44°

$$x + 5 = -2 \qquad \text{Given equation.}$$
$$x + 5 - 5 = -2 - 5 \qquad \text{Subtract 5 from both sides.}$$
$$x = -7 \qquad \text{Simplify.}$$

The average summer temperature now is -7 degrees Fahrenheit. If this temperature goes up 5 degrees in the next decade, then -7 plus 5 equals -2. The solution checks.

PROGRESS CHECK 9   Based on other data, the meteorologist in Example 9 predicts that the average winter temperature at the South Pole will go up 6 degrees in the next decade to reach -63 degrees Fahrenheit. What is the average winter temperature now?

Progress Check Answer

9. -69 degrees

EXERCISES 2.2

In Exercises 1–12, classify each equation as a conditional equation, an identity, or a false equation.

1. $x - 8 = 10$
2. $x - 10 = 8$
3. $x + 1 = 9$
4. $x + 9 = 1$
5. $18 + x = 0$
6. $15 + x = 0$
7. $8y - 7y = y$
8. $4x + 3x = 7x$
9. $3(x + 1) = 3x + 3$
10. $4(x - 2) = 4x - 8$
11. $x + 1 = x - 1$
12. $3x + 4 = 3x$

In Exercises 13–24, determine if the indicated number is a solution of the given equation.

13. $x - 8 = 10;\ 18$
14. $x - 10 = 8;\ 18$
15. $x + 1 = 9;\ 10$
16. $x + 9 = 1;\ -9$
17. $18 - 2x = 0;\ 9$
18. $15 - 3x = 0;\ 5$
19. $-x = 4;\ -4$
20. $-2x = 12;\ -6$
21. $x - 1 = 2x - 2;\ -1$
22. $1 + 3x = 1 - 3x;\ 1$
23. $\frac{1}{2}y + \frac{1}{3} = \frac{2}{5};\ \frac{1}{3}$
24. $\frac{1}{2}y + \frac{1}{3} = \frac{1}{3}y + \frac{5}{9};\ \frac{4}{3}$

In Exercises 25–54, solve each equation and check the solution.

25. $x - 3 = 4$
26. $x - 4 = 3$
27. $x - 3.5 = 1.2$
28. $x - 5.3 = 2.1$
29. $0 = x - 2$
30. $0 = y - 8$
31. $x - 1 = -3$
32. $x - 3 = -1$
33. $-7 = y - 7$
34. $-4 = y - 4$
35. $x - \frac{1}{3} = \frac{1}{2}$
36. $x - \frac{1}{4} = \frac{1}{5}$
37. $x - 3\frac{3}{4} = 1\frac{7}{8}$
38. $\frac{1}{2} = x - 3\frac{1}{3}$
39. $y + 2 = 7.8$
40. $y + 3.1 = 9.1$
41. $y + 1 = 1$
42. $x + 5 = 5$
43. $-2 = 4 + y$
44. $3 = 5 + y$
45. $x + \frac{1}{3} = 3$
46. $x + \frac{1}{4} = 4$
47. $-\frac{5}{3} + x = -\frac{1}{8}$
48. $\frac{3}{4} = x + \frac{7}{8}$
49. $-1 = x + 0.7$
50. $0 = x + 0.4$
51. $x + (-3) = -5$
52. $x + (-1.9) = -2.9$
53. $-2\frac{1}{8} = x + 2\frac{1}{8}$
54. $-\frac{27}{5} = x + \frac{27}{5}$

In Exercises 55–66, show the equation is either an identity or a false equation.

55. $3x + 6x = 9x$
56. $x + x = 2x$
57. $5x = x + 4x$
58. $3x = x + 2x$
59. $x + x - 3 = 2x - 3$
60. $2x + 5x = 3x + 4x$
61. $2y - 4 = -4 + 2y$
62. $7 - 2y = -2y + 7$
63. $3x + 4 = 3x$
64. $5x - 2 = 5x$
65. $x + 2 = x - 2$
66. $x = x + 3$

In Exercises 67–70, we have given a word problem and an equation that describes the relationship in the problem. Solve the equation and answer the question.

67. In triangle ABC the measure of angle A is $27°$ and the measure of angle B is $135°$. Find the measure of angle C. Use the equation $A + B + C = 180$.

68. In triangle DEF angle E measures three times as great as angle F, which measures $25°$. Find the measure of angle D. Use the equation $D + E + F = 180$.

69. Scientists predict that a major nuclear war would cause the average temperature of the earth to drop 15 degrees, because the debris would block the sun for such a long time. They state that there is a part of the globe where the average temperature would fall to -8 degrees because of this. What is the average temperature there now? Use the equation $x - 15 = -8$.

70. One day the Dow-Jones index on Wall Street fell 29.73 points to reach 2,675.62. What was the index before it fell? Use the equation $x - 29.73 = 2,675.62$.

In Exercises 71–76, solve each problem by first setting up an appropriate equation. (*Note:* These exercises are similar to the word problems in Exercises 67–70.)

71. If the air pressure in a tire reaches 50.5 pounds per square inch (psi), it will blow out. The pressure is now 38.7 psi. How much more can it rise before it blows out?

72. The two equal sides of an isosceles triangle are each 10.6 cm. The perimeter is 25.2 cm. Find the length of the third side.

73. Between March and May in a certain year the consumer price index in Miami rose 1.1 points to reach 120.9. What was the value of the index in March of that year?

74. A family has a budget of $3,500 per month. One-quarter goes for housing, $1,000 goes for food, and $528 is for car expenses. How much is left for other expenses?

75. The population of a city rose by 49,000 one year to reach 1 million people. What was the population at the beginning of the year?

76. The value of a share of stock in a company dropped $1.13 to reach $14.47. What was it worth before the drop?

THINK ABOUT IT

1. When you slice a pizza, the angles at the center sum to 360°. If you cut it into six 52° slices and one more left over, what will be the measure of the angle for that last piece? Show a sketch and write an appropriate equation.
2. Explain in words what the equation $x + 1 = x$ means; then explain why it must be a false equation.
3. What are equivalent equations? Explain, and give an example.
4. What is the difference between a conditional equation and an identity? Explain, and give an example of each.
5. Is the number π a solution to this equation: $\dfrac{3\pi}{4} + \dfrac{x}{4} = \pi$?

REMEMBER THIS

1. Simplify $\dfrac{3x}{3}$.
2. Simplify $\dfrac{-7x}{-7}$.
3. Simplify $2(\frac{1}{2}x)$.
4. Simplify $-4(-\frac{1}{4}x)$.
5. Simplify $-1(-x)$.

6. What is the product of -1 and $-6x$?
7. Write the formula for the area of a triangle.
8. Use the distributive law to rewrite $-(x + 2)$.
9. Represent 6 percent of x algebraically.
10. Write an algebraic expression to represent a number increased by 10 percent of itself.

2.3 The Division and Multiplication Properties of Equality

For selling a house, the real estate agent gets 6 percent of the sales price as a commission A couple wants to sell their house and end up with $150,000 after they pay the commission. Find the required selling price of the house by letting x represent the unknown and solving the equation $x - 0.06x = 150.000$.
(See Example 6.)

OBJECTIVES

1 Solve linear equations by applying the division and multiplication properties of equality.

2 Solve equations of the form $-x = n$.

3 Solve linear equations in which the first step is to combine the terms.

4 Solve applied problems by solving linear equations.

1 We have learned in Section 2.2 that we can generate equivalent equations by adding or subtracting the same number on both sides of an equation. In a similar way, we can multiply or divide both sides of an equation by the same number (as long as that number is not zero). First, consider the **division property of equality.**

> **Division Property of Equality**
>
> If A, B, and C are algebraic expressions and C does not represent zero, then the equations
> $$A = B \qquad \text{and} \qquad \frac{A}{C} = \frac{B}{C}$$
> are equivalent.

The requirement that C should not be zero is important, because division by zero is not defined. This property can be used to solve linear equations of the form $cx = n$ with $c \neq 0$, as shown in Example 1.

EXAMPLE 1   Solve the equation $3x = -21$.

Solution   Because x is multiplied by 3, we shall divide both sides of the equation by 3 to isolate the x.

$$3x = -21 \qquad \text{Given equation.}$$
$$\frac{3x}{3} = \frac{-21}{3} \qquad \text{Divide both sides by 3.}$$
$$1x = -7 \qquad \text{Simplify.}$$
$$x = -7 \qquad \text{Simplify.}$$

So, -7 is the solution to the equation.

We check the solution by replacing x by -7 in the original equation.

$$3x = -21 \qquad \text{Original equation.}$$
$$3(-7) \overset{?}{=} -21 \qquad \text{Replace } x \text{ by } -7.$$
$$-21 \overset{\checkmark}{=} -21 \qquad \text{Simplify.}$$

Since $-21 = -21$ is a true statement, the solution checks and the solution set is $\{-7\}$.

Caution   It is not helpful when solving $3x = -21$ to *subtract* 3 on both sides of the equation, because $3x - 3$ does not equal x.

PROGRESS CHECK 1   Solve the equation $5x = -105$.

Certain equations, especially those involving fractions, are best approached by multiplying both sides of the equation by the same nonzero number. We may do this by applying the **multiplication property of equality.**

Multiplication Property of Equality

If A, B, and C are algebraic expressions and C does not represent zero, then the equations

$$A = B \quad \text{and} \quad CA = CB$$

are equivalent.

We require that C not equal zero because multiplying both sides of an equation by zero results in the identity $0 = 0$, even if the original equation is not an identity. This outcome violates the principle of writing simpler *equivalent* equations at each step.

Note in the next two examples that when we apply the multiplication property, we take advantage of the fact that the product of a number and its reciprocal is 1.

EXAMPLE 2   Solve the equation $\dfrac{2}{3}x = -8$.

Solution   The coefficient of x is $\frac{2}{3}$ and the reciprocal of $\frac{2}{3}$ is $\frac{3}{2}$. Therefore, we can isolate x by multiplying both sides of the equation by $\frac{3}{2}$.

Progress Check Answer

1. $\{-21\}$

$$\frac{2}{3}x = -8 \qquad \text{Given equation.}$$

$$\frac{3}{2} \cdot \frac{2}{3}x = \frac{3}{2} \cdot (-8) \qquad \text{Multiply both sides by } \frac{3}{2}.$$

$$1 \cdot x = \frac{-24}{2} \qquad \text{Simplify.}$$

$$x = -12 \qquad \text{Simplify.}$$

We check in the usual way.

$$\frac{2}{3}x = -8 \qquad \text{Original equation.}$$

$$\frac{2}{3}(-12) \stackrel{?}{=} -8 \qquad \text{Replace } x \text{ by } -12.$$

$$-8 \stackrel{\checkmark}{=} -8 \qquad \text{Simplify.}$$

The solution checks and the solution set is $\{-12\}$.

PROGRESS CHECK 2   Solve the equation $\frac{3}{4}x = 24$.

EXAMPLE 3   Solve the equation $\dfrac{x}{7} = \dfrac{3}{11}$.

Solution   Since division by 7 produces the same result as multiplication by $\frac{1}{7}$, the equations

$$\frac{x}{7} = \frac{3}{11} \qquad \text{and} \qquad \frac{1}{7}x = \frac{3}{11}$$

are equivalent. To isolate x, we multiply both sides by 7 (the reciprocal of the coefficient of x).

$$\frac{1}{7}x = \frac{3}{11} \qquad \text{Replace } \frac{x}{7} \text{ by } \frac{1}{7}x.$$

$$7 \cdot \frac{1}{7}x = 7 \cdot \frac{3}{11} \qquad \text{Multiply both sides by 7.}$$

$$x = \frac{21}{11} \qquad \text{Simplify.}$$

We verify that $\frac{21}{11}$ solves the equation as follows.

$$\frac{x}{7} = \frac{3}{11} \qquad \text{Original equation.}$$

$$\frac{21/11}{7} \stackrel{?}{=} \frac{3}{11} \qquad \text{Replace } x \text{ by } 21/11.$$

$$\frac{21}{11} \cdot \frac{1}{7} \stackrel{?}{=} \frac{3}{11} \qquad \text{Change division by 7 to multiplication by } \frac{1}{7}.$$

$$\frac{3}{11} \stackrel{\checkmark}{=} \frac{3}{11} \qquad \text{Simplify.}$$

The solution set is $\{\frac{21}{11}\}$.

Note   We can also reason in this example that since x is divided by 7, we can isolate x by undoing this division and multiplying both sides of the equation by 7.

PROGRESS CHECK 3   Solve the equation $\dfrac{x}{3} = -\dfrac{4}{11}$.

Progress Check Answers

2. $\{32\}$

3. $\left\{-\dfrac{12}{11}\right\}$

2 In our future work in solving equations we sometimes come to an equivalent equation that has the form $-x = n$. We *cannot* stop here. Instead, we multiply both sides of the equation by -1 to obtain $x = -n$, as shown in the next example.

EXAMPLE 4   Solve the equation $4 - x = 6$.

Solution   Suppose we begin the usual way.

$$4 - x = 6 \qquad \text{Given equation.}$$
$$4 - x - 4 = 6 - 4 \qquad \text{Subtract 4 from both sides.}$$
$$-x = 2 \qquad \text{Simplify.}$$

Now since $-x = 2$ is equivalent to $-1x = 2$, we isolate x by multiplying both sides by -1.

$$(-1)(-x) = (-1)(2) \qquad \text{Multiply both sides by } -1.$$
$$x = -2 \qquad \text{Simplify.}$$

Finally, check that -2 is the solution of the equation.

$$4 - x = 6 \qquad \text{Original equation.}$$
$$4 - (-2) \overset{?}{=} 6 \qquad \text{Replace } x \text{ by } -2.$$
$$6 \overset{\checkmark}{=} 6 \qquad \text{Simplify.}$$

The solution checks and the solution set is $\{-2\}$.

PROGRESS CHECK 4   Solve the equation $3 - x = -3$. ⌐

3 When solving equations, we often must combine like terms before applying equality properties.

EXAMPLE 5   Solve the equation $4y - 5y = 20$.

Solution   We start by combining the two like terms on the left-hand side of the equation to get just one term containing y.

$$4y - 5y = 20 \qquad \text{Given equation.}$$
$$-y = 20 \qquad \text{Combine like terms.}$$
$$(-1)(-y) = (-1)(20) \qquad \text{Multiply both sides by } -1.$$
$$y = -20 \qquad \text{Simplify.}$$

Now check the solution.

$$4y - 5y = 20 \qquad \text{Original equation.}$$
$$4(-20) - 5(-20) \overset{?}{=} 20 \qquad \text{Replace } y \text{ by } -20.$$
$$-80 + 100 \overset{?}{=} 20 \qquad \text{Simplify.}$$
$$20 \overset{\checkmark}{=} 20 \qquad \text{Simplify.}$$

The solution set is $\{-20\}$.

PROGRESS CHECK 5   Solve the equation $2x - 5x = -12$. ⌐

4 The section-opening problem is a good example of an applied problem that involves solving an equation by first combining like terms.

EXAMPLE 6   Solve the problem in the section introduction on page 84.

Solution   The given equation is $x - 0.06x = 150,000$, with x representing the selling price of the house. The equation is derived by recognizing that the commission is 6

Progress Check Answers
4. $\{6\}$

5. $\{4\}$

percent of the selling price, so

$$x \quad - \quad 0.06x \quad = \quad 150{,}000$$

We solve the equation as follows.

$$0.94x = 150{,}000 \qquad \text{Combine like terms.}$$

$$\frac{0.94x}{0.94} = \frac{150{,}000}{0.94} \qquad \text{Divide both sides by 0.94.}$$

$$x = 159{,}574.47 \qquad \text{Simplify.}$$

The selling price should be $159,574.47. Of course, in practice the couple should round the number up a little, to, say, $159,600. Then

$$\text{commission} = 0.06(\$159{,}600) = \$9{,}576,$$
$$\text{proceeds from sale} = \$159{,}600 - \$9{,}576 = \$150{,}024.$$

PROGRESS CHECK 6   In a vacation community the real estate salespeople get a 12 percent commission on rentals. An owner wants to rent out a vacation house for one month and end up with $2,000 income. Find the required rental price for the house by letting x represent the unknown and solving the equation $x - 0.12x = 2{,}000$. Round the answer to the nearest $10.

Progress Check Answer

6. $\{\$2{,}270\}$

EXERCISES 2.3

In Exercises 1–84, solve the equation for the variable and check the solution.

1. $3x = -21$
2. $4x = -24$
3. $5x = 15$
4. $6x = 18$
5. $0.2x = 0.08$
6. $0.3x = 0.9$
7. $42 = -7x$
8. $42 = -6x$
9. $-13.95 = -3.1x$
10. $-2.25 = -1.5y$
11. $2x = 5$
12. $3x = 7$
13. $1 = 5y$
14. $3 = 4x$
15. $2.6x = 0$
16. $0.4x = 0.1$
17. $3x = -2.7$
18. $7x = -3.5$
19. $-2.6 = -2.6x$
20. $-1.234x = 0$
21. $\dfrac{x}{3} = 4$
22. $\dfrac{x}{2} = 11$
23. $\dfrac{3x}{4} = 3$
24. $\dfrac{5x}{6} = 10$
25. $\frac{2}{3}x = \frac{4}{9}$
26. $\frac{5}{3}y = \frac{15}{21}$
27. $-\frac{4}{3}y = 1$
28. $-\frac{5}{2}x = 1$
29. $0 = -\frac{8}{11}x$
30. $0 = -\frac{9}{5}x$
31. $-\frac{1}{2}y = -5$
32. $-\frac{1}{3}x = -2$
33. $-2x = -\frac{3}{4}$
34. $-3x = -\frac{4}{5}$
35. $\frac{11}{5}x = -\frac{11}{5}$
36. $\frac{7}{6}x = \frac{14}{6}$
37. $3 - x = 1$
38. $5 - x = 3$
39. $2 - x = 2$
40. $3 - x = 0$
41. $3 = 4 - x$
42. $1 = 1 - x$
43. $2 - y = -1$
44. $1 - y = -3$
45. $1 - x = 3$
46. $5 - x = 8$
47. $0 - x = 5$
48. $2 - x = -7$
49. $\frac{1}{2} - x = \frac{1}{2}$
50. $\frac{2}{3} - x = \frac{5}{3}$
51. $\frac{1}{4} - x = \frac{3}{8}$
52. $\frac{1}{3} - x = \frac{1}{2}$

53. $-0.1 = 0.3 - y$
54. $4.1 = 1.4 - x$
55. $\frac{2}{3} = \frac{1}{5} - x$
56. $-5 = -\frac{5}{2} - x$
57. $-x + 1 = 3.8$
58. $-y + 2.4 = 2.5$
59. $-y - \frac{2}{3} = \frac{3}{4}$
60. $-x - \frac{3}{4} = -\frac{4}{5}$
61. $3x - 4x = -7$
62. $2x - 3x = 10$
63. $5x - 2x = -12$
64. $2x - 8x = 18$
65. $-1 = 5y - 6y$
66. $-2 = 16y - 17y$
67. $4 = 5y - 7y$
68. $-1 = 12y - 16y$
69. $3x - 4x = 0$
70. $0 = 2y - 3y$
71. $x - 5x = 0$
72. $0 = 2y - 9y$
73. $-4x + 3x = 6$
74. $-2x + x = 3$
75. $-4y + y = 6$
76. $-5y + 2y = 5$
77. $0.1x - 1.1x = 1$
78. $2.5x - 3.5x = -4.1$
79. $1.4y - 3.4y = 12$
80. $2.8y - 5.8y = -20.1$
81. $\dfrac{3x}{2} - \dfrac{5x}{2} = 2$
82. $\dfrac{4x}{3} - \dfrac{7x}{3} = \dfrac{1}{3}$
83. $\dfrac{3y}{5} - \dfrac{4y}{5} = 1$
84. $\dfrac{y}{3} - \dfrac{7y}{3} = 2$

In Exercises 85–96, tell what operations should be performed on the first expression to give the second expression. (Answers may vary.)

85. $\dfrac{x}{5}$; x
86. $\dfrac{x}{3}$; x
87. $-x$; x
88. $-3x$; x
89. $\dfrac{y}{a}$; y
90. $\dfrac{y}{c}$; y
91. $\dfrac{1x}{2}$; x
92. $-\dfrac{3x}{4}$; x
93. $\dfrac{ax}{b}$; x
94. $\dfrac{cy}{d}$; y

95. $y - \dfrac{a}{b}$; y **96.** $x + \dfrac{b}{c}$; x

In Exercises 97–100, we have given a word problem and an equation that can be used to solve it. Solve the equation and answer the question.

97. When an art gallery sells a painting, the gallery gets a 15 percent commission, and the artist gets the rest. What should the selling price be so that the artist will get $2,500? If x stands for the selling price, the equation to solve is $x - 0.15x = 2,500$.

98. The agent for a sports star gets 10 percent of the star's annual income. How much does the star have to earn so that the star's own share will be $1 million? Use the equation $x - 0.10x = 1,000,000$, where x stands for the star's earnings.

99. The total amount (including a 6 percent sales tax) for the sale of an automobile was $24,183.90. How much did the auto cost before the tax was added? Use $x + 0.06x = 24,183.90$, where x is the price of the car.

100. A consumer price index rose 4 percent one year and ended up at 147.80. What was it the year before? Solve $x + 0.04x = 147.80$, where x is the previous year's value. Round your answer to two decimal places.

In Exercises 101–104, solve each problem by first setting up an appropriate equation. (*Note:* These exercises are similar to the word problems in Exercises 97–100.)

101. In a certain region there was a recession and only 528 new houses were started, which represented a 12 percent decrease from the previous year. How many houses were built the previous year?

102. Factory equipment depreciates in value each year. In one factory the value of equipment dropped 17 percent during the year. At the end of the year the owners reported to the IRS that their equipment was worth $6.2 million. How much was it worth one year earlier? Round the answer to one decimal place.

103. A charity spends 5 percent of the money it collects for administrative expenses. One year these expenses amounted to $500,000. How much did the charity collect in the first place?

104. An ecologist states that in one year $\frac{1}{20}$ of a rain forest was cut down. That amounted to 12,000 acres. How large was the forest before the trees were cut?

THINK ABOUT IT

In Exercises 1–4, set up an equation and solve it to answer the question. Round answers sensibly.

1. An auctioneer sells property and takes a 20 percent commission. The owner gets the rest. How much would the property have to sell for so that the owner would get $5,000 after the commission is paid?

2. Upon moderate exercise an adult's heart rate often increases 40 percent. If a heart rate increases 40 percent and ends up at 120 beats per minute, what was it to start with?

3. An author gets a 15 percent royalty on the sales of a book. How many dollars of book sales would net the author $60,000 in income?

4. An inventor gets a commission of one-quarter of 1 percent (0.25 percent, written 0.0025) every time one invention is used by an automobile manufacturer in a new car. What amount of new car sales will bring in $10,000 for the inventor? About how many cars must be sold if the average car price is $15,000?

5. Why do we say that division by zero is undefined? Why isn't the answer to 7 divided by 0 equal to 0?

REMEMBER THIS

1. Use the distributive property to multiply $5(x - 1)$.
2. Use the distributive property to multiply $-2(3x - 4)$.
3. Multiply $\frac{2}{3}(6 + 9x)$.
4. Use the distributive property to rewrite $-(5 - x)$
5. What is the product of 3 and $2x - 7$?
6. What is the sum of 3 and $2x - 7$?

7. Write the formula for the volume of a rectangular box.

8. Write the reciprocal of -3.
9. Write an algebraic expression to represent 4 less than one-half of a number.
10. Write an algebraic expression for the square of the sum of x and 8.

2.4   A General Approach to Solving Linear Equations

Debbie is offered two jobs selling exercise equipment. In one store she would be paid $100 per week plus a 10 percent commission on each sale. In the other she gets a straight 15 percent commission. At a certain point the second job begins to pay more than the first. Find this point by letting x stand for the amount of sales at which both jobs pay the same and solving the equation $100 + 0.10x = 0.15x$. (See Example 6.)

O B J E C T I V E S

1 Apply a general procedure to solve *any* linear equation.

2 Classify certain equations as false equations or identities.

3 Solve applied problems by solving linear equations.

1   To this point, we have used the addition or subtraction properties of equality to solve equations of the form $x + c = n$, and the multiplication or division properties to solve $cx = n$. In this final section on solving linear equations we combine our previous methods to formulate a general procedure to solve any linear equation.

Solution of Linear Equations

1. Simplify each side of the equation if necessary. This step involves mainly combining like terms, sometimes preceded by using the distributive property to remove parentheses.
2. Use the addition or subtraction properties of equality, if necessary, to write an equivalent equation of the form $cx = n$. To accomplish this, write equivalent equations with all terms involving the unknown on one side of the equation and all constant terms on the other side.
3. Use the multiplication or division properties of equality to solve $cx = n$ if $c \neq 1$. The result of this step will read $x = $ number.
4. Check the solution by substituting it in the original equation.

Consider Example 1 carefully. All four steps are needed in the solution.

EXAMPLE 1   Solve the equation $5(x - 1) = 2x + 1$.

Solution   We follow the steps outlined.

1. Simplify the left side of the equation by removing parentheses.

$$5(x - 1) = 2x + 1 \quad \text{Given equation.}$$
$$5x - 5 = 2x + 1 \quad \text{Distributive property.}$$

2. Obtain an equation of the form $cx = n$ by using the addition or subtraction properties of equality.

$$5x - 2x - 5 = 2x - 2x + 1 \quad \text{Subtract } 2x \text{ from both sides.}$$
$$3x - 5 = 1 \quad \text{Simplify.}$$
$$3x - 5 + 5 = 1 + 5 \quad \text{Add 5 to both sides.}$$
$$3x = 6 \quad \text{Simplify.}$$

3. Obtain an equation of the form $x = $ number by using the multiplication or division properties of equality.

$$\frac{3x}{3} = \frac{6}{3} \quad \text{Divide both sides by 3.}$$
$$x = 2 \quad \text{Simplify.}$$

4. Check the solution in the original equation.

$$5(x - 1) = 2x + 1 \quad \text{Original equation.}$$
$$5(2 - 1) \stackrel{?}{=} 2(2) + 1 \quad \text{Replace } x \text{ by 2.}$$
$$5(1) \stackrel{?}{=} 4 + 1 \quad \text{Simplify.}$$
$$5 \stackrel{\checkmark}{=} 5 \quad \text{Simplify.}$$

The solution set is $\{2\}$.

PROGRESS CHECK 1   Solve the equation $7(x + 2) = 5x - 4$.   ⌐

To solve certain linear equations, it may not be necessary to use all of the steps outlined (as illustrated in Example 2). Checking the solution, however, is always recommended at this stage because it helps to catch any mistakes.

EXAMPLE 2   Solve the equation $4x + 7 = 3x - 2$.

Solution   We cannot simplify either side of the equation, so we proceed to step 2 and obtain an equation of the form $cx = n$.

$$
\begin{array}{ll}
4x + 7 = 3x - 2 & \text{Given equation.} \\
4x - 3x + 7 = 3x - 3x - 2 & \text{Subtract } 3x \text{ from both sides.} \\
x + 7 = -2 & \text{Simplify.} \\
x + 7 - 7 = -2 - 7 & \text{Subtract 7 from both sides.} \\
x = -9 & \text{Simplify.}
\end{array}
$$

The equation $x = -9$ is already in the form $x =$ number, so we omit step 3 and check that -9 is the solution.

$$
\begin{array}{ll}
4x + 7 = 3x - 2 & \text{Original equation.} \\
4(-9) + 7 \overset{?}{=} 3(-9) - 2 & \text{Replace } x \text{ by } -9. \\
-36 + 7 \overset{?}{=} -27 - 2 & \text{Simplify.} \\
-29 \overset{\checkmark}{=} -29 & \text{Simplify.}
\end{array}
$$

The solution checks and the solution set is $\{-9\}$.

PROGRESS CHECK 2   Solve the equation $6x - 5 = 5x - 6$.   ⌐

In the next example, note that we choose in step 2 to write an equivalent equation with the unknown on the right side, because this choice leads to the simple form $n = x$. Remember that you may work toward the form $cx = n$ or the form $n = cx$. In only a few cases will one form be more beneficial than the other.

EXAMPLE 3   Solve the equation $3x + 2(x - 5) = 6x + 2$

Solution   Use the solution steps, except that step 3 is unnecessary.

Step 1
$$
\begin{array}{ll}
3x + 2(x - 5) = 6x + 2 & \text{Given equation.} \\
3x + 2x - 10 = 6x + 2 & \text{Distributive property.} \\
5x - 10 = 6x + 2 & \text{Combine like terms.}
\end{array}
$$
Step 2
$$
\begin{array}{ll}
5x - 5x - 10 = 6x - 5x + 2 & \text{Subtract } 5x \text{ from both sides.} \\
-10 = x + 2 & \text{Simplify.} \\
-10 - 2 = x + 2 - 2 & \text{Subtract 2 from both sides.} \\
-12 = x & \text{Simplify.}
\end{array}
$$
Step 4   Check.
$$
\begin{array}{ll}
3x + 2(x - 5) = 6x + 2 & \text{Original equation.} \\
3(-12) + 2(-12 - 5) \overset{?}{=} 6(-12) + 2 & \text{Replace } x \text{ by } -12. \\
-36 - 34 \overset{?}{=} -72 + 2 & \text{Simplify.} \\
-70 \overset{\checkmark}{=} -70 & \text{Simplify.}
\end{array}
$$

The solution set is $\{-12\}$.

Progress Check Answers
1. $\{-9\}$
2. $\{-1\}$

PROGRESS CHECK 3   Solve the equation $2(x + 7) = 4(x - 3) - x$.   ⌐

When a linear equation contains fractions, it is usually best to remove fractions first by multiplying both sides of the equation by the least common denominator, or LCD. The resulting equation will not contain fractions and can be solved in the usual four-step way.

EXAMPLE 4   Solve the equation $\frac{1}{8}x + \frac{1}{2} = \frac{3}{4}x$.

Solution   Since the denominators are 8, 2, and 4, the LCD is 8. First, multiply both sides of the equation by 8 to clear fractions, and then apply the four-step procedure.

	$\frac{1}{8}x + \frac{1}{2} = \frac{3}{4}x$	Given equation.
	$8(\frac{1}{8}x + \frac{1}{2}) = 8(\frac{3}{4}x)$	Multiply both sides by 8.
Step 1	$x + 4 = 6x$	Simplify.
Step 2	$x - x + 4 = 6x - x$	Subtract x from both sides.
Step 3	$4 = 5x$	Simplify.
	$\dfrac{4}{5} = \dfrac{5x}{5}$	Divide both sides by 5.
	$\dfrac{4}{5} = x$	Simplify.

Step 4   Check.

$\frac{1}{8}x + \frac{1}{2} = \frac{3}{4}x$	Original equation.
$\frac{1}{8}(\frac{4}{5}) + \frac{1}{2} \overset{?}{=} \frac{3}{4}(\frac{4}{5})$	Replace x by $\frac{4}{5}$.
$\frac{1}{10} + \frac{1}{2} \overset{?}{=} \frac{3}{5}$	Simplify.
$\frac{3}{5} \overset{\checkmark}{=} \frac{3}{5}$	Simplify.

The solution set is $\{\frac{4}{5}\}$.

PROGRESS CHECK 4   Solve the equation $\frac{2}{3}x - \frac{5}{12} = \frac{5}{6}x$.   ⌐

2   Certain equations can be identified as false equations or identities by applying the first two steps in our general approach to solving linear equations. The solution set of a false equation is ∅, the empty set, and the solution set of an identity is the set of all real numbers.

EXAMPLE 5   Solve the equation $4 - 2(x + 1) = 5 - 2x$.

Solution   Simplify the left side of the equation, and then use the addition property of equality.

Step 1	$4 - 2(x + 1) = 5 - 2x$	Given equation.
	$4 - 2x - 2 = 5 - 2x$	Distributive property.
	$2 - 2x = 5 - 2x$	Simplify.
Step 2	$2 - 2x + 2x = 5 - 2x + 2x$	Add $2x$ to both sides.
	$2 = 5$	Simplify.

Since $2 = 5$ is never true, the original equation has no solution and the solution set is ∅.

Caution   In this example, be careful to apply the distributive law first to remove parentheses. A common student error is to incorrectly rewrite $4 - 2(x + 1)$ as $2(x + 1)$.

PROGRESS CHECK 5   Solve the equation $2 - (3x - 1) = -3(x - 1)$.   ⌐

3   With the methods from this section we can solve the equation in the application problem that opens this section.

EXAMPLE 6   Solve the problem in the section introduction on page 89.

Solution   We are asked to solve $100 + 0.10x = 0.15x$, where x stands for the amount of sales at which both jobs pay the same. This equation is derived as follows.

		Pay at store 1		*Pay at store 2*
100	plus	10 percent of sales	equals	15 percent of sales
↓	↓		↓	
100	+	$0.10x$	=	$0.15x$

We cannot simplify either side of the equation, so begin at step 2.

Step 2 $100 + 0.10x - 0.10x = 0.15x - 0.10x$ Subtract 0.10x from both sides.

$100 = 0.05x$ Simplify.

Step 3 $\dfrac{100}{0.05} = \dfrac{0.05x}{0.05}$ Divide both sides by 0.05.

$2{,}000 = x$ Simplify.

Step 4   Check.

$100 + 0.10x = 0.15x$ Original equation.

$100 + 0.10(2{,}000) \overset{?}{=} 0.15(2{,}000)$ Replace x by 2,000.

$100 + 200 \overset{?}{=} 300$ Simplify.

$300 \overset{\checkmark}{=} 300$ Simplify.

The check shows that at \$2,000 in sales both jobs pay \$300. For sales above \$2,000 the straight 15 percent commission pays more. For sales less than \$2,000 the first job pays more.

PROGRESS CHECK 6   Redo the problem in Example 6, assuming that one store pays \$150 per week plus a 12 percent commission while the second store pays \$120 per week plus a 16 percent commission. Use $150 + 0.12x = 120 + 0.16x$ for the revised equation.

Progress Check Answers

6. $\{750\}$; when sales equals \$750, both jobs pay the same.

EXERCISES 2.4

In Exercises 1–72, solve the equation, and then check the solution.

1. $2(x + 4) = 9$
2. $3(x + 2) = 5$
3. $5(x - 1) = 12$
4. $3(x - 2) = 10$
5. $2x + 3 = 5$
6. $3x + 2 = 5$
7. $4x - 3 = -1$
8. $2x - 1 = 0$
9. $2 = -3(x - 1)$
10. $0 = -4(2 - x)$
11. $\frac{1}{2}(3x + 4) = 5$
12. $\frac{2}{3}(5y - 6) = 1$
13. $5(x + 1) = 2x - 1$
14. $3(x - 2) = 4x + 3$
15. $2(x - 4) = x - 8$
16. $4(x + 3) = x - 10$
17. $3(4 + x) = 5 - 5x$
18. $2(5 - x) = 3 - 3x$
19. $5(2x + 8) = 8x$
20. $7 + 11x = 3(5x + 2)$

21. $1 - 8y = 2(5 - 3y)$
22. $2 - 7y = 3(1 - 2y)$
23. $\frac{3}{4}(4x - 8) = x - 6$
24. $\frac{5}{3}(12 + 3x) = 20 + x$
25. $4x + 7 = 3x + 1$
26. $5x - 5 = 4x + 4$
27. $5y + 1 = 3y + 1$
28. $5y + 1 = 7 + 3y$
29. $y = 4 - 3y$
30. $7y = 2 + 3y$
31. $3x - 2 = x - 2$
32. $4x - 1 = x - 4$
33. $12x - 5 = 9x + 5$
34. $10x - 3 = 5x + 12$
35. $\frac{x}{3} - 1 = \frac{x}{4} + 1$
36. $\frac{x}{4} - 2 = \frac{x}{5} + 2$
37. $2(x + 1) = 3x + 1$
38. $2(x - 3) = 5x - 1$
39. $2x + 3(x - 4) = 7x - 18$
40. $3x - 2(x - 9) = 4 + 2x$
41. $3(x + 2) = 7(x - 2) - 2x$

42. $4(x - 3) = 6(x + 1) - 3$
43. $x + 8 = 3x + 4$
44. $6 - 7x = x - 2$
45. $3x + 2(4 - x) = 4(x - 3)$
46. $4x - (3 - x) = 5(3 + 2x)$
47. $x - (x - 1) = 1 - (1 - x)$
48. $2 - (2 - x) = x - (x - 3)$
49. $-3(x + 1) = 2x - 1$
50. $-3(x - 1) = 1 - x$
51. $-2(x + 4) = x - 8$
52. $-2(x + 4) = x$
53. $-4x + 7 = 3x + 1$
54. $-3x + 1 = 4x + 1$
55. $3 - 2(x + 1) = 2 - 5x$
56. $5 - 3(x - 2) = 3x + 6$
57. $5 - 2(3x - 4) - x = 3 - 2x$
58. $4 - 3(6x - 5) = 3(2 - 5x)$
59. $\frac{1}{2} - \frac{1}{3}(6x - 3) = \frac{1}{5}(5 - 50x)$
60. $\frac{3}{2} - \frac{1}{2}(4 + 2x) = \frac{4}{3}(3x + 9)$
61. $\frac{x}{2} + \frac{1}{3} = \frac{x}{5}$

62. $\frac{3}{4}x + \frac{1}{2} = \frac{7}{4}x$

63. $\frac{2x}{3} + \frac{12}{5} = \frac{x}{2} - \frac{5}{6}$

64. $\frac{4x}{3} + \frac{5}{2} = \frac{x}{2} - \frac{5}{6}$

65. $\frac{5}{2}(x + 4) = \frac{1}{2}x - 8$

66. $\frac{3}{2}(x - 1) = \frac{1}{2}(x + 2)$

67. $-\frac{1}{3}(3x + 1) = 4 - 2x$

68. $-\frac{1}{4}(x + \frac{2}{3}) = 5$

69. $\frac{3}{4}(x - 1) - \frac{1}{2}x = 0$

70. $\frac{5}{3} - \frac{1}{3}(9x + 5) = x$

71. $\frac{4}{5} + \frac{1}{5}(10x - 4) = 2$

72. $\frac{3}{7} + \frac{4}{7}(1 - 7x) = x - 1$

In Exercises 73–84, determine if the equation is a false equation or an identity.

73. $3(x + 1) = 3x + 1$

74. $2(3 + x) = 5 + 2x$

75. $2(x - 3) = 2x - 6$

76. $5(2x + 1) = 2 + (10x + 3)$

77. $\frac{1}{2}(4x + 6) = 3 + 2x$

78. $\frac{2}{3}(2x - 1) = \frac{4}{3}(x - \frac{1}{2})$

79. $-3(1 - 2x) = 6(x - 1)$

80. $-2(\frac{1}{2}x + 1) = 2 - x$

81. $4 + 3(5x - 1) = 2(6x + 5) + 3x$

82. $3(x + 1) + 2(x + 3) = 5x + 9$

83. $\frac{x}{2} + \frac{x}{2} = x$

84. $\frac{3x}{5} + 1 = \frac{1}{5}(3x + 10)$

In Exercises 85–93, we have given a word problem and an equation to solve it. Solve the problem.

85. One sales job pays $160 per week plus a 10 percent commission on sales. A second pays $130 per week plus a 12 percent commission. At a certain point of sales the second job pays more than the first. Determine this point by solving the equation $160 + 0.10x = 130 + 0.12x$, where x stands for the amount of sales.

86. A waiter can be paid $300 per week or paid $100 per week plus 15 percent of the cost of all the meals the waiter serves. To the nearest dollar, how many dollars worth of meals would have to be served so that the second scheme paid more? Find out by solving $300 = 100 + 0.15x$, where x stands for the cost of the meals.

87. For what value of x are the perimeters of the two rectangles in the figure equal? Use the formula $P = 2w + 2\ell$ and solve $2(\frac{1}{4}x) + 2(x - \frac{1}{2}) = 2(\frac{1}{2}x) + 2(\frac{1}{2}x)$.

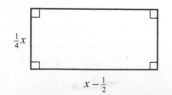

88. For what value of x are the perimeters of the two triangles in the figure equal? Solve $x + (x + 1) + (x + 2) = 2(x - 1) + 2(x - 1) + 2(x - 1)$.

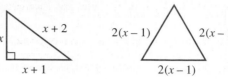

89. Show that the perimeters of these two rectangles cannot be equal by solving $2x + 2(x + 1) = 2x + 2x$.

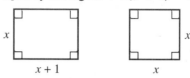

90. Show that the perimeters of these two triangles cannot be equal by solving $x + (x + 1) + (x + 2) = x + x + x$.

91. Half of a certain number equals two more than the number itself. How can this be? Let x stand for the number and find out by solving $\frac{1}{2}x = x + 2$. Check that your answer makes sense.

92. Alex gets paid $100 less per week than Bernie, but if he gets a 20 percent raise, he will catch up. Let x stand for Alex's pay, and solve $x + 0.20x = x + 100$ to find out their salaries.

93. John gets paid $100 more per week than Jim. If Jim gets a 20 percent raise and John takes a 4 percent cut, their salaries will be the same. What are their salaries now? Let x represent Jim's salary and $x + 100$ stand for John's salary. Then solve $x + 0.20x = (x + 100) - 0.04(x + 100)$.

94. Test your skills on the following problem from the *Mathematical Puzzles of Sam Loyd*, a two-volume series edited by Martin Gardner and published by Dover Publications (© 1960).

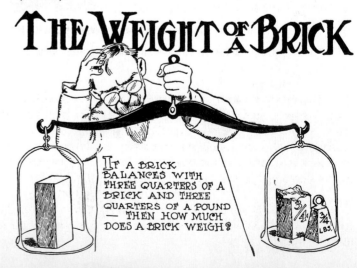

THE WEIGHT OF A BRICK

IF A BRICK BALANCES WITH THREE QUARTERS OF A BRICK AND THREE QUARTERS OF A POUND — THEN HOW MUCH DOES A BRICK WEIGH?

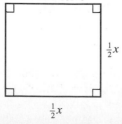

THINK ABOUT IT

In Exercises 1 and 2, solve the problem by writing an appropriate equation and solving it.

1. One sales job pays $150 per week plus a 5 percent commission on sales. A second pays $120 per week plus a 7 percent commission. Find the point at which both jobs pay the same. What is the pay at that point?

2. If you add 5 to a certain number, you will get the same answer as if you multiplied it by 3. What is the number?

3. If an equation has the form $ax + b = c$, what is the first step in solving for x? What is the second step?

4. When an equation contains fractions, why is it often a good idea to multiply both sides by the LCD of the fractions?

5. Make up a word problem that corresponds to this equation: $400 + 0.10x = 0.15x$.

REMEMBER THIS

1. What operation on the expression $ab + x$ would result in the expression x?

2. What operation on the expression cdx would result in the expression x?

3. Solve $3 = 4x$.

4. Write an expression to represent half of 5 more than a certain number.

5. Rewrite $(b + c)d$ using the distributive property.

6. Write the formula for the circumference of a circle.

7. If the product of $8 + x$ and 5 equals zero, what is x?

8. If the sum of $8 + x$ and 5 equals zero, what is x?

9. Simplify $(-1)100$.

10. Which is greater, $3^2 + 4^2$ or $(3 + 4)^2$?

2.5 Formulas and Literal Equations

For a science project a student must convert many temperature measurements from degrees Fahrenheit to degrees Celsius. In a reference book the student finds the formula $F = \frac{9}{5}C + 32$ that can be used to make the conversions, but inefficiently. Solve this formula for C in terms of F to obtain an efficient formula for the student's needs. (See Example 4.)

OBJECTIVES

1 Find the value of a variable in a formula when given values for the other variables.

2 Solve a given formula or literal equation for a specified variable.

1 To analyze relationships, we often use **literal equations,** which are equations that contain two or more letters. The letters may represent any mix of variables and constants. Common examples of literal equations are formulas, such as the simple-interest formula $I = Prt$, and general forms of equations such as $ax + b = 0$. Because literal equations and formulas are types of equations, we may use the equation-solving techniques developed in this chapter to analyze them.

EXAMPLE 1   The formula $I = Prt$ gives the simple interest I if P dollars are invested for t years at an annual interest rate r. Find the value of t if $I = \$420$, $P = \$1,200$, and $r = 0.07$.

Solution   Substitute the given values into the formula and then solve for t.

$$I = Prt$$ Given formula.
$$420 = 1,200(0.07)t$$ Substitute the given values.
$$420 = 84t$$ Simplify.
$$\frac{420}{84} = \frac{84t}{84}$$ Divide both sides by 84.
$$5 = t$$ Simplify.

Check the solution by replacing each letter by its value. You will get the identity $420 = 420$, proving that $t = 5$ is correct.

PROGRESS CHECK 1   The formula $V = \ell wh$ gives the volume of a rectangular solid with length ℓ, width w, and height h. Find the width of a rectangular packaging carton if $\ell = 12$ in., $h = 8$ in., and $V = 1,056$ in.³   ⌐

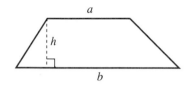

a

h

b

Figure 2.4

EXAMPLE 2   The formula $A = \frac{1}{2}h(a + b)$ gives the area of a trapezoid with height h and bases a and b, as shown in Figure 2.4. Find the value of b if $h = 9$ cm, $a = 11$ cm, and $A = 108$ cm².

Solution   After substituting the given values in the formula, we solve for b by first multiplying both sides by 2 to clear fractions.

$$A = \tfrac{1}{2}h(a + b)$$ Given formula.
$$108 = \tfrac{1}{2}(9)(11 + b)$$ Substitute the given values.
$$216 = 9(11 + b)$$ Multiply both sides by 2.
$$216 = 99 + 9b$$ Distributive property.
$$117 = 9b$$ Subtract 99 from both sides.
$$13 = b$$ Divide both sides by 9.

Since $\frac{1}{2}(9)(11 + 13) = \frac{1}{2}(9)(24) = 108$, the solution checks and $b = 13$ cm.

Note   If h is an even number, then it is easier to clear fractions in this equation by taking one-half of h.

PROGRESS CHECK 2   Find the value of a in the formula $A = \frac{1}{2}h(a + b)$ if $h = 8$ cm, $b = 17$ cm, and $A = 92$ cm².   ⌐

2   A formula is in an efficient form for a particular problem when it is possible to merely substitute the given values and perform the indicated operations (often by calculator). For example, the formula $P = 4s$ gives the perimeter P of a square in terms of the length of its side s. This formula is easy to use when we know s and need to find P. However, since the formula is not efficient if we know P and wish to find s, we may solve for s as follows.

$$P = 4s$$ Given equation.
$$\frac{P}{4} = \frac{4s}{4}$$ Divide both sides by 4.
$$\frac{P}{4} = s$$ Simplify.

Thus $P = 4s$ is equivalent to $s = P/4$. Now we may select the more useful version of the formula, depending on the given information.

Progress Check Answers

1. $w = 11$ in.

2. $a = 6$ cm

EXAMPLE 3   The area of a triangle is given by $A = \frac{1}{2}bh$, where b measures the base and h is the height. Solve this formula for h.

Solution   To solve for h, we need to isolate h. First, multiply both sides by 2 to clear fractions; then divide by b to undo the product bh.

$$A = \tfrac{1}{2}bh \qquad \text{Given formula.}$$
$$2A = 2(\tfrac{1}{2})bh \qquad \text{Multiply both sides by 2.}$$
$$2A = bh \qquad \text{Simplify.}$$
$$\frac{2A}{b} = \frac{bh}{b} \qquad \text{Divide both sides by } b.$$
$$\frac{2A}{b} = h \qquad \text{Simplify.}$$

The formula $h = 2A/b$ gives h in terms of A and b.

PROGRESS CHECK 3   Solve $V = \tfrac{1}{3}Bh$ for B.   ⌐

EXAMPLE 4   Solve the problem in the section introduction on page 95.

Solution   We will solve for C so that the form of the result is the common formula for converting from degrees Fahrenheit to degrees Celsius.

$$F = \tfrac{9}{5}C + 32 \qquad \text{Given formula.}$$
$$F - 32 = \tfrac{9}{5}C + 32 - 32 \qquad \text{Subtract 32 from each side.}$$
$$F - 32 = \tfrac{9}{5}C \qquad \text{Simplify.}$$
$$\tfrac{5}{9}(F - 32) = \tfrac{5}{9}(\tfrac{9}{5})C \qquad \text{Multiply both sides by } \tfrac{5}{9}.$$
$$\tfrac{5}{9}(F - 32) = C \qquad \text{Simplify.}$$

Thus, the student should use the formula $C = \tfrac{5}{9}(F - 32)$ to make the temperature conversions.

Note   In this example it is also logical to clear fractions as a first step by multiplying both sides by 5. This approach leads to the result

$$C = \frac{5F - 160}{9}.$$

When rearranging formulas, there may be several acceptable answer forms and you may need to consult with your instructor to see if your answer is correct.

PROGRESS CHECK 4   Solve $A = P + Prt$ for t.   ⌐

In Chapters 6 and 7 we will sometimes find it useful to solve equations of the form $ax + by = C$ for y. Example 5 illustrates this type of problem.

EXAMPLE 5   Solve $7x - y = 2$ for y.

Solution   We isolate y as follows.

$$7x - y = 2 \qquad \text{Given equation.}$$
$$7x - 7x - y = 2 - 7x \qquad \text{Subtract } 7x \text{ from both sides.}$$
$$-y = 2 - 7x \qquad \text{Simplify.}$$
$$-1(-y) = -1(2 - 7x) \qquad \text{Multiply both sides by } -1.$$
$$y = -2 + 7x \qquad \text{Simplify.}$$

The answer may be written as $y = -2 + 7x$ or $y = 7x - 2$.

PROGRESS CHECK 5   Solve $3x - y = -7$ for y.   ⌐

In our final example we begin with the general form of a linear equation and solve this equation for x. Note that this example proves that each linear equation has exactly one solution, $-b/a$.

Progress Check Answers

3. $B = \dfrac{3V}{h}$

4. $t = \dfrac{A - P}{Pr}$

5. $y = 3x + 7$

EXAMPLE 6   Solve $ax + b = 0$ with $a \neq 0$ for x.

Solution   Isolate x in the usual way.

$$ax + b = 0 \qquad \text{Given equation.}$$
$$ax + b - b = 0 - b \qquad \text{Subtract } b \text{ from both sides.}$$
$$ax = -b \qquad \text{Simplify.}$$
$$\frac{ax}{a} = \frac{-b}{a} \qquad \text{Divide both sides by } a.$$
$$x = -\frac{b}{a} \qquad \text{Simplify.}$$

Progress Check Answer

6. $x = \dfrac{n + c}{b}$

The restriction $a \neq 0$ is needed because $-b/0$ is undefined.

PROGRESS CHECK 6   Solve $bx - c = n$ with $b \neq 0$ for x.

EXERCISES 2.5

In Exercises 1–36, find the value of the indicated variable in each formula.

1. $d = rt$, $d = 130$, $r = 65$. Find t.
2. $d = rt$, $d = 165$, $r = 55$. Find t.
3. $V = \ell wh$, $V = 100$, $\ell = 4$, $w = 5$. Find h.
4. $V = \ell wh$, $V = 240$, $\ell = 10$, $w = 6$. Find h.
5. $I = Pr$, $I = 50$, $r = 0.08$. Find P.
6. $I = Pr$, $I = 75$, $r = 0.06$. Find P.
7. $P = 4s$, $P = 15$. Find s.
8. $P = 4s$, $P = 32$. Find s.
9. $E = IR$, $E = 9$, $R = 2$. Find I.
10. $E = IR$, $E = 36$, $R = 12$. Find I.
11. $F = ma$, $F = 6$, $m = 12$. Find a.
12. $F = ma$, $F = 3$, $m = 21$. Find a.
13. $C = 2\pi r$, $C = 6.28$, $\pi = 3.14$. Find r.
14. $C = 2\pi r$, $C = 314$, $\pi = 3.14$. Find r.
15. $A = P + Prt$, $A = 1{,}000$, $P = 500$, $r = 0.10$. Find t.
16. $A = P + Prt$, $A = 550$, $P = 500$, $t = 2$. Find r.
17. $5x - 2y = 10$, $x = 12$. Find y.
18. $5x - 2y = 10$, $x = 24$. Find y.
19. $3x + 2y = -8$, $x = 4$. Find y.
20. $3x + 2y = -6$, $x = 3$. Find y.
21. $A = \frac{1}{2}h(a + b)$, $A = 4$, $h = 1$, $a = 2$. Find b.
22. $A = \frac{1}{2}h(a + b)$, $A = 4$, $h = 2$, $a = 3$. Find b.
23. $F = \frac{9}{5}C + 32$, $F = 212$. Find C.
24. $F = \frac{9}{5}C + 32$, $F = 32$. Find C.
25. $A = 2ah - \frac{2}{3}a^2$, $A = 12$, $a = 3$. Find h.
26. $A = 2ah - \frac{2}{3}a^2$, $A = 24$, $a = 3$. Find h.
27. $F = \dfrac{GMm}{d^2}$, $F = 1$, $d = 1$, $G = 1$, $M = 5$. Find m.
28. $F = \dfrac{GMm}{d^2}$, $F = 2$, $d = 2$, $G = 2$, $M = 5$. Find m.
29. $A = \frac{1}{2}bh$, $A = \frac{1}{3}$, $b = 4$. Find h.
30. $A = \frac{1}{2}bh$, $A = \frac{7}{3}$, $b = \frac{3}{4}$. Find h.
31. $E = hv$, $E = \frac{1}{7}$, $h = \frac{3}{14}$. Find v.

32. $E = hv$, $E = \frac{1}{5}$, $h = \frac{1}{15}$. Find v.
33. $E = \frac{1}{2}mv^2$, $E = 0$, $v = 197.27$. Find m.
34. $E = \frac{1}{2}mv^2$, $E = 0$, $v = 0.0123$. Find m.
35. $X = \mu + z\sigma$, $X = 7.32$, $\mu = 4.08$, $\sigma = 1.1$. Find z.
36. $X = \mu + z\sigma$, $X = -3$, $\mu = -5$, $\sigma = 0.5$. Find z.

In Exercises 37–54, solve each formula for the variable indicated.

37. $d = rt$ for r.
38. $F = ma$ for a.
39. $P = 2(\ell + w)$ for ℓ.
40. $C = 5(a + b + c)$ for a.
41. $V = \ell wh$ for ℓ.
42. $V = \frac{1}{2}abh$ for h.
43. $V = \frac{1}{3}a^2h$ for h.
44. $V = \frac{4}{3}\pi r^3$ for r^3.
45. $X = \mu + z\sigma$ for z.
46. $y = mx + b$ for m.
47. $V = \pi r^2 h$ for h.
48. $C = 10s^2h$ for h.
49. $A = \dfrac{a + b}{2}$ for a.
50. $A = \dfrac{a + b + c}{3}$ for b.
51. $a^2 + b^2 = c^2$ for a^2.
52. $x^3 + y^3 = z^3$ for x^3.
53. $A = \frac{1}{2}h(a + b)$ for b.
54. $S = \frac{1}{2}n(a + \ell)$ for ℓ.

In Exercises 55–66, solve for y.

55. $3x + 2y = 11$
56. $4x + 3y = 12$
57. $2x - y = 0$
58. $5x - y = 0$
59. $-2x + 2y = 6$
60. $-3x + 3y = 6$
61. $-0.5x - 0.4y = 1$
62. $0.4x - 0.5y = 0$
63. $-x - 2y = -1$
64. $-3x - 2y = 5$
65. $\dfrac{x + y}{5} = 7$

66. $\dfrac{x + y}{3} = 0$

67. The formula for converting kilometers (k) to miles (m) is $m = 0.6214k$.

   a. Solve the formula for k.

   b. The radius of the earth is about 4,000 mi. Express this distance in kilometers.

68. The speed of sound in air is about 1,100 ft/second. This is given as $d = 1{,}100t$, where d is the distance in feet and t is the time in seconds.

   a. Solve this formula for t.

   b. How long does it take the sound of the bat hitting the ball to reach an outfielder 350 ft away from home plate?

69. The area of an ellipse is given by $A = \pi ab$, where a and b are as shown in the figure.

   a. Solve for a.

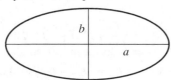

b. If the area is 100 cm² and b is 10 cm, find a.

70. The volume of a pyramid with height h and a square base of side s is given by $V = \frac{1}{3}s^2h$.

   a. Solve the formula for h.

   b. If the volume is 54 in.³ and $s = 2$ in., find h.

71. A teacher gives different weights to three exams by averaging them according to the formula $A = 0.20g_1 + 0.30g_2 + 0.50g_3$.

   a. Solve the formula for g_3 (grade 3).

   b. A student ended up with a weighted average of 83 after getting a 75 for grade 1 and an 80 for grade 2. What was the grade on exam 3?

72. An equation which approximately relates "normal" weight to height for adult males is $h = (w + 220)/5.5$, where w is the weight in pounds and h is the height in inches.

   a. Solve the formula for w.

   b. What is the "normal" weight for a man who is 5 ft 8 in. tall?

THINK ABOUT IT

1. A student solves $x - y = 4$ for y and obtains $y = -4 + x$. However, the text answer key reads $y = x - 4$. Is the student's answer also correct? Explain.

2. A student solves $x - y = 1$ for y and obtains $y = 1 - x$. However, the text answer key reads $y = x - 1$. Is the student's answer also correct? Explain.

3. The formulas $A = \frac{1}{2}bh$ and $A = bh/2$ are equivalent. Why?

4. In Exercise 69 above a formula for the area of an ellipse is given. What happens to that formula if a and b are equal?

What happens to the picture if a and b are equal?

5. Sunlight travels at about 186,000 mi/second. This is expressed as $d = 186{,}000t$, where t is in seconds and d is in miles.

   a. Solve this equation for t.

   b. The distance from the earth to the sun is about 93 million mi. Use the equation in part **a** to find out how long it takes light from the sun to reach the earth. This is how long it would take you to reach the sun if you could travel at the speed of light.

REMEMBER THIS

1. Show that -1 is a solution to the equation $3x^2 + 3x = 0$.

2. Show that $2x + 5 = 2(x + 3) - 1$ is an identity.

3. Simplify $5 - (3 - 1)$.

4. Combine like terms: $3x + 2y - 4x + 9y$.

5. Which property of numbers is expressed as $a(b + c) = ab + ac$?

6. If x stands for an unknown number, what stands for

   a. 4 times that number?

   b. 4 more than that number?

   c. 4 less than that number?

7. Find the average of 92, 86, 97, and 83.

8. Which of these numbers are integers: $\{1, \frac{1}{2}, 0, -\frac{1}{3}, -8\}$?

9. True or false?

   a. 0 is an even integer.

   b. 37 and 38 are consecutive integers.

10. True or false? $(2 + 3)^2 = 2^2 + 3^2$.

2.6 Solving Word Problems

Suppose a student's average grade on four tests must be at least 89.5 to receive a grade of A. If a student has three test grades of 92, 86, and 97, find the lowest grade for the last test that results in a grade of A. (See Example 3.)

OBJECTIVES

1 Solve word problems by translating phrases and setting up and solving equations.

2 Prove certain statements about integers.

1 To this point, we have learned to solve linear equations and we have been developing the ability to translate verbal phrases to mathematical statements. In this section we combine both of these skills and discuss the solution to several basic types of word problems. The following steps are recommended by both mathematics and reading specialists as a general approach to solving word problems.

To Solve a Word Problem

1. **Read the problem several times.** The first reading is a preview and is done quickly to obtain a general idea of the problem. The objective of the second reading is to determine exactly what you are asked to find. Write this down. Finally, read the problem carefully and note what information is given. If possible, display the given information in a sketch or chart.
2. **Let a variable represent an unknown quantity** (which is usually the quantity you are asked to find). Write down precisely what the variable represents. If there is more than one unknown, represent these unknowns in terms of the original variable.
3. **Set up an equation** that expresses the relationship between the quantities in the problem.
4. **Solve the equation.**
5. **Answer the question.**
6. **Check the answer** by interpreting the solution in the context of the word problem.

EXAMPLE 1   You need to cut a piece of string that is 85 in. long into two parts so that one piece is 4 times longer than the other. How long should you make each piece?

Solution   We need to find the length of two pieces of string that satisfy a given condition. If we let

$$x = \text{length of the shorter piece,}$$
$$\text{then} \quad 4x = \text{length of the longer piece.}$$

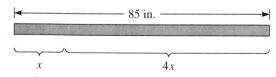

Figure 2.5

Now, illustrate the given information, as in Figure 2.5, and *set up an equation.*

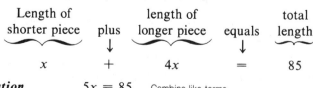

Length of shorter piece	plus	length of longer piece	equals	total length
x	$+$	$4x$	$=$	85

Solve the Equation $5x = 85$ Combine like terms.
                          $x = 17$ Divide both sides by 5.

Answer the Question   The shorter piece is 17 in. long. The longer piece is 4 times longer, so it is 68 in. long.

Check the Answer   Because 68 in. is 4 times longer than 17 in., and 17 in. plus 68 in. equals 85 in., the solution checks.

Progress Check Answer
1. 3 ft and 9 ft

PROGRESS CHECK 1   A 12-ft piece of lumber must be cut into two pieces so that one piece is 3 times longer than the other. How long are the pieces?

EXAMPLE 2   The total cost of a magazine and a Sunday newspaper is $3.75. If the newspaper costs $1.25 less than the magazine, find the cost of the magazine.

Solution   We need to find the cost of the magazine. If x = cost of the magazine, then $x - 1.25$ = cost of the Sunday newspaper.

Set up an Equation

Cost of magazine	plus	cost of newspaper	equals	total cost
x	$+$	$x - 1.25$	$=$	3.75

Solve the Equation

$$2x - 1.25 = 3.75 \qquad \text{Combine like terms.}$$
$$2x = 5.00 \qquad \text{Add 1.25 to both sides.}$$
$$x = 2.50 \qquad \text{Divide both sides by 2.}$$

Answer the Question   The magazine costs $2.50.

Check the Answer   If the magazine costs $2.50, then the Sunday newspaper costs $2.50 - $1.25 = $1.25. The solution checks because the newspaper costs $1.25 less than the magazine and the total cost of both items is $3.75.

PROGRESS CHECK 2   A bottle and a cork together cost $1.10. The cork costs $1 less than the bottle. How much does the bottle cost?

EXAMPLE 3   Solve the problem in the section introduction on page 99.

Solution   We are asked to find the lowest grade for the last test that results in a grade of A. If we let x = grade on last test, then the average grade (the sum of the grades divided by 4) is

$$\frac{92 + 86 + 97 + x}{4}.$$

The cutoff average for an A is 89.5, so *set up the equation*.

$$\frac{92 + 86 + 97 + x}{4} = 89.5$$

Solve the Equation

$$92 + 86 + 97 + x = 358 \qquad \text{Multiply both sides by 4.}$$
$$275 + x = 358 \qquad \text{Simplify.}$$
$$x = 83 \qquad \text{Subtract 275 from both sides.}$$

Answer the Question   The lowest grade for the last test that results in an A is 83.

Check the Answer   The average of 92, 86, 97, and 83 is $(92 + 86 + 97 + 83)/4 = 358/4 = 89.5$. The solution checks.

PROGRESS CHECK 3   Suppose a student's average grade on six tests must be at least 59.5 to pass a course. If a student has five test grades of 67, 45, 72, 51, and 55, find the lowest grade for the last test that will result in a passing grade.

In the remaining examples we consider word problems that help us build proficiency in translating verbal phrases into mathematical statements.

EXAMPLE 4   Six less than 5 times a number is 8 more than 4 times the number. What is the number?

Solution   We need to find a number that satisfies a given condition. Let x represent the unknown number; then *set up the equation*.

Progress Check Answers
2. $1.05
3. 67

$$\underbrace{\text{6 less than}\atop\text{5 times a number}} \quad \underset{\downarrow}{\text{is}} \quad \underbrace{\text{8 more than}\atop\text{4 times the number}}$$

$$5x - 6 \qquad\quad = \qquad\quad 4x + 8$$

Solve the Equation

$$x - 6 = \ 8 \qquad \text{\small Subtract } 4x \text{ from both sides.}$$
$$x = 14 \qquad \text{\small Add 6 to both sides.}$$

Answer the Question   The number is 14.

Check the Answer   First, 6 less than 5 times 14 is 6 less than 70, which is 64. Then, 8 more than 4 times 14 is 8 more than 56, which is 64. The results are equal, so the solution checks.

PROGRESS CHECK 4   Nine more than 7 times a number is 7 less than 9 times the number. What is the number?   ⌐

For the final two examples we need the following definitions.

$$\text{Integers} = \{ \ldots , -3, -2, -1, 0, 1, 2, 3, \ldots \}$$
$$\text{Even integers} = \{ \ldots , -6, -4, -2, 0, 2, 4, 6, \ldots \}$$
$$\text{Odd integers} = \{ \ldots , -5, -3, -1, 1, 3, 5, \ldots \}$$

Note that consecutive integers differ by 1. Consecutive even integers as well as consecutive odd integers differ by 2.

EXAMPLE 5   The sum of two consecutive even integers is 158. Find the integers.

Solution   If x represents the first even integer, then $x + 2$ represents the next even integer, and *the equation is*

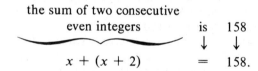

$$\underbrace{\text{the sum of two consecutive}\atop\text{even integers}} \quad \underset{\downarrow}{\text{is}} \quad \underset{\downarrow}{158}$$

$$x + (x + 2) \qquad\quad = \qquad\quad 158.$$

Solve the Equation

$$2x + 2 = 158 \qquad \text{\small Combine like terms.}$$
$$2x = 156 \qquad \text{\small Subtract 2 from both sides.}$$
$$x = 78 \qquad \text{\small Divide both sides by 2.}$$

Answer the Question   The smaller integer is 78, and the next consecutive even integer is 80.

Check the Answer   The sum of 78 and 80 is 158, and they are consecutive even integers, so the solution checks.

PROGRESS CHECK 5   The sum of two consecutive odd integers is 124. Find the integers.   ⌐

2   With a little modification we can use the skills we have developed to prove some statements about integers.

EXAMPLE 6   Show that the sum of any three consecutive integers is equal to 3 times the middle integer.

Solution   We are asked to prove a statement about consecutive integers. If we let x represent the first integer, then

$$x + 1 = \text{second consecutive integer,}$$
$$x + 2 = \text{third consecutive integer.}$$

Progress Check Answers
4. 8
5. 61 and 63

Now set up an equation.

$$\underbrace{x + (x + 1) + (x + 2)}_{\substack{\text{The sum of three} \\ \text{consecutive integers}}} \quad \underset{\downarrow}{=} \quad \underbrace{3(x + 1)}_{\substack{\text{3 times the} \\ \text{middle integer}}}$$

To prove the statement, we show that the statement is an identity.

$$x + (x + 1) + (x + 2) = 3(x + 1) \qquad \text{Derived equation.}$$
$$3x + 3 = 3(x + 1) \qquad \text{Combine like terms.}$$
$$3x + 3 = 3x + 3 \qquad \text{Distributive property.}$$

The resulting equation, $3x + 3 = 3x + 3$, is true for all replacements of x, so any three consecutive integers possess the stated property.

PROGRESS CHECK 6   Show that the sum of any five consecutive integers is equal to 5 times the middle integer.

Progress Check Answer

6. The equation leads to $5x + 10 = 5x + 10$.

EXERCISES 2.6

In Exercises 1–30, solve each problem by following the steps outlined in this section.

1. You want to divide $500 between two people so that one of them gets 3 times as much as the other. How should you split the money?

2. If you divide a set of 642 baseball cards into two piles so that one pile contains twice as many cards as the other, how many cards will be in each pile?

3. A jacket costs $40 more than a pair of pants. Together, they cost $120. How much does the jacket cost?

4. The paperback edition of a book costs $10 less than the hardback edition. It will cost you a total of $31.90 to buy one of each. What is the cost of the paperback edition?

5. A 10-ft piece of wood is cut into two pieces, where one is 9 ft shorter than the other. How long is each piece of wood?

6. In a high school class of 200 people there were 50 more males than females. How many males were in the class?

7. A 10-ft piece of wood is cut into two pieces, where one is 9 times as long as the other. How long is each?

8. The sum of a patient's systolic and diastolic blood pressures is 170, and there is a 50-point difference between the two. Find the systolic blood pressure, which is the higher one.

9. A 12-ft board is sawed into three smaller pieces. The longest piece is 2.5 ft longer than the smallest one and 2 ft longer than the middle-sized one. Find the length of the middle-sized piece.

10. You want to saw a 22-ft piece of lumber into three different-sized pieces so that the middle one is twice the length of the shortest one, and the longest is 4 times the middle one. How long should each piece be?

11. The combined floor space of three warehouses is 50,000 ft². The two smaller ones have the same area, and the largest one is 3 times the size of each of those.
   a. Find the floor space for each warehouse.

   b. Can a shipment which takes up 12,000 ft² of floor space fit in one of the smaller warehouses?

12. Three full containers in a chemistry lab hold a total of 2,409 milliliters (ml) of a solution. The two larger ones hold the same volume, which is 5 times the volume of the smallest one. What is the volume of the smallest container?

13. On a joint income tax return A's income was $\frac{1}{4}$ of B's income, and their total income was $40,000. How much was A's income?

14. A person's weight on the moon is $\frac{1}{6}$ of his or her earth weight. Suppose the sum of a person's moon and earth weight is 210 lb. What does that person weigh on the earth?

15. If your grades on three tests are 70, 70, and 70, what grade on the fourth test will bring your average up to 74.5?

16. If your grades on the first two tests are both 65, what would it take to bring the average up to 69.5 with one more test?

17. A company had sales in January of $33,000 and February sales of $34,500. The sales force will get a bonus if the average sales for the first quarter (January, February, March) is $35,000 or more. What is the least amount of sales for March that will result in a bonus?

18. A robot-operated assembly line is judged to be not working satisfactorily if it produces an average of less than 3,000 finished pieces per hour. When this happens, the line must be shut down and repaired. For the first 4 hours of one day it has turned out 3,010, 3,006, 3,002, and 2,998 pieces, respectively. What is the least number of pieces it could turn out for the fifth hour and still not be shut down?

19. Three less than 4 times a number is 5 more than 6 times the number. What is the number?

20. Six more than 5 times a number is 4 less than 3 times the number. What is the number?

21. Two numbers add to 444. Half the larger one is 85.5 more than twice the smaller one. Find the two numbers.

22. The sum of two numbers is 789. Half the larger one is 105 less than the smaller one. What is the smaller number?

23. Half of a number equals 6 less than the number. What is the number?

24. Half a number equals 6 more than the number. What is the number?

25. The sum of three consecutive integers is 102. What are the integers?
26. The sum of two consecutive odd integers is 236. What are the integers?
27. A baseball player has 115 hits in 300 at bats for a percentage of $115/300 = .383$. If the player has 100 more at bats in the season, how many hits are needed to bring the percentage up to .400?
28. A baseball player has 61 hits in 200 at bats for a percentage of .305. If the player has 40 more at bats in the season, what are the fewest number of hits that will keep the average in the "300's"?
29. The votes are being counted in a two-way contest between A and B. Right now, A has 530 out of the first 1,000 votes counted. There are 230 more votes to be counted. What's the worst A can do on these outstanding votes and still win? (To win, A needs more than 50 percent of the total votes.)
30. A film processor charges $2 for developing and 19¢ per print for a roll of 36 negatives. Joe dropped off a roll of 36 negatives, and the bill came to $7.13. How many negatives were not printable?

In Exercises 31–36, write and solve equations to prove statements about integers.

31. Show that the sum of any four consecutive integers is 6 more than 4 times the smallest one.
32. Show that the sum of any four consecutive integers is 6 less than 4 times the largest one.
33. Show that the sum of two consecutive integers is 1 less than twice the larger one and also 1 more than twice the smaller one.
34. Show that the sum of three consecutive integers is 3 more than 3 times the smallest one and 3 less than 3 times the largest one.
35. Show that the sum of three consecutive integers cannot be 100.
36. Show that the sum of three consecutive odd integers cannot be 91.

THINK ABOUT IT

1. The equation $x + 2 = 7$ can be translated into the sentence, "If 2 is added to a certain number, the result is 7." Translate each of these equations into a sentence. There are many correct translations.
   a. $x - 1 = -4$ b. $3y = 14$ c. $2x + 4 = 5$
2. In a word problem x represents one quantity and $10 - x$ represents another. What can you say about the sum of these quantities?
3. Make up a word problem involving a piece of string cut into two pieces which could be solved by the equation $10 - x = 2x$.

The last two exercises are so-called age problems. Problems like these, though they are quite artificial, have a long tradition in books written to teach algebra. Both problems are taken from Davies' *Algebra*, published in 1859 by the A. S. Barnes Company of New York. Try to solve them.

4. James is three years older than John; and one-sixth of James' age is equal to one-fifth of John's. How old are they?
5. Nancy's age is 3 times Eliza's; one-half Nancy's plus one-third of Eliza's is equal to the difference of their ages diminished by 1. What is the age of each?

REMEMBER THIS

1. Draw a sketch of a rectangle and write the formula for its perimeter.

2. What is the measure of a right angle?
3. If x stands for a certain number, what expression stands for 8 percent of that number?
4. What expression represents 10,000 more than half of a number?

5. Solve the equation $d = rt$ for r.

6. Rewrite the expression $0.06(12 - x)$ by applying the distributive property.
7. Evaluate the expression $x = \frac{a}{b}$ if $a = -0.12$ and $b = -0.03$.
8. Find the product of 3 and $\frac{2}{3}$.
9. Find the sum of 3 and $\frac{2}{3}$.
10. Simplify $10 - 0.01(500 - 200x)$.

2.7   Additional Applications of Equations

On a video display an air traffic controller notices two planes 100 mi apart and flying toward each other on a collision course. One plane is flying 500 mi/hour; the other is flying 300 mi/hour. How much time is there for the controller to prevent a crash? (See Example 5.)

1 Solve problems involving geometric figures.

2 Solve problems involving percentages.

3 Solve problems about uniform motion.

4 Solve problems about liquid mixtures.

To develop skill in problem solving, it is useful to build a supply of model problems that you understand. Then, when a new situation arises, you may find that it is similar to a problem you already know how to solve, and you will quickly know some approaches that might work. To build this base, we now consider word problems that involve geometric figures, percentages, uniform motion, and liquid mixtures.

Geometric Problems

1 To solve problems about geometric figures, we often need perimeter, area, or angle measure formulas. Section 1.3 contains the formulas you need here.

EXAMPLE 1 A piece of molding is 75 in. long. You wish to use it to make a rectangular frame that is twice as long as it is wide. What are the dimensions of the frame?

Solution To find the dimensions of the frame, let

$$x = \text{width},$$
$$\text{so} \quad 2x = \text{length}.$$

Now, illustrate the given information as in Figure 2.6.

Because the molding is 75 in. long, the perimeter of the rectangle is 75 in., and we can *set up an equation* by using $P = 2\ell + 2w$, the formula for the perimeter of a rectangle.

$$P = 2\ell + 2w$$
$$75 = 2(2x) + 2(x) \quad \text{Replace } P \text{ by 75, } \ell \text{ by 2x, and } w \text{ by x.}$$

Solve the Equation

$$75 = 6x \quad \text{Combine like terms.}$$
$$\tfrac{75}{6} = x \quad \text{Divide both sides by 6.}$$
$$12.5 = x \quad \text{Simplify.}$$

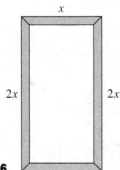

Figure 2.6

Answer the Question The width of the frame is 12.5 in. The length of the frame is $2(12.5) = 25$ in.

Check the Answer The perimeter is $2(25) + 2(12.5) = 75$ in., while 25 in. is twice as long as 12.5 in. The solution checks.

PROGRESS CHECK 1 If the perimeter of a rectangle is 40 m and the length is 3 m greater than the width, find the dimensions of the rectangle. ⌐

EXAMPLE 2 One of the two acute angles in a right triangle is 15° greater than the other. What are the measures of all the angles?

Solution By definition, a right triangle contains a right angle, so one of the angles measures 90°. To find the measures of the acute angles, let

$$x = \text{measure of the smaller acute angle},$$
$$\text{then} \quad x + 15 = \text{measure of the larger acute angle}.$$

Draw a sketch of the problem as in Figure 2.7 on page 106.

Progress Check Answer
1. $\ell = 11.5$ m, $w = 8.5$ m

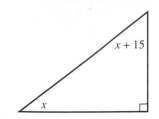

Figure 2.7

Because the sum of the angle measures in a triangle is 180°, we can *set up an equation.*

Measure of smaller acute angle	plus	measure of larger acute angle	plus	measure of right angle	equals	180
↓	↓	↓	↓	↓	↓	↓
x	$+$	$x + 15$	$+$	90	$=$	180

Solve the Equation

$$2x + 105 = 180 \quad \text{Combine like terms.}$$
$$2x = 75 \quad \text{Subtract 105 from each side.}$$
$$x = \tfrac{75}{2} \quad \text{Divide both sides by 2.}$$
$$x = 37.5 \quad \text{Simplify.}$$

Answer the Question   The angle measures are 37.5°, 52.5° (from 37.5 + 15), and 90°.

Check the Answer   Because 37.5 + 52.5 + 90 = 180, and 52.5 is 15 greater than 37.5, the solution checks.

Note   In any right triangle the sum of the measures of the acute angles is 90°. Therefore, in this example we can find x by solving

$$x + (x + 15) = 90.$$

PROGRESS CHECK 2   If one acute angle of a right triangle is 2° less than the other, what are the measures of all the angles in the triangle?   ⌐

Percentage Problems

2   To solve problems involving percentages, you may need to consult the review of percentages in the Appendix. When a problem involves annual interest, use $I = Pr$, where I represents annual interest, P represents principal, and r represents the annual interest rate.

EXAMPLE 3   You win $30,000 in a state lottery. You are advised to put some of the money in a risky investment that earns 15 percent annual interest and the rest in a safe investment that earns 7 percent annual interest. If you want the total interest to be $3,000 annually, how much should be invested at each rate?

Solution   To find the amount to invest at each rate, let

$$x = \text{amount invested at 7 percent,}$$
$$\text{then} \quad 30{,}000 - x = \text{amount invested at 15 percent.}$$

It is helpful to analyze each investment in a chart format.

Investment	Principal	Interest Rate	=	Annual Interest
Safe	x	0.07		$0.07x$
Risky	$30{,}000 - x$	0.15		$0.15(30{,}000 - x)$

Set up an equation using 3,000 as the desired amount of total interest.

Interest from safe investment	plus	interest from risky investment	equals	total interest
↓		↓	↓	↓
$0.07x$	$+$	$0.15(30{,}000 - x)$	$=$	$3{,}000$

Solve the Equation

$$0.07x + 4,500 - 0.15x = 3,000 \qquad \text{Distributive property.}$$
$$-0.08x + 4,500 = 3,000 \qquad \text{Combine like terms.}$$
$$-0.08x = -1,500 \qquad \text{Subtract 4,500 from each side.}$$
$$x = \frac{-1,500}{-0.08} \qquad \text{Divide both sides by } -0.08.$$
$$x = 18,750 \qquad \text{Simplify.}$$

Answer the Equation   You should invest $18,750 in the safe investment at 7 percent and $30,000 − $18,750, or $11,250, at 15 percent.

Check the Answer   The safe investment earns $1,312.50 (or 7 percent of $18,750), and the risky investment earns $1,687.50 (or 15 percent of $11,250). Thus the total interest is $1,312.50 + $1,687.50, or $3,000, so the solution checks.

PROGRESS CHECK 3   A $5,000 retirement contribution is split into two investments, some at 5 percent and the rest at 12 percent. If the total annual interest was $355, how much was invested at 12 percent?   ⌐

EXAMPLE 4   The total cost (including tax) for a restaurant meal is $13.50. If the tax rate on the meal is 8 percent, how much is the tax?

Solution   We are asked to find the tax. Since the tax is based on the cost of the meal, we let

$$x = \text{cost of the meal before the tax.}$$

Then the tax is 8 percent of x, so

$$0.08x = \text{tax on the meal.}$$

Now, *set up an equation.*

Cost of meal before tax	plus	tax on meal	equals	total cost
x	+	$0.08x$	=	13.50

Solve the Equation

$$1.08x = 13.50 \qquad \text{Combine like terms.}$$
$$x = \frac{13.50}{1.08} \qquad \text{Divide both sides by 1.08.}$$
$$x = 12.50 \qquad \text{Simplify.}$$

Answer the Question   The tax is $0.08x = 0.08(12.50)$, or $1.00. Note that in this problem the value of x is not the answer to the question.

Check the Answer   The meal is $12.50 and the tax is $1.00 (or 8 percent of $12.50), so the total bill is $13.50. The solution checks.

PROGRESS CHECK 4   The total cost (including tax) of a new TV set is $504.40. If the sales tax rate is 4 percent, how much is paid in taxes?   ⌐

Progress Check Answers
3. $1,500
4. $19.40

Uniform Motion Problems

3   To solve uniform motion problems, you need the formula $d = rt$, where d represents the distance traveled in time t by an object moving at a constant rate r. Analyzing the problem with both a chart and a sketch is recommended here.

EXAMPLE 5   Solve the problem in the section introduction on page 104.

Solution   We need to find how long it takes for the planes to reach each other. Let

$$t = \text{time until the planes meet.}$$

Note that both planes travel for time t. Now, analyze the problem in a chart format, as in Example 3, but in this case, use $d = rt$.

Plane	Rate (mi/hour)	Time (hours)	=	Distance (mi)
Slower plane	300	t		$300t$
Faster plane	500	t		$500t$

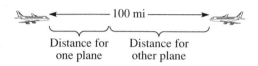

Figure 2.8

To *set up an equation*, sketch the situation as in Figure 2.8. From the sketch we can see that

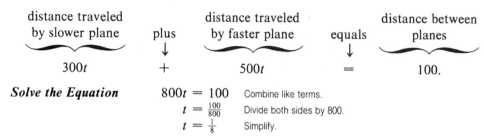

distance traveled by slower plane	plus	distance traveled by faster plane	equals	distance between planes
$300t$	$+$	$500t$	$=$	$100.$

Solve the Equation

$$800t = 100 \quad \text{Combine like terms.}$$
$$t = \tfrac{100}{800} \quad \text{Divide both sides by 800.}$$
$$t = \tfrac{1}{8} \quad \text{Simplify.}$$

Answer the Question   The planes reach each other in $\tfrac{1}{8}$ hour. Thus, the controller has $\tfrac{1}{8}$ hour, or 7.5 minutes, to prevent a collision.

Check the Answer   In $\tfrac{1}{8}$ hour the slower plane travels $300(\tfrac{1}{8}) = 37.5$ mi, while the faster plane travels $500(\tfrac{1}{8}) = 62.5$ mi. The combined distance is 100 mi, so the answer checks.

PROGRESS CHECK 5   Two satellites are 760 m apart and drifting toward each other. One drifts at a constant speed of 20 m/second, the other at 30 m/second. How long will it take for the satellites to meet?   ⌐

Liquid Mixture Problems

4   To solve problems about liquid mixtures, you need to apply the concept of percentage in the following context:

$$\begin{pmatrix} \text{percent of} \\ \text{an ingredient} \end{pmatrix} \cdot \begin{pmatrix} \text{amount of} \\ \text{solution} \end{pmatrix} = \begin{pmatrix} \text{amount of} \\ \text{the ingredient} \end{pmatrix}.$$

For example, the amount of acid in 10 liters of a solution that is 30 percent acid is $0.30(10)$, or 3 liters. Once again, a chart is recommended for this type of problem.

EXAMPLE 6   A machine shop has two large containers that are each filled with a mixture of oil and gasoline. Container A contains 3 percent oil (and 97 percent gasoline). Container B contains 6 percent oil (and 94 percent gasoline). How much of each should be used to obtain 12 qt of a new mixture that contains 5 percent oil?

Solution   To find the correct mixture, let

$$x = \text{amount used from container A,}$$
$$\text{so} \quad 12 - x = \text{amount used from container B.}$$

As recommended, we analyze the problem with a chart.

Solution	Percent Oil	Amount of Solution (qt)	=	Amount of Oil (qt)
container A	3	x		$0.03x$
container B	6	$12 - x$		$0.06(12 - x)$
New Solution	5	12		$0.05(12)$

To *set up an equation*, we reason that the amount of oil in the new solution is the sum of the amounts contributed by the solutions from containers A and B.

Amount of oil from container A	plus	amount of oil from container B	equals	amount of oil in new solution
$0.03x$	+	$0.06(12 - x)$	=	$0.05(12)$

Solve the Equation

$$0.03x + 0.72 - 0.06x = 0.60 \qquad \text{Simplify.}$$
$$-0.03x + 0.72 = 0.60 \qquad \text{Combine like terms.}$$
$$-0.03x = -0.12 \qquad \text{Subtract 0.72 from both sides.}$$
$$x = \frac{-0.12}{-0.03} \qquad \text{Divide both sides by } -0.03.$$
$$x = 4 \qquad \text{Simplify.}$$

Answer the Question   Mix 4 qt from container A with $12 - 4$, or 8 qt, from container B.

Check the Answer   The new solution contains $4 + 8 = 12$ qt. Also, 4 qt from container A contains $0.03(4) = 0.12$ qt of oil, while 8 qt from container B contains $0.06(8) = 0.48$ qt of oil. Thus the new mixture contains 0.60 qt of oil in 12 qt of mixture. Because $0.60/12 = 0.05 = 5$ percent, the new mixture does contain 5 percent oil, and the solution checks.

PROGRESS CHECK 6   A chemist has two acid solutions, one 30 percent acid and the other 70 percent acid. How much of each must be used to obtain 10 liters of a solution that is 41 percent acid?   ⌐

Progress Check Answer

6. 7.25 liters of 30 percent solution and 2.75 liters of 70 percent solution

EXERCISES 2.7

In Exercises 1–30, draw and label an appropriate sketch if none is given. Then, write and solve an equation to answer the question.

1. A rectangle is twice as long as it is wide. Its perimeter is 10 in. What are the dimensions of the rectangle?
2. A rectangle is three times as long as it is wide. Its perimeter is 15 cm. What are its dimensions?
3. The length of a rectangle is 3 in. more than the width and the perimeter is 38 in. Find the length and width.
4. The width of a rectangle is 2 ft less than the length. The perimeter is 50 ft. Find the length and width.
5. In an isosceles triangle with perimeter equal to 100 in. the base is 42 in. Find the length of the two equal sides.
6. In an isosceles triangle with perimeter equal to $\frac{3}{5}$ cm the base is $\frac{1}{10}$ cm. Find the length of the two equal sides.
7. In an isosceles triangle the two equal sides are each 1.6 cm. The perimeter is 5.7 cm. How long is the base?

8. In an isosceles triangle the two equal sides are each $\frac{1}{3}$ ft. The perimeter is $\frac{5}{3}$ ft. How long is the base?
9. An isosceles triangle is drawn with the sides twice as long as the base. The perimeter is 36 in. What are the dimensions of the triangle?
10. An isosceles triangle is drawn with the sides 4 times as long as the base. The perimeter is 99 in. How long are the two equal sides?
11. In a triangle the middle side is 1 in. longer than the shortest side and 1 in. shorter than the longest side. The perimeter is 10.5 in. How long is the shortest side? (*Hint:* Let x represent the length of the middle side.)
12. In a triangle the middle side is 2 in. longer than the shortest side and 3 in. shorter than the longest side. The perimeter is 22 in. Find the dimensions of the triangle. (*Hint:* Let x represent the length of the middle side.)

13. The perimeter of this figure is 90 cm. Find the lengths of all the sides.

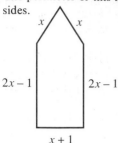

14. The perimeter of this figure is 20 cm. Find the lengths of all the sides.

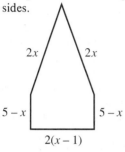

15. Find the lengths of all the sides if the perimeter is 14 ft.

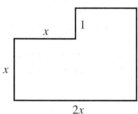

16. Find the lengths of all the sides if the perimeter is 20 ft.

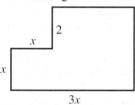

17. In a right triangle the measure of one of the acute angles is twice the measure of the other. What is the measure of the larger acute angle?

18. In a right triangle the measure of one of the acute angles is $3\frac{1}{2}$ times the measure of the other. What are the measures of all the angles?

19. In a right triangle one of the acute angles is one-third of the other. What are the measures of all the angles?

20. In a right triangle one of the acute angles is one-fourth of the other. What is the measure of the smaller acute angle?

21. In a right triangle one acute angle is 15° more than the other. Find the measure of the larger acute angle.

22. In a right triangle one acute angle is 40° less than the other. Find the measures of all the angles.

23. In a triangle the measure of the middle angle is 10° less than the measure of the largest and 10° more than the measure of the smallest. Find the measures of all the angles.

24. In a triangle the measure of the middle angle is 20° less than the largest and 20° more than the smallest. Find the measure of the largest angle.

25. Find the measure in degrees of all the angles.

26. Find the measure in degrees of all the angles.

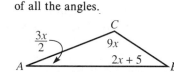

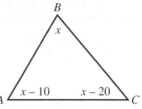

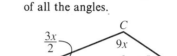

27. Find the measure in degrees of all the angles.

28. Find the measure in degrees of all the angles.

29. The measures of all the angles of a triangle are three consecutive integers. Find the measure of the smallest angle.

30. The sum of all the angles of a quadrilateral is 360°. Show that the measures of the angles cannot be four consecutive integers.

31. By writing and solving an equation, show that these labels cannot be correct for the three angle measures in a triangle: angle A: $3x - 3$; angle B: $2 - x$; angle C: $1 - x$.

32. By writing and solving an equation, show that these labels cannot be correct for the three angle measures in a triangle: angle A: $2x + 52$; angle B: $-5(x + 20)$; angle C: $3(6 - x)$.

Exercises 33–48 involve percentages. For problems involving annual interest, use the formula $I = Pr$ as explained in this section of the text.

33. Ten thousand dollars will be split into two accounts and deposited for a year. One account earns 10 percent annual interest; the other earns 6 percent. How much should be put into each account so that the total interest earned is $750?

34. The board members of an employees pension plan have $10.2 million to invest. They will split it into two accounts, one earning 10 percent annual interest, the other earning 15 percent. How much should be put into each account so that the total earnings are $1,250,000?

35. One amount of money was invested for a year at 6 percent annual interest. One thousand dollars more than that was invested at 10 percent. Altogether, both accounts earned $4,100. How much was invested in each account?

36. One amount of money was invested for a year at 8 percent annual interest. Two thousand dollars more than that was invested at 10 percent. Altogether, both accounts earned $1,100. How much was invested at 10 percent?

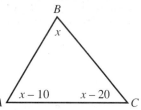

37. A certain amount of money was invested for a year at 6 percent annual interest. Twice as much was invested at 3 percent. Together, the accounts earned $1,200. How much was invested at 3 percent?

38. A certain amount of money was invested for a year at 6 percent annual interest. Three times as much was invested at 4 percent. Together, the accounts earned $1,800. How much was invested in each account?

39. A computer monitor costs $600. With the tax it comes to $627. What is the tax rate?

40. If the tax on an $18 purchase is $1.17, what is the tax rate?

41. The full price for a computer system is $2,304.75, including the 5 percent sales tax. What is the price of the system before the tax? What was the amount of tax paid?

42. A special deal offers a set of audio speakers, including 8.25 percent sales tax and $75 shipping, for $711.51. What is the price of the system not including tax and shipping?

43. The population of a city grows 2 percent in one year to reach 5.1 million people. What was the population at the beginning of the year? How many people were added to the population during the year?

44. The population of a city declines 2 percent in one year and falls to 245,000 people. How many people left the city during the year?

45. The value of a piece of factory equipment depreciates 8 percent every year. If after one year the equipment is worth $15,000, how much was it worth originally? Give the answer to the nearest $100.

46. The value of a piece of real estate in a popular community increases 6 percent every year. If after one year a property is worth $150,000, what was it worth the year before? Give the answer to the nearest $100.

47. You have $12 with you when you enter a restaurant. The tax on meals is 5 percent. The tip will be 20 percent of the menu price. What is the most you can afford to spend on the meal?

48. You have $15 with you when you enter a restaurant. The tax on meals is 5 percent. You always tip 15 percent of the menu price. What is the most you can afford to spend on the meal?

Exercises 49–60 involve uniform motion and can be solved using the basic relationship $d = rt$ as explained in the text. Make a sketch and a chart for each problem.

49. Two spacecraft are approaching each other in a straight line at a constant rate with no gravitational pull of any planet. One is moving 8,000 mi/hour, and the other is moving at 12,000 mi/hour. If they are 100,000 mi apart in outer space, how long will it take them to be 1,000 mi apart, at which distance they can identify one another?

50. Two snails are crawling toward each other in a straight line at constant speed. One goes at 1 ft/hour. The other goes at 1.5 ft/hour. If they start 5 ft apart, how long will it take them to meet?

51. One hiker starts up a 25-mi mountain trail climbing at an average rate of 0.6 mi/hour. At the same time another starts down the trail at 3.4 mi/hour. How long will it take them to meet?

52. One biker leaves the end of a 10-mi bike path and cycles at an average speed of 8 mi/hour. At the same time another leaves the other end of the path riding at the same speed. How long will it take them to meet?

53. In a race two sailboats take the same course, but one averages 15 nautical miles per hour (knots) while the other averages 12.5 knots. How long will it take the faster boat to be ahead by 20 nautical miles?

54. In a hot-air balloon race one balloon floats at an average speed of 55.2 mi/hour and another at 52.7 mi/hour. If they start together and float along the same course, how long will it take the faster one to have a 5-mi lead?

55. The hero of a movie is just beginning to run along a railroad bridge at 12 mi/hour. A train is coming up from behind at a constant speed and is now $\frac{1}{2}$ mi from the bridge. The bridge is $\frac{1}{4}$ mi long. How fast should the train go so that the hero will beat it to the end of the bridge by 5 seconds? (*Hint:* To keep the units in miles and hours, represent 1 second as $\frac{1}{3600}$ hour, and work with fractions rather than decimals.)

56. The leader in a race is $\frac{1}{4}$ mi from the finish line and running at a steady speed of 10 mi/hour.
   a. Running at a constant speed, how fast does a runner have to go to close a 0.05-mi lead (this is 88 yd) just at the finish line?
   b. If the runner who is behind can run 13 mi/hour, who will win?

57. One plane takes 6 hours for a transatlantic flight. A newer plane goes 200 mi/hour faster and makes the trip in 4 hours.
   a. How fast does each travel?
   b. What is the distance between airports?

58. One plane takes 1 hour 15 minutes for a certain flight. Under the same conditions another plane goes 100 mi/hour faster and makes the trip in 1 hour. What is the distance between airports?

59. A space crew is told that a communications satellite is 1,000 mi away and drifting away at 200 mi/hour. How fast does the ship have to go to catch up to the satellite in 10 hours?

60. A space station crew realizes that an orbiting telescope 25,000 mi away has malfunctioned and is drifting away in a straight path at 250 mi/hour. How fast does a ship launched from the station have to go to catch up to the satellite in 6 hours?

Exercises 61–72 are about liquid mixtures. It will help to draw a chart to organize the information. See Example 6 in this section for an illustration.

61. One metal contains 30 percent gold by weight and the rest silver. Another contains 50 percent gold by weight and the rest silver. They will be melted down and mixed together to form a new alloy which is 35 percent gold. How much of each should be used to form 5 lb of the new alloy?

62. One metal contains 40 percent copper by weight and the rest silver. Another contains 70 percent copper by weight and the rest silver. They will be melted down and mixed together to form a new alloy which is 60 percent copper. How much of each should be used to form 120 lb of the new alloy?

63. Two mixtures, one of which is 25 percent acid and the other 40 percent acid, will be mixed to make 5 liters of a solution which is 30 percent acid. How much of each should be used?

64. Two mixtures, one of which is 5 percent acid and the other 8 percent acid, will be mixed to make 10 liters of a solution which is 7.2 percent acid. How much of each should be used?

65. How much water should be added to 1 gal of a solution which is 50 percent antifreeze to dilute it to a 25 percent solution? How much of this new solution will you end up with?

66. How much water should be added to 5 gal of a solution which is 20 percent antifreeze to dilute it to a 15 percent solution? How much of this new solution will you end up with?

67. How much oil should be added to 2 liters of a gas/oil mixture which is 10 percent oil to get a solution which is 25 percent oil?

68. How much oil should be added to 3 liters of a gas/oil mixture which is 4 percent oil to get a solution which is 10 percent oil?

69. A jeweler has 36 oz of an alloy which is 70 percent gold and 30 percent silver. How much silver should be added to make the alloy contain 35 percent silver? Give the answer to the nearest hundredth.

70. A jeweler has 40 oz of an alloy which is 80 percent gold and 20 percent silver. How much silver should be added to make the alloy contain 25 percent silver? Give the answer to the nearest hundredth.

71. In a study on acid rain scientists mix 1 gal of a pollutant into 50 gal of solution which already contains 1 percent pollutant. What will be the percentage of pollutant after they do this?

72. One quart of pure antifreeze is mixed with 3 qt of a solution which is 10 percent antifreeze. What is the percentage of antifreeze in this new mixture?

THINK ABOUT IT

1. You have 120 in. of string to shape into a rectangle with perimeter 120 in. Make a chart which shows various possibilities for the dimensions. Do they all have the same area? What dimensions appear to yield the maximum area? The minimum area?

2. The length of a rectangle is 2 in. more than twice the width. The perimeter is 4.2 in. Find the length, width, and area. What happens if the perimeter is given as 4.0 in.?

3. A chemist has two mixtures, 30 percent acid and 20 percent acid. Can they be mixed to form 2 liters of a 40 percent solution? Set up the equation for this and see what the solution tells you.

4. If a motorcycle goes 120 mi at 40 mi/hour and then 120 mi at 60 mi/hour, what is its average speed for the whole trip? Recall that average speed equals total distance divided by total time.

5. How much time do you save driving to work at 65 mi/hour rather than 55 mi/hour if the distance to work is 20 mi? 30 mi? Over what distance would the savings amount to 15 minutes?

REMEMBER THIS

1. Graph the set of prime numbers between 10 and 20.

2. True or false? $4 \geq 4$.

3. Which of these replacements for x make $3x + 1 \geq 10$ a true statement? Choose x from $\{1,2,3,4\}$.

4. Write a formula for the average of a, b, and c.

5. Solve the equation $3x + 1 = 10$.

6. Explain why $\frac{1}{2}$ is greater than -100.

7. Which is greater, $-2(3)$ or $-2(4)$?

8. Which is greater, the sum or the product of -5 and 7?

9. Write an equation which describes this statement: 4 less than 3 times a number is equal to 1 more than twice the number.

10. Find a replacement for x which makes $2x$
    a. greater than x b. equal to x
    c. less than x

2.8 Solving Linear Inequalities

A publishing company plans to produce and sell x textbooks. The cost of producing x books is given by $180{,}000 + 11x$, which represents a setup cost of $180,000 plus a variable cost of $11 per book. The publisher receives $26 on the sale of each book, so the revenue for selling x books is given by $26x$. For what values of x is revenue *greater than* cost? (See Example 9.)

OBJECTIVES

1 Specify solution sets of inequalities by using graphs and set-builder notation.

2 Solve linear inequalities by applying properties of inequalities.

3 Solve applied problems involving linear inequalities.

4 Solve compound inequalities that specify numbers between two numbers.

The inequality signs ($<, \leq, >, \geq$) often express the required relation in a problem. For instance, the publishing company in the section-opening problem clearly wants revenue to be *greater than* cost so that the textbook is profitable. Throughout this section it is essential that you remember the meanings of the inequality symbols introduced in Section 1.1.

Symbol	Meaning
$<$	less than
$\leq$	less than or equal to
$>$	greater than
$\geq$	greater than or equal to

1 An **inequality** is a statement that relates expressions by using the inequality symbols above. For example,

$$x + 1 \geq -5, \quad x > 1, \quad \text{and} \quad 0 \leq x < 10$$

are inequalities. As with equations, a solution of an inequality is a value for the variable that makes the inequality a true statement, and the set of all such solutions is called the **solution set.** Using these definitions, we say the solution set of the inequality $x > 1$ is the set of *all real numbers* greater than 1. This is an infinite set of numbers, and we need some special ways to describe such sets. One way is to use a graph on the real number line.

EXAMPLE 1 Graph the solution set of $x > 1$.

Solution Draw a number line. Place an open circle at 1; then draw an arrow from the circle to the right, as in Figure 2.9. The open circle means the value 1 is *not* included in the solution set, and the arrow specifies all real numbers greater than 1.

PROGRESS CHECK 1 Graph the solution set of $x > -2$.   ⌐

EXAMPLE 2 Graph the solution set of $x \leq 2$.

Solution To indicate that the inequality is true when x equals 2, we draw a solid dot at 2. Because we also need to specify real numbers less than 2, we draw an arrow to the left from this dot. Figure 2.10 shows the completed graph.

PROGRESS CHECK 2 Graph the solution set of $x \leq 0$.   ⌐

We often need to specify a set of numbers between two numbers. To do this, we may use a compound inequality. The inequality $2 < x < 5$ is read "2 is less than x and x is less than 5" or "x is between 2 and 5"; and the statement is true for the set of numbers graphed in Figure 2.11. Note that 2 and 5 are not included in the solution set. Example 3 shows how to include an endpoint in a solution set.

Figure 2.9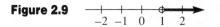

Figure 2.10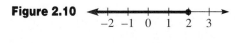

Figure 2.11

Progress Check Answers

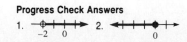

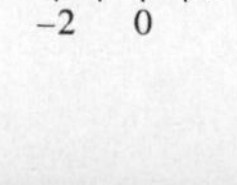

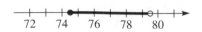

Figure 2.12

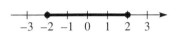

Figure 2.13

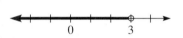

Figure 2.14

EXAMPLE 3   In a math class a student receives a grade of C+ for an average a which satisfies $74.5 \leq a < 79.5$. Graph this interval.

Solution   An average of 74.5 is included in the C+ interval while an average of 79.5 is not. To draw the graph, we connect a solid dot at 74.5 to an open circle at 79.5, as in Figure 2.12.

PROGRESS CHECK 3   In a sociology class, a grade of A corresponds to an average a which satisfies $89.5 < a \leq 100$. Graph this interval.   ⌐

A second way to specify an infinite set of numbers is to use **set-builder notation.** Using this notation, sets are written in the form

$$\{x: x \text{ has property } P\},$$

which is read "the set of all elements x such that x has property P." The colon (:) is read "such that."

EXAMPLE 4   Specify the set of numbers graphed in Figure 2.13 by using set-builder notation.

Solution   We need to specify the set of real numbers from -2 to 2, including both endpoints. So in set-builder notation we write

$$\{x: -2 \leq x \leq 2\}.$$

Note that we assume x represents a real number unless stated otherwise.

PROGRESS CHECK 4   Specify the set of numbers graphed in Figure 2.14 by using set-builder notation.   ⌐

2   By analogy to equations, a **first-degree** or **linear inequality** results if the equal sign in a linear equation is replaced by one of the inequality symbols. For example,

$$2x + 1 < 5, \qquad x \geq 1, \qquad \text{and} \qquad 5(x - 1) > 2x + 1$$

are linear inequalities. We solve linear inequalities using essentially the same methods we use to solve linear equations.

The first properties of inequality we develop relate to addition and subtraction. If we begin with a true inequality, say $4 < 8$, and add 5 to both sides, we obtain

$$4 + 5 < 8 + 5$$
$$9 < 13.$$

Note that the result $9 < 13$ is also a true statement. Similarly, if we begin with $4 < 8$ and subtract 5 from both sides, we obtain the true statement

$$4 - 5 < 8 - 5 \qquad \text{or} \qquad -1 < 3.$$

These examples suggest the following properties of inequality.

Addition and Subtraction Properties of Inequality

If A, B, and C are algebraic expressions, then the inequality $A < B$ has the same solution set as

$$A + C < B + C \qquad \text{and} \qquad A - C < B - C.$$

Similar properties may be stated for $>$, $\leq$, and $\geq$.

Progress Check Answers

3.

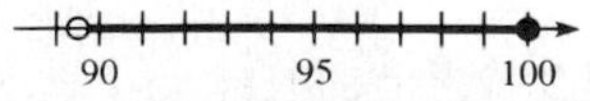

90   95   100

4. $\{x: x < 3\}$

These properties mean that in solving inequalities, you may add (or subtract) the same expression to (or from) both sides of an inequality.

EXAMPLE 5   Solve $x + 3 > -2$ and graph the solution set.

Solution   As with equations, the goal is to isolate x on one side of the inequality.

$$x + 3 > -2 \qquad \text{Given inequality.}$$
$$x + 3 - 3 > -2 - 3 \qquad \text{Subtract 3 from both sides.}$$
$$x > -5 \qquad \text{Simplify.}$$

Thus, all real numbers greater than -5 make the original inequality a true statement. The solution set is written $\{x: x > -5\}$ and is graphed in Figure 2.15.

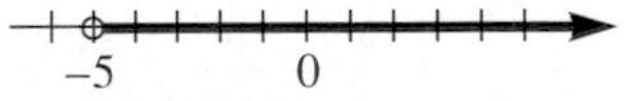

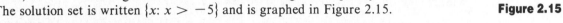

Figure 2.15

PROGRESS CHECK 5   Solve $x - 3 < -2$ and graph the solution set.   ⌐

The properties of inequality that relate to multiplication and division demand special care. To see why, notice that the true inequality $-5 < 2$ leads to a true inequality if we multiply both sides by *positive* 3.

$$-5(3) < 2(3) \qquad \text{Multiply both sides by 3.}$$
$$-15 < 6 \qquad \text{True inequality.}$$

But $-5 < 2$ leads to a false inequality if we multiply both sides by *negative* 3.

$$-5(-3) < 2(-3) \qquad \text{Multiply both sides by } -3.$$
$$15 < -6 \qquad \text{False inequality.}$$

To obtain a true statement when we multiply both sides by -3, we must reverse the direction of the inequality.

$$-5 < 2 \qquad \text{Original inequality.}$$
$$-5(-3) > 2(-3) \qquad \text{Multiply both sides by } -3 \text{ and } \textit{reverse} \text{ the inequality symbol.}$$
$$15 > -6 \qquad \text{True inequality.}$$

These examples show that the multiplication and division properties of inequality involve two cases: one for positive multipliers and one for negative multipliers.

Multiplication and Division Properties of Inequality

Case 1: C is *positive*.
If A, B, and C are algebraic expressions with $C > 0$, then the inequality $A < B$ has the same solution set as

$$AC < BC \qquad \text{and} \qquad \frac{A}{C} < \frac{B}{C}.$$

Case 2: C is *negative*.
If A, B, and C are algebraic expressions with $C < 0$, then the inequality $A < B$ has the same solution set as

$$AC > BC \qquad \text{and} \qquad \frac{A}{C} > \frac{B}{C}.$$

Similar properties may be stated for $>$, $\leq$, and $\geq$.

To this point, notice that the properties of inequality are analogous to the properties of equality, with one major exception: **If we multiply or divide both sides of an inequality by a negative number, we must reverse the direction of the inequality.**

EXAMPLE 6   Solve each inequality and graph the solution set.

a. $5x < -10$ **b.** $-5x < 10$

Progress Check Answer
5. $\{x: x < 1\}$

Solution

a. To isolate x, divide both sides by 5. We are applying the division property of inequality with a positive divisor, so the direction of the inequality remains the same.

$$5x < -10 \qquad \text{Original inequality.}$$
$$\frac{5x}{5} < \frac{-10}{5} \qquad \text{Divide both sides by 5.}$$
$$x < -2 \qquad \text{Simplify.}$$

The solution set is graphed in Figure 2.16 and is written $\{x: x < -2\}$. Note that the negative sign of the constant term -10 does not affect the direction of the inequality.

Figure 2.16

b. In this part, we isolate x by dividing both sides by -5. Because we are dividing by a negative number, we must reverse the inequality.

$$-5x < 10 \qquad \text{Original inequality.}$$
$$\frac{-5x}{-5} > \frac{10}{-5} \qquad \begin{array}{l}\text{Divide both sides by } -5 \text{ and reverse the}\\ \text{inequality symbol.}\end{array}$$
$$x > -2 \qquad \text{Simplify.}$$

The solution set is $\{x: x > -2\}$, as shown in Figure 2.17.

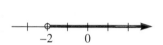

Figure 2.17

PROGRESS CHECK 6   Solve each inequality and graph the solution set.

a. $-2x \geq -10$ 　　　　　　　　　　　**b.** $3x > -9$

The combination of our previous methods allows us to solve any linear inequality. Note the similarities between the next example and Example 1 in Section 2.4.

EXAMPLE 7   Solve $5(x - 1) > 2x + 1$ and graph the solution set.

Solution   We follow similar steps to the general procedure for solving linear equations.

1. Simplify the left side of the inequality by removing parentheses.

$$5(x - 1) > 2x + 1 \qquad \text{Original inequality.}$$
$$5x - 5 > 2x + 1 \qquad \text{Distributive property.}$$

2. Write an inequality of the form $cx > n$ by using the addition or subtraction properties of inequality.

$$5x - 2x - 5 > 2x - 2x + 1 \qquad \text{Subtract } 2x \text{ from both sides.}$$
$$3x - 5 > 1 \qquad \text{Simplify.}$$
$$3x - 5 + 5 > 1 + 5 \qquad \text{Add 5 to both sides.}$$
$$3x > 6 \qquad \text{Simplify.}$$

3. Obtain an inequality of the form $x >$ number by using the multiplication or division properties of inequality.

$$\frac{3x}{3} > \frac{6}{3} \qquad \text{Divide both sides by 3.}$$
$$x > 2 \qquad \text{Simplify.}$$

4. Although we cannot check the entire solution set, because it is an infinite set of numbers, we can at least select one number greater than 2 and check that our result is reasonable. Picking 3 yields the following.

$$5(x - 1) > 2x + 1 \qquad \text{Original inequality.}$$
$$5(3 - 1) > 2(3) + 1 \qquad \text{Replace } x \text{ with 3.}$$
$$10 > 7 \qquad \text{True inequality.}$$

Progress Check Answers

6. (a) $\{x: x \leq 5\}$ 　　　　　(b) $\{x: x > -3\}$

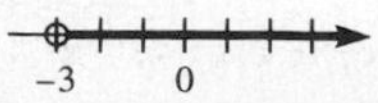

The check confirms that 3 is a solution, so $x > 2$ is reasonable. The solution set is $\{x: x > 2\}$, as shown in Figure 2.18.

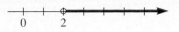

Figure 2.18

PROGRESS CHECK 7   Solve $7(x + 2) > 5x - 4$ and graph the solution set. ⌐

We may write $4 < 8$ as $8 > 4$, and vice versa. In general, the inequality

$$a > b \qquad \text{is equivalent} \qquad \text{to } b < a.$$

This equivalence means that we may interchange the left and right side of an inequality provided we also reverse the inequality symbol. We use this reversal in the next example because $x > 1$ is easier to visualize than $1 < x$.

EXAMPLE 8   Solve $4 - x < 3x$ and graph the solution set.

Solution   We cannot simplify either side of the inequality, so we begin by applying the addition property.

$$
\begin{aligned}
4 - x &< 3x && \text{Original inequality.} \\
4 - x + x &< 3x + x && \text{Add } x \text{ to both sides.} \\
4 &< 4x && \text{Simplify.}
\end{aligned}
$$

Now, divide both sides by 4 and maintain the direction of the inequality.

$$
\begin{aligned}
\frac{4}{4} &< \frac{4x}{4} && \text{Divide both sides by 4.} \\
1 &< x && \text{Simplify.}
\end{aligned}
$$

Finally, the answer is easier to visualize if we interchange sides and reverse the inequality symbol.

$$x > 1$$

Substituting any number greater than 1 in the original inequality will show that our answer is reasonable. The solution set is $\{x: x > 1\}$, as graphed in Figure 2.19.

Figure 2.19

PROGRESS CHECK 8   Solve $6 - 2x > x$ and graph the solution set. ⌐

3   We now consider a business application of inequalities.

EXAMPLE 9   Solve the problem in the section introduction on page 112.

Solution   The publisher plans to produce and sell x textbooks. We need to find values of x, so that

$$
\underbrace{\text{revenue}}_{26x} \qquad \underbrace{\text{is greater than}}_{>} \qquad \underbrace{\text{cost}}_{180{,}000 + 11x}.
$$

Solve the Inequality

$$
\begin{aligned}
15x &> 180{,}000 && \text{Subtract } 11x \text{ from both sides.} \\
x &> 12{,}000 && \text{Divide both sides by 15.}
\end{aligned}
$$

Answer the Question   Revenue is greater than cost when the publisher produces and sells more than 12,000 textbooks.

Check the Answer   The revenue from the sale of 12,000 books is $\$26(12{,}000) = \$312{,}000$, while the cost of producing 12,000 books is $\$180{,}000 + \$11(12{,}000) = \$312{,}000$. Therefore, 12,000 books is the break-even point, and revenue is greater than cost after this point. Note also that only *integers* make sense as solutions.

PROGRESS CHECK 9   Redo the problem in Example 9 assuming that the publisher's cost is given by $161{,}000 + 9x$ and the revenue is given by $23x$. ⌐

Progress Check Answers

7. $\{x: x > -9\}$

8. $\{x: x < 2\}$

9. Break-even point is 11,500 books.

4 The next two examples show how to solve a compound inequality that specifies a set of numbers between two numbers.

EXAMPLE 10 Solve $-3 < 6 - 2x < 3$ and graph the solution set.

Solution To solve this type of inequality, we isolate *x in the middle.*

$-3 < 6 - 2x < 3$	Original inequality.
$-3 - 6 < 6 - 6 - 2x < 3 - 6$	Subtract 6 from each member.
$-9 < -2x < -3$	Simplify.
$\dfrac{-9}{-2} > \dfrac{-2x}{-2} > \dfrac{-3}{-2}$	Divide each member by -2 and reverse the inequality symbols.
$4.5 > x > 1.5$	Simplify.

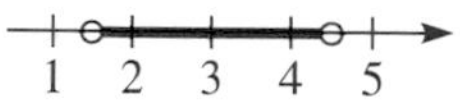

A check using any number between 1.5 and 4.5 will show this result is reasonable. The solution set is graphed in Figure 2.20. Because $1.5 < x < 4.5$ is equivalent to $4.5 > x > 1.5$ and is more natural to write, we specify the solution set as $\{x: 1.5 < x < 4.5\}$.

Figure 2.20

PROGRESS CHECK 10 Solve $0 \le 2 - 5x \le 9$ and graph the solution set. ⌐

EXAMPLE 11 In a math class an average from 74.5 up to but not including 79.5 results in a grade of C+. A student's first three test grades are 74, 72, and 81. With one test left, find all possible grades on the last test that result in a final grade of C+.

Solution Let x represent the grade on the last exam. To result in a grade of C+, x must satisfy

$$74.5 \quad \le \quad \text{average of grades} \quad < \quad 79.5$$
$$\downarrow \quad \downarrow \qquad \qquad \qquad \qquad \downarrow \quad \downarrow$$
$$74.5 \quad \le \quad \frac{74 + 72 + 81 + x}{4} \quad < \quad 79.5.$$

Solve the Inequality

$74.5 \le \dfrac{227 + x}{4} < 79.5$	Simplify.
$298 \le 227 + x < 318$	Multiply each member by 4.
$71 \le x < 91$	Subtract 227 from each member.

Answer the Question Any grade on the last test from 71 up to but not including 91 results in a grade of C+

Check the Answer If the last grade is 71, then the average of the test grades is $(74 + 72 + 81 + 71)/4 = 74.5$, the lowest average that produces a grade of C+. If the last test grade is 91, the average is $(74 + 72 + 81 + 91)/4 = 79.5$, the excluded endpoint of the C+ interval. The solution checks.

PROGRESS CHECK 11 An average from 84.5 up to but not including 89.5 earns a B+. If your first three grades are 83, 88, and 87, find all possible grades on the fourth exam that will result in a grade of B+. ⌐

Progress Check Answers

10. $\{x: -\frac{7}{5} \le x \le \frac{2}{5}\}$

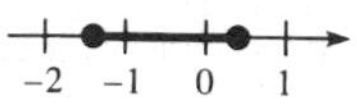

11. Any grade from 80 up to but not including 100

EXERCISES 2.8

In Exercises 1–18, graph the solution set of the given inequality.

1. $x > 2$

2. $x > 2.5$

3. $x < -1$

4. $x < -0.5$

5. $x \ge 3$

6. $x \ge -2.5$

7. $x \ge 0$

8. $x \le -3$

9. $x < \frac{5}{4}$

10. $x > -\frac{13}{2}$

11. $x \le -3\frac{1}{4}$

12. $x \ge 1\frac{1}{2}$

13. $2 < x < 3$

14. $-4 < x < 1.5$

15. $0 \le x \le 1$

16. $\frac{3}{2} \le x < \frac{7}{2}$

17. $-\frac{13}{4} < x < -\frac{1}{4}$

18. $-3 \le x < 0$

In Exercises 19–30, specify the set of numbers shown in the graph by using set-builder notation.

19.
1

20.
2

21.
3

22.
0

23.
1   3

24.
−3   3

25.
−5   −1

26.
−7   2

27.
$-\frac{13}{2}$

28.
100.38

29.
0

30.
1

In Exercises 31–54, solve the inequality and graph the solution set.

31. $x + 2 > -3$

32. $x - 2 > -3$

33. $x - 8 \le 0$

34. $x + 4 \ge -4$

35. $2x - 3 < x + 1$

36. $3x + 1 > 2x - 3$

37. $4x \ge 3x - 1$

38. $5x \le 4x + 4$

39. $3x < 12$

40. $4x < 0$

41. $-2x < 14$

42. $-7x > 21$

43. $-12 < 2x$

44. $8 \le \frac{4}{3}x$

45. $-\frac{1}{2}x + 2 < 3$

46. $-\frac{1}{4}x - 3 \ge -1$

47. $4(x - 3) > 2x$

48. $2(x - 1) \le 4x - 2$

49. $3(2 - x) < 0$

50. $-2(2 + x) > 5$

51. $x \ge 2(3 + x)$

52. $-2x > -(x - 2)$

53. $1 - x > 2 - x$
54. $3 - 2x < 4 - 2x$

In Exercises 55–66, solve the compound inequality and graph the solution set.

55. $-1 < x + 2 < 1$

56. $-3 \le x - 2 < 3$

57. $-2 < 4x + 7 \le 2$

58. $-4 \le 3x - 1 \le 4$

59. $0 \le 3 - x \le 1$

60. $1 < 2 - x < 2$

61. $-1 < 4 - 2x < 0$

62. $-3 \le 5 - 4x < 13$

63. $\frac{1}{4} < x - 2 < \frac{3}{4}$

64. $\frac{2}{3} \le x + 1 \le \frac{5}{3}$

65. $\frac{1}{2} \le \frac{x}{3} < \frac{2}{5}$

66. $\frac{-3}{2} < \frac{-x}{3} < -166$

In Exercises 67–76, solve the problem by writing and solving an appropriate inequality.

67. The cost of producing x widgets is $200,000 plus $14 per widget. The manufacturer sells the widgets for $28. For what values of x does the manufacturer make a profit?

68. An electrician charges a flat fee of $50 plus $18 an hour for a job. How many hours must a job take for the electrician to make more than $500?

69. Normal weight in pounds for children of a certain age and height is 60 to 79, inclusive. This interval is represented as $60 \le p \le 79$. Use the formula $p = 2.205k$ to express this interval in kilograms.

70. According to the directions on the can, a gallon of paint will cover from 175 to 425 ft² of surface. Express this inequality in terms of square yards by using the relationship $f = 9y$, where f stands for square feet and y stands for square yards.

71. An advertisement for a Canadian ski vacation says that the temperature will be anywhere from -15 to -5 degrees Celsius. This is represented by $-15 \le C \le -5$. Find the corresponding range in Fahrenheit by using the formula $C = \frac{5}{9}(F - 32)$.

72. For a chemistry experiment to be successful the temperature must be maintained between 32.5 and 33.5 degrees Fahrenheit. This is written $32.5 < F < 33.5$. Convert this to a range of Celsius temperatures by using the formula $F = \frac{9}{5}C + 32$.

73. In a math class an average of from 70.0 up to but not including 74.5 results in a grade of C. A student's first four test grades are 69, 67, 72, and 74. Find all possible grades (x) on the fifth test that will result in a grade of C.

74. A student got a grade of 85 on each of the first four tests. If the average for five tests is at least 89.5, the student will get an A. Show that the student cannot do it even with a perfect score on the last test. Assume that the highest possible grade on the last test is 100.

75. The expense of running a program is $1,000 plus $240 per person enrolled. If the minimum budget is $8,000 and the maximum budget is $9,500, how many people can enroll in the program? Remember that only positive integers make sense as answers.

76. A computer service bills its users $25 per month plus $6 for every hour they are connected to the service. A user proposes a minimum monthly budget of $150 and a maximum budget of $250 for this service. With this budget how many hours of connect time can the user get?

THINK ABOUT IT

1. Why do you reverse the inequality when you multiply both sides of an inequality by a negative number?

2. Can the square of a positive number be smaller than the number itself? Can the square of a negative number be less than the number itself?

3. Show that the inequality $x + 1 < x$ has no solutions. Explain why this makes sense by translating the inequality into an English sentence.

4. Solve for x: $x - 5 < 10 < x + 8$. (*Hint:* As a first step, subtract x from all three members of the inequality.)

5. Solve for x: $x - 5 < 10 < 2x + 1$. (*Hint:* Write this as two separate inequalities, and solve each one. Then describe the set of numbers that are in both solution sets.)

REMEMBER THIS

1. Evaluate $4x^3$ when (a) $x = 5$ and (b) $x = -4$.

2. Compute: (a) $(-7)(7)$; (b) $7(-7)$; (c) $-(7)(7)$.

3. Compute: (a) $(5) - (5)$; (b) $(5)(-5)$.

4. What is one-third of one-third?

5. Are any of these equal? $(2 + 3)^3$, $2^3 + 3^3$, $(2 \cdot 3)^3$.

6. Write an expression for 8 less than 5 times the square of a number.

7. Write an expression for the square of the sum of the two numbers m and n.

8. Write an expression for the sum of the squares of the two numbers m and n.

9. Solve $C = 2\pi r$ for r.

10. Four times a number is the same as 5 more than twice the number. What is the number?

Chapter 2   SUMMARY

OBJECTIVES CHECKLIST   Specific chapter objectives are summarized below along with numbered example problems from the text that should clarify the objectives. If you do not understand any objectives or do not know how to do the selected problems, then restudy the material.

2.1 Can you:

1. Identify the terms and the numerical coefficients of the terms in an algebraic expression?
   a. Identify the terms in $x^2 - 3x + 4$. [Example 1b]
   b. Specify the numerical coefficient in $-x$. [Example 2d]

2. Identify like terms and unlike terms?
   Identify any like terms in $5 + 7x - 3x + 4$. [Example 3b]

3. Simplify algebraic expressions by combining like terms?
   Simplify $y - 11y$. [Example 4b]

4. Simplify algebraic expressions by removing parentheses and combining like terms?
   Simplify $4(x + 1) - 6(x + 2)$. [Example 7a]

2.2 Can you:

1. Classify an equation as a conditional equation, an identity, or a false equation?
   Classify $x + 1 = x$. [Example 1b]

2. Determine if a number is a solution of an equation?
   Determine if -1 is a solution of $5 - x = 4$. [Example 2b]

3. **Solve linear equations by applying the addition or subtraction properties of equality?**
   a. Solve the equation $x - 3 = 8$. [Example 3]
   b. Solve the equation $y + 8 = -1$. [Example 5]

4. **Use the addition or subtraction properties to show that certain equations are identities or false equations?**
   Show that $x + 1 = x$ is a false equation. [Example 7]

5. **Solve applied problems by solving linear equations?**
   A rise in temperature of 5 degrees will result in a temperature of -2 degrees. What is the original temperature? [Example 9]

2.3 Can you:
   1. **Solve linear equations by applying the division and multiplication properties of equality?**
   a. Solve the equation $3x = -21$. [Example 1]
   b. Solve the equation $\dfrac{x}{7} = \dfrac{3}{11}$. [Example 3]

   2. **Solve equations of the form $-x = n$?**
   Solve the equation $4 - x = 6$. [Example 4]

   3. **Solve linear equations in which the first step is to combine the terms?**
   Solve the equation $4y - 5y = 20$. [Example 5]

   4. **Solve applied problems by solving linear equations?**
   By solving $x - 0.06x = 150,000$, find the selling price for a house which nets \$150,000 for the seller after the agent takes a 6 percent commission. [Example 6]

2.4 Can you:
   1. **Apply a general procedure to solve *any* linear equation?**
   Solve the equation $5(x - 1) = 2x + 1$. [Example 1]

   2. **Classify certain equations as false equations or identities?**
   Solve the equation $4 - 2(x + 1) = 5 - 2x$. [Example 5]

   3. **Solve applied problems by solving linear equations?**
   One job pays \$100 per week plus a 10 percent commission. Another pays a straight 15 percent commission. At what point does the second job begin to pay more than the first? Solve $100 + 0.10x = 0.15x$. [Example 6]

2.5 Can you:
   1. **Find the value of a variable in a formula when given values for the other variables?**
   Using $A = \frac{1}{2}h(a + b)$, find the value of b if $h = 9$ cm, $a = 11$ cm, and $A = 108$ cm². [Example 2]

   2. **Solve a given formula or literal equation for a specified variable?**
   Solve $F = \frac{9}{5}C + 32$ for C. [Example 4]

2.6 Can you:
   1. **Solve word problems by translating phrases and setting up and solving equations?**
   You need to cut a piece of string that is 85 in. long into two parts so that one piece is 4 times longer than the other. How long should you make each piece? [Example 1]

   2. **Prove certain statements about integers?**
   Show that the sum of any three consecutive integers is equal to 3 times the middle integer. [Example 6]

2.7 Can you:
   1. **Solve problems involving geometric figures?**
   A piece of molding is 75 in. long. You wish to use it to make a rectangular frame that is twice as long as it is wide. What are the dimensions of the frame? [Example 1]

2. **Solve problems involving percentages?**
   You win $30,000 in a state lottery. You are advised to put some of the money in a risky investment that earns 15 percent annual interest and the rest in a safe investment that earns 7 percent annual interest. If you want the total interest to be $3,000 annually, how much should be invested at each rate?   [Example 3]

3. **Solve problems about uniform motion?**
   An air traffic controller sees that two planes 100 mi apart are flying toward each other, one at 500 mi/hour and one at 300 mi/hour. How much time is there to prevent a crash?   [Example 5]

4. **Solve problems about liquid mixtures?**
   One mixture of gasoline and oil contains 3 percent oil and another contains 6 percent oil. How much of each should be used to obtain 12 qt of a new mixture that contains 5 percent oil?   [Example 6]

2.8 **Can you:**
1. **Specify solution sets of inequalities by using graphs and set-builder notation?**
   a. Graph the solution set of $x \leq 2$.   [Example 2]
   b. Specify the set of numbers graphed by using set-builder notation.   [Example 4]

   $$\begin{array}{ccccccc} + & \bullet & + & + & + & \bullet & + \rightarrow \\ -3 & -2 & -1 & 0 & 1 & 2 & 3 \end{array}$$

2. **Solve linear inequalities by applying properties of inequalities?**
   Solve $5(x - 1) > 2x + 1$ and graph the solution set.   [Example 7]

3. **Solve applied problems involving linear inequalities?**
   The cost of producing x books is $180,000 + 11x$, which represents a setup cost of $180,000 plus a variable cost of $11 per book. The publisher receives $26 for each sale, so the revenue is $26x$ for selling x books. For what values of x is revenue greater than cost?   [Example 9]

4. **Solve compound inequalities that specify numbers between two numbers.**
   Solve $-3 < 6 - 2x < 3$ and graph the solution set.   [Example 10]

KEY TERMS

Conditional equations (2.2)	Like terms (2.1)	Set-builder notation (2.8)
Equivalent equations (2.2)	Linear equations (2.2)	Solution set (2.2)
False equation (2.2)	Linear inequality (2.8)	Terms (2.1)
Identity (2.2)	Literal equation (2.5)	Unlike terms (2.1)
Inequalities (2.8)	Numerical coefficient (2.1)	

KEY CONCEPTS AND PROCEDURES

Section	Key Concepts or Procedures to Review
2.1	■ Only like terms can be combined. We combine like terms by combining their numerical coefficients using the distributive property.
2.2	■ Addition and subtraction properties of equality: If A, B, and C are algebraic expressions, then the equation $A = B$ is equivalent to $$A + C = B + C \quad \text{and} \quad A - C = B - C$$ ■ Methods to determine if an equation is a conditional equation, an identity, or a false equation

Section	Key Concepts or Procedures to Review
2.3	■ Multiplication and division properties of equality: If A, B, and C are algebraic expressions, and C does not represent zero, then the equation $A = B$ is equivalent to $$\frac{A}{{}^{,}C} = \frac{B}{C} \quad \text{and} \quad CA = CB$$
2.4	■ General procedures to solve linear equations 1. Simplify each side of the equation if necessary. 2. Use the addition and subtraction properties of equality, if necessary, to write an equivalent equation of the form $cx = n$. 3. Use the multiplication or division properties of equality to solve $cx = n$ if $c \neq 1$. 4. Check the solution by substituting it in the original equation.
2.5	■ The equation-solving techniques of this chapter apply to solving formulas and literal equations for a specified variable.
2.6	■ Guidelines to solving a word problem 1. Read the problem several times. 4. Solve the equation. 2. Let a variable represent an unknown. 5. Answer the question. 3. Set up an equation. 6. Check the answer in the word problem.
2.7	■ For Word Problems Involving Use

	For Word Problems Involving	Use
2.7	Geometric figures	Formulas from Section 1.3
	Percentage	$I = Pr$ annual interest = principal · rate
	Uniform motion	$d = rt$ distance = rate · time
	Liquid mixtures	Ingredient amount = ingredient percent times solution amount

2.8	■ Addition and subtraction properties of inequality: If A, B, and C are algebraic expressions, then the inequality $A < B$ has the same solution set as $$A + C < B + C \quad \text{and} \quad A - C < B - C$$ ■ Multiplication and division properties of equality: If A, B, and C are algebraic expressions, then 1. $A < B$ has the same solution set as $AC < BC$ and $\dfrac{A}{C} < \dfrac{B}{C}$, if $C > 0$ 2. $A < B$ has the same solution set as $AC > BC$ and $\dfrac{A}{C} > \dfrac{B}{C}$, if $C < 0$ ■ $a < b$ is equivalent to $b > a$.

CHAPTER 2 REVIEW EXERCISES

2.1

1. Identify the terms in $4 - 5y - 6y^2$.

2. Specify the numerical coefficient of $\dfrac{x}{4}$.

3. Identify any sets of like terms in $3x^3 + 4x + 6x^3$.
4. Simplify by combining like terms: $5 + 4x - x$.

5. Find the perimeter of the given figure.

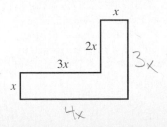

6. Simplify: $3(6x - 5) - 3(5 - 6x)$
7. Simplify: $3y^2 - 5(2y + 3y^2)$.
8. The length of a rectangle is 6 times the width. Find a simple formula for the perimeter. Use x for the width.

2.2

9. Classify each equation as a conditional equation, an identity, or a false equation.
   a. $2x - 9 = 11$
   b. $5x + 4x = 9x$
   c. $x + 2 = x - 2$
10. Is 10 a solution of $x + 3 = 7$?
11. Solve and check: $x - 4 = 5$.
12. Solve and check: $y + 3 = 8.9$.
13. Show that $x + 2x = 3x$ is an identity.
14. Show that $6x - 6 = 6x$ is a false equation.
15. A family has $2,400 in its monthly budget. One-quarter goes to rent, one-third goes for food, and $120 goes for a car payment. How much is left for other expenses? Use the equation $\frac{1}{4}(2,400) + \frac{1}{3}(2,400) + 120 + x = 2,400$.

2.3

16. Solve and check: $4x = -28$.
17. Solve and check: $\dfrac{x}{2} = 7$.
18. Solve and check: $-y = 67$.
19. Solve and check: $2 - y = -5$.
20. Solve and check: $5x - 6x = 9$.
21. The sales price of a car plus the 6 percent tax came to $15,741. How much did the car cost before the tax was figured in? Use $x + 0.06x = 15,741$, where x is the sales price.
22. One year after it was purchased, a farm tractor depreciated in value 8 percent, to $29,440. How much was the purchase price? Use $x - 0.08x = 29,440$.

2.4

23. Solve and check: $3(x + 5) = 10$.
24. Solve and check: $6(x + 2) = 3x$.
25. Solve and check: $y = 5 - 6y$.
26. Solve and check: $2x - 3(x + 4) = x + 2$.
27. Show that $3(4x + 2) = 12x + 2$ is a false equation.
28. Half of a certain number is 5 less than twice the number. Find the number. Solve $\frac{1}{2}x = 2x - 5$.

2.5

29. Given $d = rt$, $d = 150$, and $r = 45$, find t.
30. Given $A = P + Prt$, $A = 2,375$, $P = 1,250$, and $r = 0.09$, find t.
31. Given $E = \frac{1}{2}mv^2$, $v = 10$, and $E = 1,000$, find m.
32. Solve $3x + 4y = -10$ for y.
33. Solve $P = 2(\ell + w)$ for w.
34. The speed of sound in water is about 4,800 ft/second. Since $d = rt$, we can represent this as $d = 4,800t$, where d is the distance the sound travels in feet and t is the time elapsed in seconds. Solve for t. Then find the time it takes for the call of a whale to travel 1 mi (1 mi $= 5,280$ ft).

2.6

35. A 9-ft piece of wood is sawed into two pieces, where one is 8 ft shorter than the other. Find the length of the shorter piece.
36. One person's income was half of another's and the total of both incomes was $38,100. What was the smaller of the two incomes?
37. Five less than 3 times a number is 6 more than the number. Find the number.
38. Show that the sum of any five consecutive integers is 10 more than 5 times the smallest one.

2.7

39. The length of a rectangle is 6 yd longer than its width, and the perimeter is 24 yd. Find the area of the rectangle.
40. In a right triangle one acute angle is 45° larger than the other. Find the measure of the smaller acute angle.
41. A person took out two loans totaling $8,000 for a year. The first loan was at an annual interest rate of 10 percent, and the second was at 9 percent. The total interest charged for both loans was $770 for the year. How much was borrowed at 9 percent?
42. A student spent 20 percent of a grant on books and 25 percent of the grant on a computer printer, for a total of $540. How much did the printer cost?
43. Two bikers are cycling toward one another along a highway. Each one is going 10 mi/hour. How long will it take them to meet if they are now 15 mi apart?
44. A truck leaves a rest area and is going down an interstate highway averaging 45 mi/hour. Twenty minutes later a car leaves the same rest area following the truck but averaging 55 mi/hour. If neither vehicle stops, how long will it take the car to catch up with the truck? (*Hint:* Use $\frac{1}{3}$ hour to represent 20 minutes.)
45. A jar contains a quart of liquid made up of 50 percent olive oil and 50 percent wine vinegar. How much oil should be added so that the mixture is only 40 percent vinegar?
46. One mixture is 3 percent acid and another is 5 percent acid. How much of each should be used to make 50 ml of a new mixture which is 3.4 percent acid?

2.8

47. Graph the solution set of $x < 3.5$.
48. Graph the solution set of $-\frac{1}{2} < x \le \frac{1}{2}$.
49. For $5(x + 1) > 0$, solve the inequality and graph the solution set.
50. For $-2 \le 3x - 4 < 1$, solve the inequality and graph the solution set.
51. The cost of manufacturing n wooden puzzles is estimated at $50,000 (setup costs) plus $4 per puzzle. The anticipated selling price is $12 per piece. For what values of n would the manufacturer make a profit? Suppose the manufacturer can reasonably expect to sell 5,000 puzzles. Is this a good business situation?

CHAPTER 2 TEST

1. Identify any sets of like terms in $1 + 2x + 3x^3 - 5x$. Explain why they are like terms.

2. Simplify: $x - 2(x - 3)$.
3. Draw and label a sketch and then write an expression for the perimeter of a rectangle where the length is twice the width. Let x represent the width of the rectangle.

4. Is -1 a solution of $x - 1 = 2x$? Explain how you can tell.

In Questions 5–9, solve the equation and check the solution.

5. $-3 = 3 + y$
6. $\dfrac{4x}{5} = 8$
7. $4x - 5x = -6$
8. $13x - 6 = 10x + 6$
9. $\dfrac{x}{3} + \dfrac{1}{4} = \dfrac{x}{5}$

10. Solve this inequality and graph the solution set: $-2 < 4 - 5x < 0$.

11. Solve $4x + 3y = 12$ for y.

12. What is a false equation? Show that $3x + 4 = 3x - 4$ is a false equation. What is the solution set?

13. One-third of a number equals 3 more than the number. Find the number by solving $\frac{1}{3}x = x + 3$.
14. Use $A = P + Prt$ to find t if $A = 2,000$, $P = 1,000$, and $r = 0.12$.

15. One piece of mail weighs 13 oz more than another, and the total weight for both is 20 oz. Find the weight of each.
16. The average of three numbers is -5. Two of the numbers are 2 and -8. Find the third number.
17. Find the measure in degrees of all the angles in this triangle. Use the fact that the sum of the measures must be 180°.

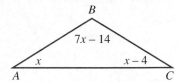

18. Together, two bonds totaling $2,000 earned $106 interest in one year. One earned interest at the 5 percent rate and one earned interest at 6.5 percent. How much was invested in each bond? How much interest did each bond earn for the year?

19. Two very long, straight conveyor belts run side by side. One travels at 1 ft/second, and the other travels at 0.3 ft/second. If two objects start next to each other, one on each belt, how long will it take before the one on the slower belt is 7.98 ft behind the faster one?

20. A potter has 1,000 grams (g) of a ceramic glaze mix, of which 4 percent is a "secret" ingredient. How much more of this ingredient should be added to bring it up to 5 percent of the mix? (Round your answer to the nearest tenth of a gram.)

CUMULATIVE TEST 2

1. Compute: $\frac{7}{4} + \frac{3}{8} - \frac{51}{40}$.
2. Graph the set of odd integers from -3 to 3.

3. Evaluate $\dfrac{3x^2 + 2}{3 + 2x^2}$ when $x = 4$.

4. Evaluate $2,000(1 + 0.05)^8$ using a calculator. Round to 2 decimal places.
5. Compute the volume of a cylinder whose radius and height are each 3 in. Use $\pi = 3.14$.
6. Evaluate $|x| - |-x|$ when $x = 5$.
7. Evaluate $-2 + [3 - 4(5 - 1)]$.
8. Write down expressions for the area and perimeter. Then evaluate these expressions when $x = -1$.

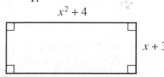

9. The unemployment rate in a state went from 5.1 percent to 4.5 percent. Express this as a percent change. Round to the nearest hundredth of a percent.
10. Rewrite $-(3 - 3x + 4y)$ without parentheses.

11. Simplify $3x - 4(x - 1)$.
12. Is -2 a solution of $3x - 1 = x - 5$? Explain how you can tell.
13. Solve and check: $12x - 7 = 7 - 12x$.

14. Solve and check: $\dfrac{x}{5} - \dfrac{3}{4} = 1$.

15. Solve and check: $-3 < 1 - 2x < -1$.
16. One-fourth of a number equals 4 more than the number. Find the number.

17. Solve for c: $a = \dfrac{c + b}{d}$.

18. Find the measure in degrees of the smallest angle in this triangle.

19. How much water should be added to a gallon of solution which is 80 percent antifreeze to make a solution which is 60 percent antifreeze?
20. Use $A = P + Prt$ to find r when $P = 1,500$, $t = 3$, and $A = 1,860$.

3 Exponents and Polynomials

The volume of a sphere of radius r is given by the formula $V = \frac{4}{3}\pi r^3$. Determine a formula for the volume in terms of the diameter d. (See Example 7 of Section 3.1.)

THE ABILITY to apply the definitions and properties of exponents is a crucial skill in algebra. In this chapter we show how to deal with exponents that are integers and how to use these concepts to work with scientific notation and perform operations with polynomials. As you will see, polynomials are basically algebraic expressions with terms of the form

$$(\text{real number})x^{\text{nonnegative integer}},$$

so typical polynomial terms look like $5x^3$ and $7x$. In many areas of mathematics it is common to analyze problems by using polynomial expressions.

3.1 Product and Power Properties of Exponents

1 Apply positive integer exponent definitions.

2 Apply the product property of exponents.

3 Apply the power properties of exponents.

4 Simplify exponential expressions with literal exponents.

1 Recall from Chapter 1 that a positive integer exponent is a shortcut way of expressing a repeated factor. That is,

$$5^3 \quad \text{is shorthand for} \quad 5 \cdot 5 \cdot 5,$$
$$(-4)^2 \quad \text{is shorthand for} \quad (-4) \cdot (-4),$$

a^n, where n is a positive integer, is shorthand for

$$\underbrace{a \cdot a \cdot a \cdot a \cdots a.}_{n \text{ factors}}$$

Thus by a^n, where n is a positive integer, we mean to use a as a factor n times. The expression a^n is read "the **nth power of a,**" and a^n is called an **exponential expression** with **base** a and **exponent** n. For the first power of a, remember that $a^1 = a$.

EXAMPLE 1   Write each expression using exponents. In each case, identify the base and the exponent.

a. $3 \cdot 3 \cdot 3 \cdot 3$ **b.** $\left(\frac{2}{3}\right) \cdot \left(\frac{2}{3}\right)$

c. $(-2) \cdot (-2) \cdot (-2)$ **d.** $-2 \cdot 2 \cdot 2$

Solution   Apply the definition of a positive integer exponent.

a. $3 \cdot 3 \cdot 3 \cdot 3 = 3^4$. The base is 3 and the exponent is 4.

b. $\left(\frac{2}{3}\right) \cdot \left(\frac{2}{3}\right) = \left(\frac{2}{3}\right)^2$. The base is $\frac{2}{3}$ and the exponent is 2. Note that parentheses are needed to show that the base is $\frac{2}{3}$.

c. $(-2) \cdot (-2) \cdot (-2) = (-2)^3$. The base is -2 and the exponent is 3. Once again, parentheses are needed to identify the base.

d. $-2 \cdot 2 \cdot 2 = (-1)2 \cdot 2 \cdot 2 = (-1)2^3$, which is the negative of 2^3, so $-2 \cdot 2 \cdot 2 = -2^3$. The base is 2 and the exponent is 3.

Caution   Be careful to distinguish between the form $(-a)^n$, as in part **c,** and the form $-a^n$, as in part **d.** In the expression $(-a)^n$ the parentheses indicate that the base is $-a$. The absence of parentheses in $-a^n$ means that the expression symbolizes the negative of a^n, or $(-1)a^n$.

PROGRESS CHECK 1   Write each expression using exponents. In each case, identify the base and the exponent.

a. $2 \cdot 2 \cdot 2 \cdot 2 \cdot 2$ **b.** $-7 \cdot 7$

c. $(-7) \cdot (-7)$ **d.** $\left(\frac{1}{9}\right) \cdot \left(\frac{1}{9}\right) \cdot \left(\frac{1}{9}\right)$ ⌐

EXAMPLE 2   Evaluate each expression. Check your result with a calculator.

a. $3^4 - 4^3$ **b.** -4^3 **c.** $5^2 \cdot 5^4$

Solution   Apply the definition of a positive integer exponent. For the calculator check, use the power function key $\boxed{y^x}$, as discussed in Section 1.2.

Progress Check Answers

1. (a) 2^5; 2; 5   (b) -7^2; 7; 2   (c) $(-7)^2$; -7; 2
(d) $\left(\frac{1}{9}\right)^3$; $\frac{1}{9}$; 3

a. $3^4 = 3 \cdot 3 \cdot 3 \cdot 3 = 81$, and $4^3 = 4 \cdot 4 \cdot 4 = 64$, so $3^4 - 4^3 = 81 - 64 = 17$.

Calculator Check   3 $\boxed{y^x}$ 4 $\boxed{-}$ 4 $\boxed{y^x}$ 3 $\boxed{=}$ $\boxed{\qquad 17}$

b. As discussed in Example 1, -4^3 equals the negative of 4^3, so $-4^3 = -64$. We use the change sign key $\boxed{+/-}$ in the check.

Calculator Check   4 $\boxed{y^x}$ 3 $\boxed{=}$ $\boxed{+/-}$ $\boxed{\qquad -64}$

c. $5^2 = 5 \cdot 5 = 25$, and $5^4 = 5 \cdot 5 \cdot 5 \cdot 5 = 625$, so $5^2 \cdot 5^4 = 25 \cdot 625 = 15{,}625$. Use the square key $\boxed{x^2}$ to compute 5^2 in the check.

Calculator Check   5 $\boxed{x^2}$ $\boxed{\times}$ 5 $\boxed{y^x}$ 4 $\boxed{=}$ $\boxed{\quad 15625}$

PROGRESS CHECK 2   Evaluate each expression. Check your result with a calculator.

a. $6^3 + 5^6$ **b.** -3^5 **c.** $4^7 \cdot 4^2$

2   The first rule of exponents we develop concerns the product of two exponential expressions with the same base. For instance, reconsider the product $5^2 \cdot 5^4$ in Example 2c. We can reason that

$$5^2 \cdot 5^4 = \underbrace{(5 \cdot 5)}_{2 \text{ factors}} \cdot \underbrace{(5 \cdot 5 \cdot 5 \cdot 5)}_{4 \text{ factors}} = \underbrace{5 \cdot 5 \cdot 5 \cdot 5 \cdot 5 \cdot 5}_{6 \text{ factors}} = 5^6.$$

Then by calculator $5^6 = 15{,}625$. Note that the exponent 6 on the resulting product is the sum of the original exponents (2 and 4). This example illustrates the product property of exponents.

Product Property of Exponents

If m and n are positive integers and a is a real number, then

$$a^m \cdot a^n = a^{m+n}.$$

Remember that this property applies only to exponential expressions with the *same base*. When computing a product this way, you can see we are really just *counting* the number of times a appears as a factor.

EXAMPLE 3   Use the product property of exponents and simplify each expression.

a. $3^4 \cdot 3^2$ **b.** $x^3 \cdot x^5$
c. $(-2) \cdot (-2)^6$ **d.** $a^7 a^4 a$

Solution

a. $3^4 \cdot 3^2 = 3^{4+2} = 3^6 = 729$
b. $x^3 \cdot x^5 = x^{3+5} = x^8$
c. Because $(-2) = (-2)^1$, we have $(-2) \cdot (-2)^6 = (-2)^1 \cdot (-2)^6 = (-2)^{1+6}$
$= (-2)^7 = -128$.
d. The three factors have a common base, so add all three exponents.

$$a^7 a^4 a = a^{7+4+1} = a^{12}$$

PROGRESS CHECK 3   Simplify each expression.

a. $10^2 \cdot 10^3$ **b.** $y^6 y^4$
c. $(-4)^4 \cdot (-4)$ **d.** $b^3 b^5 b$

Progress Check Answers
2. (a) 15,841   (b) -243 (c) 262,144
3. (a) $10^5 = 100{,}000$ (b) y^{10}
(c) $(-4)^5 = -1{,}024$ (d) b^9

EXAMPLE 4   Find (a) the product and (b) the sum of $4x^3$ and $5x^3$ in simplest form.

Solution

a. To find the product, we *multiply* the expressions.

$$4x^3 \cdot 5x^3 = (4 \cdot 5)(x^3 \cdot x^3) \qquad \text{Reorder and group.}$$
$$= (4 \cdot 5)x^6 \qquad \text{Product property of exponents.}$$
$$= 20x^6 \qquad \text{Simplify.}$$

b. To find the sum, we *add* the expressions. Simplify by combining like terms.

$$4x^3 + 5x^3 = (4 + 5)x^3 \qquad \text{Distributive property.}$$
$$= 9x^3 \qquad \text{Simplify.}$$

PROGRESS CHECK 4 Find (a) the product and (b) the sum of $-2x^4$ and $7x^4$ in simplest form. ⌐

3 A second property of exponents concerns raising to a power an expression that contains exponents. For example, what is the result when 5^2 is raised to the fourth power?

$$(5^2)^4 \text{ means } 5^2 \cdot 5^2 \cdot 5^2 \cdot 5^2 = 5^{2+2+2+2} = 5^8$$

Note that the exponent 8 in the result is the product of the original exponents (2 and 4). This example illustrates the power-to-a-power property of exponents.

Power-to-a-Power Property of Exponents

If m and n are positive integers and a is a real number, then

$$(a^m)^n = a^{mn}.$$

To raise an exponential expression to a power, use the same base and multiply the exponents.

EXAMPLE 5 Simplify each expression.

a. $(10^3)^3$ **b.** $(x^3)^4$ **c.** $[(-3)^4]^2$

Solution Apply the power-to-a-power property of exponents.

a. $(10^3)^3 = 10^{3 \cdot 3} = 10^9$ or $1,000,000,000$ (1 billion)
b. $(x^3)^4 = x^{3 \cdot 4} = x^{12}$
c. $[(-3)^4]^2 = (-3)^8 = 6,561$

PROGRESS CHECK 5 Simplify each expression.

a. $(3^4)^4$ **b.** $(y^6)^5$ **c.** $[(-4)^2]^3$ ⌐

We often raise products and quotients to powers. First, consider the result when $3 \cdot 4$ is raised to the third power.

$$(3 \cdot 4)^3 = (3 \cdot 4)(3 \cdot 4)(3 \cdot 4) \qquad \text{Definition of positive integer exponent.}$$
$$= (3 \cdot 3 \cdot 3)(4 \cdot 4 \cdot 4) \qquad \text{Reorder and regroup.}$$
$$= 3^3 \cdot 4^3 \qquad \text{Definition of positive integer exponent.}$$
$$= 27 \cdot 64 \qquad \text{Simplify.}$$
$$= 1,728 \qquad \text{Simplify.}$$

Thus $(3 \cdot 4)^3 = 3^3 \cdot 4^3 = 1,728$. To check this result observe that $(3 \cdot 4)^3 = 12^3$, which also equals $1,728$.

Next, note that

$$\left(\frac{3}{4}\right)^3 = \left(\frac{3}{4}\right)\left(\frac{3}{4}\right)\left(\frac{3}{4}\right) = \frac{3 \cdot 3 \cdot 3}{4 \cdot 4 \cdot 4} = \frac{3^3}{4^3}.$$

Progress Check Answers
4. (a) $-14x^8$ (b) $5x^4$
5. (a) $3^{16} = 43,046,721$ (b) y^{30}
(c) $(-4)^6 = 4,096$

These examples illustrate the following power properties.

> ### Power Properties of Products and Quotients
>
> If n is a positive integer and a and b are real numbers, then
>
> $$(ab)^n = a^n b^n \qquad \text{and} \qquad \left(\frac{a}{b}\right)^n = \frac{a^n}{b^n} \qquad b \neq 0.$$

The next example involves the three power properties of exponents considered in this section.

EXAMPLE 6   Simplify each expression.

a. $(6x)^3$
b. $(-3x^4y)^2$
c. $\left(\dfrac{-a^3}{7x^5}\right)^2$

Solution   Apply the necessary power properties.

a. $(6x)^3 = 6^3 x^3 = 216x^3$
b. $(-3x^4y)^2 = (-3)^2(x^4)^2 y^2 = 9x^8 y^2$
c. $\left(\dfrac{-a^3}{7x^5}\right)^2 = \dfrac{(-1)^2(a^3)^2}{7^2(x^5)^2} = \dfrac{a^6}{49x^{10}}$

PROGRESS CHECK 6   Simplify each expression.

a. $(5y)^2$
b. $(-2ax^4)^3$
c. $\left(\dfrac{-x^2}{4y^3}\right)^4$

EXAMPLE 7   Solve the problem in the chapter introduction on page 126.

Solution   Because the radius is one-half of the diameter, we replace r by $d/2$ and simplify.

$$V = \frac{4}{3}\pi r^3 \qquad \text{Given formula.}$$

$$= \frac{4}{3}\pi\left(\frac{d}{2}\right)^3 \qquad \text{Replace } r \text{ by } \frac{d}{2}.$$

$$= \frac{4}{3}\pi\left(\frac{d^3}{2^3}\right) \qquad \text{Quotient-to-a-power property.}$$

$$= \frac{1}{6}\pi d^3 \qquad \text{Simplify.}$$

Thus, the volume of a sphere in terms of the diameter is given by $V = \frac{1}{6}\pi d^3$.

PROGRESS CHECK 7   The surface area of a sphere of diameter d is given by $A = \pi d^2$. Determine a formula for the area in terms of the radius r.

In Example 8 we first apply power properties to clear parentheses; then we use the product property of exponents to simplify the result.

EXAMPLE 8   Simplify the expressions.

a. $(y^2)^4(y^5)^3$
b. $(3a^5x)^2(-2a^3x^3)^4$

Solution

a. $(y^2)^4(y^5)^3$
   $= y^8 y^{15}$
   $= y^{8+15}$
   $= y^{23}$

b. $(3a^5x)^2(-2a^3x^3)^4$
   $= 3^2 a^{10} x^2 (-2)^4 a^{12} x^{12}$
   $= 3^2(-2)^4 a^{10+12} x^{2+12}$
   $= 144a^{22}x^{14}$

PROGRESS CHECK 8   Simplify the expressions.

a. $(b^3)^3(b^2)^5$ **b.** $(-5xy^2)^3(2x^2y)^2$ ⌐

4   Exponents may appear that represent variables or arbitrary constants that may be replaced by positive integers. To simplify such expressions, apply the appropriate properties of exponents and then simplify the resulting exponent in the usual algebraic way.

EXAMPLE 9   Simplify the expressions. Assume n is a positive integer.

a. $2^n \cdot 2^{n+1}$ **b.** $(x^{n+1})^2$

Solution

a. $2^n \cdot 2^{n+1} = 2^{n+(n+1)}$ Product property of exponents.

$\qquad\qquad = 2^{2n+1}$ Combine like terms in the exponent.

b. $(x^{n+1})^2 = x^{(n+1)\cdot 2}$ Power-to-a-power property.

$\qquad\qquad = x^{2n+2}$ Distributive property in the exponent.

PROGRESS CHECK 9   Simplify the expressions. Assume n is a positive integer.

a. $5^{2n} \cdot 5^n$ **b.** $(y^{2n-1})^3$ ⌐

Progress Check Answers

8. (a) b^{19} (b) $-500x^7y^8$

9. (a) 5^{3n} (b) y^{6n-3}

EXERCISES 3.1

In Exercises 1–18, write each expression using exponents. In each case, identity the base and the exponent.

1. $4 \cdot 4 \cdot 4 \cdot 4 \cdot 4$

2. $5 \cdot 5 \cdot 5$

3. $(\frac{3}{4}) \cdot (\frac{3}{4}) \cdot (\frac{3}{4})$

4. $(\frac{1}{8}) \cdot (\frac{1}{8})$

5. $(-3) \cdot (-3) \cdot (-3) \cdot (-3)$

6. $(-4) \cdot (-4) \cdot (-4)$

7. $-3 \cdot 3 \cdot 3 \cdot 3$

8. $-10 \cdot 10 \cdot 10 \cdot 10 \cdot 10$

9. $(a+b)(a+b)(a+b)$

10. $(x-y)(x-y)(x-y)(x-y)$

11. $\dfrac{1}{xxx}$ 12. $\dfrac{5}{(-x)(-x)(-x)}$

13. $(1.4x)(1.4x)$

14. $(0.08p)(0.08p)(0.08p)$

15. $(3-x)(3-x)(3-x)$

16. $(4+2x)(4+2x)$

17. $\dfrac{1}{(x+y)(x+y)}$

18. $\dfrac{1}{(a-b)(a-b)}$

In Exercises 19–30, evaluate each expression. Then check your result with a calculator.

19. $2^3 + 3^2$ 20. $2^1 - 1^2$ 21. $0^2 - 2^3$

22. $0^3 - 3^2$ 23. $5^3 \cdot 4^2$ 24. $2^3 \cdot 3^2$

25. -2^3 26. -3^2 27. $(12-2)^3$

28. $(9-3)^2$ 29. $12 - 2^3$ 30. $9 - 3^2$

In Exercises 31–54, simplify the given expressions by using the product property of exponents.

31. $3^5 \cdot 3^4$ 32. $2^5 \cdot 2^3$

33. $(-3)^4(-3)^2(-3)^3$

34. $(-4)^3(-4)^2(-4)^2$

35. $(-2)^3(-2)$ 36. $(-5)^2(-5)$

37. x^4x^5 38. y^3y^3 39. $a^2a^2a^2$

40. $b^3b^3b^3$ 41. $a^3b^2a^4b^5$ 42. $x^2y^2x^3y$

43. $(2x)(5x)$ 44. $(3a)(7a)$ 45. $(3a^2)(4a^5)$

46. $(-5x^2)(7x^3)$ 47. $(2x)(-3x^2)(4x^3)$

48. $(3a^2)(3a^2)(3a^2)$ 49. $2^n \cdot 2^3$

50. $3^a \cdot 3^5$ 51. $2^n \cdot 2^{6+n}$

52. $3^n \cdot 3^{5-n}$ 53. x^ax^a 54. y^by^{3b}

In Exercises 55–60, find (a) the product and (b) the sum of the given expressions. Express the answer in simplest form.

55. $3x^3$ and $4x^3$ 56. $2x^5$ and $3x^5$

57. $-3x^2$ and $-5x^2$

58. $-2x^3$ and $-4x^3$

59. $-x^2$ and x^2 60. $-3x^4$ and $3x^4$

In Exercises 61–88, use the power properties of exponents to simplify each expression.

61. $(2^3)^4$ 62. $(3^2)^3$

63. $[(-2)^3]^2$ 64. $[(-3)^2]^2$

65. $[(\frac{1}{2})^3]^2$ 66. $[(\frac{2}{3})^3]^3$

67. $(a^2)^3$ 68. $(b^3)^2$ 69. $[(-a)^2]^4$

70. $[(-a)^3]^3$ 71. $-(3^4)^2$ 72. $-(2^3)^5$

73. $(3x)^2$ 74. $(4y)^2$ 75. $(3xy^2)^3$

76. $(2x^2y)^4$ 77. $(-xy)^3$ 78. $(-2x^2y)^3$

79. $\left(\dfrac{3a^2}{4x}\right)^2$ 80. $\left(\dfrac{2a^3}{3b^2}\right)^3$ 81. $\left(\dfrac{-2x^2}{-3y}\right)^2$

82. $\left(\dfrac{-3a^3}{4b^2}\right)^3$ 83. $\left(\dfrac{-a^2b}{2c}\right)^3$ 84. $\left(\dfrac{3a^2b}{-c}\right)^2$

85. $(5^{n+1})^2$ 86. $(4^{n+2})^3$ 87. $(4^{3n})^{2n}$

88. $(3^n)^{n+1}$

In Exercises 89–100, simplify the given expressions by applying any necessary power properties and the product property of exponents.

89. $(x^2)^3(x^3)^4$ 90. $(x^2)^2(x^3)^3$

91. $(2x)^2(3x)^2$

92. $(5y)^3(4y)^2$

93. $(3x^2)^3(2x^3)^2$

94. $(3a)^4(3a^4)^2$

95. $(7a^5)^2(-5a^5)$

96. $(2x)^3(-3x)$

97. $(4a^2x)^3(-2ax^2)^4$

98. $(-x)^4(3ax)^2$

99. $(2ax^2)^n(2x)^n$

100. $(3a^2x)^n(3ax)^{2n}$

In Exercises 101–106, to solve the equation for x, first simplify by using the product property of exponents; next use the fact that if $a^n = a^m$, then n must equal m.

101. $4^2 \cdot 4^3 = 4^x$

102. $5^3 \cdot 5 = 5^x$

103. $7^x \cdot 7^2 = 7^{11}$

104. $8^{4x} \cdot 8^4 = 8^{12}$

105. $6^{2x} \cdot 6^{3x} = 6^5$

106. $3^x 3^{3x} = 3^8$

107. A mathematician is using a formula $f = mnd^2$ but would prefer a formula that used t instead of d. If it is true that $d = \frac{1}{2}t$, rewrite the formula in terms of t.

108. A mathematician is using a formula $s = ax^2$ but would prefer a formula that used y instead of x. If it is true that $x = 3y$, rewrite the formula in terms of y.

109. The volume of a cylinder is $V = \pi r^2 h$, where r is the radius of the circular base. Determine a formula for V in terms of the diameter d.

110. Rewrite the formula $V = 8\pi r^3$ in terms of d if $r = d/2$.

111. Rewrite the formula for the area of rectangle ($A = \ell w$) if both ℓ and w are equal to $5x$.

112. Rewrite the formula for the volume of a rectangular box ($V = \ell wh$) if ℓ, w, and h are each equal to $3x$.

In Exercises 113–116, write an expression for the area of the given figure. Simplify the expression using the properties of exponents.

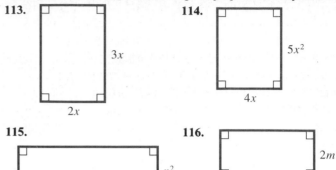

113.

114.

115.

116.

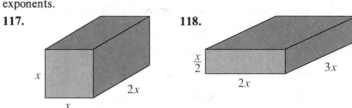

In Exercises 117–118, write an expression for the volume of the given figure. Simplify the expression using the properties of exponents.

117.

118.

THINK ABOUT IT

1. 2^{2^2} can be evaluated as $(2^2)^2 = 4^2$ or as $(2)^{(2^2)} = 2^4$.
   a. Show that these are both equal to 16.
   b. Is it true that 3^{3^3} gives the same answer as $(3^3)^3$ and as $(3)^{(3^3)}$?

2. a. When you raise a proper fraction like $\frac{2}{3}$ to higher and higher powers, does the answer get smaller and smaller or larger and larger?
   b. What happens when you raise an improper fraction to higher and higher powers?
   c. What happens when you raise the number 1 to higher and higher powers?

   d. Why can you not tell which is a larger quantity, x^2 or x^3?

3. What does *common* mean when you say that two exponential expressions have a common base?

4. What are the usual names for these numbers?
   a. 10^3
   b. 10^6
   c. 10^9
   d. 10^{12}

5. The formula to change feet into inches is $I = 12f$. The formula for changing square feet into square inches, therefore, is $I^2 = (12f)^2$, or $I^2 = 144f^2$. So 1 ft² equals 144 in.². Find a formula for converting cubic feet into cubic inches. How many cubic inches are there in 1 ft³?

REMEMBER THIS

1. Evaluate $3x^2$ and $(3x)^2$ when $x = 5$.

2. Write an expression for "3 more than 3 times a number."

3. Evaluate $a + b^2$ and $(a + b)^2$ when $a = 9$ and $b = 1$. 10; 100

4. Solve the inequality $x - (-3) < -2$ and graph the solution set.

5. Solve the inequality $-3x + 2 > 5$ and graph the solution set.

6. One amount of money was invested for a year at 7 percent annual interest. Two thousand dollars more than that was invested at 10 percent. The total interest earned was $285. How much was invested in each account?

7. Two cars start out on an interstate highway together, one at 55 mi/hour and one at 65 mi/hour. How much time must pass before the slower car is 4 mi behind the faster one?

8. Solve the formula $F = ma$ for m.

9. Find the shaded area.

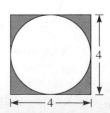

10. Calculate $2^7 - 2^6$.

3.2 Integer Exponents and the Quotient Property of Exponents

The annual budget of the U.S. government is currently between \$1 and \$2 trillion. One trillion may be written as 1,000,000,000,000, or 10^{12}, while 1 million may be written as 1,000,000, or 10^6. How many times does the government have to spend a million dollars in order to spend a trillion dollars? (See Example 5.)

OBJECTIVES

1 Evaluate expressions with a zero exponent.

2 Evaluate expressions with a negative integer exponent.

3 Apply the quotient property of exponents.

4 Apply combinations of exponent definitions and properties.

5 Simplify exponential expressions with literal exponents that denote integers.

6 Evaluate integer exponents on a calculator.

To this point, the only numbers we have used as exponents have been positive integers $\{1, 2, 3, \ldots\}$. It is a goal of algebra to be able to use any real number as an exponent. In this section we will give meaning to exponents that are negative integers or zero. The determining principle is that any rules that work for positive integer exponents should continue to work with zero and negative integer exponents.

1 We start by considering the product property of exponents.

$$a^m \cdot a^n = a^{m+n}$$

If we apply this property to the product of 5^2 and 5^0, we have

$$5^2 \cdot 5^0 = 5^{2+0} = 5^2.$$

When we multiply 5^2 by 5^0, the result is 5^2. Thus 5^0 must equal 1, and we make the following definition.

Zero Exponent

If a is a nonzero real number, then

$$a^0 = 1.$$

EXAMPLE 1 Evaluate each expression. Assume $x \neq 0$.

a. $2^0 + 3^0$ **b.** $(2 + 3)^0$ **c.** $(2x)^0$ **d.** $2x^0$

Solution Apply the zero exponent definition.

a. $2^0 = 1$ and $3^0 = 1$, so $2^0 + 3^0 = 1 + 1 = 2.$

b. $(2 + 3)^0 = 5^0 = 1$. Alternatively, the parentheses indicate that a nonzero number is raised to the zero power, so the result is 1. Note $(2 + 3)^0 \neq 2^0 + 3^0.$

c. Once again, the parentheses indicate $(2x)^0 = 1$.

d. Because $2x^0 = 2^1 \cdot x^0$, we have $2x^0 = 2(1) = 2$.

PROGRESS CHECK 1 Evaluate each expression. Assume $x \neq 0$.

a. $4^0 + (-4)^0$ **b.** $(7 - 2)^0$ **c.** $(-3x)^0$ **d.** $-3x^0$

2 To obtain a definition for exponents that are negative integers, consider the result of applying the product property of exponents to $5^2 \cdot 5^{-2}$.

$$5^2 \cdot 5^{-2} = 5^{2 + (-2)} = 5^0 = 1$$

When we multiply 5^2 by 5^{-2}, the result is 1. Therefore, 5^{-2} is the *reciprocal* of 5^2, or $5^{-2} = 1/5^2$. In general, our previous laws of exponents may be extended by making the following definition.

Negative Exponent

If a is a nonzero real number and n is an integer, then

$$a^{-n} = \frac{1}{a^n}.$$

This definition means that a^{-n} is the **reciprocal** of a^n.

EXAMPLE 2 Evaluate each expression.

a. 2^{-3} **b.** $(-2)^{-3}$ **c.** $3^{-1} + 2^{-1}$ **d.** $(\frac{2}{3})^{-4}$

Solution Use the negative exponent definition and the fraction principles from Section 1.1.

a. $2^{-3} = \dfrac{1}{2^3} = \dfrac{1}{8}$ **b.** $(-2)^{-3} = \dfrac{1}{(-2)^3} = \dfrac{1}{-8} = -\dfrac{1}{8}$

c. $3^{-1} + 2^{-1} = \dfrac{1}{3^1} + \dfrac{1}{2^1} = \dfrac{2}{6} + \dfrac{3}{6} = \dfrac{5}{6}$

d. $\left(\dfrac{2}{3}\right)^{-4} = \dfrac{1}{(\frac{2}{3})^4} = \dfrac{1}{\frac{16}{81}} = 1 \cdot \dfrac{81}{16} = \dfrac{81}{16}$

Caution The sign of the exponent does not tell you anything about the sign of the resulting expression. For instance, in this example

$$2^{-3} \text{ is positive; } (-2)^{-3} \text{ is negative.}$$

Negative exponents indicate reciprocals that may or may not be negative numbers.

PROGRESS CHECK 2 Evaluate each expression.

a. 10^{-1} **b.** $(-10)^{-1}$ **c.** $2^{-2} - 3^{-2}$ **d.** $(\frac{3}{5})^{-3}$

Example 2d shows the evaluation of a fraction raised to a negative power. Such evaluations may be done more easily by using

$$\left(\frac{a}{b}\right)^{-n} = \left(\frac{b}{a}\right)^n \qquad a, b \neq 0.$$

We can verify this property from our previous methods.

$$\left(\frac{a}{b}\right)^{-n} = \frac{1}{(a/b)^n} = \frac{1}{a^n/b^n} = 1 \cdot \frac{b^n}{a^n} = \frac{b^n}{a^n} = \left(\frac{b}{a}\right)^n$$

In the next example we redo Example 2d using this property.

EXAMPLE 3   Evaluate $(\frac{2}{3})^{-4}$.

Solution

$$\left(\frac{2}{3}\right)^{-4} = \left(\frac{3}{2}\right)^{4} \qquad \left(\frac{a}{b}\right)^{-n} = \left(\frac{b}{a}\right)^{n}$$

$$= \frac{3^4}{2^4} \qquad \text{Quotient-to-a-power property.}$$

$$= \frac{81}{16} \qquad \text{Simplify.}$$

PROGRESS CHECK 3   Evaluate $(\frac{3}{5})^{-3}$.

3   To multiply exponential expressions with the same base, we *add exponents*. So it is natural to expect that we divide exponential expressions with the same base by *subtracting exponents*. To see that $a^m/a^n = a^{m-n}$, consider the following three cases.

Case	Example Using Positive Integer Exponent Definition	Example Using $a^m/a^n = a^{m-n}$
$m > n$	$\dfrac{5^6}{5^2} = \dfrac{5 \cdot 5 \cdot 5 \cdot 5 \cdot 5 \cdot 5}{5 \cdot 5} = 5^4$	$\dfrac{5^6}{5^2} = 5^{6-2} = 5^4$
$m = n$	$\dfrac{5^2}{5^2} = \dfrac{5 \cdot 5}{5 \cdot 5} = 1$	$\dfrac{5^2}{5^2} = 5^{2-2} = 5^0 = 1$
$m < n$	$\dfrac{5^2}{5^6} = \dfrac{5 \cdot 5}{5 \cdot 5 \cdot 5 \cdot 5 \cdot 5 \cdot 5} = \dfrac{1}{5^4}$	$\dfrac{5^2}{5^6} = 5^{2-6} = 5^{-4} = \dfrac{1}{5^4}$

Thus, with the aid of the definition for negative integer and zero exponents we may state the quotient property of exponents.

Quotient Property of Exponents

If m and n are integers and a is a nonzero real number,

$$\frac{a^m}{a^n} = a^{m-n}.$$

EXAMPLE 4   Simplify and write the result using only positive exponents. Assume $x \neq 0$.

a. $\dfrac{x^7}{x^4}$ b. $\dfrac{3^2}{3^{-2}}$ c. $\dfrac{4^2 x^3}{4^5 x^3}$ d. $\dfrac{10^{-5} x^2}{10^{-6} x^3}$

Solution   Apply the quotient property of exponents.

a. $\dfrac{x^7}{x^4} = x^{7-4} = x^3$ b. $\dfrac{3^2}{3^{-2}} = 3^{2-(-2)} = 3^4 = 81$

c. $\dfrac{4^2 x^3}{4^5 x^3} = 4^{2-5} x^{3-3} = 4^{-3} x^0 = \dfrac{1}{4^3} \cdot 1 = \dfrac{1}{64}$

d. $\dfrac{10^{-5} x^2}{10^{-6} x^3} = 10^{-5-(-6)} x^{2-3} = 10^1 x^{-1} = 10 \cdot \dfrac{1}{x} = \dfrac{10}{x}$

Progress Check Answer

3. $\dfrac{125}{27}$

PROGRESS CHECK 4   Simplify and write the result using only positive integers. Assume $y \neq 0$.

a. $\dfrac{2^5}{2^3}$

b. $\dfrac{y^{-4}}{y^4}$

c. $\dfrac{5^3 y^5}{5^3 y^6}$

d. $\dfrac{3^{-1} y}{3 y^7}$

EXAMPLE 5   Solve the problem in the section introduction on page 133.

Solution   To help us understand the problem, we will restate the question with simple numbers: How many times do you have to spend $2 in order to spend $8? The answer is obviously 4, and the procedure is to divide 8 by 2. Thus, to solve the given problem, divide a trillion by a million.

$$\frac{10^{12}}{10^6} = 10^{12-6} = 10^6 \quad \text{or} \quad 1{,}000{,}000$$

The government must spend a million dollars a million times in order to spend a trillion dollars. In other words, a trillion is a million millions.

PROGRESS CHECK 5   One billion may be written as 1,000,000,000, or 10^9. How many billions make up a trillion?

4   For the next example, remember that we have defined integer exponents so that all previous properties of exponents continue to apply.

EXAMPLE 6   Simplify and write the result using only positive exponents. Assume $x \neq 0$ and $y \neq 0$.

a. $x^{-3} \cdot x^{-2}$

b. $(5x^{-3})^{-2}$

c. $\left(\dfrac{3x}{4}\right)^{-3}$

d. $\left(\dfrac{y^2}{y^{-1}}\right)^3$

Solution

a. $x^{-3} \cdot x^{-2} = x^{-3+(-2)}$ Product property of exponents.

$= x^{-5}$ Simplify the exponent.

$= \dfrac{1}{x^5}$ Negative exponent definition.

b. $(5x^{-3})^{-2} = 5^{-2}(x^{-3})^{-2}$ Product-to-a-power property.

$= 5^{-2}x^6$ Power-to-a-power property.

$= \dfrac{1}{5^2} \cdot x^6$ Negative exponent definition.

$= \dfrac{x^6}{25}$ Simplify.

c. $\left(\dfrac{3x}{4}\right)^{-3} = \left(\dfrac{4}{3x}\right)^3$ $\left(\dfrac{a}{b}\right)^{-n} = \left(\dfrac{b}{a}\right)^n$.

$= \dfrac{4^3}{3^3 x^3}$ Quotient- and product-to-a-power properties.

$= \dfrac{64}{27x^3}$ Simplify.

d. $\left(\dfrac{y^2}{y^{-1}}\right)^3 = (y^{2-(-1)})^3$ Quotient property of exponents.

$= (y^3)^3$ Simplify.

$= y^9$ Power-to-a-power property.

Note   Alternative methods are possible when simplifying expressions with negative exponents. For instance, in part **d** we can simplify as follows.

$$\left(\frac{y^2}{y^{-1}}\right)^3 = \frac{(y^2)^3}{(y^{-1})^3} = \frac{y^6}{y^{-3}} = y^{6-(-3)} = y^9$$

With practice it becomes easier to spot efficient paths to a solution.

Progress Check Answers

4. (a) 4   (b) $\dfrac{1}{y^8}$ (c) $\dfrac{1}{y}$ (d) $\dfrac{1}{9y^6}$

5. 1,000

PROGRESS CHECK 6 Simplify and write the result using only positive exponents. Assume $x \neq 0$ and $b \neq 0$.

a. $x^2 \cdot x^{-5}$ b. $(2b^{-1})^{-3}$ c. $\left(\dfrac{4x^2}{5}\right)^{-2}$ d. $\left(\dfrac{b^{-2}}{b^{-3}}\right)^4$

5 Exponential expressions with literal exponents that represent integers are simplified in Example 7.

EXAMPLE 7 Simplify the expression. Assume that $x \neq 0$ and n is an integer.

a. $\dfrac{x^{2n+2}}{x^2}$ b. $\dfrac{3}{3^{-n}}$ c. $(x^{n-1})^{-2}$

Solution

a. Apply the quotient property of exponents.

$$\dfrac{x^{2n+2}}{x^2} = x^{2n+2-2} = x^{2n}$$

b. Note that $3 = 3^1$ and apply the quotient property of exponents.

$$\dfrac{3}{3^{-n}} = 3^{1-(-n)} = 3^{1+n}$$

c. Apply the power-to-a-power property of exponents.

$$(x^{n-1})^{-2} = x^{(n-1)(-2)} = x^{-2n+2}$$

PROGRESS CHECK 7 Simplify the expression. Assume that $x \neq 0$ and n is an integer.

a. $\dfrac{x^{2n}}{x^n}$ b. $\dfrac{5^n}{5}$ c. $(x^{2-n})^{-1}$

6 Integer exponents are usually evaluated on a calculator with the power function key $\boxed{y^x}$ in the usual way. For example, 5^{-3} is computed as

$$5 \boxed{y^x} 3 \boxed{+/-} \boxed{=} \boxed{\quad 0.008}.$$

An exponent of -1 is best computed with the reciprocal key; so 4^{-1} is simply

$$4 \boxed{1/x} \boxed{\quad 0.25}.$$

As discussed in Section 1.7, if your calculator does not permit usage of the power key $\boxed{y^x}$ with a negative base, you will need to enter a positive base and then decide the correct sign yourself. Thus you would compute $(-5)^{-3}$ by evaluating $(5)^{-3}$ as above. And then you would decide that the correct answer is negative 0.008, because a negative base raised to an odd power is negative.

EXAMPLE 8 Evaluate $(1+i)^{-n}$ if $i = 0.06$ and $n = 10$.

Solution We can evaluate $(1+0.06)^{-10}$ on a calculator as follows.

$$1 \boxed{+} .06 \boxed{=} \boxed{y^x} 10 \boxed{+/-} \boxed{=} \boxed{0.5583948}$$

Note The formula $(1+i)^{-n}$ tells a financial analyst how much one dollar n years from now is worth today, assuming an annual growth rate of i. For instance, the result of this example indicates that a dollar 10 years from now is worth about 56 cents today, assuming a 6 percent annual growth rate. Equivalently, 56 cents today will grow to about one dollar in 10 years (assuming the same rate of growth).

Progress Check Answers

6. (a) $\dfrac{1}{x^3}$ (b) $\dfrac{b^3}{8}$ (c) $\dfrac{25}{16x^4}$ (d) b^4

7. (a) x^n (b) 5^{n-1} (c) x^{-2+n}

8. 0.3768895

PROGRESS CHECK 8 Evaluate $(1+i)^{-n}$ if $i = 0.05$ and $n = 20$.

EXERCISES 3.2

In Exercises 1–36, evaluate the given expression. Assume $x \neq 0$.

1. $2^0 + 5^0 + 6^0$ **2.** $1^0 + 2^0 + 3^0$ **3.** $(3 + 4)^0$

4. $(5 + 6)^0$ **5.** $(7 - 2)^0$ **6.** $(6 - 3)^0$

7. $6^0 \cdot 4$ **8.** $5^0 \cdot 8$ **9.** $(2^3)^0$

10. $(4^0)^3$ **11.** $(3x)^0$ **12.** $(4x)^0$

13. $(3x + 8)^0$, $x \neq -\frac{8}{3}$ **14.** $(2x + 9)^0$, $x \neq -\frac{9}{2}$ **15.** $3x^0$

16. $4x^0$ **17.** $-5x^0$ **18.** $-6x^0$

19. 3^{-2} **20.** 3^{-3} **21.** $3^{-2} + 3^{-3}$

22. $4^{-2} + 4^{-1}$ **23.** $(-3)^{-1}$ **24.** $(-3)^{-2}$

25. $4^{-2} \cdot 16$ **26.** $5^{-2} \cdot 25$ **27.** $(\frac{1}{4})^{-2}$

28. $(\frac{1}{3})^{-2}$ **29.** $(\frac{3}{4})^{-1}$ **30.** $(\frac{4}{5})^{-1}$

31. $(\frac{2}{3})^{-2} + (\frac{3}{2})^{-2}$ **32.** $(\frac{3}{4})^{-2} + (\frac{4}{3})^{-2}$

33. $(\frac{3}{7})^{-2} \cdot (\frac{7}{3})^{-3}$ **34.** $(\frac{5}{8})^{-3} \cdot (\frac{8}{5})^{-4}$

35. $5^{-2}(25x)$ **36.** $3^{-2}(18x)$

In Exercises 37–42, find (a) the sum and (b) the product of the quantities given.

37. 3^0 and 4^0 **38.** 4^0 and 5^0

39. $(\frac{3}{4})^{-1}$ and $\frac{3}{4}$ **40.** $(\frac{2}{3})^{-1}$ and $\frac{2}{3}$

41. 10^{-2} and 10^2 **42.** 10^{-1} and 10^1

In Exercises 43–54, without calculating, determine if the given quantity is a negative or positive number.

43. $(-2)^2$ **44.** $(-2)^3$ **45.** 3^{-3}

46. 3^{-4} **47.** -2^5 **48.** -2^6

49. $(\frac{1}{3})^{-2}$ **50.** $(\frac{1}{4})^{-3}$ **51.** $(-2)^5$

52. $(-2)^6$ **53.** $(-1,324)^{520}$ **54.** $(-5,281)^{6,401}$

In Exercises 55–84, simplify the given expression and write the result using only positive exponents. Assume that $x \neq 0$ and n is an integer.

55. $\dfrac{x^6}{x^5}$ **56.** $\dfrac{x^8}{x^5}$ **57.** $\dfrac{x^6}{x^8}$

58. $\dfrac{x^3}{x^7}$ **59.** $\dfrac{2^3}{2^{-3}}$ **60.** $\dfrac{5^2}{5^{-2}}$

61. $\dfrac{3^5 x^2}{3^3 x^2}$ **62.** $\dfrac{4^2 x}{4^3 x^2}$ **63.** $\dfrac{x^{-3}}{x^{-3}}$

64. $\dfrac{x^{-4}}{x^{-4}}$ **65.** $\dfrac{3^5 \cdot 7^4}{3^3 \cdot 7^6}$ **66.** $\dfrac{4^2 \cdot 5^6}{4^4 \cdot 5^5}$

67. $\dfrac{3xy^2}{5x^2 y}$ **68.** $\dfrac{4x^3 y}{8xy^3}$ **69.** $\dfrac{2^3 \cdot 3^4 x^2 y^3}{2 \cdot 3^2 x}$

70. $\dfrac{3^4 \cdot 5^2 xy^4}{3^2 \cdot 5^3 x^2 y^2}$ **71.** $\dfrac{4^{-2} y^{-3}}{4y^{-2}}$ **72.** $\dfrac{5^{-3} x^{-2}}{5x^{-3}}$

73. $\dfrac{10^{-3} x^2 y^{-3}}{10^{-5} x^3 y^5}$ **74.** $\dfrac{10^{-4} x^{-2} y^{-4}}{10^{-4} x^{-2} y^{-6}}$ **75.** $x^{-3} x^{-4}$

76. $y^{-1} y^{-2}$ **77.** $(3x)^{-2}(4x^{-3})$ **78.** $(4x)^{-2}(5x^{-1})$

79. $(4x^{-2})^{-3}$ **80.** $(5x^{-1})^{-4}$ **81.** $\left(\dfrac{2x}{3}\right)^{-2}$

82. $\left(\dfrac{3}{2x}\right)^{-2}$ **83.** $\left(\dfrac{x^{-2}}{x}\right)^3$ **84.** $\left(\dfrac{x^{-3}}{x^{-2}}\right)^2$

In Exercises 85–90, simplify these expressions. Assume n is an integer.

85. $\dfrac{x^{3n+4}}{x^n}$ **86.** $\dfrac{x^{2n-5}}{x^{2n}}$

87. $\dfrac{2}{2^{-n}}$ **88.** $\dfrac{3^7}{3^{7-n}}$

89. $(x^{n+3})^{-2}$ **90.** $(x^{2-n})^{-3}$

In Exercises 91–102, use a calculator to evaluate the given expressions. Round decimals to three places.

91. $(1 + i)^{-n}$ if $i = 0.05$ and $n = 10$

92. $(1 + i)^{-n}$ if $i = 0.07$ and $n = 15$

93. $4,000(1 + i)^{-n}$ if $i = 0.08$ and $n = 12$
This gives how much \$4,000 _____ years from now is worth today, assuming an annual growth rate of _____ .

94. $5,000(1 + i)^{-n}$ if $i = 0.10$ and $n = 5$
This gives how much \$5,000 _____ years from now is worth today, assuming an annual growth rate of _____ .

95. 1.34^{-4} **96.** 3.96^{-4}

97. $(-2.079)^{-2}$ **98.** $(-4.086)^{-3}$

99. Divide 10^{-7} by 10^{-8}.   **100.** Divide 10^{15} by 10^{12}.

101. Divide 1 trillion (10^{12}) by 10 million (10^7).

102. Divide 100 million (10^8) by 1,000 (10^3).

103. One million is 10^6, and 1 billion is 10^9. How many millions make up a billion?

104. One googol is 10^{100} and 1 million is 10^6. How many millions make up a googol? This answer is so big you will have to leave it as an exponential expression.

105. The frequency of FM radio stations is about 10^8 hertz. The frequency of AM stations is about 10^6 hertz. How many times higher is the frequency of FM than AM?

106. Some of the smallest protozoa (single-celled animals) are about $2 \cdot 10^{-2}$ mm long, and some of the largest protozoa are about 2 mm long. How many times larger are the largest ones than the smallest ones?

107. The speed of light is about 300,000,000 m/second (300 million meters per second). How long does it take light to travel a distance of 0.3 m? Use the relationship

$$\text{time} = \frac{\text{distance}}{\text{speed}},$$

and express the speed of light as $3 \cdot 10^8$ and the distance as $3 \cdot 10^{-1}$. The answer approximates the time it takes for the light from your bed lamp to reach your pillow.

THINK ABOUT IT

1. Why do we say "assume $x \neq 0$" for an expression like $7x^{-3}$?
2. What is the reciprocal of $(-\frac{3}{4})^{-8}$?
3. Why is it not correct to say "Any number raised to the zero power equals 1"?

4. Is it always true that a negative number raised to a positive integer power gives a negative result?
5. Why is it true that any negative number raised to an even power gives a positive result?

REMEMBER THIS

1. Write this expression using exponents: $4 \cdot 4 \cdot 4 \cdot 4 \cdot 4 \cdot 4 \cdot 5 \cdot 5$.

2. Evaluate $3^5 - 5^3$.
3. Evaluate $(-3)^3$.
4. Simplify $x^2 x^3 x^5$.
5. Find the product of $5x^3$ and $-3x$.
6. Find $[(\frac{1}{2})^2]^3$.

7. Write an expression for the sum of the squares of x and y.

8. Rewrite this as an addition expression: $77 - (-18c)$.
9. Find the volume of a cylinder with radius 4 in. and height 6 in.

10. Solve the equation $-8x + 7 = 5 - x$.

3.3   Scientific Notation

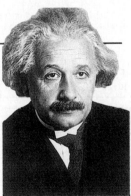

The largest hydrogen bomb ever exploded (Soviet Union, 1961) released 2.4×10^{17} joules of energy, the equivalent of 57 million tons of TNT. Use Einstein's formula $E = mc^2$ to find m and determine how much matter was used to create this energy. If you take c, the speed of light, equal to 3×10^8 m/second, then the units for mass will be kilograms (kg). (See Example 5.)

OBJECTIVES

1. Convert a number from standard notation to scientific notation.

2. Convert a number from scientific notation to standard notation.

3. Compute products and quotients using scientific notation.

4. Use the scientific notation capabilities of a calculator and solve applied problems.

1   Many numbers that appear in scientific and technical work are either very large or very small. For example, a microwave oven produces energy with a frequency of 2,450,000 cycles per second, which translates into one cycle in about 0.000000408 second. To work effectively with such numbers, we often write them in scientific notation form. A positive number N is expressed in **scientific notation** when it is written in the form

$$N = m \times 10^k,$$

where $1 \leq m < 10$ and k is an integer. For example,

$$2,450,000 = 2.45(1,000,000) = 2.45 \times 10^6$$
$$0.000000408 = 4.08(1/10,000,000) = 4.08 \times 10^{-7}$$

Note that the effect of multiplying a number by 10^6 is to move the decimal point 6 places to the right, while the effect of multiplying a number by 10^{-7} is to move the decimal point 7 places to the left. This observation leads to the following procedure for converting a number from standard notation to scientific notation.

To Convert to Scientific Notation

1. Immediately after the first nonzero digit of the number, place an apostrophe (').
2. Starting at the apostrophe, count the number of places to the decimal point. If you move to the right, your count is expressed as a positive number. If you move to the left, the count is negative.
3. The apostrophe indicates the position of the decimal in the factor between 1 and 10; the count represents the exponent to be used in the factor, which is a power of 10.

The following chart illustrates how this procedure is used. The direction of the counting is indicated by the arrow.

Number	=	Number from 1 to 10	×	Power of 10
6'95,990,000.	=	6.9599	×	10^8
0.0003'7	=	3.7	×	10^{-4}
5'.2	=	5.2	×	10^0

EXAMPLE 1   The speed of light is about 186,000 mi/second and it takes light about 0.0000054 second to travel 1 mi. Write these numbers in scientific notation.

Solution   Follow the procedure outlined above.

$$1'86,000. = 1.86 \times 10^5$$

$$0.000005'4 = 5.4 \times 10^{-6}$$

PROGRESS CHECK 1   Write each number in scientific notation.
a. 1,223,000,000   **b.** 0.65

2   To convert a number from scientific notation to standard notation, just perform the indicated multiplication. Move the decimal point to the right when the power of 10 has a positive exponent, and move it to the left when the power of 10 has a negative exponent. Insert, as needed, zeros that are required to indicate the position of the decimal point.

EXAMPLE 2   Express each number in standard notation.
a. 6.91×10^5 **b.** 4.518×10^{-3}

Solution

a. To multiply by 10^5, we move the decimal point 5 places to the right.

$$6.91 \times 10^5 = 6.91000. = 691,000$$

b. To multiply by 10^{-3}, we move the decimal point 3 places to the left.

$$4.518 \times 10^{-3} = 0.004.518 = 0.004518$$

Progress Check Answers
1. (a) 1.223×10^9 (b) 6.5×10^{-1}
2. (a) 0.000075   (b) 114,000,000

PROGRESS CHECK 2   Express each number in standard notation.
a. 7.5×10^{-5} **b.** 1.14×10^8

3 One of the important advantages of scientific notation is that it is easy to multiply and divide numbers expressed this way. Note that the procedure shown in the following example is effective because the product and quotient properties of exponents are easily applied to numbers in scientific notation form.

EXAMPLE 3 Perform the indicated operations and express the result in standard notation.

a. $(3 \times 10^{-4})(6 \times 10^7)$

b. $\dfrac{0.085}{5,000}$

Solution

a.
$$
\begin{aligned}
(3 \times 10^{-4})(6 \times 10^7) &= (3 \cdot 6)(10^{-4}10^7) & \text{Reorder and regroup.} \\
&= (3 \cdot 6) \times 10^{-4 + 7} & \text{Product property of exponents.} \\
&= 18 \times 10^3 & \text{Simplify.} \\
&= 18,000 & \text{Standard notation.}
\end{aligned}
$$

b.
$$
\begin{aligned}
\dfrac{0.085}{5,000} &= \dfrac{8.5 \times 10^{-2}}{5 \times 10^3} & \text{Scientific notation.} \\
&= \dfrac{8.5}{5} \times \dfrac{10^{-2}}{10^3} & \dfrac{ac}{bd} = \dfrac{a}{b} \cdot \dfrac{c}{d}. \\
&= \dfrac{8.5}{5} \times 10^{-2-3} & \text{Quotient property of exponents.} \\
&= 1.7 \times 10^{-5} & \text{Simplify.} \\
&= 0.000017 & \text{Standard notation.}
\end{aligned}
$$

PROGRESS CHECK 3 Perform the indicated operations and express the result in standard notation.

a. $(7 \times 10^5)(2 \times 10^{-1})$

b. $\dfrac{0.00081}{300}$

4 Scientific calculators are programmed to work with scientific notation. To enter a number in scientific notation, first enter the significant digits of the number from 1 to 10, press $\boxed{\text{EE}}$ or $\boxed{\text{EXP}}$, and finally, enter the exponent of the power of 10. For example, to enter 8.5×10^{-2}, press

$$8.5 \;\boxed{\text{EE}}\; 2 \;\boxed{+/-}\;.$$

The display looks like

$$\boxed{8.5 \qquad -02},$$

with the exponent appearing on the right in the display. Read the owner's manual for your calculator to learn its scientific notation capabilities. Here are two common features you need to keep in mind.

1. If a calculation results in an answer with too many digits for display, the calculator automatically displays the answer in scientific notation.

$$\text{Example: } 47 \;\boxed{y^x}\; 5 \;\boxed{=}\; \boxed{2.2935 \quad 08}$$

The display shown here is common for a calculator with an eight-digit display. Note that the result shows only five significant digits followed by the current exponent. However, the exact value of 47^5 is 229,345,007 and the calculator carries this answer internally. Check your owner's manual to determine the accuracy you can expect in displayed and internal values.

2. Entries in standard and scientific notation may be mixed in the same problem.

$$\text{Example: } (7.6 \times 10^3) + 127$$
$$7.6 \;\boxed{\text{EE}}\; 3 \;\boxed{+}\; 127 \;\boxed{=}\; \boxed{7.727 \quad 03}$$

Progress Check Answers

3. (a) 140,000   (b) 0.0000027

Some calculators would display the result in this example in standard notation.

With the aid of the calculator we can now solve some applied problems that involve very large or small numbers.

EXAMPLE 4 If one red blood cell contains about 270,000,000 hemoglobin molecules, about how many molecules of hemoglobin are there in 25,000 red blood cells?

Solution $270{,}000{,}000 = 2.7 \times 10^8$ and $25{,}000 = 2.5 \times 10^4$. Find the product of these numbers on the calculator as follows.

$$2.7 \boxed{EE} \; 8 \; \boxed{\times} \; 2.5 \boxed{EE} \; 4 \; \boxed{=} \; \boxed{6.75 \quad 12}$$

Thus, there are about 6.75×10^{12} hemoglobin molecules in 25,000 red blood cells. This is 6.75 trillion molecules.

PROGRESS CHECK 4 The speed of light is about 186,000 mi/second. About how far does light travel in 86,400 seconds (which is one day)? ⌟

EXAMPLE 5 Solve the problem in the section introduction on page 139.

Solution First, solve for m using the given information.

$E = mc^2$	Given formula.
$2.4 \times 10^{17} = m(3 \times 10^8)^2$	Replace E by 2.4×10^{17} and c by 3×10^8.
$\dfrac{2.4 \times 10^{17}}{(3 \times 10^8)^2} = m$	Divide by $(3 \times 10^8)^2$.

Now, compute the quotient on a calculator as follows.

$$2.4 \boxed{EE} \; 17 \; \boxed{\div} \; 3 \boxed{EE} \; 8 \; \boxed{x^2} \; \boxed{=} \; \boxed{2.6667 \quad 00}$$

Thus, to the nearest tenth, about 2.7 kg of matter were used. Since 1 kg is about 2.2 lb, this explosion was generated from about 6 lb of matter!

Progress Check Answers
4. 1.60704×10^{10} mi   5. 1.0889 kg

PROGRESS CHECK 5 Use the information in Example 5 and determine how much matter is needed to create an explosion that releases 9.8×10^{16} joules of energy. ⌟

EXERCISES 3.3

In Exercises 1–12, convert the given number into scientific notation.

1. 4,235,000,000
2. 23,400,000
3. 345,000
4. 341,333,000,000
5. 0.00000568
6. 0.0002347
7. 0.0008402
8. 0.00539
9. 1,000
10. 10,000
11. 0.01
12. 0.0001

In Exercises 13–24, convert the given number into standard notation.

13. 1.35×10^3
14. 8.64×10^6
15. 4.01×10^2
16. 5.601×10^2
17. 10^{-3}
18. 10^{-1}
19. 5.3028×10^{-1}
20. 5.3028×10^{-3}
21. 9.807×10^{-2}
22. 1.001×10^{-1}
23. 1.2×10^9
24. 1.2×10^{-9}

In Exercises 25–30, a number is mentioned in standard notation. Write it in scientific notation.

25. In 1988 the federal government collected about $401,181,000,000 in individual income taxes.

26. In 1988 the federal government collected about $94,195,000,000 in corporate income taxes.

27. In 1988 about 40.4 billion shares of stock were sold on the New York Stock Exchange.

28. In 1988 about 2.52 billion shares of stock were sold on the American Stock Exchange.

29. The average distance from Mercury to the sun is about 36,000,000 mi.

30. The average distance from the earth to the sun is about 92,900,000 mi.

In Exercises 31–36, a number is mentioned in scientific notation. Write it in standard notation.

31. A mole is a quantity (like a dozen is a quantity, which is equal to 12) which equals 6.022×10^{23} things.

32. The mass of one electron is about 9.1×10^{-31} kg.

33. A copper penny contains about 2.4×10^{21} copper atoms.

34. In empty space light travels at about 3×10^8 m/second.

35. The half-life of uranium 238 is about 4.5×10^9 years.

36. The average distance a molecule of oxygen travels in air before it collides with another molecule is 1.3×10^{-7} m.

In Exercises 37–48, use scientific notation while doing the indicated operation and then express the result in standard notation.

37. $(4 \times 10^{-3})(7 \times 10^6)$
38. $(2 \times 10^3)(3 \times 10^{-5})$
39. $(2.3 \times 10^5)(4.0 \times 10^{-5})$
40. $(5.46 \times 10^{22})(1.00 \times 10^{-22})$
41. $(5.61 \times 10^{-2})(1.1 \times 10^{-1})$
42. $(7.34 \times 10^{-2})(3.47 \times 10^{-2})$

43. $\dfrac{3.66 \times 10^4}{1.22 \times 10^2}$

44. $\dfrac{9.99 \times 10^{15}}{3.33 \times 10^{12}}$

45. $\dfrac{7.5 \times 10^{-12}}{2.5 \times 10^{-14}}$

46. $\dfrac{4.2 \times 10^{-12}}{2.1 \times 10^{-10}}$

47. $\dfrac{270}{0.009}$

48. $\dfrac{0.00015}{500}$

In Exercises 49–54, it will be easier to use a scientific calculator and to enter some quantities in scientific notation.

49. Sunlight travels at about 186,000 mi/second. How far does it travel in one year? This distance is called one light-year. (One year is about 31,536,000 seconds.) Use the formula $d = rt$.

50. The formula for the surface area of a sphere is $S = 4\pi r^2$, where r is the radius of the sphere. The radius of the earth is about 6,400 km. Find the surface area of the earth and express it in scientific notation. The units will be square kilometers. Show two decimal places.

51. Energy greater than 4.5 electron volts (eV) is enough to break up water molecules and therefore harm living organisms. The energy of radiation is calculated by the formula $e = hf$, where h is Planck's constant (4.14×10^{-15}) and f is the frequency (in hertz) of the radiation. Ultraviolet light has a frequency of 1.5×10^{15} hertz. Compute the energy of ultraviolet light (which is a part of sunlight) and see why too much sun is dangerous to your health.

52. The speed of sound in air is about 1,100 ft/second. How long does it take the sound of the bat hitting the ball to reach an outfielder 350 ft away from home plate? Write your answer in scientific notation. Use the formula $d = rt$.

53. The distance from earth to the moon is about 3.844×10^8 m. Radio waves travel at the speed of light, 300 million m/second. If a person sends a radio message from the moon to the earth, about how long does it take to get there? Use the formula $d = rt$.

54. A table titled "Planets—Physical Data" includes these entries.

	Mass (10^{24} kg)
Mercury	0.3302
Earth	1.0834
Jupiter	1899.728

   a. What is the mass of the earth in kilograms?
   b. How many times more massive is Jupiter than the earth?

THINK ABOUT IT

1. From 1987 to 1988 defense spending in the United States increased from 274×10^9 to 282×10^9.
   a. How much was the increase in dollars?
   b. What percentage of the 1987 figure was this increase?
2. How many times bigger than 2.34×10^4 is 2.34×10^5?
3. What number is 100 times bigger than 4.81×10^{21}?
4. A black hole in space is a star that has such a strong gravitational force pulling from its center that not even light can escape. Light begins to escape at a distance from the center called the Schwarzschild radius. This radius (r) depends on the mass (M) of the black hole, in accordance with the formula $r = 1.48 \times 10^{-27}M$. When mass is in kilograms, the radius is in meters. What would be the mass of a black hole with a Schwarzschild radius of 1 m?
5. Why do astronomers and nuclear physicists use scientific notation?

REMEMBER THIS

1. Which is greater, 2^{-3} or 3^{-2}?
2. Which is greater, 4^{-2} or 2^{-4}?
3. Simplify $\dfrac{x^3}{x^4}$. Write the answer using positive exponents.
4. Simplify $[(1^2)^3 - (3^2)^1]^0$.
5. Some numbers of the form $2^n - 1$ are prime. Test the formula for $n = 1, 2, 3, 4$. For which of these values do you get a prime number?
6. Solve the equation $\dfrac{2x}{3} - \dfrac{1}{2} = x$.
7. Solve this equation for a: $b - c = ad$.
8. Graph the solution set for this inequality: $-3 < x + 2 < 3$.
9. Five more than 3 times a number equals 1 less than 4 times the number. Find the number.
10. Simplify this expression: $\dfrac{(3x^2)^3 - (2x^3)^2}{23x}$.

3.4 Addition and Subtraction of Polynomials

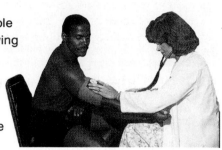

A statistician used a procedure called multiple regression to discover a formula for showing the relationship between age (x) and average systolic blood pressure (y) for the patients in a medical research project. The formula was $y = 113.41 + 0.088x + 0.01x^2$. What is the predicted systolic blood pressure for someone who is 19 years old? (See Example 4.)

OBJECTIVES

1 Identify polynomials.

2 State the degree of a polynomial and identify monomials, binomials, and trinomials.

3 Evaluate polynomials.

4 Add polynomials.

5 Subtract polynomials.

1 An important type of algebraic expression is called a polynomial. Examples of polynomials are

$$9x^2 + 24x + 16, \qquad \tfrac{4}{3}\pi r^3, \qquad \text{and} \qquad a^2 - b^2,$$

and a **polynomial** is defined as an algebraic expression which may be written as a finite sum of terms that contain only nonnegative integer exponents on the variables. Symbolically, a term in a polynomial in a single variable, say x, may be written in the form

$$ax^n,$$

where n is a nonnegative integer and a is a real number constant. The following example will help you to identify polynomials.

EXAMPLE 1 Which of the following algebraic expressions are polynomials?

a. $x^2 - 3x + 4$ **b.** $d/2$ **c.** $5/t$

Solution

a. Recall from Section 2.1 that those parts of an algebraic expression separated by plus ($+$) signs are called terms of the expression. Therefore, $x^2 - 3x + 4$, which may be written as $x^2 + (-3x) + 4$, is an algebraic expression with three terms: x^2, $-3x$, and 4. Because $-3x = -3x^1$ and $4 = 4x^0$, each term contains only nonnegative integer exponents on the variable x. Thus, $x^2 - 3x + 4$ is a polynomial.

b. $d/2$ may be written as $\tfrac{1}{2}d$ and is an algebraic expression with one term. In this term the exponent on d is the nonnegative integer 1. Thus, $d/2$ is a polynomial.

c. $5/t$ may be written as $5t^{-1}$. Since the exponent on the variable is a negative integer, $5/t$ is *not* a polynomial.

PROGRESS CHECK 1 Which of the following algebraic expressions are polynomials?

a. $-81/x^4$ **b.** $x^4 - 81$ **c.** $-x^4/81$ ⌐

2 We often work with polynomials that contain one term, two terms, or three terms, so it is useful to have names for these types of expressions. A polynomial with just one term is called a **monomial;** one that contains exactly two terms is called a **binomial;** and one with exactly three terms is a **trinomial.** The **degree of a monomial** is the number of variable factors in the term. Therefore, the degree of a monomial is given by the sum of the exponents on the variables in the term. The **degree of a polynomial** is the same as the degree of its highest monomial term. Thus the degree of $9x^3y^3$ is 6, the degree of $18xy$ is 2, and the degree of $9x^3y^3 + 18xy$ is 6, since $9x^3y^3$ is the highest-degree term.

EXAMPLE 2   Give the degree of each polynomial and identify polynomials that are monomials, binomials, or trinomials.

a. $4x^5 - 9x^3$ **b.** -8 **c.** $18x^2y + 2x^3y + 40xy$

Solution

a. $4x^5 - 9x^3$ is a polynomial with two terms, so it is a binomial. The degree of $4x^5$ is 5 and the degree of $-9x^3$ is 3, because the terms contain 5 variable factors and 3 variable factors, respectively. Note that if a term contains only one variable, then the degree of the term is its exponent. Since $4x^5$ is the highest-degree term, $4x^5 - 9x^3$ is a polynomial of degree 5.

b. -8 is a monomial because it is a polynomial with one term. Since -8 may be written as $-8x^0$, it is a polynomial of degree 0.

c. $18x^2y + 2x^3y + 40xy$ is a polynomial with three terms, so it is a trinomial. The degree of the polynomial is 4, because the degree of $2x^3y$ is 4 and the degree of each of the remaining terms is less than 4.

Note   It is convenient to arrange the terms of a polynomial so that the degrees of the terms are in descending order. For instance, we usually write the polynomial in part **c** as

$$2x^3y + 18x^2y + 40xy.$$

This arrangement makes it easy to identify the degree of the polynomial and to spot the absence of certain lower-degree terms.

PROGRESS CHECK 2   Give the degree of each polynomial and identify polynomials that are monomials, binomials, or trinomials.

a. $x^2 - 5x^4 + x$ **b.** $6x^2y^3 + 9xy^2$ **c.** 2 ⌐

⌐**3** ⌐   One basic skill is to evaluate a polynomial for some known value of a variable. To do this, replace all occurrences of a variable by its given value and then perform the indicated operations.

EXAMPLE 3   Evaluate $5x^3 - 7x^2 + 4x - 11$ for $x = 4$. Check the result on a calculator.

Solution   Substitute 4 for x and simplify.

$$5x^3 - 7x^2 + 4x - 11 = 5(4)^3 - 7(4)^2 + 4(4) - 11$$
$$= 5 \cdot 64 - 7 \cdot 16 + 16 - 11$$
$$= 213$$

Calculator Check   Use the power key $\boxed{y^x}$ to compute 4^3 and the square key $\boxed{x^2}$ to compute 4^2.

$5\ \boxed{\times}\ 4\ \boxed{y^x}\ 3\ \boxed{-}\ 7\ \boxed{\times}\ 4\ \boxed{x^2}\ \boxed{+}\ 4\ \boxed{\times}\ 4\ \boxed{-}\ 11\ \boxed{=}\ \boxed{213}$

Instead of keying in a number several times, it can be useful to enter x just once by using the memory keys.

$4\ \boxed{\text{STO}}\ 5\ \boxed{\times}\ \boxed{\text{RCL}}\ \boxed{y^x}\ 3\ \boxed{-}\ 7\ \boxed{\times}\ \boxed{\text{RCL}}\ \boxed{x^2}\ \boxed{+}\ 4\ \boxed{\times}\ \boxed{\text{RCL}}\ \boxed{-}\ 11\ \boxed{=}\ \boxed{213}$

Progress Check Answers

2. (a) 4, trinomial   (b) 5, binomial   (c) 0, monomial

PROGRESS CHECK 3   Evaluate $3x^5 - 10x^3 - x^2 + 7$ for $x = 5$. Check the result on a calculator. ⌐

When we apply a formula, we are often just evaluating a polynomial. To illustrate this, we now consider the section-opening problem.

EXAMPLE 4   Solve the problem in the section introduction on page 144.

Solution   Substitute 19 for x in the given formula and simplify.

$$y = 113.41 + 0.088x + 0.01x^2 \qquad \text{Given formula.}$$
$$= 113.41 + 0.088(19) + 0.01(19)^2 \qquad \text{Replace } x \text{ by 19.}$$
$$= 118.692 \qquad \text{Simplify on calculator.}$$

Thus, the predicted systolic blood pressure is about 119.

PROGRESS CHECK 4   Use the formula in this example to predict the systolic blood pressure for someone who is 50 years old. ⌐

4   To add polynomials, recall from Section 2.1 that the distributive property indicates that we may combine like terms by combining their numerical coefficients and that only like terms may be combined. **Like terms** are terms that have exactly the same variables with the same exponents, while the constant factor in a term is called the **numerical coefficient** of the term. For example,

$$7x^3 + 2x^3 = (7 + 2)x^3 = 9x^3,$$
$$5ab - 2ab = (5 - 2)ab = 3ab,$$
$$2x^2 + 3x \qquad \text{does not simplify.}$$

Because polynomial addition often involves the sum of many terms, it is common to display the problem by aligning like terms in columns, as shown in the following example.

EXAMPLE 5   Add $7x^2 - 4x + 2$ and $-5x^2 + 9x - 6$.

Solution   The given problem may be written as follows.

$$\begin{array}{r} 7x^2 - 4x + 2 \\ \text{Add:} \quad -5x^2 + 9x - 6 \end{array}$$

Now, add $7x^2$ and $-5x^2$; then add $-4x$ and $9x$; and finally, add 2 and -6. The result is $2x^2 + 5x - 4$, as displayed below.

$$\begin{array}{r} 7x^2 - 4x + 2 \\ \text{Add:} \quad \underline{-5x^2 + 9x - 6} \\ 2x^2 + 5x - 4 \end{array}$$

Thus, $2x^2 + 5x - 4$ is the sum of the given polynomials.

Caution   At this point, students sometimes *mistakenly* add exponents when adding polynomials, because we have often added exponents in the first three sections of this chapter. Remember that

$$7x^3 + 2x^3 = (7 + 2)x^3 = 9x^3.$$

In the next section of the chapter when we explain how to *multiply* polynomials, we will add exponents. For instance,

$$(7x^3)(2x^3) = (7 \cdot 2)(x^3 \cdot x^3) = 14x^6.$$

PROGRESS CHECK 5   Add $-8x^2 + x - 3$ and $2x^2 - 7x + 6$. ⌐

Progress Check Answers

3. 8,107
4. 142.81 ≈ 143
5. $-6x^2 - 6x + 3$

To add polynomials, we sometimes need to remove parentheses that group certain terms together. Parentheses are removed by applying the distributive property, as discussed in Section 1.9. Note that the problems in Example 5 and Example 6 are equivalent to one another.

EXAMPLE 6   Simplify $(7x^2 - 4x + 2) + (-5x^2 + 9x - 6)$.

Solution   Because the grouping symbols are preceded by a plus sign or no symbol, we may think of the given problem as

$$1(7x^2 - 4x + 2) + 1(-5x^2 + 9x - 6).$$

So when parentheses are removed, we obtain

$$7x^2 - 4x + 2 - 5x^2 + 9x - 6.$$

Now, proceed as follows.

$$
\begin{aligned}
7x^2 - 4x &+ 2 - 5x^2 + 9x - 6 \\
&= (7x^2 - 5x^2) + (-4x + 9x) + (2 - 6) \qquad \text{Reorder and regroup.} \\
&= 2x^2 + 5x - 4 \qquad\qquad\qquad\qquad\qquad\;\; \text{Combine like terms.}
\end{aligned}
$$

Thus, the given expression simplifies to $2x^2 + 5x - 4$.

PROGRESS CHECK 6   Simplify $(-2x^2 - x + 1) + (-x^2 + 2x - 6)$.   ⌐

5   To subtract polynomials, we use the subtraction definition $a - b = a + (-b)$. In arithmetic we can subtract b from a by adding to a the negative of b. In terms of polynomials the negative of a polynomial is obtained by changing the signs on all of the terms. For example, the negative of $x^2 - 2x - 5$ is written $-(x^2 - 2x - 5)$ and simplifies by the distributive property as shown in Section 1.9.

$$
\begin{aligned}
-(x^2 - 2x - 5) &= -1(x^2 - 2x - 5) \\
&= -x^2 + 2x + 5
\end{aligned}
$$

Thus, $-x^2 + 2x + 5$ is the negative of $x^2 - 2x - 5$, and we can obtain the result by just changing the signs on all of the terms. Our discussion leads to the following method for polynomial subtraction.

To Subtract Polynomials

To subtract two polynomials, change the signs on all of the terms in the polynomial that is being subtracted, and then add the polynomials.

EXAMPLE 7   Subtract $x^2 - 2x - 5$ from $3x^2 - 4x + 1$.

Solution   The given problem may be written as

$$(3x^2 - 4x + 1) - (x^2 - 2x - 5).$$

According to the stated subtraction procedure, change the signs on all of the terms in $x^2 - 2x - 5$ and then add the polynomials.

$$
\begin{aligned}
(3x^2 - 4x + 1) - (x^2 - 2x - 5) &= (3x^2 - 4x + 1) + (-x^2 + 2x + 5) \\
&= (3x^2 - x^2) + (-4x + 2x) + (1 + 5) \\
&= 2x^2 - 2x + 6
\end{aligned}
$$

PROGRESS CHECK 7   Subtract $5x^2 - 2x + 3$ from $2x^2 - 7x - 1$.   ⌐

Progress Check Answers
6. $-3x^2 + x - 5$
7. $-3x^2 - 5x - 4$

As with polynomial addition, you may find it helpful to arrange a polynomial subtraction with like terms aligned vertically.

EXAMPLE 8 Subtract $5x^4 - x^3 + 7x$ from $3x^4 + x^2 - 6x$.

Solution Align like terms vertically and display the problem as follows.

$$\text{Subtract:} \quad \begin{array}{l} 3x^4 + x^2 - 6x \\ \underline{5x^4 - x^3 + 7x} \end{array}$$

Now, change the signs on all of the terms in $5x^4 - x^3 + 7x$ and then add.

$$\text{Add:} \quad \begin{array}{l} 3x^4 + x^2 - 6x \\ \underline{-5x^4 + x^3 - 7x} \\ -2x^4 + x^3 + x^2 - 13x \end{array}$$

Thus, the answer is $-2x^4 + x^3 + x^2 - 13x$.

PROGRESS CHECK 8 Subtract $7x^3 + x^2 - 3$ from $10x^3 + 5x - 9$. ⌐

EXERCISES 3.4

In Exercises 1–12, identify which of the given algebraic expressions are polynomials.

1. $x^2 - 3x + 5$
2. $x^3 + x - 2$
3. $4x^3 + 2x^2 - 3x - 5$
4. $4 + x^3$
5. $\dfrac{x}{4} + 3$
6. $\dfrac{2x}{3} - x^2$
7. $\dfrac{5}{x}$
8. $x^2 - \dfrac{4}{x}$
9. $4x^{-2}$
10. $-2x$
11. $-3/x^2$
12. $x^3 + 2x - 4x^{-3}$

In Exercises 13–24, give the degree of the polynomial and identify any polynomials that are monomials, binomials, or trinomials.

13. $5x^4 + 3x^2$
14. $-2x^5 + x$
15. $x^3 - x^2 + x$
16. $x^5 - x - 2$
17. $10x - 21x^2 - 3$
18. $1 - x^5$
19. 2
20. $27x^3$
21. $x - 3x^2 + 2 - 5x^4$
22. $1 + 3x^3 - 2x^2 + \frac{5}{2}x$
23. $5xy - 6xy^2 + 7xy^3$
24. $x^3 + 3x^2y + 3xy^2 + y^3$

In Exercises 25–36, evaluate the given polynomial for the given value of x; then check the result on a calculator.

25. $3x - 12$; $x = 2$
26. $14 - 3x$; $x = 9$
27. $2.7x^2 + 4.1x - 7$; $x = 0$
28. $-3.2x^2 - 4x + 11.3$; $x = 0$
29. $4x^3 - 3x^2$; $x = -3$
30. $-5x^3 - 4x$; $x = -3$
31. $\frac{1}{2}x^2 + \frac{1}{4}x + \frac{1}{2}$; $x = \frac{1}{2}$
32. $\frac{1}{2}x^2 + \frac{1}{4}x + \frac{1}{2}$; $x = -\frac{1}{2}$
33. $x^4 - x^3 + x^2 - x + 1$; $x = 1$
34. $x^4 - x^3 + x^2 - x + 1$; $x = -1$
35. $5x^2 - 3x - 3.45$; $x = -2.5$
36. $-2x^3 + 3x^2 - 0.864$; $x = 1.2$

In Exercises 37–42, answer the question by evaluating the given polynomial.

37. When an object is dropped and falls straight down, the distance it has fallen after t seconds is given by the formula $y = 16t^2$,

where y is measured in feet. How far will an object fall in 1 second? In 2 seconds? In 3 seconds?

38. When an object is dropped and falls straight down, it goes faster and faster. Its speed after t seconds is given by the formula $s = 32t$, where s is measured in feet per second. How fast is the object going after 1 second? After 2 seconds? After 3 seconds?

39. In a physics laboratory an object is moved along a straight line marked like a number line. Its position is given by the formula $x = 3t^2 - t^3$, where x is in feet and t is in seconds. At what position on the line is it when $t = 0$ seconds? When t is 2 seconds? When t is 3 seconds?

40. In a physics laboratory an object is moved along a straight line marked like a number line. Its velocity is given by the formula $v = 6t - 3t^2$, where v is in feet per second and t is in seconds. When v is positive, it is moving toward higher numbers on the line. When v is negative, it is moving toward lower numbers. Find its velocity and direction when $t = 0$ seconds, 1 second, 2 seconds, and 3 seconds.

41. If there are n people in a room and everyone shakes hands with everyone else, then the total number of handshakes is given by the polynomial $\dfrac{n^2}{2} - \dfrac{n}{2}$. How many handshakes are there if the number of people present is 2? 3? 4? 5? 20?

42. The eighteenth-century Swiss mathematician Leonhard Euler wrote a polynomial expression which produces prime numbers when evaluated for all integers from 0 to 39. The formula is $n^2 + n + 41$. Test the formula by trying a few values for n; try $n = 0$, 1, and 5. Then show that the formula does not yield a prime number for $n = 40$.

In Exercises 43–54, find the sum of the given polynomials.

43. $6x^2 - 5x + 3$; $-4x^2 + 8x - 5$
44. $-2x^2 + 3x - 1$; $2x^2 + x - 2$
45. $-8x^3 + 4x^2 - 3x + 1$; $-5x^3 - 4x^2 - x - 5$
46. $8x^3 - 4x^2 - 3x + 1$; $-8x^3 - 4x^2 + 3x + 1$

47. $3x^2 - 1; -3x^3 + 5x + 1$
48. $3x^3 - x; -3x^3 + 5x^2 + 1$
49. $7x + 1; 4 - 5x; x - 2$
50. $-3x + 1; 3 - 5x; x - 4$
51. $x^2 + x + 1; 2x - 2; 3x^2 - x$
52. $x^2 - x - 1; 2x + 2; 3x^2 + x$
53. $\dfrac{x^2}{2} - \dfrac{x}{3} + \dfrac{3}{4}; \dfrac{3x^2}{4} - \dfrac{x}{2} - \dfrac{1}{8}$
54. $\dfrac{3x^2}{2} - \dfrac{x}{3} - \dfrac{3}{4}; -\dfrac{3x^2}{4} + \dfrac{x}{3} - \dfrac{1}{8}$

In Exercises 55–66, find the difference of the given polynomials. In each case, subtract the second polynomial from the first one.

55. $-8x + 3; 8x - 5$
56. $7x - 2; -3x - 2$
57. $2x^2 - 2x - 1; -2x^2 - 2x + 3$
58. $x^2 - x - 4; x^2 - x + 4$
59. $-8x^3 + 4x^2 - 3x + 1; -5x^3 - 4x^2 - x - 5$
60. $8x^3 - 4x^2 - 3x + 1; 8x^3 - 4x^2 + 3x + 1$
61. $3x^2 - 1; -3x^3 + 5x + 1$
62. $3x^3 - x; -3x^2 + 5x^2 + 1$
63. $-x^2 + x + 1; 2x - 2$
64. $x^2 + x + 1; 2x - 2$
65. $\dfrac{x^2}{2} - \dfrac{x}{3} + \dfrac{3}{4}; \dfrac{3x^2}{4} - \dfrac{x}{3} - \dfrac{1}{8}$
66. $\dfrac{3x^2}{2} - \dfrac{x}{3} - \dfrac{3}{4}; -\dfrac{3x^2}{4} + \dfrac{x}{3} - \dfrac{1}{8}$

In Exercises 67–78, simplify the given expressions.

67. $(6x^2 - 3x + 3) + (-5x^2 + 8x - 5)$
68. $(-7x^2 - 5x + 3) + (5x^2 + 3x - 1)$
69. $(8x^2 - 4x + 2) - (-5x^2 + 3x - 7)$
70. $(-9x^2 - 7x + 5) - (5x^2 + 5x - 5)$
71. $(x^3 - x^2 - x - 1) - (x^3 + x^2 + x + 1)$
72. $(x^3 - 2x^2 - 3x - 4) - (x^3 + 2x^2 + 3x + 4)$
73. $(x - 2x^2 + 1) + (3x^2 - 9 - 4x)$
74. $(7x - 8x^2 + 1) + (-3x^2 - 1 - 2x)$
75. $(1 + 2x^3 - 3x) - (1 - 4x)$
76. $(-x^2 + x^3) - (x + x^2)$
77. $\left(\dfrac{x^2}{5} - \dfrac{x}{3} + \dfrac{3}{2}\right) + \left(\dfrac{3x^2}{4} - \dfrac{x}{5} - \dfrac{1}{6}\right)$
78. $\left(\dfrac{3x^2}{5} - \dfrac{x}{3} - \dfrac{3}{2}\right) - \left(-\dfrac{3x^2}{4} + \dfrac{x}{5} - \dfrac{1}{8}\right)$

In Exercises 79–84, find a polynomial expression for the perimeter of the given figure.

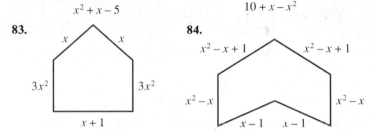

79. $3x^2 - 4$, $x + 8$
80. $x^2 + x + 1$, $x^2 - x + 3$
81. x, $x^2 - 2x - 1$, $x^2 + x - 5$
82. x^2, $8 - x^2$, $10 + x - x^2$
83. x, x, $3x^2$, $3x^2$, $x + 1$
84. $x^2 - x + 1$, $x^2 - x + 1$, $x^2 - x$, $x^2 - x$, $x - 1$, $x - 1$

In Exercises 85–90, solve the given equations for x.

85. $x^2 + 3x - 2 = x^2 - x + 6$
86. $3x^2 + 5x - 3 = 3x^2 - x + 6$
87. $(x^2 + 2x - 1) - (x^2 + 3x - 1) = 0$
88. $(4x^2 + 3x - 1) - (4x^2 + 2x + 1) = 0$
89. $(x - 2x^2 + 3) - (7x - 2x^2 - 3) = 0$
90. $(8x - x^2 + 3) - (7x - x^2 - 3) = 2$

THINK ABOUT IT

For Exercises 1 and 2, assume that x is the only variable in the polynomials mentioned.

1. If a term in a polynomial contains no variables, but is just a constant numerical value (like the 3 in $x^2 + 2x + 3$), then it is called "the constant term." Explain why, when you evaluate a polynomial for $x = 0$, the result is just the constant term. Evaluate $2.1973x^3 - 3.421x^2 + \pi x - 5$ when $x = 0$.

2. Explain why, when you evaluate a polynomial for $x = 1$, the result is just the sum of the coefficients. Find the value of $x^3 - 2x^2 + 3x - 2$ when $x = 1$.

3. For some polynomials you get the same value when you substitute a number and its negative, and this happens no matter what number you substitute. Such a polynomial is called "even." Which of these polynomials are even? Make up another even polynomial yourself.
    a. $x^2 + 5$ b. $x^4 - 1$ c. $x^4 + x^2 + 1$ d. $x^3 + 2$

4. Sometimes polynomials are used to approximate other, more complex formulas. The polynomial given below gives a reasonable approximation for $\sqrt{x}$ when x is a number near 1.

$$1 + \dfrac{x - 1}{2} - \dfrac{(x - 1)^2}{8}$$
$$+ \dfrac{(x - 1)^3}{16} - \dfrac{5(x - 1)^4}{128} + \dfrac{7(x - 1)^5}{256}$$

   a. Use the polynomial to estimate $\sqrt{1.1}$ (by replacing x with 1.1), and then check the answer against your calculator's square root key. Use eight decimal places.
   b. Use the polynomial to estimate $\sqrt{2}$ (by replacing x with 2), and then check the answer against your calculator's square root key.
   c. Can you tell from the calculations in parts a and b why the polynomial works better when the number is closer to 1?

5. Look up the prefixes mono-, bi-, tri-, and poly- in the dictionary. What would be a good name for a polynomial with four terms?

REMEMBER THIS

1. Multiply 3.46×10^{55} by 2×10^{65}.
2. Divide 6.66×10^{101} by 3.33×10^{99}.
3. Simplify $3.8 \times 10^{-2} + 6.2 \times 10^{-2}$.
4. Simplify $\dfrac{24x^2y^3}{8x^2y}$.
5. True or false?   **a.** $2^4 = 4^2$ **b.** $1^0 = 0^1$
6. Which of these are undefined? **a.** 0^{-2} **b.** $(-2)^0$
7. Sunlight travels at 186,000 mi/second, and the sun is 93 million mi from the earth. Express each of these numbers in scientific notation, and then compute the time it takes for sunlight to reach the earth.
8. Which of these are equal?   **a.** $|(-2) + 5|$ **b.** $|2 + (-5)|$ **c.** $|5 - (-2)|$ **d.** $|5 - 2|$
9. The negative of a number equals 3 more than half the number. Find the number.
10. The perimeter of a right triangle is 12, and the lengths of its sides are three consecutive integers. Find the area.

3.5 Multiplication of Polynomials

An airline offers a charter flight at a fare of $300 per person if 100 passengers sign up. For each passenger above 100, the fare for each passenger is reduced by $2. Because revenue = (number of passengers) · (fare per passenger) the total revenue for the flight is given by $(100 + x)(300 - 2x)$, where x represents the number of passengers above 100. Perform the indicated multiplication and write a polynomial expression for the total revenue. (See Example 5.)

OBJECTIVES

|1| Multiply a monomial by a monomial.

|2| Multiply a monomial by a polynomial with more than one term.

|3| Multiply any two polynomials.

|4| Expand powers of a binomial.

|1| The simplest polynomial multiplication involves the product of two monomials. To compute this type of product, simply follow our methods from Section 3.1. For instance, we repeat below the solution to Example 4a from that section.

$$4x^3 \cdot 5x^3 = (4 \cdot 5)(x^3 \cdot x^3) \qquad \text{Reorder and regroup.}$$
$$= (4 \cdot 5)x^{3+3} \qquad \text{Product property of exponents.}$$
$$= 20x^6 \qquad \text{Simplify.}$$

In words, to multiply two monomials, first multiply the numerical coefficients and then multiply the variable factors. Apply the product property of exponents to the product of two exponential expressions with the same base.

EXAMPLE 1   Find each product.

  a. $(-4y^5)(7y^2)$ **b.** $(-5x^2)(\frac{2}{5}y^3)$ **c.** $(2x^2y)(3x^4y^3)$

Solution   Multiply the numerical coefficients and multiply the variable factors.

  a. $(-4y^5)(7y^2) = (-4 \cdot 7)(y^5 \cdot y^2) = -28y^{5+2} = -28y^7$
  b. $(-5x^2)(\frac{2}{5}y^3) = (-5 \cdot \frac{2}{5})(x^2 \cdot y^3) = -2x^2y^3$
  c. $(2x^2y)(3x^4y^3) = (2 \cdot 3)(x^2 \cdot x^4)(y \cdot y^3) = 6x^{2+4}y^{1+3} = 6x^6y^4$

PROGRESS CHECK 1   Find each product.

a. $(-3x^4)(-4x^3)$ **b.** $(-\frac{2}{3}x)(9y^2)$ **c.** $(5a^3b^2)(6a^4b)$ ⌟

2   In the general procedure for multiplying polynomials, we use the properties of exponents and the distributive property in the forms

$$a(b + c) = ab + ac \quad \text{and} \quad (a + b)c = ac + bc,$$

or in an extended form of the distributive property, such as

$$a(b + c + \cdots + n) = ab + ac + \cdots + an.$$

Example 2 shows the procedure for multiplying a monomial and a polynomial with more than one term.

EXAMPLE 2   Find each product.

a. $-5x(3x^2 - 4x + 2)$ **b.** $(4x + 5y)2x$

Solution   Apply the distributive property to remove parentheses, and then simplify by performing the indicated multiplications. Once again, we frequently switch between addition and subtraction by using the subtraction definition, $a - b = a + (-b)$.

a. $-5x(3x^2 - 4x + 2) = (-5x)(3x^2) + (-5x)(-4x) + (-5x)(2)$
$$= -15x^3 + 20x^2 - 10x$$
b. $(4x + 5y)2x = (4x)(2x) + (5y)(2x)$
$$= 8x^2 + 10xy$$

PROGRESS CHECK 2   Find each product.

a. $2x^2(x^2 + 3x - 4)$ **b.** $(7y - x)5y$ ⌟

3   If both factors in the multiplication contain more than one term, the distributive property must be used more than once. For example, no matter what expression is inside the parentheses,

$$(\blacksquare)(x + 3) \quad \text{equals} \quad (\blacksquare)x + (\blacksquare)3.$$

Thus,

$$(x + 4)(x + 3) \quad \text{equals} \quad (x + 4)x + (x + 4)3.$$

Using the distributive property the second time, we obtain $(x + 4)x + (x + 4)3 = x^2 + 4x + 3x + 12 = x^2 + 7x + 12$. Therefore,

$$(x + 4)(x + 3) = x^2 + 7x + 12.$$

The rectangles shown in Figure 3.1 give a geometric interpretation of the product $(x + 4)(x + 3)$ in terms of area.

Figure 3.1

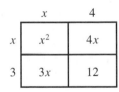

EXAMPLE 3   Find each product.

a. $(3x - 1)(5x + 2)$ **b.** $(2x^2 - 5x + 1)(x - 3)$

Solution   Use the distributive property twice and simplify.

a. $(3x - 1)(5x + 2) = (3x - 1)5x + (3x - 1)2$
$$= 15x^2 - 5x + 6x - 2$$
$$= 15x^2 + x - 2$$
b. $(2x^2 - 5x + 1)(x - 3) = (2x^2 - 5x + 1)x + (2x^2 - 5x + 1)(-3)$
$$= 2x^3 - 5x^2 + x - 6x^2 + 15x - 3$$
$$= 2x^3 - 11x^2 + 16x - 3$$

PROGRESS CHECK 3   Find each product.

a. $(4x - 2)(3x + 1)$ **b.** $(5x^2 - 2x + 4)(x - 2)$ ⌟

Progress Check Answers
1. (a) $12x^7$ (b) $-6xy^2$ (c) $30a^7b^3$
2. (a) $2x^4 + 6x^3 - 8x^2$ (b) $35y^2 - 5xy$
3. (a) $12x^2 - 2x - 2$ (b) $5x^3 - 12x^2 + 8x - 8$

Notice from the examples that applying the distributive property twice is equivalent to multiplying each term of the first factor by each term of the second factor. Therefore, in practice we can multiply polynomials as follows.

> **To Multiply Polynomials**
>
> To multiply two polynomials, multiply each term of one polynomial by each term of the other polynomial, and then combine like terms.

EXAMPLE 4   Find the product of $x^2 - 7x + 5$ and $2x + 4$.

Solution   We multiply each term of $x^2 - 7x + 5$ by each term of $2x + 4$ and combine like terms.

$$(2x + 4)(x^2 - 7x + 5)$$
$$= 2x(x^2) + 2x(-7x) + 2x(5) + 4(x^2) + 4(-7x) + 4(5)$$
$$= 2x^3 - 14x^2 + 10x + 4x^2 - 28x + 20$$
$$= 2x^3 - 10x^2 - 18x + 20$$

Note that the commutative property of multiplication allows us to write the factors in either order when we write the multiplication problem.

PROGRESS CHECK 4   Find the product of $2x^2 + 6x - 4$ and $3x + 1$.   ⌐

EXAMPLE 5   Solve the problem in the section introduction on page 150.

Solution   Multiply each term of $100 + x$ by each term of $300 - 2x$ and combine like terms.

$$(100 + x)(300 - 2x) = 100(300) + 100(-2x) + x(300) + x(-2x)$$
$$= 30,000 - 200x + 300x - 2x^2$$
$$= 30,000 + 100x - 2x^2$$

The total revenue for the charter flight is given by $30,000 + 100x - 2x^2$, where x represents the number of passengers above 100.

PROGRESS CHECK 5   An airline offers a charter flight at a fare of $400 per person if 90 passengers sign up. For each passenger above 90, the fare for each passenger is reduced by $3. The total revenue for the flight is given by $(90 + x)(400 - 3x)$, where x represents the number of passengers above 90. Perform the indicated multiplication and write a polynomial expression for the total revenue.   ⌐

When a polynomial multiplication involves the sum of many terms, it is often useful to display the problem by aligning like terms in columns. In the next example we redo the problem in Example 4 using this type of vertical arrangement.

EXAMPLE 6   Find the product of $x^2 - 7x + 5$ and $2x + 4$.

Solution   Once again, multiply each term of $x^2 - 7x + 5$ by $2x$ and then each term of $x^2 - 7x + 5$ by 4. Arrange like terms in the products under each other; then add.

$$
\begin{array}{r}
x^2 - 7x + 5 \\
2x + 4 \\
\hline
2x^3 - 14x^2 + 10x \\
4x^2 - 28x + 20 \\
\hline
2x^3 - 10x^2 - 18x + 20
\end{array}
$$

Add:

This line equals $2x(x^2 - 7x + 5)$.
This line equals $4(x^2 - 7x + 5)$.

The product is $2x^3 - 10x^2 - 18x + 20$, as expected.

PROGRESS CHECK 6 Find the product of $2x^2 + 6x - 4$ and $3x + 1$ using a vertical arrangement. ⌐

4 The next two examples show how to raise a binomial to a power.

EXAMPLE 7 Expand $(3x - 5)^2$.

Solution $(3x - 5)^2$ is short for $(3x - 5)(3x - 5)$, and to expand $(3x - 5)^2$ means to multiply this expression.

$$(3x - 5)(3x - 5) = 3x(3x) + 3x(-5) + (-5)(3x) + (-5)(-5)$$
$$= 9x^2 - 15x - 15x + 25$$
$$= 9x^2 - 30x + 25$$

The expanded form of $(3x - 5)^2$ is $9x^2 - 30x + 25$.

Caution Remember to multiply each term of the first factor by each term of the second factor to square a binomial. Note that $(3x - 5)^2$ does *not* equal the sum $(3x)^2 + (-5)^2$, because this answer leaves out the middle term, $-30x$. The square of a binomial is discussed further in the next section.

PROGRESS CHECK 7 Expand $(4x - 3)^2$. ⌐

EXAMPLE 8 Expand $(x + h)^3$.

Solution $(x + h)^3$ is short for $(x + h)(x + h)(x + h)$. To multiply this expression, first multiply $(x + h)(x + h)$, as follows.

$$(x + h)(x + h) = x(x) + x(h) + h(x) + h(h)$$
$$= x^2 + 2xh + h^2$$

Now, find the product of this result and the third factor of $x + h$.

$$
\begin{array}{ll}
x^2 + 2xh + h^2 & \\
\underline{\quad\quad\; x + h} & \\
x^3 + 2x^2h + xh^2 & \text{This line equals } x(x^2 + 2xh + h^2). \\
\text{Add:} \quad \underline{\quad\; x^2h + 2xh^2 + h^3} & \text{This line equals } h(x^2 + 2xh + h^2). \\
x^3 + 3x^2h + 3xh^2 + h^3 &
\end{array}
$$

In expanded form $(x + h)^3$ equals $x^3 + 3x^2h + 3xh^2 + h^3$.

PROGRESS CHECK 8 Expand $(a - b)^3$. ⌐

Progress Check Answers
6. $6x^3 + 20x^2 - 6x - 4$
7. $16x^2 - 24x + 9$
8. $a^3 - 3a^2b + 3ab^2 - b^3$

EXERCISES 3.5

In Exercises 1–24, find the product of the given polynomials.

1. $(2x)(3x)$
2. $(5y)(6y)$
3. $-4d(5d^2)$
4. $-7t^3(t)$
5. $3x^2(-2x^2)$
6. $4y^5(-3y^5)$
7. $(-3x^3)(-5x^2)$
8. $(-x^5)(-9x)$
9. $2x(3y^2)$
10. $6y(4x^3)$
11. $(-7y^2)(\frac{1}{7}y)$
12. $(-5x^4)(-\frac{3}{5}x^2)$
13. $3x(2x + 1)$
14. $x(7x + 2)$
15. $-3x(x + 2)$
16. $(2 + 5t)(-5t)$
17. $-x(x^2 - 7x - 9)$
18. $-3y(2y^2 - y)$
19. $(1 + 2s - 3s^2)2s^2$
20. $(1 - x^3 - x^5)x^2$
21. $ab(a^2 + ab + b^2)$
22. $cd(c^3 - 3c^2d + 3cd^2 - d^3)$
23. $-3x^2y(2x - 3xy + 5y)$
24. $-xy^2(x^2 + 2xy - y^2)$

In Exercises 25–48, find the product of the given binomials.

25. $(x + 1)(x + 2)$
26. $(x + 3)(x + 4)$
27. $(2a + 1)(2a + 3)$
28. $(b + 5)(3b + 1)$
29. $(x + 3)(x - 3)$
30. $(2x - 1)(2x + 1)$
31. $(7y - 6)(y - 2)$
32. $(x - 1)(2x - 3)$
33. $(4 - 3x)^2$
34. $(-3 - 8x)^2$
35. $(6x + 2)(3x - 1)$
36. $(4x - 8)(3x + 6)$
37. $(3x + 5)(3x + 5)$
38. $(6x - 2)(3x - 1)$
39. $(a + 2)(c + 3)$
40. $(a + 1)(b - 1)$

41. $(3x - y)(x + 3y)$
42. $(x - 5y)(5x - y)$
43. $(-x + y)(y - x)$
44. $(-2x - 3y)(4y + 5x)$
45. $(2x + y)(3x - y)$
46. $(x + y)(5x + 5y)$
47. $(x + 1)(x^3 - 1)$
48. $(y^2 - 3)(y^2 - 4)$

In Exercises 49–60, find the given product.

49. $(x + 1)(x^2 + x + 1)$
50. $(y - 1)(y^2 + y + 1)$
51. $(3y^2 - y + 1)(y - 1)$
52. $(5x^2 - 3x - 2)(x - 2)$
53. $(2x + 1)(4x^2 - 2x + 1)$
54. $(2x + 1)(4x^2 - 2x - 1)$
55. $(a + b)(c + d + e)$
56. $(a - b)(c - d - e)$
57. $(x + y)(x^2 + xy + y^2)$
58. $(x - y)(x^2 + xy - y^2)$
59. $(2x + 3y)(x^2 - 5xy + 7y^2)$
60. $(4x - 3y)(2x^2 - 3xy - y^2)$

In Exercises 61–72, find the given product.

61. $(x^2 + x + 1)(x^2 + 2x + 3)$
62. $(3x^2 - 2x + 3)(x^2 + 4x - 5)$
63. $x(x + 1)(x + 2)$
64. $3y(2y - 3)(y + 2)$
65. $(y + 1)(y - 2)(y + 3)$
66. $(x - 2)(x - 2)(3x + 5)$
67. $(a + 2)^3$ 68. $(b - 3)^3$
69. $(x - 1)(x + 1)^2$
70. $(2x - 5)(x - 3)^2$
71. $(x + 1)^4$
72. $(3x - 2)^4$

In Exercises 73–76, find a polynomial expression for the area of the given figure.

73. Rectangle: length $= 3x + 9$; width $= x - 7$
74. Square: side $= 3x - 4$
75. Triangle: base $= 2x + 1$; height $= x$
76. Circle: radius $= 3x - 10$

In Exercises 77–78, find a polynomial expression for the volume of the given figure.

77. Rectangular solid: length $= x + 1$; width $= x$; height $= x - 1$
78. Rectangular solid: length $= 2x + 1$; width $= 2x - 1$; height $= 3x$

In Exercises 79–82, solve the word problem by carrying out the indicated multiplication.

79. A manufacturer plans to introduce a new model of computer at a list price of $4,000. The sales department assumes that for each $100 the price is cut, sales will go up by 500 units. At $4,000 it estimates selling 10,000 units. The formula for estimated revenue then is $(4,000 - 100x)(10,000 + 500x)$, where x represents the number of $100 price cuts. Multiply the two factors and write a polynomial expression for the total estimated revenue. What is the selling price when $x = 1$? What is the estimated revenue when $x = 1$?

80. A manufacturer has been selling about 2,000 lawn tractors per year for $1,700. The sales department assumes that for each $100 the price is raised, sales will fall by 25 units. The formula $(1,700 + 100x)(2,000 - 25x)$ gives the estimated revenue, where x represents the number of $100 price increases. Multiply the two factors and write a polynomial expression for the total estimated revenue. What is the selling price when $x = 1$? What is the estimated revenue when $x = 1$?

81. In many applications of probability p and $1 - p$ represent two positive fractions whose sum is 1. Often, the product of these numbers is important.
    a. Write a polynomial expression for the product of p and $1 - p$.
    b. Show that the product is higher when $p = \frac{1}{2}$ than it is when $p = \frac{9}{10}$.

82. If you make a rectangle with a piece of string that is 50 in. long, the sketch can be labeled as shown.

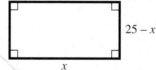

    a. Write an expression for the perimeter and show that it simplifies to 50 even when x is not known.

    b. Find a polynomial expression for the area.
    c. Complete the table shown.

length	width	area
x	$25 - x$	
6	19	
12		
13		
18		

    d. Guess what value for x would give the maximum area.

THINK ABOUT IT

1. Show that $(3x + 0)(4x + 0)$ is the same as $3x(4x)$.
2. A geometric representation of $(a + b)^2$ is shown. This corresponds to $(a + b)(a + b) = a^2 + 2ab + b^2$. Draw a similar three-dimensional sketch to illustrate $(a + b)^3$, and see if you can figure out the polynomial expression for $(a + b)^3$ from the sketch.

	a	b
a	a^2	ab
b	ab	b^2

3. For three consecutive integers the square of the middle one is always one more than the product of the first and last. Show this algebraically in two ways, and then decide if one way is easier than the other.
   a. Call the three numbers x, $x + 1$, and $x + 2$. Show that $(x + 1)^2$ is 1 more than $x(x + 2)$.
   b. Call the three numbers $x - 1$, x, and $x + 1$. Show that x^2 is 1 more than $(x - 1)(x + 1)$.
   c. Using $x = 3$, show that the result can be seen visually by comparing the sketch of a rectangle with sides x and $x + 2$ to the sketch of a square with side $x + 1$.

4. If you multiply a polynomial of degree 2 times a polynomial of degree 3, what degree will the product polynomial have? Show this by an example.

5. Show that you cannot say in general what degree polynomial you will get if you add or subtract two polynomials of degree 2.
   a. Make up an example where their sum has degree 2.
   b. Make up an example where their sum has degree 1.

REMEMBER THIS

1. Which of these expressions are polynomials?
   a. $\dfrac{1}{x}$ b. $x^2 + 4$ c. 5   d. $3x + \sqrt{x}$

2. Give the degree of each of these polynomials.
   a. $2x^3 + x - 1$ b. $25 - x^2$ c. $x + 12$ d. $xy + x$

3. Evaluate $4x^2 - 2x - 1$ when $x = 0, \frac{1}{2}, -\frac{1}{2}$.

4. Add $x^2 - 3x - 10$ and $4 + 2x - 3x^2$.

5. Subtract $5x^2 - 2x - 8$ from $5x^2 + 2x - 5$.

6. Convert 0.000485 into scientific notation.

7. Evaluate $E = mc^2$ when $m = 2.3 \times 10^5$ and $c = 3 \times 10^8$.

8. Simplify $\dfrac{4^{-1}x^2y^{-2}z}{4^{-2}x^3y^4z^{-1}}$.

9. Four more than one-third of a number equals 6 less than half the number. Find the number.

10. How much acid should be added to 800 ml of solution which is 8 percent acid to make a solution which is 10 percent acid?

3.6 Products of Binomials

Consider the square with side length $a + b$ shown in the figure.
a. Algebraically, calculate the area $(a + b)^2$.
b. Find the area by adding the areas of regions 1, 2, 3, and 4.
c. Does $(a + b)^2 = a^2 + b^2$? (See Example 6.)

OBJECTIVES

1 Use the FOIL method to multiply binomials.

2 Multiply binomials that are the sum and difference of the same two terms.

3 Use the formulas for the square of a binomial.

4 Multiply binomials and then solve linear equations.

Mental shortcuts and formulas are available for multiplying two binomials. Because this type of product occurs often, it is important that you learn the efficient procedures shown in this section.

1 According to our general procedure, to multiply two binomials, we multiply each term of one binomial by each term of the other binomial and then combine like terms. A simple method for doing this is shown below. This method is called the FOIL method and the letters F, O, I, and L denote the products of the first, outer, inner, and last terms, respectively.

$$\overset{L}{\overset{F}{(a + b)}(c + d)} = ac + ad + bc + bd \quad F + O + I + L$$

To multiply using the FOIL method, add the following products.

1. Multiply the **First** terms: ac.
2. Multiply the **Outer** terms: ad.
3. Multiply the **Inner** terms: bc.
4. Multiply the **Last** terms: bd.

In the following examples, note that the outer product and the inner product are often like terms and are therefore combined in such cases.

EXAMPLE 1   Multiply using the FOIL method: $(x + 5)(x + 2)$.

Solution

$$(x + 5)(x + 2) = x(x) + x(2) + 5(x) + 5(2)$$
$$= x^2 + 2x + 5x + 10$$
$$= x^2 + 7x + 10$$

Thus $(x + 5)(x + 2) = x^2 + 7x + 10$.

Note   After you have become proficient with the FOIL method, the work displayed above is done *mentally* and you can often just write down the answer. Work toward this goal.

PROGRESS CHECK 1   Multiply by the FOIL method: $(x + 3)(x + 7)$.   ⌐

EXAMPLE 2   Multiply using the FOIL method: $(2x - 5)(3x - 1)$.

Solution

$$(2x - 5)(3x - 1) = 2x(3x) + 2x(-1) + (-5)(3x) + (-5)(-1)$$
$$= 6x^2 - 2x - 15x + 5$$
$$= 6x^2 - 17x + 5$$

Thus, $(2x - 5)(3x - 1) = 6x^2 - 17x + 5$.

PROGRESS CHECK 2   Multiply by the FOIL method: $(4x - 3)(5x - 2)$.   ⌐

EXAMPLE 3   Multiply $(5x - y)(2x + 3y)$. Use the FOIL method.

Solution

$$(5x - y)(2x + 3y) = 5x(2x) + 5x(3y) + (-y)(2x) + (-y)(3y)$$
$$= 10x^2 + 15xy - 2xy - 3y^2$$
$$= 10x^2 + 13xy - 3y^2$$

Thus, $(5x - y)(2x + 3y) = 10x^2 + 13xy - 3y^2$.

PROGRESS CHECK 3   Multiply $(7x + 2y)(3x - 4y)$.   ⌐

EXAMPLE 4   Multiply $(3x + 5)(3x - 5)$. Use the FOIL method.

Solution

$$
\begin{aligned}
(3x + 5)(3x - 5) &= \overset{F}{3x(3x)} + \overset{O}{3x(-5)} + \overset{I}{5(3x)} + \overset{L}{5(-5)} \\
&= 9x^2 - 15x + 15x - 25 \\
&= 9x^2 - 25
\end{aligned}
$$

Thus, $(3x + 5)(3x - 5) = 9x^2 - 25$.

PROGRESS CHECK 4   Multiply $(2x - 7)(2x + 7)$ by the FOIL method. ⌐

2 Note in Example 4 that the factors in $(3x + 5)(3x - 5)$ differ only in the operation between $3x$ and 5, while the result of the multiplication is $(3x)^2$ and 5^2 with a minus sign between them. In general,

$$(a + b)(a - b) = a^2 - ab + ba - b^2 = a^2 - b^2,$$

so we may use the following special product formula.

Product of the Sum and Difference of Two Expressions

The product of the sum and difference of two terms is the square of the first term minus the square of the second term.

$$(a + b)(a - b) = a^2 - b^2$$

EXAMPLE 5   Find each product. Use $(a + b)(a - b) = a^2 - b^2$.

a. $(x + 6)(x - 6)$ **b.** $(2 + 3x)(2 - 3x)$ **c.** $(5x - 4y)(5x + 4y)$

Solution

a. Substitute x for a and 6 for b in the given formula.

$$
\begin{aligned}
(a + b)(a - b) &= a^2 - b^2 \\
(x + 6)(x - 6) &= x^2 - 6^2 \\
&= x^2 - 36
\end{aligned}
$$

b. Here $a = 2$ and $b = 3x$, so

$$
\begin{aligned}
(2 + 3x)(2 - 3x) &= 2^2 - (3x)^2 \\
&= 4 - 9x^2.
\end{aligned}
$$

c. Replace a by $5x$ and b by $4y$. The product is the square of the first term minus the square of the second term.

$$
\begin{aligned}
(5x - 4y)(5x + 4y) &= (5x)^2 - (4y)^2 \\
&= 25x^2 - 16y^2
\end{aligned}
$$

PROGRESS CHECK 5   Find each product. Use $(a + b)(a - b) = a^2 - b^2$.

a. $(x + 4)(x - 4)$ **b.** $(5 + 2x)(5 - 2x)$ **c.** $(6x - 7y)(6x + 7y)$ ⌐

3 Another important special product is the square of a binomial. To develop this formula, we now solve the problem which opened this section.

Progress Check Answers
4. $4x^2 - 49$
5. (a) $x^2 - 16$ (b) $25 - 4x^2$ (c) $36x^2 - 49y^2$

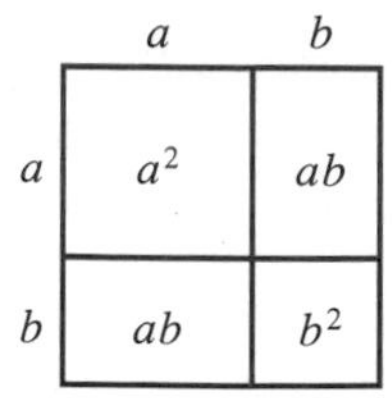

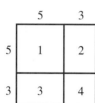

Figure 3.2

Example 6 Solve the problem in the section introduction on page 155.

Solution See Figure 3.2.

a. Algebraically, $(a + b)^2 = (a + b)(a + b)$, and we can multiply this expression by the FOIL method.

$$\begin{aligned} & \quad \text{F} + \text{O} + \text{I} + \text{L} \\ (a + b)(a + b) &= a^2 + ab + ba + b^2 \\ &= a^2 + 2ab + b^2 \end{aligned}$$

Thus, the square is $(a + b)^2$, which equals $a^2 + 2ab + b^2$.

b. The areas of regions 1, 2, 3, and 4 are a^2, ab, ba, and b^2, respectively, as shown in Figure 3.2. Adding these areas shows that the area of the square is given by $a^2 + 2ab + b^2$.

c. From both an algebraic and a geometric viewpoint, it is clear that **$(a + b)^2$ does not equal $a^2 + b^2$.**

Figure 3.3

PROGRESS CHECK 6 Consider the square in Figure 3.3, with side length $5 + 3$.

a. Arithmetically, calculate the area $(5 + 3)^2$.

b. Find the area by adding the areas of regions 1, 2, 3, and 4.

c. Does $(5 + 3)^2 = 5^2 + 3^2$?

Example 6 showed that $(a + b)^2 = a^2 + 2ab + b^2$. If we switch the sign between a and b from $+$ to $-$, then $(a - b)^2 = (a - b)(a - b) = a^2 - ab - ba + b^2 = a^2 - 2ab + b^2$. Thus, we can square a binomial by applying the following formulas.

Square-of-a-Binomial Formulas

To square a binomial, use

$$\begin{aligned} (a + b)^2 &= a^2 + 2ab + b^2 \\ \text{or} \quad (a - b)^2 &= a^2 - 2ab + b^2. \end{aligned}$$

The square of a binomial is the square of the first term, plus or minus twice the product of the two terms, plus the square of the second term.

EXAMPLE 7 Find each product. Use the square-of-a-binomial formulas.

a. $(x + 3)^2$ **b.** $(4x - 5)^2$ **c.** $(3x + 5y)(3x + 5y)$

Solution

a. Use the formula for $(a + b)^2$ with $a = x$ and $b = 3$.

$$\begin{aligned} (a + b)^2 &= a^2 + 2ab + b^2 \\ (x + 3)^2 &= (x)^2 + 2(x)(3) + (3)^2 \\ &= x^2 + 6x + 9 \end{aligned}$$

b. Use the formula for $(a - b)^2$ with $a = 4x$ and $b = 5$.

$$\begin{aligned} (a - b)^2 &= a^2 - 2ab + b^2 \\ (4x - 5) &= (4x)^2 - 2(4x)(5) + (5)^2 \\ &= 16x^2 - 40x + 25 \end{aligned}$$

c. Because $(3x + 5y)(3x + 5y) = (3x + 5y)^2$, the product is the square of the first term, plus twice the product of the two terms, plus the square of the second term.

Progress Check Answers

6. (a) 64 (b) 64 (c) No

$$(a + b)^2 = a^2 + 2ab + b^2$$
$$(3x + 5y)^2 = (3x)^2 + 2(3x)(5y) + (5y)^2$$
$$= 9x^2 + 30xy + 25y^2$$

PROGRESS CHECK 7 Find each product. Use the formulas for the square of a binomial.

a. $(x + 4)^2$ **b.** $(3x - 7)^2$ **c.** $(y - 5x)(y - 5x)$ ⌐

4 The procedures for solving a linear equation together with the methods for multiplying binomials enable us to solve the problem in the next example.

EXAMPLE 8 The difference in area between the rectangle and the square shown in Figure 3.4 is 63 in.². Find the length of the rectangle.

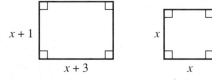

Figure 3.4

Solution
Set up an equation, using $A = \ell w$, $A = s^2$, and the given information.

Area of rectangle	minus	area of square	equals	63
⌣	↓	⌣	↓	↓
$(x + 3)(x + 1)$	−	x^2	=	63

Solve the Equation

$$x^2 + x + 3x + 3 - x^2 = 63 \quad \text{FOIL multiplication.}$$
$$4x + 3 = 63 \quad \text{Combine like terms.}$$
$$4x = 60 \quad \text{Subtract 3 from both sides.}$$
$$x = 15 \quad \text{Divide both sides by 4.}$$

Answer the Question   The length of the rectangle is 18 in., because the length is 3 more than x.

Check the Answer   For the square, $A = 15^2 = 225$. For the rectangle, $A = 18(16) = 288$. Because $288 - 225 = 63$, the solution checks.

PROGRESS CHECK 8 The length and the width of a rectangle are given by $x + 5$ and $x + 2$, respectively, and the side length of a square is given by x. Find the width of the rectangle if the area of the rectangle exceeds the area of the square by 80 in.².   ⌐

Progress Check Answers
7. (a) $x^2 + 8x + 16$ (b) $9x^2 - 42x + 49$
(c) $y^2 - 10xy + 25x^2$
8. Width = 12 in.

EXERCISES 3.6

In Exercises 1–24, find the product using the FOIL method. Try to calculate the middle term mentally.

1. $(x + 1)(x + 3)$ **2.** $(x + 3)(x + 4)$
3. $(w - 2)(w + 4)$ **4.** $(s - 5)(s + 6)$
5. $(x + 3)(x - 5)$ **6.** $(x + 7)(x - 9)$
7. $(y - 8)(y - 1)$ **8.** $(y - 6)(y - 10)$
9. $(2x + 3)(3x + 1)$
10. $(4x - 1)(x + 1)$
11. $(4x + 6)(3x - 2)$
12. $(8t - 12)(3t - 2)$
13. $(3 - 7c)(5 + 3c)$
14. $(1 - a)(2 + a)$
15. $(\frac{1}{2}x + 4)(\frac{1}{2}x + 6)$
16. $(\frac{1}{4}y + 4)(\frac{1}{4}y - 8)$
17. $(-2 - z)(4 - 2z)$

18. $(-3 + 2z)(-4 - 3z)$
19. $(x + 0.1)(x - 0.2)$
20. $(p - 0.1)(p - 0.9)$
21. $(2x + 7y)(x - 9y)$
22. $(8x - 8y)(x + y)$
23. $(\frac{1}{2}a - b)(a + \frac{1}{2}b)$
24. $(\frac{1}{4}g + \frac{1}{2}h)(\frac{1}{2}g - \frac{1}{4}h)$

In Exercises 25–36, find the product by using the pattern $(a + b)(a - b) = a^2 - b^2$.

25. $(x + 1)(x - 1)$ **26.** $(y + 2)(y - 2)$
27. $(3w + 5)(3w - 5)$ **28.** $(7n + 8)(7n - 8)$
29. $(2x + 3y)(2x - 3y)$
30. $(9x + y)(9x - y)$
31. $(4x - 3u)(4x + 3u)$
32. $(5f - h)(5f + h)$

33. $\left(\dfrac{3r}{4} + \dfrac{1}{4}\right)\left(\dfrac{3r}{4} - \dfrac{1}{4}\right)$

34. $\left(\dfrac{2x}{3} + \dfrac{5}{9}\right)\left(\dfrac{2x}{3} - \dfrac{5}{9}\right)$

35. $(x - 0.5)(x + 0.5)$

36. $(0.1y - 0.2x)(0.1y + 0.2x)$

In Exercises 37–48, find the product by using the pattern for the square of a binomial.

37. $(x + 4)^2$

38. $(x - 3)^2$

39. $(3y - 5)^2$

40. $(7y + 1)^2$

41. $(2 + 6k)^2$

42. $(3 - 4c)^2$

43. $(2x + 0.2)^2$

44. $(10v - 0.1)^2$

45. $(3s - 2r)^2$

46. $(a + 5b)^2$

47. $(\frac{1}{4}j + 2k)^2$

48. $(\frac{1}{2}x - y)^2$

In Exercises 49–54, fill in the blank to make a true statement.

49. $(x - 5)$ _____ $= x^2 - 25$

50. $(x + 3)$ _____ $= x^2 - 9$

51. $(m - n)$ _____ $= m^2 - n^2$

52. $(2w + 1)$ _____ $= 4w^2 - 1$

53. $(x + y)$ _____ $= x^2 + 2xy + y^2$

54. $(3s - 5r)$ _____ $= 9s^2 - 30sr + 25r^2$

55. Find the difference in area between a square with side equal to x and a rectangle with sides $x - 1$ and $x + 1$. Check your answer by looking at the case for $x = 5$. What happens for $x \leq 1$?

56. Find the difference in area between a square with side equal to x and a rectangle with sides $x - 2$ and $x + 2$. Check your answer by looking at the case for $x = 8$. What happens for $x \leq 2$?

57. The area of the rectangle shown is 10 more than the area of the square. Find the dimensions of both figures.

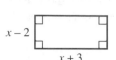

58. The area of the rectangle shown is 10 less than the area of the square. Find the dimensions of both figures.

59. Find the area of the small square if the shaded area equals 21 square units.

60. Find the radius of the larger circle shown if the shaded area equals 30π square units.

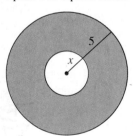

THINK ABOUT IT

1. What do the letters in FOIL stand for? Would the answer be the same for the LIFO method? [Try it for $(a + b)(c + d)$.] How many different ways can you arrange the letters {F,O,I,L}? Which properties assure us that the answer will be the same for each arrangement?

2. The expression $(3a + b)^2$ is the square of a binomial. The figure shown corresponds to this expression. Show the pieces inside the square have a total area equal to $9a^2 + 6ab + b^2$, which is what you get when you evaluate $(3a + b)^2$ by the methods of this chapter.

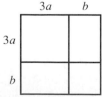

3. In general, $(x + n)^2 \neq x^2 + n^2$, because there should be a term in the middle for $2xn$. But can you find any particular value for n where they do turn out to be equal?

4. Use the FOIL method to multiply 29×31 by writing the product as $(30 - 1)(30 + 1)$. Now multiply 28×32 mentally using this same trick.

5. Compute 19^2 mentally by thinking of it as $(20 - 1)^2$.

REMEMBER THIS

1. Rewrite these expressions using the distributive property.
   a. $2x(y + w)$
   b. $(2 + x)(y + w)$
2. Find an expression for the shaded area.

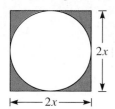

3. Find the product: $(x + 2)(x^2 + 3x + 5)$.
4. Evaluate the polynomial $x^2 - x + 1$ when $x = -1, 0,$ and 1.

5. Which of these are binomials?
   a. $2x^2$ **b.** $2 + x^2$ **c.** $x^2 + x + 2$ **d.** 5
6. Graph the numbers in the set $\{-9, -3, 0, 3, 9\}$ that make $x^2 - 9$ equal to zero.

7. Subtract $3x^2 - 2x + 5$ from $4x^2 - 2x - 6$.
8. What number is 1,000 times greater than 3.45×10^{16}?

9. How much silver must be added to 6 lb of gold to have an alloy which is one-quarter silver?
10. Solve the equation: $\dfrac{x + 4}{9} = \dfrac{x - 2}{2}$.

3.7 Division of Polynomials

When solving the formula $A = \frac{1}{2}h(a + b)$ for b in Exercise 53 in Section 2.5, a student obtains the answer $b = \dfrac{2A}{h} - a$.

However, the answer given in the text answer key is

$$b = \frac{2A - ah}{h}.$$

Is the student's answer also correct?
(See Example 5.)

OBJECTIVES

1 Divide a polynomial by a monomial.

2 Divide a polynomial by a binomial.

1 To complete the discussion of the four basic operations with polynomials, we now consider division of polynomials. In the simplest case we divide a monomial by another monomial by applying the quotient property of exponents as shown in Section 3.2. Because division by 0 is undefined, assume throughout this section that we exclude any value for a variable that results in a zero denominator.

EXAMPLE 1 Find each quotient. Write the result using only positive exponents.

a. $\dfrac{28x^6}{4x^2}$ **b.** $\dfrac{6y^2}{3y^3}$ **c.** $\dfrac{9x^2y}{-12x^2y^4}$

Solution   Divide the numerical coefficients and divide the variable factors.

a. $\dfrac{28x^6}{4x^2} = \dfrac{28}{4}x^{6-2} = 7x^4$

b. $\dfrac{6y^2}{3y^3} = \dfrac{6}{3}y^{2-3} = 2y^{-1} = \dfrac{2}{y}$

c. $\dfrac{9x^2y}{-12x^2y^4} = \dfrac{9}{-12}x^{2-2}y^{1-4} = -\dfrac{3}{4}x^0y^{-3}$

$\qquad\qquad = -\dfrac{3}{4} \cdot 1 \cdot \dfrac{1}{y^3} = -\dfrac{3}{4y^3}$

Note   Whereas addition, subtraction, and multiplication of two polynomials always produces another polynomial, note in this example that the quotient of two polynomials may or may not be another polynomial.

PROGRESS CHECK 1   Find each quotient. Write the result using only positive exponents.

a. $\dfrac{15y^5}{5y^3}$
b. $\dfrac{32x}{4x^4}$
c. $\dfrac{6xy^3}{-16x^3y^3}$

To divide a polynomial with more than one term by a monomial, recall that $a \div b = a(1/b)$. Thus, we may divide $a + c$ by b as follows.

$$(a + c) \div b = (a + c) \cdot \dfrac{1}{b} \qquad \text{Use } a \div b = a\left(\dfrac{1}{b}\right).$$

$$= a \cdot \dfrac{1}{b} + c \cdot \dfrac{1}{b} \qquad \text{Distributive property.}$$

$$= \dfrac{a}{b} + \dfrac{c}{b} \qquad \text{Simplify.}$$

The result shows a fast and practical way to perform this type of division.

To Divide a Polynomial by a Monomial

To divide a polynomial by a monomial, divide each term of the polynomial by the monomial. In symbols,

$$\dfrac{a + c}{b} = \dfrac{a}{b} + \dfrac{c}{b}, \qquad b \neq 0.$$

From a different viewpoint, note that starting with Section 1.1 we have often used

$$\dfrac{a}{b} + \dfrac{c}{b} = \dfrac{a + c}{b}, \qquad b \neq 0,$$

to add fractions. So you may think of the procedure for dividing by a monomial as the reverse process of adding fractions with a common denominator.

EXAMPLE 2   Divide $5x^4 + 12x^2 + 2x$ by $2x$.

Solution   Divide *each* term of $5x^4 + 12x^2 + 2x$ by $2x$ and simplify.

$$\dfrac{5x^4 + 12x^2 + 2x}{2x} = \dfrac{5x^4}{2x} + \dfrac{12x^2}{2x} + \dfrac{2x}{2x}$$

$$= \dfrac{5}{2}x^3 + 6x + 1$$

Note The answer to a division problem can be checked by multiplying the answer by the divisor. In this case

$$2x(\tfrac{5}{2}x^3 + 6x + 1) = 5x^4 + 12x^2 + 2x,$$

so the answer checks.

PROGRESS CHECK 2   Divide $6x^3 + 2x^2 + 3x$ by $3x$, and check the answer. ⌟

EXAMPLE 3   Divide $\dfrac{4x^3 - 8x^2 + 5x}{4x^2}$.

Solution   Divide each term in the numerator by $4x^2$ and simplify.

$$\frac{4x^3 - 8x^2 + 5x}{4x^2} = \frac{4x^3}{4x^2} - \frac{8x^2}{4x^2} + \frac{5x}{4x^2} = x - 2 + \frac{5}{4x}$$

PROGRESS CHECK 3   Divide $\dfrac{10x^4 + 5x^2 - 4x}{5x^2}$. ⌟

EXAMPLE 4   Divide $\dfrac{-20x^2y^2 - 25x^2y}{-10xy}$.

Solution   Because of the negative sign in the denominator, we choose to express the subtraction in the numerator as an addition expression, using the relationship $a - b = a + (-b)$.

$$\frac{-20x^2y^2 - 25x^2y}{-10xy} = \frac{-20x^2y^2}{-10xy} + \frac{-25x^2y}{-10xy} = 2xy + \frac{5}{2}x$$

PROGRESS CHECK 4   Divide $\dfrac{6x^2y - 8xy^2}{-4xy}$. ⌟

EXAMPLE 5   Solve the problem in the section introduction on page 161.

Solution   Divide each term of $2A - ah$ by h and simplify.

$$\frac{2A - ah}{h} = \frac{2A}{h} - \frac{ah}{h} = \frac{2A}{h} - a$$

Thus, the student's answer is also correct.

Caution   A common student error when dividing $2A - ah$ by h is to divide h only into the term containing h and incorrectly answer $2A - a$. Because of the distributive property, remember to divide *each* term of the polynomial by the monomial in such divisions.

PROGRESS CHECK 5   When solving the formula $S = \tfrac{1}{2}n(a + \ell)$ for ℓ, a student obtains the answer $\ell = \dfrac{2S}{n} - a$. If the answer in the text key is $\ell = \dfrac{2S - na}{n}$, is the student's answer also correct? ⌟

2   We divide a polynomial by a binomial in a manner similar to long division in arithmetic. To illustrate, consider the division of 169 by 12 shown below, and note the names of the key components.

$$
\begin{array}{r}
14 \leftarrow \text{quotient} \\
\text{divisor} \rightarrow 12\overline{)169} \leftarrow \text{dividend} \\
\underline{12} \\
49 \\
\underline{48} \\
1 \leftarrow \text{remainder}
\end{array}
$$

Progress Check Answers

2. $2x^2 + \frac{2}{3}x + 1$

3. $2x^2 + 1 - \dfrac{4}{5x}$

4. $-\dfrac{3}{2}x + 2y$

5. Yes

The result may be expressed as

$$\frac{169}{12} = 14 + \frac{1}{12}.$$

And in general, we may answer in the form

$$\frac{\text{dividend}}{\text{divisor}} = \text{quotient} + \frac{\text{remainder}}{\text{divisor}}.$$

Now, dividing 12 into 169 is like dividing $x + 2$ into $x^2 + 6x + 9$, where x stands for 10. As we carry out the division, see how the steps compare. First, set up the problem as follows. Note that both polynomials must be written with descending powers.

$$x + 2 \overline{)x^2 + 6x + 9}$$

Divide the first term of the dividend by the first term of the divisor to obtain the first term of the quotient.

$$\begin{array}{r} x \\ x + 2 \overline{)x^2 + 6x + 9} \end{array}$$

Next, multiply the entire divisor by the first term of the quotient and subtract this product from the dividend.

$$\begin{array}{r} x \\ x + 2 \overline{)x^2 + 6x + 9} \\ \text{Subtract:} \quad \underline{x^2 + 2x } \\ 4x + 9 \end{array}$$

Use the remainder as the new dividend and repeat the above procedure until the remainder is of lower degree than the divisor.

$$\begin{array}{r} x + 4 \\ x + 2 \overline{)x^2 + 6x + 9} \\ \text{Subtract:} \quad \underline{x^2 + 2x } \\ 4x + 9 \\ \text{Subtract:} \quad \underline{4x + 8} \\ 1 \end{array}$$

Thus, the quotient is $x + 4$, the remainder is 1, and we may answer

$$\frac{x^2 + 6x + 9}{x + 2} = x + 4 + \frac{1}{x + 2}.$$

We now summarize the procedure.

Long Division of Polynomials

1. Arrange the terms of the dividend and the divisor with descending powers. If a lower power is absent in the dividend, write 0 as its coefficient.
2. Divide the first term of the dividend by the first term of the divisor to obtain the first term of the quotient.
3. Multiply the entire divisor by the first term of the quotient and subtract this result from the dividend.
4. Use the remainder as the new dividend and repeat the above procedure until the remainder is of lower degree than the divisor.

A summary cannot do justice to the systematic procedure described above. Careful consideration of Examples 6–8 should help clarify the process.

EXAMPLE 6   Divide $2x^2 - 3x + 5$ by $x - 3$.

Solution   The division is performed below. As shown in color, the key divisions are that $2x^2$ divided by x is $2x$, and $3x$ divided by x is 3.

$$\frac{2x^2}{x} = 2x \qquad \frac{3x}{x} = 3$$

$$
\begin{array}{r}
2x + 3 \\
x - 3 \overline{)\, 2x^2 - 3x + 5}
\end{array}
$$

Subtract: $2x^2 - 6x$ This line is $2x(x-3)$.

$3x + 5$

Subtract: $3x - 9$ This line is $3(x-3)$.

14

The quotient is $2x + 3$, the remainder is 14, and we may write

$$\frac{2x^2 - 3x + 5}{x - 3} = 2x + 3 + \frac{14}{x - 3}.$$

Note   Check that dividend = (divisor)(quotient) + remainder. In this example,

$$(x - 3)(2x + 3) + 14 = 2x^2 + 3x - 6x - 9 + 14$$
$$= 2x^2 - 3x + 5.$$

Because the dividend is $2x^2 - 3x + 5$, the answer checks.

PROGRESS CHECK 6   Divide $3x^2 - 2x - 6$ by $x - 2$, and check your answer.   ⌐

EXAMPLE 7   Divide $\dfrac{x^3 - 1}{x - 1}$.

Solution   The dividend is a polynomial of degree 3 with no second-degree or first-degree terms. In the division below, note that $0x^2$ and $0x$ are inserted so that like terms align vertically.

$$
\begin{array}{r}
x^2 + x + 1 \\
x - 1 \overline{)\, x^3 + 0x^2 + 0x - 1}
\end{array}
$$

$\dfrac{x^3}{x} = x^2, \dfrac{x^2}{x} = x,$ and $\dfrac{x}{x} = 1.$

Subtract: $x^3 - x^2$ This line is $x^2(x-1)$.

$x^2 + 0x - 1$

Subtract: $x^2 - x$ This line is $x(x-1)$.

$x - 1$

Subtract: $x - 1$ This line is $1(x-1)$.

0

In the answer we do not include a zero remainder, so

$$\frac{x^3 - 1}{x - 1} = x^2 + x + 1.$$

PROGRESS CHECK 7   Divide $\dfrac{x^3 + 1}{x + 1}$.   ⌐

EXAMPLE 8   Divide $\dfrac{2x^4 - 8x^3 - x^2 + 7x - 12}{2x^2 + 5}$.

Solution   Be careful to align like terms, as shown in the following division.

Progress Check Answers

6. $3x + 4 + \dfrac{2}{x - 2}$

7. $x^2 - x + 1$

b. Assuming that the population of the United States is about 250 million, how many dollars per person is this amount? Write the given numbers in scientific notation before you do the division.

c. If you spent $500 billion at the rate of $1,000 per second, how long would it take to spend all of it?

6. Evaluate (a) 2^{-1};   (b) 10^{-2};   (c) $(\frac{1}{3})^{-3}$;   (d) 16^0.

7. Evaluate $(a^b)^c$ when $a = 5$, $b = 3$, and $c = 2$. Is this equal to $(a^c)^b$?

8. Graph the solution to the inequality $-3x + 5 > 8$.

9. The total cost of a restaurant meal and the 4 percent tax was $9.31. What was the cost of the meal before tax?

10. Solve the equation $x(r + s) = t$ for x.

Chapter 3   SUMMARY

OBJECTIVES CHECKLIST   Specific chapter objectives are summarized below along with numbered example problems from the text that should clarify the objectives. If you do not understand any objectives or do not know how to do the selected problems, then restudy the material.

3.1 Can you:

1. **Apply positive integer exponent definitions?**
   Write $(-2) \cdot (-2) \cdot (-2)$ using exponents. Identify the base and the exponent.   [Example 1c]

2. **Apply the product property of exponents?**
   Use the product property of exponents and simplify $x^3 \cdot x^5$.   [Example 3b]

3. **Apply the power properties of exponents?**
   Simplify $(x^3)^4$.   [Example 5b]

4. **Simplify exponential expressions with literal exponents?**
   Simplify $2^n \cdot 2^{n+1}$. Assume n is a positive integer.   [Example 9a]

3.2 Can you:

1. **Evaluate expressions with a zero exponent?**
   Evaluate $(2 + 3)^0$.   [Example 1b]

2. **Evaluate expressions with a negative integer exponent?**
   Evaluate 2^{-3}.   [Example 2a]

3. **Apply the quotient property of exponents?**
   Simplify $\dfrac{4^2 x^3}{4^5 x^3}$.   [Example 4c]

4. **Apply combinations of exponent definitions and properties?**
   Simplify $(5x^{-3})^{-2}$ and write the result using only positive exponents. Assume $x \neq 0$.   [Example 6b]

5. **Simplify exponential expressions with literal exponents that denote integers?**
   Simplify $\dfrac{x^{2n+2}}{x^2}$. Assume that $x \neq 0$ and n is an integer.   [Example 7a]

6. **Evaluate integer exponents on a calculator?**
   Evaluate $(1 + i)^{-n}$ if $i = 0.06$ and $n = 10$.   [Example 8]

3.3 Can you:

1. **Convert a number from standard notation to scientific notation?**
   The speed of light is about 186,000 mi/second and it takes light about 0.0000054 second to travel 1 mi. Write these numbers in scientific notation.   [Example 1]

2. **Convert a number from scientific notation to standard notation?**
   Express 6.91×10^5 in standard notation.   [Example 2a]

3. **Compute products and quotients using scientific notation?**
Perform the indicated operations and express the result in standard notation:
$(3 \times 10^{-4})(6 \times 10^{7})$. [Example 3a]

4. **Use the scientific notation capabilities of a calculator and solve applied problems?**
If one red blood cell contains about 270,000,000 hemoglobin molecules, about how many
molecules of hemoglobin are there in 25,000 red blood cells? [Example 4]

3.4 **Can you:**

1. **Identify polynomials?**
Which of the following algebraic expressions are polynomials?
a. $x^2 - 3x + 4$ **b.** $d/2$ **c.** $5/t$ [Example 1]

2. **State the degree of a polynomial and identify monomials, binomials, and trinomials?**
Give the degree of this polynomial and identify it as a monomial, binomial, or trinomial:
$4x^5 - 9x^3$. [Example 2a]

3. **Evaluate polynomials?**
A statistician used a procedure called multiple regression to discover a formula for showing
the relationship between age (x) and average systolic blood pressure (y) for the patients in a
medical research project. The formula was $y = 113.41 + 0.088x + 0.01x^2$. What is the
predicted systolic blood pressure for someone who is 19 years old? [Example 4]

4. **Add polynomials?**
Simplify $(7x^2 - 4x + 2) + (-5x^2 + 9x - 6)$. [Example 6]

5. **Subtract polynomials?**
Subtract $x^2 - 2x - 5$ from $3x^2 - 4x + 1$. [Example 7]

3.5 **Can you:**

1. **Multiply a monomial by a monomial?**
Find the product: $(-4y^5)(7y^2)$. [Example 1a]

2. **Multiply a monomial by a polynomial with more than one term?**
Find the product: $-5x(3x^2 - 4x + 2)$. [Example 2a]

3. **Multiply any two polynomials?**
Find the product of $x^2 - 7x + 5$ and $2x + 4$. [Example 4]

4. **Expand powers of a binomial?**
Expand $(3x - 5)^2$. [Example 7]

3.6 **Can you:**

1. **Use the FOIL method to multiply binomials?**
Multiply $(5x - y)(2x + 3y)$. Use the FOIL method. [Example 3]

2. **Multiply binomials that are the sum and difference of the same two terms?**
Find the product: $(x + 6)(x - 6)$. Use $(a + b)(a - b) = a^2 - b^2$. [Example 5a]

3. **Use the formulas for the square of a binomial?**
Find the product $(4x - 5)^2$. Use the square-of-a-binomial formula. [Example 7b]

4. **Multiply binomials and then solve linear equations?**
The difference in area between the rectangle and the square in the figure is 63 in.^{2}. Find the
length of the rectangle. [Example 8]

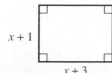

3.7 **Can you:**

1. **Divide a polynomial by a monomial?**
   Divide $5x^4 + 12x^2 + 2x$ by $2x$.   [Example 2]

2. **Divide a polynomial by a binomial?**
   Divide $2x^2 - 3x + 5$ by $x - 3$.   [Example 6]

KEY TERMS

Base (3.1)	Exponential expression (3.1)	Polynomial (3.4)
Binomial (3.4)	FOIL method (3.6)	Quotient (3.7)
Degree of a monomial (3.4)	Like terms (3.4)	Reciprocal (3.2)
Degree of a polynomial (3.4)	Monomial (3.4)	Remainder (3.7)
Dividend (3.7)	nth power of a number (3.1)	Scientific notation (3.3)
Divisor (3.7)	Numerical coefficient (3.4)	Trinomial (3.4)
Exponent (3.1)		

KEY CONCEPTS AND PROCEDURES

Section	Key Concepts or Procedures to Review
3.1	■ Product property of exponents: If m and n are positive integers and a is a real number, then $a^m \cdot a^n = a^{m+n}$.
	■ Power-to-a-power property of exponents: If m and n are positive integers and a is a real number, then $(a^m)^n = a^{mn}$.
	■ Power properties of products and quotients: If n is a positive integer and a and b are real numbers, then $$(ab)^n = a^n b^n \quad \text{and} \quad \left(\frac{a}{b}\right)^n = \frac{a^n}{b^n} \quad b \neq 0$$
3.2	■ Zero exponent: If a is a nonzero real number, then $a^0 = 1$.
	■ Negative exponent: If a is a nonzero real number and n is an integer, then $a^{-n} = \dfrac{1}{a^n}$.
	■ Quotient property of exponents: If m and n are integers and a is a nonzero real number, then $\dfrac{a^m}{a^n} = a^{m-n}$.
3.3	■ To convert to scientific notation
	1. Immediately after the first nonzero digit of the number, place an apostrophe (').
	2. Starting at the apostrophe, count the number of places to the decimal point. If you move to the right, your count is expressed as a positive number. If you move to the left, the count is negative.
	3. The apostrophe indicates the position of the decimal in the factor between 1 and 10; the count represents the exponent to be used in the factor, which is a power of 10.
	■ Scientific calculators are programmed to work with scientific notation. To enter a number in scientific notation, first enter the significant digits of the number from 1 to 10, press $\boxed{\text{EE}}$ or $\boxed{\text{EXP}}$, and finally, enter the exponent of the power of 10. For example, to enter 8.5×10^{-2}, press 8.5 $\boxed{\text{EE}}$ 2 $\boxed{+/-}$.
3.4	■ To add polynomials, combine like terms by combining their numerical coefficients.
	■ To subtract polynomials: To subtract two polynomials, change the signs on all of the terms in the polynomial that is being subtracted, and then add the polynomials.
3.5	■ To multiply two monomials, first multiply the numerical coefficients and then multiply the variable factors.
	■ To multiply polynomials: To multiply two polynomials, multiply each term of one polynomial by each term of the other polynomial, and then combine like terms.
3.6	■ To multiply two binomials of the form $(a + b)(c + d)$ using the FOIL method, add the following products.
	1. Multiply the First terms: ac.

Section	Key Concepts or Procedures to Review

2. Multiply the Outer terms: *ad*.

3. Multiply the Inner terms: *bc*.

4. Multiply the Last terms: *bd*.

■ Product of the sum and difference of two expressions: The product of the sum and difference of two terms is the square of the first term minus the square of the second term.
$$(a + b)(a - b) = a^2 - b^2$$

■ Square-of-a-binomial formulas: To square a binomial, use
$$(a + b)^2 = a^2 + 2ab + b^2$$
or $$(a - b)^2 = a^2 - 2ab + b^2$$

The square of a binomial is the square of the first term, plus or minus twice the product of the two terms, plus the square of the second term.

3.7

■ To divide a polynomial by a monomial: To divide a polynomial by a monomial, divide each term of the polynomial by the monomial. In symbols,
$$\frac{a + c}{b} = \frac{a}{b} + \frac{c}{b} \qquad b \neq 0$$

■ Long division of polynomials

1. Arrange the terms of the dividend and the divisor with descending powers. If a lower power is absent in the dividend, write 0 as its coefficient.

2. Divide the first term of the dividend by the first term of the divisor to obtain the first term of the quotient.

3. Multiply the entire divisor by the first term of the quotient and subtract this result from the dividend.

4. Use the remainder as the new dividend and repeat the above procedure until the remainder is of lower degree than the divisor.

CHAPTER 3 REVIEW EXERCISES

3.1

1. Write $a \cdot a \cdot a \cdot a$ using exponents.
2. Evaluate $3^2 - 2^3$.
3. Simplify $(4a^2)(5a^3)$.
4. Simplify $-(x^3)^2$.
5. If the length of a rectangle is $3x$ and the width is $2x^2$, find and simplify an expression for the area.

3.2

6. Evaluate $3^0 + (2 - 4)^0$.
7. Evaluate $(-2)^{-1}$.
8. Simplify $\frac{x^7}{x^9}$. Write the result using only positive exponents.
9. Simplify $y^{-2}y^{-3}$ and write the result using only positive exponents.
10. Simplify $\frac{3}{3^{-m}}$.
11. Use a calculator to evaluate $(1 + i)^{-n}$ if $i = 0.03$ and $n = 30$. Round the answer to two decimal places.
12. How many times bigger than 10^{21} is 10^{22}?

3.3

13. Convert 3,211,000 to scientific notation.
14. Convert -2.5×10^{-2} to standard notation.
15. Express the answer in standard notation: $(5 \times 10^{-3})(3 \times 10^5)$.

16. If the speed of light is 186,000 mi/second, how many *feet* does light travel in 1 second? There are 5,280 ft per mile. Give the answer in scientific notation.

3.4

17. Which of these expressions are polynomials?
a. $5 + 1/x$ **b.** 7 **c.** $3x^2 + 8x - 4$
18. What is the degree of this polynomial: $x^2 - 2x + x^3 - 5$?
19. Evaluate $4x^2 - 3x - 2$ when $x = -1$.
20. The distance y (in feet) a freely falling object falls in t seconds is given by the formula $y = 16t^2$. How far will an object fall in 10 seconds?
21. Find the sum of these two polynomials: $3x^2 - 5x + 2$ and $-x^2 - 3x - 2$.
22. Subtract $2x^2 - 3x + 1$ from $4x + 8$.
23. Simplify $(2x^2 - 2x + 2) - (3x - 4) + (x^2 + 1)$.

3.5

24. Find the product: $(m^2 - 4m - 5)2m^2$.
25. Find the product: $(6y + 1)(2y - 3)$.
26. Find the product: $(x + 1)(x^2 - x + 1)$.

3.6

27. Use the FOIL method to find the product: $(2c - 3)(c - 2)$.

28. Find the product: $(3w + 4)(3w - 4)$. Use the special product pattern $(a + b)(a - b) = a^2 - b^2$.

29. Expand this expression using the special pattern for the square of a binomial: $(3v - 2)^2$.

30. Find the radius of the larger circle if the shaded area equals 32π square units.

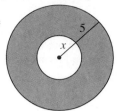

3.7

31. Find the quotient: $\dfrac{9xy^3}{36x^2y^2}$.

32. Find the quotient: $\dfrac{4x^2 - 2x}{2x}$.

33. Solve $y = \frac{3}{4}(x + 2)$ for x.

34. Divide $3x^2 + 11x + 10$ by $x + 2$.

CHAPTER 3 TEST

1. Write this expression using exponents:
$$\frac{1}{(2x - 4)(2x - 4)}.$$

2. Evaluate -3^4.

3. Find (a) the sum and (b) the product of $3y^2$ and $4y^2$.

4. Evaluate 5^{-3}.

5. Simplify $\left(\dfrac{3x}{2}\right)^{-2}$ and write the result using only positive exponents.

6. Simplify $\dfrac{2^3 x^2 y^4}{2x^3 y^{-2}}$.

7. One red blood cell contains about 270 million hemoglobin molecules. About how many hemoglobin molecules are there in 50,000 red blood cells? Give the answer in scientific notation.

8. Use scientific notation to perform the given operation; then express the answer in standard notation:
$(2.35 \times 10^{-12})(2.35 \times 10^{14})$.

9. Use scientific notation to write the number that is 100 times larger than 2.38×10^{31}.

10. Give the degree of this polynomial: $2x^2 - 3x^3 + x - 7$.

11. At a family reunion everyone hugs everyone else. The total number of hugs is given by the formula $\dfrac{n^2}{2} - \dfrac{n}{2}$, where n is the number of people at the reunion. If there are 20 people at the reunion, how many hugs are there?

12. Simplify $(2x^2 - 3x + 5) - (2x - 4) + (x^2 + x)$.

13. Find the product: $3m(2m - 4)$.

14. Find the product: $(b - 4)(3b + 1)$.

15. Find the product: $(x + 2)(x^2 + x - 1)$.

16. Find the product: $(5w - 4)(5w + 4)$.

17. Find the product: $(x - 1)^2$.

18. Find the difference in area between a square with side x and a rectangle with sides $x + 3$ and $x - 3$. Assume $x > 3$.

19. Find the quotient: $\dfrac{3h^3 + 2h^2 + h}{h}$.

20. Divide $3x^2 + 2x - 8$ by $x + 2$.

CUMULATIVE TEST 3

1. Evaluate $4 - 2(3 + 4^2)$.

2. Write an algebraic expression which represents "8 percent of the price p."

3. Which of these is equal to zero, the sum of -4 and -4 or the difference of -4 and -4?

4. Write an addition expression that is equivalent to $6 - (-5)$.

5. What is the reciprocal of -5?

6. Evaluate the formula $V = -300t + 9,000$ when $t = 9$.

7. Divide $\frac{3}{4}$ by 5.

8. Write an expression which is equivalent to $-(3c - 5)$, but without parentheses.

9. Simplify $3(x - 1) + 2(x + 1)$. Combine like terms.

10. Solve $3x - 2 = 5$.

11. Why is this equation called an identity: $3x - 3 = 3(x - 1)$?

12. You have 64 oz of flour, and you must split it into three batches, where the heaviest batch is twice the weight of the middle one and the middle one is twice the weight of the lightest one.

Translate these relationships into an equation, where x represents the weight of the lightest batch.

13. Two planes are 50 mi apart and flying toward one another. One is flying at 200 mi/hour and the other at 300 mi/hour. If they stay on this course, how many minutes will elapse before they crash?

14. Solve $-3x + 2 < 5$. Give the solution set in set-builder notation and graph it.

15. Simplify $a^5 \cdot a^3 \cdot a$.

16. Simplify $(2x^2)^3(3x)^2$.

17. Evaluate $2^{-1} + 1^{-2} - (\frac{3}{2})^0$.

18. Simplify $(3x^2 - 4x + 2) - (x^2 - x + 2)$.

19. Multiply $(2x - 1)(x^2 - 2x + 3)$.

20. The side of a square is represented by $2n + 3$.
   a. Find a trinomial expression for its area. Give the degree of the trinomial.
   b. Find a binomial expression for its perimeter. Give the degree of the binomial.

4 Factoring

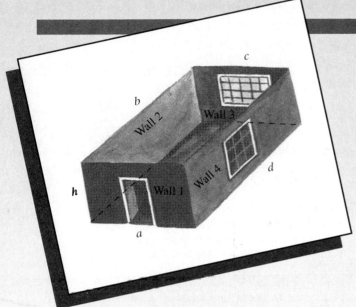

To paint the walls in a room, we may need to approximate the square footage to be covered. If subtractions for doors and windows are disregarded, then a common guide for determining this square footage is to measure (in feet) the height of a wall and multiply by the distance around the room. Show why this procedure works in the following way, using the given figure.

a. Write an expression for the total square footage of wall space by adding the square footage of each of the four walls.

b. Factor out the common factor in the expression in part **a** and interpret the result in terms of the above procedure for estimating the wall space in a room. (See Example 5 of Section 4.1.)

IN ALGEBRA we often need to transform an expression into a form that is more useful for a particular problem. A key technique for obtaining many useful forms is factoring, which is the reverse of multiplication. With our work from Chapter 3 on multiplying polynomials as a basis, this chapter first discusses a variety of procedures for factoring polynomials and then applies these factoring methods to analyze equations that extend beyond linear equations.

4.1 Factoring Polynomials with Common Factors

OBJECTIVES

1 Find the greatest common factor of a set of terms.

2 Factor out the greatest common factor from a polynomial.

3 Factor by grouping.

In Sections 3.5 and 3.6 we considered how to multiply polynomials and change a product into a sum. For instance, we saw that

$$8(x + y) = 8x + 8y.$$

$$\underbrace{8(x + y)}_{\text{product}} = \underbrace{8x + 8y}_{\text{sum}}.$$

Each of the expressions that is to be multiplied in the product is called a **factor;** each of the expressions that is to be added in the sum is called a **term.** Thus, in the above example, 8 and $x + y$ are factors, and $8x$ and $8y$ are terms. The process of factoring reverses the process of multiplication, as shown below.

To change a product into a sum, we multiply.

$$8(x + y) \xrightarrow{\text{multiplying}} 8x + 8y$$

To change a sum into a product, we factor.

$$8(x + y) \xleftarrow{\text{factoring}} 8x + 8y$$

Both processes are important, because in different situations one form may be more useful than the other.

1 The first method that should be used in factoring a sum is to attempt to find the greatest factor that is common to each of the terms. For example, consider this table in the margin. Notice that in each case we identify the *greatest* common factor that will divide into each term in the sum. Therefore, although 2 is a common factor of $8x + 8y$, a preferable common factor is 8, because 8 is the largest number that divides both terms. To identify the greatest common factor, or GCF, it is helpful to first review a procedure from arithmetic for finding the GCF of a set of numbers.

Polynomial	Greatest Common Factor
$8x + 8y$	8
$12x + 18$	6
$2x^2 + (-4x)$ or $2x^2 - 4x$	$2x$

To Find the GCF

1. Write each number as a product of prime factors.
2. The greatest common factor (GCF) is the product of all the *common* prime factors, with each prime number appearing the *fewest* number of times it appears in any one of the factorizations. If there are no common prime factors, the GCF is 1.

EXAMPLE 1 Find the greatest common factor of 120 and 36.

Solution Express 120 and 36 in prime factored form using the method shown in Section 1.1.

$$120 = 2 \cdot 2 \cdot 2 \cdot 3 \cdot 5 = 2^3 \cdot 3 \cdot 5$$
$$36 = 2 \cdot 2 \cdot 3 \cdot 3 = 2^2 \cdot 3^2$$

The prime number 2 appears in both factorizations, and the fewest number of times it appears is twice. Similarly, the prime number 3 appears in both factorizations, and the fewest number of times it appears is once. The prime number 5 is not a common factor. Thus,

$$GCF = 2^2 \cdot 3 = 12.$$

The largest number that divides both 120 and 36 is 12.

PROGRESS CHECK 1   Find the GCF of 90 and 72.   ⌐

Just as the GCF of $2^3 \cdot 3 \cdot 5$ and $2^2 \cdot 3^2$ is $2^2 \cdot 3$, the greatest common factor of x^3yz and x^2y is x^2y. With respect to variables, the GCF is the product of all common variable factors, with each variable appearing the *fewest* number of times it appears in any one of the factorizations.

EXAMPLE 2   Find the GCF of $15x^3$, $5x^4$, and $10x^2$.

Solution   The GCF of 15, 5, and 10 is 5. The GCF of x^3, x^4, and x^2 is x^2. Thus, the GCF of the three terms is $5x^2$.

PROGRESS CHECK 2   Find the GCF of $16y^5$, $20y$, and $12y^3$.   ⌐

2   After we identify the greatest common factor of all the terms in a polynomial, we may use the distributive property to factor it out. For instance, the GCF of the terms in $6x^2 + 8x$ is $2x$, and we factor it out as follows.

$$6x^2 + 8x = 2x \cdot 3x + 2x \cdot 4 \quad \textit{Think:} \; \frac{6x^2}{2x} = 3x; \; \frac{8x}{2x} = 4.$$

$$= 2x(3x + 4) \qquad \text{Distributive property.}$$

Thus, $2x(3x + 4)$ represents $6x^2 + 8x$ in factored form. Because factoring reverses multiplying, it is recommended that answers be checked through multiplication. In this case

$$2x(3x + 4) = 2x \cdot 3x + 2x \cdot 4 = 6x^2 + 8x,$$

so the result checks.

We now summarize the above procedure.

To Factor Out the GCF

1. Find the GCF of the terms in the polynomial.
2. Express each term in the polynomial as a product with the GCF as one factor.
3. Factor out the GCF using the distributive property.
4. Check the answer through multiplication.

EXAMPLE 3   Factor out the greatest common factor.

  a. $8x + 12$ **b.** $5x^4 - 15x^3 + 10x^2$ **c.** $27x^4y + 12x^2y^3$

Solution

a. The GCF of $8x$ and 12 is 4.

$$8x + 12 = 4 \cdot 2x + 4 \cdot 3 \quad \textit{Think:} \; \frac{8x}{4} = 2x; \; \frac{12}{4} = 3.$$

$$= 4(2x + 3) \qquad \text{Distributive property.}$$

The answer $4(2x + 3)$ checks, because multiplying 4 and $2x + 3$ gives $8x + 12$.

Progress Check Answers
1. 18
2. $4y$

b. The GCF of $5x^4$, $-15x^3$, and $10x^2$ is $5x^2$.

$$5x^4 - 15x^3 + 10x^2$$

$$= 5x^2(x^2) + 5x^2(-3x) + 5x^2(2) \qquad \frac{5x^4}{5x^2} = x^2; \frac{-15x^3}{5x^2} = -3x; \frac{10x^2}{5x^2} = 2$$

$$= 5x^2(x^2 - 3x + 2) \qquad\qquad \text{Distributive property.}$$

Check the answer $5x^2(x^2 - 3x + 2)$ through multiplication.

c. The GCF of $27x^4y$ and $12x^2y^3$ is $3x^2y$.

$$27x^4y + 12x^2y^3$$

$$= 3x^2y(9x^2) + 3x^2y(4y^2) \qquad \frac{27x^4y}{3x^2y} = 9x^2; \frac{12x^2y^3}{3x^2y} = 4y^2$$

$$= 3x^2y(9x^2 + 4y^2) \qquad\qquad \text{Distributive property.}$$

Check the answer $3x^2y(9x^2 + 4y^2)$ through multiplication.

PROGRESS CHECK 3   Factor out the greatest common factor.

a. $15x - 10$ **b.** $8y^5 + 4y^3 - 12y$ **c.** $20x^5y^2 + 32x^3y^4$ ⌐

The next example shows two special situations that you need to consider carefully.

EXAMPLE 4   Factor out the greatest common factor.

a. $x^3 + x$ **b.** $x(x + 5) + 2(x + 5)$

Solution

a. The GCF of x^3 and x is x.

$$x^3 + x = x \cdot x^2 + x \cdot 1 \qquad \text{Note that } x = x \cdot 1.$$

$$= x(x^2 + 1) \qquad\qquad \text{Distributive property.}$$

When one of the terms is the GCF, it is helpful to express this term as GCF $\cdot$ 1 before applying the distributive property, because the 1 is needed in the answer. Checking that multiplying your answer produces the initial polynomial is especially valuable here.

b. The greatest common factor is the binomial $x + 5$. Factoring out the GCF using the distributive property gives

$$\overset{\text{GCF}}{x(x + 5)} + 2(x + 5) = (x + 5)(x + 2).$$

The answer $(x + 5)(x + 2)$ may be converted to $x(x + 5) + 2(x + 5)$ through multiplication, so the answer checks.

PROGRESS CHECK 4   Factor out the greatest common factor.

a. $12x^2 - 4x$ **b.** $x(x - 3) + 4(x - 3)$ ⌐

The chapter-opening problem (see Example 5) shows a case in which factoring out a common factor changes an expression to a more useful form. The associated "Progress Check" exercise gives another illustration of the special situation considered in Example 4a.

EXAMPLE 5   Solve the problem in the chapter introduction on page 173.

Solution

a. The walls are rectangular, so we may use $A = \ell w$ to find the square footage for each wall. Then add to find the total square footage.

	wall 1		wall 2		wall 3		wall 4
Total square footage =	$h \cdot a$	+	$h \cdot b$	+	$h \cdot c$	+	$h \cdot d$

b. Factoring out the common factor h from the sum in part **a** gives

$$\text{total square footage} = h(\overbrace{a + b + c + d}^{\substack{\text{distance} \\ \text{around room}}}).$$

This form shows that the total square footage is simply the height of a wall multiplied by the distance around the room.

PROGRESS CHECK 5   If the profit on the sale of an item is to be p percent of the cost c, then a formula for the selling price s is

$$s = c + \frac{p}{100} \cdot c$$

Factor the expression on the right side of this equation to obtain another useful version of this formula.   ⌐

3   In Example 4b we factored out a common factor that was a binomial. This situation often occurs when we factor a polynomial with four terms by a method called **factoring by grouping.** For example,

$$ax + ay + 2x + 2y$$

can be factored by rewriting the expression as

$$(ax + ay) + (2x + 2y).$$

If we factor out the common factor in each group, we have

$$a(x + y) + 2(x + y).$$

Now $x + y$ is a common binomial factor (obtaining such a factor is the goal here). Factoring out this common factor gives

$$(x + y)(a + 2)$$

as factored form for the original polynomial. You should check here that an alternative grouping of the original expression, like

$$(ax + 2x) + (ay + 2y),$$

also achieves the same result.

EXAMPLE 6   Factor by grouping.

a. $5x^2 + 10x + 4x + 8$ **b.** $x^2 - 4x - 3x + 12$

Solution

a. $5x^2 + 10x + 4x + 8$

$$
\begin{aligned}
&= \overbrace{(5x^2 + 10x)}^{\text{common factor: } 5x} + \overbrace{(4x + 8)}^{\text{common factor: } 4} & \text{Group terms with common factors.} \\
&= 5x(x + 2) + 4(x + 2) & \text{Factor in each group.} \\
&= (x + 2)(5x + 4) & \text{Factor out the common binomial factor.}
\end{aligned}
$$

b. Note in the following solution that we factor out -3 instead of 3 from the grouping on the right to reach the goal of obtaining a common binomial factor.

$$
\begin{aligned}
& x^2 - 4x - 3x + 12 \\
&= (x^2 - 4x) + (-3x + 12) & \text{Group terms with common factor.} \\
&= x(x - 4) - 3(x - 4) & \text{Factor in each group.} \\
&= (x - 4)(x - 3) & \text{Factor out the common binomial factor.}
\end{aligned}
$$

PROGRESS CHECK 6   Factor by grouping.

a. $2x^2 + 8x + 3x + 12$ **b.** $x^2 - 5x - 2x + 10$ ⌐

Progress Check Answers

5. $s = c\left(1 + \dfrac{p}{100}\right)$

6. (a) $(2x + 3)(x + 4)$ (b) $(x - 2)(x - 5)$

EXERCISES 4.1

In Exercises 1–24, find the greatest common factor (GCF) of the given expressions.

1. 120, 42
2. 60, 84
3. 100, 10
4. 18, 90
5. 5, 7
6. 27, 16
7. 120, 144, 168
8. 216, 162, 126
9. $2^3 \cdot 3^2, 2^2 \cdot 3 \cdot 5$
10. $3 \cdot 5^2 \cdot 7, 5 \cdot 7^3 \cdot 11$
11. $12x^2, 16x, 4x^3$
12. $6x^5, 9x^4, 12x^3$
13. $10y, 20y^4, 25y^3$
14. $3y, 2y^2, y^3$
15. $2x^2y, 4xy^2$
16. $4x^5y^6, 6x^5y^4$
17. $3xy, 4yz, 7xz$
18. $2x^2, 4y^2, 6z^2$
19. $4x, x$
20. $12y, 24y^2$
21. $3(x + y), 7(x + y)$
22. $a(x - 3), b(x - 3)$
23. $(w - s)(x + y), (q - r)(x + y)$
24. $(x + 2)(x - 4), (x + 5)(x - 4)$
25. If we factor $3x$ from $6x^2$, what is the remaining factor?
26. If we factor $4x^2$ from $12x^5$, what is the remaining factor?
27. If we factor $12xy$ from $-36x^2y$, what is the remaining factor?
28. If we factor x^2y from $-6x^2y$, what is the remaining factor?
29. If we factor $3x$ from $3x^2 + 3x$, what is the remaining factor?
30. If we factor $15x^3$ from $60x^4 - 15x^3$, what is the remaining factor?

In Exercises 31–54, factor out the GCF from the given expressions. Check your answer by multiplication.

31. $2x - 6$
32. $3x + 18$
33. $4xz + 8uv$
34. $3xy - 5xz$
35. $3x^4 - 9x^3 + 6x^2$
36. $12y^3 + 8y - 10y^5$
37. $6y + 5xy - 30x^2y$
38. $12x^2 - 15x^3 + 19x^5$
39. $36x^3y^2 + 24x^2y^3$
40. $36x^4y^3 - 54x^2y^4$
41. $2xy + y$
42. $s + 8st$
43. $x^4 - x$
44. $y^3 + y^2$
45. $16y^2 - 4y$
46. $15x^3 + 5x^2$
47. $3(b + c) + d(b + c)$
48. $(a - b)x + (a - b)y$
49. $x(x + 7) - 2(x + 7)$
50. $y(y - 1) + 5(y - 1)$
51. $2x(x + 4) + (x + 4)$
52. $3n(n + 1) - (n + 1)$
53. $3x(3x - 5) - 5(3x - 5)$
54. $(2x + y)y + 2x(2x + y)$

In Exercises 55–60, you can do the given computation mentally if you first factor the GCF.

55. $13 \cdot 53 + 13 \cdot 47$
56. $29 \cdot 7 + 29 \cdot 3$
57. $75 \cdot 116 - 75 \cdot 16$
58. $23 \cdot 14 - 13 \cdot 14$
59. $21.8 \cdot 19 + 21.8 \cdot 3 - 21.8 \cdot 22$
60. $6 \cdot 34.6 - 5 \cdot 34.6$

In Exercises 61–72, factor by grouping. If necessary, first rearrange terms.

61. $3x^2 + 6x + 5x + 10$
62. $x^2 + 3x + 2x + 6$
63. $y^2 - 3y + 2y - 6$
64. $5y^2 - 2y + 10y - 4$
65. $2x^2 + 6 + 4x + 3x$
66. $5 + 14x^2 + 10x + 7x$
67. $4x^2 + 5xy + 8xy + 10y^2$
68. $x^2 + 3xy + 3xy + 9y^2$
69. $a^2 - ab + ab - b^2$
70. $c^2 + cd + cd + d^2$
71. $8a^2 - b^2 + 4ab - 2ab$
72. $6a^2 + 6b^2 + 9ab + 4ab$

In Exercises 73–78, each word problem leads to a factorable expression. Factor the expression and interpret the new expression.

73. If you unfold the paper box shown, you can see that the surface area is the sum of six areas. Write an expression for this surface area in factored form.

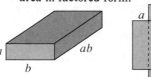
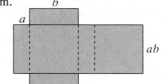

74. One formula for the volume of the wedge shown in the diagram is $V = \dfrac{\ell_1 ab}{3} + \dfrac{\ell_2 ab}{6}$. Express both terms with a common denominator and then factor to find another useful version of the formula.

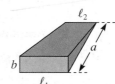

75. A shopper bought two items each at a 15 percent discount. An expression for the total amount of the discount on these purchases is $0.15p_1 + 0.15p_2$. Factor this expression and interpret the result. (p_1 and p_2 represent the two "original" prices.)

76. A shopper bought three items each at a 20 percent discount. An expression for the total amount paid for these items is $0.80p_1 + 0.80p_2 + 0.80p_3$. Factor this expression and interpret the result. (p_1, p_2, and p_3 represent the three "original" prices.)

77. If a storekeeper raises a price 20 percent and then reduces this new price by 20 percent, this transaction can be represented as $(x + 0.20x) - 0.20(x + 0.20x)$, where x represents the original price. Factor out the GCF, $x + 0.20x$, simplify, and interpret the result.

78. If a storekeeper reduces a price by 50 percent and then increases this new price by 50 percent, this transaction can be represented as $(x - 0.50x) + 0.50(x - 0.50x)$, where x represents the original price. Factor out the GCF, $x - 0.50x$, simplify, and interpret the result. Is the final price more or less than the original price?

THINK ABOUT IT

1. a. Explain each word in "greatest common factor" of two positive integers.
   b. What would be meant by the "smallest common factor" of any two positive integers?
2. Find two expressions whose greatest common factor is
   a. 7   **b.** $4x^2$
3. If the GCF for two positive integers is 1, they are said to be "relatively prime." Find two composite numbers which are relatively prime. Recall that a composite number is an integer greater than 1 that is not prime.
4. True or false? The GCF for an odd number and an even number must be 1. Justify your answer.
5. An automatic procedure for finding the GCF of two positive integers was given by Euclid (about 300 B.C.) in Book VII of his

Elements. This procedure is known as "Euclid's algorithm." It is basically a series of divisions, and in each step you divide the previous divisor by the newest remainder. The GCF is the last nonzero remainder. As an illustration, we use it to find the GCF of 12 and 20. We start by dividing 20 by 12, and we continue until the remainder is 0.

Step 1: $20 = 12 \times 1 + 8$
Step 2: $12 = 8 \times 1 + 4$
Step 3: $8 = 4 \times 2 + 0$

Conclusion: The GCF of 20 and 12 is 4.

Use the algorithm to find the GCF of 182 and 210.

REMEMBER THIS

1. Find the product of $2x$ and x.
2. Find the product of $2x$ and $3x$ and $4x$.
3. Multiply $(x + 2)(x + 4)$.
4. True or false? $(x + 3)^2 = x^2 + 9$ for all values of x.
5. Multiply $-3(x - 4)$.
6. Divide $2x^2 - 3x + 5$ by $x - 4$.
7. Simplify $\dfrac{30x^3y^5}{6x^5y}$.

8. Find the length of a rectangle whose area is $x^2 - 1$ and whose width is $x - 1$.
9. Solve the equation $\dfrac{x - 1}{2} = 7 + x$.
10. What is the result (in scientific notation) of dividing 1.08×10^{11} by 100?

4.2 Factoring $ax^2 + bx + c$ with $a = 1$

A gardener wants to use 100 ft of fencing to form a rectangular enclosure with an area of 400 ft². The dimensions for this enclosure may be found by solving the equation

$$x^2 - 50x + 400 = 0.$$

Factor the left side of this equation. (See Example 3.)

OBJECTIVES

1 Factor $ax^2 + bx + c$ with $a = 1$.

2 Identify prime polynomials.

3 Factor out the GCF and then factor trinomials.

1   In Section 3.6 we discussed the FOIL method for multiplying binomials. For instance,

$$(x + 2)(x + 4) = x^2 + 4x + 2x + 8$$
$$= x^2 + 6x + 8$$

Note that when we multiply binomials of the form $x + m$ and $x + n$, the result is a trinomial of the form $x^2 + (m + n)x + mn$. We can verify this observation: By FOIL multiplication

$$(x + m)(x + n) = x^2 + nx + mx + mn,$$

and factoring out the common factor x from the middle two terms gives

$$(x + m)(x + n) = x^2 + (m + n)x + mn.$$

This equation provides a model for factoring trinomials of the form $ax^2 + bx + c$ when $a = 1$. Examples 1–5 explain how to apply this factoring model. Note in these examples that when we factor a trinomial with integer coefficients, we will be interested only in binomial factors with integer coefficients.

EXAMPLE 1   Factor $x^2 + 7x + 10$.

Solution   Match the given trinomial to our factoring model.

$$x^2 + \underbrace{(m + n)}x + \underbrace{mn} = (x + m)(x + n)$$
$$x^2 + \quad 7x \quad + 10 = (x + ?)(x + ?)$$

The question marks represent two integers m and n such that $mn = 10$ and $m + n = 7$. Thus, the main problem is to find two integers whose product is 10 and whose sum is 7. Because the product and the sum are positive, we need consider only positive possibilities for m and n.

Possible Factor Pairs of 10	Associated Sums
$1 \cdot 10 = 10$	$1 + 10 = 11$
$2 \cdot 5 = 10$	$2 + 5 = 7$

The combination of 2 and 5 satisfies the required conditions and produces the middle term of $7x$, as shown below.

$$(x + 2)(x + 5)$$
$$5x + 2x = 7x$$

Thus, $x^2 + 7x + 10 = (x + 2)(x + 5)$. We may also write $(x + 5)(x + 2)$ for this factorization, noting that we may always reorder factors according to the commutative property of multiplication.

PROGRESS CHECK 1   Factor $x^2 + 9x + 8$.

EXAMPLE 2   Factor $y^2 - 13y + 12$.

Solution   Any variable may replace x in our factoring model. In this case the variable is y, so the factorization fits the form $(y + m)(y + n)$, where m and n are two integers whose product is 12 and whose sum is -13. Because the product is positive and the sum is negative, we need consider only negative values for m and n.

Possible Factor Pairs of 12	Associated Sums
$(-1)(-12) = 12$	$-1 + (-12) = -13$
$(-2)(-6) = 12$	$-2 + (-6) = -8$
$(-3)(-4) = 12$	$-3 + (-4) = -7$

The required integers are -1 and -12, so

$$y^2 - 13y + 12 = (y - 1)(y - 12).$$

Check the factorization through multiplication.

PROGRESS CHECK 2   Factor $t^2 - 11t + 18$. ⌐

Examples 1 and 2 illustrate that when we factor $ax^2 + bx + c$ with $a = 1$ into $(x + m)(x + n)$, m and n have the same sign if the constant term c is positive. The common sign matches the sign of b in the middle term in the trinomial. We can use this insight to solve the section-opening problem.

EXAMPLE 3   Solve the problem in the section introduction on page 179.

Solution   We need to factor $x^2 - 50x + 400$. Because the constant term is positive and the coefficient of the middle term is negative, we look for *two negative integers* whose product is 400 and whose sum is -50.

Possible Factor Pairs of 400	Associated Sums
$(-1)(-400) = 400$	$-1 + (-400) = -401$
$(-2)(-200) = 400$	$-2 + (-200) = -202$
$(-4)(-100) = 400$	$-4 + (-100) = -104$
$(-5)(-80) = 400$	$-5 + (-80) = -85$
$(-8)(-50) = 400$	$-8 + (-50) = -58$
$(-10)(-40) = 400$	$-10 + (-40) = -50$
$(-16)(-25) = 400$	$-16 + (-25) = -41$
$(-20)(-20) = 400$	$-20 + (-20) = -40$

The required integers are -10 and -40, so

$$x^2 - 50x + 400 = 0 \quad \text{is equivalent to} \quad (x - 10)(x - 40) = 0.$$

In this form the equation is easy to solve, as you will see in Section 4.6.

Note   In practice, we only need to consider possible factor pairs until a pair meets the requirements. We will continue to list all possibilities in the examples so that you can see the choices.

PROGRESS CHECK 3   A rectangular enclosure is to have an area of 450 ft^2 and a perimeter of 90 ft. The dimensions for the enclosure may be found by solving the equation $x^2 - 45x + 450 = 0$. Factor the left side of this equation. ⌐

We now consider the case when the constant term c in the trinomial to be factored is negative.

EXAMPLE 4   Factor $x^2 + 2x - 35$.

Solution   The factorization fits the form $(x + m)(x + n)$, where m and n are two integers whose product is -35 and whose sum is 2. Because the product is negative, m and n must have opposite signs.

Possible Factor Pairs of -35	Associated Sums
$1(-35) = -35$	$1 + (-35) = -34$
$-1(35) = -35$	$-1 + 35 = 34$
$5(-7) = -35$	$5 + (-7) = -2$
$-5(7) = -35$	$-5 + 7 = 2$

The integers -5 and 7 are the required pair, so

$$x^2 + 2x - 35 = (x - 5)(x + 7).$$

PROGRESS CHECK 4   Factor $x^2 - 4x - 12$.   ⌐

The next example shows how to extend our current methods to factor similar trinomials involving two variables.

EXAMPLE 5   Factor $x^2 - ax - 6a^2$.

Solution   The factorization fits the form $(x + m)(x + n)$, where m and n are two *polynomials* whose product is $-6a^2$ and whose sum is $-a$. Because of the negative sign in the product, m and n must have opposite signs.

Possible Factor Pairs of $-6a^2$	Associated Sums
$a(-6a) \ = -6a^2$	$a + (-6a) = -5a$
$-a(6a) \ = -6a^2$	$-a + 6a \ = 5a$
$2a(-3a) = -6a^2$	$2a + (-3a) = -a$
$-2a(3a) = -6a^2$	$-2a + 3a \ = a$

The required polynomials are $2a$ and $-3a$. Note below how this combination produces the middle term of $-ax$.

$$(x + 2a)(x - 3a)$$

$$-3ax + 2ax = -ax$$

Thus, $x^2 - ax - 6a^2 = (x + 2a)(x - 3a)$.

PROGRESS CHECK 5   Factor $x^2 - 3bx - 10b^2$.   ⌐

2   Many trinomials cannot be factored if we use only integer coefficients in the factorization. For example, we will try to factor $x^2 + 2x + 6$. Because the constant term is positive and the coefficient of the middle term is positive, we look for two positive integers whose product is 6 and whose sum is 2.

Possible Factor Pairs of 6	Associated Sums
$1 \cdot 6 = 6$	$1 + 6 = 7$
$2 \cdot 3 = 6$	$2 + 3 = 5$

Neither possibility has a sum of 2, so $x^2 + 2x + 6$ cannot be factored using integer coefficients. A polynomial that cannot be factored into two polynomial factors of positive degree with integer coefficients is called a **prime polynomial** over the set of integers.

EXAMPLE 6   Is $y^2 + 3y - 6$ a prime polynomial? If no, then write the polynomial in factored form.

Solution   We look for two integers whose product is -6 and whose sum is 3.

Possible Factor Pairs of -6	Associated Sums
$1(-6) = -6$	$1 + (-6) = 5$
$-1(6) = -6$	$-1 + 6 \ = -5$
$2(-3) = -6$	$2 + (-3) = -1$
$-2(3) = -6$	$-2 + 3 \ = 1$

Progress Check Answers

4. $(x - 6)(x + 2)$ 5. $(x + 2b)(x - 5b)$

By considering the possibilities above, we find that no pair of integers satisfies the requirements. Thus, $y^2 + 3y - 6$ is a prime polynomial.

PROGRESS CHECK 6   Is $y^2 - 12y - 20$ a prime polynomial? If no, then write the polynomial in factored form. ⌐

3   The directions in factoring problems often use the phrase "factor completely." This expression directs us to continue factoring until the polynomial contains no factors of two or more terms that can be factored again. These directions emphasize that we often need to apply more than one factoring procedure, as shown in the next example. You should assume that "factor" and "factor completely" are equivalent directions.

EXAMPLE 7   Factor completely $3x^3 - 6x^2 - 24x$.

Solution   First, look for common factors. The GCF is $3x$, so factor it out.

$$3x^3 - 6x^2 - 24x = 3x(x^2 - 2x - 8)$$

Then $x^2 - 2x - 8$ factors into $(x + 2)(x - 4)$ by the methods of this section, so

$$3x^3 - 6x^2 - 24x = 3x(x + 2)(x - 4).$$

Remember to include the common factor $3x$ in your answer. Forgetting this factor is a common student error. Once again, a check that multiplying your answer produces the original polynomial is valuable here.

Note   As a general strategy, it is recommended that you factor out any common factors as your first factoring procedure.

PROGRESS CHECK 7   Factor completely $5y^4 - 15y^3 - 50y^2$. ⌐

Progress Check Answers
6. It is prime.   7. $5y^2(y - 5)(y + 2)$

EXERCISES 4.2

In Exercises 1–12, fill in the blank to show a correct factorization. Check your answer by multiplication.

1. $x^2 + 12x + 35 = (x + 7)(x + \underline{\quad})$
2. $s^2 + 7s + 12 = (s + \underline{\quad})(s + 3)$
3. $x^2 - 12x + 36 = (x - 6)(x - \underline{\quad})$
4. $n^2 - 2n + 1 = (n - \underline{\quad})(n - 1)$
5. $y^2 + y - 56 = (y - 7)(y \underline{\quad} 8)$
6. $x^2 + 4x - 5 = (x \underline{\quad} 1)(x + 5)$
7. $x^2 - 5x - 24 = (x \underline{\quad} 8)(x \underline{\quad} 3)$
8. $r^2 - 6r - 16 = (r \underline{\quad} 2)(r \underline{\quad} 8)$
9. $t^2 + 2t - 80 = (t \underline{\quad} \underline{\quad})(t \underline{\quad} 8)$
10. $x^2 + 8x - 20 = (x \underline{\quad} 2)(x \underline{\quad} \underline{\quad})$
11. $t^2 - 3t - 54 = (t \underline{\quad} 6)(t \underline{\quad} \underline{\quad})$
12. $y^2 - 10y - 24 = (y \underline{\quad} \underline{\quad})(y \underline{\quad} 2)$

In Exercises 13–42, factor the given expression by using the model $x^2 + (m + n)x + mn = (x + m)(x + n)$.

13. $x^2 + 3x + 2$
14. $x^2 + 5x + 6$
15. $y^2 + 7y + 12$
16. $y^2 + 8y + 15$
17. $t^2 + 6t + 5$
18. $t^2 + 12t + 11$
19. $x^2 - 9x + 20$
20. $x^2 - 12x + 20$
21. $y^2 - 8y + 12$
22. $y^2 - 13y + 12$
23. $t^2 - 9t + 18$
24. $t^2 - 11t + 18$
25. $x^2 + 3x - 40$
26. $x^2 + 2x - 8$
27. $y^2 + y - 12$
28. $y^2 + 4y - 21$
29. $t^2 + 5t - 14$
30. $t^2 + 6t - 40$
31. $x^2 - 7x - 8$
32. $x^2 - 7x - 18$
33. $y^2 - 8y - 20$
34. $y^2 - y - 20$
35. $t^2 - 6t - 72$
36. $t^2 - 5t - 50$
37. $x^2 + 3bx + 2b^2$
38. $x^2 + 5ax + 6a^2$
39. $y^2 - 9cy + 20c^2$
40. $t^2 - 11at + 18a^2$
41. $z^2 - 2dz - 15d^2$
42. $x^2 + 3ax - 54a^2$

In Exercises 43–54, factor completely. (*Hint:* Factor out the GCF as a first step.)

43. $2x^2 + 6x + 4$
44. $3x^2 + 15x + 18$
45. $4y^3 + 28y^2 + 48y$
46. $2y^3 + 16y^2 + 30y$
47. $3x^4 - 27x^3 + 60x^2$
48. $4x^4 - 48x^3 + 80x^2$
49. $2t^5 - 10t^4 - 28t^3$
50. $y^3 - y^2 - 20y$
51. $ax^2 - 3ax + 2a$
52. $b^2x^2 - 2b^2x - 15b^2$
53. $x^2(x + 3) + 6x(x + 3) + 9(x + 3)$
54. $y^2(y - 5) - 10y(y - 5) + 25(y - 5)$

In Exercises 55–60, fill in the blank in such a way that the resulting trinomial can be factored. Then factor it. There is more than one correct answer.

55. $x^2 + 5x + \underline{\hspace{1cm}}$
56. $x^2 + 7x + \underline{\hspace{1cm}}$
57. $y^2 + 2y - \underline{\hspace{1cm}}$
58. $y^2 + 3y - \underline{\hspace{1cm}}$
59. $a^2 - 3a + \underline{\hspace{1cm}}$
60. $a^2 - 4a + \underline{\hspace{1cm}}$

In Exercises 61–72, show, by testing all possibilities, that these trinomials cannot be factored and are therefore prime polynomials over the integers.

61. $x^2 + x + 1$ **62.** $x^2 + x + 2$
63. $x^2 + 2x + 4$ **64.** $x^2 + 8x + 6$
65. $y^2 - 2y + 7$ **66.** $y^2 - 2y + 5$
67. $t^2 + 3t - 6$ **68.** $t^2 + 3t - 3$
69. $s^2 - 4s - 8$ **70.** $s^2 - 4s - 16$
71. $y^2 - 10y - 12$ **72.** $y^2 - 4y - 18$

In Exercises 73–78, decide if the expression is a prime polynomial. If it is not, then factor it.

73. $x^2 + 3x + 4$ **74.** $x^2 + 3x - 4$
75. $y^2 - y - 6$ **76.** $y^2 - y + 6$
77. $x^2 + 12x - 24$
78. $x^2 - 20x + 51$

In Exercises 79–84, we interpret factoring geometrically. The trinomial to be factored represents the total area of a rectangle. The factors are the length and the width. In each exercise we have started the procedure.

a. Find the missing part of the expressions for the length and width.
b. Compute the areas of the remaining cells.
c. Then show that the sum of all the areas gives the original trinomial.

79. Factor $x^2 + 5x + 6$.

	x	$+$	?
x	x^2		?
$+$			
?	?		6

80. Factor $x^2 + 11x + 24$.

	x	$+$	?
x	x^2		?
$+$			
?	?		24

81. Factor $d^2 + 15d + 56$.

	d	$+$	?
d	d^2		?
$+$			
?	?		56

82. Factor $r^2 + 25r + 156$.

	r	$+$	?
r	r^2		?
$+$			
?	?		156

83. Factor $x^2 + 9ax + 8a^2$.

	x	$+$	?
x	x^2		?
$+$			
?	?		$8a^2$

84. Factor $x^2 + 11bx + 18b^2$.

	x	$+$	?
x	x^2		?
$+$			
?	?		$18b^2$

In Exercises 85–90, an equation is given which can be used to solve a word problem. One step in the solution is to factor the left side.

85. The perimeter of the rectangle shown is 50 ft. Its area is 156 ft². The dimensions can be found by solving the equation $x^2 - 25x + 156 = 0$. Factor the left side of this equation.

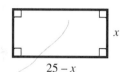

86. The area of the isosceles triangle shown is 12 in.². The dimensions can be found by solving $x^2 - 11x + 24 = 0$. Factor the left side of this equation.

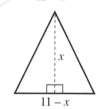

87. A ball is thrown down from a roof 160 ft high with an initial velocity of 48 ft/second. To find the time it takes to hit the ground, you solve $16t^2 + 48t - 160 = 0$, where t is the time in seconds. Factor the left side of this equation.

88. A thin sheet of metal is 12 in.². The corners will be cut out as shown and the sides folded up to make an open box. To find out what size corners to cut out to get the box with maximum volume, you can solve $12x^2 - 96x + 144 = 0$. Factor the left side of this equation.

89. The sum of the odd integers from 1 to a certain number m equals 36. This number m can be found by solving the equation $m^2 + 2m - 4(36) + 1 = 0$. Simplify the left-hand side of the equation and then factor it.

90. The sum of all the integers from 1 to a certain number n equals 210. This number n can be found by solving the equation $n^2 + n - 2(210) = 0$. Simplify the left side of this equation and then factor it.

THINK ABOUT IT

1. Why is $x^2 - 4x + 6$ called a trinomial?
2. **a.** If $x^2 + bx + 7$ is factorable, what are the only two possible values of b?
   b. If $x^2 + bx + c$ is factorable and c is a prime number, what are the only two possible values of b?
3. True or false? If the sign of the constant term of a factorable trinomial is positive, then both binomial factors must have plus signs before their constant terms. Explain.

4. A trinomial of the form $ax^2 + bx + c$ is factorable over the integers if and only if $b^2 - 4ac$ is a perfect square. Use this rule to show that $x^2 + 10x + 5$ cannot be factored but that $x^2 + 10x + 16$ can be factored.
5. If you know one factor of a trinomial, you can find the other factor by division. Find the second factor of $x^2 - 2x - 143$ given that one factor is $x + 11$. Then complete the identity $x^2 - 2x - 143 = (x + 11)(\qquad)$.

REMEMBER THIS

1. Multiply $(3x + 2)(x + 3)$.
2. Multiply $3x(6x + 1)(x - 2)$.
3. Find the GCF of $15x^3$, $5x^5$, and $10x^2$.
4. Factor out the greatest common factor: $27x^3y + 15x^2y^2$.
5. Factor by grouping: $5x^2 + 10x + 3x + 6$.
6. Solve the inequality: $-2 < 6 - 3x < 2$.
7. Part of a $5,000 bank deposit was invested at a 6.5 percent annual interest rate and the rest at 8 percent. If the total annual interest was $377.50, how much was invested at 8 percent?

8. Solve the equation: $4x + 7 = 2x - 3$.
9. What is the sum of the additive inverse of 5 and the multiplicative inverse of $\frac{1}{5}$?
10. Evaluate $\dfrac{(x - 1)(x - 3)}{(1 - x)(3 - x)}$ if $x = -2$.

4.3   Factoring $ax^2 + bx + c$ with $a \neq 1$

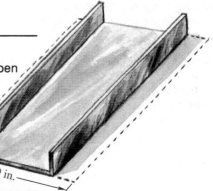

A long strip of galvanized sheet metal 9 in. wide is to be shaped into an open gutter by bending up the edges to form a gutter of cross-sectional area 9 in.². The two possible heights (x) of this gutter may be found by solving the equation

$$2x^2 - 9x + 9 = 0.$$

Factor the left side of this equation. (See Example 3.)

OBJECTIVES

1 Factor $ax^2 + bx + c$ with $a \neq 1$ by reversing FOIL.

2 Factor $ax^2 + bx + c$ with $a \neq 1$ by the *ac* method.

When $a \neq 1$, the trinomial $ax^2 + bx + c$ is harder to factor. The reason for this difficulty can be seen in the following product.

$$(px + m)(qx + n) = (pq)x^2 + (pn + qm)x + mn$$

Note that the factors for the last term, m and n, do not now necessarily add up to the middle coefficient, which has a more complex form. In this section we will show two methods for factoring these more complicated trinomials. In different situations one method may be easier than the other.

1 **Method 1: Reversing FOIL**

In factoring by this method, the key observation is that the middle term in the trinomial to be factored is the sum of the outer product and the inner product from the FOIL multiplication. Also, note in the examples that when we factor first terms with positive coefficients like $3x^2$, we prefer to use $3x$ and x, although $-3x$ and $-x$ are possible. Choosing positive coefficients permits us to carry over the sign restrictions for m and n discussed in the previous section.

EXAMPLE 1   Factor $3x^2 + 19x + 6$.

Solution   The factorization fits the form $(px + m)(qx + n)$. By reversing FOIL, the first term $3x^2$ is the result of multiplying the first terms in the FOIL method. Thus, a possible factorization begins

$$3x^2 + 19x + 6 = (3x + ?)(x + ?).$$

The last term is the result of multiplying the last terms in the FOIL method. Since $6 = (6)(1)$ and $(3)(2)$, we have four possibilities. We eliminate negative possibilities for m and n because the middle term, $19x$, has a positive coefficient.

$$3x^2 + 19x + 6 \overset{?}{=} \begin{array}{l} (3x + 6)(x + 1) \\ (3x + 1)(x + 6) \\ (3x + 3)(x + 2) \\ (3x + 2)(x + 3) \end{array}$$

The middle term is the sum of the inner and outer products.

$$(3x + 6)(x + 1) \qquad (3x + 1)(x + 6)$$
$$3x + 6x = 9x \qquad 18x + x = 19x$$

$$(3x + 3)(x + 2) \qquad (3x + 2)(x + 3)$$
$$6x + 3x = 9x \qquad 9x + 2x = 11x$$

Thus, $3x^2 + 19x + 6 = (3x + 1)(x + 6)$ is the correct choice.

PROGRESS CHECK 1   Factor $5x^2 + 14x + 8$.   ⌐

In Example 1 we did not really need to test the products $(3x + 6)(x + 1)$ and $(3x + 3)(x + 2)$ as possible factorizations of $3x^2 + 19x + 6$. The reason is that 3 is a common factor of $3x + 6$ in the first case and $3x + 3$ in the second, so

$$(3x + 6)(x + 1) = 3(x + 2)(x + 1)$$
$$(3x + 3)(x + 2) = 3(x + 1)(x + 2).$$

These factorizations imply that 3 is a common factor of the original trinomial, which is not the case. Therefore, you should only test factorizations that are possible, according to the following guideline.

Requirement for Binomial Factors

If the terms of a trinomial have no common integer factors (except 1 and -1), then the same is true for both of its binomial factors.

EXAMPLE 2   Factor $6x^2 - 7x + 2$.

Solution   The first term, $6x^2$, is the result of multiplying the first terms in the FOIL method. Thus,

$$6x^2 - 7x + 2 \overset{?}{=} \begin{array}{c} (6x + ?)(x + ?) \\ \text{or} \\ (3x + ?)(2x + ?). \end{array}$$

The last term, 2, is the result of multiplying the last term in the FOIL method. There is only one possibility for 2, $(-1)(-2)$, since we eliminate $(1)(2)$ because the coefficient of the middle integer is negative.

$$6x^2 - 7x + 2 \overset{?}{=} \begin{array}{l} (6x - 1)(x - 2) \\ (6x - 2)(x - 1) \\ (3x - 1)(2x - 2) \\ (3x - 2)(2x - 1) \end{array}$$

By the requirement for binomial factors, eliminate the products $(6x - 2)(x - 1)$ and $(3x - 1)(2x - 2)$, because they have binomial factors with a common factor of 2 (but 2 is not a common factor of $6x^2 - 7x + 2$). Test the two remaining possibilities.

$$\begin{array}{cc} (6x - 1)(x - 2) & (3x - 2)(2x - 1) \\ -12x - x = -13x & -3x - 4x = -7x \end{array}$$

Thus, $6x^2 - 7x + 2 = (3x - 2)(2x - 1)$.

PROGRESS CHECK 2   Factor $4x^2 - 9x + 2$.   ⌐

EXAMPLE 3   Solve the problem in the section introduction on page 185.

Solution   We need to factor $2x^2 - 9x + 9$. The first term is $2x^2$, so a possible factorization begins $(2x + ?)(x + ?)$. Because the constant term is positive and the coefficient of the middle term is negative, we look for negative factors of 9, so we use $(-1)(-9)$ or $(-3)(-3)$.

$$2x^2 - 9x + 9 \overset{?}{=} \begin{array}{l} (2x - 1)(x - 9) \\ (2x - 9)(x - 1) \\ (2x - 3)(x - 3) \end{array}$$

The middle term, $-9x$, is the sum of the inner and the outer products.

$$\begin{array}{c} (2x - 3)(x - 3) \\ -6x - 3x = -9x \end{array}$$

Therefore, the equation $2x^2 - 9x + 9 = 0$ may be written as

$$(2x - 3)(x - 3) = 0.$$

PROGRESS CHECK 3   If the sheet metal is 25 cm wide and the desired cross-sectional area is 75 cm², then the two possible heights (x) of a gutter designed as in Example 3 may be found by solving the equation

$$2x^2 - 25x + 75 = 0.$$

Factor the left side of this equation.   ⌐

EXAMPLE 4   Factor completely $18x^3 - 33x^2 - 6x$.

Solution   First, look for common factors. The GCF is $3x$, so factor it out.

$$18x^2 - 33x^2 - 6x = 3x(6x^2 - 11x - 2)$$

Now, suppose we try $(6x - 1)(x + 2)$ as a factorization for $6x^2 - 11x - 2$.

$$\begin{array}{c} (6x - 1)(x + 2) \\ 12x - x = 11x \end{array}$$

The desired middle term, $-11x$, is the opposite of the above result, so we adjust here by merely switching signs on the factors of 2, from $(-1)(2)$ to $(1)(-2)$.

$$(6x + 1)(x - 2)$$
$$-12x + x = -11x$$

Thus, $6x^2 - 11x - 2 = (6x + 1)(x - 2)$, so the complete factorization is

$$18x^3 - 33x^2 - 6x = 3x(6x + 1)(x - 2).$$

PROGRESS CHECK 4   Factor completely $18x^4 - 50x^3 - 12x^2$.   ⌐

To factor $ax^2 + bx + c$ when a is negative, we recommend that you first factor out -1; then proceed as in our previous examples.

EXAMPLE 5   Factor $-2x^2 + x + 1$.

Solution   First, factor out -1 to make the coefficient of the squared term positive.

$$-2x^2 + x + 1 = -1(2x^2 - x - 1)$$

Now, $2x^2 - x - 1 = (2x + 1)(x - 1)$, so the complete factorization is

$$-2x^2 + x + 1 = -1(2x + 1)(x - 1)   \quad \text{or} \quad   -(2x + 1)(x - 1).$$

PROGRESS CHECK 5   Factor $-5x^2 + 7x + 6$.   ⌐

2   **Method 2: *ac* method**

The *ac* method for factoring $ax^2 + bx + c$ may involve less trial and error than our previous approach, particularly if you cannot *mentally* eliminate many of the possibilities. Example 6 explains in detail the steps for this method. Note that the *ac* method relies on factoring by grouping, which was discussed in Section 4.1.

EXAMPLE 6   Factor $3x^2 + 11x + 6$ by the *ac* method.

Solution   Follow the steps below.

Step 1   **Find two integers whose product is *ac* and whose sum is *b*.**
In the trinomial $3x^2 + 11x + 6$, we have $a = 3$, $b = 11$, and $c = 6$. Thus, $ac = 3 \cdot 6 = 18$. We look for two integers whose product is 18 and whose sum is 11. Because the product and the sum are both positive, we need only consider positive factor pairs of 18.

Possible Factor Pairs of 18	Associated Sums
$1 \cdot 18 = 18$	$1 + 18 = 19$
$2 \cdot 9 = 18$	$2 + 9 = 11$
$3 \cdot 6 = 18$	$3 + 6 = 9$

The required integers are 2 and 9.

Step 2   **Replace *b* by the sum of the two integers from step 1 and then distribute *x*.**

$$3x^2 + 11x + 6$$
$$= 3x^2 + (9 + 2)x + 6 \quad \text{Replace 11 by 9 + 2.}$$
$$= 3x^2 + 9x + 2x + 6 \quad \text{Distributive property.}$$

Step 3   **Factor by grouping.**

$$3x^2 + 9x + 2x + 6$$
$$= (3x^2 + 9x) + (2x + 6) \quad \text{Group terms with common factors.}$$
$$= 3x(x + 3) + 2(x + 3) \quad \text{Factor in each group.}$$
$$= (x + 3)(3x + 2) \quad \text{Factor out the common binomial factor.}$$

Thus, $3x^2 + 11x + 6 = (x + 3)(3x + 2)$. Check this result through multiplication.

Note   We may also replace 11 by $2 + 9$ and factor by grouping as follows.

$$3x^2 + 11x + 6 = 3x^2 + 2x + 9x + 6$$
$$= x(3x + 2) + 3(3x + 2)$$
$$= (3x + 2)(x + 3)$$

The answers differ only in the order of the factors, and both answers are correct.

PROGRESS CHECK 6   Factor $5x^2 + 9x + 4$ by the *ac* method.   ⌟

If we reconsider the product model given at the beginning of this section, we can see the basis for the *ac* method.

$$\underset{\displaystyle (px \;+\; m)(qx \;+\; n) \;=\;}{}\; \overset{ax^2}{\underset{\downarrow}{(pq)x^2}} \;+\; \overset{bx}{\underset{\downarrow}{(pn + qm)x}} \;+\; \overset{c}{\underset{\downarrow}{mn}}$$

$$ac = (pq)(mn) = (pn)(qm)$$

This diagram shows that the trial-and-error part involves finding two integers (in our notation they are pn and qm) whose product is ac and whose sum is b. If such a pair does not exist, then the trinomial is a prime polynomial.

In the next example reversing FOIL is difficult because there are many possible factor pairs for both $12x^2$ and -18. In such cases the systematic approach of the *ac* method is especially helpful.

EXAMPLE 7   Factor $12x^2 + 19x - 18$.

Solution   Follow the steps given in Example 6.

Step 1   For this trinomial, observe that $a = 12$, $b = 19$, and $c = -18$. Therefore, $ac = 12(-18) = -216$. We look for two integers whose product is -216 and whose sum is 19. With a little trial and error, we find that

$$(-8)(27) = -216 \qquad \text{and} \qquad -8 + 27 = 19,$$

so the required integers are -8 and 27.

Step 2   Replace 19 by $-8 + 27$ and then distribute x.

$$12x^2 + 19x - 18$$
$$= 12x^2 + (-8 + 27)x - 18 \qquad \text{Replace 19 by } -8 + 27.$$
$$= 12x^2 - 8x + 27x - 18 \qquad \text{Distributive property.}$$

Step 3   Factor by grouping.

$$(12x^2 - 8x) + (27x - 18)$$
$$= 4x(3x - 2) + 9(3x - 2) \qquad \text{Factor in each group.}$$
$$= (3x - 2)(4x + 9) \qquad \text{Factor out the common binomial factor.}$$

Thus, $12x^2 + 19x - 18 = (3x - 2)(4x + 9)$.

PROGRESS CHECK 7   Factor $12x^2 - x - 20$.   ⌟

Progress Check Answers
6. $(5x + 4)(x + 1)$
7. $(3x - 4)(4x + 5)$

EXERCISES 4.3

In Exercises 1–6, choose the correct factorization.

1. $2x^2 + 7x + 5 =$
   a. $(2x + 5)(x + 1)$ b. $(2x + 1)(x + 5)$

2. $3x^2 - 8x + 5 =$
   a. $(x - 5)(3x - 1)$ b. $(x - 1)(3x - 5)$

3. $3x^2 - 13x + 14 =$
   a. $(3x - 14)(x - 1)$ b. $(3x - 1)(x - 14)$
   c. $(3x - 7)(x - 2)$ d. $(3x - 2)(x - 7)$
4. $4x^2 + 12x + 5 =$
   a. $(4x + 5)(x + 1)$ b. $(4x + 1)(x + 5)$
   c. $(2x + 5)(2x + 1)$
5. $5x^2 - 9x - 2 =$
   a. $(5x - 1)(x + 2)$ b. $(5x + 1)(x - 2)$
   c. $(5x + 2)(x - 1)$ d. $(5x - 2)(x + 1)$
6. $6x^2 - 5x - 6 =$
   a. $(3x - 2)(2x + 3)$ b. $(3x + 2)(2x - 3)$
   c. $(6x - 1)(x + 6)$ d. $(x - 6)(6x + 1)$

In Exercises 7–12, complete the factorization of the given trinomial. Check your answer by multiplication.

7. $2x^2 + 5x + 3 = (2x + \underline{\quad})(x + \underline{\quad})$
8. $3x^2 - 16x + 5 = (3x - \underline{\quad})(x - \underline{\quad})$
9. $3x^2 + 16x - 12 = (3x \underline{\quad} 2)(x \underline{\quad} 6)$
10. $16x^2 + 6x - 1 = (\underline{\quad} x + 1)(\underline{\quad} x - 1)$
11. $12x^2 - 23x - 24 = (3x \underline{\quad} \underline{\quad})(4x \underline{\quad} \underline{\quad})$

12. $6x^2 - x - 15 = (\underline{\quad} x - 5)(\underline{\quad} x + 3)$

13. Match the correct form of the factors to the given trinomial. Assume all variables represent positive integers. There may be more than one correct answer.

   1. $ax^2 + bx + c$ a. $(px + n)(qx - m)$
   2. $ax^2 - bx + c$ b. $(px - n)(qx - m)$
   3. $ax^2 + bx - c$ c. $(px + n)(qx + m)$
   4. $ax^2 - bx - c$ d. $(px - n)(qx + m)$

14. Judging by the arrangement of plus and minus signs, which of these forms cannot be correct choices for a factorization of $24x^2 - 172x - 60$?
   a. $(12x - 4)(2x - 15)$ b. $(12x + 4)(2x + 15)$
   c. $(12x + 4)(2x - 15)$

In Exercises 15–68, factor the trinomial. Check your answer by multiplication.

15. $2x^2 + 5x + 2$ 16. $2x^2 + 3x + 1$
17. $3x^2 + 4x + 1$ 18. $3x^2 + 5x + 2$
19. $4x^2 + 9x + 2$ 20. $5x^2 + 11x + 2$
21. $2x^2 + 11x + 14$ 22. $3x^2 + 10x + 7$
23. $4x^2 + 11x + 6$ 24. $4x^2 + 13x + 10$
25. $20x^2 + 49x + 30$
26. $10x^2 + 21x + 9$
27. $5x^2 - 6x + 1$
28. $5x^2 - 16x + 3$
29. $5x^2 - 16x + 12$
30. $4x^2 - 12x + 5$
31. $12x^2 - 7x + 1$
32. $9x^2 - 24x + 7$
33. $14x^2 - 25x + 6$
34. $8x^2 - 33x + 4$
35. $18x^2 - 25x + 8$
36. $15x^2 - 49x + 24$
37. $10x^2 - 33x + 20$
38. $9x^2 - 18x + 8$
39. $3x^2 + 2x - 5$
40. $3x^2 + 14x - 5$

41. $5x^2 + 54x - 11$
42. $7x^2 + 6x - 1$
43. $5x^2 + 7x - 6$
44. $4x^2 + 12x - 7$
45. $8x^2 + 22x - 13$
46. $11x^2 + x - 10$
47. $4x^2 + 4x - 35$
48. $15x^2 + 19x - 10$
49. $16x^2 + 6x - 27$
50. $9x^2 + 6x - 8$
51. $2x^2 - x - 3$
52. $2x^2 - 9x - 5$
53. $3x^2 - 8x - 3$
54. $3x^2 - 4x - 7$
55. $17x^2 - 32x - 4$
56. $6x^2 - 37x - 13$
57. $12x^2 - 16x - 11$
58. $5x^2 - 9x - 18$
59. $6x^2 - 39x - 21$
60. $2x^2 - 5x - 12$
61. $36x^2 - 34x - 30$
62. $36x^2 - 37x - 70$
63. $2x^2 + 3xy - 14y^2$
64. $25x^2 + 55xy - 12y^2$
65. $12a^2 - 35ab + 25b^2$
66. $8a^2 + 15ab - 2b^2$
67. $24c^2 - 50cd - 9d^2$
68. $6c^2 + 13cd - 5d^2$

In Exercises 69–88, factor the given trinomial by first factoring out any common factors. If the coefficient of x^2 is negative, take out a negative factor. Check your answer by multiplication.

69. $-x^2 + x + 2$
70. $-x^2 - 2x + 15$
71. $-y^2 - 10y - 24$
72. $-y^2 + 15y - 54$
73. $3c^2 + 3c - 6$
74. $5d^2 + 25d + 30$
75. $12x^3 - 2x^2 - 30x$
76. $60x^3 + 28x^2 - 16x$
77. $-28x^2 - 26x + 24$
78. $-9x^2 - 24x - 15$
79. $-6x^4 - 12x^3 - 6x^2$
80. $-30x^4 - 35x^3 + 15x^2$
81. $16x^2y - 10xy - 6y$
82. $24a^2b^2 + 10ab^2 - 4b^2$
83. $6a^2b - 5ab^2 + b^3$
84. $12m^3 + 17m^2n + 6mn^2$
85. $6a^3bc - 5a^2b^2c - 4ab^3c$
86. $20x^3y^2 + 9x^2y^3 - 20xy^4$
87. $2x^2(x + 1) + 4x(x + 1) + 2(x + 1)$
88. $3y^2(y - 1) - 6y(y - 1) + 3(y - 1)$

Exercises 89–94 involve just the first steps of factoring trinomials of the form $ax^2 + bx + c$ by the ac method. For each exercise:
a. Write down b and ac.
b. Find two integers whose sum is b and whose product is ac.
89. $6x^2 + 11x + 3$
90. $24x^2 + 34x + 5$

EX/

a.

Sol

a.

91. $12x^2 - 31x + 20$
92. $6x^2 - 17x + 12$
93. $6x^2 + 7x - 24$
94. $20x^2 - x - 30$

In Exercises 95–106, factor the trinomial using the *ac* method. If possible, first factor out the GCF. Check your answer by multiplication.

95. $2x^2 + 11x + 9$
96. $3x^2 + 16x + 20$
97. $10x^2 + x - 2$
98. $16x^2 - 8x - 3$
99. $8x^2 + 26x + 15$
100. $10x^2 + 19x + 6$
101. $9x^2 - 18x + 8$
102. $16x^2 - 32x + 15$
103. $30y^2 + 11y - 30$
104. $18y^2 - 3y - 10$
105. $12x^2 - 18x - 12$
106. $24x^2 - 28x - 12$

b.

In Exercises 107–112, an equation is given which can be used to solve a word problem. One step in the solution is to factor the left side.

107. The perimeter of the rectangle shown is 8 in. and its area is $3\frac{3}{4}$ in.^{2}. The dimensions can be found by solving the equation $4x^2 - 16x + 15 = 0$. Factor the left side of this equation.

c.

$4 - x$

108. The perimeter of the right triangle shown is 18 in. and its area is 13.5 in.^{2}. The dimensions can be found by solving the equation $2x^2 + 3x - 54 = 0$. Factor the left side of this equation.

PI

a.

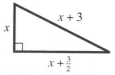

$x + 3$

x

$x + \frac{3}{2}$

109. Two posts are 10 ft apart, as shown. One is 3 ft high and one is 5 ft high. A stake will be put into the ground between them and cables will be attached to it from the top of each post. How far from the shorter post should the stake be placed so that the least amount of cable can be used? This can be found by solving $4x^2 + 45x - 225 = 0$. Factor the left side of this equation.

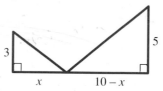

3 5

x $10 - x$

110. There are two numbers for which it is true that the number plus twice its square equals 21. One of these numbers is 3. The other can be found by solving $2x^2 + x - 21 = 0$. Factor the left side of this equation.

111. A 5-ft length of wire will be bent as shown into three sides of a rectangle. There are two ways to do this so that the area of the rectangle is 3 ft^2. These lengths can be found by solving $2x^2 - 5x + 3 = 0$. Factor the left side of this equation.

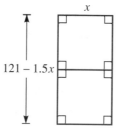

x

$5 - 2x$

112. Using 242 ft of fence to enclose two corrals as shown, an area of 1,820 ft^2 can be enclosed in two different ways. What are the two sets of dimensions? They can be found by solving $3x^2 - 242x + 3,640 = 0$. Factor the left side of this equation.

x

$121 - 1.5x$

2

Ar
sq

N
th
R

THINK ABOUT IT

1. Why is the *ac* method for factoring trinomials called "*ac*"?
2. When you rewrite a sum like $3x^2 + 12x$ as the product $3x(x + 4)$, this step is called *factoring*. Why? What does this step have to do with the noun *factor*?
3. The trinomial $ax^2 + bx + c$ will be factorable over the integers if and only if $b^2 - 4ac$ is a perfect square. Use this test to show that $2x^2 + 3x + 1$ is factorable but $2x^2 + 3x - 1$ is not.
4. One way to factor $-2x^2 + x + 1$ is to first factor out -1.

Another way is to rearrange the trinomial as $1 + x - 2x^2$ and to then factor it in the form $(? + x)(? - x)$. Try it both ways and compare your results.
5. By multiplying each expression, show that they are all equal.
  a. $-(2x^2 - x - 1)$ **b.** $-(2x + 1)(x - 1)$
  c. $(-2x - 1)(x - 1)$ **d.** $(2x + 1)(-x + 1)$
  e. $(2x + 1)(1 - x)$

REMEMBER THIS

K

1. Find $(3x - 2)^2$ by multiplying $(3x - 2)(3x - 2)$.
2. Find $(x + 1)^3$ by multiplying $(x + 1)(x + 1)(x + 1)$.

3. Multiply $(3x - 5)(3x + 5)$.
4. True or false? $(x + 1)^3 = x^3 + 1$ for all values of x.

5.
6.
7.
8.

EXAMPLE 2   Factor each difference of squares.

a. $x^2 - 25$ **b.** $9y^2 - 64x^2$ **c.** $4x^6 - t^4$

Solution

a. Substitute x for a and 5 for b in the given model.

$$a^2 - b^2 = (a + b)(a - b)$$
$$x^2 - 25 = x^2 - 5^2 = (x + 5)(x - 5)$$

b. Because $9y^2 = (3y)^2$ and $64x^2 = (8x)^2$, replace a by $3y$ and b by $8x$ in the factoring model.

$$a^2 - b^2 = (a + b)(a - b)$$
$$9y^2 - 64x^2 = (3y)^2 - (8x)^2 = (3y + 8x)(3y - 8x)$$

c. Note that $4x^6 - t^4 = (2x^3)^2 - (t^2)^2$, and follow the factoring formula for a difference of squares.

$$a^2 - b^2 = (a + b)(a - b)$$
$$4x^6 - t^4 = (2x^3)^2 - (t^2)^2 = (2x^3 + t^2)(2x^3 - t^2)$$

PROGRESS CHECK 2   Factor each difference of squares.

a. $x^2 - 100$ **b.** $25t^2 - 36c^2$ **c.** $x^8 - 9y^6$ ⌐

In the next two examples we will need to factor more than once in order to factor completely. Remember that, if possible, you should factor out any common factors as your first factoring procedure.

EXAMPLE 3   Factor completely.   **a.** $2x^3 - 2x$ **b.** $x^4 - 16$

Solution

a. First, look for common factors. The GCF is $2x$, so factor it out.

$$2x^3 - 2x = 2x(x^2 - 1)$$

Now, factor $x^2 - 1$ using the difference-of-squares model.

$$2x^3 - 2x = 2x(x + 1)(x - 1)$$

b. There are no common factors, so factor $x^4 - 16$ as a difference of squares.

$$x^4 - 16 = (x^2 + 4)(x^2 - 4)$$

Then $x^2 + 4$ does not factor, but $x^2 - 4$ factors into $(x + 2)(x - 2)$. Thus,

$$x^4 - 16 = (x^2 + 4)(x + 2)(x - 2).$$

Caution   In Example 3b note that $x^2 + 4$ is a prime polynomial. In general, the sum of squares $a^2 + b^2$ cannot be factored and is a prime polynomial.

PROGRESS CHECK 3   Factor completely.

a. $x - xy^2$ **b.** $16y^4 - 1$ ⌐

EXAMPLE 4   Solve the problem in the section introduction on page 192.

Solution   We are asked to factor $\pi R^2 - \pi r^2$ completely.

$$\pi R^2 - \pi r^2 = \pi(R^2 - r^2) \qquad \text{Factor out } \pi.$$
$$= \pi(R + r)(R - r) \qquad \text{Difference of squares.}$$

Thus, $A = \pi(R + r)(R - r)$ is a common formula for the area of an annulus.

PROGRESS CHECK 4   A ball is dropped from a roof 144 ft high. The required time in seconds for the ball to hit the ground may be found by solving the equation $144 - 16t^2 = 0$. Factor the left side of this equation. ⌐

3   **Sum or difference of two cubes**

The product of $a + b$ and $a^2 - ab + b^2$ is $a^3 + b^3$, as shown below.

$$(a + b)(a^2 - ab + b^2) = a^3 - a^2b + ab^2 + a^2b - ab^2 + b^3$$
$$= a^3 + b^3.$$

In a similar way,

$$(a - b)(a^2 + ab + b^2) = a^3 + a^2b + ab^2 - a^2b - ab^2 - b^3$$
$$= a^3 - b^3.$$

By reversing these special products, we obtain factoring models for the sum and the difference of two cubes.

Factoring Models for Sums and Differences of Cubes

$$a^3 + b^3 = (a + b)(a^2 - ab + b^2)$$
$$a^3 - b^3 = (a - b)(a^2 + ab + b^2)$$

The idea is to identify appropriate replacements for a and b in the expression to be factored; then substitute in these formulas.

EXAMPLE 5   Factor.

a. $x^3 + 8$

b. $27y^3 - 1$

Solution

a. Use the formula for the sum of two cubes. Replace a with x and b with 2 (because $8 = 2^3$).

$$a^3 + b^3 = (a + b)(a^2 - ab + b^2)$$
$$x^3 + 8 = x^3 + 2^3 = (x + 2)[x^2 - x(2) + 2^2]$$
$$= (x + 2)(x^2 - 2x + 4)$$

b. To factor $27y^3 - 1$, use the formula for the difference of two cubes. Because $27y^3 = (3y)^3$ and $1 = 1^3$, replace a with $3y$ and b with 1.

$$a^3 - b^3 = (a - b)(a^2 + ab + b^2)$$
$$27y^3 - 1 = (3y)^3 - 1^3 = (3y - 1)[(3y)^2 + (3y)(1) + 1^2]$$
$$= (3y - 1)(9y^2 + 3y + 1)$$

Caution   The trinomial factors $a^2 - ab + b^2$ and $a^2 + ab + b^2$ in these factoring models are prime polynomials. Students often waste time attempting to factor these trinomials, which are not factorable.

PROGRESS CHECK 5   Factor.

a. $y^3 + 1$

b. $8x^3 - 27$ ⌐

The next example points out once again that we often need to apply more than one factoring procedure to factor completely. Also, to factor polynomials as in Example 6b, keep in mind that powers of variables that are multiples of 3, like x^9 and y^6, are perfect cubes.

Progress Check Answers

4. $16(3 + t)(3 - t) = 0$

5. (a) $(y + 1)(y^2 - y + 1)$

(b) $(2x - 3)(4x^2 + 6x + 9)$

EXAMPLE 6   Factor completely.

a. $3x^3 - 24$ **b.** $x^9y + 125y^7$

Solution

a. First, factor out the common factor 3, and then apply the factoring model for the difference of two cubes.

$$3x^3 - 24 = 3(x^3 - 8) \qquad \text{Factor out 3.}$$
$$= 3(x - 2)(x^2 + 2x + 4) \qquad \text{Difference of cubes.}$$

b. First, factor out the common factor y.

$$x^9y + 125y^7 = y(x^9 + 125y^6)$$

Because $x^9 = (x^3)^3$ and $125y^6 = (5y^2)^3$, factor $x^9 + 125y^6$ by applying the model for the sum of two cubes, substituting x^3 for a and $5y^2$ for b.

$$x^9y + 125y^7 = y(x^3 + 5y^2)[(x^3)^2 - (x^3)(5y^2) + (5y^2)^2]$$
$$= y(x^3 + 5y^2)(x^6 - 5x^3y^2 + 25y^4)$$

Progress Check Answers
6. (a) $2(x + 5)(x^2 - 5x + 25)$
(b) $y^2(x^5 - 2y^3)(x^{10} + 2x^5y^3 + 4y^6)$

PROGRESS CHECK 6   Factor completely.

a. $2x^3 + 250$ **b.** $x^{15}y^2 - 8y^{11}$

EXERCISES 4.4

1. Which of the given expressions equals $(x + 1)^2$?
   a. $x^2 + 1$ **b.** $x^2 + x + 1$ **c.** $x^2 + 2x + 1$
2. Which of the given expressions equals $(x + 3)^2$?
   a. $x^2 + 9$ **b.** $x^2 + 6x + 9$ **c.** $x^2 + 3x + 6$
3. Which of the given expressions equals $(y - 2)^2$?
   a. $y^2 + 4y + 4$ **b.** $y^2 - 4y + 4$ **c.** $y^2 - 4$
4. Which of the given expressions equals $(y - 4)^2$?
   a. $y^2 - 8y + 16$ **b.** $y^2 - 4y + 8$ **c.** $y^2 + 16$
5. Which of the given expressions equals $(3x + 5y)^2$?
   a. $3x^2 + 15xy + 5y^2$ **b.** $9x^2 + 15xy + 25y^2$
   c. $9x^2 + 30xy + 25y^2$ **d.** $9x^2 + 25y^2$
6. Which of the given expressions equals $(4a - 3b)^2$?
   a. $16a^2 - 9b^2$ **b.** $16a^2 + 9b^2$
   c. $16a^2 + 24ab + 9b^2$ **d.** $16a^2 - 24ab + 9b^2$

In Exercises 7–12, indicate whether or not the given expression is a perfect square.

7. $x^2 + 4$ 8. $y^2 + 9$
9. $y^2 + 6y + 9$ 10. $x^2 + x + 1$
11. $4a^2 + 9ab + 16b^2$ 12. $a^2 + 6ab + 9b^2$

In Exercises 13–18 the given expressions are perfect square trinomials which match the pattern $a^2 + 2ab + b^2$ or the pattern $a^2 - 2ab + b^2$. In each exercise, indicate what a and b stand for.

13. $x^2 + 4x + 4$
14. $x^2 - 6x + 9$
15. $16x^2 - 40xy + 25y^2$
16. $9n^2 + 24mn + 16m^2$
17. $s^2 + 12st + 36t^2$
18. $49w^2 - 14wv + v^2$

In Exercises 19–30, fill in the blanks to make a perfect square trinomial.

19. $x^2 +$ _____ $+ 16$ 20. $x^2 +$ _____ $+ 25$

21. $y^2 -$ _____ $+ 4$ 22. $y^2 -$ _____ $+ 9$
23. $16x^2 +$ _____ $+ 4y^2$ 24. $9x^2 +$ _____ $+ y^2$
25. $4a^2 -$ _____ $+ 25b^2$ 26. $49c^2 -$ _____ $+ 36d^2$
27. $9a^4 +$ _____ $+ b^6$ 28. $m^8 +$ _____ $+ n^{10}$
29. $4x^2y^8 -$ _____ $+ z^4$ 30. $9x^4y^6 -$ _____ $+ z^2$

In Exercises 31–42, factor the perfect square polynomials.

31. $x^2 + 4x + 4$ 32. $x^2 - 4x + 4$
33. $y^2 - 10y + 25$ 34. $y^2 + 10y + 25$
35. $9a^2 + 24a + 16$ 36. $9b^2 - 24a + 16$
37. $36x^2 - 12x + 1$ 38. $y^2 + 14y + 49$
39. $w^2 + 2wb + b^2$ 40. $v^2 - 2vf + f^2$
41. $4r^2 - 12rs + 9s^2$
42. $16m^2 + 72mn + 81n^2$

In Exercises 43–48, the given expression is a difference of squares of the form $a^2 - b^2$. In each exercise, indicate what a and b stand for.

43. $9x^2 - 16$ 44. $25y^2 - 9$
45. $9f^2 - 25g^2$ 46. $64m^2 - 121k^2$
47. $4x^6 - 49y^8$ 48. $100x^{10} - 81y^4$

In Exercises 49–60, factor the difference of squares.

49. $x^2 - 1$ 50. $9 - y^2$
51. $81 - s^2$ 52. $x^2 - 100$
53. $64a^2 - b^2$ 54. $144d^2 - h^2$
55. $49z^2 - 4w^2$ 56. $81x^2 - 25y^2$
57. $x^4 - y^6$ 58. $s^{10} - t^4$
59. $36f^8 - k^2$ 60. $r^2 - 25s^{12}$

In Exercises 61–66, each polynomial is the sum or difference of cubes matching one of the patterns $a^3 + b^3$ or $a^3 - b^3$. In each exercise, indicate what a and b stand for.

61. $x^3 - 8$ 62. $27 + y^3$

1 In this chapter we have shown many different methods of factoring. At this point it is useful to summarize all of our factoring models and to state a general strategy to help you systematically factor a variety of polynomials.

Summary of Factoring Models

1. $ab + ac = a(b + c)$ — Common factor
2. $x^2 + (m + n)x + mn = (x + m)(x + n)$ — Trinomial ($a = 1$)
3. $(pq)x^2 + (pn + qm)x + mn = (px + m)(qx + n)$ — General trinomial
4. $a^2 + 2ab + b^2 = (a + b)^2$ — Perfect square trinomial
5. $a^2 - 2ab + b^2 = (a - b)^2$ — Perfect square trinomial
6. $a^2 - b^2 = (a + b)(a - b)$ — Difference of squares
7. $a^3 + b^3 = (a + b)(a^2 - ab + b^2)$ — Sum of cubes
8. $a^3 - b^3 = (a - b)(a^2 + ab + b^2)$ — Difference of cubes

With the aid of these models you may use the following steps as a guideline for factoring polynomials.

Guidelines for Factoring a Polynomial

1. Factor out any common factors (if present) as the first factoring procedure.
2. Check for factorizations according to the number of terms in the polynomial.

   Two terms: Look for a difference of squares or cubes or for a sum of cubes. Then apply model 6, 7, or 8, if applicable. Remember that $a^2 + b^2$ is a prime polynomial.

   Three terms: Is the trinomial a perfect square trinomial? If yes, use model 4 or 5. If no, and the coefficient of the squared term is 1, try to use model 2. If no, and the coefficient of the squared term is not 1, use FOIL reversal or the *ac* method.

   Four terms: Try factoring by grouping.
3. Make sure that no factors of two or more terms can be factored again.

Remember that you can always check your factoring by multiplying your answer.

EXAMPLE 1 Factor completely $5x^2 - 40x + 60$.

Solution First, look for common factors. The GCF is 5, so factor it out.

$$5x^2 - 40x + 60 = 5(x^2 - 8x + 12)$$

Now, try to factor $x^2 - 8x + 12$, which has three terms. We rule out a perfect square trinomial and use factoring model 2 (because $a = 1$), to obtain

$$5x^2 - 40x + 60 = 5(x - 2)(x - 6).$$

This result represents the complete factorization because neither binomial factor can be factored further.

PROGRESS CHECK 1 Factor completely $4x^2 - 16x + 16$.

Progress Check Answer
1. $4(x - 2)^2$

---

63. $27 + 8x^3$
64. $64 - 27x^3$
65. $64x^{15} - 27y^{12}$
66. $125y^6 - x^9$

In Exercises 67–78, factor the given sum or difference of cubes.

67. $a^3 + c^3$
68. $m^3 - n^3$
69. $d^3 - 27$
70. $s^3 + 64$
71. $8x^3 - 1$
72. $27y^3 + 1$
73. $8j^3 + 27k^3$
74. $27m^3 - 125n^3$
75. $125g^9 + h^9$
76. $a^6 - 1{,}000b^{15}$
77. $216x^6 - 343y^9$
78. $343x^{15} - 64y^{12}$

In Exercises 79–102, factor completely the given polynomial. First, try to factor out the GCF. If the polynomial cannot be factored, then write "prime."

79. $16x^2 - 25y^2$
80. $36f^2 - 49g^2$
81. $x^2 + 4$
82. $25d^2 + 1$
83. $h^2 + 10h + 25$
84. $w^2 - 14w + 49$
85. $t^2 + 8t + 12$
86. $r^2 - 3r - 40$
87. $x^3 + 64$
88. $y^3 - 8$
89. $z^4 - 81c^2$
90. $25j^2 - 16k^4$
91. $4x^2 - 20xy + 25y^2$
92. $9a^2 + 12ab + 4b^2$
93. $1{,}000s^4 - 1{,}000s^2$
94. $45d^3 - 5d$
95. $81x^4 - 16y^4$
96. $z^4 - 1$
97. $x^2 + 2x + 25$
98. $x^2 - 2x - 9$
99. $ax^2 - 2a^2x + a^3$
100. $2u^3v + 4u^2v^2 + 2uv^3$
101. $16cx^3 - 128c^4$
102. $a^2bx^3 + a^2b^4$

103. The sketch shown consists of three areas whose sum is $a^2 + 2ab + b^2$.
   a. Factor this polynomial to find another expression for this area.
   b. Show that the three pieces can be cut up and rearranged to make a square whose dimensions correspond to the two factors from part **a.**

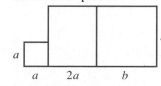

104. The shaded area shown is $a^2 - b^2$.
   a. Factor this polynomial to find another expression for this area.

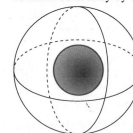

   b. Show that the shaded area can be cut up and rearranged into a rectangle whose length and width are the two factors found in part **a.**

105. An object is dropped from a platform 100 ft off the ground. The time it takes the object to hit the ground may be found by solving the equation $100 - 16t^2 = 0$. Perform the first step in finding the solution, which is to factor the left side of this equation.

106. In a physics experiment an object is set in motion along a straight track marked as shown in the sketch. It starts at position -25; and when it reaches the flag at 0, it is going 40 ft/second. It then decelerates at the constant rate of 8 ft/second every second. To calculate when it will reach the flag at 75, you can solve $75 = -25 + 40t - \frac{1}{2}8t^2$. This can be rewritten as $4t^2 - 40t + 100 = 0$. Perform the first step in solving this equation by factoring the left side.

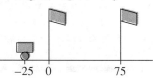

$-25 \quad 0 \qquad 75$

107. A golf ball is made with a smaller spherical core inside, as shown in the figure. If the radius of the golf ball is R and the radius of the core is r, then the part of the ball which is made up of the outer material is given by the difference of the two volumes of the two spheres, $V = \frac{4}{3}\pi R^3 - \frac{4}{3}\pi r^3$. Express this formula differently by factoring it completely.

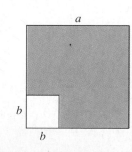

108. A metal piece of a lock consists of a smaller cube (side $= s$) mounted on top of a larger cube (side $= b$). The total weight of this piece is given by $W = 8s^3 + 27b^3$, where 8 and 27 are the respective densities of the small and large cubes. Express this formula differently by factoring it completely.

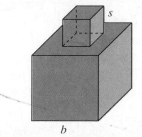

THINK ABOUT IT

1. Since $36 - 16$ is a difference of squares, it must be equal to $(6 - 4)(6 + 4)$. Show that this is so.
2. Why are trinomials of the form $a^2 + 2ab + b^2$ called perfect square trinomials? Fill in the blanks to make a perfect square trinomial. **a.** $x^2 + 4x + $ _____ **b.** $x^2 + 6x + $ _____
3. The numerical coefficients in $a + b$ are 1, 1. The numerical coefficients in $a^2 + 2ab + b^2$, which is the expansion of $(a + b)^2$, are 1, 2, 1. Furthermore, the numerical coefficients in $a^3 + 3a^2b + 3ab^2 + b^3$, which is the expansion of $(a + b)^3$, are 1, 3, 3, 1.
   a. Find the numerical coefficients in the expansion of $(a + b)^4$.
   b. Do you see a pattern in these sets of coefficients? Can you guess what the numerical coefficients would be for the expansion of $(a + b)^5$?
4. In the text we have restricted factoring to the domain of integers, so that we say $3x + 4y$ cannot be factored. But if you allow rational coefficients, then you *can* factor it. For example, you can factor out a 3 and get $3x + 4y = 3(x + \frac{4}{3}y)$.
   a. What would you get if you factored out 4 from $3x + 4y$?
   b. Factor out 10 from $3x + 4y$.
   c. Factor out x from $3x + 4y$.

5. A smaller square with side x is centered inside a larger one with side y, as shown in the figure.

   a. Find an expression for the shaded area and express it in factored form.
   b. The little square is now moved to the corner of the larger one, as shown in this figure. Find an expression for the total shaded area by adding the areas of the three shaded regions. Why must this expression be equivalent to the answer in part **a?**

REMEMBER THIS

1. True or false? x^{10} is a perfect square.
2. True or false? $x^2 + 9x + 9$ is a perfect square trinomial.
3. True or false? $27x^3 + 1$ is factorable because it is a sum of two cubes.
4. True or false? **a.** $\dfrac{a + x}{b + x}$ simplifies to $\dfrac{a}{b}$ if $x \neq 0$.
   b. $\dfrac{ax}{bx}$ simplifies to $\dfrac{a}{b}$ if $x \neq 0$.
5. Solve $y = 4nx - 5c$ for x.

6. Find the quotient: $\dfrac{14x - x^2 + 21x^3}{-7x^2}$. Write the answer using only positive exponents.
7. Find the product $(x^2 - 7x + 5)(x - 4)$.
8. Evaluate $10^{-1} + 10^{-2} + 10^{-3}$.
9. Evaluate $(-3)^{-3}$.
10. If neither a nor b equals zero, then $(a + b)^0$ is (equal to, less than, greater than) $a^0 + b^0$.

4.5   General Factoring Strategy

The total surface area of a cylinder is given by

$$A = 2\pi r^2 + 2\pi rh.$$

Write a common alternative formula by factoring the right side of this equation. (See Example 4.)

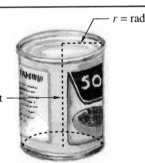
r = radius
h = height

OBJECTIVES

1 Apply a general strategy to systematically factor polynomials.

EXAMPLE 2   Factor completely $5x^2 + 38x - 16$.

Solution   There are no common factors to factor out and the polynomial is not a perfect square trinomial. Because the coefficient of the squared term is not 1, we use FOIL reversal or the *ac* method to obtain

$$5x^2 + 38x - 16 = (5x - 2)(x + 8).$$

This result expresses the complete factorization.

PROGRESS CHECK 2   Factor completely $4x^2 - 20x + 21$.

EXAMPLE 3   Factor completely $x^4 - y^4$.

Solution   There are no common factors to factor out and the polynomial is a difference that contains two terms. Look for a difference of squares or a difference of cubes. Because $x^4 - y^4$ is a difference of squares,

$$x^4 - y^4 = (x^2 + y^2)(x^2 - y^2).$$

Now $x^2 + y^2$ is a prime polynomial, but $x^2 - y^2$ is a difference of squares. So

$$x^4 - y^4 = (x^2 + y^2)(x + y)(x - y)$$

represents the complete factorization.

PROGRESS CHECK 3   Factor completely $1 - x^4$.

EXAMPLE 4   Solve the problem in the section introduction on page 198.

Solution   We are asked to factor $2\pi r^2 + 2\pi rh$. First, factor out $2\pi r$, the GCF.

$$2\pi r^2 + 2\pi rh = 2\pi r(r + h)$$

The binomial, $r + h$, does not factor, so the requested formula is

$$A = 2\pi r(r + h).$$

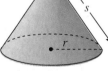

Figure 4.1

PROGRESS CHECK 4   The total surface area of a right circular cone, as shown in Figure 4.1, is given by $A = \pi r^2 + \pi rs$. Write an alternative formula by factoring the right side of this equation.

EXAMPLE 5   Factor completely $x^2 + 5x + ax + 5a$.

Solution   There are no common factors to factor out and the polynomial contains four terms. Try factoring by grouping.

$$x^2 + 5x + ax + 5a = x(x + 5) + a(x + 5)$$
$$= (x + 5)(x + a)$$

PROGRESS CHECK 5   Factor completely $y^2 + ay + by + ab$.

EXAMPLE 6   Factor completely $x^6 - 1$.

Solution   There are no common factors to factor out. The polynomial contains two terms, and we find that $x^6 - 1$ is both a difference of squares and a difference of cubes. In such cases, apply the difference-of-squares model first.

$$x^6 - 1 = (x^3)^2 - 1^2 = (x^3 + 1)(x^3 - 1)$$

Now, factor $x^3 + 1$ by the sum-of-cubes model and $x^3 - 1$ by the difference-of-cubes model.

$$x^6 - 1 = (x + 1)(x^2 - x + 1)(x - 1)(x^2 + x + 1)$$

At this point, no factors can be factored further.

Progress Check Answers
2. $(2x - 7)(2x - 3)$
3. $(1 - x)(1 + x)(1 + x^2)$
4. $A = \pi r(r + s)$
5. $(y + a)(y + b)$

PROGRESS CHECK 6   Factor completely $y^6 - 64$.

Progress Check Answer
6. $(y - 2)(y + 2)(y^2 - 2y + 4)(y^2 + 2y + 4)$

EXERCISES 4.5

In Exercises 1–6, assign as many of these labels as are correct to each expression.

a. Trinomial
b. Perfect square trinomial
c. Difference of squares
d. Sum of cubes
e. Difference of cubes

1. $x^2 + 2x + 1$
2. $y^2 + 2y + 2$
3. $n^3 - 125$
4. $27x^3 + 125y^6$
5. $x^2 + 1$
6. $v^2 - 25$

In Exercises 7–36, factor the given expression completely. Check your answer by multiplication.

7. $6x^2 - 48x + 72$
8. $y^2 + 4y - 77$
9. $5x^2 - 20x + 20$
10. $3b^2 - 12bc + 12c^2$
11. $7x^2 - 54x - 16$
12. $9y^2 + 50y - 24$
13. $8x^2 - 34x + 21$
14. $12x^2 + 35x + 25$
15. $16 - y^4$
16. $4q^4 - r^6$
17. $2cr^2t + 2crs$
18. $\frac{2}{3}cR^3 - \frac{2}{3}cr^3$
19. $x^2 + 6x + ax + 6a$
20. $5z^2 + 15z + 6z + 18$
21. $x - bx + 1 - b$
22. $w^3 - aw^2 + uw^2 - uaw$
23. $x^6 - 64y^6$
24. $16n^4 - a^4$
25. $9x^2 + 66xy + 121y^2$
26. $4s^2 + 4s + 4$
27. $8x^3 + 27$
28. $1 - t^3$
29. $x^6 - 4$
30. $z^6 + 4$
31. $x^2 + 9$
32. $x^2 - 5$
33. $g^2 + 2g + 3$
34. $4h^2 + 40h + 100$
35. $343u^2 - 49u - 392$
36. $125v^3 + 25v^2 - 100v$

37. Find an expression for the surface area and express it in factored form. (*Hint:* The top and bottom are squares and the four sides are identical trapezoids.)

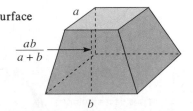

38. Find an expression for the surface area and express it in factored form. (*Hint:* The bottom is a square and the four sides are identical triangles.)

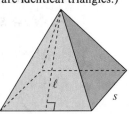

In Exercises 39–40, find an expression for the shaded area and express it in factored form.

39.

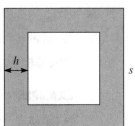

40.

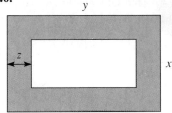

In Exercises 41 and 42, find the shaded volume and express it in factored form.

41.

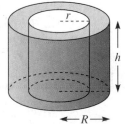

42.

THINK ABOUT IT

1. Use the fact that $1 + 2 + 3 + \cdots + 100 = 5{,}050$ to find the sum of $2 + 4 + 6 + \cdots + 200$. (*Hint:* Factor the GCF.) What is the sum of $3 + 6 + 9 + \cdots + 300$?

2. In Example 6 of this section we factored $x^6 - 1$ completely as $(x + 1)(x - 1)(x^2 - x + 1)(x^2 + x + 1)$ by first treating $x^6 - 1$ as a difference of squares.

a. Try the problem again as a difference of cubes.
b. Show that the two solutions are the same by showing that $(x^2 - x + 1)(x^2 + x + 1) = x^4 + x^2 + 1$.

3. Show that $x + 1$ is a factor of $x^5 + 1$ by finding the quotient $x + 1 \overline{)x^5 + 1}$. This means that $x^5 + 1$ can be factored into $(x + 1)(\ ?\ )$.

4. $x - 1$ can be written as $(x - 1)(1)$.
$x^2 - 1$ can be written as $(x - 1)(x + 1)$.
$x^3 - 1$ can be written as $(x - 1)(x^2 + x + 1)$.
Can $x^4 - 1$ be written as $(x - 1)(x^3 + x^2 + x + 1)$?
How can you factor $x^5 - 1$?

5. The volume of a cone is given by $V = \frac{1}{3}\pi r^2 h$, where h is the height and r is the radius. Use this formula to find expressions for the volumes of the shaded solids. Factor if possible.

(a)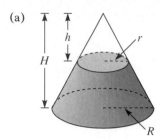

(b)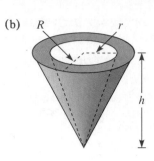

REMEMBER THIS

1. If $7x = 0$, then what number does x stand for?
2. True or false? If ab equals 0, then one or both of the letters must stand for 0.
3. True or false? If ab equals 5, then one or both of the letters must stand for 5.
4. Translate into an algebraic expression: "The product of three consecutive integers equals zero."
5. Evaluate $(x - 3)(x + 4)$ when x equals 3, -4, and 0.
6. Evaluate $7x(x - 1)$ when x equals 0, 1, and 7.

7. Evaluate $2^{-3} + (-3)^2$.
8. Which is smaller, 3^{-4} or 4^{-3}?
9. Solve $n(a + b) = c$ for n.
10. A waiter can be paid $220 per week plus 8 percent of the cost of the meals served, or $80 per week plus 15 percent of the cost of the meals served. At how many dollars worth of meals does the second scheme catch up to the first?

4.6 Solving Equations by Factoring

The height (y) of a projectile that is shot directly up from the ground with an initial velocity of 80 ft/second is given by the formula

$$y = 80t - 16t^2,$$

where t is the elapsed time in seconds. For what value(s) of t is the projectile 96 ft off the ground? (See Example 5.)

OBJECTIVES

1 Solve certain quadratic equations using factoring.

2 Solve applied problems involving formulas that lead to quadratic equations.

3 Solve other equations using factoring.

1 Our factoring methods may be applied to analyze certain equations that extend beyond first-degree or linear equations. Moving up one degree, we now define the next higher-degree polynomial equation.

Definition of Quadratic Equation

A **second-degree** or **quadratic equation** is an equation that can be written in the form

$$ax^2 + bx + c = 0,$$

where a, b, and c are real numbers, with $a \neq 0$.

Some examples of quadratic equations are

$$6x^2 - 5x + 1 = 0, \qquad y^2 = 4y, \qquad \text{and} \qquad (t - 2)^2 = 1.$$

To show how factoring can sometimes be used to solve a quadratic equation, let us try to find replacements for x that make $x^2 - 2x = 8$ a true statement. First, it will be necessary to rewrite the equation so that one side is 0.

$$x^2 - 2x = 8 \qquad \text{Original equation.}$$
$$x^2 - 2x - 8 = 0 \qquad \text{Subtract 8 from both sides.}$$

Now, factor the nonzero side of the equation.

$$(x - 4)(x + 2) = 0$$

When the product of two factors is zero, we have a special situation, as outlined in the zero product principle.

Zero Product Principle

For any numbers a and b,

$$\text{if } ab = 0, \text{ then } a = 0 \text{ or } b = 0.$$

In words, the principle states that when the product of two factors is 0, then at least one of the factors is 0. By applying this principle in our example, we obtain $(x - 4)(x + 2) = 0$ implies

$$x - 4 = 0 \qquad \text{or} \qquad x + 2 = 0.$$

Here we have two linear equations that may be solved as in Chapter 2.

$$x - 4 = 0 \qquad \text{or} \qquad x + 2 = 0$$
$$x = 4 \qquad \text{or} \qquad x = -2$$

Finally, the following check shows that both 4 and -2 are solutions.

$$x^2 - 2x = 8 \qquad\qquad x^2 - 2x = 8$$
$$4^2 - 2(4) \stackrel{?}{=} 8 \qquad (-2)^2 - 2(-2) \stackrel{?}{=} 8$$
$$16 - 8 \stackrel{?}{=} 8 \qquad\qquad 4 + 4 \stackrel{?}{=} 8$$
$$8 \stackrel{\checkmark}{=} 8 \qquad\qquad\qquad 8 \stackrel{\checkmark}{=} 8$$

Thus, the solution set of $x^2 - 2x = 8$ is $\{4, -2\}$.

Example 1 shows a concise format for the procedure just described.

EXAMPLE 1   Solve $3x^2 = 15x$.

Solution   Follow this format.

$$3x^2 = 15x \qquad \text{Original equation.}$$
$$3x^2 - 15x = 0 \qquad \text{Rewrite the equation so one side is 0.}$$
$$3x(x - 5) = 0 \qquad \text{Factor the nonzero side.}$$
$$3x = 0 \quad \text{or} \quad x - 5 = 0 \qquad \text{Set each factor equal to 0.}$$
$$x = \frac{0}{3} \qquad\qquad x = 5 \qquad \text{Solve each linear equation.}$$
$$x = 0$$

Check each possible solution by substituting it in the original equation.

$$3x^2 = 15x \qquad\qquad 3x^2 = 15x$$
$$3(0)^2 \overset{?}{=} 15(0) \qquad 3(5)^2 \overset{?}{=} 15(5)$$
$$0 \overset{\checkmark}{=} 0 \qquad\qquad 75 \overset{\checkmark}{=} 75$$

Thus, the solution set is $\{0,5\}$.

Note   As long as no mistakes are made along the way, each "possible solution" from the above procedure will always check. The check is needed only to catch any mistakes.

PROGRESS CHECK 1   Solve $2x^2 = 18x$.

To summarize, quadratic equations may be solved as follows, provided we can factor the polynomial (step 2). Chapter 9 will show alternative approaches that apply when we cannot factor.

Factoring Method for Solving Quadratic Equations

1. If necessary, change the form of the equation so that one side is 0.
2. Factor the nonzero side of the equation.
3. Set each factor equal to 0 and obtain the solution(s) by solving the resulting equations.
4. Check each solution by substituting it in the original equation.

Example 2 shows that it is possible for a quadratic equation to have only one solution. This is why we wrote "solution(s)" in step 3.

EXAMPLE 2   Solve $y^2 + 4 = 4y$.

Solution   Follow the steps given.

$$y^2 + 4 = 4y \qquad \text{Original equation.}$$
$$y^2 - 4y + 4 = 0 \qquad \text{Rewrite the equation so one side is 0.}$$
$$(y - 2)(y - 2) = 0 \qquad \text{Factor the nonzero side.}$$
$$y - 2 = 0 \quad \text{or} \quad y - 2 = 0 \qquad \text{Set each factor equal to 0.}$$
$$y = 2 \qquad\qquad y = 2 \qquad \text{Solve each linear equation.}$$

The solution, 2, checks, because $2^2 + 4 = 4(2)$ is a true statement, and the solution set is $\{2\}$.

PROGRESS CHECK 2   Solve $y^2 + 9 = -6y$.

To solve certain quadratic equations, it may not be necessary to use all of the steps outlined, as shown in the next two examples.

EXAMPLE 3   Solve $(4x + 1)(x - 5) = 0$.

Solution   One side is already 0 and the nonzero side is in factored form. Therefore, simply set each factor equal to 0, and solve the resulting linear equations.

$$(4x + 1)(x - 5) = 0 \qquad \text{Original equation.}$$
$$4x + 1 = 0 \quad \text{or} \quad x - 5 = 0 \qquad \text{Set each factor equal to 0.}$$
$$4x = -1 \qquad\qquad x = 5 \qquad \text{Solve each linear equation.}$$
$$x = -\tfrac{1}{4}$$

$$\text{Check} \qquad (4x + 1)(x - 5) = 0 \qquad\qquad (4x + 1)(x - 5) = 0$$

$$[4(-\tfrac{1}{4}) + 1](-\tfrac{1}{4} - 5) \overset{?}{=} 0 \qquad [4(5) + 1](5 - 5) \overset{?}{=} 0$$

$$(0)(-\tfrac{21}{4}) \overset{?}{=} 0 \qquad\qquad (21)(0) \overset{?}{=} 0$$

$$0 \overset{\checkmark}{=} 0 \qquad\qquad\qquad 0 \overset{\checkmark}{=} 0$$

Thus, the solution set is $\{-\tfrac{1}{4}, 5\}$.

Caution   The equation $(4x + 1)(x - 5) = 0$ is written in a form that allows us to immediately apply the zero product principle. We do *not* need to multiply $4x + 1$ and $x - 5$ as a first step here, which would only complicate the solution. However, remember that you must have 0 on one side of the equation in order to apply the *zero* product principle. So for equations like

$$(x - 1)(x + 2) = 10,$$

you must multiply $x - 1$ and $x + 2$ as a first step.

PROGRESS CHECK 3   Solve $(3x - 4)(x + 3) = 0$. ⌐

EXAMPLE 4   Solve $6x^2 - 5x + 1 = 0$.

Solution   One side is already 0, so begin by factoring the nonzero side.

$$6x^2 - 5x + 1 = 0 \qquad \text{Original equation.}$$
$$(3x - 1)(2x - 1) = 0 \qquad \text{Factor the nonzero side.}$$
$$3x - 1 = 0 \quad \text{or} \quad 2x - 1 = 0 \qquad \text{Set each factor equal to 0.}$$
$$3x = 1 \qquad\qquad 2x = 1 \qquad \text{Solve each linear equation.}$$
$$x = \tfrac{1}{3} \qquad\qquad x = \tfrac{1}{2}$$

$$\text{Check} \qquad 6x^2 - 5x + 1 = 0 \qquad\qquad 6x^2 - 5x + 1 = 0$$

$$6(\tfrac{1}{3})^2 - 5(\tfrac{1}{3}) + 1 \overset{?}{=} 0 \qquad 6(\tfrac{1}{2})^2 - 5(\tfrac{1}{2}) + 1 \overset{?}{=} 0$$

$$\tfrac{2}{3} - \tfrac{5}{3} + \tfrac{3}{3} \overset{?}{=} 0 \qquad\qquad \tfrac{3}{2} - \tfrac{5}{2} + \tfrac{2}{2} \overset{?}{=} 0$$

$$0 \overset{\checkmark}{=} 0 \qquad\qquad\qquad 0 \overset{\checkmark}{=} 0$$

Thus, the solution set is $\{\tfrac{1}{3}, \tfrac{1}{2}\}$.

PROGRESS CHECK 4   Solve $6x^2 - x - 2 = 0$. ⌐

2   To analyze problems that involve formulas containing second-degree polynomials, we sometimes need to solve quadratic equations. To illustrate this, we now consider the section-opening problem.

EXAMPLE 5   Solve the problem in the section introduction on page 202.

Solution   First, replace y by 96 in the given formula.

$$y = 80t - 16t^2$$
$$96 = 80t - 16t^2$$

Progress Check Answers

3. $\{\tfrac{4}{3}, -3\}$ 4. $\{\tfrac{2}{3}, -\tfrac{1}{2}\}$

Now, rewrite the equation so that one side is 0 and the coefficient of the squared term is positive. Then, factor the nonzero side.

$$16t^2 - 80t + 96 = 0$$
$$16(t^2 - 5t + 6) = 0$$
$$16(t - 3)(t - 2) = 0$$

Because the constant factor, 16, cannot be 0, we need only set equal to 0 the two factors that contain a variable.

$$t - 3 = 0 \quad \text{or} \quad t - 2 = 0$$
$$t = 3 \qquad\qquad t = 2$$

Both solutions check. The projectile attains a height of 96 ft after 2 seconds (on the way up) and again when 3 seconds have elapsed (on its way down).

PROGRESS CHECK 5   For what value(s) of t is the projectile considered in this example 64 ft off the ground?   ⌐

3   The factoring method associated with the zero product principle may be applied to solve higher-degree equations provided we can factor the polynomial on the nonzero side of the equation.

EXAMPLE 6   Solve each equation.

a. $(5x - 1)(x + 4)(x - 7) = 0$ b. $x^3 = 9x$

Solution

a. As in Example 3, one side is already 0 and the nonzero side is in factored form. Therefore, simply set each factor equal to 0 and solve the resulting equation.

$(5x - 1)(x + 4)(x - 7) = 0$	Original equation.
$5x - 1 = 0 \quad \text{or} \quad x + 4 = 0 \quad \text{or} \quad x - 7 = 0$	Set each factor equal to 0.
$x = \frac{1}{5} \qquad\qquad x = -4 \qquad\qquad x = 7$	Solve each linear equation.

All three solutions check (verify this), and the solution set is $\{\frac{1}{5}, -4, 7\}$.

b. First rewrite the equation so that one side is 0.

$$x^3 = 9x$$
$$x^3 - 9x = 0$$

Now, factor completely.

$$x(x^2 - 9) = 0$$
$$x(x + 3)(x - 3) = 0$$

Setting each factor equal to 0 and solving gives

$$x = 0 \quad \text{or} \quad x + 3 = 0 \quad \text{or} \quad x - 3 = 0$$
$$x = -3 \qquad\qquad x = 3.$$

All three solutions check, because $0^3 = 9(0)$, $(-3)^3 = 9(-3)$, and $3^3 = 9(3)$ are all true statements. The solution set is $\{0, -3, 3\}$.

PROGRESS CHECK 6   Solve each equation.

a. $(2x + 1)(2x - 1)(x - 1) = 0$ b. $y^3 = 25y$ ⌐

EXERCISES 4.6

In Exercises 1–6, indicate whether or not the given equation is a quadratic equation. If it is not quadratic, explain why.

1. $x^2 - 3x + 3 = 0$

2. $x^3 - 3x^2 + 4x - 2 = 0$
3. $y^2 = 3y + 3$
4. $w - 4 = 0$
5. $t - 5^2 = 0$
6. $(x - 4)^2 = x + 2$

In Exercises 7–12, find any values for the variable that will make the given product equal to zero.

7. $2x$
8. $3(x - 4)$
9. $4y(y - 5)$
10. $(y - 2)(y + 3)$
11. $(x + 1)(x - 1)x$
12. $(2x - 5)(3x + 1)$

In Exercises 13–48, solve the given quadratic equation by the factoring method. Check each possible solution.

13. $x^2 - 5x + 6 = 0$
14. $x^2 - 11x + 28 = 0$
15. $y^2 - 6y + 9 = 0$
16. $y^2 + 12y + 36 = 0$
17. $35a^2 - a - 6 = 0$
18. $3a^2 + 16a - 99 = 0$
19. $3x^2 - 12x = 0$
20. $25x^2 + 5x = 0$
21. $4x^2 = 20x$
22. $4x^2 = 18x$
23. $3x^2 = 11x$
24. $x^2 = 10x$
25. $y^2 + 16 = 8y$
26. $y^2 + 9 = -6y$
27. $b^2 - 10b = -9$
28. $c^2 + 8c = 20$
29. $x(x - 5) = 24$
30. $x(x + 2) = 3$
31. $2n(n - 3) = n^2 - 8$
32. $3n(n + 2) = 2n^2 + 160$
33. $3x^2 + 46 = 4x^2 + 10$
34. $x^2 - 1 = 2x^2 - 26$
35. $p^2 = -p$
36. $10q^2 = -5q$
37. $q^2 = 9 - 8q$
38. $p^2 = 2p(p + 7)$
39. $(x - 1)(x - 2) = 2$
40. $(x + 3)(x - 4) = -6$
41. $3y^2 + 6y = 2(3y + 54)$
42. $4y^2 - y = y(y - 1)$
43. $(t - 3)^2 = 25$
44. $(t + 2)^2 = 16$
45. $(2s + 5)^2 = 81$
46. $(3s - 1)^2 = 1$
47. $(r - 2)^2 = r + 4$
48. $3(r + 1)^2 = 2r + 10$

In Exercises 49–54, solve the equation by using the zero product principle.
49. $(x - 3)(x + 4)(2x - 3) = 0$
50. $(y + 1)(4y + 3)(3y - 4) = 0$
51. $x^3 - x^2 - 2x = 0$
52. $3y^3 - 15y^2 + 18y = 0$
53. $16x^3 = 49x$
54. $25x^4 = 9x^2$

55. Based on a study of many volunteers, a research psychologist worked out a statistical formula (called a regression model) which relates anxiety level (x) to performance (y). Both of these were measured by special tests invented by the psychologist. The formula used was $y = 10 + 10x - x^2$. This formula shows that as anxiety increases, so does performance, for a while; then with increasing anxiety, performance deteriorates. The maximum possible value of y according to this formula is 35. For what value of anxiety will y be 35?

56. The formula $P = EI - rI^2$ can be used to describe the power output of an electrical generator. In this formula P stands for power output in kilowatts, E is voltage, r is resistance in ohms, and I is the current in amperes. Notice that the formula is quadratic in terms of I. Solve the equation for I given that $E = 20$ volts, $r = 1$ ohm, and $P = 75$ kilowatts. There are two answers.

57. The equation for the motion of an object falling to the earth in the absence of air resistance is $s = 16t^2$, where s is the distance in feet and t is the time of the fall in seconds. Use the formula to answer these questions.
   a. Find the time it takes an object to fall 16 ft. (Note that negative solutions are not sensible for these applications.)
   b. Find the time it takes an object to fall 64 ft.
   c. Does it take 4 times as long to fall 64 ft as it takes to fall 16 ft?

58. A formula that describes the motion of a projectile which is shot straight up from the ground and then falls to earth is $d = v_0t - 16t^2$, where v_0 is the initial velocity, t is the elapsed time, and d is the distance above the ground. Suppose the initial velocity is 160 ft/second. Find the duration of the flight by setting $v_0 = 160$ and $d = 0$ and solving for t.

59. The number of diagonals in an n-sided polygon is given by $d = (n^2 - 3n)/2$.
   a. Use the formula to show that a triangle has no diagonals, and that a square has two diagonals.
   b. A certain polynomial has 20 diagonals. How many sides does it have?

60. In a formal meeting of diplomats each person must shake hands with each other person. The formula for the total number of handshakes is $h = n(n - 1)/2$, where n is the number of people present.
   a. How many handshakes will there be if six people are present?
   b. At one meeting there were 66 handshakes. How many people were present?

61. Mathematical texts (on clay tablets) which are more than 4,000 years old have been found in the Middle East. Here is a problem from one of these tablets that a student solved before 2000 B.C. If the area of a square is diminished by the side, the result is 870. Find the side of the square. You can answer this by solving the equation $s^2 - s = 870$. Only one of the two solutions will make sense. Why?

62. Here is a problem from an American algebra textbook from the year 1859. Two merchants each sold the same kind of stuff; the second sold 3 yd more of it than the first, and together they received $35. The first said to the second, "I would have received $24 for your stuff." The other replied, "And I should have received $12\frac{1}{2}$ for yours." How many yards did each of them sell? If you let x equal the number of yards sold by the first merchant, then $x + 3$ is the number sold by the second, and the equation $x\left(\dfrac{24}{x + 3}\right) + (x + 3)\left(\dfrac{12.5}{x}\right) = 35$ describes the situation. Multiply both sides of the equation by $x(x + 3)$ to get $24x^2 + 12.5(x + 3)^2 = 35x(x + 3)$, and solve this equation to find the answer. (*Hint:* Before you factor, divide both sides by the coefficient of x^2.)

THINK ABOUT IT

1. Second-degree equations are called quadratic because *to quadrate* means "to make square." Any quadratic equation can be rewritten into the form $(x + a)^2 = b$, which is the square of a binomial equal to a constant.
   a. Show that $x^2 + 4x - 5 = 0$ is equivalent to $(x + 2)^2 = 9$ by solving the first equation and then showing that both of its solutions are also solutions of the second equation.
   b. Show that by multiplying out the left-hand side of the second equation and collecting all terms on the left-hand side, you get the first equation.

2. a. Solve $3x^2 = 15x$ by the method of this chapter.
   b. Solve the equation by dividing both sides by x. Notice that the solution $x = 0$ is lost this way. Thus you should avoid dividing both sides of an equation by a variable whose value is not known.

3. What is the zero product principle?

4. Why is it not correct to solve $(x - 2)(x - 7) = 5$ by setting each factor equal to 5? Show that if you try to solve the equation this way, you will get incorrect solutions.

5. By taking advantage of the zero product principle, make up an equation which has three solutions, 1, 2, and 3.

REMEMBER THIS

1. Translate into algebraic symbols: The sum of two consecutive integers is -21.
2. The perimeter of a rectangle is 3 times the length, and the width is 3 in. Find the area.
3. Write an expression for the area of this triangle.

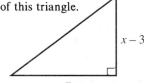

$x - 3$

x

4. True or false? $3^2 + 4^2 = 5^2$.
5. True or false? $1^2 + 2^2 = 3^2$.
6. Factor $x^2 - 25$. Is this a difference of two squares?
7. Is $x^2 + 10x + 25$ a perfect square trinomial?
8. Graph the solution set of $2x - 2 > 8$.
9. Simplify $\dfrac{(x^2y)^3}{(x^3y)^2}$.
10. Simplify $1 - [x - 3(x - 2)]$.

4.7 Additional Applications

A manufacturer wants to convert a square piece of cardboard into an uncovered box by cutting out from each of its four corners a square piece 2 in. on a side and then turning up the flaps. If the volume of the box is to be 98 in.³, what should be the side length of the original piece of cardboard? (See Example 4.)

OBJECTIVES

Solve word problems that lead to quadratic equations and involve the following:

1 Number relations

2 Geometric figures

3 The Pythagorean theorem

1 The applications in this section all require us to set up and then solve a quadratic equation. You should review the general approach to solving word problems given in Section 2.6 before considering these problems.

EXAMPLE 1   The product of two consecutive integers is 132. Find the integers.

Solution   We need to find two integers that satisfy certain conditions. As discussed in Section 2.6, if x represents the first integer, then $x + 1$ represents the next consecutive integer. Now, *set up an equation.*

The product of two consecutive integers is 132.
$$x(x + 1) \qquad\qquad = 132$$

Solve the Equation

$x^2 + x = 132$	Remove parentheses.
$x^2 + x - 132 = 0$	Subtract 132 from both sides.
$(x - 11)(x + 12) = 0$	Factor the nonzero side.
$x - 11 = 0 \quad\text{or}\quad x + 12 = 0$	Set each factor equal to 0.
$x = 11 \qquad\qquad x = -12$	Solve each linear equation.

Answer the Question   There are two pairs of integers that answer the question. If x is 11, then $x + 1$ is 12; and when x is -12, then $x + 1$ is -11. One solution is 11 and 12, and a second solution is -12 and -11.

Check the Answer   11 and 12 are consecutive integers and $11(12) = 132$. Also, -12 and -11 are consecutive integers and $-12(-11) = 132$. Both solutions check.

PROGRESS CHECK 1   The product of two consecutive integers is 72. Find the integers. ⌐

2   To solve problems about geometric figures, we often need the perimeter, area, and volume formulas given in Section 1.3.

EXAMPLE 2   The area of a square is 12 more than the perimeter. Find the side length for the square when this length is measured in meters.

Solution   We need to find the side length, which we represent by s. From Section 1.3 the area and perimeter formulas for a square are

$$A = s^2 \qquad \text{and} \qquad P = 4s.$$

Now, *set up an equation.*

The area is 12 more than the perimeter.
$$s^2 \qquad = \qquad 4s + 12$$

Solve the Equation

$s^2 - 4s - 12 = 0$	Rewrite so one side is 0.
$(s - 6)(s + 2) = 0$	Factor the nonzero side.
$s - 6 = 0 \quad\text{or}\quad s + 2 = 0$	Set each factor equal to 0.
$s = 6 \qquad\qquad s = -2$	Solve each linear equation.

Answer the Question   Because a square cannot have a negative side length, reject -2 as a solution to the problem. The side length is 6 m.

Check the Answer   If $s = 6$, then $A = s^2$ and $P = 4s$ give $A = 36$ and $P = 24$. Because 36 is 12 more than 24, the solution checks.

PROGRESS CHECK 2   The area of a square is 32 more than the perimeter. Find the side length of the square if the length is measured in feet. ⌐

Progress Check Answers
1. 8, 9 and $-8, -9$
2. 8 ft

EXAMPLE 3   The area of a triangle is 20 ft². If the height is 3 less than the length of the base, find the height of the triangle.

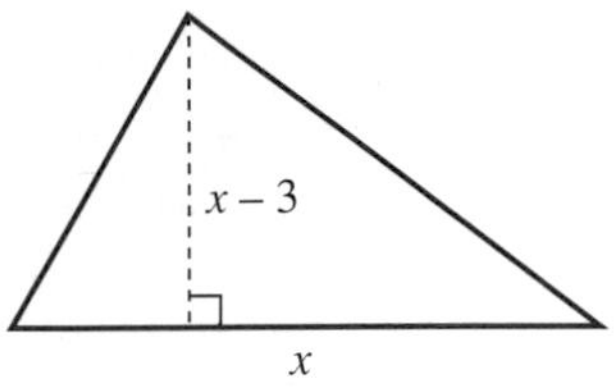

Solution   We need to find the height, which is defined in terms of the base. Therefore, let

$$x = \text{length of the base}$$
$$\text{so} \quad x - 3 = \text{height}.$$

Now, illustrate the given information, as in Figure 4.2, and use the formula for the area of a triangle (from Section 1.3) to *set up an equation.*

$$A = \tfrac{1}{2}bh$$
$$20 = \tfrac{1}{2}x(x - 3) \quad \text{Replace } A \text{ by 20, } b \text{ by } x, \text{ and } h \text{ by } x - 3.$$

Solve the Equation

$40 = x(x - 3)$	Multiply both sides by 2.
$40 = x^2 - 3x$	Remove parentheses.
$0 = x^2 - 3x - 40$	Subtract 40 from both sides.
$0 = (x - 8)(x + 5)$	Factor the nonzero side.
$x - 8 = 0 \quad \text{or} \quad x + 5 = 0$	Set each factor equal to 0.
$x = 8 \qquad\qquad x = -5$	Solve each linear equation.

Answer the Question   Because the length of the base cannot be negative, reject $x = -5$. So the length of the base is 8 ft. Then the height is $8 - 3 = 5$ ft.

Check the Answer   If $b = 8$ and $h = 5$, then $A = \tfrac{1}{2}(8)(5) = 20$. The solution checks.

PROGRESS CHECK 3   The area of a triangle is 18 ft². If the height is 5 less than the length of the base, find the height of the triangle.   ⌐

EXAMPLE 4   Solve the problem in the section introduction on page 208.

Solution   Let x represent the side length of the original piece of cardboard. When a 2-in. square is cut from each corner and the flaps turned up, then the cardboard square converts to a cardboard box with dimensions as shown in Figure 4.3. Now, use the formula for the volume of a rectangular solid to *set up an equation.*

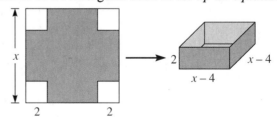

Figure 4.3

$$V = \ell wh$$
$$98 = (x - 4)(x - 4)(2) \quad \text{Replace } V \text{ by 98, } \ell \text{ by } x - 4, w \text{ by } x - 4, \text{ and } h \text{ by 2.}$$

Solve the Equation

$49 = (x - 4)(x - 4)$	Divide both sides by 2.
$49 = x^2 - 8x + 16$	Multiply.
$0 = x^2 - 8x - 33$	Subtract 49 from each side.
$0 = (x - 11)(x + 3)$	Factor the nonzero side.
$x - 11 = 0 \quad \text{or} \quad x + 3 = 0$	Set each factor equal to 0.
$x = 11 \qquad\qquad x = -3$	Solve each linear equation.

Figure 4.2 (labels: $x - 3$, x)

Answer the Question   The side length, which must be positive, is 11 in.

Check the Answer   If the side length of the original piece of cardboard is 11 in., then the length, width, and height of the resulting box are 7, 7, and 2 in., respectively. Because $V = (7 \text{ in.})(7 \text{ in.})(2 \text{ in.}) = 98 \text{ in.}^3$, the solution checks.

PROGRESS CHECK 4   In the above problem, assume that a square piece 3 in. on a side is cut from each corner and that the volume of the box is to be 243 in.³. What should be the side length of the original piece of cardboard?   ⌐

3   An important geometric figure for analyzing problems is a **right triangle,** which is a triangle that contains a 90° angle. The side opposite the 90° angle is called the **hypotenuse,** and it is always the longest side of the triangle. The other two sides are called the **legs** of the triangle. The Pythagorean theorem provides a formula that relates the lengths of the hypotenuse and the legs.

Pythagorean Theorem

In a right triangle with legs of length a and b and hypotenuse of length c,

$$c^2 = a^2 + b^2.$$

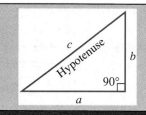

In words, the square of the length of the hypotenuse is equal to the sum of the squares of the lengths of the two legs.

EXAMPLE 5   How far from the base of a building should a 13-ft ladder be placed so that it reaches 12 ft up on the building?

Solution   Let x represent the distance from the base of the building to the bottom of the ladder. Now, illustrate the solution as in Figure 4.4, and use the Pythagorean theorem to *set up an equation.*

$$c^2 = a^2 + b^2$$
$$13^2 = x^2 + 12^2 \qquad \text{Replace } c \text{ by 13, } a \text{ by } x, \text{ and } b \text{ by 12.}$$

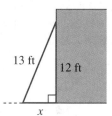

Figure 4.4

Solve the Equation

$169 = x^2 + 144$	Evaluate powers.
$0 = x^2 - 25$	Subtract 169 from each side.
$0 = (x + 5)(x - 5)$	Factor the nonzero side.
$x + 5 = 0 \quad \text{or} \quad x - 5 = 0$	Set each factor equal to 0.
$x = -5 \qquad\qquad x = 5$	Solve each linear equation.

Answer the Question   Only the positive solution is meaningful here. The bottom of the ladder should be placed 5 ft from the base of the building.

Check the Answer   Because $13^2 = 5^2 + 12^2$ is a true statement, placing the ladder 5 ft from the building forms a right triangle that satisfies the given conditions. The solution checks.

PROGRESS CHECK 5   How far from the base of a building should a 17-ft ladder be placed so that it reaches 15 ft up the side of the building?   ⌐

EXAMPLE 6   The length of a rectangle is 2 more than the width, and a diagonal measures 10 ft. Find the dimensions of the rectangle.

Solution   To find the dimensions, let
$$x = \text{width}$$
so $$x + 2 = \text{length.}$$

Progress Check Answers
4. 15 in.
5. 8 ft

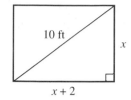

Figure 4.5

Now, sketch the situation, as in Figure 4.5, and note that a diagonal may be used to form a right triangle inside the rectangle. We may then use the Pythagorean theorem to *set up an equation.*

$$c^2 = a^2 + b^2$$
$$10^2 = (x + 2)^2 + x^2 \quad \text{Replace } c \text{ by 10, } a \text{ by } x + 2 \text{, and } b \text{ by } x.$$

Solve the Equation

$$
\begin{array}{ll}
100 = x^2 + 4x + 4 + x^2 & \text{Expand powers.} \\
0 = 2x^2 + 4x - 96 & \text{Rewrite so one side is 0.} \\
0 = 2(x^2 + 2x - 48) & \text{Factor out 2.} \\
0 = 2(x - 6)(x + 8) & \text{Factor } x^2 + 2x - 48. \\
x - 6 = 0 \quad \text{or} \quad x + 8 = 0 & \text{Set each nonconstant factor equal to 0.} \\
\quad\ x = 6 \qquad\qquad x = -8 & \text{Solve each linear equation.}
\end{array}
$$

Answer the Question   Reject the negative solution. The width is 6 ft and the length is $6 + 2 = 8$ ft.

Check the Answer   Because $10^2 = 8^2 + 6^2$ is a true statement, a rectangle that measures 8 ft by 6 ft satisfies the given conditions.

Progress Check Answer

6. 9 m by 12 m

PROGRESS CHECK 6   The length of a rectangle is 3 more than the width, and a diagonal measures 15 m. Find the dimensions of the rectangle.   ⌐

EXERCISES 4.7

Exercises 1–12 concern number relationships.

1. The product of two consecutive integers is 0. What are the integers?
2. The product of two consecutive integers is 156. Find the integers.
3. The sum of the squares of two consecutive integers is 181. What are the integers?
4. The square of the sum of two consecutive integers is 121. What are the integers?
5. One number is 2.5 more than another and their product is 9. Find the numbers.
6. The difference between two numbers is 5, and their product is 104. What are the numbers?
7. Find three consecutive integers for which the sum of the squares of the two smaller ones equals the square of the largest one.
8. The difference between the squares of two consecutive integers is 25. Find the integers.
9. The difference between two numbers is 6. The product of these two numbers is 9 less than their average. Find the numbers.
10. The sum of two numbers is 3, and their product is -10. Find the numbers. (*Hint:* If the sum of two numbers is 3, then one can be called x and the other $3 - x$.)
11. The sum of twice a number and the square of the number is 1 less than the square of 1 more than the number. Find the number.
12. The difference between two numbers is 6. The product of these two numbers is 9 less than the square of their average. Find the numbers.

Exercises 13–24 all deal with geometry. Remember to eliminate all solutions which do not make sense in the context of the problem.

13. The area of a square is half the perimeter. Find the dimensions of the square. Find the area and the perimeter. Assume that the side is measured in meters.
14. The area of a square is twice the perimeter. Find the dimensions of the square and its area and perimeter. Assume that the side is measured in inches.
15. The sides of a square were increased by 1, and as a result, the area increased by 15. What were the original dimensions? Assume that the side is measured in feet.
16. The sides of a square were decreased by 1, and as a result, the area decreased by 17. What were the original dimensions? Assume that the side is measured in centimeters.
17. The area of a triangle is 24 in.² and the height is 2 in. more than the base. Find the base of the triangle.
18. The base of a triangle is 5 more than the height. The area is 15 more than the base. Find the base of the triangle.
19. A rectangular piece of paper has length double the width. Find the dimensions of the paper so that if a 1-in. square is cut from each corner, it can be folded into a box with volume equal to 84 in.³.
20. A rectangular piece of paper has length double the width. Find the dimensions of the paper so that if a 2-in. square is cut from each corner, it can be folded into a box with volume equal to 96 in.³.
21. The longer base of a trapezoid is 3 times the shorter base, and the height is 1 more than the shorter base. The area is 24 cm². Find the height. Recall that the area of a trapezoid is the product of the height and the average of the bases.

22. The shorter base of a trapezoid is 3 less than the height, and the longer base is 5 more than the height. The area is 30 in.². Find the longer base.

23. The shaded area equals 3 times the unshaded area. Find the side length of the larger square.

24. Show that if the side of the outer square is 3 times the side of the inner square, then the shaded area is 8 times the unshaded area.

Exercises 25–36 all make use of the Pythagorean theorem.

25. In a right triangle one leg is 3 in. longer than the other and 3 in. shorter than the hypotenuse. How long is the hypotenuse?

26. In a right triangle one leg is 7 cm longer than the other and 1 cm shorter than the hypotenuse. Find the shorter leg.

27. Hindu religious writings from before 500 B.C. indicate that a right triangle with one side of 15 units and a hypotenuse of 39 was used in constructing a temple. Find the other side of this triangle.

28. Several sets of measurements for right triangles were found on a Babylonian cuneiform tablet dating from before 1600 B.C. One set had a hypotenuse equal to 18,541 and one side equal to 13,500. Find the other side.

29. A square has area equal to 50 in.². Find the diagonal.

30. A square has area equal to 18 in.². Find the diagonal.

31. **a.** Show that a triangle with sides 3, 4, and 5 must be a right triangle because the numbers satisfy the Pythagorean theorem.
   b. Show that a triangle with sides 4, 5, and 6 cannot be a right triangle because the numbers do not satisfy the Pythagorean theorem.

32. **a.** Show that a triangle with sides 5, 12, and 13 must be a right triangle because the numbers satisfy the Pythagorean theorem.
   b. Show that a triangle with sides 6, 13, and 14 cannot be a right triangle because the numbers do not satisfy the Pythagorean theorem.

33. A person can drive from A to X to B at 55 mi/hour or from A to B at 40 mi/hour. Which way takes less time?

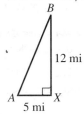

34. A person can drive from A to X to B at 55 mi/hour or from A to B at 40 mi/hour. Which way takes less time?

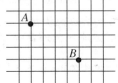

35. Find the distance between points A and B according to this scaled grid. The side of each small square represents a length of 10 m.

36. Find the distance between points A and B according to this scaled grid. The side of each small square represents a length of 5 m.

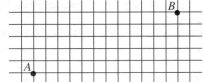

THINK ABOUT IT

1. In a right triangle the hypotenuse is 4.5 cm longer than one leg and 1 cm longer than the other. Find the shorter leg. (*Hint:* Multiply both sides of the equation by a number which will make all the coefficients integers.)

2. If three integers a, b, and c, satisfy $a^2 + b^2 = c^2$, they are called a Pythagorean triple and can be the sides of a right triangle. Here is one procedure that produces all Pythagorean triples.

Step 1. Find any two positive integers u and v, where
   (i) u and v have a GCF of 1
   (ii) $u > v$
   (iii) one of u and v is even and the other is odd

Step 2. Let $a = 2uv$. Let $b = u^2 - v^2$. Let $c = u^2 + v^2$. Then a, b, and c are a Pythagorean triple.
   Find the Pythagorean triple that results when
   a. $u = 2$ and $v = 1$
   b. $u = 3$ and $v = 2$
   c. $u = 4$ and $v = 1$

3. There are an unlimited number of Pythagorean triples (see "Think About It" Exercise 2). But there is no set of positive integers that satisfies $a^3 + b^3 = c^3$. Can you find the set of positive integers for a, b, and c that makes $a^3 + b^3$ come out *closest* to c^3?

4. a. Find an expression for the area of a square whose diagonal is d.
   b. Find the shaded area.

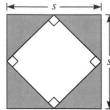

5. What does the Pythagorean theorem say about the areas of the three squares in this figure?

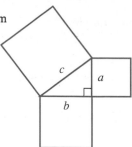

REMEMBER THIS

1. True or false? $\frac{3}{0}$ is undefined.

2. True or false? $\frac{3}{5}$ is a rational number.

3. Show that $\frac{3}{5} = \frac{4.5}{7.5}$ by showing that $3(7.5) = 5(4.5)$.

   Does $\frac{3}{5} = \frac{9}{25}$?

4. Evaluate $\frac{7}{x-4}$ for $x = -3$, 5, and 4.

5. True or false? $\frac{7}{x}$ can never be equal to zero.

6. Solve $x^2 + 2x - 3 = 0$.

7. Simplify $\frac{x^2 + x - 2}{x^2 + 2x - 3}$.

8. Factor $8x^3 + y^3$.

9. Find the product: $(2x + 1)(x^2 - 2x - 3)$.

10. Evaluate $-(x^2 - 2y)$ if $x = -1$ and $y = -3$.

Chapter 4   SUMMARY

OBJECTIVES CHECKLIST   Specific chapter objectives are summarized below along with numbered example problems from the text that should clarify the objectives. If you do not understand any objectives or do not know how to do the selected problems, then restudy the material.

4.1 Can you:

1. Find the greatest common factor of a set of terms?
Find the GCF of $15x^3$, $5x^4$, and $10x^2$.   [Example 2]

2. Factor out the greatest common factor from a polynomial?
Factor out the GCF from $5x^4 - 15x^3 + 10x^2$.   [Example 3b]

3. Factor by grouping?
Factor by grouping: $5x^2 + 10x + 4x + 8$.   [Example 6a]

4.2 Can you:

1. Factor $ax^2 + bx + c$ with $a = 1$?
Factor $x^2 + 7x + 10$.   [Example 1]

2. Identify prime polynomials?
Is $y^2 + 3y - 6$ a prime polynomial?   [Example 6]

3. Factor out the GCF and then factor trinomials?
Factor completely $3x^3 - 6x^2 - 24x$.   [Example 7]

4.3 Can you:

1. Factor $ax^2 + bx + c$ with $a \neq 1$ by reversing FOIL?
Factor $3x^2 + 19x + 6$.   [Example 1]

2. Factor $ax^2 + bx + c$ with $a \neq 1$ by the ac method?
Factor $3x^2 + 11x + 6$ by the ac method.   [Example 6]

4.4 Can you:

1. **Identify and factor perfect square trinomials?**
   Factor $9x^2 - 12xy + 4y^2$.   [Example 1b]

2. **Factor the difference of two squares?**
   Factor $x^2 - 25$.   [Example 2a]

3. **Factor the sum or the difference of two cubes?**
   Factor $27y^3 - 1$.   [Example 5b]

4.5 Can you:

1. **Apply a general strategy to systematically factor polynomials?**
   Factor completely $5x^2 - 40x + 60$.   [Example 1]
   Factor completely $x^6 - 1$.   [Example 6]

4.6 Can you:

1. **Solve certain quadratic equations using factoring?**
   Solve $y^2 + 4 = 4y$.   [Example 2]

2. **Solve applied problems involving formulas that lead to quadratic equations?**
   The height (y) of a projectile is given by the formula $y = 80t - 16t^2$, where t is the elapsed
   time in seconds. For what value(s) of t is the projectile 96 ft off the ground?   [Example 5]

3. **Solve other equations using factoring?**
   Solve $(5x - 1)(x + 4)(x - 7) = 0$.   [Example 6a]

4.7 Can you solve word problems that lead to quadratic equations and involve the following?

1. **Number relations**
   The product of two consecutive integers is 132. Find the integers.   [Example 1]

2. **Geometric figures**
   The area of a square is 12 more than the perimeter. Find the side length for the square
   when this length is measured in meters.   [Example 2]

3. **The Pythagorean theorem**
   How far from the base of a building should a 13-ft ladder be placed so that it reaches 12 ft
   up on the building?   [Example 5]

KEY TERMS

Annulus (4.4)	Hypotenuse (4.7)	Quadratic equation (4.6)
Factor (4.1)	Legs of right triangle (4.7)	Right triangle (4.7)
Factoring (4.1)	Perfect square trinomial (4.4)	Second-degree equation (4.6)
Factoring by grouping (4.1)	Prime polynomial (4.2)	Term (4.1)

KEY CONCEPTS AND PROCEDURES

Section	Key Concepts or Procedures to Review
4.1	■ To find the GCF of a set of numbers

1. Write each number as a product of prime factors.

2. The greatest common factor (GCF) is the product of all the *common* prime factors, with each prime
   number appearing the *fewest* number of times it appears in any one of the factorizations. If there are no
   common prime factors, the GCF is 1.

■ With respect to variables, the GCF is the product of all common variable factors, with each variable
appearing the fewest number of times it appears in any one of the factorizations.

Section	Key Concepts or Procedures to Review
	■ To factor out the GCF from the terms of a polynomial
	1. Find the GCF of the terms in the polynomial.
	2. Express each term in the polynomial as a product with the GCF as one factor.
	3. Factor out the GCF using the distributive property.
	4. Check the answer through multiplication.
4.2	■ Factoring model for trinomials of the form $ax^2 + bx + c$ when $a = 1$ $$(x + m)(x + n) = x^2 + (m + n)x + mn$$
4.3	■ In factoring by reversing FOIL, the key observation is that the middle term in the trinomial to be factored is the sum of the outer product and the inner product from the FOIL multiplication.
	■ If the terms of a trinomial have no common integer factors (except 1 and -1), then the same is true for both of its binomial factors.
	■ In factoring $ax^2 + bx + c$ by the ac method
	1. Find two integers whose product is ac and whose sum is b.
	2. Replace b by the sum of the two integers from step 1.
	3. Factor by grouping.
4.4	■ Factoring models for perfect square trinomials $$a^2 + 2ab + b^2 = (a + b)^2 \qquad a^2 - 2ab + b^2 = (a - b)^2$$
	■ Factoring model for a difference of squares $$a^2 - b^2 = (a + b)(a - b)$$
	■ Factoring models for sums and differences of cubes $$a^3 + b^3 = (a + b)(a^2 - ab + b^2) \qquad a^3 - b^3 = (a - b)(a^2 + ab + b^2)$$
4.5	■ Guidelines for factoring a polynomial
	1. Factor out any common factors (if present).
	2. Check for factorizations according to the number of terms in the polynomial.
	Two terms: Look for a difference of squares or cubes or for a sum of cubes. Remember that $a^2 + b^2$ is a prime polynomial.
	Three terms: Is the trinomial a perfect square trinomial? If not, use the appropriate model, depending on whether or not $a = 1$. If $a \neq 1$, use FOIL reversal or the ac method.
	Four terms: Try factoring by grouping.
	3. Make sure that no factors of two or more terms can be factored again.
4.6	■ Zero product principle: For any numbers a and b, if $ab = 0$, then $a = 0$ or $b = 0$.
	■ Factoring method for solving quadratic equations
	1. If necessary, change the form of the equation so that one side is 0.
	2. Factor the nonzero side of the equation.
	3. Set each factor equal to 0 and obtain the solution(s) by solving the resulting equations.
	4. Check each solution by substituting it in the original equation.
4.7	■ Pythagorean theorem: In a right triangle with legs of length a and b and hypotenuse of length c, $$c^2 = a^2 + b^2.$$
	■ Many word problems lead to quadratic equations. But remember to check all solutions to see if they make sense as answers to the word problem.

CHAPTER 4 REVIEW EXERCISES

4.1

1. Find the GCF of $12x^2$, $16x$, and $4x^3$.
2. If we factor $3x$ from $3x^2 + 3x$, what is the remaining factor?
3. Factor out the GCF from $8x^2 - 8y$.
4. Factor by grouping: $x^2 - 3x + 2x - 6$.
5. To find the area of a trapezoid, you can split it into two triangles as shown. Write the formula for the sum of the areas of the two triangles; then factor to get the formula for the area of the trapezoid.

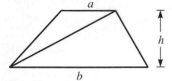

4.2

6. Factor $a^2 + 3a + 2$.
7. Factor $x^2 - 6x - 72$.
8. Factor $3y^4 - 27y^3 + 60y^2$.
9. Show that $x^2 + x + 3$ is a prime polynomial over the integers.

4.3

10. Complete the factorization
    $3y^2 - 16y + 5 = (3y - ?)(y - ?)$.
11. Factor $3y^2 + 5y + 2$.
12. Factor $24x^2 + 23x - 12$.
13. Factor completely $-9y^2 - 24y - 15$.
14. Find two integers whose sum is b and whose product is ac:
    $6x^2 - 5x - 6$.
15. Factor by the ac method: $6x^2 - 5x - 6$.

4.4

16. Which of the given expressions equals $(a + 2)^2$?
    a. $a^2 + 4$ **b.** $a^2 + 2a + 4$ **c.** $a^2 + 4a + 4$
17. Fill in the blank to make a perfect square trinomial:
    $x^2 + \_\_\_ + 9$.
18. Factor $x^2 - 10x + 25$.
19. Factor $4a^2 + 20a + 25$.
20. Factor $9x^2 - 25$.
21. Factor $x^3 + 1$.

4.5

For Exercises 22–25, factor where possible.

22. $49 - x^2$
23. $x^2 + 25$
24. $x^2 + 5x - 14$
25. $a^3 + 8b^3$
26. Find a formula for the shaded area and express it in factored form.

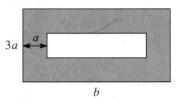

4.6

27. Find any values for x that will make the product equal to 0:
    $x(x - 2)(x + 3)$.
28. Solve by factoring: $x^2 + 3x - 4 = 0$.
29. Solve $5x^2 - 15x = 0$.
30. Solve $x(x - 5) = -6$.
31. Solve $6x^2 - x - 2 = 0$.
32. A manager of a clothing store can buy a certain suit at $20 wholesale. She estimates that if she prices it at $100, she will sell 50 of them, and that for each $5 *increase* in price she will sell 1 *less* suit. Letting n be the number of $5 increases, the formula $I = 5,000 + 150n - 5n^2$ gives the resulting income. By trying different values of n, she realizes that the maximum possible income is $6,125. What price will generate this income? To find out, let $I = \$6,125$ and solve the resulting equation.

4.7

33. One number is 2.5 more than another, and their product is 26. Find the number.
34. The perimeter of a square is 8 times the area. Find the dimensions of the square and its area and perimeter. Assume that the side is measured in feet.
35. The diagonal of a rectangle is 15 in. and the length is 3 in. longer than the width. Find the area of the rectangle.

CHAPTER 4 TEST

1. Find the greatest common factor of $24x^4$, $8x^3$, and $20x^2$.
2. Factor out the GCF: $6a^4 - 18a^3 + 12a^2$.

For Problems 3–13, factor the given expression completely. If the expression cannot be factored, write "prime polynomial."

3. $x^2 + 7x + 12$
4. $y^2 - 8y + 12$
5. $x^2 + 3x - 18$
6. $x^2 - 6x - 16$
7. $y^2 + 3y - 6$
8. $6x^3 + 12x^2 + 6x$
9. $10x^2 - 3x - 4$
10. $a^2 + 6a + 9$
11. $16c^2 - 9$
12. $8x^2 - 8$
13. $a^3 - 8$
14. Factor by grouping: $3x^2 + 6x + 2x + 4$.
15. Use the ac method to factor $3x^2 + 20x + 12$. Show all steps, including factoring by grouping.

For Problems 16–18, solve the given equation.

16. $5x^2 = 15x$
17. $x(2x + 1)(3x - 4) = 0$

5.1 Simplification of Rational Expressions

O B J E C T I V E S

1 Find all values of a variable which make a rational expression undefined.

2 Determine whether a pair of fractions are equivalent.

3 Express rational expressions in lowest terms.

1 In the chapter-opening problem the expression for the percent of profit

$$\frac{s - 2}{s}$$

is an algebraic fraction. The numerator is the polynomial $s - 2$, and the denominator is the polynomial s. This type of fraction is called a rational expression.

Rational Expression

A **rational expression** is an expression of the form

$$\frac{A}{B},$$

where A and B are polynomials with $B \neq 0$.

Some other examples of rational expressions are

$$\frac{5}{7}, \quad \frac{x^3 + 1}{2}, \quad \frac{7y - 4}{y^2 - 1}, \quad \text{and} \quad \frac{xy}{y + x}.$$

The above definition states that the polynomial in the denominator cannot be 0, because division by 0 is undefined (as discussed in Section 1.8). Thus, it is important to note any values for a variable which lead to a zero denominator. Rational expressions are meaningless for such values. They will be defined, however, for all other values of the variable, unless some restrictions apply because of the context of the problem.

EXAMPLE 1 Find all values of the variable which make each rational expression undefined.

a. $\dfrac{7}{3x - 4}$ **b.** $\dfrac{y}{y^2 - 1}$ **c.** $\dfrac{x}{x^2 + 1}$

Solution

a. Set the denominator $3x - 4$ equal to 0 and solve.

$$3x - 4 = 0$$
$$3x = 4$$
$$x = \frac{4}{3}$$

The only number that makes $\dfrac{7}{3x - 4}$ undefined is $\dfrac{4}{3}$.

b. Set the denominator $y^2 - 1$ equal to 0 and solve as in Section 4.6.

$$y^2 - 1 = 0$$
$$(y + 1)(y - 1) = 0$$
$$y + 1 = 0 \qquad \text{or} \qquad y - 1 = 0$$
$$y = -1 \qquad\qquad\qquad y = 1$$

The given expression is undefined for -1 and for 1.

c. For all x, x^2 is greater than or equal to 0, and the denominator $x^2 + 1$ is greater than or equal to 1. Thus, no replacements for x lead to a denominator of 0, so there are no values which make the expression undefined.

PROGRESS CHECK 1   Find all values of the variable which make each rational expression undefined.

a. $\dfrac{x - 6}{5x + 2}$

b. $\dfrac{1}{x^2 - 2x - 8}$

c. $\dfrac{x^2 - 1}{2}$

EXAMPLE 2   Solve the problem in the chapter introduction on page 219.

Solution

a. Replace s by 2.50 in the expression given for the percent of profit. If $s = 2.50$, then

$$\frac{s - 2}{s} = \frac{2.50 - 2}{2.50} = \frac{0.50}{2.50} = 0.2, \text{ or 20 percent.}$$

The percent of profit (based on the selling price) is 20%.

b. If the denominator is set equal to 0, the nonnegative value of s for which the profit is undefined is $s = 0$.

PROGRESS CHECK 2   A sporting goods store sells a can of tennis balls for $3. If this item cost the store c dollars, then the percent of profit based on the cost is given by $\dfrac{3 - c}{c}$.

a. Find the percent of profit if the cost is $2.

b. For what nonnegative value of c is the percent of profit undefined?

2   Because different quotients have the same value, there are many fractions that are equivalent. For example, $\frac{1}{2}$ and $\frac{4}{8}$ are equivalent fractions; although they appear in different forms, they have the same value. The following principle provides a simple method for determining if two fractions are equivalent.

Equivalence of Fractions

Let a, b, c, and d be real numbers, with $b \neq 0$ and $d \neq 0$. Then

$$\frac{a}{b} = \frac{c}{d} \text{ if } ad = bc; \quad \text{and} \quad \frac{a}{b} \neq \frac{c}{d} \text{ if } ad \neq bc.$$

EXAMPLE 3   Determine whether each pair of fractions are equivalent.

a. $\dfrac{1}{2}, \dfrac{2}{4}$

b. $\dfrac{-2}{3}, \dfrac{2}{-3}$

c. $\dfrac{8}{36}, \dfrac{1}{4}$

d. $\dfrac{3 \cdot 2}{4 \cdot 2}, \dfrac{3}{4}$

Solution   Use the equivalence-of-fractions principle and compare cross products.

a. Because $1(4) = 2(2)$, $\dfrac{1}{2}$ and $\dfrac{2}{4}$ are equivalent fractions.

b. Because $(-2)(-3) = 3(2)$, $\dfrac{-2}{3}$ and $\dfrac{2}{-3}$ are equivalent fractions.

c. Because $8(4) \neq 36(1)$, $\dfrac{8}{36}$ and $\dfrac{1}{4}$ are not equivalent fractions.

d. Because $(3 \cdot 2)4 = (4 \cdot 2)3$, $\dfrac{3 \cdot 2}{4 \cdot 2}$ and $\dfrac{3}{4}$ are equivalent fractions.

Progress Check Answers

1. (a) $-\dfrac{2}{5}$ (b) 4, -2 (c) None

2. (a) 50 percent   (b) 0

PROGRESS CHECK 3 Determine whether each pair of fractions are equivalent.

a. $\dfrac{6}{9}, \dfrac{2}{3}$ **b.** $\dfrac{3}{8}, \dfrac{6}{14}$ **c.** $\dfrac{-4}{-5}, \dfrac{4}{5}$ **d.** $\dfrac{3\cdot 5}{3\cdot 6}, \dfrac{5}{6}$

3 Although there are many equivalent fractional forms, we usually need to express a fraction in lowest terms. For example, recall from Section 1.1 that $\frac{6}{9}$ simplifies to $\frac{2}{3}$ as follows.

$$\frac{6}{9} = \frac{2\cdot 3}{3\cdot 3} = \frac{2\cdot \overset{1}{\cancel{3}}}{3\cdot \underset{1}{\cancel{3}}} = \frac{2}{3}$$

In general, to express a fraction in lowest terms, we divide out all nonzero factors that are common to the numerator and the denominator, according to the fundamental principle of fractions.

Fundamental Principle of Fractions

If a, b, and k are real numbers, with $b \neq 0$ and $k \neq 0$, then

$$\frac{ak}{bk} = \frac{a}{b}.$$

Note that the fundamental principle can be established from the criteria for the equality of two fractions, since $(ak)b$ is equal to $(bk)a$. The next example reviews the entire procedure that was used in Section 1.1 to express a fraction in lowest terms.

EXAMPLE 4 Express $\frac{18}{45}$ in lowest terms.

Solution Express 18 as $2 \cdot 3 \cdot 3$ and 45 as $5 \cdot 3 \cdot 3$. Then apply the fundamental principle of fractions.

$$\frac{18}{45} = \frac{2\cdot 3\cdot 3}{5\cdot 3\cdot 3} \qquad \text{Write 18 and 45 in prime factored form.}$$
$$= \frac{2}{5} \qquad \text{Divide out the common factor } 3\cdot 3 \text{ according to the fundamental principle.}$$

In lowest terms, $\frac{18}{45}$ is expressed as $\frac{2}{5}$.

PROGRESS CHECK 4 Express $\frac{20}{28}$ in lowest terms.

Rational expressions may be reduced to lowest terms using the same methods. First, however, let us restate the fundamental principle in terms of rational expressions.

Fundamental Principle of Rational Expressions

If A, B, and K are polynomials, with $B \neq 0$ and $K \neq 0$, then

$$\frac{AK}{BK} = \frac{A}{B}.$$

Now, Example 5 explains in detail how to reduce a rational expression.

EXAMPLE 5 Express $\dfrac{2x + 6}{7x + 21}$ in lowest terms.

Solution First, factor completely the numerator and the denominator in an attempt to find a factor that is common to both.

$$\frac{2x + 6}{7x + 21} = \frac{2(x + 3)}{7(x + 3)}$$

Since the numerator and the denominator both contain the factor $x + 3$, we can divide out this common factor according to the fundamental principle, provided $x \neq -3$.

$$\frac{2\overset{1}{\cancel{(x + 3)}}}{7\underset{1}{\cancel{(x + 3)}}} = \frac{2}{7}$$

In lowest terms $\dfrac{2x + 6}{7x + 21} = \dfrac{2}{7}$, provided that $x \neq -3$.

PROGRESS CHECK 5 Express $\dfrac{3x - 15}{8x - 40}$ in lowest terms. ⌐

Example 5 shows that reducing a rational expression to lowest terms is essentially a two-step procedure.

To Reduce Rational Expressions

1. Factor completely the numerator and the denominator of the fraction.
2. Divide out nonzero factors that are common to the numerator and the denominator according to the fundamental principle.

EXAMPLE 6 Express each rational expression in lowest terms.

a. $\dfrac{5x + 10}{x^2 + x - 2}$

b. $\dfrac{y^2 - 1}{(y - 1)^2}$

Solution Factor and use the fundamental principle.

a. $\dfrac{5x + 10}{x^2 + x - 2} = \dfrac{5(x + 2)}{(x - 1)(x + 2)} = \dfrac{5}{x - 1}$ for $x \neq 1$ or -2

b. $\dfrac{y^2 - 1}{(y - 1)^2} = \dfrac{(y + 1)(y - 1)}{(y - 1)(y - 1)} = \dfrac{y + 1}{y - 1}$ for $y \neq 1$

PROGRESS CHECK 6 Express each rational expression in lowest terms.

a. $\dfrac{x^2 - 9}{4x + 12}$

b. $\dfrac{x^2 - 5x + 4}{x^2 - 3x - 4}$ ⌐

In Example 6 the restrictions $x \neq 1$ or -2 and $y \neq 1$ are necessary because we may divide out only nonzero factors, since division by 0 is undefined. With the understanding that such restrictions always apply, we will not continue to list them from this point on. In the next example consider carefully the ideas discussed in the note.

EXAMPLE 7 Express $\dfrac{x}{x^2 + x}$ in lowest terms.

Solution Once again, factor and apply the fundamental principle.

$$\frac{x}{x^2 + x} = \frac{x}{x(x + 1)} = \frac{1}{x + 1}$$

Progress Check Answers

5. $\frac{3}{8}$

6. (a) $\dfrac{x - 3}{4}, x \neq -3$ (b) $\dfrac{x - 1}{x + 1}, x \neq -1, 4$

Note   In this example the common factor x is the entire numerator, so x must be replaced by 1 when we divide out x.

$$\frac{x}{x(x+1)} = \frac{\overset{1}{\cancel{x}}}{\underset{1}{\cancel{x}}(x+1)} = \frac{1}{x+1}$$

Also, note that we may divide out common *factors,* not common terms. Often students make the error of dividing out terms, as shown below.

$$\frac{x}{x^2+x} = \frac{\overset{1}{\cancel{x}}}{x^2+\underset{1}{\cancel{x}}} = \frac{1}{x^2+1} \qquad \textbf{WRONG}$$

Remember that factors are the components in a product, and terms are the components in a sum.

PROGRESS CHECK 7   Express $\dfrac{y}{y^2-y}$ in lowest terms. ⌐

To reduce rational expressions, it is important to note what happens when we reverse the order in which numbers are written in addition and subtraction. For addition, $a+b$ is equal to $b+a$. However, for subtraction, $a-b$ is the opposite of $b-a$. Thus,

$$\frac{a+b}{b+a} = 1 \qquad \text{while} \qquad \frac{a-b}{b-a} = \frac{-1(b-a)}{b-a} = -1.$$

Both of these principles are used in the next example.

EXAMPLE 8   Express each rational expression in lowest terms.

a. $\dfrac{x^2+ax}{a^2+ax}$ **b.** $\dfrac{1-x}{x-1}$ **c.** $\dfrac{y-x}{x^2-y^2}$

Solution

a. Reduce as before, noting that $x+a=a+x$.

$$\frac{x^2+ax}{a^2+ax} = \frac{x(x+a)}{a(a+x)} = \frac{x}{a}$$

b. The quotient is -1, as shown below. In general, the quotient of two (nonzero) opposites is always -1.

$$\frac{1-x}{x-1} = \frac{-1(x-1)}{x-1} = -1$$

c. x^2-y^2 factors into $(x+y)(x-y)$, and $y-x$ is the opposite of $x-y$, so

$$\frac{y-x}{x^2-y^2} = \frac{-1(x-y)}{(x+y)(x-y)} = \frac{-1}{x+y}.$$

Note   Recall from Section 1.8 that the placement of a negative sign in a fraction is arbitrary. For example, for the answer in part **c,**

$$\frac{-1}{x+y} = -\frac{1}{x+y} = \frac{1}{-(x+y)}.$$

In general, for polynomials A and B, $B \neq 0$,

$$-\frac{A}{B} = \frac{-A}{B} = \frac{A}{-B}.$$

The form $\dfrac{A}{-B}$, however, is rarely used to express a quotient.

PROGRESS CHECK 8   Express each rational expression in lowest terms.

a. $\dfrac{y - 5}{5 - y}$

b. $\dfrac{1 - x^2}{x - 1}$

c. $\dfrac{y + x}{x^2 + 2xy + y^2}$

Progress Check Answers

8. (a) -1 (b) $(-1)(1 + x)$ (c) $\dfrac{1}{x + y}$

EXERCISES 5.1

In Exercises 1–12, find all values of the variable which make each rational expression undefined.

1. $\dfrac{8}{x - 2}$

2. $\dfrac{5}{x - 3}$

3. $\dfrac{2}{x}$

4. $\dfrac{3 + y}{y}$

5. $\dfrac{y}{3y + 6}$

6. $\dfrac{2x}{x + 2}$

7. $\dfrac{3}{x^2 + 5}$

8. $\dfrac{2z}{3z^2 + 5}$

9. $\dfrac{1}{(x - 2)^2}$

10. $\dfrac{x}{2}$

11. $\dfrac{5y}{y^2 - 9}$

12. $\dfrac{7}{y^2 + 2y - 3}$

In Exercises 13–18, determine whether the given pair of fractions are equivalent.

13. $\dfrac{6}{8}, \dfrac{9}{12}$

14. $\dfrac{10}{4}, \dfrac{15}{6}$

15. $\dfrac{3}{5}, \dfrac{9}{25}$

16. $\dfrac{2}{3}, \dfrac{4}{9}$

17. $\dfrac{0}{1}, \dfrac{0}{2}$

18. $\dfrac{-2}{-5}, \dfrac{4}{10}$

Yes or no? The first fraction can be simplified to equal the second fraction.

19. $\dfrac{3 + 1}{3 + 5}; \dfrac{1}{5}$

20. $\dfrac{10 - 5}{11 - 5}; \dfrac{10}{11}$

21. $\dfrac{3x}{3 + x}; \dfrac{x}{1 + x}$

22. $\dfrac{4y}{4 + y}; \dfrac{4}{5}$

23. Which of these are equivalent to $\frac{1}{2}$? Assume the denominator does not equal 0.

a. $\dfrac{x}{2x}$

b. $\dfrac{x}{x + 2}$

c. $\dfrac{2}{4x}$

d. $\dfrac{2x}{4}$

e. $\dfrac{1 + x}{2 + x}$

24. Which of these are equivalent to $\frac{2}{3}$?

a. $\dfrac{2y}{3x}$

b. $\dfrac{2y}{3y}$

c. $\dfrac{2xy}{3xy}$

d. $\dfrac{2 - xy}{3 - xy}$

e. $\dfrac{2/y}{3/y}$

In Exercises 25–72, express the given fraction in lowest terms. If it is already in lowest terms, say so. Assume that denominators do not equal 0.

25. $\dfrac{27}{36}$

26. $\dfrac{11}{99}$

27. $\dfrac{1,010}{4,040}$

28. $\dfrac{1,212}{2,323}$

29. $\dfrac{33c^3}{11c}$

30. $\dfrac{25n^2}{5n}$

31. $\dfrac{4}{4x}$

32. $\dfrac{ab}{abx}$

33. $\dfrac{-2ab}{4a^2b}$

34. $\dfrac{14a^3b^2}{-7ab}$

35. $\dfrac{-8(y - 2)}{2(y - 2)}$

36. $\dfrac{5(x + 3)}{-10(x + 3)}$

37. $\dfrac{3x - 3}{5x - 5}$

38. $\dfrac{8y + 16}{2y + 4}$

39. $\dfrac{3(x - 3)(x + 4)}{5(x - 3)}$

40. $\dfrac{(x - 1)(y + 2)^2}{5(x - 1)(y + 2)}$

41. $\dfrac{3x - 12}{x - 4}$

42. $\dfrac{5x + 10}{x + 2}$

43. $\dfrac{3x - 12}{x^2 - 9x + 20}$

44. $\dfrac{7x + 14}{x^2 - 5x - 14}$

45. $\dfrac{y - 1}{y^2 - 1}$

46. $\dfrac{4x + 8}{4x^2 - 16}$

47. $\dfrac{x^2 + 4x + 4}{x^2 + 5x + 6}$

48. $\dfrac{y^2 - 7y + 12}{y^2 - 6y + 9}$

49. $\dfrac{c^2 - c - 6}{c^2 - 2c - 8}$

50. $\dfrac{d^2 + 4d - 5}{d^2 + 5d - 6}$

51. $\dfrac{x^2}{x^2 + x}$

52. $\dfrac{y^2 - xy}{xy}$

53. $\dfrac{x}{x + 1}$

54. $\dfrac{3y}{y + 3}$

55. $\dfrac{x^2 - y^2}{x + y}$

56. $\dfrac{x^2 - y^2}{x - y}$

57. $\dfrac{2x^2 + 6x + 4}{2x^2 + 8x + 6}$

58. $\dfrac{3x^2 + 15x + 18}{3x^2 + 9x}$

59. $\dfrac{Y - 1}{1 - Y}$

60. $\dfrac{a^2 - b^2}{b^2 - a^2}$

61. $\dfrac{y + 1}{y - 1}$

62. $\dfrac{a + 2b}{b + 2a}$

63. $\dfrac{x - y}{y^2 + xy - 2x^2}$

64. $\dfrac{5(x - y)}{2y^2 + xy - 3x^2}$

65. $\dfrac{x^2 - bx}{b^2 - bx}$

66. $\dfrac{4ax - ay}{by - 4bx}$

67. $\dfrac{a^3 + b^3}{a^2 - b^2}$

68. $\dfrac{a^3 - b^3}{a^4 - b^4}$

69. $\dfrac{m^3 - n^3}{m - n}$

70. $\dfrac{8m^3 + 64n^3}{m^3 + 8n^3}$

71. $\dfrac{2c + 2d + c^2 + cd}{cd + d^2}$

72. $\dfrac{x^2 + xy}{ax + ay + bx + by}$

73. To find the percentage change for the cost of an item, evaluate

$$\dfrac{C_n - C_0}{C_0} \times 100,$$

where C_n is the new cost and C_0 is the old cost.

a. Find the percent change for the price of an automobile that rises from $15,000 to $15,600.

b. For what value of C_0 is this expression not defined?

c. An item used to be free but now costs $1. What percent increase is that?

74. a. Find the percentage decrease for the cost of a calculator that falls from $20 to $15. (*Hint:* Use the formula in Exercise 73.)

b. For what value of C_0 is the formula not defined?

c. An item used to cost $1 but now is free. What percent decrease is that?

THINK ABOUT IT

1. a. True or false? $-(2x)$, $(-2)x$, and $2(-x)$ are all equal.
b. Check part **a** using $x = 5$ and then $x = -6$.

2. True or false?
a. If you square the numerator and denominator of a proper fraction, the new fraction is equivalent to the original one.
b. If you add 3 to the numerator and denominator of a proper fraction, the new fraction is equivalent to the original one.
c. If you multiply the numerator and denominator of a proper fraction by 3, the new fraction is equivalent to the original one.

3. Show that $\frac{4}{6}$ and $\frac{6}{9}$ are equivalent fractions in three different ways.
a. Rewrite both using the LCD, which is 18.
b. Reduce them both to lowest terms.
c. Show that the cross products are equal.

4. When we say that $\frac{2x + 6}{7x + 21}$ reduces to $\frac{2}{7}$, we mean that for al values of x for which the fraction is defined, $\frac{2x + 6}{7x + 21}$ will be equivalent to $\frac{2}{7}$.
a. Show that this is true for $x = 0$, 1, and -1.
b. Show that this is not true for $x = -3$.

5. Explain why $\frac{3}{3x}$ does not reduce to x.

REMEMBER THIS

1. Multiply $\frac{2}{3} \cdot \frac{4}{5}$.

2. Multiply $\frac{1}{8}(-16)(4x)$.

3. Multiply $7 \cdot \frac{2}{5}$.

4. Divide x^a by x^{a-1}.

5. True or false? $\dfrac{x^2 - 1}{2}$ is undefined when $x = -1$.

6. Find an expression for the area of this trapezoid.

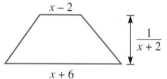

7. Reduce to lowest terms: $\dfrac{5x + 10}{x^2 + x - 2}$.

8. Solve $x^2 = x + 2$.

9. Which property allows you to rewrite $3(x + y)$ as $3x + 3y$?

10. True or false? $\dfrac{a^5}{a} = \dfrac{a^{-1}}{a^{-5}}$. (Assume that $a \neq 0$.)

5.2 Multiplication and Division of Rational Expressions

If the chance of winning on a certain wager in roulette is s/n, then the odds in favor of winning are given by

$$\frac{s}{n} \div \frac{n - s}{n}.$$

Express this division in lowest terms. (See Example 5.)

OBJECTIVES

1 Multiply rational expressions.

2 Divide rational expressions.

1　In arithmetic we know that the product of two or more fractions is the product of their numerators divided by the product of their denominators. For example,

$$\frac{2}{3} \cdot \frac{4}{5} = \frac{2 \cdot 4}{3 \cdot 5} = \frac{8}{15},$$

$$\frac{2}{3} \cdot \frac{3}{4} \cdot \frac{4}{5} = \frac{2 \cdot \overset{1}{\cancel{3}} \cdot \overset{1}{\cancel{4}}}{\underset{1}{\cancel{3}} \cdot \underset{1}{\cancel{4}} \cdot 5} = \frac{2}{5}.$$

Note from the last example that if the same factor appears in both the numerator and the denominator, it is usually easier to divide out this factor before multiplying. These methods are also used to multiply rational expressions.

Multiplication of Rational Expressions

If $\dfrac{A}{B}$ and $\dfrac{C}{D}$ are rational expressions, with $B \neq 0$ and $D \neq 0$, then

$$\frac{A}{B} \cdot \frac{C}{D} = \frac{AC}{BD}.$$

To express products in lowest terms, factoring and the fundamental principle are usually required, as shown in the following examples.

EXAMPLE 1　Multiply and express each answer in lowest terms.

a. $6x^2 \cdot \dfrac{5}{3x^3}$

b. $\dfrac{2x}{5y} \cdot \dfrac{10y^2}{8x^3}$

Solution

a. $6x^2 \cdot \dfrac{5}{3x^3} = \dfrac{6x^2}{1} \cdot \dfrac{5}{3x^3}$　　　Write $6x^2$ as $\dfrac{6x^2}{1}$.

$= \dfrac{6x^2 \cdot 5}{1 \cdot 3x^3}$　　　Multiply fractions.

$= \dfrac{2 \cdot 3 \cdot x \cdot x \cdot 5}{3 \cdot x \cdot x \cdot x}$　　　Write in prime factored form.

$= \dfrac{10}{x}$　　　Write in lowest terms, using the fundamental principle.

b. $\dfrac{2x}{5y} \cdot \dfrac{10y^2}{8x^3} = \dfrac{2x \cdot 10y^2}{5y \cdot 8x^3}$　　　Multiply fractions.

$= \dfrac{2 \cdot x \cdot 2 \cdot 5 \cdot y \cdot y}{5 \cdot y \cdot 2 \cdot 2 \cdot 2 \cdot x \cdot x \cdot x}$　　　Write in prime factored form.

$= \dfrac{y}{2x^2}$　　　Write in lowest terms, using the fundamental principle.

PROGRESS CHECK 1　Multiply and express each answer in lowest terms.

a. $\dfrac{7}{5x^5} \cdot 20x^2$

b. $\dfrac{27x^3}{12y^2} \cdot \dfrac{4y}{15x^3}$

Progress Check Answers

1. (a) $\dfrac{28}{x^3}$　　(b) $\dfrac{3}{5y}$

EXAMPLE 2   Find (in lowest terms) the product of $\dfrac{6x + 18}{x^2}$ and $\dfrac{x}{x^2 - 9}$.

Solution   First, factor completely all numerators and denominators. Then use the multiplication definition, and express the result in lowest terms.

$$\frac{6x + 18}{x^2} \cdot \frac{x}{x^2 - 9} = \frac{6(x + 3)}{x \cdot x} \cdot \frac{x}{(x + 3)(x - 3)} \qquad \text{Factor completely.}$$

$$= \frac{6(x + 3) \cdot x}{x \cdot x(x + 3)(x - 3)} \qquad \text{Multiply fractions.}$$

$$= \frac{6}{x(x - 3)} \qquad \begin{array}{l} \text{Write in lowest terms, using} \\ \text{the fundamental principle.} \end{array}$$

Note that multiplications like $x(x - 3)$ do not have to be multiplied out when expressing answers to these problems.

PROGRESS CHECK 2   Find (in lowest terms) the product of $\dfrac{x^2 - x - 12}{x}$ and $\dfrac{x^2}{x - 4}$.

EXAMPLE 3   Multiply $\dfrac{t^2 + 3t - 4}{t^2 - 3t - 4} \cdot \dfrac{t^2 - 5t + 4}{t^2 + 5t + 4}$. Express the answer in lowest terms.

Solution   Proceed as in Example 2.

$$\frac{t^2 + 3t - 4}{t^2 - 3t - 4} \cdot \frac{t^2 - 5t + 4}{t^2 + 5t + 4}$$

$$= \frac{(t + 4)(t - 1)}{(t - 4)(t + 1)} \cdot \frac{(t - 4)(t - 1)}{(t + 4)(t + 1)} \qquad \text{Factor completely.}$$

$$= \frac{(t + 4)(t - 1)(t - 4)(t - 1)}{(t - 4)(t + 1)(t + 4)(t + 1)} \qquad \text{Multiply fractions.}$$

$$= \frac{(t - 1)^2}{(t + 1)^2} \qquad \begin{array}{l} \text{Write in lowest terms, using} \\ \text{the fundamental principle.} \end{array}$$

PROGRESS CHECK 3   Multiply $\dfrac{y^2 - 3y + 2}{y^2 - 4y + 4} \cdot \dfrac{y^2 - 6y + 5}{y^2 - 2y + 1}$. Express the answer in lowest terms.

2   To divide two rational expressions, recall from Section 1.1 that to divide two fractions, we invert the fraction by which we are dividing to find its reciprocal, and then we multiply. For example,

$$\frac{2}{3} \div \frac{5}{9} = \frac{2}{3} \cdot \frac{9}{5} = \frac{6}{5}.$$

This procedure is also used to divide rational expressions.

Division of Rational Expressions

If $\dfrac{A}{B}$ and $\dfrac{C}{D}$ are rational expressions, with $B \neq 0$, $C \neq 0$, and $D \neq 0$, then

$$\frac{A}{B} \div \frac{C}{D} = \frac{A}{B} \cdot \frac{D}{C} = \frac{AD}{BC}$$

Progress Check Answers

2. $x(x + 3)$

3. $\dfrac{y - 5}{y - 2}$

EXAMPLE 4   Divide: $\dfrac{35}{x^2} \div \dfrac{7}{x}$. Express the answer in lowest terms.

Solution   Convert to multiplication and then proceed as before.

$$\frac{35}{x^2} \div \frac{7}{x} = \frac{35}{x^2} \cdot \frac{x}{7} \qquad \text{Multiply } \frac{35}{x^2} \text{ by the reciprocal of } \frac{7}{x}.$$

$$= \frac{5 \cdot 7 \cdot x}{x \cdot x \cdot 7} \qquad \text{Multiply and factor.}$$

$$= \frac{5}{x} \qquad \begin{array}{l}\text{Express in lowest terms, using}\\ \text{the fundamental principle.}\end{array}$$

PROGRESS CHECK 4   Divide: $\dfrac{12}{x^4} \div \dfrac{8}{x^2}$. Express the answer in lowest terms. ⌟

EXAMPLE 5   Solve the problem in the section introduction on page 226.

Solution   Simplify the expression given for the odds in favor of winning.

$$\frac{s}{n} \div \frac{n-s}{n} = \frac{s}{n} \cdot \frac{n}{n-s} \qquad \text{Division definition.}$$

$$= \frac{s \cdot n}{n(n-s)} \qquad \text{Multiply fractions.}$$

$$= \frac{s}{n-s} \qquad \begin{array}{l}\text{Express in lowest terms, using}\\ \text{the fundamental principle.}\end{array}$$

If the chance of winning is $\dfrac{s}{n}$, the odds in favor of winning are $\dfrac{s}{n-s}$. In the context of odds, this answer is usually read as "s to $n-s$."

PROGRESS CHECK 5   If the chance of winning is $\dfrac{s}{n}$, then the odds *against* winning are given by $\dfrac{n-s}{n} \div \dfrac{s}{n}$. Express this division in lowest terms. ⌟

EXAMPLE 6   Divide: $\dfrac{(y-1)^2}{3y} \div \dfrac{y^2-1}{15y}$. Express the answer in lowest terms.

Solution

$$\frac{(y-1)^2}{3y} \div \frac{y^2-1}{15y} = \frac{(y-1)^2}{3y} \cdot \frac{15y}{y^2-1} \qquad \text{Division by definition.}$$

$$= \frac{(y-1)(y-1)}{3y} \cdot \frac{3 \cdot 5 \cdot y}{(y+1)(y-1)} \qquad \text{Factor completely.}$$

$$= \frac{(y-1)(y-1) \cdot 3 \cdot 5 \cdot y}{3 \cdot y(y+1)(y-1)} \qquad \text{Multiply fractions.}$$

$$= \frac{5(y-1)}{y+1} \qquad \begin{array}{l}\text{Write in lowest terms, using the}\\ \text{fundamental principle.}\end{array}$$

PROGRESS CHECK 6   Divide: $\dfrac{(t+4)^2}{4t} \div \dfrac{t^2-16}{16t^2}$. Express the answer in lowest terms.

EXAMPLE 7   Divide: $\dfrac{1-x}{y^2-y} \div \dfrac{x^2-1}{y^2-2y+1}$. Express the answer in lowest terms.

Solution   Note that the following solution contains an extra step, because $1-x$ and $x-1$ are opposites, so we may divide out the factors as shown.

Progress Check Answers

4. $\dfrac{3}{2x^2}$

5. $\dfrac{n-s}{s}$

6. $\dfrac{4t(t+4)}{t-4}$

$$\frac{1-x}{y^2-y} \div \frac{x^2-1}{y^2-2y+1} = \frac{1-x}{y^2-y} \cdot \frac{y^2-2y+1}{x^2-1}$$ Division definition.

$$= \frac{1-x}{y(y-1)} \cdot \frac{(y-1)(y-1)}{(x+1)(x-1)}$$ Factor completely.

$$= \frac{(1-x)(y-1)(y-1)}{y(y-1)(x+1)(x-1)}$$ Multiply fractions.

$$= \frac{-1(x-1)(y-1)(y-1)}{y(y-1)(x+1)(x-1)}$$ Replace $1-x$ by $-1(x-1)$.

$$= \frac{-1(y-1)}{y(x+1)} \text{ , or } \frac{1-y}{y(x+1)}$$ Write in lowest terms, using the fundamental principle.

Progress Check Answer

7. $\dfrac{-xy}{x+1}$

PROGRESS CHECK 7 Divide: $\dfrac{x^2+x}{y-1} \div \dfrac{x^2+2x+1}{y-y^2}$. Express the answer in lowest terms.

EXERCISES 5.2

In Exercises 1–24, multiply the given expressions and express each answer in lowest terms. Assume denominators are not equal to 0.

1. $15x^2 \cdot \dfrac{6}{5x^4}$

2. $\dfrac{10}{3y} \cdot 9y^5$

3. $11x \cdot \dfrac{3}{121x}$

4. $\dfrac{-2}{10y^2} \cdot 5y^2$

5. $\dfrac{a}{b} \cdot \dfrac{b}{a}$

6. $\dfrac{5}{c} \cdot \dfrac{c}{b}$

7. $\dfrac{3x}{40y^3} \cdot \dfrac{5y^3}{6x^2}$

8. $\dfrac{4x}{24y^3} \cdot \dfrac{6y^2}{7x^3}$

9. $\dfrac{3}{x} \cdot \dfrac{x^2}{4} \cdot \dfrac{8x}{3}$

10. $\dfrac{2}{y^2} \cdot \dfrac{y^3}{5} \cdot \dfrac{15}{2y^5}$

11. $\dfrac{2x+3}{x^2} \cdot \dfrac{x}{2x+3}$

12. $\dfrac{3y-5}{y} \cdot \dfrac{y^3}{3y-5}$

13. $\dfrac{8x-12}{x+1} \cdot \dfrac{x+1}{2x-3}$

14. $\dfrac{12x+15}{x+1} \cdot \dfrac{3x+3}{4x+5}$

15. $\dfrac{x^2+3x+2}{x+2} \cdot \dfrac{1}{x^2+4x+3}$

16. $\dfrac{y^2+3y-4}{y+4} \cdot \dfrac{2}{y-1}$

17. $\dfrac{t-2}{s^2} \cdot \dfrac{s}{2-t}$

18. $\dfrac{w-v}{w} \cdot \dfrac{w^5}{v-w}$

19. $\dfrac{y}{3x-y} \cdot \dfrac{y-3x}{x}$

20. $\dfrac{a}{2cx-cy} \cdot \dfrac{by-2bx}{ab}$

21. $\dfrac{t^2-3t+2}{t^2+t-2} \cdot \dfrac{t^2+5t+6}{t^2+t-6}$

22. $\dfrac{s^2+3s-10}{s^2-3s-10} \cdot \dfrac{s^2-2s-8}{s^2+2s-8}$

23. $\dfrac{4a^2+4a-8}{a^2+4a-21} \cdot \dfrac{a^2+2a-15}{4a^2+8a-12}$

24. $\dfrac{3c^2+6c+3}{2c^2-10c+12} \cdot \dfrac{2c^2-12c+18}{3c^2+9c+6}$

In Exercises 25–48, do the given division and express the answer in lowest terms. Assume denominators are not equal to 0.

25. $\dfrac{45}{x^2} \div \dfrac{9}{x}$

26. $\dfrac{12}{x^3} \div x$

27. $\dfrac{b}{x} \div \dfrac{b}{x^2}$

28. $\dfrac{x}{7} \div \dfrac{x^2}{49}$

29. $\dfrac{x+y}{n} \div \dfrac{x-y}{n}$

30. $\dfrac{3x+y}{n+m} \div \dfrac{x+y}{n+m}$

31. $\dfrac{20}{n} \div \dfrac{5}{n}$

32. $\dfrac{n}{5} \div \dfrac{n}{10}$

33. $\dfrac{3x}{4y} \div \dfrac{4x}{3y}$

34. $\dfrac{5x}{6y} \div \dfrac{6y}{5x}$

35. $\dfrac{(x-3)^2}{x} \div \dfrac{x^2-9}{2x}$

36. $\dfrac{5(x-1)^2}{x} \div \dfrac{x^2-1}{5x^2}$

37. $\dfrac{(t+5)^2}{3t} \div \dfrac{t^2-25}{3t}$

38. $\dfrac{(s+3)^2}{5} \div \dfrac{s^2-9}{5s}$

39. $\dfrac{2-y}{x^2+2x} \div \dfrac{y^2-4}{x^2+x-2}$

40. $\dfrac{3-y}{x^2+3x} \div \dfrac{y^2-9}{x^2+x-6}$

41. $\dfrac{a+b}{c-d} \div \dfrac{a^2+2ab+b^2}{c^2-d^2}$

42. $\dfrac{d-c}{d+c} \div \dfrac{c^2-2cd+d^2}{c^2+2cd+d^2}$

43. $\dfrac{x^2+4x+3}{x^2-4x+3} \div \dfrac{x^2+5x+6}{x^2-5x+6}$

44. $\dfrac{x^2+8x+15}{x^2-8x+15} \div \dfrac{x^2+2x-15}{x^2-4x-5}$

45. $\dfrac{y^2+5y}{3y^2-y} \div \dfrac{2y^2+13y+15}{3y^3-y^2}$

46. $\dfrac{y^2-2y}{6y^2+3y} \div \dfrac{y^2-4y+4}{3y}$

47. $\dfrac{3x-3a}{cy+bc} \div \dfrac{cx-ac}{dy-bd}$

48. $\dfrac{ab-ac}{ab+bc} \div \dfrac{bd-cd}{a^2-c^2}$

49. The ratio of the area of a square of side s to the area of the inscribed circle is given by $s^2 \div \pi(s/2)^2$. Express this ratio in lowest terms.

50. The ratio of the volume of a cube of side s to the inscribed sphere is given by $s^3 \div (4/3)\pi(s/2)^3$. Express this ratio in lowest terms.

THINK ABOUT IT

1. A student made a terrible mistake and got the right answer when reducing $\frac{16}{64}$ to $\frac{1}{4}$ by crossing out the 6's. Can you find another fraction that will reduce weirdly like this?

2. When you reduce a fraction like $\frac{3a}{4a}$ by dividing out the a's, why does it make sense to replace the a's by 1's? Would it be correct to replace the a's with 0's?

3. Prove that it is wrong to reduce $\frac{11 + 4}{1 + 4}$ by dividing out the 4's.

   Prove it by showing that the original fraction is not equivalent to the new fraction.

4. Archimedes (287–212 B.C.) was one of the greatest mathematicians of all time. Among his most notable achievements is the discovery of formulas for the volumes and surface areas of cylinders and spheres. He made a most remarkable observation about a cylinder and its inscribed sphere (see the sketch). He proved that the ratio of the two volumes is exactly the same as the ratio of the two surface areas. Use the given formulas to find out what he discovered. (*Hint:* In the formulas that involve h, express h in terms of r, so that r is the only variable.)

   Volume of a cylinder $= \pi r^2 h$
   Volume of a sphere $= \frac{4}{3}\pi r^3$
   Surface area of a
   $\quad$ cylinder $= 2\pi rh + 2\pi r^2$
   Surface area of a sphere $= 4\pi r^2$

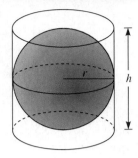

5. **a.** Calculate $\frac{1}{2} \cdot \frac{2}{3} \cdot \frac{3}{4}$.
   b. If you multiply all the fractions indicated from $\frac{1}{2}$ to $\frac{99}{100}$, what will the product be?
   $$\frac{1}{2} \cdot \frac{2}{3} \cdot \frac{3}{4} \cdot \frac{4}{5} \cdots \frac{99}{100}$$
   c. If you continue until $\frac{999}{1,000}$, what will the product be?
   d. Will the product equal 0 if you go far enough?

REMEMBER THIS

1. Add $\frac{2}{3} + \frac{4}{3}$.
2. True or false? If you multiply the numerator and denominator of $\frac{5}{8}$ by -1, the resulting fraction is still equivalent to $\frac{5}{8}$.
3. Find the sum of $\frac{3}{5}$ and $\frac{3}{-5}$.
4. Simplify $\dfrac{x^2 - (x^2 - 2x + 2)}{x - 1}$.
5. Solve for a: $3ab - c = c$.

6. True or false? $\dfrac{a^5}{a} = \dfrac{a^{-2}}{a^{-6}}$.
7. Simplify $\dfrac{(3x)^2}{3x^2}$.
8. If the area of this rectangle equals 18 cm², what is the perimeter?

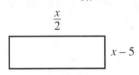

9. Graph the solution set of $-2x > 8$.
10. Write an algebraic expression for the sum of the squares of two consecutive integers.

5.3 Addition and Subtraction of Rational Expressions with the Same and with Opposite Denominators

f the chance of success in a business venture is $\dfrac{s}{n}$, then the chance of not

obtaining success is $\dfrac{n - s}{n}$. Find the sum of these probabilities. That is, add

$$\frac{s}{n} + \frac{n - s}{n} .$$

Express the result in lowest terms. (See Example 3.)

OBJECTIVES

1 Add and subtract rational expressions with the same denominator.

2 Add and subtract rational expressions with opposite denominators.

1   Recall from Section 1.1 that the sum (or difference) of two or more fractions that have the same denominator is given by the sum (or difference) of the numerators divided by the common denominator. For example,

$$\frac{1}{5} + \frac{2}{5} = \frac{1+2}{5} = \frac{3}{5},$$

$$\frac{9}{7} - \frac{4}{7} = \frac{9-4}{7} = \frac{5}{7}.$$

We can add or subtract rational expressions in the same way.

Addition or Subtraction of Rational Expressions

If $\dfrac{A}{B}$ and $\dfrac{C}{B}$ are rational expressions, with $B \neq 0$, then

$$\frac{A}{B} + \frac{C}{B} = \frac{A+C}{B} \quad \text{and} \quad \frac{A}{B} - \frac{C}{B} = \frac{A-C}{B}.$$

EXAMPLE 1   Write as a single fraction in lowest terms.

a. $\dfrac{4x}{9} + \dfrac{x}{9}$

b. $\dfrac{2}{y+7} - \dfrac{5}{y+7}$

Solution

a. $\dfrac{4x}{9} + \dfrac{x}{9} = \dfrac{4x+x}{9}$ Sum of the numerators divided by the common denominator.

$= \dfrac{5x}{9}$ Add in the numerator.

b. $\dfrac{2}{y+7} - \dfrac{5}{y+7} = \dfrac{2-5}{y+7}$ Difference of the numerators divided by the common denominator.

$= \dfrac{-3}{y+7}$ Subtract in the numerator.

PROGRESS CHECK 1   Write as a single fraction in lowest terms.

a. $\dfrac{6}{7t} - \dfrac{1}{7t}$

b. $\dfrac{x}{x-5} + \dfrac{3x}{x-5}$ ⌟

Any numerators with more than one term should be enclosed in parentheses when applying the addition or subtraction definition for rational expressions. In particular, this practice will help you to avoid the common student error discussed in the caution in Example 2.

EXAMPLE 2   Write as a single fraction in lowest terms.

a. $\dfrac{x^2+3x}{x^2+1} + \dfrac{2x^2-5x}{x^2+1}$

b. $\dfrac{4x+1}{2x-6} - \dfrac{2x+4}{2x-6}$

Solution

a. $\dfrac{x^2+3x}{x^2+1} + \dfrac{2x^2-5x}{x^2+1} = \dfrac{(x^2+3x)+(2x^2-5x)}{x^2+1}$ Add the fractions.

$= \dfrac{3x^2-2x}{x^2+1}$ Simplify in the numerator.

Progress Check Answers

1. (a) $\dfrac{5}{7t}$ (b) $\dfrac{4x}{x-5}$

b. $\dfrac{4x + 1}{2x - 6} - \dfrac{2x + 4}{2x - 6} = \dfrac{(4x + 1) - (2x + 4)}{2x - 6}$ Subtract the fractions.

$\qquad\qquad\qquad\qquad = \dfrac{4x + 1 - 2x - 4}{2x - 6}$ Remove parentheses.

$\qquad\qquad\qquad\qquad = \dfrac{2x - 3}{2x - 6}$ Combine like terms.

Caution   In Example 2b, note that we must subtract the entire numerator of the fraction on the right. Often students make the error of subtracting only the first term, as shown below.

$$\dfrac{4x + 1}{2x - 6} - \dfrac{2x + 4}{2x - 6} = \dfrac{4x + 1 - 2x + 4}{2x - 6} \qquad \textbf{WRONG}$$

Using parentheses as shown above will help you avoid this mistake.

PROGRESS CHECK 2   Write as a single fraction in lowest terms.

a. $\dfrac{4x^2 - 7}{x^2 + x} + \dfrac{3x^2 + 1}{x^2 + x}$ **b.** $\dfrac{5x - 6}{x - 1} - \dfrac{7x - 1}{x - 1}$ ⌐

In the next two examples additional steps are necessary in order to express the answer in lowest terms.

EXAMPLE 3   Solve the problem in the section introduction on page 231.

Solution   Simplify the expression given for the sum of the probabilities.

$$\dfrac{s}{n} + \dfrac{n - s}{n} = \dfrac{s + (n - s)}{n}$$ Add the fractions.

$$= \dfrac{n}{n}$$ Simplify in the numerator.

$$= 1$$ Express in lowest terms.

The sum of the given probabilities is 1. This represents a general rule in probability: The probability that an event will occur added to the probability that it will not occur give a sum of 1.

PROGRESS CHECK 3   The expected profit of a gambler who wagers x dollars in a certain (fair) game is given by

$$6x\left(\dfrac{s}{n}\right) + (-x)\left(\dfrac{6s}{n}\right) \text{ dollars.}$$

Simplify this expression to lowest terms.   ⌐

EXAMPLE 4   Add $\dfrac{3x + 4}{x^2 - 1} + \dfrac{-2x - 3}{x^2 - 1}$. Express the result in lowest terms.

Solution

$$\dfrac{3x + 4}{x^2 - 1} + \dfrac{-2x - 3}{x^2 - 1} = \dfrac{(3x + 4) + (-2x - 3)}{x^2 - 1}$$ Add the fractions.

$$= \dfrac{x + 1}{x^2 - 1}$$ Simplify in the numerator.

$$= \dfrac{x + 1}{(x + 1)(x - 1)}$$ Factor completely.

$$= \dfrac{1}{x - 1}$$ Express in lowest terms, using the fundamental principle.

PROGRESS CHECK 4   Add $\dfrac{4x-5}{x^2+x-2}+\dfrac{7-3x}{x^2+x-2}$. Express the result in lowest terms.

2   To add or subtract rational expressions when one denominator is the opposite of the other, choose either fraction and multiply the numerator and the denominator of this fraction by -1. Because opposite expressions differ only by a factor of -1, we may now write both expressions with the same denominator and combine them by our previous methods.

EXAMPLE 5   Write as a single fraction in lowest terms.

a. $\dfrac{7}{y}-\dfrac{5}{-y}$

b. $\dfrac{4x+5}{3-2x}+\dfrac{1-x}{2x-3}$

Solution   For the common denominator we usually choose to make the coefficient of the highest power of the variable a positive number.

a. $\dfrac{7}{y}-\dfrac{5}{-y}=\dfrac{7}{y}-\dfrac{5(-1)}{-y(-1)}$ Fundamental principle.

$=\dfrac{7}{y}-\dfrac{-5}{y}$ Perform the -1 multiplications.

$=\dfrac{7-(-5)}{y}$ Subtract the fractions.

$=\dfrac{12}{y}$ Simplify.

b. $\dfrac{4x+5}{3-2x}+\dfrac{1-x}{2x-3}=\dfrac{(4x+5)(-1)}{(3-2x)(-1)}+\dfrac{1-x}{2x-3}$ Fundamental principle.

$=\dfrac{-4x-5}{2x-3}+\dfrac{1-x}{2x-3}$ Perform the -1 multiplications.

$=\dfrac{(-4x-5)+(1-x)}{2x-3}$ Add the fractions.

$=\dfrac{-5x-4}{2x-3}$ Simplify.

Note   In Example 5b it is also sensible to obtain the same denominator by multiplying by -1 in the numerator and the denominator of the fraction on the right. Then,

$$\dfrac{4x+5}{3-2x}+\dfrac{1-x}{2x-3}=\dfrac{4x+5}{3-2x}+\dfrac{-1+x}{3-2x}=\dfrac{5x+4}{3-2x}.$$

The two answers are equivalent (show this) and both answers are correct.

PROGRESS CHECK 5   Write as a single fraction in lowest terms.

a. $\dfrac{3}{-x}-\dfrac{3}{x}$

b. $\dfrac{t+1}{t-1}+\dfrac{2-3t}{1-t}$

EXAMPLE 6   Add $\dfrac{y^2}{y-2}+\dfrac{4}{2-y}$. Express the answer in lowest terms.

Solution

$\dfrac{y^2}{y-2}+\dfrac{4}{2-y}=\dfrac{y^2}{y-2}+\dfrac{4(-1)}{(2-y)(-1)}$ Fundamental principle.

$=\dfrac{y^2}{y-2}+\dfrac{-4}{y-2}$ Perform the -1 multiplications.

$$= \frac{y^2 - 4}{y - 2} \qquad \text{Add the fractions.}$$

$$= \frac{(y + 2)(y - 2)}{y - 2} \qquad \text{Factor completely.}$$

$$= y + 2 \qquad \text{Express in lowest terms.}$$

PROGRESS CHECK 6 Add $\dfrac{y^2}{y - 3} + \dfrac{2y + 3}{3 - y}$. Express the result in lowest terms.

Progress Check Answer

⌐ 6. $y + 1$

EXERCISES 5.3

In Exercises 1–18, perform the addition or subtraction and express the answer in lowest terms.

1. $\dfrac{5x}{7} + \dfrac{2x}{7}$

2. $\dfrac{12y}{5} - \dfrac{7y}{5}$

3. $\dfrac{3}{x + 4} - \dfrac{5}{x + 4}$

4. $\dfrac{6}{x - 2} + \dfrac{4}{x - 2}$

5. $\dfrac{x}{x + 1} + \dfrac{1}{x + 1}$

6. $\dfrac{y}{y - 4} - \dfrac{4}{y - 4}$

7. $\dfrac{y^2 + y}{y^2 + 3} - \dfrac{y - 3}{y^2 + 3}$

8. $\dfrac{y^2 + x}{y^2 - 1} - \dfrac{x + 1}{y^2 - 1}$

9. $\dfrac{n}{m + n} - \dfrac{m - n}{m + n}$

10. $\dfrac{1 - x}{n} - \dfrac{x - x^2}{n}$

11. $\dfrac{5 - x^2}{x + 5} + \dfrac{x^2 - 5}{x + 5}$

12. $\dfrac{3x + 1}{3x + 2} + \dfrac{3x + 3}{3x + 2}$

13. $\dfrac{x^2}{x - 1} - \dfrac{2x - 1}{x - 1}$

14. $\dfrac{x^2 + 1}{x + 1} + \dfrac{2x}{x + 1}$

15. $\dfrac{2x}{9} - \dfrac{x - 1}{9} - \dfrac{x - 2}{9}$

16. $\dfrac{1}{x^2 + 2x} - \dfrac{1 - 3x}{x^2 + 2x} - \dfrac{2x}{x^2 + 2x}$

17. $\dfrac{x^2 - 3x - 1}{x^2 + 3x + 2} - \dfrac{x^2 - 4x - 2}{x^2 + 3x + 2}$

18. $\dfrac{y^2 + 5y + 1}{y^2 - 7y + 10} - \dfrac{y^2 + 6y - 1}{y^2 - 7y + 10}$

In Exercises 19–24, write the given fraction as the sum or difference of two fractions.

19. $\dfrac{x + 1}{y}$

20. $\dfrac{x - 5}{y}$

21. $\dfrac{y^2 + y}{x}$

22. $\dfrac{3x^2 - 4x}{2y}$

23. $\dfrac{x + 1}{x^2 + 1}$

24. $\dfrac{y - 5}{y^2 - 5}$

In Exercises 25–30, which choice is the negative of the given expression? Recall that the sum of an expression and its negative is zero.

25. $2x - 1$
  a. $1 - 2x$ **b.** $-2x - 1$ **c.** $2x + 1$

26. $3x + 4$
  a. $-3x - 4$ **b.** $-3x + 4$ **c.** $3x - 4$

27. $y^2 - 9$
  a. $9 - y^2$ **b.** $-y^2 - 9$ **c.** $y^2 + 9$

28. $x^2 + 25$
  a. $x^2 - 25$ **b.** $25 - x^2$ **c.** $-x^2 - 25$

29. $n^2 + 3n + 1$
  a. $n^2 - 3n - 1$ **b.** $-n^2 - 3n - 1$ **c.** $-n^2 + 3n + 1$

30. $x^2 - x - 1$
  a. $1 + x - x^2$ **b.** $x^2 + x + 1$ **c.** $-x^2 - x - 1$

In Exercises 31–48, perform the addition or subtraction and express the answer in lowest terms.

31. $\dfrac{6}{y} + \dfrac{5}{-y}$

32. $\dfrac{x}{3} + \dfrac{8x}{-3}$

33. $\dfrac{1}{2x} - \dfrac{1}{-2x}$

34. $\dfrac{1}{5y^2} - \dfrac{4}{-5y^2}$

35. $\dfrac{t - 1}{t} + \dfrac{t - 1}{-t}$

36. $\dfrac{w - 2}{w} - \dfrac{2 - w}{-w}$

37. $\dfrac{x + 2}{2x - 1} + \dfrac{3 - x}{1 - 2x}$

38. $\dfrac{3x}{3x - 2} + \dfrac{2}{2 - 3x}$

39. $\dfrac{y^2}{y - 3} + \dfrac{9}{3 - y}$

40. $\dfrac{z^2}{z - 5} + \dfrac{25}{5 - z}$

41. $\dfrac{2x^2}{2x - 1} + \dfrac{1 - x}{1 - 2x}$

42. $\dfrac{4x^2}{3x - 2} + \dfrac{x^2 - x + 2}{2 - 3x}$

43. $\dfrac{2t^2 + t + 1}{3t - 2} - \dfrac{t^2 + 3t - 5}{2 - 3t}$

44. $\dfrac{2t^2 + 5t - 3}{4t - 3} - \dfrac{2t^2 + 4t - 6}{3 - 4t}$

45. $\dfrac{y - y^2}{y^2 - 3y - 4} - \dfrac{y^2 - 4}{4 + 3y - y^2}$

46. $\dfrac{2}{y^2 - 4} - \dfrac{y}{4 - y^2}$

47. $\dfrac{x}{x^2 - x - 1} + \dfrac{x - 1}{1 + x - x^2}$

48. $\dfrac{x^2}{x^2 - 2x - 1} + \dfrac{x^2 - x}{1 + 2x - x^2}$

In Exercises 49–54, find an expression for the perimeter. Express it in lowest terms.

49. $\dfrac{3}{x}$

$\dfrac{1}{x}$

50. $\dfrac{3}{5x}$

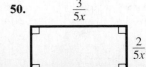

$\dfrac{2}{5x}$

51. $\dfrac{2-x}{x+1}$

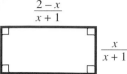

$\dfrac{x}{x+1}$

52. $\dfrac{x^2}{x+2}$

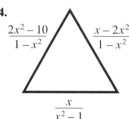

$\dfrac{1-x^2}{x+2}$

53.

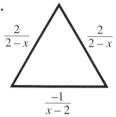

$\dfrac{2}{2-x}$ $\dfrac{2}{2-x}$

$\dfrac{-1}{x-2}$

54.

$\dfrac{2x^2-10}{1-x^2}$ $\dfrac{x-2x^2}{1-x^2}$

$\dfrac{x}{x^2-1}$

55. $(1-w)\dfrac{a}{c} + w\left(\dfrac{b}{c}\right)$ is the weighted average of $\dfrac{a}{c}$ and $\dfrac{b}{c}$. Express this average as a single fraction by performing the given addition.

56. In a probability exercise concerning a game where you will either win or lose c dollars, the expected value of the game is $\dfrac{x}{n}(c) + \dfrac{n-x}{n}(-c)$. The fraction $\dfrac{x}{n}$ represents the probability you will win, and $\dfrac{n-x}{n}$ represents the probability you will lose.

  a. Simplify the expression for the expected value.

  b. Calculate the expected value when $x = 1$ and $n = 2$.

  c. Calculate the expected value when $x = 2$ and $n = 3$.

THINK ABOUT IT

1. Two expressions are negatives of each other if their sum is 0. What is the negative of $x^2 - x + 1$?

2. Which of these sums are equal to 0?

  a. $\dfrac{3}{x} + \dfrac{3}{-x}$ **b.** $\dfrac{3}{x} + \dfrac{-3}{x}$ **c.** $\dfrac{3}{x} + \dfrac{-x}{3}$

3. a. Simplify $\dfrac{1}{2 + \dfrac{1}{2}}$.   **b.** Simplify $\dfrac{a}{x + \dfrac{a}{2}}$.

4. Why is $\dfrac{3x+4}{2x-6}$ called a "rational" expression?

5. What expression added to $\dfrac{x-2}{x+3}$ will equal 0?

REMEMBER THIS

1. Find the LCD for $\frac{3}{8}$ and $\frac{3}{20}$.

2. Simplify $\frac{1}{12} - \frac{2}{15} + \frac{1}{24}$.

3. Divide $18x^2y^3$ by $6x^2y$.

4. True or false? $y - 1$ is a factor of both $3y - 3$ and $y^2 - 1$.

5. Multiply and express the answer in lowest terms: $\dfrac{3x}{5y} \cdot \dfrac{10y^2}{6x^3}$.

6. Divide: $\dfrac{1-x}{y^2-y} \div \dfrac{x^2-1}{y^2-2y+1}$.

7. How much antifreeze should be added to 4 gal of a solution which is 80 percent antifreeze to get a 90 percent solution?

8. Solve $5x(x-1)(2x+5) = 0$.

9. Simplify $\dfrac{2x^2y^{-4}}{2^{-1}(x^{-1}y)^{-3}}$.

10. True or false? No matter what real number x stands for, $3x$ will be greater than x.

<table>
<tr><td>**5.4**</td><td>**Addition and Subtraction of Rational Expressions with Unlike Denominators**</td></tr>
</table>

When solving the formula $A = \frac{1}{2}h(a + b)$ for b in Exercise 53 of Section 2.5, a student obtains the answer $b = \dfrac{2A}{h} - a$. However, the answer given in the text answer key is

$$b = \dfrac{2A - ah}{h}.$$

Perform the subtraction in $\dfrac{2A}{h} - a$ to show that the student's answer is also correct. (See Example 6.)

1 Find the least common denominator (LCD).

2 Add and subtract rational expressions with unlike denominators.

1   For the remainder of this chapter, we will need to find an expression that is exactly divisible by the denominators in a given problem. Our work will be simpler if we find the smallest possible common denominator, called the **least common denominator,** or **LCD.** Example 1 reviews the procedure given in Section 1.1 for finding the LCD of fractions in arithmetic.

EXAMPLE 1   Find the least common denominator for $\frac{5}{18}$ and $\frac{11}{42}$.

Solution   To find the LCD, first express 18 and 42 as products of prime factors.

$$18 = 2 \cdot 3 \cdot 3 \quad \text{and} \quad 42 = 2 \cdot 3 \cdot 7$$

The LCD is the product of all the different prime factors, with each prime number appearing the greatest number of times it appears in any one factorization. Thus,

$$\text{LCD} = 2 \cdot 3 \cdot 3 \cdot 7 = 126.$$

PROGRESS CHECK 1   Find the least common denominator for $\frac{17}{12}$ and $\frac{7}{30}$.   ⌐

To find the least common denominator for rational expressions, we use a similar procedure.

> ### To Find the LCD
>
> 1. Factor completely each denominator.
> 2. The LCD is the product of all the different factors, with each factor appearing the greatest number of times it appears in any one factorization.

This procedure is illustrated in Examples 2 and 3.

EXAMPLE 2   Find the least common denominator for $\dfrac{5}{6x^2y}$ and $\dfrac{4}{9xy^3}$.

Solution   First, factor completely each denominator.

$$6x^2y = 2 \cdot 3 \cdot x \cdot x \cdot y$$
$$9xy^3 = 3 \cdot 3 \cdot x \cdot y \cdot y \cdot y$$

The LCD will contain the factors 2, 3, x, and y. In any factorization the greatest number of times that 2 appears is once, that 3 and that x appear is twice, and that y appears is three times. Thus,

$$\text{LCD} = 2 \cdot 3 \cdot 3 \cdot x \cdot x \cdot y \cdot y \cdot y = 18x^2y^3.$$

PROGRESS CHECK 2   Find the LCD for $\dfrac{1}{4xy^2}$ and $\dfrac{7}{6x^2y}$.   ⌐

EXAMPLE 3   Find the LCD for $\dfrac{y}{3y - 3}$ and $\dfrac{2}{y^2 - 1}$.

Solution   Factor each denominator completely.

$$3y - 3 = 3(y - 1)$$
$$y^2 - 1 = (y + 1)(y - 1)$$

The least common denominator will contain the factors 3, $y + 1$, and $y - 1$. The greatest number of times each factor appears in any one factorization is once. Thus,

$$LCD = 3(y + 1)(y - 1).$$

PROGRESS CHECK 3   Find the LCD for $\dfrac{1}{x^2 + 4x}$ and $\dfrac{x}{3x + 12}$. ⌐

2   Now we consider an important application of the LCD. Recall from Section 1.1 that when fractions have unlike denominators, we change them into equivalent fractions with the same denominator before we add or subtract them. Example 4 reviews the complete process shown earlier.

EXAMPLE 4   Add $\frac{5}{18} + \frac{11}{42}$. Express the result in lowest terms.

Solution   As shown in Example 1 of this section, the LCD here is 126. Because $126 = 18 \cdot 7$ and $126 = 42 \cdot 3$, we may add the fractions as shown below.

$$\frac{5}{18} + \frac{11}{42} = \frac{5 \cdot 7}{18 \cdot 7} + \frac{11 \cdot 3}{42 \cdot 3} \qquad \text{Fundamental principle.}$$

$$= \frac{35}{126} + \frac{33}{126} \qquad \text{Perform the multiplications.}$$

$$= \frac{68}{126} \qquad \text{Add the fractions.}$$

$$= \frac{34}{63} \qquad \text{Express in lowest terms.}$$

PROGRESS CHECK 4   Add $\frac{7}{30} + \frac{17}{18}$. Express the result in lowest terms. ⌐

The same method is also used to add or subtract rational expressions, and we may summarize the procedure as follows.

To Add or Subtract Rational Expressions

1. Completely factor each denominator and find the LCD.
2. For each fraction, obtain an equivalent fraction by applying the fundamental principle and multiplying the numerator and the denominator of the fraction by the factors of the LCD that are not contained in the denominator of that fraction.
3. Add or subtract the numerators and divide this result by the common denominator.
4. Express the answer in lowest terms.

With the understanding that final results should always be expressed in lowest terms (as indicated in step 4), we will not continue to state these directions from this point on.

EXAMPLE 5   Add $\dfrac{3}{2x} + \dfrac{5}{4}$.

Solution   First, find the LCD. The denominator $2x$ is $2 \cdot x$, while 4 factors as $2 \cdot 2$. The greatest number of times that 2 appears is twice, and that x appears is once; so the LCD is $2 \cdot 2 \cdot x$, or $4x$. Because $4x = 2x \cdot 2$ and $4x = 4 \cdot x$, we have the following addition.

$$\frac{3}{2x} + \frac{5}{4} = \frac{3 \cdot 2}{2x \cdot 2} + \frac{5 \cdot x}{4 \cdot x} \qquad \text{Fundamental principle.}$$

$$= \frac{3 \cdot 2 + 5 \cdot x}{4x} \qquad \text{Add the fractions.}$$

$$= \frac{6 + 5x}{4x} \text{, or } \frac{5x + 6}{4x} \qquad \text{Simplify in the numerator.}$$

PROGRESS CHECK 5   Add: $\dfrac{4}{3x} + \dfrac{2}{9}$.

EXAMPLE 6   Solve the problem in the section introduction on page 236.

Solution   Write a as $a/1$, so the expression becomes

$$\frac{2A}{h} - \frac{a}{1} .$$

The LCD is h, so the subtraction is as follows.

$$\frac{2A}{h} - \frac{a}{1} = \frac{2A}{h} - \frac{a \cdot h}{1 \cdot h} \qquad \text{Fundamental principle.}$$

$$= \frac{2A - ah}{h} \qquad \text{Subtract the fractions.}$$

This result shows that the student's answer is also correct.

PROGRESS CHECK 6   When solving the formula $S = \frac{1}{2}n(a + \ell)$ for ℓ, a student obtains the answer $\ell = \dfrac{2S}{n} - a$. Subtract on the right side of this equation to show that the student's answer is equivalent to $\ell = \dfrac{2S - na}{n}$, which is the solution in the text answer key.

EXAMPLE 7   Subtract $\dfrac{5}{6x^2y} - \dfrac{4}{9xy^3}$.

Solution   The LCD is $18x^2y^3$ (as explained in Example 2).

$$\frac{5}{6x^2y} - \frac{4}{9xy^3} = \frac{5(3y^2)}{6x^2y(3y^2)} - \frac{4(2x)}{9xy^3(2x)} \qquad \text{Fundamental principle.}$$

$$= \frac{5(3y^2) - 4(2x)}{18x^2y^3} \qquad \text{Subtract the fractions.}$$

$$= \frac{15y^2 - 8x}{18x^2y^3} \qquad \text{Simplify in the numerator.}$$

PROGRESS CHECK 7   Subtract $\dfrac{1}{4xy^2} - \dfrac{7}{6x^2y}$.

EXAMPLE 8   Add $\dfrac{x}{x - 4} + \dfrac{5}{x}$.

Solution   Neither denominator factors, so the LCD is $x(x - 4)$, which is the product of the denominators.

$$\frac{x}{x - 4} + \frac{5}{x} = \frac{x(x)}{(x - 4)(x)} + \frac{5(x - 4)}{x(x - 4)} \qquad \text{Fundamental principle.}$$

$$= \frac{x(x) + 5(x - 4)}{x(x - 4)} \qquad \text{Add the fractions.}$$

$$= \frac{x^2 + 5x - 20}{x(x - 4)} \qquad \text{Simplify in the numerator.}$$

Progress Check Answers

5. $\dfrac{12 + 2x}{9x}$

6. $\dfrac{2S}{n} - \dfrac{na}{n} = \dfrac{2S - na}{n}$

7. $\dfrac{3x - 14y}{12x^2y^2}$

PROGRESS CHECK 8   Add $\dfrac{y}{y-3} + \dfrac{2}{y}$.

EXAMPLE 9   Subtract $\dfrac{y}{y-1} - \dfrac{2}{y^2-1}$.

Solution   First, by noting that $y^2 - 1$ factors as $(y+1)(y-1)$, we determine that the LCD is $(y+1)(y-1)$. Then, we subtract as follows and note that extra steps are needed to express the answer in lowest terms.

$$\frac{y}{y-1} - \frac{2}{y^2-1}$$

$$= \frac{y}{y-1} - \frac{2}{(y+1)(y-1)} \qquad \text{Factor completely.}$$

$$= \frac{y(y+1)}{(y-1)(y+1)} - \frac{2}{(y-1)(y+1)} \qquad \text{Fundamental principle.}$$

$$= \frac{y(y+1) - 2}{(y+1)(y-1)} \qquad \text{Subtract the fractions.}$$

$$= \frac{y^2 + y - 2}{(y+1)(y-1)} \qquad \text{Simplify in the numerator.}$$

$$= \frac{(y+2)(y-1)}{(y+1)(y-1)} \qquad \text{Factor in the numerator.}$$

$$= \frac{y+2}{y+1} \qquad \text{Express in lowest terms.}$$

PROGRESS CHECK 9   Subtract $\dfrac{2t}{t^2-1} - \dfrac{1}{t+1}$.

EXAMPLE 10   Add $\dfrac{x+1}{x^2-x-6} + \dfrac{3x}{x^2-6x+9}$.

Solution   First, find the LCD. The denominator $x^2 - x - 6$ factors as $(x-3)(x+2)$, while $x^2 - 6x + 9$ factors as $(x-3)(x-3)$. The greatest number of times that $x - 3$ appears is twice, and that $x + 2$ appears is once. Thus, the LCD is $(x-3)(x-3)(x+2)$.

$$\frac{x+1}{x^2-x-6} + \frac{3x}{x^2-6x+9}$$

$$= \frac{x+1}{(x-3)(x+2)} + \frac{3x}{(x-3)(x-3)} \qquad \text{Factor completely.}$$

$$= \frac{(x+1)(x-3)}{(x-3)(x+2)(x-3)} + \frac{3x(x+2)}{(x-3)(x-3)(x+2)} \qquad \text{Fundamental principle.}$$

$$= \frac{(x+1)(x-3) + 3x(x+2)}{(x-3)(x-3)(x+2)} \qquad \text{Add the fractions.}$$

$$= \frac{x^2 - 3x + x - 3 + 3x^2 + 6x}{(x-3)(x-3)(x+2)} \qquad \begin{array}{l}\text{Remove parentheses}\\\text{in numerator.}\end{array}$$

$$= \frac{4x^2 + 4x - 3}{(x-3)(x-3)(x+2)} \text{, or } \frac{4x^2 + 4x - 3}{(x-3)^2(x+2)} \qquad \text{Simplify.}$$

Although the numerator $4x^2 + 4x - 3$ factors as $(2x+3)(2x-1)$, there are no common factors to divide out, so the above result represents lowest terms.

PROGRESS CHECK 10   Add $\dfrac{x-1}{x^2-4x+4} + \dfrac{2x}{x^2+x-6}$.

Progress Check Answers

8. $\dfrac{y^2 + 2y - 6}{y(y-3)}$

9. $\dfrac{1}{t-1}$

10. $\dfrac{3x^2 - 2x - 3}{(x+3)(x-2)^2}$

EXERCISES 5.4

In Exercises 1–12, find the least common denominator for the given fractions.

1. $\dfrac{5}{12}, \dfrac{1}{40}$ **2.** $\dfrac{5}{24}, \dfrac{7}{54}$ **3.** $\dfrac{1}{x}, \dfrac{2}{y}$

4. $\dfrac{3}{a}, \dfrac{4}{b}$ **5.** $\dfrac{5}{3xy}, \dfrac{1}{2x}$ **6.** $\dfrac{2}{15ab}, \dfrac{3}{5b}$

7. $\dfrac{7}{6x^2y^3}, \dfrac{3}{5x^3y}$ **8.** $\dfrac{1}{8ab^4}, \dfrac{2}{3a^3b^2}$

9. $\dfrac{x}{4x - 8}, \dfrac{x}{5x - 10}$ **10.** $\dfrac{y}{y + 5}, \dfrac{5}{3y + 15}$

11. $\dfrac{1}{y^2 - 9}, \dfrac{1}{3y + 9}$

12. $\dfrac{3}{n^2 - 25}, \dfrac{5}{2n - 10}$

In Exercises 13–42, express the result in lowest terms.

13. $\frac{5}{12} - \frac{1}{40}$ **14.** $\frac{5}{24} + \frac{7}{54}$

15. $\dfrac{1}{x} + \dfrac{2}{5}$ **16.** $\dfrac{3}{a} - \dfrac{9}{7a}$

17. $\dfrac{1}{ab} + \dfrac{2}{3b}$ **18.** $\dfrac{c}{2a} - \dfrac{1}{ab}$

19. $\dfrac{4}{3x^2y} - \dfrac{3}{4xy}$ **20.** $\dfrac{6}{5ab^3} + \dfrac{5}{6ab}$

21. $\dfrac{x}{x + 1} + \dfrac{3}{2x + 2}$ **22.** $\dfrac{4}{x - 3} - \dfrac{3x}{2x - 6}$

23. $\dfrac{x}{3x + 3} - \dfrac{1}{4x + 4}$ **24.** $\dfrac{3}{2y - 2} + \dfrac{y}{3y - 3}$

25. $\dfrac{2}{y} + \dfrac{3}{y + 1}$ **26.** $\dfrac{1}{x} - \dfrac{x}{x - 5}$

27. $\dfrac{3}{x} - \dfrac{2}{x - 2}$ **28.** $\dfrac{1}{y - 7} + \dfrac{4}{3y}$

29. $\dfrac{x}{x - 8} + \dfrac{x}{x + 8}$ **30.** $\dfrac{1}{a - 1} + \dfrac{1}{a + 1}$

31. $\dfrac{1}{y + 4} - \dfrac{1}{y}$ **32.** $\dfrac{1}{c - 2} - \dfrac{3}{3c}$

33. $\dfrac{d + 2}{d + 1} + \dfrac{d + 1}{d + 2}$

34. $\dfrac{d - 2}{d - 1} + \dfrac{d - 1}{d - 2}$

35. $\dfrac{x}{x + 3} - \dfrac{4}{x + 4}$

36. $\dfrac{y}{2y + 1} - \dfrac{1}{y + 2}$

37. $\dfrac{y}{y^2 - 4} + \dfrac{1}{y + 2}$ **38.** $\dfrac{x}{x^2 - 4} - \dfrac{1}{x + 2}$

39. $\dfrac{2}{x^2 + 3x + 2} - \dfrac{1}{x^2 + 2x + 1}$

40. $\dfrac{3x}{x^2 + 3x + 2} - \dfrac{6}{x^2 + 4x + 4}$

41. $\dfrac{x + 2}{x^2 - x - 2} + \dfrac{x - 1}{x^2 - 2x - 3}$

42. $\dfrac{x}{x^2 + x - 6} + \dfrac{x + 1}{x^2 + 2x - 8}$

In Exercises 43–46, find the perimeter of the given figure. Express it as a fraction in lowest terms.

43.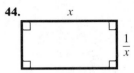

44.

45.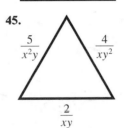

46.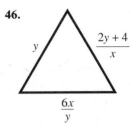

47. When solving the formula $V = \frac{1}{2}hd(a + b)$ for a, a student gets $a = \dfrac{2V}{hd} - b$. The answer in the text is $a = \dfrac{2V - bhd}{hd}$.

Perform the subtraction in $\dfrac{2V}{hd} - b$ to show that the student's answer is correct.

48. When solving $V = \frac{1}{6}hd(a + b)$ for b, a student gets $b = \dfrac{6V}{hd} - a$. The answer in the text is $b = \dfrac{6V - ahd}{hd}$.

Perform the subtraction in $\dfrac{6V}{hd} - a$ to show that the student's answer is correct.

THINK ABOUT IT

1. The average of 3 and 5 is 4. Is the average of $\frac{1}{3}$ and $\frac{1}{5}$ equal to $\frac{1}{4}$?

   a. Compute the average of $\dfrac{1}{3}$ and $\dfrac{1}{5}$.

   b. Compute the average of $\dfrac{1}{a}$ and $\dfrac{1}{b}$.

2. Simplify:

   a. $\dfrac{1}{x} + \dfrac{1}{2x}$ **b.** $\dfrac{1}{x} + \dfrac{1}{2x} + \dfrac{1}{4x}$ **c.** $\dfrac{1}{x} + \dfrac{1}{2x} + \dfrac{1}{4x} + \dfrac{1}{8x}$

   d. Can you guess what the sum will be if you extend the expression so that it has 5 terms? 10 terms?

3. The Egyptians in ancient times expressed most fractional quantities as sums of unit fractions, which are fractions with numerators equal to 1. For example, they expressed $\frac{2}{7}$ as $\frac{1}{4} + \frac{1}{28}$.

   a. Show that $\dfrac{1}{x} + \dfrac{1}{y} = \dfrac{x+y}{xy}$.

   b. Let $x = 4$ and $y = 28$ to show that the Egyptians were correct.

   c. Let $x = 2$ and $y = 14$ to find another quantity that can be expressed as the sum of two unit fractions.

   d. Can you find two unit fractions with different denominators whose sum equals $\frac{2}{9}$?

4. a. Show that $\dfrac{1}{10} + \dfrac{2}{10^2} + \dfrac{3}{10^3}$ equals $\dfrac{123}{1,000}$, or 0.123.

   b. A fraction which is the sum of fractions with denominators which are powers of the same positive integer is called a *radix fraction* for the *base* of the integer. Therefore $\dfrac{1}{10} + \dfrac{2}{10^2} + \dfrac{3}{10^3}$ is a radix fraction for base 10.

   Simplify $\dfrac{1}{b} + \dfrac{2}{b^2} + \dfrac{3}{b^3}$, a radix fraction in base b.

5. Express the sum as a single fraction.

   a. $1 + \dfrac{1}{x}$

   b. $1 + \dfrac{1}{x} + \dfrac{1}{x^2}$

   c. $1 + \dfrac{1}{x} + \dfrac{1}{x^2} + \dfrac{1}{x^3}$

   d. Can you guess what the sum will look like if you extend the expression so that the sum has 5 terms? 10 terms?

REMEMBER THIS

1. Simplify $(2 - \frac{3}{10}) - (\frac{4}{5} + 1)$.

2. Multiply $4\left(a + \dfrac{b}{4}\right)$.

3. Multiply $ad\left(\dfrac{1}{a} - \dfrac{3}{d}\right)$.

4. Simplify $\dfrac{1}{a/b}$.

5. What is one-half of the sum of two-thirds and three-fourths?

6. Add $\dfrac{3x}{4-y} + \dfrac{2x}{y-4}$.

7. Simplify $\dfrac{1}{x^2 + x - 2} + \dfrac{1}{3 - 2x - x^2}$.

8. Solve for c: $b =$

9. Graph the solution set of $x(x + 1) = 12$.

10. Find the length of a rectangle if its width is 12 cm and its diagonal is 20 cm.

5.5   Complex Fractions

In photography the focal length f of a lens is given by

$$f = \dfrac{1}{\dfrac{1}{d} + \dfrac{1}{a}},$$

where d is the distance from some object to the lens and a is the distance of its image from the lens. Rewrite this formula by simplifying the expression on the right side of this equation. (See Example 2.)

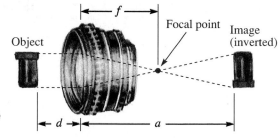

Object   Focal point   Image (inverted)

OBJECTIVES

1   Simplify complex fractions.

1   In the section-opening problem the expression on the right side of the focal length formula,

$$\dfrac{1}{\dfrac{1}{d} + \dfrac{1}{a}},$$

is an example of a complex fraction. A **complex fraction** is a fraction in which the numerator or the denominator, or both, involves fractions. The diagram below shows what is meant by the numerator and the denominator of a complex fraction.

$$\dfrac{1}{\dfrac{1}{d} + \dfrac{1}{a}}$$

Numerator of complex fraction
Primary fraction bar
Denominator of complex fraction

Two methods for simplifying a complex fraction are explained with an arithmetic illustration in Example 1.

EXAMPLE 1   Simplify the complex fraction $\dfrac{2 - \dfrac{3}{10}}{\dfrac{4}{5} + 1}$.

Solution   Consider how both methods show that the fraction simplifies to $\frac{17}{18}$.

Method 1   First, obtain a single fraction in the numerator and in the denominator of the complex fraction.

$$\dfrac{2 - \dfrac{3}{10}}{\dfrac{4}{5} + 1} = \dfrac{\dfrac{2 \cdot 10}{1 \cdot 10} - \dfrac{3}{10}}{\dfrac{4}{5} + \dfrac{1 \cdot 5}{1 \cdot 5}} = \dfrac{\dfrac{20}{10} - \dfrac{3}{10}}{\dfrac{4}{5} + \dfrac{5}{5}} = \dfrac{\dfrac{17}{10}}{\dfrac{9}{5}}$$

Then, because a/b is equivalent to $a \div b$, we get

$$\dfrac{\dfrac{17}{10}}{\dfrac{9}{5}} = \dfrac{17}{10} \div \dfrac{9}{5} = \dfrac{17}{10} \cdot \dfrac{5}{9} = \dfrac{17 \cdot 5}{10 \cdot 9} = \dfrac{17}{18} .$$

Method 2   First, find the LCD of all the fractions in the numerator and the denominator of the complex fraction.

$$\text{LCD of } \dfrac{2}{1}, \dfrac{3}{10}, \dfrac{4}{5}, \text{ and } \dfrac{1}{1} \text{ is 10.}$$

Then, multiply the numerator and the denominator of the complex fraction by this LCD and simplify the result.

$$\dfrac{2 - \dfrac{3}{10}}{\dfrac{4}{5} + 1} = \dfrac{10\left(2 - \dfrac{3}{10}\right)}{10\left(\dfrac{4}{5} + 1\right)} = \dfrac{10 \cdot 2 - 10 \cdot \dfrac{3}{10}}{10 \cdot \dfrac{4}{5} + 10 \cdot 1} = \dfrac{20 - 3}{8 + 10} = \dfrac{17}{18}$$

PROGRESS CHECK 1   Simplify $\dfrac{4 - \dfrac{1}{3}}{\dfrac{5}{6} + 3}$. Use both methods of Example 1.   ⌐

Use Example 1 as a basis for understanding the following procedures for simplifying complex fractions that contain rational expressions.

Progress Check Answer

1. $\dfrac{22}{23}$

Methods to Simplify Complex Fractions

Method 1 (Obtain single fractions and divide): Obtain single fractions in both the numerator and the denominator of the complex fraction. Then divide by multiplying by the reciprocal of the denominator.

Method 2 (Multiply using the LCD): Find the LCD of all the fractions that appear in the numerator and the denominator of the complex fraction. Then multiply the numerator and the denominator of the complex fraction by the LCD and simplify the results.

In Example 2 we will once again contrast the two methods.

EXAMPLE 2 Solve the problem in the section introduction on page 242.

Solution The given formula may be expressed as $f = \dfrac{ad}{a + d}$, as shown below.

Method 1 To obtain single fractions, we need only add in the denominator.

$$\frac{1}{\dfrac{1}{d} + \dfrac{1}{a}} = \frac{1}{\dfrac{a}{da} + \dfrac{d}{ad}} = \frac{1}{\dfrac{a + d}{ad}}$$

Then divide by multiplying by the reciprocal of the denominator.

$$1 \div \frac{a + d}{ad} = \frac{1}{1} \cdot \frac{ad}{a + d} = \frac{ad}{a + d}$$

Method 2 The LCD of $1/1$, $1/d$, and $1/a$ is ad. Therefore, we multiply in the numerator and the denominator of the complex fraction by ad and then simplify.

$$\frac{1}{\dfrac{1}{d} + \dfrac{1}{a}} = \frac{ad(1)}{ad\left(\dfrac{1}{d} + \dfrac{1}{a}\right)} = \frac{ad(1)}{ad\left(\dfrac{1}{d}\right) + ad\left(\dfrac{1}{a}\right)} = \frac{ad}{a + d}$$

PROGRESS CHECK 2 The harmonic mean M of two numbers a and b is given by

$$M = \frac{2}{\dfrac{1}{a} + \dfrac{1}{b}}.$$

Rewrite the formula by simplifying the expression on the right side of the equation.

Selecting an efficient method for simplifying will depend on your strengths and the particular problem. However, the next two examples include some guidelines for selecting a method that you should consider.

Progress Check Answer

2. $M = \dfrac{2ab}{a + b}$

EXAMPLE 3 Simplify the complex fraction $\dfrac{\dfrac{2}{a + b}}{\dfrac{1}{a} + \dfrac{1}{b}}$.

Solution   The LCD, which is $ab(a + b)$, is relatively complex, and the numerator is already a single fraction. Therefore, we will choose method 1 and begin by adding in the denominator.

$$\frac{\dfrac{2}{a + b}}{\dfrac{1}{a} + \dfrac{1}{b}} = \frac{\dfrac{2}{a + b}}{\dfrac{b}{ab} + \dfrac{a}{ba}} = \frac{\dfrac{2}{a + b}}{\dfrac{b + a}{ab}}$$

Then, divide by multiplying by the reciprocal of the denominator.

$$\frac{2}{a + b} \div \frac{b + a}{ab} = \frac{2}{a + b} \cdot \frac{ab}{b + a} = \frac{2ab}{(a + b)^2}$$

PROGRESS CHECK 3   Simplify the complex fraction $\dfrac{\dfrac{1}{y} + \dfrac{1}{x}}{\dfrac{y}{y + x}}$.

EXAMPLE 4   Simplify the complex fraction $\dfrac{\dfrac{1}{3} - \dfrac{2}{y}}{\dfrac{5}{6} + \dfrac{1}{y^2}}$.

Solution   The LCD of $\frac{1}{3}$, $2/y$, $\frac{5}{6}$, and $1/y^2$ is $6y^2$. The LCD is relatively simple, and method 1 would require us to obtain single fractions in both the numerator and the denominator. Therefore, we select method 2.

$$\frac{\dfrac{1}{3} - \dfrac{2}{y}}{\dfrac{5}{6} + \dfrac{1}{y^2}} = \frac{6y^2\left(\dfrac{1}{3} - \dfrac{2}{y}\right)}{6y^2\left(\dfrac{5}{6} + \dfrac{1}{y^2}\right)} \qquad \text{Multiply the numerator and the denominator by } 6y^2.$$

$$= \frac{6y^2\left(\dfrac{1}{3}\right) - 6y^2\left(\dfrac{2}{y}\right)}{6y^2\left(\dfrac{5}{6}\right) + 6y^2\left(\dfrac{1}{y^2}\right)} \qquad \text{Distributive property.}$$

$$= \frac{2y^2 - 12y}{5y^2 + 6} \qquad \text{Simplify.}$$

PROGRESS CHECK 4   Simplify the complex fraction $\dfrac{\dfrac{1}{2} + \dfrac{3}{x}}{\dfrac{7}{4} - \dfrac{5}{x^2}}$.

Example 5 points out once again that we always need to check that final results are expressed in lowest terms.

EXAMPLE 5   Simplify the complex fraction $\dfrac{\dfrac{y}{x} - 1}{\dfrac{y^2}{x^2} - 1}$.

Progress Check Answers

3. $\dfrac{(x + y)^2}{xy^2}$

4. $\dfrac{2x^2 + 12x}{7x^2 - 20}$

Solution   The LCD is simply x^2, and we will use method 2 for the same reasons that were given in Example 4.

$$\frac{\dfrac{y}{x} - 1}{\dfrac{y^2}{x^2} - 1} = \frac{x^2\left(\dfrac{y}{x} - 1\right)}{x^2\left(\dfrac{y^2}{x^2} - 1\right)}$$ Multiply the numerator and the denominator by x^2.

$$= \frac{x^2\left(\dfrac{y}{x}\right) - x^2 \cdot 1}{x^2\left(\dfrac{y^2}{x^2}\right) - x^2 \cdot 1}$$ Distributive property.

$$= \frac{xy - x^2}{y^2 - x^2}$$ Simplify.

$$= \frac{x(y - x)}{(y + x)(y - x)}$$ Factor completely.

$$= \frac{x}{y + x}$$ Write in lowest terms.

PROGRESS CHECK 5   Simplify the complex fraction $\dfrac{\dfrac{1}{25} - \dfrac{1}{n^2}}{\dfrac{1}{5} - \dfrac{1}{n}}$.

EXERCISES 5.5

In Exercises 1–12, show that the fraction simplifies to the given result by two methods.

  a. Obtain single fractions in the numerator and the denominator and then divide.

  b. Multiply the numerator and the denominator of the complex fraction by the LCD of all the fractions in the numerator and the denominator.

1. Show that $\dfrac{3 - \dfrac{4}{5}}{\dfrac{3}{10} + 2}$ equals $\dfrac{22}{23}$.

2. Show that $\dfrac{1 - \dfrac{3}{8}}{2 + \dfrac{1}{2}}$ equals $\dfrac{1}{4}$.

3. Show that $\dfrac{5 + \dfrac{3}{2}}{2 - \dfrac{1}{3}}$ equals $\dfrac{39}{10}$.

4. Show that $\dfrac{3 + \dfrac{1}{4}}{6 - \dfrac{1}{5}}$ equals $\dfrac{65}{116}$.

5. Show that $\dfrac{\dfrac{1}{4} + \dfrac{1}{5}}{\dfrac{1}{2} + \dfrac{1}{3}}$ equals $\dfrac{27}{50}$.

6. Show that $\dfrac{\dfrac{1}{5} + \dfrac{1}{6}}{\dfrac{1}{3} + \dfrac{1}{4}}$ equals $\dfrac{22}{35}$.

7. Show that $\dfrac{a - \dfrac{1}{b}}{a + \dfrac{1}{b}}$ equals $\dfrac{ab - 1}{ab + 1}$.

8. Show that $\dfrac{x + \dfrac{1}{y}}{x + \dfrac{2}{y}}$ equals $\dfrac{xy + 1}{xy + 2}$.

9. Show that $\dfrac{a - \dfrac{1}{b}}{a - \dfrac{1}{2b}}$ equals $\dfrac{2ab - 2}{2ab - 1}$.

10. Show that $\dfrac{5 + \dfrac{1}{a}}{10 + \dfrac{2}{a}}$ equals $\dfrac{1}{2}$.

11. Show that $\dfrac{a + \dfrac{b}{a}}{b + \dfrac{a}{b}}$ equals $\dfrac{a^2 b + b^2}{ab^2 + a^2}$.

12. Show that $\dfrac{a + \dfrac{b}{2}}{b + \dfrac{a}{3}}$ equals $\dfrac{6a + 3b}{6b + 2a}$.

In Exercises 13–36, simplify the complex fraction.

13. $\dfrac{3}{\dfrac{1}{a} + \dfrac{2}{b}}$

14. $\dfrac{5}{\dfrac{3}{a} - \dfrac{6}{b}}$

15. $\dfrac{7}{\dfrac{3}{a} + \dfrac{5}{2a}}$

16. $\dfrac{x}{\dfrac{3}{x} - \dfrac{1}{2x}}$

17. $\dfrac{5}{\dfrac{1}{a} - \dfrac{3}{a^2}}$

18. $\dfrac{8}{\dfrac{4}{n} + \dfrac{2}{n^2}}$

19. $\dfrac{\dfrac{3}{x+y}}{\dfrac{1}{x}+\dfrac{2}{y}}$

20. $\dfrac{\dfrac{1}{x-y}}{\dfrac{1}{x}-\dfrac{1}{y}}$

21. $\dfrac{\dfrac{2}{x}+\dfrac{3}{y}}{\dfrac{6}{xy}}$

22. $\dfrac{\dfrac{3}{2x}-\dfrac{2}{3y}}{\dfrac{1}{xy}}$

23. $\dfrac{\dfrac{x}{y}-\dfrac{y}{x}}{\dfrac{x-y}{x+y}}$

24. $\dfrac{\dfrac{a+b}{a-b}}{\dfrac{1}{a}+\dfrac{1}{b}}$

25. $\dfrac{\dfrac{x}{y}+\dfrac{1}{x+y}}{\dfrac{2}{x+y}}$

26. $\dfrac{\dfrac{2}{a-b}-\dfrac{1}{a}}{\dfrac{a+b}{a-b}}$

27. $\dfrac{\dfrac{2}{3}+\dfrac{1}{n}}{\dfrac{3}{4}-\dfrac{1}{n^2}}$

28. $\dfrac{1-\dfrac{1}{m^2}}{\dfrac{1}{2}+m}$

29. $\dfrac{\dfrac{2}{1+n}}{\dfrac{1}{n^2}+\dfrac{1}{n}}$

30. $\dfrac{\dfrac{1}{m+n}+\dfrac{1}{m-n}}{\dfrac{1}{m^2-n^2}}$

31. $\dfrac{\dfrac{a}{b}-2}{\dfrac{a^2}{b^2}-4}$

32. $\dfrac{1-\dfrac{x^2}{4y^2}}{1+\dfrac{x}{2y}}$

33. $\dfrac{\dfrac{a}{a+b}-1}{\dfrac{b}{a+b}-1}$

34. $\dfrac{\dfrac{ab}{a-b}-a}{\dfrac{ab}{a-b}-b}$

35. $\dfrac{\dfrac{c}{d+1}-2}{2-\dfrac{c}{d+1}}$

36. $\dfrac{\dfrac{1}{2x+4}-x}{2x-\dfrac{1}{x+2}}$

37. Use the definition of the harmonic mean of two numbers given in "Progress Check" 2 to show that the harmonic mean of 6 and 12 is 8. The word *harmonic* is related to its use in music. If a guitar string is 12 units long, and if it is shortened to length 8, the new tone (called the "fifth") is in "harmony" with the original. Similarly, if it is shortened to length 6, the new tone (called the "octave") is in harmony. Also, because a cube has 6 faces, 8 vertices, and 12 edges, it is called a *harmonic body*.

38. The harmonic mean M of three numbers a, b, and c is given by

$$M=\dfrac{3}{\dfrac{1}{a}+\dfrac{1}{b}+\dfrac{1}{c}}.$$

Simplify the expression on the right side to get another version of the formula.

THINK ABOUT IT

1. A complex fraction approximation for π is given by

$$3+\dfrac{1}{7+\frac{1}{16}}.$$

Evaluate this expression and compare it with π (which equals 3.141592654 to nine decimal places).

2. Consider these expressions.

$$\dfrac{1}{2+\frac{1}{2}},\quad \dfrac{1}{2+\dfrac{1}{2+\frac{1}{2}}},\quad \text{and} \quad \dfrac{1}{2+\dfrac{1}{2+\dfrac{1}{2+\frac{1}{2}}}}$$

  a. Show that they simplify to $\frac{2}{5}$, $\frac{5}{12}$, and $\frac{12}{29}$, respectively.

  b. If you keep going like this, the expression is called a *continued fraction*. (The term *continued fraction* was introduced by the English mathematician John Wallis in 1695.) There is a pattern to the fractions in part **a**. The next fraction is $\frac{29}{70}$. What comes after that?

  c. If you express each fraction as a decimal and add 1 to each, you may see that the answers are getting closer and closer to $\sqrt{2}$. Find an approximation to $\sqrt{2}$ on a calculator and see how these answers compare.

3. Simplify each of these complex fractions.

  a. $\dfrac{1}{2+\frac{1}{4}}$
  b. $\dfrac{1}{2+\dfrac{1}{4+\frac{1}{8}}}$
  c. $\dfrac{1}{2+\dfrac{1}{4+\dfrac{1}{8+\frac{1}{16}}}}$

4. If you inscribe a square in a right triangle as shown, the side of the square will be *half* the harmonic mean of the two legs of the triangle. (See "Progress Check" 2.)

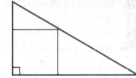

  a. Show that a 4-unit square can be inscribed in a right triangle with legs 6 and 12. Show that all three triangles are similar by showing that the sides are in proportion.

  b. What is the length of the side of the square inscribed in a 3–4–5 right triangle?

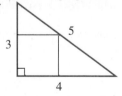

5. Using the formula from the previous problem, find the ratio of the area of the square to the area of right triangle ABC.

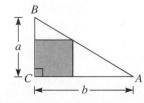

REMEMBER THIS

1. Multiply $\dfrac{3x}{8} - \dfrac{1}{4}$ by 8.

2. What is the LCD of $\dfrac{4}{x}$, 1, and $\dfrac{5}{2x}$?

3. Multiply $30R\left(\dfrac{1}{R} + \dfrac{1}{10}\right)$.

4. Prove that $\dfrac{1{,}010}{2{,}323} = \dfrac{10}{23}$ by showing that the cross products, 1,010(23) and 10(2,323), are equal.

5. Subtract $\dfrac{2A}{h} - a$ to get one fraction.

6. Simplify $\dfrac{x}{x-1} + \dfrac{x}{1-x}$.

7. The area of the rectangle is 22 ft². What is the radius of the circle?

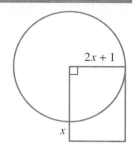

8. Simplify $\dfrac{3x^{-2}y^3z^{-4}}{3x^{-2}x^{-4}y^{-3}z^{-2}}$.

9. The sum of three consecutive integers is -117. Find the largest of the three integers.

10. Factor $2a^2 + 3a + 8a + 12$ by grouping.

5.6   Solving Equations Containing Rational Expressions

I f a resistor of R_1 ohms is connected in parallel to a resistor of R_2 ohms, as shown in the diagram, then the total resistance R is given by

$$\frac{1}{R} = \frac{1}{R_1} + \frac{1}{R_2}.$$

Find R_1 when $R = 6$ ohms and $R_2 = 10$ ohms. (See Example 3.)

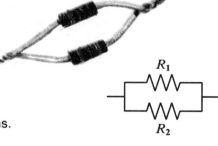

OBJECTIVES

1 Solve equations that contain rational expressions.

2 Solve formulas that contain rational expressions for a specified variable.

1 Many applied problems involve equations or formulas that contain rational expressions. The usual procedure for solving these equations is to remove the fractions by multiplying both sides of the equation by the least common denominator, or LCD. The resulting equation will not contain fractions and in many cases may be solved by methods we have already discussed.

EXAMPLE 1   Solve $\dfrac{3x}{8} - \dfrac{1}{4} = \dfrac{x}{2}$.

Solution   Multiply both sides of the equation by the LCD, 8.

$$8\left(\frac{3x}{8}-\frac{1}{4}\right)=8\left(\frac{x}{2}\right) \quad \text{Multiply both sides by 8.}$$

$$8\left(\frac{3x}{8}\right)-8\left(\frac{1}{4}\right)=8\left(\frac{x}{2}\right) \quad \text{Distributive property.}$$

$$3x-2=4x \quad \text{Simplify.}$$

$$-2=x \quad \text{Subtract } 3x \text{ from both sides.}$$

Now, replace x by -2 in the original equation to check the solution.

$$\frac{3(-2)}{8}-\frac{1}{4}\overset{?}{=}\frac{-2}{2}$$

$$\frac{-3}{4}-\frac{1}{4}\overset{?}{=}-1$$

$$-1\overset{\checkmark}{=}-1$$

The solution checks, and the solution set is $\{-2\}$.

PROGRESS CHECK 1   Solve $\dfrac{x}{5}+1=\dfrac{x+2}{10}$.

EXAMPLE 2   Solve $\dfrac{4}{x}-1=\dfrac{5}{2x}$.

Solution   We first remove fractions by multiplying both sides of the equation by the LCD, which is $2x$. Note that this step requires the restriction that $x\neq 0$ to ensure that we are multiplying both sides of the equation by a nonzero number.

$$2x\left(\frac{4}{x}-1\right)=2x\left(\frac{5}{2x}\right) \quad \text{Multiply both sides by } 2x.$$

$$2x\left(\frac{4}{x}\right)-2x(1)=2x\left(\frac{5}{2x}\right) \quad \text{Distributive property.}$$

$$8-2x=5 \quad \text{Simplify.}$$

$$-2x=-3 \quad \text{Subtract 8 from both sides.}$$

$$x=\frac{3}{2} \quad \text{Divide both sides by } -2 \text{ and simplify.}$$

To check the solution, replace x by $\frac{3}{2}$ in the original equation.

$$\frac{4}{\frac{3}{2}}-1\overset{?}{=}\frac{5}{2(\frac{3}{2})}$$

$$\frac{8}{3}-1\overset{?}{=}\frac{5}{3}$$

$$\frac{5}{3}\overset{\checkmark}{=}\frac{5}{3}$$

Thus, the solution set is $\{\frac{3}{2}\}$.

PROGRESS CHECK 2   Solve $\dfrac{5}{x}-3=\dfrac{5}{4x}$.

EXAMPLE 3   Solve the problem in the section introduction on page 248.

Solution   Replace R by 6 and R_2 by 10 in the given formula. Then solve for R_1 by first multiplying both sides of the equation by the LCD, $30R_1$. In the symbols R_1 and

R_2 the numbers 1 and 2 are called **subscripts.** In this case R_1 and R_2 are used to denote two resistor values. Note that R, R_1, and R_2 are *not* like terms.

$$\frac{1}{R} = \frac{1}{R_1} + \frac{1}{R_2} \qquad \text{Given formula.}$$

$$\frac{1}{6} = \frac{1}{R_1} + \frac{1}{10} \qquad \text{Replace } R \text{ by 6 and } R_2 \text{ by 10.}$$

$$30R_1\left(\frac{1}{6}\right) = 30R_1\left(\frac{1}{R_1} + \frac{1}{10}\right) \qquad \text{Multiply both sides by } 30R_1.$$

$$30R_1\left(\frac{1}{6}\right) = 30R_1\left(\frac{1}{R_1}\right) + 30R_1\left(\frac{1}{10}\right) \qquad \text{Distributive property.}$$

$$5R_1 = 30 + 3R_1 \qquad \text{Simplify.}$$

$$2R_1 = 30 \qquad \text{Subtract } 3R_1 \text{ from both sides.}$$

$$R_1 = 15 \qquad \text{Divide both sides by 2.}$$

Because $\frac{1}{6} = \frac{1}{15} + \frac{1}{10}$ is a true statement, the solution checks, and R_1 is a resistor of 15 ohms.

PROGRESS CHECK 3   Use the formula in Example 3 to find R_2 if $R = 4$ ohms and $R_1 = 8$ ohms.   ⌐

EXAMPLE 4   Solve $\dfrac{6}{y+2} = \dfrac{2}{y}$.

Solution   The LCD is $y(y+2)$, which is the product of the denominators. Assume, then, that $y \neq 0$ and $y \neq -2$.

$$y(y+2)\left(\frac{6}{y+2}\right) = y(y+2)\left(\frac{2}{y}\right) \qquad \text{Multiply both sides by } y(y+2).$$

$$6y = 2(y+2) \qquad \text{Simplify.}$$

$$6y = 2y + 4 \qquad \text{Distributive property.}$$

$$4y = 4 \qquad \text{Subtract } 2y \text{ from both sides.}$$

$$y = 1 \qquad \text{Divide both sides by 4.}$$

The solution checks because $\dfrac{6}{1+2} = \dfrac{2}{1}$ is a true statement, and the solution set is $\{1\}$.

PROGRESS CHECK 4   Solve $\dfrac{1}{x} = \dfrac{3}{x-4}$.   ⌐

In Example 4 the equation fit the form

$$\frac{A}{B} = \frac{C}{D}.$$

In such cases BD is a common denominator. Multiplying both sides of the equation by BD, with $B \neq 0$ and $D \neq 0$, gives

$$BD \cdot \frac{A}{B} = BD \cdot \frac{C}{D}$$

$$AD = BC.$$

This equation is easily produced through cross multiplication, as shown in Figure 5.1. And the following property is often useful.

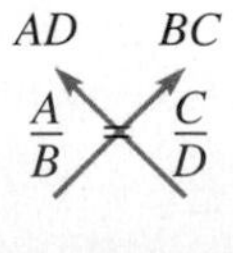

$$AD \qquad BC$$

$$\frac{A}{B} \diagdown\!\!\!\!\!\diagup \frac{C}{D}$$

Figure 5.1

Cross Multiplication Property

If $\dfrac{A}{B}$ and $\dfrac{C}{D}$ are rational expressions, with $B \neq 0$ and $D \neq 0$, then

$$\frac{A}{B} = \frac{C}{D} \qquad \text{implies} \qquad AD = BC.$$

Cross multiplication is most effective when BD is the least common denominator (as in Example 4), because cross multiplying in such cases is equivalent to multiplying both sides of the equation by the LCD.

EXAMPLE 5   Solve $\dfrac{x-1}{x+1} = \dfrac{x-5}{x-2}$.

Solution   The equation fits the form $A/B = C/D$, and the LCD is $(x+1)(x-2)$, which is the product of the denominators. Therefore, use cross multiplication, as shown below.

$$\frac{x-1}{x+1} = \frac{x-5}{x-2} \qquad \text{Given equation.}$$
$$(x-1)(x-2) = (x+1)(x-5) \qquad \text{Cross multiply, } x \neq -1, x \neq 2.$$
$$x^2 - 3x + 2 = x^2 - 4x - 5 \qquad \text{Multiply.}$$
$$-3x + 2 = -4x - 5 \qquad \text{Subtract } x^2 \text{ from both sides.}$$
$$x + 2 = -5 \qquad \text{Add } 4x \text{ to both sides.}$$
$$x = -7 \qquad \text{Subtract 2 from both sides.}$$

Because

$$\frac{-7-1}{-7+1} = \frac{-8}{-6} = \frac{4}{3} \qquad \text{and} \qquad \frac{-7-5}{-7-2} = \frac{-12}{-9} = \frac{4}{3},$$

the solution checks. The solution set is $\{-7\}$.

PROGRESS CHECK 5   Solve $\dfrac{x+2}{x-3} = \dfrac{x-4}{x+1}$. ⌐

EXAMPLE 6   Solve $\dfrac{2}{y^2-1} = \dfrac{3}{y^2-y}$.

Solution   First, by noting that $y^2 - 1$ factors as $(y+1)(y-1)$ and $y^2 - y$ factors as $y(y-1)$, we determine that the LCD is $y(y+1)(y-1)$. Although the equation fits the form $A/B = C/D$, the LCD is simpler than the product of the denominators, so it is more complicated to cross multiply. Instead, we begin by multiplying both sides by $y(y+1)(y-1)$, assuming $y \neq 0$, $y \neq -1$, and $y \neq 1$.

$$y(y+1)(y-1)\left[\frac{2}{(y+1)(y-1)}\right] = y(y+1)(y-1)\left[\frac{3}{y(y-1)}\right] \qquad \text{Multiply both sides by the LCD.}$$
$$2y = 3(y+1) \qquad \text{Simplify.}$$
$$2y = 3y + 3 \qquad \text{Distributive property.}$$
$$-y = 3 \qquad \text{Subtract } 3y \text{ from both sides.}$$
$$y = -3 \qquad \text{Multiply both sides by } -1.$$

Because

$$\frac{2}{(-3)^2-1} = \frac{2}{8} = \frac{1}{4} \qquad \text{and} \qquad \frac{3}{(-3)^2-(-3)} = \frac{3}{12} = \frac{1}{4},$$

the solution checks. The solution set is $\{-3\}$.

Progress Check Answer
5. $\{1\}$

PROGRESS CHECK 6   Solve $\dfrac{4}{t^2 + 2t} = \dfrac{3}{t^2 - 4}$.

In the next two examples, when the equation is cleared of fractions, the result will be a quadratic equation that can be solved by the methods of Section 4.6.

EXAMPLE 7   Solve $x - \dfrac{4}{x} = 3$.

Solution   Multiply both sides of the equation by the LCD, x.

$$x\left(x - \frac{4}{x}\right) = x \cdot 3 \qquad \text{Multiply both sides by } x.$$

$$x \cdot x - x\left(\frac{4}{x}\right) = x \cdot 3 \qquad \text{Distributive property.}$$

$$
\begin{aligned}
x^2 - 4 &= 3x && \text{Simplify.} \\
x^2 - 3x - 4 &= 0 && \text{Subtract } 3x \text{ from both sides.} \\
(x - 4)(x + 1) &= 0 && \text{Factor the nonzero side.} \\
x - 4 = 0 \quad \text{or} \quad x + 1 &= 0 && \text{Set each factor equal to 0.} \\
x = 4 \qquad\qquad x &= -1 && \text{Solve each linear equation.}
\end{aligned}
$$

Because $4 - \dfrac{4}{4}$ and $-1 - \dfrac{4}{-1}$ both simplify to 3, both solutions check. The solution set is $\{4, -1\}$.

PROGRESS CHECK 7   Solve $y - \dfrac{8}{y} = 2$.

The next example shows that the methods of this section may lead to **extraneous solutions,** which are solutions that do not check in the *original* equation. We always discard extraneous solutions.

EXAMPLE 8   Solve $1 = \dfrac{3}{x - 2} - \dfrac{12}{x^2 - 4}$.

Solution   Begin by factoring $x^2 - 4$ as $(x + 2)(x - 2)$ and determining that the LCD is $(x + 2)(x - 2)$. Then multiply both sides of the equation by the LCD $(x + 2)\,(x - 2)$, assuming $x \neq -2$ and $x \neq 2$.

$$(x + 2)(x - 2) \cdot 1 = (x + 2)(x - 2)\left(\frac{3}{x - 2} - \frac{12}{x^2 - 4}\right) \qquad \text{Multiply both sides by the LCD.}$$

$$(x + 2)(x - 2) \cdot 1 = (x + 2)(x - 2)\left(\frac{3}{x - 2}\right) - (x + 2)(x - 2)\left(\frac{12}{x^2 - 4}\right) \qquad \text{Distributive property.}$$

$$
\begin{aligned}
(x + 2)(x - 2) &= 3(x + 2) - 12 && \text{Simplify.} \\
x^2 - 4 &= 3x + 6 - 12 && \text{Remove parentheses.} \\
x^2 - 3x + 2 &= 0 && \text{Rewrite the equation so one side is 0.} \\
(x - 2)(x - 1) &= 0 && \text{Factor.} \\
x - 2 = 0 \quad \text{or} \quad x - 1 &= 0 && \text{Set each factor equal to 0.} \\
x = 2 \qquad\qquad x &= 1 && \text{Solve each equation.}
\end{aligned}
$$

The following check verifies that 1 is a solution and confirms that 2 is extraneous, because division by 0 is undefined.

Check

$$1 \overset{?}{=} \frac{3}{2 - 2} - \frac{12}{2^2 - 4} \qquad\qquad 1 \overset{?}{=} \frac{3}{1 - 2} - \frac{12}{1^2 - 4}$$

$$1 \neq \frac{3}{0} - \frac{12}{0} \qquad\qquad\qquad 1 \overset{?}{=} -3 - (-4)$$

$$1 \overset{\checkmark}{=} 1$$

Thus, 2 is an extraneous solution. The solution set is $\{1\}$.

PROGRESS CHECK 8   Solve $1 = \dfrac{4}{x^2 - 1} - \dfrac{2}{x - 1}$.   ⌐

2   Recall from Section 2.5 that we sometimes need to convert a formula to a form that is more efficient for a particular problem. The methods of this section are used to rearrange formulas that contain fractions.

EXAMPLE 9   In statistics the z score formula is $z = \dfrac{x - \mu}{\sigma}$, where x is a raw score, μ is the mean, and σ is the standard deviation. Solve this formula for the raw score x.

Solution   For this equation, multiplying both sides by σ is equivalent to cross multiplying.

$$z = \frac{x - \mu}{\sigma} \qquad \text{Given formula.}$$

$$\sigma z = x - \mu \qquad \text{Multiply both sides by } \sigma.$$

$$\mu + \sigma z = x \qquad \text{Add } \mu \text{ to both sides.}$$

The formula $x = \mu + \sigma z$ gives x in terms of μ, σ, and z.

Note   The symbol μ is lowercase *mu*, the Greek letter for *m*, while the symbol σ is lowercase *sigma*, the Greek letter for *s*.

PROGRESS CHECK 9   Solve $t = \dfrac{v - v_0}{a}$ for v.   ⌐

EXAMPLE 10   Solve the formula in the section introduction, $\dfrac{1}{R} = \dfrac{1}{R_1} + \dfrac{1}{R_2}$, for R_1.

Solution   First, multiply both sides of the equation by the LCD, RR_1R_2; then simplify.

$$RR_1R_2\left(\frac{1}{R}\right) = RR_1R_2\left(\frac{1}{R_1} + \frac{1}{R_2}\right) \qquad \text{Multiply both sides by } RR_1R_2.$$

$$RR_1R_2\left(\frac{1}{R}\right) = RR_1R_2\left(\frac{1}{R_1}\right) + RR_1R_2\left(\frac{1}{R_2}\right) \qquad \text{Distributive property.}$$

$$R_1R_2 = RR_2 + RR_1 \qquad \text{Simplify.}$$

To solve for R_1, rewrite the equation so that both terms containing R_1 are isolated on one side of the equation.

$$R_1R_2 - RR_1 = RR_2 \qquad \text{Subtract } RR_1 \text{ from both sides.}$$

Finally, factor out the common factor R_1 and proceed in the usual way.

$$R_1(R_2 - R) = RR_2 \qquad \text{Factor out } R_1.$$

$$R_1 = \frac{RR_2}{R_2 - R} \qquad \text{Divide both sides by } R_2 - R.$$

PROGRESS CHECK 10   The total capacitance C of a circuit containing two capacitances C_1 and C_2 in series is given by $\dfrac{1}{C} = \dfrac{1}{C_1} + \dfrac{1}{C_2}$. Solve the formula for C_2.   ⌐

Progress Check Answers

8. $\{-3\}$

9. $v = v_0 + at$

10. $C_2 = \dfrac{CC_1}{C_1 - C}$

EXERCISES 5.6

In Exercises 1–18, solve the given equation. Assume all denominators with variables are not equal to 0.

1. $\dfrac{5x}{6} - \dfrac{1}{3} = \dfrac{x}{2}$

2. $\dfrac{4y}{3} - \dfrac{1}{6} = \dfrac{2y}{3}$

3. $\dfrac{x}{4} + 3 = \dfrac{x - 1}{8}$

4. $\dfrac{3y}{10} + 2 = \dfrac{2y - 3}{5}$

5. $\dfrac{y}{4} - \dfrac{11}{3} = \dfrac{3y + 5}{6}$

6. $\dfrac{5x}{12} - \dfrac{1}{4} = \dfrac{2 - x}{8}$

7. $\dfrac{3}{x} - 2 = \dfrac{4}{3x}$

8. $\dfrac{1}{2x} + 1 = \dfrac{1}{x}$

9. $\dfrac{1}{y} + \dfrac{2}{3} = 1$

10. $\dfrac{2}{y} + \dfrac{5}{9} = 1$

11. $\dfrac{5}{n} - \dfrac{1}{2} = 2$

12. $3 + \dfrac{7}{n} = 17$

13. $\dfrac{5}{a} + \dfrac{3}{2} = \dfrac{1}{3a}$

14. $\dfrac{2}{3b} - \dfrac{3}{4} = \dfrac{1}{6b}$

15. $\dfrac{5}{3n} + \dfrac{2}{3} = \dfrac{5+n}{2n}$

16. $\dfrac{m-8}{3m} - \dfrac{1}{m} = \dfrac{m-19}{5m}$

17. $\dfrac{x-5}{5} = x + 3$

18. $\dfrac{x+1}{6} = x - 4$

In Exercises 19–48, solve the given equations. Check for extraneous solutions.

19. $\dfrac{5}{y+1} = \dfrac{10}{3y}$

20. $\dfrac{2}{x} = \dfrac{3}{x-10}$

21. $\dfrac{1}{x-5} = \dfrac{3}{x-15}$

22. $\dfrac{4}{x+1} = \dfrac{8}{2-x}$

23. $\dfrac{2}{x+1} = \dfrac{3}{x+1}$

24. $\dfrac{1}{x+1} = \dfrac{2}{2x+2}$

25. $\dfrac{x-1}{x+3} = \dfrac{x-4}{x-3}$

26. $\dfrac{x+2}{x+4} = \dfrac{x-2}{x-1}$

27. $\dfrac{x+2}{x-3} = \dfrac{x-3}{x+2}$

28. $\dfrac{x-5}{x+4} = \dfrac{x+4}{x-5}$

29. $\dfrac{4x-1}{2x-1} = \dfrac{6x+2}{3x+2}$

30. $\dfrac{3x-1}{x+1} = \dfrac{6x}{2x+2}$

31. $\dfrac{2}{x^2+x} = \dfrac{1}{x^2-1}$

32. $\dfrac{4}{x^2-x} = \dfrac{2}{x^2-1}$

33. $\dfrac{3}{t^2+3t} = \dfrac{21}{t^2-3t}$

34. $\dfrac{2}{t^2-2t} = \dfrac{10}{t^2+2t}$

35. $\dfrac{1}{r^2-r} = \dfrac{2}{r-1}$

36. $\dfrac{1}{r^2+r} = \dfrac{3}{r+1}$

37. $x - \dfrac{6}{x} = 1$

38. $x + \dfrac{3}{x} = 4$

39. $2y - \dfrac{1}{3y} = \dfrac{5}{3}$

40. $3y - \dfrac{2}{3y} = -\dfrac{7}{3}$

41. $1 = \dfrac{2}{x+3} + \dfrac{12}{x^2-9}$

42. $1 = \dfrac{1}{x+2} + \dfrac{4}{x^2-4}$

43. $\dfrac{1}{x-1} + \dfrac{1}{3} = \dfrac{4}{x^2-1}$

44. $\dfrac{3}{x+1} - \dfrac{1}{8} = \dfrac{5}{x^2-1}$

45. $\dfrac{3}{x-5} - \dfrac{11}{x^2-25} = 2$

46. $\dfrac{2}{x-1} - \dfrac{4}{x^2-1} = -2$

47. $\dfrac{2}{x-1} - \dfrac{4}{x^2-1} = 1$

48. $\dfrac{3}{x-2} - \dfrac{12}{x^2-4} = 1$

In Exercises 49–66, solve for the given variable.

49. $\dfrac{1}{a} = \dfrac{1}{b} - \dfrac{1}{c}$: If $a = 5$ and $b = 4$, solve for c.

50. $\dfrac{1}{x} = \dfrac{2}{y} - \dfrac{3}{z}$: If $x = 2$ and $z = -5$, solve for y.

51. $\dfrac{1}{x} = \dfrac{3y}{y+1} + \dfrac{3z}{z+2}$:

If $y = 1$ and $z = -1$, solve for x.

52. $\dfrac{2a-1}{a} = \dfrac{b}{b-1} + \dfrac{c}{c-2}$:

If $b = -1$ and $c = -3$, solve for a.

53. $\dfrac{1}{x} = \dfrac{1}{y} + \dfrac{1}{z}$: If $x = 3$ and $y = 3$, solve for z.

54. $\dfrac{1}{a} + \dfrac{2}{b} = \dfrac{3}{c}$: If $a = \dfrac{1}{2}$ and $c = \dfrac{3}{2}$, solve for b.

55. $\dfrac{y-1}{x} = m$: Solve for y.

56. $\dfrac{x+2}{y} = m$: Solve for x.

57. $s = \dfrac{kr}{\ell}$: Solve for r.

58. $h = \dfrac{kE}{V}$: Solve for V.

59. $A = \dfrac{a+b}{2}$: Solve for a.

60. $A = \dfrac{a-b}{2}$: Solve for b.

61. $S = \dfrac{a}{1-r}$: Solve for r.

62. $D = \dfrac{1-p}{p}$: Solve for p.

63. $\dfrac{1}{x+1} = y$: Solve for x.

64. $\dfrac{a}{b+c} = d$: Solve for b.

65. $\dfrac{1}{a} = \dfrac{1}{b} + \dfrac{1}{c}$: Solve for a.

66. $\dfrac{1}{a} = \dfrac{1}{b} - \dfrac{1}{c}$: Solve for c.

67. For what value of x will the perimeter of the rectangle equal the area?

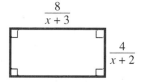

68. For what value of x will the perimeter of the rectangle equal its area?

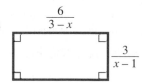

69. If the total cost of producing x units of a certain product consists of paying \$800 in fixed costs and \$5 per unit, then the *average* cost per unit, A, is given by $A = \dfrac{5x + 800}{x}$. How many units should be produced for the average cost to be \$10 per unit?

70. In Exercise 69, how many units should be produced for the average cost to be \$9 per unit?

THINK ABOUT IT

1. Solve $\dfrac{3}{x^2 + 3x + 2} + \dfrac{x + 5}{x^2 + 5x + 6} = \dfrac{x + 7}{x^2 + 4x + 3}$.

2. Solve $\dfrac{x + 5}{x^2 - 1} - \dfrac{4}{x^2 - x - 2} = \dfrac{x + 1}{x^2 - 3x + 2}$.

3. Solve $\dfrac{x - 1}{x^2 + x} = \dfrac{x^2 - x}{x + 1}$.

4. Positive integers a, b, and c are in "harmonic proportion" if $\dfrac{a}{c} = \dfrac{a - b}{b - c}$. Solve for b.

5. The Greek mathematician Diophantus, who worked in Alexandria, Egypt, is credited with originating an algebraic style of notation for equations, which before then were solved by referring to geometric diagrams. Little is known about him; he probably lived in the third century A.D., but a math puzzle about his life remains in the form of a poem. Can you put the information into an equation and find out how long Diophantus lived?

"Here lies Diophantus." The wonder behold—
Through art algebraic, the stone tells how old:
"God gave him his boyhood one-sixth of his life,
One-twelfth more as youth while whiskers grew rife;
And then yet one-seventh ere marriage begun;
In five years there came a bouncing new son.
Alas, the dear child of master and sage
Met fate at just half his dad's final age.
Four years yet his studies gave solace from grief;
Then leaving the scenes earthly he, too, found relief."

This is the version of the poem given in the article "The History of Algebra" by John Baumgart, which is in the book *Historical Topics for the Mathematics Classroom*, published by the NCTM in 1989.

REMEMBER THIS

1. What number is one and one-half times 7?
2. True or false?
   a. 1.6 ft is the same as 1 ft, 6 in.
   b. 1.6 hours is the same as 1 hour, 6 minutes.
3. A recipe which serves 8 is cut to serve 4 by proportionately reducing all the ingredients. The original recipe calls for $\frac{3}{4}$ teaspoon cayenne pepper. How much cayenne is in the new recipe?
4. Show that if you keep the height the same but double the base of a rectangle, then you double the area. (Divide the larger area by the smaller.)

   $h\;\boxed{}$ $h\;\boxed{}$
   $\quad\;\;x$ $\quad\;\;\;2x$

5. Simplify $\dfrac{1}{\dfrac{2}{c} - \dfrac{3}{d}}$.

6. Subtract $\dfrac{4}{5x^2y} - \dfrac{3}{8xy^2}$.

7. Subtract $\dfrac{4x - 5}{x^2 + 3x + 2} - \dfrac{3x - 7}{x^2 + 3x + 2}$.

8. Two cyclists leave point A together. After 3 hours the faster one has gone 18 mi, is 6 mi ahead, and stops to wait for her friend. How long will she have to wait?

9. Solve $x^2 - 4x + 16 = 4x + 9$.

10. Solve $3(x^2 + 1) - 3x = 3x^2 - 3(x + 1)$.

5.7   Ratio, Proportion, and Work Problems

A pool player wishes to make a ball at A strike the cushion at P to hit another ball at B, as shown in the figure. The ball will rebound off the cushion at the same angle at which it strikes. Use the dimensions given in the figure and find the length of line segment CP to determine the location of point P. (See Example 4.)

OBJECTIVES

1   Express a ratio in lowest terms.

2   Set up and solve proportion problems.

3   Set up and solve work problems.

4   Set up and solve uniform motion problems.

The applications in this section require us to set up and solve equations that contain rational expressions. Once again, it will be useful to take into account the general approach to solving word problems given in Section 2.6.

1 A **ratio** is a comparison of two quantities by division. For example, suppose the length of a room is 15 ft and the width of the room is 12 ft. Then the ratio of the length to the width is expressed in either of the following ways (which are read "15 to 12").

1. *Fractional form:* Write $\frac{15}{12}$.
2. *Colon form:* Write 15:12.

For ease of interpretation a ratio should usually be expressed in lowest terms. Thus $\frac{15}{12}$ would become $\frac{5}{4}$, and 15:12 would become 5:4.

It is important to remember that when you express ratios in many applications, the ratios will be more sensible when the quantities are measured in the same units. For instance, the ratio of the length of a 1-ft ruler to the length of a 1-yd stick is not 1:1 but 1:3, because there are 3 ft in a yard.

EXAMPLE 1   Express in lowest terms the ratio of a time interval measured at 30 seconds to another interval measured at 5 minutes.

Solution   Since there are 60(5) = 300 seconds in 5 minutes, the ratio is

$$\frac{30 \text{ seconds}}{5 \text{ minutes}} = \frac{30 \text{ seconds}}{300 \text{ seconds}} = \frac{1}{10}.$$

This ratio may also be expressed as 1:10.

PROGRESS CHECK 1   Express in lowest terms the ratio of 2 dollars to 75 cents.

2 A **proportion** is a statement that two ratios are equal. For example, the ratios $\frac{4}{5}$ and $\frac{8}{10}$ are equal and form a proportion that may be written as $\frac{4}{5} = \frac{8}{10}$ (read: 4 is to 5 as 8 is to 10) or as 4:5 = 8:10. When we work with proportions, it is easier to write the ratios as fractions and then use the techniques we have developed for solving an equation.

EXAMPLE 2   Fifty gallons of water flow through a feeder pipe in 20 minutes. At the same rate, how many gallons of water will flow through the pipe in 32 minutes?

Solution   If we let x gal represent the unknown amount of water and *set up two equal ratios* that compare like measurements, we have the proportion

$$\frac{20 \text{ minutes}}{32 \text{ minutes}} = \frac{50 \text{ gal}}{x \text{ gal}}.$$

Now, *solve the equation* by disregarding the units and cross multiplying.

$$20x = 32(50) \qquad \text{Cross multiply.}$$
$$x = \frac{32(50)}{20} \qquad \text{Divide both sides by 20.}$$
$$x = 80 \qquad \text{Simplify.}$$

Answer the Question   Eighty gallons of water will flow through the feeder pipe in 32 minutes.

Check the Answer   In the context of the word problem, 50 gal in 20 minutes is a rate of 2.5 gal/minute; and 80 gal in 32 minutes is also a rate of 2.5 gal/minute. The solution checks.

Note   It is possible to set up the proportion in other ways, as long as both sides of the equation express similar ratios. For instance, as suggested in the check, a logical proportion in this example would be

$$\frac{50 \text{ gal}}{20 \text{ minutes}} = \frac{x \text{ gal}}{32 \text{ minutes}}.$$

PROGRESS CHECK 2   Fifty-four gallons of oil flow through a pipe in 12 minutes. At the same rate, how many gallons of oil will flow through the pipe in 28 minutes?

Similar triangles are triangles that have the same shape but not necessarily the same size. We know that two triangles are similar if they have the same angle measures. And **an important property of similar triangles is that the lengths of corresponding sides are in proportion.** We will use this property in the next two examples.

EXAMPLE 3   A yardstick casts a shadow 4 ft long at the same time that a tree casts a shadow of 52 ft. What is the height of the tree?

Solution   Let x represent the height of the tree, and illustrate the given information as in Figure 5.2, using 3 ft for 1 yd. The right angles and the angles at which the sun's

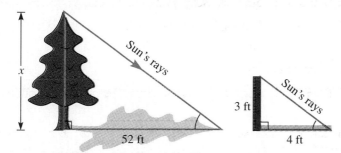

Figure 5.2

rays hit the ground have the same measures, so the triangles are similar. Thus, corresponding side lengths are in proportion, so

$$\frac{x}{3} = \frac{52}{4}.$$

Solve the Equation

$$3\left(\frac{x}{3}\right) = 3\left(\frac{52}{4}\right) \qquad \text{Multiply both sides by 3.}$$
$$x = 39 \qquad \text{Simplify.}$$

Answer the Question   The height of the tree is 39 ft.

Check the Answer   If $x = 39$ ft, then corresponding side lengths are in proportion, since $\frac{39}{3} = \frac{52}{4}$. The solution checks.

PROGRESS CHECK 3   A yardstick casts a shadow 5 ft long at the same time that a building casts a shadow of 130 ft. What is the height of the building?

EXAMPLE 4   Solve the problem in the section introduction on page 255.

Solution   Let x represent the length of segment CP, and consider the simplified sketch of the problem in Figure 5.3. Note that if $CD = 4$ ft and $CP = x$, then $DP = 4 - x$. Because the ball rebounds off the cushion at the same angle at which it strikes,

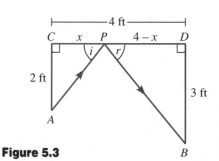

Figure 5.3

Progress Check Answers
2. 126 gal
3. 78 ft

$i = r$, so triangles ACP and BDP are similar and corresponding side lengths are in proportion. Thus,

$$\frac{x}{4-x} = \frac{2}{3} \qquad \frac{CP}{DP} = \frac{AC}{BD}$$

Solve the Equation

$3x = 2(4 - x)$ Cross multiply.

$3x = 8 - 2x$ Distributive property.

$5x = 8$ Add 2x to both sides.

$x = \frac{8}{5}$ Divide both sides by 5.

Answer the Question   The length of segment CP is $\frac{8}{5}$, or 1.6, ft.

Check the Answer   If $CP = 1.6$ ft, then $DP = 2.4$ ft, and corresponding side lengths are in proportion, because $1.6/2.4 = 2/3$ is a true statement. The solution checks.

PROGRESS CHECK 4   Redo the problem in Example 4 given that $AC = 3$ ft, $CD = 5$ ft, and $BD = 1$ ft. ⌐

3   When two or more people or machines work together, then finding the time needed to complete a job is usually referred to as a **work problem.** The key concept for analyzing such problems is assuming a constant work rate: **If a job requires t units of time to complete, then $1/t$ of the job is completed in 1 unit of time.** For example, if a student can type a term paper in 5 hours, then (assuming a constant typing rate) the student types 1/5 of the paper for each hour of typing. Example 5 shows how to solve a typical work problem.

EXAMPLE 5   Coin sorter A can process a sack of coins in 20 minutes; sorter B can process the same sack in 30 minutes. How long would it take the two machines working together to process the sack of coins?

Solution   Let x represent the number of minutes required to process the coins with both machines operating. Then apply the key concept outlined above.

Sorter A: This sorter processes the coins in 20 minutes, so it processes 1/20 of the coins in 1 minute.

Sorter B: This sorter processes the coins in 30 minutes, so it processes 1/30 of the coins in 1 minute.

Together: The sorters together process the coins in x minutes, so they process $1/x$ of the sack in 1 minute.

Now, *set up an equation* as follows.

Part done by sorter A in 1 minute		part done by sorter B in 1 minute		part done by both sorters in 1 minute
$\dfrac{1}{20}$	$+$	$\dfrac{1}{30}$	$=$	$\dfrac{1}{x}$

Solve the Equation

$$60x\left(\frac{1}{20} + \frac{1}{30}\right) = 60x\left(\frac{1}{x}\right) \qquad \text{Multiply both sides by 60x.}$$

$$60x\left(\frac{1}{20}\right) + 60x\left(\frac{1}{30}\right) = 60x\left(\frac{1}{x}\right) \qquad \text{Distributive property.}$$

$3x + 2x = 60$ Simplify.

$5x = 60$ Combine like terms.

$x = 12$ Divide both sides by 5.

Answer the Question   Together, the two sorters process the sack in 12 minutes.

Check the Answer   In 12 minutes sorter A does $\frac{1}{20} \cdot 12$ of the job while sorter B does $\frac{1}{30} \cdot 12$ of the job. Because $\frac{12}{20} + \frac{12}{30} = \frac{36}{60} + \frac{24}{60} = 1$, the whole job is completed. The solution checks.

PROGRESS CHECK 5   A lawn can be mowed in 40 minutes with a riding mower and in 120 minutes with a self-propelled mower. How long will it take to mow the lawn using the two mowers together?   ⌐

4   In Section 2.7 we solved certain uniform motion problems in a chart format, using the formula

$$\text{distance} = \text{rate} \cdot \text{time}, \quad \text{or} \quad d = rt.$$

For the uniform motion problem in the next example, it is useful to rewrite this formula as

$$\text{time} = \frac{\text{distance}}{\text{rate}}, \quad \text{or} \quad t = \frac{d}{r},$$

when setting up the chart to analyze the problem.

EXAMPLE 6   A motorboat can travel 30 mi downstream (with the current) in the same time it can travel 15 mi upstream (against the current). If the current of the river is 3 mi/hour, what is the boat's rate in still water?

Solution   Let x represent the boat's rate in still water. Then

$$x + 3 = \text{rate downstream (add the current)}$$
$$\text{and} \quad x - 3 = \text{rate upstream (subtract the current)}.$$

Now, analyze the problem in a chart format, using $t = d/r$.

Direction	Distance	÷	Rate	=	Time
Downstream	30		$x + 3$		$\dfrac{30}{x + 3}$
Upstream	15		$x - 3$		$\dfrac{15}{x - 3}$

To *set up an equation,* use the given condition that both trips require the same amount of time.

$$\underbrace{\text{Time downstream}}_{\dfrac{30}{x + 3}} \quad \text{equals} \quad \underbrace{\text{time upstream.}}_{\dfrac{30}{x - 3}}$$

$$\frac{30}{x + 3} = \frac{30}{x - 3}$$

Solve the Equation

$30(x - 3) = 15(x + 3)$	Cross multiply.
$30x - 90 = 15x + 45$	Distributive property.
$15x - 90 = 45$	Subtract 15x from both sides.
$15x = 135$	Add 90 to both sides.
$x = 9$	Divide both sides by 15.

Answer the Question   The boat's rate in still water is 9 mi/hour.

Check the Answer   If $x = 9$, then the trip downstream takes $\frac{30}{12}$, or 2.5, hours, while the trip upstream takes $\frac{15}{6}$, or 2.5, hours. The times are the same, and the solution checks.

Progress Check Answer

5. 30 minutes

Progress Check Answer

6. 10 mi/hour

PROGRESS CHECK 6 A motorboat can travel 18 mi downstream in the same time that it can travel 12 mi upstream. If the current of the river is 2 mi/hour, what is the boat's rate in still water? ⌐

EXERCISES 5.7

In Exercises 1–6, express the given ratio in lowest terms. Express both values in the same units first.

1. A time interval measured at 10 seconds to another measured at 1 minute
2. A time interval of 10 minutes to another of 90 seconds
3. A length of 1 yd to a length of 15 in.
4. A length of 1 ft to a length of 2 yd
5. A price of $3\frac{1}{2}$ dollars to a price of 35 cents
6. A cost of 10 dollars to a cost of 25 cents
7. The probability that a card drawn at random from a deck is an ace is the same as the ratio of the number of aces in the deck to the total number of cards. There are four of each value of card in a deck; the values are 2 through 10, ace, king, queen, and jack. Find the probability of drawing an ace.
8. The probability that the ball on a roulette wheel lands in a red slot is the ratio of the number of red slots to the total number of slots. On a roulette wheel there are 18 red, 18 black, and 2 green slots. Find the probability that the ball lands in a red slot.
9. In the United States in 1988 there were about 3,913,000 babies born out of a total population of about 246 million people. About how many babies is this for every 1,000 people in the country? This is called the annual birth rate (per 1,000 population) for 1988.
10. In the United States in 1988 there were about 2,171,000 deaths out of a total population of about 246 million. About how many deaths is this for every 100,000 people? This is called the annual death rate (per 100,000 population) for 1988.

In Exercises 11–26, solve the given problem by writing and solving a proportion.

11. A well provides 4 gal of water per minute. At this rate, how long will it take to provide 130 gal of water?
12. A well provides 20 gal of water in 5 minutes. At this rate, how many gallons will it provide in 44 minutes?
13. A recipe for hot cereal calls for 2 cups of water to $\frac{1}{3}$ cup of dry oats. Maintaining this ratio, how many cups of oats would be used with 3 cups of water?
14. A recipe for rice uses $1\frac{2}{3}$ cups of water for 1 cup of rice. How much water goes with $1\frac{1}{2}$ cups of rice?
15. a. Show that 1.5 ft equals 18 in. by solving the proportion $\dfrac{1\ \text{ft}}{12\ \text{in.}} = \dfrac{1.5\ \text{ft}}{x\ \text{in.}}$.
    b. How many inches are in 1.6 ft?
16. a. Show that 1.5 hours equals 90 minutes by solving the proportion $\dfrac{1\ \text{hour}}{60\ \text{minutes}} = \dfrac{1.5\ \text{hours}}{x\ \text{minutes}}$.
    b. How many minutes are there in 1.6 hours?

17. The scale on a road map is 1 in. to 150 mi.
    a. On this map cities A and B are 3.2 in. apart. What is the distance between them in miles?
    b. If you drive at an average speed of 50 mi/hour, what time should you leave A to arrive at B before 5 P.M.?
18. The scale on a map is 1 in. to 180 mi. If two cities are 300 mi apart, what is the distance between them on the map?
19. a. If an idling car uses 40 oz of gasoline in 50 minutes, how long must it idle to use 1 gal of gas? (One gallon is 128 oz.)
    b. If this car idles 30 minutes per day, about how many days does it take to idle away $15 worth of gas? Assume gas costs $1.50 per gallon.
20. Suppose the car in Exercise 19 is tuned up so that it uses 25 oz of gasoline in 50 minutes when it idles. Now, how long must it idle to use 1 gal of gas?
21. The two triangles shown are similar. Find the lengths of sides *DE* and *EF*.

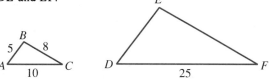

22. The two triangles shown are similar. Find the lengths of sides *DE* and *EF*.

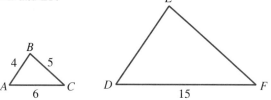

23. A tree casts a 12-ft shadow at the same time a yardstick casts a 24-in. shadow. Find the height of the tree.
24. A little after noon a tall building casts a 4-ft shadow. At the same time an 8-ft pole casts a 4-in. shadow. Find the height of the building.
25. The chart shows the breakdown of the responses to a survey in which 100 people were asked if they were in favor of a new tax.

	Men	Women	Total
Yes	x	$20 - x$	20
No	$40 - x$	$40 + x$	80
Total	40	60	100

a. How many men and how many women were in the survey?

b. How many people said they were in favor of the new tax? How many were opposed?

c. The fraction of men who said yes is $x/40$. The fraction of women who said yes is $(20 - x)/60$. For what value of x are these fractions equal? What is the value of the fraction?

26. The chart shows the breakdown of the responses to a survey in which 200 voters were asked if they were in favor of extending the school year by one month.

	Men	**Women**	**Total**
Yes	x	$120 - x$	120
No	$105 - x$	$x - 25$	80
Total	105	95	200

a. How many men and how many women were in the survey?

b. How many people said they were in favor of extending the school year? How many were opposed?

c. For what value of x will the percent of men and women who say yes be the same? What is the value of that percent?

For Exercises 27 and 28, use this information. When a light beam strikes a flat mirror, the angle of reflection (α) is the same as the angle of incidence (β); see the sketch. This means that the two triangles shown are similar.

27. a. Using the data in the figure, find the distance (x) from the mirror to the point where the beam is reflected on the wall.

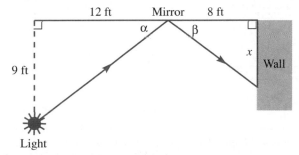

b. Find the total distance that the beam traveled.

28. Using the data in the figure, where should the light strike the mirror so that it will reflect off the mirror and hit the target?

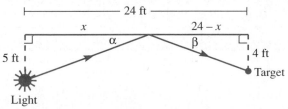

Exercises 29–34 are work problems, which can be solved by the methods of Example 5 in this section.

29. One robot assembly line can complete a full order in 4 hours, while a second one can complete a full order in 6 hours. How

long will it take to complete a full order if the machines are started at the same time and run together?

30. A mail-processing machine can sort 20,000 pieces of mail in 1 hour. A newer one can sort 20,000 pieces of mail in 40 minutes.

a. If they operate together, how long will it take them to sort 20,000 pieces of mail?

b. If they operate together, how long will it take them to sort 10,000 pieces of mail?

31. A pump can empty a tank in 100 minutes. Show algebraically that two such pumps working together will cut the time in half.

32. A milk-processing machine can fill 20 1-qt containers per minute.

a. How long will it take two of these machines working together to fill 20 1-qt containers?

b. How long will it take two of these machines working together to fill 500 1-qt containers?

33. An old photocopy machine can make 20 copies a minute, while the new one can make 80 copies a minute.

a. How long will it take the old machine to make 800 copies?

b. How long will it take the new machine to make 800 copies?

c. How long will it take both machines working together to make 800 copies?

34. One tape-duplicating machine can make 200 copies in 10 minutes, while another can make 100 copies in 10 minutes. How long will it take both working together to make 200 copies? (*Hint:* First, determine how long it will take each machine alone to make 200 copies.)

Exercises 35–40 are uniform motion problems. They can be solved by the methods of Example 6 in this section.

35. A motorboat can travel 20 mi downstream in the same time it can travel 10 mi upstream.

a. If the current of the river is 2 mi/hour, what is the boat's rate in still water?

b. How long would it take the boat to travel 16 mi upstream?

c. About how long would it take to make a round-trip of 8 mi each way if it took 15 minutes to unload cargo before turning around?

36. A motorboat can travel 18 mi downstream in the same time it can travel 12 mi upstream.

a. If the current of the river is 1 mi/hour, what is the boat's rate in still water?

b. How long would it take the boat to travel 12 mi downstream?

c. About how long would it take to make a round-trip of 5 miles each way if it took half an hour to make repairs before turning around?

37. An experimental light plane, powered by the pilot's pedaling, flew a distance of $\frac{1}{2}$ mi against a 2-mi/hour wind. With that same wind behind it, it flew 1 mi in the same time.

a. What is the speed of the plane in still air?

b. How far could it go in 6 minutes with a 4-mi/hour wind behind it?

38. An improved version of the plane from Exercise 37 flew 1 mile with a 3-mi/hour wind pushing it in the same time it took to fly 0.6 mi against that wind. What is the speed of the plane in still air?

39. A bicycle racer travels for a set time 10 mi along a level desert road moving the pedals at a constant rate against a 10-mi/hour head wind. Then the cyclist turns around and travels for that same time with the wind pushing behind. On the return trip the cyclist covers 30 mi. Cycling at this rate, how long would it take the cyclist to cover 50 mi with a 5-mi/hour wind pushing from behind?

40. A bicycle rider travels 5 mi along a level road pedaling at a constant rate against a 5-mi wind. Then the cyclist turns around and travels for the same length of time with the wind pushing from behind and is able to cover 15 mi. At this rate, how long would it take the cyclist to cover 15 mi going against a 4-mi/hour head wind?

THINK ABOUT IT

1. A well provides 5 gal of water per minute. At that rate, how long will it take to fill a circular swimming pool of diameter 10 yd to a depth of 4 ft? There are about 7.5 gal/ft³ of water.

2. In an emergency three pumps are started to empty a large vat. The fastest one could complete the job alone in 3 hours, the next in 4 hours, and the slowest in 6 hours. Can the three of them together get the job done in less than an hour?

3. A commuter airplane flies from a New York airport to a Vermont airport in 1.25 hours flying against the wind. If it were flying the other way with the wind behind it, the flight would take 1 hour. Show that the plane speed is nine times the wind speed.

4. **a.** If 1 cat can catch 1 rat in 1 day, how many rats can 1 cat catch in 10 days?
   b. How many rats can 2 cats catch in 10 days?

5. Bob was 24 when his daughter was born.
   a. What is the ratio of his age to hers when she is 12 years old?
   b. What is the ratio when she is 24 years old?
   c. What is the ratio when she is 36 years old?
   d. What happens to this ratio as she grows older?
   e. How long will it take before they are the same age?

REMEMBER THIS

1. Evaluate $2x - 1$ for $x = -1, 0,$ and 1.

2. Use the formula $y = 3x + 4$ to find y when $x = -\frac{4}{3}$.

3. If $x = 1$ and $y = 2$, is it true that $3x - 2y = 1$?

4. Solve $3x + 4y + 5 = 0$ for y.

5. Find the perimeter of this rectangle if the area equals 96 m².

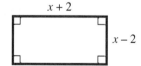

$x + 2$

$x - 2$

6. Factor completely: $18x^2 - 8y^2$.

7. Find an expression for the area of this triangle.

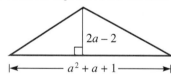

$2a - 2$

$a^2 + a + 1$

8. Simplify $2^1 + 2^0 + 2^{-1}$.

9. Simplify $1 - [2 - (3 - 4)]$.

10. One number is 2 more than another, and 9 times their product equals -5. Find the numbers. (There are two such pairs.)

Chapter 5   SUMMARY

OBJECTIVES CHECKLIST   Specific chapter objectives are summarized below along with numbered example problems from the text that should clarify the objectives. If you do not understand any objectives or do not know how to do the selected problems, then restudy the material.

5.1 **Can you:**

1. Find all values of a variable which make a rational expression undefined?

Find all values of y which make $\dfrac{y}{y^2 - 1}$ undefined.

[Example 1b]

2. **Determine whether a pair of fractions are equivalent?**

Determine whether $\dfrac{-2}{3}$ and $\dfrac{2}{-3}$ are equivalent. [Example 3b]

3. **Express rational expressions in lowest terms?**

Express $\dfrac{5x + 10}{x^2 + x - 2}$ in lowest terms. [Example 6a]

5.2 **Can you:**

1. **Multiply rational expressions?**

Multiply and express the product in lowest terms: $\dfrac{2x}{5y} \cdot \dfrac{10y^2}{8x^3}$. [Example 1b]

2. **Divide rational expressions?**

Divide: $\dfrac{1 - x}{y^2 - y} \div \dfrac{x^2 - 1}{y^2 - 2y + 1}$. Express the answer in lowest terms. [Example 7]

5.3 **Can you:**

1. **Add and subtract rational expressions with the same denominator?**

Write as a single fraction in lowest terms: $\dfrac{4x + 1}{2x - 6} - \dfrac{2x + 4}{2x - 6}$. [Example 2b]

2. **Add and subtract rational expressions with opposite denominators?**

Write as a single fraction in lowest terms: $\dfrac{4x + 5}{3 - 2x} + \dfrac{1 - x}{2x - 3}$. [Example 5b]

5.4 **Can you:**

1. **Find the least common denominator (LCD)?**

Find the least common denominator for $\dfrac{5}{6x^2y}$ and $\dfrac{4}{9xy^3}$. [Example 2]

2. **Add and subtract rational expressions with unlike denominators?**

Add: $\dfrac{x}{x - 4} + \dfrac{5}{x}$. [Example 8]

5.5 **Can you:**

1. **Simplify complex fractions?**

Simplify the complex fraction $\dfrac{\dfrac{1}{3} - \dfrac{2}{y}}{\dfrac{5}{6} + \dfrac{1}{y^2}}$. [Example 4]

5.6 **Can you:**

1. **Solve equations that contain rational expressions?**

Solve $\dfrac{4}{x} - 1 = \dfrac{5}{2x}$. [Example 2]

2. **Solve formulas that contain rational expressions for a specified variable?**

Solve the formula $z = \dfrac{x - \mu}{\sigma}$ for x. [Example 9]

5.7 **Can you:**

1. **Express a ratio in lowest terms?**

Express in lowest terms the ratio of a time interval measured at 30 seconds to another interval measured at 5 minutes. [Example 1]

2. Set up and solve proportion problems?

Fifty gallons of water flow through a feeder pipe in 20 minutes. At the same rate, how many gallons of water will flow through the pipe in 32 minutes? [Example 2]

3. Set up and solve work problems?

Coin sorter A can process a sack of coins in 20 minutes; sorter B can process the same sack in 30 minutes. How long would it take the two machines working together to process the sack of coins? [Example 5]

4. Set up and solve uniform motion problems?

A motorboat can travel 30 mi downstream (with the current) in the same time it can travel 15 mi upstream (against the current). If the current of the river is 3 mi/hour, what is the boat's rate in still water? [Example 6]

KEY TERMS

Complex fraction (5.5)	Proportion (5.7)	Similar triangles (5.7)
Extraneous solutions (5.6)	Ratio (5.7)	Subscripts (5.6)
Least common denominator (5.4)	Rational expression (5.1)	Work problem (5.7)

KEY CONCEPTS AND PROCEDURES

Section	Key Concepts or Procedures to Review

5.1

- Rational expression: A rational expression is an expression of the form $\dfrac{A}{B}$, where A and B are polynomials, with $B \neq 0$.

- Equivalence of fractions: Let a, b, c, and d be real numbers, with $b \neq 0$ and $d \neq 0$. Then

$$\frac{a}{b} = \frac{c}{d} \quad \text{if} \quad ad = bc; \qquad \text{and} \qquad \frac{a}{b} \neq \frac{c}{d} \quad \text{if} \quad ad \neq bc$$

- Fundamental principle of fractions: If a, b, and k are real numbers, with $b \neq 0$ and $k \neq 0$, then

$$\frac{ak}{bk} = \frac{a}{b}$$

- Fundamental principle of rational expressions: If A, B, and K are polynomials, with $B \neq 0$ and $K \neq 0$, then

$$\frac{AK}{BK} = \frac{A}{B}$$

- To reduce rational expressions

  1. Factor completely the numerator and the denominator of the fraction.

  2. Divide out nonzero factors that are common to the numerator and the denominator according to the fundamental principle.

- $\dfrac{a + b}{b + a} = 1$, while $\dfrac{a - b}{b - a} = \dfrac{-1(b - a)}{b - a} = -1$

5.2

- Multiplication of rational expressions: If $\dfrac{A}{B}$ and $\dfrac{C}{D}$ are rational expressions, with $B \neq 0$ and $D \neq 0$, then

$$\frac{A}{B} \cdot \frac{C}{D} = \frac{AC}{BD}$$

- Division of rational expressions: If $\dfrac{A}{B}$ and $\dfrac{C}{D}$ are rational expressions, with $B \neq 0$, $C \neq 0$, and $D \neq 0$, then

$$\frac{A}{B} \div \frac{C}{D} = \frac{A}{B} \cdot \frac{D}{C} = \frac{AD}{BC}$$

Section	Key Concepts or Procedures to Review
5.3	■ Addition or subtraction of rational expressions: If $\dfrac{A}{B}$ and $\dfrac{C}{B}$ are rational expressions, with $B \neq 0$, then $$\frac{A}{B} + \frac{C}{B} = \frac{A + C}{B} \quad \text{and} \quad \frac{A}{B} - \frac{C}{B} = \frac{A - C}{B}$$

Section	Key Concepts or Procedures to Review

5.4	■ To find the LCD
	  1. Factor completely each denominator.
	  2. The LCD is the product of all the different factors, with each factor appearing the greatest number of times it appears in any one factorization.
	■ To add or subtract rational expressions
	  1. Completely factor each denominator and find the LCD.
	  2. For each fraction, obtain an equivalent fraction by applying the fundamental principle and multiplying the numerator and the denominator of the fraction by the factors of the LCD that are not contained in the denominator of that fraction.
	  3. Add or subtract the numerators and divide this result by the common denominator.
	  4. Express the answer in lowest terms.

Section	Key Concepts or Procedures to Review
5.5	■ Methods to simplify complex fractions *Method 1* (Obtain single fractions and divide): Obtain single fractions in both the numerator and the denominator of the complex fraction. Then divide by multiplying by the reciprocal of the denominator. *Method 2* (Multiply using the LCD): Find the LCD of all the fractions that appear in the numerator and the denominator of the complex fraction. Then multiply the numerator and the denominator of the complex fraction by the LCD and simplify the results.

Section	Key Concepts or Procedures to Review
5.6	■ Cross multiplication property If $\dfrac{A}{B}$ and $\dfrac{C}{D}$ are rational expressions, with $B \neq 0$ and $D \neq 0$, then $$\frac{A}{B} = \frac{C}{D} \quad \text{implies} \quad AD = BC$$

Section	Key Concepts or Procedures to Review
5.7	■ The ratio of a to b can be expressed in fractional form, such as a/b, or in colon form, as in $a{:}b$. Fractional form is more convenient when working with proportions. ■ In similar triangles the lengths of corresponding sides are in proportion. ■ In the solution of work problems, if a job requires t units of time to complete, then $1/t$ of the job is completed in 1 unit of time.

CHAPTER 5 REVIEW EXERCISES

5.1

1. Find all values of y which make this rational expression undefined: $\dfrac{y}{2y + 6}$.

2. Determine whether these two fractions are equivalent: $\frac{3}{4}$ and $\frac{9}{16}$.

3. Express in lowest terms: $\dfrac{x^2 + 6x + 9}{x^2 + 5x + 6}$.

5.2

4. Multiply and express the product in lowest terms:
$$\frac{8x - 12}{3x + 6} \cdot \frac{x + 2}{2x - 3}.$$

5. Divide and express the answer in lowest terms:
$$\frac{2 - y}{x^2 + 3x} \div \frac{y^2 - 4}{x^2 - 3x - 18}.$$

5.3

6. Add the two fractions and express the answer in lowest terms:
$$\frac{x^2}{x + 3} + \frac{2x - 3}{x + 3}.$$

7. Subtract the two fractions and express the answer in lowest terms: $\dfrac{x^2}{x + 3} - \dfrac{2x - 3}{x + 3}.$

8. Perform the calculation and express the result in lowest terms:
$$\frac{2x}{2x-3} + \frac{3}{3-2x}.$$

9. Find an expression for the perimeter and write it in lowest terms.

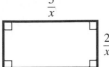

$\frac{5}{x}$

$\frac{2}{x}$

5.4

10. Find the least common denominator for $\frac{1}{y^2-4}$ and $\frac{1}{2y+4}$.

11. Combine fractions and express the result in lowest terms:
$$\frac{x}{x+2} - \frac{5}{x+3}.$$

12. In solving a literal equation, a student gets an answer $c = \frac{24b}{n} - 5d$. The answer in the text is $c = \frac{24b - 5dn}{n}$.
Perform the subtraction in the first answer to show that the student's answer is equivalent to the one in the text.

5.5

13. Show that $\dfrac{4 - \dfrac{5}{6}}{\dfrac{1}{3} + 4} = \dfrac{19}{26}$.

  a. Use the method where you first obtain single fractions in the numerator and the denominator.

  b. Use the method where you multiply numerator and denominator by the LCD of all the fractions.

14. Simplify the complex fraction $\dfrac{\dfrac{x-y}{n}}{\dfrac{1}{x} - \dfrac{1}{y}}$.

5.6

15. Solve $\dfrac{1}{x} + \dfrac{3}{4} = 2$.

16. Solve $\dfrac{1}{t^2+3t} = \dfrac{-1}{5t^2-15t}$.

17. Solve $A = \dfrac{a+b+c}{3}$ for b.

5.7

18. Express the given ratio in lowest terms: a price of 9 dollars to a price of 75 cents.

19. A recipe for rice calls for $1\frac{3}{4}$ cups of water for 1 cup of rice. How much water goes with $1\frac{1}{2}$ cups of rice?

20. At a direct-mail business one job consists of folding a quantity of printed ads and inserting them into envelopes. One of the machines can do this job in 20 minutes, while another can do it in 10 minutes. If the two machines work together, how long will it take for this task?

21. A motorboat can travel 10 mi downstream in the same time that it takes to travel 5 mi upstream. The current of the river is 2 mi/hour.
  a. What is the boat's speed in still water?
  b. How long will it take this boat to go 24 mi downstream?

CHAPTER 5 TEST

1. For what value of x is the expression $\dfrac{4}{x+1}$ undefined?

2. Express $\dfrac{4x-4}{6x-6}$ in lowest terms. Assume the denominator does not equal 0.

3. Multiply and express the product in lowest terms:
$$\frac{x-y}{x} \cdot \frac{x^3}{y-x}.$$

4. Divide and express the quotient in lowest terms:
$$\frac{m-n}{m^3n^2} \div \frac{m^2-2mn+n^2}{m^2n^3}.$$

5. Combine fractions and express the result in lowest terms:
$$\frac{3+x}{x-5} + \frac{2-x}{x-5}.$$

6. Combine fractions and express the result in lowest terms:
$$\frac{x}{x-8} - \frac{8}{x-8}.$$

7. Combine fractions and express the result in lowest terms:
$$\frac{x^2}{x-5} + \frac{25}{5-x}.$$

8. Combine fractions and express the result in lowest terms:
$$\frac{x}{x+3} - \frac{6}{x+2}.$$

9. Combine fractions and express the result in lowest terms:
$$\frac{x^2}{x-5} - \frac{3x+10}{x-5}.$$

10. Simplify the given complex fraction: $\dfrac{2}{\dfrac{1}{a} + \dfrac{1}{b}}$

11. Simplify the given complex fraction: $\dfrac{\dfrac{a}{b} + 3}{\dfrac{a^2}{b^2} - 9}$.

12. Solve $\dfrac{3}{y} + \dfrac{2}{7} = 1$.

13. Solve $x - \dfrac{2}{x} = 1$.

14. Solve for r: $S = \dfrac{5}{1 - r}$. $r =$

15. **a.** For what value of x will the perimeter equal the area?

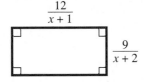

$\dfrac{12}{x+1}$

$\dfrac{9}{x+2}$

    b. What are the dimensions of the rectangle?

16. Fifty gallons of oil flow through a pipe in 2.5 minutes. At the same rate, how many gallons of oil will flow through the pipe in 8 minutes?

17. One printing press can produce 1,000 copies of a newspaper in 2 hours, while another takes 1.5 hours for that task. If they both run together, can they produce 1,000 copies of the paper in less than an hour?

18. Choose all fractions in this list which are equivalent to x/y.

    a. $\dfrac{2x}{2y}$ **b.** $\dfrac{x^2}{y^2}$ **c.** $\dfrac{x+2}{y+2}$ **d.** $\dfrac{x-2}{y-2}$ **e.** $\dfrac{\frac{x}{2}}{\frac{y}{2}}$

19. In multiplying $\dfrac{t^2 + 3t - 4}{t^2 - 3t - 4} \cdot \dfrac{t^2 - 5t + 4}{t^2 + 5t + 4}$, explain why it is a good idea to factor each trinomial first.

20. Express the ratio of the perimeter of a square to its area in lowest terms. Assume the square has side of length s.

CUMULATIVE TEST 5

1. Compute $(-4)(3)(-2)$.
2. Rewrite $-(3x - 4)$ without parentheses.
3. True or false? If x represents a negative number, then x^2 represents a positive number.
4. Choose the relation $\{<, =, >\}$ that makes this statement true:
   $4 - 4 \_\_\_\_ 4(-4)$.
5. In the expression $\dfrac{x^2}{2} + \dfrac{x}{3} + \dfrac{1}{4}$, what is the coefficient of x^2?
6. Solve $5 - y = 6$.
7. Show that $3x + 2 = 3(x - 5)$ is a false equation.
8. A kitchen container holds 64 oz of a mix which is half oil and half vinegar. How much oil (to the nearest tenth of an ounce) should be added to get a mix which is 55 percent oil and 45 percent vinegar.
9. Evaluate $3^{-2} + 3^{-1} + 3^0$. Give the answer as an improper fraction.
10. Simplify $(9x^2 + 3x - 2) - (x^2 - 3x + 1)$.

11. Find the product $(2x - 3)(2x + 3)$.
12. Solve $y = \dfrac{x + 3}{5}$ for x.
13. Factor $b^2 - b - 12$.
14. Factor $4y^2 - 9$.
15. Solve $3x(x - 2)(2x + 3) = 0$.
16. The perimeter of a square is 1.75 more than the area. Find the perimeter.
17. For what value(s) of x is $\dfrac{x}{x^2 + 1}$ undefined?
18. Express in lowest terms: $\dfrac{x^2 + x - 2}{x^2 + 2x - 3}$.
19. Solve $\dfrac{1}{x} - \dfrac{3}{5} = 3$.
20. If one machine can complete a job in 1 hour and another can complete it in 3 hours, how long should it take to complete the job if both machines operate at the same time?

Graphing Linear Equations and Inequalities

If an economy car rents for $59 a week plus 10¢ per mile, then the weekly cost C is given by

$$C = 0.10m + 59,$$

where m is the mileage for the week. Fill in the missing component in each of the following ordered pairs, which are of the form (m,C), so that the pair is a solution of this equation.

a. (90,   )   **b.** (,90)

(See Example 3 of Section 6.1.)

ANALYZING RELATIONSHIPS using graphs and equations with two variables is one of the most useful considerations in mathematics. In this chapter we will begin this study by developing a system that enables us to draw a geometric picture (graph) that is a line corresponding to a certain type of equation with two variables. Then the graph can help us to visualize the relationship between the variables, while the power methods of algebra can be used to analyze graphs.

6.1   Ordered Pairs and Their Graphs

OBJECTIVES

1 Determine if an ordered pair is a solution of an equation.

2 Complete ordered pairs so that they are solutions of an equation.

3 Graph ordered pairs and determine coordinates of points.

1 Equations like $C = 0.10m + 59$, $\ell + w = 50$, and $y = 2x - 2$ express a relationship between two variables. Solutions of such equations are not single numbers but pairs of numbers that show corresponding values of the variables. For example, consider carefully the following chart, in which we generate some solutions to the equation $y = 2x - 1$.

If x Equals	Then $y = 2x - 1$	Thus, a Solution Is
2	$2(2) - 1 = 3$	$x = 2, y = 3$
1	$2(1) - 1 = 1$	$x = 1, y = 1$
0	$2(0) - 1 = -1$	$x = 0, y = -1$
-1	$2(-1) - 1 = -3$	$x = -1, y = -3$
-2	$2(-2) - 1 = -5$	$x = -2, y = -5$

In mathematical notation **ordered pairs** are used to abbreviate the type of solution contained in this chart. For example, the solution "$x = 2, y = 3$" is written

$$(2,3).$$

value of x ⟋   ⟍ corresponding value of y

Note that because the x value is listed first, the order of the numbers in the pair is significant. In the equation $y = 2x - 1$, (2,3) is an ordered pair that makes the equation a true statement, so (2,3) is a solution of this equation. However, (3,2) means "$x = 3, y = 2$," and this ordered pair is not a solution of the equation $y = 2x - 1$, because $2 = 2(3) - 1$ is a false statement.

EXAMPLE 1 Determine if the ordered pair is a solution of the equation.

a. $2x - y = 4$; $(1,-2)$ **b.** $y = -3x + 2$; $(-1,1)$

Solution

a. $(1,-2)$ means $x = 1, y = -2$. Then we have

$$2x - y = 4 \quad \text{Given equation.}$$
$$2(1) - (-2) \overset{?}{=} 4 \quad \text{Replace } x \text{ by 1 and } y \text{ by } -2.$$
$$4 \overset{\checkmark}{=} 4. \quad \text{Simplify.}$$

Since $4 = 4$ is a true statement, $(1,-2)$ is a solution of $2x - y = 4$.

b. $(-1,1)$ means $x = -1, y = 1$.

$$y = -3x + 2$$
$$1 \overset{?}{=} -3(-1) + 2$$
$$1 \neq 5$$

Since $1 = 5$ is a false statement, $(-1,1)$ is not a solution of $y = -3x + 2$.

For the given equation, note that $(1,-1)$ *is* a solution. So we must always be careful to replace x and y in the agreed-on manner for ordered-pair notation.

PROGRESS CHECK 1   Determine if the ordered pair is a solution of the equation.

a. $5x - y = 2$; $(0,-2)$ **b.** $y = -2x + 5$; $(-1,5)$ ⌐

2 To *find* ordered-pair solutions to an equation in two variables, we usually substitute specific values for one variable and then find the corresponding values of the other variable, as shown in the next two examples.

EXAMPLE 2   Fill in the missing component in each of the following ordered pairs so that the pair is a solution of the equation $2x - y = 4$.

a. (,6)   **b.** (0,)

Solution

a. To complete the ordered pair (,6), find x when y is 6 in the given equation.

$$\begin{aligned} 2x - y &= 4 && \text{Given equation.} \\ 2x - 6 &= 4 && \text{Replace } y \text{ by 6.} \\ 2x &= 10 && \text{Add 6 to both sides.} \\ x &= 5 && \text{Divide both sides by 2.} \end{aligned}$$

Thus, $(5,6)$ is a solution of $2x - y = 4$.

b. To complete the ordered pair (0,), find y when x is 0 in the given equation.

$$\begin{aligned} 2x - y &= 4 && \text{Given equation.} \\ 2(0) - y &= 4 && \text{Replace } x \text{ by 0.} \\ -y &= 4 && \text{Simplify.} \\ y &= -4 && \text{Multiply both sides by } -1. \end{aligned}$$

Thus, $(0,-4)$ is a solution of $2x - y = 4$.

PROGRESS CHECK 2   Fill in the missing component in each of the following ordered pairs so that the pair is a solution of the equation $3x - y = 5$.

a. (,4)   **b.** (0,)   ⌐

Until now, all ordered pairs have taken the form (x,y). Many applied problems involve letters besides x and y, and in such cases the form of the ordered pairs will depend on the particular application. This is illustrated by the chapter-opening problem.

EXAMPLE 3   Solve the problem in the chapter introduction on page 268.

Solution   The problem states that ordered pairs are of the form (m,C). This form is used because mileage is determined first, and then cost is based on mileage.

a. To complete (90,), find C when m is 90 in the given equation.

$$\begin{aligned} C &= 0.10m + 59 && \text{Given equation.} \\ C &= 0.10(90) + 59 && \text{Replace } m \text{ by 90.} \\ C &= 68 && \text{Simplify.} \end{aligned}$$

Thus, $(90,68)$ is a solution of $C = 0.10m + 59$. In practical terms, the weekly cost is \$68 if the rental car is driven 90 mi.

b. To complete (,90) find m when C is 90 in the given equation.

$$\begin{aligned} C &= 0.10m + 59 && \text{Given equation.} \\ 90 &= 0.10m + 59 && \text{Replace } C \text{ by 90.} \\ 31 &= 0.10m && \text{Subtract 59 from both sides.} \\ 310 &= m && \text{Divide both sides by 0.10.} \end{aligned}$$

Progress Check Answers

1. (a) Yes   (b) No
2. (a) (3,4)   (b) (0,−5)

Thus, if the cost is $90, then the mileage for the week is 310 mi, and (310,90) is a solution of the given equation.

PROGRESS CHECK 3   If a minivan rents for $99 a week plus 15¢ per mile, then the weekly cost C in terms of the mileage m for the week is given by $C = 0.15m + 99$. Complete the following ordered pairs of the form (m,C) so that the pairs are solutions of this equation.

a. (200,   )   **b.** (,165)

3   We can represent geometrically an ordered pair of real numbers by using the **Cartesian (or rectangular) coordinate system.** This system was devised by the French mathematician and philosopher René Descartes and is formed from the intersection of two real number lines at right angles. The values for x (or the first component in the ordered pairs) are represented on a horizontal number line. The values for y (or the second component in the ordered pairs) are represented on a vertical number line. These two lines are called the **x-axis** and the **y-axis,** and they intersect at their common zero point, which is called the **origin** (see Figure 6.1).

Any ordered pair can be represented as a point in this coordinate system. The first component indicates the distance of the point to the right or left of the vertical axis. The second component indicates the distance of the point above or below the horizontal axis. These components are called the **coordinates** of the point, and the point is called the **graph** of the ordered pair.

Figure 6.1

EXAMPLE 4   Graph the following ordered pairs.

a. (4,2)   **b.** (−3,−4)

Solution

a. To graph (4,2), start at the origin and go 4 units to the right and 2 units up, as shown in Figure 6.2(a).

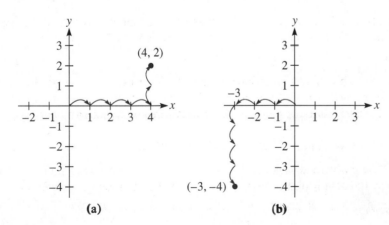

(a)   (b)

Figure 6.2

b. To graph (−3,−4), start at the origin and go 3 units to the left and 4 units down, as shown in Figure 6.2(b).

PROGRESS CHECK 4   Graph the following ordered pairs.

a. (3,4)   **b.** (−1,−2)

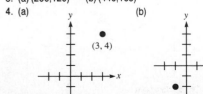

The Cartesian coordinate system divides the plane into four regions called **quadrants.** The quadrant in which both x and y are positive is designated the first quadrant. The remaining quadrants are labeled in a counterclockwise direction. Figure 6.3 shows the name of each quadrant as well as the sign of x and y in that quadrant.

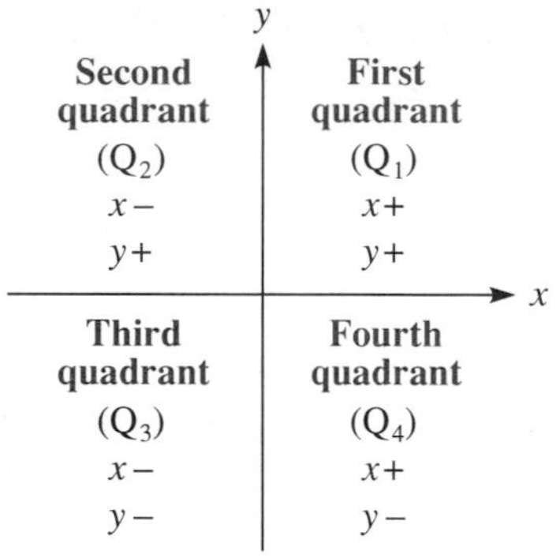

Second quadrant (Q_2) $x-$ $y+$	First quadrant (Q_1) $x+$ $y+$
Third quadrant (Q_3) $x-$ $y-$	Fourth quadrant (Q_4) $x+$ $y-$

Figure 6.3

EXAMPLE 5 Graph the following ordered pairs and indicate the quadrant location of each point.

a. $(-2,3)$ **b.** $(4,-3)$ **c.** $(2,0)$

Solution See Figure 6.4.

a. To graph $(-2,3)$, start at the origin and go 2 units to the left and 3 units up. The graph $(-2,3)$ lies in quadrant 2 (or Q_2).
b. To graph $(4,-3)$, start at the origin and go 4 units to the right and 3 units down. The graph $(4,-3)$ lies in quadrant 4 (or Q_4).
c. To graph $(2,0)$, start at the origin and go 2 units to the right and 0 units up, which means $(2,0)$ lies on the x-axis. Any point that lies on either axis is not located in a quadrant.

PROGRESS CHECK 5 Graph the following ordered pairs and indicate the quadrant location of each point.

a. $(2,-1)$ **b.** $(0,-3)$ **c.** $(-1,4)$

We often need to interpret a graph by reading the coordinates of points in the graph.

EXAMPLE 6 Approximate (use integers) the ordered pair corresponding to the points in Figure 6.5.

Solution See Figure 6.6.

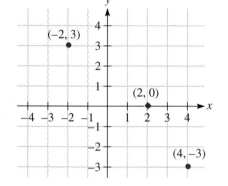

Figure 6.4

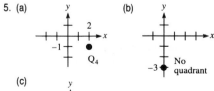

Progress Check Answers
5. (a)   (b)
(c)

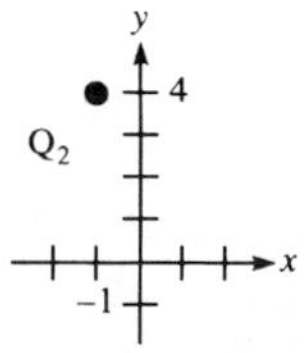

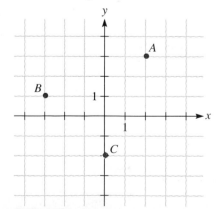

Figure 6.5

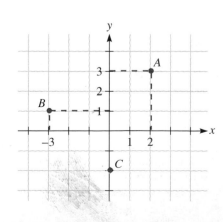

Figure 6.6

a. Point A lines up with 2 on the x-axis and 3 on the y-axis. Thus, the ordered pair corresponding to point A is $(2,3)$.

b. Point B lines up with -3 on the x-axis and 1 on the y-axis, so $(-3,1)$ is the corresponding ordered pair for point B.

c. The x-coordinate of any point on the y-axis is 0. Because point C is located on the y-axis at -2, point C is represented as $(0,-2)$.

PROGRESS CHECK 6 Approximate (use integers) the ordered pair corresponding to the points in Figure 6.7.

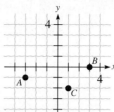

Figure 6.7

Progress Check Answer

6. $A(-3,-1)$, $B(3,0)$, $C(1,-2)$

EXERCISES 6.1

In Exercises 1–6, determine if the given ordered pair is a solution to the equation.

1. $3x + y = 0$; $(1,-3)$
2. $2x - 3y = 13$; $(5,-1)$
3. $y = -2x + 4$; $(0,4)$
4. $y = 2x - 5$; $(-1,7)$
5. $2x - y = 2$; $(1,4)$
6. $-x + 8y = 7$; $(-1,-1)$

In Exercises 7–12, fill in the missing component of the given ordered pairs so that each pair is a solution of the given equation.

7. $3x + y = 0$
  a. $(6, \quad)$
  b. $(\quad,6)$
9. $x - 3y = 6$
  a. $(0, \quad)$
  b. $(\quad,0)$
11. $y = 2x - 5$
  a. $(1, \quad)$
  b. $(\quad,-9)$

8. $2x - 3y = 13$
  a. $(5, \quad)$
  b. $(\quad,5)$
10. $4x + 3y = -24$
  a. $(0, \quad)$
  b. $(\quad,0)$
12. $y = -x + 3$
  a. $(-4, \quad)$
  b. $(\quad,0)$

13. The charge for photo developing is \$3.50 per roll plus \$0.25 per print. The total cost is then given by $C = 3.50 + 0.25n$, where n is the number of printable negatives. Complete the given ordered pairs of the form (n,C) and interpret their meaning.
  a. $(10, \quad)$
  b. $(\quad,10)$

14. The pay for a laborer at a construction site is \$10 for lunch and travel plus \$7 per hour for labor. The daily pay is given by the equation $P = 10 + 7h$, where h is the number of hours worked. Complete the given ordered pairs of the form (h,P) and interpret their meaning.
  a. $(8, \quad)$
  b. $(\quad,55.5)$

15. The charge for a mail order of computer shareware disks is \$3 per disk plus \$4 shipping. The total cost is given by the equation $C = 3n + 4$, where n is the number of disks ordered. Complete the given ordered pairs of the form (n,C) and interpret their meaning.
  a. $(13, \quad)$
  b. $(\quad,13)$

16. The rental rate for a recording studio is \$100 plus \$50 per hour. The total cost is given by $C = 100 + 50h$, where h is the number of hours the studio is rented. Complete the given ordered pairs of the form (h,C) and interpret their meaning.
  a. $(5, \quad)$
  b. $(\quad,500)$

In Exercises 17–24, graph the given ordered pairs and indicate the quadrant for each.

17. **a.** $(1,1)$
  b. $(2,2)$
  c. $(-3,-3)$

18. **a.** $(1,2)$
  b. $(3,4)$
  c. $(-3,-2)$

19. **a.** $(1,0)$
  b. $(0,2)$
  c. $(-3,0)$

20. **a.** $(0,4)$
  b. $(-1,0)$
  c. $(0,-3)$

21. **a.** $(3,-2)$
  b. $(-4,5)$
  c. $(0,0)$

22. **a.** $(5,-7)$
  b. $(-3,1)$
  c. $(-1,-1)$

23. **a.** $(\frac{1}{2},5)$
  b. $(1.3,3.1)$
  c. $(-2.1,1.2)$

24. **a.** $(2,-\frac{7}{3})$
  b. $(-0.3,-3.0)$
  c. $(-\frac{3}{2},\frac{2}{3})$

In Exercises 25–30, give an approximate ordered pair which corresponds to each point. (Use integers.)

25.

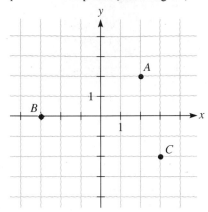

26.

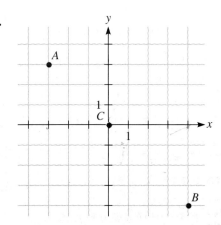

27.

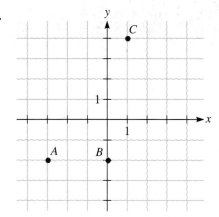

28.

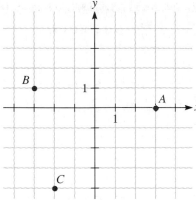

29.

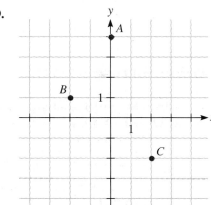

30.

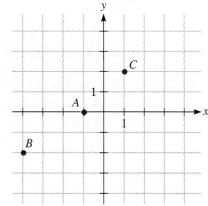

31. Which of these points is *highest* on the graph?
  a. (1,8) **b.** (8,1) **c.** (−400,−100)
32. Which of these points is furthest to the *left?*
  a. (0,1) **b.** (1,0) **c.** (3,−5)
33. Which of these points is furthest to the *right?*
  a. (−4,5) **b.** (−2,6) **c.** (0,1)
34. Which of these points is *lowest* on the graph?
  a. (−5,−1) **b.** (−4,−2) **c.** (−3,−3)

THINK ABOUT IT

1. Three distinct points are on the same vertical line. Which of these statements must be true?
  a. They all have the same second component.
  b. They all have the same first component.
  c. They all have different second components.
  d. They all have different first components.

2. Point (m,n) is in quadrant 2. Point (r,s) is in quadrant 3. Which of these statements must be true?
   a. m must be a negative number.
   b. s must be a negative number.
   c. $m > r$.
   d. $n > s$.

3. Point (p,q) is such that $p + q = 1$.
   a. In what quadrant(s) can this point be?
   b. In what quadrant(s) can it *not* be?
   c. Explain your conclusion.

4. Point (x,y) is such that $xy = 1$. In what quadrant(s) can this point be? In what quadrant(s) can it *not* be? Explain your conclusion.

5. Plot these four points: $(1,\frac{1}{2})$, $(2,\frac{1}{4})$, $(3,\frac{1}{8})$, $(4,\frac{1}{16})$.
   a. If you continue this sequence, what is the next point?
   b. Is each new point lower than the previous point?
   c. If the sequence is continued *forever*, will all the points still be in quadrant 1?
   d. What is the second component for the hundredth point?

REMEMBER THIS

1. Evaluate $y = 2x + 1$ when $x = 2, 0, -2$.
2. True or false? In the formula $y = 2x + 5$, when you increase x by 1, then y increases by 2. (Try it for $x = 1, 2, 3$.)
3. Solve $3x - 4y - 5 = 0$ for y.
4. Graph the solution set of $-3 \le 2x + 1 < 3$.
5. What value of x will make y equal to 0 in the equation $-3x + 2y = 12$?
6. In Northern Ireland between 1969 and 1989, 2,690 people out of a total population of about 1.5 million were killed in political violence. What would this number be proportionately equivalent to in the United States, where the total population at that time was about 225 million?
7. Solve $\dfrac{1}{x} + \dfrac{3}{2x} = 10$.
8. Simplify $\dfrac{\dfrac{1}{a} - b}{\dfrac{1}{b} + a}$.
9. Solve $2x^2 - 3x - 5 = 0$.
10. Express 45.8 billion in scientific notation.

6.2   Graphing Linear Equations

A video store needs to stock multiple copies of a recent hit movie. A formula for the cost C of buying n video tapes if each tape cost $50 is $C = 50n$. Graph the equation. (See Example 3.)

OBJECTIVES

1 Graph linear equations by plotting ordered-pair solutions.

2 Graph linear equations by using intercepts.

3 Graph $x = a$ or $y = b$.

1 Using the Cartesian coordinate system, we can represent any ordered pair of real numbers by a particular point in this system. This enables us to draw a geometric picture (graph) of an equation in two variables.

> **Graph**
>
> The **graph** of an equation in two variables is the set of all points in a coordinate system that correspond to ordered-pair solutions of the equation.

There are many techniques associated with determining the graph of an equation in two variables. One method is to simply choose values for one variable and then determine a list of ordered-pair solutions. By plotting enough of these solutions, we may be able to establish a trend and then complete the graph by following the established pattern. We will use this method in Example 1.

EXAMPLE 1   Graph the equation $y = 2x + 1$.

Solution   Begin by establishing integer values for x from, say, 2 to -2 and making a list of ordered-pair solutions.

If $x =$	Then $y = 2x + 1$ Is	Thus, the Ordered-Pair Solutions Are
2	$2(2) + 1 = 5$	$(2,5)$
1	$2(1) + 1 = 3$	$(1,3)$
0	$2(0) + 1 = 1$	$(0,1)$
-1	$2(-1) + 1 = -1$	$(-1,-1)$
-2	$2(-2) + 1 = -3$	$(-2,-3)$

Now, graph these ordered pairs, as in Figure 6.8, and note that these points all lie on a straight line. In fact, the graph of $y = 2x + 1$ is the straight line shown in Figure 6.9. Every ordered pair that satisfies $y = 2x + 1$ corresponds to a point on this line, and every point on the line corresponds to an ordered pair that satisfies $y = 2x + 1$.

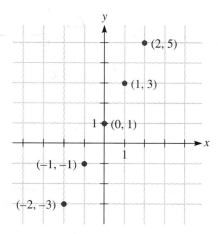

Figure 6.8

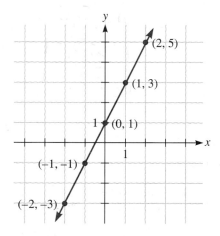

Figure 6.9

PROGRESS CHECK 1   Graph the equation $y = 2x - 1$.

We cannot possibly list all the solutions of most equations in two variables because they form an infinite set of ordered pairs. Determining how many and which points to plot can be a difficult decision. Therefore, we often try to determine the basic shape of the graph from the form of the equation. Throughout this chapter we will focus on linear equations, which means that the equation and the graph will take the following forms.

Progress Check Answer

1.

Linear Equations in Two Variables

A **linear equation in two variables** is an equation that can be written in the general form

$$Ax + By = C,$$

where A, B, and C are real numbers, with A and B not both zero. The graph of a linear equation in two variables is a straight line.

For example, we know that the equation in Example 1 graphs as a straight line because $y = 2x + 1$ is a linear equation (in two variables). In general form, this equation is written $-2x + y = 1$. Other examples of linear equations are

$$y = -x - 1, \qquad 5x - 2y = 10, \qquad y = 2, \qquad \text{and} \qquad x = -3.$$

Once we identify an equation as a linear equation, we can draw its graph by finding any two distinct points in the graph and then drawing a line through them. In practice, it is recommended that a third point in the graph be determined as a checkpoint. If you plot three points and they are not on a straight line, then there is an error in your work.

EXAMPLE 2   Graph $y = -x - 1$.

Solution   The equation $y = -x - 1$ is a linear equation (which may be written in general form as $x + y = -1$). Therefore, the graph of $y = -x - 1$ is a straight line. We graph this line as outlined above by first finding three distinct points in the graph. We will find these points by arbitrarily letting x equal 2, 0, and -2.

If $x =$	Then $y = -x - 1$ Is	Thus, the Ordered-Pair Solutions Are
2	$-(2) - 1 = -3$	$(2, -3)$
0	$-(0) - 1 = -1$	$(0, -1)$
-2	$-(-2) - 1 = 1$	$(-2, 1)$

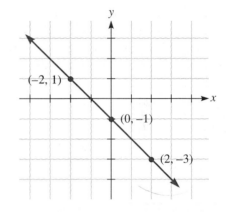

Figure 6.10

Now, plot the three points and draw the line passing through them, as shown in Figure 6.10.

PROGRESS CHECK 2   Graph $y = -x - 3$. ⌐

Graphing linear equations that arise from applied problems will often require a few adjustments to our current methods, as illustrated by the section-opening problem.

EXAMPLE 3   Solve the problem in the section introduction on page 275.

Solution   From the context of the problem, $C = 50n$ is a linear equation that is meaningful only for nonnegative values of n. Thus, the graph of $C = 50n$ is a line that starts at the origin and passes through the points listed in the table below, as shown in Figure 6.11.

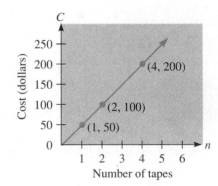

Figure 6.11

If $n =$	Then $C = 50n$ Is	Thus, the Ordered-Pair Solutions Are
1	$(50)1 = 50$	$(1, 50)$
2	$(50)2 = 100$	$(2, 100)$
4	$(50)4 = 200$	$(4, 200)$

Note that it would not be practical to draw this graph with both variables scaled in the same manner. Thus, the two axes are scaled differently in this graph, as are many graphs that represent physical data. Technically, the graph should consist only of points that correspond to nonnegative *integer* values of n, since the video store may purchase only a whole number of tapes. But in common practice, we connect the points and show a linear graph to obtain a better visual image of the relationship.

PROGRESS CHECK 3   It costs a company \$10 to manufacture one unit of a particular product. Therefore, the total cost C of manufacturing n units of this product is given by $C = 10n$. Graph this equation. ⌐

Progress Check Answers

2.   3.

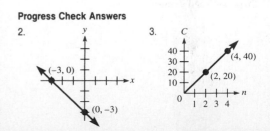

EXAMPLE 4   Graph $2x - 5y = 10$.

Solution   The linear equation is given in general form, so it is not solved for either y or x. In such cases the easiest way to find two distinct points is usually to first let $x = 0$ and find y; then let $y = 0$ and find x.

Let $x = 0$.

$$2x - 5y = 10$$
$$2(0) - 5y = 10$$
$$-5y = 10$$
$$y = -2$$

Let $y = 0$.

$$2x - 5y = 10$$
$$2x - 5(0) = 10$$
$$2x = 10$$
$$x = 5$$

Thus $(0, -2)$ and $(5, 0)$ are solutions.

Now, obtain a third point (or checkpoint) by assigning a convenient number to one of the variables, say $y = 1$, and solving for x.

$$2x - 5y = 10$$
$$2x - 5(1) = 10$$
$$2x - 5 = 10$$
$$2x = 15$$
$$x = \tfrac{15}{2}, \text{ or } 7.5$$

Finally, graph the three ordered pairs and draw the straight line determined by the three points, as shown in Figure 6.12.

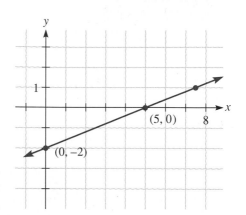

Figure 6.12

PROGRESS CHECK 4   Graph $3x - 2y = 6$.   ⌐

2   Graphing linear equations using intercepts

In Example 4 when we let $x = 0$ and found y, note in Figure 6.12 that we found the point where the graph crosses the y-axis. We call this point the **y-intercept.** Similarly, the point where the graph crosses the x-axis is called the **x-intercept.** So $(5, 0)$ is the x-intercept for the graph of $2x - 5y = 10$. Because the x-intercept always has a y-coordinate of 0, and the y-intercept always has an x-coordinate of 0, we may find intercepts as follows.

> ### To Find Intercepts
>
> To find the x-intercept $(a, 0)$, let $y = 0$ and solve for x. To find the y-intercept $(0, b)$, let $x = 0$ and solve for y.

Graphing linear equations by drawing a line through the intercepts is especially useful when the linear equation is given in general form, not solved for y or x.

EXAMPLE 5   Graph $3x + 2y = -6$ by using the x- and y-intercepts.

Solution   To find the x-intercept, let $y = 0$ and solve for x.

$3x + 2y = -6$	Given equation.
$3x + 2(0) = -6$	Replace y by 0.
$3x = -6$	Simplify.
$x = -2$	Divide both sides by 3.

Thus, the x-intercept is $(-2, 0)$.

To find the y-intercept, let $x = 0$ and solve for y.

$3x + 2y = -6$	Given equation.
$3(0) + 2y = -6$	Replace x by 0.
$2y = -6$	Simplify.
$y = -3$	Divide both sides by 2.

Progress Check Answer

4.

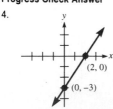

The y-intercept is $(0, -3)$.

To obtain a checkpoint, we arbitrarily let $x = 2$, which leads to the solution $(2, -6)$. The graph in Figure 6.13 is produced by plotting the intercepts and the checkpoint and then drawing the line determined by these points.

Note   Because the x-coordinate of the y-intercept is always 0, it is conventional to take this fact for granted and define the y-intercept as the y-coordinate of the point where the graph intersects the y-axis. For example, this definition means that the y-intercept of the graph in Figure 6.13 is -3. Since writing $(0, -3)$ for the y-intercept serves as a useful reminder that the x-coordinate is 0, we will continue to use our previous definition and write out the entire ordered pair throughout this text. Similar remarks are true for the x-intercept.

PROGRESS CHECK 5   Graph $x + 2y = -4$ by using the x- and y-intercepts. ⌐

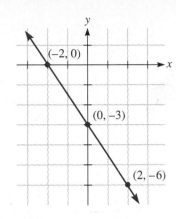

Figure 6.13

3   Graphing horizontal and vertical lines

When a linear equation may be written as simply $x = a$ or $y = b$, then the graph will be either a vertical line or a horizontal line, respectively, as shown in the next two examples.

EXAMPLE 6   Graph $y = 2$.

Solution   The linear equation $y = 2$ is equivalent to $0x + y = 2$, which means that $y = 2$ for all values of x. Thus, a few ordered-pair solutions are

$$(1,2), \quad (0,2), \quad \text{and} \quad (-3,2).$$

And in general, all solutions fit the form $(x, 2)$, with x being any real number. By plotting the three points above and drawing the line that passes through them, we obtain the graph in Figure 6.14. **Note that the graph is a horizontal line and is parallel to the x-axis.**

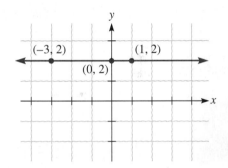

Figure 6.14

PROGRESS CHECK 6   Graph $y = -3$. ⌐

Example 7   Graph $x = -3$.

Solution   The linear equation $x = -3$ is equivalent to $x + 0y = -3$, which means $x = -3$ for all values of y. Thus, all points in the graph have an x-coordinate of -3. Three arbitrary points in the graph correspond to $(-3, 2)$, $(-3, 0)$, and $(-3, -3)$, and Figure 6.15 shows the graph of $x = -3$. **Note that the graph is a vertical line and is parallel to the y-axis.**

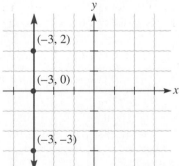

Figure 6.15

PROGRESS CHECK 7   Graph $x = 4$. ⌐

You can use examples 6 and 7 as a basis for understanding the general principles that follow. These principles enable you to quickly graph the special cases $x = a$ and $y = b$, and you should always check to see if a linear equation fits one of these special forms first.

Progress Check Answers

5.

6.

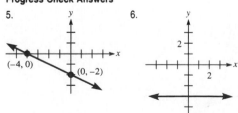

7.

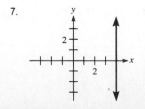

> **To Graph $x = a$ and $y = b$**
>
> 1. The graph of the linear equation $x = a$ is a vertical line that contains the point $(a,0)$. If $a = 0$, then the equation is $x = 0$, and the line is the y-axis.
> 2. The graph of the linear equation $y = b$ is a horizontal line that contains the point $(0,b)$. If $b = 0$, then the equation is $y = 0$ and the line is the x-axis.

EXERCISES 6.2

In Exercises 1–12, graph the given equation by plotting ordered-pair solutions.

1. $y = 3x + 2$

2. $y = 2x + 3$

3. $y = x - 1$

4. $y = x - 2$

5. $y = -x + 2$

6. $y = -x - 2$

7. $y = -3x - 1$

8. $y = -4x + 4$

9. $2x + 3y = 12$

10. $3x + 4y = -12$

11. $3x - y = -6$

12. $x - 2y = 4$

In Exercises 13–24, graph the given equation by using the intercepts.

13. $x + y = 5$

14. $x + y = 6$

15. $x - y = 3$

16. $x - y = 4$

17. $2x + 3y = -6$

18. $3x + 5y = -15$

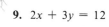

19. $-3x + y = 3$

20. $-2x + 3y = 12$

34. $x = 0$

35. $3y - 1 = 5$

36. $2x + 1 = -7$

37. If the cost of each CD is $9, then an equation for the cost of n CDs is $C = 9n$. Graph the equation. Label both axes with a variable and a description.

21. $x = y + 1$

22. $x = y + 3$

23. $y = 2x - 10$

24. $y = 2x + 10$

38. If each bus fare costs $1.50, then the cost of m bus fares is given by $C = 1.50m$. Graph the equation. Label both axes with a variable and a description.

In Exercises 25–36, graph the given equation. Indicate whether the graph is a horizontal line or a vertical line.

25. $y = 3$

26. $y = 4$

27. $y = -1$

39. An order of computer shareware disks costs $3 per disk plus $4 shipping per order.
  a. Find an equation for the total cost (C) in terms of the number (n) of disks ordered.
  b. Graph the equation.

28. $y = -2$

29. $x = 2$

30. $x = 3$

40. The cost of a car rental is $80 per week plus $0.20 per mile.
  a. Find an equation for the total cost (C) for one week in terms of the miles (m) driven.
  b. Graph the equation.

31. $x = -1$

32. $x = -2$

33. $y = 0$

THINK ABOUT IT

1. The cost of an audiocassette is $6 and the cost of a CD is $12. You wish to spend exactly $60 for a combination of these. An equation which describes this is $6x + 12y = 60$, where x is the number of cassettes purchased and y is the number of CDs purchased. Graph this equation. Find and interpret several points on the line, excluding the intercepts. Note that only nonnegative integers make sense for x and y.
2. Find a point on the line $2x + 3y = 6$ for which the x and y components are equal. Show that your answer is correct by substituting the value of the components into the equation.
3. Graph these two equations on the same set of axes.

$$3x - 4y = 2$$
$$x + y = 3$$

a. See if you can tell where the 2 lines intersect. Find the coordinates of the point of intersection.
b. Show that the point of intersection is on both lines by showing that its coordinates satisfy *both* equations.
c. How many points can satisfy both of these equations?
4. For a certain line the x-intercept is $(5,0)$ and the y-intercept is $(0,-2)$.
a. Find an equation for this line.
b. Show that the coordinates of both intercepts satisfy the equation.
5. Find an equation of a line whose y-intercept is 3 and which has no x-intercept.

REMEMBER THIS

1. Evaluate the formula $\dfrac{y_2 - y_1}{x_2 - x_1}$ when $x_1 = -1, y_1 = 2, x_2 = 3,$ and $y_2 = -1$.

2. If you draw a line through $(1,-1)$ and $(7,-1)$, will you get a horizontal line or a vertical line?
3. A quadrilateral has _____ sides.
4. Which number has the larger absolute value, -3 or -5?

5. Is $(1,-2)$ a solution of $y + 4 = 2x$?
6. Simplify $28x - 4x[4 + 3(1 - x)]$.
7. Find the volume of a cylinder whose radius is 6 in. and whose height is 4 ft.
8. Multiply $(3y + 2)(y^2 - y + 1)$.
9. Simplify $\dfrac{(4x^2)^{-1}(2x)^{-2}}{(3x)^{-1}(4x^2)^{-2}}$.
10. Solve for n: $n^2 + 9 = 6n$.

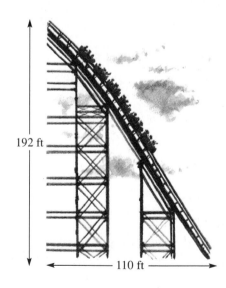

192 ft

110 ft

6.3   The Slope of a Line

The Magnum roller coaster in Sandusky, Ohio, is the world's tallest and steepest coaster. On this coaster riders drop a vertical distance of 192 ft (a 20-story drop) over a horizontal distance of 110 ft. Find the slope (steepness) of this portion of the coaster. (See Example 5.)

OBJECTIVES

1 Find and interpret the slope of a line.

2 Determine if lines are parallel, perpendicular, or neither by using slope.

3 Interpret line graphs.

1 The steepness of a roller coaster drop, the grade of a roadway, and the pitch of a roof serve as concrete illustrations of the mathematical concept of slope. The slope of a line is a measure of its steepness or inclination with respect to the horizontal axis. To define this measure, we consider the line in Figure 6.16 that contains the arbitrary points (x_1,y_1) and (x_2,y_2).* To find the slope, calculate the vertical change $y_2 - y_1$, called the **rise,** and divide this number by the horizontal change $x_2 - x_1$, called the **run.**

*Recall from Section 5.6 that in the symbols $x_1, x_2, y_1,$ and y_2 the numbers 1 and 2 are called **subscripts.** Subscript notation serves many useful identification purposes.

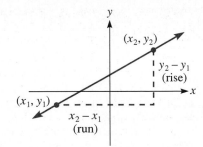

Figure 6.16

Slope of a Line

If (x_1, y_1) and (x_2, y_2) are any two distinct points on a line, with $x_1 \neq x_2$, then the slope m of the line is

$$m = \frac{\text{rise}}{\text{run}} = \frac{\text{change in } y}{\text{change in } x} = \frac{y_2 - y_1}{x_2 - x_1}.$$

EXAMPLE 1   Find the slope of the line through (1,2) and (3,3).

Solution   If we label (1,2) as point 1, then $x_1 = 1$, $y_1 = 2$, $x_2 = 3$, and $y_2 = 3$. The slope formula now gives

$$m = \frac{\text{rise}}{\text{run}} = \frac{y_2 - y_1}{x_2 - x_1} = \frac{3 - 2}{3 - 1} = \frac{1}{2}.$$

The slope is $\frac{1}{2}$, which means that y increases 1 unit for each 2-unit increase in x, as shown in Figure 6.17.

Note   The slope is unaffected by the way we label the points. In this example if we label (3,3) as point 1, then $x_1 = 3$, $y_1 = 3$, $x_2 = 1$, and $y_2 = 2$, so

$$m = \frac{y_2 - y_1}{x_2 - x_1} = \frac{2 - 3}{1 - 3} = \frac{-1}{-2} = \frac{1}{2}.$$

PROGRESS CHECK 1   Find the slope of the line through (2,1) and (5,3).

EXAMPLE 2   Find the slope of the line through $(-1,2)$ and $(3,-1)$.

Solution   We let $x_1 = -1$, $y_1 = 2$, $x_2 = 3$, and $y_2 = -1$, and we substitute in the slope formula.

$$m = \frac{\text{rise}}{\text{run}} = \frac{y_2 - y_1}{x_2 - x_1} = \frac{-1 - 2}{3 - (-1)} = \frac{-3}{4}.$$

A slope of $-\frac{3}{4}$ indicates that as x increases 4 units, y decreases 3 units, as shown in Figure 6.18.

PROGRESS CHECK 2   Find the slope of the line through $(-3,3)$ and $(0,-2)$.

When the slope formula is applied to lines that are horizontal or vertical, then special cases occur, as shown in the next two examples.

EXAMPLE 3   Find the slope of the line through $(-2,3)$ and (4,3).

Solution   Figure 6.19 shows that the line through the given points is horizontal. Applying the slope formula gives

$$m = \frac{\text{rise}}{\text{run}} = \frac{y_2 - y_1}{x_2 - x_1} = \frac{3 - 3}{4 - (-2)} = \frac{0}{6} = 0.$$

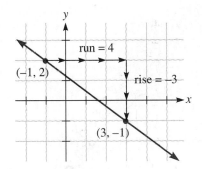

Figure 6.17

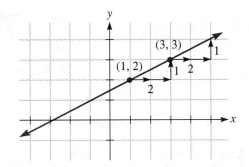

Figure 6.18

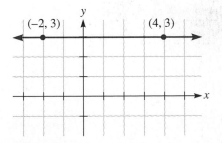

Figure 6.19

Progress Check Answers
1. $m = \frac{2}{3}$
2. $m = -\frac{5}{3}$

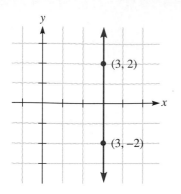

Figure 6.20

The slope of every horizontal line is 0, because the numerator of the slope ratio (the change in y) is always 0.

PROGRESS CHECK 3   Find the slope of the line through (6,1) and (0,1).   ⌐

EXAMPLE 4   Find the slope of the line through $(3, -2)$ and $(3, 2)$.

Solution   Figure 6.20 shows that the line through the given points is vertical. The slope formula is not meaningful here, since

$$m = \frac{\text{rise}}{\text{run}} = \frac{y_2 - y_1}{x_2 - x_1} = \frac{2 - (-2)}{3 - 3} = \frac{4}{0},$$

and m is undefined.

  The slope of every vertical line is undefined, because the denominator of the slope ratio (the change in x) is always zero.

PROGRESS CHECK 4   Find the slope of the line through $(-1, 1)$ and $(-1, -1)$.   ⌐

  Examples 1–4 have illustrated the cases in which the slope of a line is positive, negative, zero, or undefined. Consider carefully Figure 6.21, which summarizes our results and shows what to look for in each of these cases.

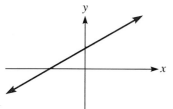

Slope is positive.
Line is higher on the right.
y increases as x increases.

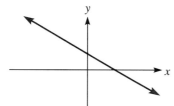

Slope is negative.
Line is lower on the right.
y decreases as x increases.

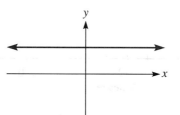

Slope is zero.
Line is horizontal.
y remains constant as x increases.

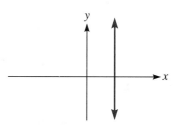

Slope is undefined.
Line is vertical.
x remains constant as y increases.

Figure 6.21

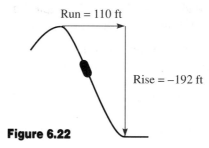

Figure 6.22

  In Examples 5 and 6 we interpret the meaning of slope in applied problems.

EXAMPLE 5   Solve the problem in the section introduction on page 282.

Solution   To measure the steepness of this roller coaster fall, note that the drop is almost linear and use the slope formula. Represent the rise by -192, as shown in Figure 6.22, because the coaster is falling in this portion of the ride.

$$m = \frac{\text{rise}}{\text{run}} = \frac{-192}{110} = \frac{-96}{55} = -1.75$$

The slope shows that the coaster falls about 1.75 ft per horizontal foot. In fact, this drop is about a 60° plunge at more than 70 mi/hour!

PROGRESS CHECK 5   Find the slope of a roof that rises 4 ft vertically through a horizontal distance of 24 ft. ⌐

If the relationship between two variables x and y graphs as a line, then the slope of the line measures the **rate of change** of y with respect to x. We will use this interpretation in the next example.

EXAMPLE 6   The weekly cost of renting a compact car (y) is $80 if the weekly mileage (x) is 100 mi and is $91 if the weekly mileage is 200 mi. If the relation between x and y graphs as a line, calculate and interpret the slope.

Solution   Figure 6.23 shows the graph of the relation between x and y. We let $x_1 = 100$, $y_1 = 80$, $x_2 = 200$, and $y_2 = 91$, and we substitute in the slope formula.

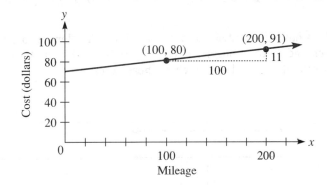

Figure 6.23

$$m = \frac{\text{change in } y}{\text{change in } x} = \frac{y_2 - y_1}{x_2 - x_1} = \frac{91 - 80}{200 - 100} = \frac{11}{100} = 0.11$$

Because the unit for y is dollars and the unit for x is miles, the slope describes dollars per mile. Therefore, a slope of 0.11 means that the mileage charge for the rental is 11 cents per mile.

PROGRESS CHECK 6   Redo the problem in Example 6 assuming that the cost is $76 if the weekly mileage is 100 mi and is $82 if the weekly mileage is 150 miles. ⌐

2   Slope is an important concept when we use algebra to compare lines. Two lines in a plane that never intersect are called **parallel lines.** Intuitively, it is easy to understand that parallel lines have the same slope, or inclination. Two lines that meet at right angles are called **perpendicular lines.** It can be shown that if the product of the slopes of two lines is -1, then the lines are perpendicular. In the following summary of these two properties, note that vertical lines are excluded, because the slope of vertical lines is undefined.

Parallel and Perpendicular Lines

1. Two nonvertical lines are parallel if the slopes are equal.
2. Two nonvertical lines are perpendicular if the product of their slopes is -1.

When applying these properties in the next example, you will need to know that a **parallelogram** is a quadrilateral in which both pairs of opposite sides are parallel, while a **rectangle** is a parallelogram in which all four angles are right angles.

Progress Check Answers
5. $\frac{1}{6}$
6. Mileage charge is 12 cents per mile.

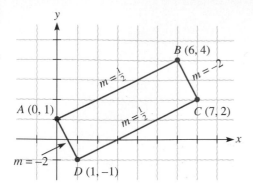

Figure 6.24

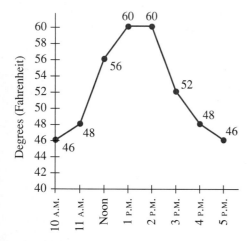

Figure 6.25

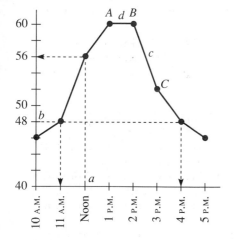

Figure 6.26

Progress Check Answers

7. (a) $m_{AB} = m_{CD} = \frac{1}{3}$; $m_{BC} = m_{AD} = -3$

(b) Because $\frac{1}{3}(-3) = -1$, the intersecting sides are perpendicular.

EXAMPLE 7 A quadrilateral has vertices at $A(0,1)$, $B(6,4)$, $C(7,2)$, and $D(1,-1)$.

  a. Show that quadrilateral $ABCD$ is a parallelogram.
  b. Show that parallelogram $ABCD$ is a rectangle.

Solution Quadrilateral $ABCD$ is shown in Figure 6.24.

  a. First, compute the slope of each of the four sides.

$$m_{AB} = \frac{4 - 1}{6 - 0} = \frac{3}{6} = \frac{1}{2}$$

$$m_{BC} = \frac{2 - 4}{7 - 6} = \frac{-2}{1} = -2$$

$$m_{CD} = \frac{2 - (-1)}{7 - 1} = \frac{3}{6} = \frac{1}{2}$$

$$m_{AD} = \frac{-1 - 1}{1 - 0} = \frac{-2}{1} = -2$$

Each pair of opposite sides is parallel, because $m_{AB} = m_{CD} = \frac{1}{2}$, and $m_{BC} = m_{AD} = -2$. Thus $ABCD$ is a parallelogram.

  b. All four angles are right angles, because

$$m_{AB} \cdot m_{BC} = m_{AB} \cdot m_{AD} = m_{CD} \cdot m_{BC} = m_{CD} \cdot m_{AD} = \tfrac{1}{2}(-2) = -1.$$

Thus, parallelogram $ABCD$ is a rectangle.

PROGRESS CHECK 7 A quadrilateral has vertices at $A(0,2)$, $B(6,4)$, $C(7,1)$, and $D(1,-1)$.

  a. Show that quadrilateral $ABCD$ is a parallelogram.
  b. Show that quadrilateral $ABCD$ is a rectangle.

3 The concepts of this chapter are useful for interpreting a common type of graph called a **line graph.** In such graphs points are plotted that correspond to data in a relationship, and the points are then connected with line segments. The resulting picture very effectively shows changes and trends in the relationship.

EXAMPLE 8 The line graph in Figure 6.25 shows the results of hourly outdoor temperature readings from 10 A.M. to 5 P.M. on a certain day. Answer the following questions using this graph.

  a. What is the outdoor temperature at noon?
  b. When was the outdoor temperature 48 degrees?
  c. Between which two hourly readings was the temperature decrease the greatest?
  d. Between what two hourly readings was there no change in temperature?

Solution See Figure 6.26.

  a. Locate noon on the horizontal axis and draw a vertical line through noon that intersects the graph. Then drawing a horizontal line from this intersection point to the vertical axis shows that the temperature at noon was 56 degrees.
  b. Locate 48 degrees on the vertical axis and draw a horizontal line through 48 degrees and the graph. Drawing vertical lines from *both* intersection points to the horizontal axis shows that the temperature was 48 degrees at 11 A.M. and 4 P.M.
  c. The line segment with the negative slope that is *largest in absolute value* corresponds to the greatest decrease in hourly temperatures. In Figure 6.26 line segment BC shows the largest drop (an hourly fall of 8 degrees), so the temperature decrease was greatest between the hourly readings at 2 P.M. and 3 P.M.

d. A line segment with zero slope corresponds to no change in hourly temperature. A zero slope implies a horizontal line, so by considering line segment *AB* in Figure 6.26, we determine that there was no change in temperature between the hourly readings at 1 P.M. and 2 P.M.

PROGRESS CHECK 8 Answer the following questions using the line graph in Figure 6.25.

a. What was the outdoor temperature at 3 P.M.?

b. When was the outdoor temperature 46 degrees?

c. Between which two hourly readings was the temperature increase the greatest?

EXERCISES 6.3

In Exercises 1–12, find the slope of the line which passes through the given points. Sketch the line. In each case, give the rise and the run.

1. (1,3) and (2,4) **2.** (2,5) and (3,7) **3.** (−2,0) and (0,4)

4. (−3,1) and (0,4) **5.** (1,2) and (2,−1)

6. (2,3) and (4,−5) **7.** (−2,−3) and (1,5)

8. (−3,−1) and (3,1) **9.** (0,0) and (5,0)

10. (−2,5) and (3,5) **11.** (5,1) and (5,−2)

12. (−2,4) and (−2,−1)

In Exercises 13–24, find the missing coordinate so that the line connecting the points has the required slope. Check your work by a sketch.

13. (1,1), (2,y); $m = 2$ **14.** (2,2), (3,y); $m = 3$
15. (0,0), (1,y); $m = -3$ **16.** (2,0), (3,y); $m = -1$
17. (−2,−3), (−1,y); $m = 4$ **18.** (−3,−2), (−2,y); $m = 5$
19. (−3,2), (−2,y); $m = -2$
20. (−3,0), (−2,y); $m = -2$
21. (5,1), (7,y); $m = 3$ **22.** (3,0), (5,y); $m = 2$
23. (0,0), (−3,y); $m = 2$ **24.** (0,0), (−2,y); $m = 1$
25. A wheelchair ramp rises 1 ft through a horizontal distance of 6 ft. Find the absolute value of the slope of the ramp.
26. A ladder is leaning against a wall so that the foot of the ladder is 3 ft from the wall and the top of the ladder hits the wall 8 ft above the ground. Find the absolute value of the slope of the ladder.
27. Two points on a horizontal line are connected. The distance between them is 4 units. Find the rise, run, and slope as you go from left to right.
28. Two points on a vertical line are connected. The distance between them is 3 units. Find the rise, run, and slope as you go from the lower point to the higher one.

29. A ski trail drops 50 ft over a distance of 50 ft. Find the slope of the trail over this segment.

30. An A-frame house has the cross section shown in the figure. Find the magnitude of the slope of the sidewall.

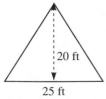

25 ft

31. The weekly cost y of renting a car is $100 when the mileage x is 250 mi and $150 when the mileage is 500 mi. Assume that the relation between x and y graphs as a line; then calculate and interpret the slope. Show a sketch with labeled axes. Find and interpret the y-intercept.

32. A printing shop charges $20 for 500 copies of a flyer and $25 for 750 copies. If the relation between the number of copies x and cost y graphs as a line, calculate and interpret the slope. Sketch the line. Find and interpret the y-intercept.

33. A car starts a trip with 15 gal of gas x and the trip meter y at 0. At the end of the trip it has 4 gal of gas and the trip meter is at 231 mi. If the relation between x and y graphs as a line, calculate and interpret the slope. Show a sketch with labeled axes. What would the y-intercept tell you in this problem?

34. An object is released from a helicopter. Its initial velocity is 0, but as it falls, it speeds up. After 2 seconds its speed is 64 ft/second. If the relation between speed y and elapsed time x graphs as a line, find the slope. What does the slope tell you about the rate at which the speed increases? Graph the line.

35. A laundry charges $36.25 to come to a house and clean a 5- × 7-ft rug. It charges $91.00 to come and clean a 9- × 12-ft carpet. The relationship between the *area* of the rug x and the price y can be graphed as a straight line. Calculate and interpret the slope. Show a sketch with labeled axes.

36. A goldsmith charges $100 for a shipment of 1 oz of a gold alloy and $140 to ship 1.5 oz. If the relation between price y and weight x graphs as a line, find and interpret the slope.

37. A quadrilateral has vertices at $A(1,2)$, $B(5,4)$, $C(6,2)$, and $D(2,0)$.
   a. Sketch $ABCD$.
   b. Show that $ABCD$ is a parallelogram.

   c. Show that $ABCD$ is a rectangle.

38. A quadrilateral has vertices at $A(-1,5)$, $B(5,8)$, $C(6,6)$, and $D(0,3)$.
   a. Sketch $ABCD$.
   b. Show that $ABCD$ is a parallelogram.

   c. Show that $ABCD$ is a rectangle.

39. A quadrilateral has vertices at $A(0,1)$, $B(-1,4)$, $C(4,5)$, and $D(5,2)$.
   a. Sketch $ABCD$.
   b. Show that $ABCD$ is a parallelogram.

   c. Show that $ABCD$ is *not* a rectangle.

40. A quadrilateral has vertices at $A(-2,-2)$, $B(2,2)$, $C(4,-1)$, and $D(0,-5)$.
   a. Sketch $ABCD$.
   b. Show that $ABCD$ is a parallelogram.

   c. Show that $ABCD$ is *not* a rectangle.

41. Two lines are perpendicular.
   a. If the slope of one is 1, then the slope of the other is _____ .
   b. If the slope of one is -2, then the slope of the other is _____ .
   c. If the slope of one is $\frac{1}{4}$, then the slope of the other is _____ .
   d. If the slope of one is 0, then the slope of the other is _____ .
   e. If the slope of one is 0.8, then the slope of the other is _____ .

42. Two lines are perpendicular.
   a. If the slope of one is -5, then the slope of the other is _____ .
   b. If the slope of one is $\frac{3}{4}$, then the slope of the other is _____ .
   c. If the slope of one is -0.25, then the slope of the other is _____ .
   d. If the slope of one is 100, then the slope of the other is _____ .
   e. If the slope of one is undefined, then the slope of the other is _____ .

43. By using the line graph, answer the given questions. The graph shows the hourly temperature at a weather station for one day.

Degrees Fahrenheit

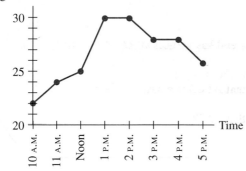

   a. What was the temperature at 10 A.M.? 4 P.M.?

   b. When was the temperature 24 degrees?
   c. Between which two hourly readings was the temperature increase the greatest?
   d. What was the temperature change between 3 P.M. and 4 P.M.?

44. The line graph shows the progress of the Dow-Jones Industrial Average for one day.

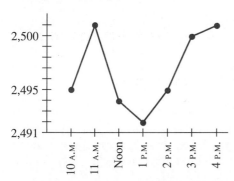

   a. At what time was the index at its lowest value of the day?

   b. Over which interval did the index change most?

   c. Over which interval did it change least?

45. A pot of water was heated on a kitchen stove and the temperature recorded every minute for 7 minutes.

Degrees Fahrenheit

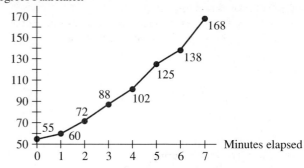

   a. Over which interval did the temperature rise most?
   b. Find the slope of a line connecting the first and last points. Interpret the slope. Round to the nearest tenth.

46. The graph shows barometric air pressure measured at various altitudes one day over one spot on the earth.

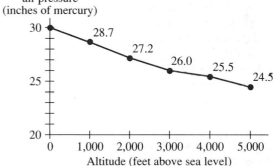

   a. What does this graph show about the relationship of atmospheric pressure and altitude?
   b. What is the barometric pressure at sea level?
   c. Find the slope of a line connecting the first and last points. What does the value of the slope tell you?

47. Using the table, create a line graph. Label the axes carefully.

Year	Population of Small Town
1950	950
1960	1,000
1970	1,400
1980	1,250
1990	1,200

   a. In which decade was there the greatest change in population?
   b. Compute the slope between the points for 1980 and 1990 and interpret it.

48. Here are United Nations estimates of world population size. Use the data to create a line graph.

Year	Population (Billions)
1950	2.5
1960	3.0
1970	3.7
1980	4.5
1990	5.3
2000 (projected)	6.3

a. In what decade is there the greatest change in population?

b. Compute the slope between 1980 and 1990 and interpret it.

THINK ABOUT IT

1. Find two points so that the line through them has slope
  a. 3   **b.** 0   **c.** $\frac{1}{2}$ **d.** $-\frac{3}{4}$ **e.** undefined

2. Since $-\dfrac{3}{4}$ is the same as $\dfrac{-3}{4}$ and $\dfrac{3}{-4}$, then a rise of -3 with a run of 4 describes the same slope as a rise of 3 and a run of -4. Show that this is correct by starting at the origin and drawing the rise and run to find a second point.

3. Why do some people say that a vertical line has an "infinite" slope? Find two points so that the line through them has slope 1 million. Is this a vertical line?

4. A graph starts at (0,0). The next point is (1,1). The next point is $(2,\frac{3}{2})$. Then $(3,\frac{7}{4})$. Then $(4,\frac{15}{8})$. What is the next point? Is each point higher than the previous one? If the sequence of points continues forever, will the points ever be higher than the line $y = 2$? Which of these shapes describes the graph?

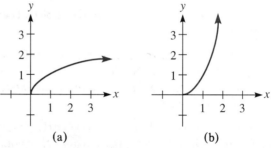

(a)   (b)

5. Find the length of the line segment between $A(1,1)$ and $B(3,3)$ by taking the rise and run as sides of a right triangle and using the Pythagorean theorem.

REMEMBER THIS

1. Show that these two equations are equivalent:
$y + 3 = -4x + 8$ and $4x + y = 5$.
(Change the first into the second by applying rules of algebra.)

2. Show that these two equations are equivalent:
$y - 5 = -\frac{1}{3}(x - 2)$ and $y - 6 = -\frac{1}{3}(x + 1)$.
(Write both in general form and show that they end up the same.)

3. Solve $ax + by = c$ for y.

4. Evaluate $y = mx + b$ when $x = 0$.

5. Graph the line $x + 3 = 0$.

6. Find both intercepts of $3x - 4y = 24$.

7. After the price of a gallon of gasoline was raised by 8 percent, the cost was 148.5 cents per gallon. What was the price before the increase?

8. The difference in length between the length and width of a rectangle is 3.6 cm and the area is 7 cm². Find the perimeter. (*Hint:* Multiply both sides of the equation by 10 to get rid of decimals.)

9. True or false? $(-3)^{-2} = (3^2)^{-1}$

10. Solve $\dfrac{1}{x - 1} + \dfrac{1}{x + 1} = \dfrac{4}{3}$.

6.4 Equations of a Line

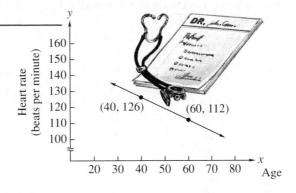

T he minimum target heart rate in beats per minute that produces a
training effect when exercising is 126 at age 40 and 112 at age 60.

  a. The relation between minimum target heart rate y and age x graphs
     as a line, as shown in the figure. Write the equation of this line
     in the form $Ax + By = C$.

  b. What is the minimum target heart rate for a 20-year-old?

  c. How old is someone with a minimum target heart rate of 133?
     (See Example 3.)

OBJECTIVES

1   Write an equation for a line given its slope and a point on the line.

2   Write an equation for a line given two points on the line.

3   Find the slope and y-intercept given an equation for the line.

4   Graph and write an equation for a line given the slope and y-intercept.

5   Determine if the graphs of two linear equations are distinct parallel lines.

There are many forms for the equation of a line. The form $Ax + By = C$ is called
the *general form*, as discussed in Section 6.2. In this section we will consider two other
forms, called the *point-slope equation* and the *slope-intercept equation*, that are often
used when working with lines.

1   Consider any nonvertical line with slope m that passes through the point (x_1, y_1)
as shown in Figure 6.27. If (x, y) represents any other point on this line, then the slope
definition gives

$$\frac{y - y_1}{x - x_1} = m.$$

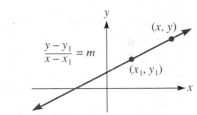

If we multiply both sides by $x - x_1$, this equation becomes

$$y - y_1 = m(x - x_1),$$

Figure 6.27

and the result is called the *point-slope* form of the equation of a line.

Point-Slope Equation

An equation of the line with slope m passing through (x_1, y_1) is

$$y - y_1 = m(x - x_1).$$

This equation is called the **point-slope** form of the equation of a line.

EXAMPLE 1   Find an equation of the line through $(2, -3)$ with slope -4. Express
the answer in the general form $Ax + By = C$.

Solution   We are given that $x_1 = 2$, $y_1 = -3$, and $m = -4$. Substitute these
numbers in the point-slope equation and simplify to the form requested.

$$y - y_1 = m(x - x_1) \qquad \text{Point-slope equation.}$$
$$y - (-3) = -4(x - 2) \qquad \text{Replace } y_1 \text{ by } -3, m \text{ by } -4, \text{ and } x_1 \text{ by 2.}$$
$$y + 3 = -4x + 8 \qquad \text{Remove parentheses.}$$
$$4x + y + 3 = 8 \qquad \text{Add } 4x \text{ to both sides.}$$
$$4x + y = 5 \qquad \text{Subtract 3 from both sides.}$$

In general form, the equation of the line is $4x + y = 5$.

PROGRESS CHECK 1   Express in general form the equation of the line through $(5, -2)$ with slope -3. ⌐

2   The point-slope equation is also useful when two points on a line are known and we need to find an equation for the line.

EXAMPLE 2   Find an equation for the line through the points $(-1,6)$ and $(2,5)$. Express the answer in the general form $Ax + By = C$.

Solution   First, we find the slope of the line.

$$m = \frac{y_2 - y_1}{x_2 - x_1} = \frac{5 - 6}{2 - (-1)} = \frac{-1}{3} = -\frac{1}{3}$$

Now, use the point-slope equation with $m = -\frac{1}{3}$ and either $(-1,6)$ or $(2,5)$ for (x_1,y_1). Using $x_1 = 2$ and $y_1 = 5$, we have the following steps.

$$y - y_1 = m(x - x_1) \qquad \text{Point-slope equation.}$$
$$y - 5 = -\tfrac{1}{3}(x - 2) \qquad \text{Replace } y_1 \text{ by 5, } m \text{ by } -\tfrac{1}{3}, \text{ and } x_1 \text{ by 2.}$$
$$3(y - 5) = -1(x - 2) \qquad \text{Multiply both sides by 3.}$$
$$3y - 15 = -x + 2 \qquad \text{Remove parentheses.}$$
$$x + 3y = 17 \qquad \text{Add } x \text{ and 15 to both sides.}$$

In general form, the equation of the line is $x + 3y = 17$.

Note   Either $(-1,6)$ or $(2,5)$ may be used for (x_1,y_1) because both points are on the line. Using $(-1,6)$ for (x_1,y_1) leads to the following equation.

$$y - 6 = -\tfrac{1}{3}[x - (-1)] \qquad \text{Replace } y_1 \text{ by 6, } m \text{ by } -\tfrac{1}{3}, \text{ and } x_1 \text{ by } -1.$$
$$3(y - 6) = -1[x - (-1)] \qquad \text{Multiply both sides by 3.}$$
$$3y - 18 = -x - 1 \qquad \text{Remove parentheses.}$$
$$x + 3y = 17 \qquad \text{Add } x \text{ and 18 to both sides.}$$

As expected, the result is the same.

PROGRESS CHECK 2   Find an equation for the line through the points $(-3,5)$ and $(1,4)$. Express the answer in the form $Ax + By = C$. ⌐

EXAMPLE 3   Solve the problem in the section introduction on page 291.

Solution

a. We are asked to find in general form the equation of the line through $(40,126)$ and $(60,112)$. As in Example 2, we first find the slope.

$$m = \frac{y_2 - y_1}{x_2 - x_1} = \frac{112 - 126}{60 - 40} = \frac{-14}{20} = -\frac{7}{10}$$

Now, use the point-slope equation with one of the points, say $(60,112)$, as follows.

$$y - y_1 = m(x - x_1) \qquad \text{Point-slope equation.}$$
$$y - 112 = -\tfrac{7}{10}(x - 60) \qquad \text{Replace } y_1 \text{ by 112, } m \text{ by } -\tfrac{7}{10}, \text{ and } x_1 \text{ by 60.}$$
$$10(y - 112) = -7(x - 60) \qquad \text{Multiply both sides by 10.}$$
$$10y - 1{,}120 = -7x + 420 \qquad \text{Remove parentheses.}$$
$$7x + 10y = 1{,}540 \qquad \text{Add } 7x \text{ and 1,120 to both sides.}$$

Progress Check Answers
1. $3x + y = 13$
2. $x + 4y = 17$

In general form the equation of the line is $7x + 10y = 1,540$.

b. To find the minimum target heart rate for a 20-year-old, replace x by 20 in the equation that defines the relation and solve for y.

$$7x + 10y = 1,540 \quad \text{Equation of the line.}$$
$$7(20) + 10y = 1,540 \quad \text{Replace } x \text{ by 20.}$$
$$10y = 1,400 \quad \text{Subtract 7(20), or 140, from both sides.}$$
$$y = 140 \quad \text{Divide both sides by 10.}$$

At age 20, the minimum target heart rate is 140 beats per minute.

c. To find the age of someone with a minimum target heart rate of 133, replace y by 133 in the equation that defines the relation and solve for x.

$$7x + 10y = 1,540 \quad \text{Equation of the line.}$$
$$7x + 10(133) = 1,540 \quad \text{Replace } y \text{ by 133.}$$
$$7x = 210 \quad \text{Subtract 10(133), or 1,330, from both sides.}$$
$$x = 30 \quad \text{Divide both sides by 7.}$$

A minimum target heart rate of 133 corresponds to a 30-year-old.

PROGRESS CHECK 3   The maximum target heart rate when exercising is 153 at age 40 and 136 at age 60.

a. Write in general form the equation of the line that defines the linear relation between maximum target heart rate y and age x.

b. What is the maximum target heart rate for a 20-year-old?

c. How old is someone with a maximum target heart rate of 119?

3   The point-slope equation is used extensively to find the equation of a line, but it is not very helpful for graphing lines or interpreting relations that graph as lines from a known equation. To develop a form that is useful in such cases, consider Figure 6.28. Note that we will use b to denote the y-coordinate of the point where the graph crosses the y-axis, so the point $(0,b)$ is the y-intercept. If we apply the point-slope equation in the case when (x_1,y_1) is the y-intercept, we have the following steps.

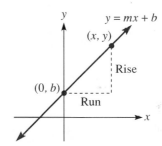

$$y - y_1 = m(x - x_1) \quad \text{Point-slope equation.}$$
$$y - b = m(x - 0) \quad \text{Replace } y_1 \text{ by } b \text{ and } x_1 \text{ by 0.}$$
$$y - b = mx \quad \text{Simplify.}$$
$$y = mx + b \quad \text{Add } b \text{ to both sides.}$$

Figure 6.28

This result is called the *slope-intercept* form of the equation of a line. When we express a linear equation in this form, we may easily find the slope and y-intercept.

Slope-Intercept Equation

The graph of the equation
$$y = mx + b$$
is a line with slope m and y-intercept $(0,b)$.

EXAMPLE 4   Find the slope and the y-intercept of the line defined by the following equations.

a. $y = 4x + 5$

b. $2x + 3y = -7$

Solution

a. The equation is given in the form $y = mx + b$, with $m = 4$ and $b = 5$. Thus:

The slope of the line is 4.

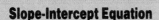

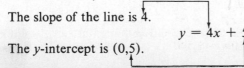
$$y = 4x + 5$$

The y-intercept is $(0,5)$.

Progress Check Answers
3. (a) $17x + 20y = 3,740$ (b) 170   (c) 80

b. First, express the equation in the form $y = mx + b$.

$$2x + 3y = -7 \qquad \text{Given equation.}$$
$$3y = -2x - 7 \qquad \text{Subtract } 2x \text{ from both sides.}$$
$$y = -\tfrac{2x}{3} - \tfrac{7}{3} \qquad \text{Divide both sides by 3.}$$

Matching the equation to the form $y = mx + b$, we conclude that

$$m = -\tfrac{2}{3} \qquad \text{and} \qquad b = -\tfrac{7}{3}.$$

Thus, the slope is $-\tfrac{2}{3}$ and the y-intercept is $(0, -\tfrac{7}{3})$.

PROGRESS CHECK 4   Find the slope and the y-intercept of the line defined by the following equations.

a. $y = -6x + 7$

b. $5x - 2y = 3$ ⌐

4 When the slope and the y-intercept are known, the line may be graphed and its equation may be found, as shown in the next example.

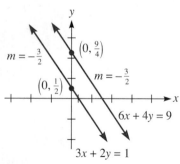

Figure 6.29

EXAMPLE 5   The slope of a line is $-\tfrac{1}{2}$ and the y-intercept is $(0,3)$.

a. Graph the line.

b. Find the equation of the line in slope-intercept form.

Solution

a. We are given the y-intercept, so one point on the line is $(0,3)$. To find another point, we may interpret a slope of $-\tfrac{1}{2}$ to mean that when x increases 2 units, y decreases 1 unit. By starting at $(0,3)$ and going 2 units to the right and 1 unit down, we obtain a second point on the line, at $(2,2)$. Drawing a line through these two points produces the graph in Figure 6.29.

b. By letting $m = -\tfrac{1}{2}$ and $b = 3$ in the slope-intercept equation $y = mx + b$, we determine that

$$y = -\tfrac{1}{2}x + 3$$

is the equation of the line in the slope-intercept form.

PROGRESS CHECK 5   The slope of a line is $-\tfrac{2}{3}$ and the y-intercept is $(0,6)$.

a. Graph the line.

b. Find the equation of the line in slope-intercept form.   ⌐

5 In Chapter 7 it will be useful to know when the graphs of a pair of linear equations are distinct parallel lines. From our work in Section 6.3, **two distinct parallel lines must have the same slope and different y-intercepts.**

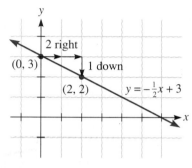

Figure 6.30

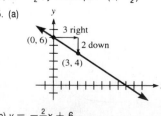
EXAMPLE 6   Determine if the graphs of $3x + 2y = 1$ and $6x + 4y = 9$ are distinct parallel lines.

Solution

$$3x + 2y = 1 \qquad\qquad 6x + 4y = 9$$
$$2y = -3x + 1 \qquad\qquad 4y = -6x + 9$$
$$y = -\tfrac{3}{2}x + \tfrac{1}{2} \qquad\qquad y = -\tfrac{6}{4}x + \tfrac{9}{4}$$
$$\qquad\qquad\qquad\qquad y = -\tfrac{3}{2}x + \tfrac{9}{4}$$

Now, determine the slope and y-intercept for each line.

Slope: $-\tfrac{3}{2}$ Slope: $-\tfrac{3}{2}$

y-intercept: $(0, \tfrac{1}{2})$ y-intercept: $(0, \tfrac{9}{4})$

The lines have the same slope and different y-intercepts, so they are distinct parallel lines, as shown in Figure 6.30.

PROGRESS CHECK 6   Determine if the graphs of $2x + y = 1$ and $-2x - y = 3$
are distinct parallel lines.   ⏌

Progress Check Answer

6. Yes; both slopes are -2 and the intercepts differ.

EXERCISES 6.4

In Exercises 1–18, find an equation of the line through the given
point which has the given slope.

	Point	Slope		Point	Slope
1.	(0,0)	1	2.	(0,0)	-2
3.	(1,2)	-3	4.	(3,4)	5
5.	$(-2,3)$	2	6.	$(-3,2)$	3
7.	$(1,-3)$	1	8.	$(2,-2)$	2
9.	(1,1)	$\frac{1}{2}$	10.	(3,1)	$\frac{2}{3}$
11.	(1,3)	$-\frac{1}{4}$	12.	(2,2)	$-\frac{3}{4}$
13.	(4,1)	0	14.	$(-3,-1)$	0
15.	(3,5)	Undefined	16.	$(-1,8)$	Undefined
17.	(0,2)	-0.3	18.	(2,0)	-0.3

In Exercises 19–36, find an equation for the line through the two
given points.

19. (0,0) and (3,3)
20. (0,0) and $(-2,2)$
21. (1,2) and (4,3)
22. (1,3) and (5,5)
23. $(-2,3)$ and (4,0)
24. $(-3,1)$ and (2,0)
25. $(-2,-3)$ and $(1,-1)$
26. $(-3,-4)$ and $(3,-2)$
27. $(-1,-2)$ and (1,2)
28. $(-2,-3)$ and (2,3)
29. (2,3) and (5,3)
30. $(-5,1)$ and (2,1)
31. (4,1) and $(4,-1)$
32. $(-2,0)$ and $(-2,5)$
33. $(\frac{1}{2},\frac{3}{2})$ and $(\frac{3}{2},\frac{5}{2})$
34. $(\frac{1}{3},\frac{2}{3})$ and $(\frac{4}{3},\frac{5}{3})$
35. $(\frac{1}{2},\frac{1}{3})$ and $(\frac{1}{4},\frac{1}{5})$
36. $(\frac{2}{3},\frac{3}{4})$ and $(\frac{4}{5},\frac{5}{6})$

In Exercises 37–48, find the slope and the y-intercept of the line
defined by the given equation.

37. $y = 3x + 2$
38. $y = 2x + 1$
39. $y = -x + 8$
40. $y = -x - 4$
41. $y = \dfrac{x}{3} + 1$
42. $y = \dfrac{5x}{2} + 3$
43. $2x + 3y = -3$
44. $3x + 2y = -1$
45. $3x - y = 2$
46. $4x - 2y = 1$
47. $5x - 3y = 6$
48. $4x - 5y = -20$

In Exercises 49–60, find the equation of the line in slope-intercept
form and graph it. The slope and the y-intercept are given.

	Slope	y-intercept
49.	1	(0,0)
50.	2	(0,0)
51.	1	(0,2)
52.	2	$(0,-3)$
53.	-3	(0,0)
54.	$-\frac{1}{2}$	(0,0)
55.	$\frac{1}{3}$	(0,3)
56.	$\frac{1}{4}$	(0,4)
57.	-2	$(0,\frac{2}{3})$

Slope   *y-intercept*

58. -3 $(0, -\frac{1}{4})$

59. 0 $(0,2)$

60. 0 $(0,-3)$

In Exercises 61–72, determine if the graphs of the two given equations are distinct parallel lines.

61. $3x + 2y = 4$
$3x + 2y = 5$
62. $x - y = 10$
$x - y = 5$
63. $2x - y = 1$
$4x - 2y = 1$
64. $3x + y = 5$
$6x + 2y = 5$
65. $3x + y = 2$
$6x + 2y = 4$
66. $x - 2y = 3$
$3x - 6y = 9$
67. $x + y = 1$
$x + 2y = 1$
68. $2x + 5y = 7$
$3x + 4y = 7$
69. $y = 4$
$y = 2$
70. $y = 3$
$2y = 4$
71. $-3x + 2y = 1$
$6x - 4y = 2$
72. $2x - 5y = 0$
$-2x + 5y = 5$

73. The charge to develop and print 36 photos is $11. The charge to develop and print 24 photos is $8. The graph of the relationship between the number of photos (x) and the price (y) is a line.
   a. Find the equation of the line in slope-intercept form.
   b. What is the charge if no photos are printable?
   c. If the price is $7, how many photos were printed?

74. To use a certain on-line information service, you pay a basic connection fee plus an hourly charge. The final cost (y) has a linear relation to the time (x). The cost of a 30-minute call is $8. The cost of a 50-minute call is $12.
   a. Find the equation of the line in slope-intercept form.
   b. What is the basic connection fee?
   c. How long can you be connected for $15?
   d. What is the *hourly* charge?

75. A train brakes and is slowing down so that every second its speed decreases by a certain amount. After 3 seconds its speed is 96 mi/hour. After 5 seconds its speed is 80 mi/hour.
   a. The relationship between speed (y) and time elapsed (x) is linear. Find an equation for this line, and graph it.
   b. How fast was the train going when the brakes were first applied?
   c. How long will it take the train to stop?

76. A rocket is fired in such a way that during part of its flight its speed increases by a constant amount every second. At the start of this interval its speed is 7 mi/second. One second later its speed is 7.2 mi/second.
   a. Find the equation in slope-intercept form of the line that relates speed (y) to time elapsed (x) over this portion of the flight.
   b. How long will it take to reach a speed of 8 mi/second?
   c. How long will it take to reach a speed of 26,000 mi per hour? (Round to the nearest tenth of a second.)

THINK ABOUT IT

1. If one line has slope 0 and another has slope undefined, are they perpendicular, parallel, or neither?
2. The cost for a car rental is $80 per week, with the first 200 mi free. After that the cost per mile is $0.20.
   a. Draw a graph of the relationship between miles driven and cost for one week.
   b. You need a combination of two equations to describe the relationship. The first equation covers miles from 0 to 200. The second covers miles over 200. Find both equations.
3. Find an equation of the line with y-intercept $(0,5)$ which is parallel to the line $6x - 2y = 5$. Sketch both lines.
4. Find an equation of the line with x-intercept $(5,0)$ which is perpendicular to the line $6x - 2y = 5$.
5. Find an equation of the line whose x-intercept is $(a,0)$ and whose y-intercept is $(0,a)$. Sketch the line if $a > 0$.

REMEMBER THIS

1. True or false? $4 \le 4$.

2. Graph the solution set of $-3x \ge x + 1$.

3. If $x = 1$ and $y = 2$, is the statement $3x - 2y > -1$ true or false?

4. What is the slope of the line through $(-2, -3)$ and $(3, 2)$?

5. Line ℓ_1 has slope -3. Line ℓ_2 is perpendicular to ℓ_1. What is the slope of ℓ_2?

6. Solve $x(2x - 1) + 1 = 2(x^2 + 3) - (x + 5)$.

7. Add $\dfrac{x^2 + 2x - 3}{x^2 + x - 2} + \dfrac{x^2 + x - 2}{x^2 - 3x + 2}$.

8. Divide $3x^2 - 2x + 1 \div x - 1$.

9. Twelve ounces of an alloy is 50 percent gold and 50 percent silver. How much silver should be added so that the alloy is 60 percent silver?

10. Simplify $\dfrac{(x^{-3}y)^2}{(-xy^{-2})^3}$.

6.5 Graphing Linear Inequalities

While on a diet, a teacher has yogurt and milk for lunch. If low-fat yogurt contains about 15 calories an ounce, skim milk contains about 10 calories an ounce, and lunch is to result in at most 180 calories, then the inequality

$$15x + 10y \le 180$$

expresses the calorie restriction for lunch, where x and y represent the number of ounces eaten of yogurt and milk, respectively. Graph this inequality for $x \ge 0$ and $y \ge 0$. (See Example 6.)

OBJECTIVES

1. Determine if an ordered pair is a solution of an inequality.

2. Graph linear inequalities.

3. Graph linear inequalities restricted to $x \ge 0$, $y \ge 0$.

1   The inequality signs ($<$, $\le$, $>$, $\ge$) often express the required relationship between two variables. For example, in the section-opening problem the phrase "at most" translates to the inequality symbol $\le$ when we describe how x and y are related. By analogy to equations, a **linear inequality** results if the equal sign in a linear equation is replaced by one of the inequality symbols. As with equations, a solution of an inequality in two variables is an ordered pair that makes the inequality a true statement.

EXAMPLE 1   Determine if $(4,0)$ is a solution of $y \le x$.

Solution $(4,0)$ means $x = 4$, $y = 0$. Then

$$y \le x \qquad \text{Given inequality.}$$
$$0 \le 4. \qquad \text{Replace } y \text{ by } 0 \text{ and } x \text{ by } 4.$$

Recall that $0 \le 4$ is true if either $0 < 4$ or $0 = 4$ is true. Because $0 < 4$ is true, $(4,0)$ is a solution of $y \le x$.

PROGRESS CHECK 1   Determine if $(0, -4)$ is a solution of $2x - 3y \le 12$.   ⌐

2   A linear inequality in two variables such as $y \le x$ has infinitely many solutions. Because $y \le x$ means

$$y < x \qquad \text{or} \qquad y = x,$$

Progress Check Answer

1. Yes

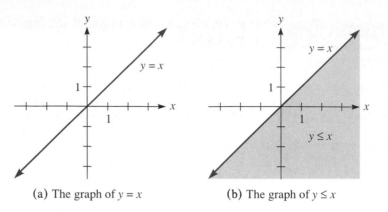

(a) The graph of $y = x$ (b) The graph of $y \leq x$

Figure 6.31

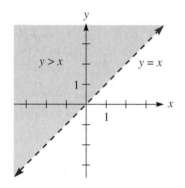

Figure 6.32

we illustrate these solutions graphically by first drawing the line $y = x$, as shown in Figure 6.31(a). Note that the line $y = x$ separates the plane into two regions. Each region consists of the set of points on one side of the line and is called a **half plane.** The solution set to $y < x$ is given by the half plane *below* $y = x$, while the solution set to $y > x$ is given by the half plane *above* $y = x$. Thus, we graph $y \leq x$ as shown in Figure 6.31(b), using shading to indicate that the half plane in the solution set is *below* $y = x$.

When the inequality symbol is either $<$ or $>$, then the points on the line are *not* part of the solution set. We indicate this by drawing the line as a dashed line, as shown in Example 2.

EXAMPLE 2   Graph $y > x$.

Solution   As discussed, the solution set to $y > x$ is given by the half plane *above* $y = x$. Figure 6.32, therefore, shows the graph of $y > x$. Note that $y = x$ is drawn as a dashed line in this figure to show that the points on the line are *not* included in the solution set.

PROGRESS CHECK 2   Graph $y < -x$.   ⌐

EXAMPLE 3   Graph $y \leq -2x + 1$.

Solution $y \leq -2x + 1$ means $y < -2x + 1$ or $y = -2x + 1$. First, graph the equation $y = -2x + 1$ as a solid line to indicate that this boundary line is included in the solution set. Then graph $y < -2x + 1$ by shading in the half plane *below* this line. Figure 6.33 shows the graph of $y \leq -2x + 1$.

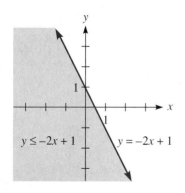

Figure 6.33

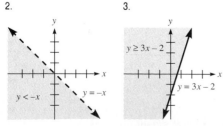

PROGRESS CHECK 3   Graph $y \geq 3x - 2$.   ⌐

Vertical lines and horizontal lines may be boundary lines when graphing linear inequalities, and the next example considers such a case.

EXAMPLE 4   Graph $x \geq 2$.

Solution   As an inequality in two variables, note that the inequality $x \geq 2$ is equivalent to $x + 0y \geq 2$. First, graph $x = 2$ and make the vertical line solid, because the given inequality symbol is $\geq$. Then graph $x > 2$ by shading in the half plane to the *right* of this line, because the *x*-coordinates of all points to the right of this line are greater than 2. Figure 6.34 shows the graph of $x \geq 2$.

PROGRESS CHECK 4 Graph $x < 3$.

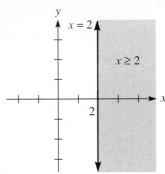

Figure 6.34

When the inequality is not solved for *y* or *x*, it is not obvious which half plane to use in the solution set. The easiest way to decide in such cases is usually to pick some **test point** that is not on the line and substitute the coordinates of this point into the inequality. If the resulting statement is true, then shade the half plane containing the test point. Otherwise, shade the half plane on the other side of the line from the test point. The origin (0,0) is a convenient point to use in this test provided the origin is not on the line.

EXAMPLE 5 Graph $2x - y > 4$.

Solution First, graph $2x - y = 4$ as a dashed line, because the given inequality symbol is $>$, which means that the points on the line are not included in the solution set. The origin is not on the line $2x - y = 4$, so substitute 0 for both *x* and *y* in the given inequality to decide which half plane to shade in the answer.

$$2x - y > 4 \quad \text{Given inequality.}$$
$$2(0) - 0 > 4 \quad \text{Replace } x \text{ by 0 and } y \text{ by 0.}$$
$$0 > 4 \quad \text{Simplify.}$$

Because $0 > 4$ is a false statement, (0,0) is not a solution; so shade the half plane not containing the origin. Figure 6.35 shows the graph of $2x - y > 4$.

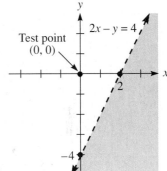

Figure 6.35

PROGRESS CHECK 5 Graph $3x - 2y < 6$.

3 In applied problems *x* and *y* may have meaning only for nonnegative numbers. In such cases the graph should be restricted to points in quadrant 1 and points on the nonnegative axes which serve as boundary lines.

EXAMPLE 6 Solve the problem in the section introduction on page 297.

Solution We need to graph $15x + 10y \leq 180$, where $x \geq 0$ and $y \geq 0$, because the number of ounces of either food cannot be negative. First, graph $15x + 10y = 180$ for $x \geq 0$ and $y \geq 0$ by determining that the intercepts are (12,0) and (0,18) and then drawing a solid line segment with these intercepts as endpoints of the segment. To graph $15x + 10y < 180$, we will choose (0,0) as a test point.

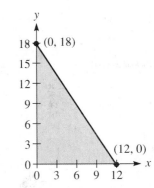

Figure 6.36

$$15x + 10y \leq 180 \quad \text{Given inequality.}$$
$$15(0) + 10(0) \leq 180 \quad \text{Replace } x \text{ by 0 and } y \text{ by 0.}$$
$$0 \leq 180 \quad \text{Simplify.}$$

Because $0 \leq 180$ is true, we shade the points in quadrant 1 on the same side of the line segment as the origin, using the nonnegative *x*- and *y*-axes as boundary lines. Figure 6.36 shows the graph of $15x + 10y \leq 180$ for $x \geq 0$ and $y \geq 0$.

PROGRESS CHECK 6 A football player who wishes to gain weight snacks on potato chips and chocolate malts. If potato chips contain about 150 calories an ounce, chocolate malts contain about 60 calories an ounce, and the snack is to result in at least 600 calories, then the inequality $150x + 60y \geq 600$ expresses the calorie requirement for this snack, where *x* and *y* represent the number of ounces eaten of potato chips and chocolate malts, respectively. Graph this inequality for $x \geq 0$ and $y \geq 0$.

In summary we have used the following procedures to graph linear inequalities.

Progress Check Answers

4. 5.

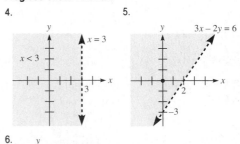

6.

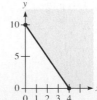

To Graph a Linear Inequality

1. Graph the linear equation that results when the inequality symbol is replaced by $=$. Draw this boundary line as a solid line if the inequality symbol is $\leq$ or $\geq$, and draw a dashed line for $<$ or $>$.
2. Shade the half plane on one side of the boundary line as follows.

   Case 1: The inequality is solved for y or x.

Inequality Begins	Shade the Half Plane
$y <$	Below the line
$y >$	Above the line
$x <$	Left of the line
$x >$	Right of the line

   Case 2: The inequality is not solved for y or x. Choose a test point that is not on the boundary line, and substitute the coordinates of this point into the inequality.

Resulting Inequality Is	Shade the Half Plane
True	Containing the test point
False	Not containing the test point

EXERCISES 6.5

In Exercises 1–12, determine if the given point is a solution of the inequality.

	Point	Inequality
1.	$(3,1)$	$y \leq x$
2.	$(3,5)$	$y \geq x$
3.	$(3,-9)$	$x - y < 10$
4.	$(0,-3)$	$x - y < 2$
5.	$(1,1)$	$3x + y \geq 4$
6.	$(1,2)$	$2x + y \leq 4$
7.	$(\frac{1}{3},-\frac{1}{2})$	$3x - 2y \leq 0$
8.	$(-\frac{3}{2},\frac{1}{3})$	$2x + 3y \leq -3$
9.	$(-3,-1)$	$y < 3x + 8$
10.	$(0,-5)$	$y > 2x - 5$
11.	$(5,6)$	$y > 2x - 8$
12.	$(-10,0)$	$y < \frac{1}{2}x + 6$

In Exercises 13–30, graph the given inequality.

13. $y < x$

14. $y > -x$

15. $y \leq 2x + 3$

16. $y \leq 3x + 2$

17. $y > x - 3$

18. $y > x - 1$

19. $x + y \leq 5$

20. $x + y \leq 2$

21. $2x + 3y \geq 6$

22. $3x - 4y \geq 12$

23. $y > 3$

24. $y < -3$

25. $x \leq 0$ **26.** $x > 0$ **27.** $y < 3x - 3$

28. $y \leq 2x - 6$ **29.** $x - \frac{1}{2}y \leq 5$ **30.** $\frac{2}{3}x - y \geq 1$

31. You need to buy some ribbon for decorating costumes. Gold ribbon costs \$3 per foot; silver costs \$2 per foot. The total amount spent on ribbon cannot be more than \$24. If x and y represent the number of feet of gold and silver ribbon purchased, respectively, then $3x + 2y \leq 24$ expresses the expense restriction.
   a. Graph this inequality, but note that both x and y cannot be negative.

   b. If you spend \$3 on gold ribbon, what is the longest piece of silver ribbon you can afford?
   c. What is the longest piece of silver ribbon you can afford to buy?
32. In a blend of ingredients for dog food one component (x) supplies 1.25 grams (g) of fat per ounce and another (y) supplies 2.5 g of fat per ounce. One can of the mixture must contain no more than 5 g of fat from these two sources. The inequality $1.25x + 2.5y \leq 5$ describes this restriction.
   a. Graph the inequality, restricting $x \geq 0$ and $y \geq 0$.

b. Which of these combinations are acceptable?

Ounces of Ingredient x	Ounces of Ingredient y
1	1
2	1
1	2

   c. If 1 oz of ingredient x is used, what is the maximum permissible amount of ingredient y?
33. For a party you will buy some sliced ham at \$8 per pound and some cheddar cheese at \$4 per pound. You wish to spend no more than \$16 on these two items. Let x and y represent the amounts of ham and cheese, respectively. Then $8x + 4y \leq 16$ represents the expense restriction.
   a. Graph the inequality, and note that x and y cannot be negative.

   b. If you decide to buy $1\frac{1}{2}$ lb of cheese, can you also buy a pound of ham?
   c. If you buy 2 lb of cheese, what is the most ham you can buy?
34. A car loan has a 10 percent annual interest rate. A loan for furniture has a 16 percent annual interest rate. You can afford over the year at most \$1,000 in interest payments for these two purchases. This restriction is given by $0.10x + 0.16y \leq 1,000$, where x is the amount of the car loan and y is the amount of the furniture loan.
   a. Graph this inequality.

   b. What is the maximum car loan you can afford?
   c. If you get a \$4,000 car loan, what is the most you can borrow for furniture?

THINK ABOUT IT

1. The line $y = x$ goes through the origin with a slope of 1. It is shown in the figure, with half the plane shaded. Which of these must be true for every point in the shaded area?
   a. $y \geq x$ **b.** $y \leq x$ **c.** $y - x \geq 0$

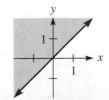

2. Find an inequality which corresponds to the shaded region. You will have to put restrictions on x and y.

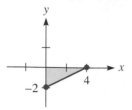

3. Find and graph an inequality which describes the following relationship. At a record store CDs cost $12.99 and cassettes cost $7.99. For gifts you want to buy a mix of cassettes and CDs. There will be a 6 percent sales tax on your purchase. You can spend no more than $100. What is the maximum number of cassettes you can buy? What is the maximum number of CDs you can buy? What purchase gives you the least change?

4. You have up to $30 to buy a combination of 30¢ and 20¢ envelopes at the post office.
   a. Find and graph an inequality which describes this restriction.
   b. You decided to use all the money and to buy an equal number of each. How many of each can you buy?

c. You decided to buy twice as many 30¢ envelopes as 20¢ envelopes. What is the maximum number of 30¢ envelopes you can buy?

5. You are going to make a diet lunch consisting of low-fat yogurt, which has 20 calories per ounce, skim milk, which has 10 calories per ounce, and chocolate chip cookies, which have 100 calories per ounce. You can eat no more than 200 calories per lunch. A graph of these restrictions is shown in the figure.

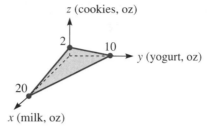

   a. Write an inequality that describes the restrictions of the diet.
   b. If you decide to eat one cookie, write an inequality that describes the resulting relationship between the milk and the yogurt.

REMEMBER THIS

1. For what value of x is the fraction $\dfrac{3-x}{x-1}$ undefined?

2. When we say that $(3,-2)$ is a solution of $3y - x + 9 = 0$, which one of the two numbers replaces the letter x?

3. True or false? Both 3 and -3 are solutions to $y^2 = 9$.

4. What two numbers satisfy $|n| = 5$?

5. Write an equation in general form for the line through $(2,5)$ with a slope equal to 3.

6. Write in slope-intercept form the equation of the line through $(-2,-3)$ and $(3,2)$.

7. What is the slope of the line $y = 4x - 5$?

8. Use the Pythagorean theorem to find the length of the hypotenuse of the triangle $A(-1,-1)$, $B(3,2)$, $C(3,-1)$.

9. Solve $2x(2x - 1)(2x + 2) = 0$.

10. One number is 2 more than another one. When you square the difference between 2 times the larger one and 4 times the smaller one, you get 25. What are the two numbers?

6.6 Functions

The distance d traveled by a free-falling body is a function of time and is given by
$$d = f(t) = 16t^2,$$
where d is measured in feet and t is the elapsed time measured in seconds. Find $f(10)$; that is, find d when $t = 10$ seconds. (See Example 8.)

OBJECTIVES

1 Find the domain and range given a set of ordered pairs.

2 Determine if a set of ordered pairs is a function.

3 Determine if an equation defines y as a function of x.

4 Find a formula that defines the functional relationship between two variables.

5 Use functional notation.

1 In this chapter we have used linear equations and linear inequalities to analyze the relationship between two variables. For instance, in the chapter-opening problem the equation $C = 0.10m + 59$ showed how the cost C of a certain car rental was related to the weekly mileage m. This equation serves as a rule that establishes a correspondence between values of m and values of C which may be written as a set of ordered pairs. It is not necessary that two variables be related by an equation for a relationship to exist, so we will now take a more general approach to analyzing relationships, and we call *any* set of ordered pairs a *relation*.

Relation Definition

A **relation** is a set of ordered pairs. The set of all first components of the ordered pairs is called the **domain** of the relation. The set of all second components is called the **range** of the relation.

EXAMPLE 1 In a certain class the correspondence between students' final averages and their final grades is given by {(80,B),(92,A),(74,C),(86,B)}. Find the domain and range of this relation.

Solution The domain, which is the set of all first components, is {80,92,74,86}. The range, which is the set of all second components, is {A,B,C}.

PROGRESS CHECK 1 Find the domain and the range of the relation {(61,D),(100,A),(38,F),(94,A)}. ⌐

EXAMPLE 2 The relation "less than" in the set {1,2,3} is defined by {(1,2),(1,3),(2,3)}. Find the domain and range of this relation.

Solution The domain of this relation is {1,2}, and the range is {2,3}. In this example, note that an ordered pair like (1,2) belongs to the relation, because 1 is less than 2.

PROGRESS CHECK 2 The relation "greater than" in the set {1,2,3} is defined by {(3,1),(3,2),(2,1)}. Find the domain and range of this relation. ⌐

2 Note in the ordered pairs in Example 1 that all the first components (which customarily represent x) are different, while this is not the case in Example 2. It is convenient when there is no duplication in the x values, because then each x value leads to *exactly one* y value. In other words, the rule involved in the relation is such that the rule produces exactly one answer whenever we use it. This discussion leads to the definition of a function, which is a special type of relation.

Function

A **function** is a relation in which no two different ordered pairs have the same first component.

EXAMPLE 3 Determine if the given relation is a function.

a. {(5,1),(6,1),(7,1)}   **b.** {(1,1),(0,0),(1,−1)}

Solution

a. This relation is a function because the first component in the ordered pairs is always different. Note that the definition of a function does not require the second components to be different.

b. This relation is not a function because the number 1 is the first component in more than one ordered pair.

PROGRESS CHECK 3   Determine if the given relation is a function.

a. $\{(-1,1),(0,0),(1,1)\}$ **b.** $\{(1,5),(1,6),(1,7)\}$ ⌐

3   Infinite sets of ordered pairs are common types of relations and functions. For example, the solution set of the linear equation $y = x + 1$ is the infinite set of ordered pairs that defines the relation,

$$\{(x,y): y = x + 1\}.$$

Furthermore, this relation is a function, because corresponding to each number x there is exactly one number y. (It is the number that is one more than x.) When y is a function of x, as just described, the value of y depends on the choice of x; so we call x the **independent variable** and y the **dependent variable.** To work with a relation or function that is an infinite set of ordered pairs, we usually focus on the rule, like $y = x + 1$, that is used to define the relation, as shown in the next two examples.

EXAMPLE 4   Does the given rule determine y as a function of x?

a. $y = x^2$ **b.** $y^2 = x$ **c.** $y < x$ **d.** $y = 2$

Solution

a. The equation $y = x^2$ determines y as a function of x, since to each real number x there corresponds exactly one number y that is the square of x.
b. The equation $y^2 = x$ does not determine y as a function of x, because it is possible for one x value to correspond to two different y values. For instance, if $x = 4$,

$$y^2 = x \qquad \text{yields} \qquad y^2 = 4$$

so $y = 2$ or $y = -2$.
c. The inequality $y < x$ does not determine y as a function of x, because to each real number x there corresponds an infinite number of y values that are less than x.
d. The equation $y = 2$, which is equivalent to $y = 2 + 0x$, determines y as a function of x, because to each real number x there corresponds exactly one y value, namely, 2.

PROGRESS CHECK 4   Does the given rule determine y as a function of x?

a. $y = |x|$ **b.** $|y| = x$ **c.** $x = 2$ **d.** $y \geq x$ ⌐

4   One of the first steps we use when analyzing a relationship is to determine a rule that defines a functional relationship between two variables. In some cases the rule may be derived from common formulas, as in the next example. The wording of the question in this example is phrased in the language associated with functions, and you will need to become familiar with this language.

EXAMPLE 5   Write a rule that is a formula or equation to define the functional relationship between the two variables.

a. Express the perimeter y of a square as a function of its side length x.
b. Express the distance y that a car going 50 mi/hour will travel as a function of t, the hours spent traveling.

Solution

a. Using $P = 4s$, the equation $y = 4x$ expresses the perimeter y as a function of the side length x.
b. Using $d = rt$, the equation $y = 50t$ expresses the distance y as a function of the time t when r equals 50.

Progress Check Answers

3. (a) Yes   (b) No
4. (a) Yes   (b) No   (c) No   (d) No

PROGRESS CHECK 5   Use the directions in Example 5.

a. Express the area A of a square as a function of its side length x.
b. Express the distance y that a plane going 400 mi/hour will travel as a function of t, the hours spent traveling.   ⌐

5   Functional notation is used extensively in mathematics. It is useful because it allows us to represent more conveniently the value of the dependent variable for a particular value of the independent variable. In this notation we use a letter like f to name a function. Then an equation like

$$y = 4x + 1 \quad \text{is written as} \quad f(x) = 4x + 1.$$

The dependent variable y is replaced by $f(x)$, with the independent variable x appearing in parentheses. The notation $f(x)$ is read as "f of x" or "f at x" and means the value of the function (the y value) corresponding to the value of x. Similarly, $f(2)$ means the value of the function when $x = 2$. If $f(x) = 4x + 1$, we find $f(2)$ by substituting 2 for x in the equation.

$$
\begin{aligned}
f(x) &= 4x + 1 & &\text{Given equation.} \\
f(2) &= 4(2) + 1 & &\text{Replace } x \text{ by 2.} \\
&= 9 & &\text{Simplify.}
\end{aligned}
$$

The result $f(2) = 9$ says that $y = 9$ when $x = 2$. Note that the notation $f(x)$ does not mean f times x.

EXAMPLE 6   If $y = f(x) = 5x - 1$, find $f(3), f(-2)$, and $f(\frac{1}{2})$.

Solution   Replace x by the number inside the parentheses and then simplify.

$$
\begin{aligned}
y_{\text{when } x = 3} &= f(3) = 5(3) - 1 = 14 \\
y_{\text{when } x = -2} &= f(-2) = 5(-2) - 1 = -11 \\
y_{\text{when } x = \frac{1}{2}} &= f(\tfrac{1}{2}) = 5(\tfrac{1}{2}) - 1 = \tfrac{5}{2} - \tfrac{2}{2} = \tfrac{3}{2}
\end{aligned}
$$

Thus, $f(3) = 14, f(-2) = -11$, and $f(\frac{1}{2}) = \frac{3}{2}$.

PROGRESS CHECK 6   If $y = f(x) = 3x - 2$, find $f(4), f(-1)$, and $f(\frac{3}{2})$.   ⌐

In functional notation it is customary to use the symbols f and x, but other symbols work just as well and are often useful. The notations $f(x) = 4x$, $g(t) = 4t$, and $h(r) = 4r$ all define exactly the same function if x, t, and r may be replaced by the same numbers.

EXAMPLE 7   If $g(x) = 3x^2 + 7x - 2$, find $g(-4)$.

Solution   To find $g(-4)$, replace *all occurrences* of x by -4 and simplify.

$$
\begin{aligned}
g(x) &= 3x^2 + 7x - 2 & &\text{Given equation.} \\
g(-4) &= 3(-4)^2 + 7(-4) - 2 & &\text{Replace } x \text{ by } -4. \\
&= 48 - 28 - 2 & &\text{Remove parentheses.} \\
&= 18 & &\text{Subtract.}
\end{aligned}
$$

PROGRESS CHECK 7   If $h(x) = 4x^2 + 9x - 5$, find $h(-3)$.   ⌐

EXAMPLE 8   Solve the problem in the section introduction on page 302.

Solution $f(10)$ gives d when $t = 10$ seconds. Using $d = f(t) = 16t^2$,

$$d_{\text{when } t = 10} = f(10) = 16(10)^2 = 1,600.$$

Thus, the distance traveled by a free-falling body in 10 seconds is 1,600 ft.

Progress Check Answers
5. (a) $A = x^2$ (b) $y = 400t$
6. $f(4) = 10$, $f(-1) = -5$, $f(\frac{3}{2}) = \frac{5}{2}$
7. $h(-3) = 4$

Progress Check Answer

8. $f(5) = 400$; a free-falling body travels 400 ft in 5 seconds.

PROGRESS CHECK 8   Use $d = f(t) = 16t^2$ and find $f(5)$. Interpret the meaning of $f(5)$ in the context of Example 8.   ⌐

EXERCISES 6.6

In Exercises 1–6, find the domain and range of the given relations.

1. The correspondence between "class rank" and the students Mary, Ellen, and Janice is given by $\{(\text{Mary},1),(\text{Ellen},2),(\text{Janice},3)\}$.

2. The correspondence between "final grade" and the students Tom, Dick, and Harry is given by $\{(\text{Tom},B),(\text{Dick},B),(\text{Harry},B)\}$.

3. The relation "east of" on the set $\{\text{Los Angeles, Chicago, New York}\}$ is given by $\{(\text{Chicago, Los Angeles}),(\text{New York, Chicago}),(\text{New York, Los Angeles})\}$.

4. The relation "harder than" on the set $\{\text{diamond, aluminum, balsa wood}\}$ is given by $\{(\text{diamond, aluminum}),(\text{diamond, balsa wood}),(\text{aluminum, balsa wood})\}$.

5. The relation "is more than the square of" on the set $\{-1,0,1,2\}$ is given by $\{(1,0),(2,0),(2,1),(2,-1)\}$.

6. The relation "is 3 more than" on the set $\{1,2,3,4,5\}$ is given by $\{(5,2),(4,1)\}$.

In Exercises 7–18, determine if the set of ordered pairs is a function.

7. $\{(2,3),(4,3),(6,3)\}$
8. $\{(-2,0),(-3,0),(-4,0)\}$
9. $\{(1,2),(1,3),(1,4)\}$
10. $\{(3,6),(3,9),(3,12)\}$
11. $\{(-2,4),(2,4)\}$
12. $\{(9,3),(9,-3)\}$
13. $\{(1,2),(2,3),(3,4)\}$
14. $\{(2,4),(4,8),(6,12)\}$
15. $\{(1,1),(2,2),(2,3),(4,4)\}$
16. $\{(1,1),(2,2),(3,3),(4,4)\}$
17. $\{(1,5)\}$
18. $\{(0,0)\}$

In Exercises 19–30, decide if the given rule determines y as a function of x.

19. $y = 2x + 3$
20. $y = 3x - 4$
21. $y = 6$
22. $y = -4$
23. $x = 1$
24. $x = -2$
25. $y \le x$
26. $y \le 2x + 1$
27. $y = 3x^2 - 1$
28. $y = 1 - x^2$
29. $x = y^2 + 1$
30. $x = y^2 - 3$

In Exercises 31–42, write a rule that is a formula or equation to define the functional relationship between the two variables.

31. Express the diameter (d) of a circle as a function of the radius (r).

32. Express the radius (r) of a circle as a function of the diameter (d).

33. Express the length of a board in inches (i) as a function of its length in feet (f).

34. Express the length of a board in feet (f) as a function of its length in inches (i).

35. Express the weight (w) of a container of liquid in terms of its volume (v) if the liquid weighs 65 lb/ft³ and the empty container weighs 5 lb.

36. Express the price (p) of a salad in terms of its weight (w) if the cost is \$1.29 per pound plus 25¢ for the container.

37. Express the area (A) of a semicircle as a function of the diameter (d).

38. Express the volume (V) of a sphere as a function of its diameter (d).

39. Express the travel time (t) in hours as a function of distance (d) for a vehicle going 55 mi/hour.

40. Express the time in hours (h) to cook a turkey as a function of weight (w) if it should cook for 15 minutes per pound.

41. A recipe for hot cereal says to use 3 cups of water to 2 cups of dry cereal. If you use the same proportions for other amounts, express the cups of dry cereal (c) as a function of the cups of water (w).

42. Express shelf space (s) in feet as a function of the number of books (b) in a library if, on average, there are 8 books per foot on the shelves.

43. If $y = f(x) = 4x + 2$, find $f(1), f(0), f(-\frac{1}{2}), f(0.1)$.

44. If $y = f(x) = 2x - 4$, find $f(0), f(2), f(-2), f(0.2)$.

45. If $y = g(x) = x^2 - 1$, find $g(-1), g(1), g(0), g(0.5)$.

46. If $y = g(x) = 4 - x^2$, find $g(2), g(-1), g(0), g(0.1)$.

47. If $y = h(x) = x^2 + x - 2$, find $h(1), h(-2), h(-1), h(0.3)$.

48. If $y = h(x) = x^2 + x - 6$, find $h(2), h(-3), h(-1), h(0.3)$.

49. If $f(x) = -4x^2$, find $f(3), f(-3), f(\frac{1}{2}), f(0.1)$.

50. If $f(x) = (-4x)^2$, find $f(3), f(-3), f(\frac{1}{2}), f(0.1)$.

51. If $g(t) = 100 + 10t - t^2$, find $g(0), g(1), g(\frac{1}{2}), g(0.2)$.

52. If $g(s) = 25 + 5s - s^2$, find $g(0), g(-5), g(\frac{1}{5}), g(0.5)$.

53. If $h(r) = \pi r^2 + r$, find $h(1), h(0), h(-\pi)$.

54. If $h(u) = (u - 4)(u + 2)$, find $h(4)$, $h(-2)$, $h(0)$.

55. The distance (d) traveled by a ball hurled down from the roof of a 100-story skyscraper is a function of time (t) and is given by $d = f(t) = 80t + 16t^2$, where d is measured in feet and t is elapsed time in seconds. Find $f(2)$ and interpret what it means.

56. The distance (d) traveled by a stone released into a vat of thick liquid is a function of time and is given by $d = f(t) = \frac{1}{2}t^2$, where d is measured in feet and t is elapsed time in seconds. Find $f(3)$ and interpret what it means.

57. A function which approximates the value V of a $95,000 house which increases in value each year at a rate of 5 percent per year is $V = f(x) = 95,000(1.05)^x$, where x is the number of years since the house was purchased.
  a. Find $f(10)$ and interpret its meaning.

  b. After how many full years of ownership is the house worth more than $200,000? (Find by trial and error.)

58. If a radioactive material has a half-life of 10 years, then half of the amount of the current amount of material disappears every 10 years. When a research lab starts with 4 lb of this material, then the approximate amount left after x years is given by $A = f(x) = 4(\frac{1}{2})^{x/10}$.
  a. Show that $f(10) = 2$. What does this mean?

  b. Find and interpret $f(30)$.

THINK ABOUT IT

1. The relation "is not more than" on the set $\{1,2,3\}$ is given by what set of ordered pairs? Is it a function?
2. Which of these is not true?
  a. All relations are functions.
  b. All functions are relations.
3. Express the length (d) of the diagonal of a square in terms of the side (s).
4. If you step out of a helicopter 25 ft in the air, the distance you fall is given by $d = f(t) = 16t^2$, where t is elapsed time in seconds and d is distance in feet.
  a. Find $f(1)$ and $f(1.25)$ and explain what they mean.
  b. How long will it take you to hit the ground?

  c. The function says that $f(2) = 64$. Explain why that is probably not a meaningful answer.
  d. The speed at which you are falling is given in feet per second by $s = g(t) = 32t$. How fast are you going when you hit the ground?
5. For each set, find an equation which describes the relation shown in the given set of ordered pairs.
  a. $\{(1,2),(2,3),(3,4)\}$
  b. $\{(1,4),(2,3),(3,2),(4,1)\}$
  c. $\{(1,1),(2,4),(3,9),(4,16)\}$
  d. $\{(1,1),(2,3),(3,5),(4,7)\}$

REMEMBER THIS

1. An item costs n dollars. The sales tax rate is 4 percent, so the tax is $0.04n$. If you double the original cost to $2n$ dollars, the tax is $0.04(2n)$. Is this double the original tax?
2. A square has side length s and area s^2. If you triple the side length to $3s$, the new area is $(3s)^2$. Is this triple the original area?
3. In the formula $y = \dfrac{36}{x}$, when $x = 2$, $y = 18$. If you double x, does the value of y also double?
4. Graph the inequality $y \leq 2x + 3$.

5. What is the slope-intercept equation of the line with slope equal to -3 and y-intercept $(0,1)$?

6. Find and interpret the *slope*. The graph shows charges for using a telephone computer service.

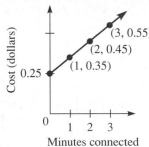

7. Which of these pairs is a solution to $3x - y = 9$, $(0,-9)$ or $(0,3)$?
8. Add $\dfrac{1}{x + 2} + \dfrac{x - 1}{x}$.

9. Solve $\dfrac{x}{x + 2} = \dfrac{x - 1}{x}$.

10. Simplify $\dfrac{\dfrac{1}{a} - \dfrac{1}{a^2}}{\dfrac{1}{b} + \dfrac{1}{b^2}}$.

6.7 Variation

The weekly earnings *E* of a part-time employee at a fast-food restaurant varies directly as the number *n* of hours worked. If an employee makes $105 for working 25 hours, how much will this employee make working 18 hours? (See Example 3.)

OBJECTIVES

Solve problems involving the following:

1 Direct variation

2 Inverse variation

3 Variation of powers of variables

1 In applied fields, particularly the sciences, the functional relationship between variables is often stated in the language of variation. The simplest type of variation occurs when one variable multiplied by a constant results in the other variable.

> **_y_ Varies Directly as _x_**
>
> The statement "*y* varies directly as *x*" means that
>
> $$y = kx,$$
>
> where *k* is a constant called the **variation constant.**

EXAMPLE 1 Write a variation equation for the given statement and identify the variation constant if it is known.

  a. The perimeter *P* of a square varies directly as the side length *s*.
  b. The sales tax *T* on a purchase varies directly as the price *p* of the item.

Solution

  a. *P* varies directly as *s* means $P = ks$. In this case we know the variation constant *k* is 4.
  b. *T* varies directly as *p* means $T = kp$. The variation constant *k* depends on the sales tax rate, which is fixed for any particular location.

PROGRESS CHECK 1 Write a variation equation for the given statement and identify the variation constant if it is known.

  a. If a car travels at a constant speed of 55 mi/hour, the distance *d* traveled varies directly as the time *t*.
  b. The property tax *T* varies directly as assessed valuation *v*.

Figure 6.37 shows the graph of the variation equation $y = kx$ for $k > 0$. Note that the graph is a straight line through the origin with slope *k*, and that as *x* increases, *y* increases. We may determine the value of *k* if one pair of values for the variables, other than (0,0), is known. The value of *k* may then be used to find other corresponding values of the variables.

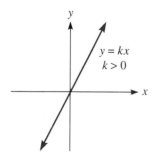

Figure 6.37

EXAMPLE 2 *y* varies directly as *x*, and $y = 4$ when $x = 3$. Find *y* when $x = 12$.

Solution   Since y varies directly as x, we have

$$y = kx.$$

To find k, replace y by 4 and x by 3.

$$4 = k3 \qquad \text{so} \qquad \tfrac{4}{3} = k$$

The variation equation is $y = \tfrac{4}{3}x$. When $x = 12$,

$$y = \tfrac{4}{3}(12) = 16.$$

Thus, $y = 16$ when $x = 12$.

PROGRESS CHECK 2 y varies directly as x, and $y = 3$ when $x = 5$. Find y when $x = 30$.

EXAMPLE 3   Solve the problem in the section introduction on page 308.

Solution   Since earnings E vary directly as the number n of hours worked,

$$E = kn.$$

To find k, replace E by 105 and n by 25.

$$105 = k25$$
$$\tfrac{105}{25} = k, \qquad \text{or} \qquad k = 4.2$$

Thus, $E = 4.2n$, which means that the hourly wage is \$4.20. When $n = 18$,

$$E = 4.2(18) = 75.6.$$

The employee makes \$75.60 for working 18 hours.

Note   You may have noticed that this problem may also be solved by using the methods of Section 5.7 and setting up a proportion like

$$\frac{105 \text{ dollars}}{25 \text{ hours}} = \frac{x \text{ dollars}}{18 \text{ hours}}.$$

The proportion method can always be used in such cases; because if y varies directly as x, then $y = kx$, so $y/x = k$. Thus, there is a constant ratio between corresponding values of x and y. For this reason the variation constant k is sometimes called the constant of proportionality, and the expression "y varies directly as x" is sometimes stated as "y is proportional to x." However, for our present purpose the variation method is more informative, because this approach emphasizes the relationship between the variables.

PROGRESS CHECK 3   The weight of an object on the moon varies directly as the weight of the object on earth. An object that weighs 102 lb on earth weighs 17 lb on the moon. If a person weighs 168 lb on the earth, how much will the person weigh on the moon?

2   Another common type of variation, called inverse variation, occurs when the products of corresponding values of the variables is a constant k.

_y_ Varies Inversely as _x_

The statement "y varies inversely as x" means $xy = k$, or

$$y = \frac{k}{x},$$

where k is a constant called the **variation constant.**

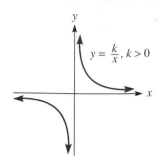

Figure 6.38

Figure 6.38 shows the graph of the variation equation $y = k/x$ for $k > 0$. Although the curve in this figure, which is called a **hyperbola,** is not among the graphs we will study in this text, we can still use the graph to better understand the concept of inverse variation. In particular, note that as x increases, y decreases; as x decreases, y increases; and the variation equation $y = k/x$ is meaningless if $x = 0$ or $y = 0$.

EXAMPLE 4 y varies inversely as x, and $y = 6$ when $x = 8$. Find y when $x = 2$.

Solution   Since y varies inversely as x, we have

$$y = \frac{k}{x}.$$

To find k, replace y by 6, x by 8, and cross multiply.

$$6 = \frac{k}{8} \qquad \text{so} \qquad 48 = k$$

Thus, $y = 48/x$. When $x = 2$, $y = 48/2 = 24$.

PROGRESS CHECK 4 y varies inversely as x, and $y = 4$ when $x = 9$. Find y when $x = 6$.   ⌐

EXAMPLE 5   If a car travels at a constant rate r, then the time t required to drive a fixed distance varies inversely as r. If it takes 13 hours to drive a certain distance at 55 mi/hour, how long will it take to drive this distance at 65 mi/hour?

Solution   Since t varies inversely as r,

$$t = \frac{k}{r}.$$

To find k, replace t by 13, r by 55, and cross multiply.

$$13 = \frac{k}{55} \qquad \text{so} \qquad 715 = k$$

Thus, $t = 715/r$. When $r = 65$,

$$t = \tfrac{715}{65} = 11.$$

It takes 11 hours to drive this distance at 65 mi/hour.

PROGRESS CHECK 5   If the area of a rectangle remains constant, the length varies inversely as the width. The length of a certain rectangle is 16 centimeters (cm) when the width is 6 cm. If the area of the rectangle remains constant, find the length when the width is 8 cm.   ⌐

3   The concept of variation extends to include both direct and inverse variation of variables raised to specific powers.

EXAMPLE 6 y varies inversely as the cube of x, and $y = 32$ when $x = \frac{1}{2}$. Find y when $x = 2$.

Solution   Since y varies inversely as x^3,

$$y = \frac{k}{x^3}.$$

Progress Check Answers
4. 6
5. 12 cm

Now, find k as follows.

$$32 = \frac{k}{\left(\frac{1}{2}\right)^3} \qquad \text{Replace } y \text{ by 32 and } x \text{ by } \frac{1}{2}.$$

$$32 = \frac{k}{\frac{1}{8}} \qquad \text{Evaluate a power.}$$

$$4 = k \qquad \text{Cross multiply.}$$

Thus, $y = 4/x^3$. When $x = 2$,

$$y = \frac{4}{2^3} = \frac{4}{8} = \frac{1}{2}.$$

PROGRESS CHECK 6 y varies inversely as the square of x, and $y = 81$ when $x = \frac{1}{3}$. Find y when $x = 3$. ⌐

EXAMPLE 7   In a storm the wind pressure P varies directly as the square of the wind velocity v, and P is a good measure of the destructive capability of the wind. How many *times* more destructive is a 150-mi/hour wind (a major hurricane) than a 50-mi/hour wind (a strong gale)?

Solution   Since P varies directly as the square of v,

$$P = kv^2.$$

If we let $v = 50$ mi/hour, then $3v = 150$ mi/hour. In other words, we are asked to find the change in P if v is tripled.

$$P = k(3v)^2 = 9kv^2$$

Because $9kv^2$ is nine times greater than kv^2, a 150-mi/hour wind is nine times more destructive than a 50-mi/hour wind.

PROGRESS CHECK 7   Use the variation statement in Example 7 and determine how many times more destructive a 120-mi/hour wind is than a 60-mi/hour wind. ⌐

Progress Check Answers
6. 1
7. 4 times more destructive

EXERCISES 6.7

In Exercises 1–6, write a variation equation for the given statement. Give the value of the variation constant if it is known.

1. The price (p) for a chunk of cheese varies directly as the weight (w).
2. The weight (w) of a solid cube varies directly as the volume (v).
3. The cost (c) of a piece of ribbon varies directly as the length (ℓ).
4. The price (p) for a carpet varies directly as the area (A).
5. If a plane flies at a constant speed of 400 mi/hour, the distance (d) covered varies directly as the time (t).
6. If the sales tax rate is 6 percent of the purchase price, then the tax (T) on a purchase varies directly as the price (P) of the item.

In Exercises 7–12, y varies directly as x.

7. $y = 5$ when $x = 4$; find y when $x = 6$.
8. $y = -5$ when $x = 2$; find y when $x = -3$.
9. $y = \frac{1}{2}$ when $x = 3$; find y when $x = 10$.
10. $y = \frac{1}{3}$ when $x = 4$; find y when $x = 13$.
11. $y = \frac{2}{3}$ when $x = \frac{2}{3}$; find y when $x = -\frac{2}{3}$.
12. $y = \frac{3}{4}$ when $x = \frac{4}{3}$; find y when $x = -\frac{8}{3}$.
13. The earnings of a baby-sitter vary directly as the number of hours worked. If the sitter charges \$16.25 for 5 hours, what is the charge for 4 hours of work?
14. A part-time employee is paid \$151.20 for 24 hours of work. If pay varies directly with the number of hours worked, what is the pay for 30 hours of work?
15. The price of 1×12 pine boards varies directly with length. An 8-ft board costs \$2.16. How much will 18 ft of this board cost?
16. The time it takes to cook a turkey varies directly with its weight. If it takes 3 hours and 36 minutes to cook a 12-lb turkey, how long will it take to cook a 20-lb turkey?
17. The weight of an object on the moon varies directly with its weight on earth. An object that weighs 96 lb on earth weighs 16 lb on the moon. How much will a person who weighs 171 lb on earth weigh on the moon?

18. The price for carpet varies directly as its *area*. If a 4- × 6-ft carpet costs $114, how much will a 3- × 5-ft carpet cost?

In Exercises 19–24, write a variation equation for the given statement.

19. The time (t) for a trip of a certain distance varies inversely as the speed (s) traveled.

20. The volume (V) of a fixed weight of gas varies inversely as the pressure (p) the gas exerts.

21. The time (t) to complete a manufacturing order varies inversely as the number (n) of machines used.

22. The number of ounces of fish (w) you can buy for a fixed amount of money varies inversely as the price per pound (p).

23. The number of times a wheel revolves (r) while it rolls a fixed distance varies inversely with the diameter of the wheel (d).

24. The frequency (f) of an electromagnetic wave varies inversely with its wavelength (w).

In Exercises 25–30, y varies *inversely* as x.

25. $y = 6$ when $x = 10$; find y when $x = 5$.

26. $y = -6$ when $x = 1$; find y when $x = 2$.

27. $y = 10$ when $x = -2.4$; find y when $x = 6$.

28. $y = \frac{1}{2}$ when $x = 3$; find y when $x = 6$.

29. $y = \frac{3}{5}$ when $x = 2$; find y when $x = 1$.

30. $y = 1$ when $x = 1$; find y when $x = -4$.

31. If a car travels at a constant rate (r), then the time (t) to drive a fixed distance varies inversely as r. If it takes 22 hours to cover a certain distance at 65 mi/hour, how long will it take at 55 mi/hour?

32. If the area of a rectangle remains constant, the length varies inversely with the width. The length of a certain rectangle is 32 cm when the width is 16 cm. Keeping the area constant, what is the length when the width is 20 cm?

33. The time it takes to fill a swimming pool varies inversely with the rate at which the water is pumped in. A certain pool can be filled in 25 hours when the water enters at 4 gal/minute. How long will it take to fill this pool if the water enters at a rate of 3 gal/minute?

34. There is an inverse relationship between the length of a lever arm and the force needed to lift a given weight using a fixed fulcrum (see the figure). A lever will be used to lift the corner of a cast-iron wood stove. If the lever arm is 2 ft long, 200 lb of force will be needed. If a person is only strong enough to exert 60 lb of force, how long does the lever arm need to be?

In Exercises 35–40, write a variation equation for the given statement.

35. Luminosity (l) varies inversely as the square of the distance (d) from the light source.

36. Gravitational attraction (f) varies inversely as the square of the distance (d) between two objects.

37. The volume (V) of a sphere varies directly as the cube of the radius (r).

38. The volume (V) of a cube varies directly as the cube of the side (s).

39. The kinetic energy (E) of a moving mass varies directly as the square of the velocity (v).

40. The angular momentum (M) of a ball whirling at the end of a rope varies directly as the square of the length (r) of the string.

41. y varies inversely as the square of x, and $y = 60$ when $x = \frac{1}{2}$. Find y when $x = 1$.

42. y varies inversely as the square of x, and $y = 5$ when $x = 2$. Find y when $x = \frac{1}{2}$.

43. y varies directly as the cube of x, and $y = 135$ when $x = 3$. Find y when $x = 4$.

44. y varies directly as the cube of x, and $y = 162.5$ when $x = 5$. Find y when $x = 10$.

45. y varies directly as the square of x, and $y = -27$ when $x = 3$. Find y when $x = -3$.

46. y varies directly as the cube of x, and $y = -81$ when $x = 3$. Find y when $x = -3$.

47. The distance (d) covered by a freely falling body varies directly as the square of the elapsed time (t). When the elapsed time is 0.5 second, the distance covered is 4 ft. What distance is covered when the elapsed time is 2.5 seconds?

48. As the moon revolves around the earth, its motion has acceleration toward the earth. The acceleration (a) varies directly with the velocity (v) of the moon in its orbit. The velocity of the moon is 1,020 meters per second (m/second), and its acceleration is 0.00273 m/second². If the moon speeded up to 2,000 m/second, what would its acceleration toward the earth be?

49. The weight (w) of a solid metal cube varies directly as the cube of the side (s).
   a. If you double the length of the side, what happens to the weight?
   b. Suppose a cube that measures 1 ft per side weighs 50 lb. What would a cube of this material weigh if it measured 2 ft per side?

50. The area (A) of a square varies directly as the square of the side (s).
   a. If you cut the side in half, what happens to the area?
   b. Compare the area of a square which is 3 cm per side to one which is 6 cm per side.

51. In physics the work (w) done in compressing a spring a distance (x) beyond its unstretched length varies directly with the square of that distance. What happens to the amount of work done if instead of compressing the spring 1 in., you compress it 3 in.?

THINK ABOUT IT

1. If y varies directly as x, does x vary directly as y?
2. If y varies directly as x, what is the value of y when x is 0?
3. If three cats can catch three rats in three days, how many rats can twelve cats catch in twelve days? (Can you solve this as a problem in direct variation?)
4. **a.** z varies directly as x and inversely as y. This is denoted by $z = \dfrac{kx}{y}$. What happens to z when x and y are both tripled?

    b. z varies directly as x and inversely as y^2. This is denoted by $z = \dfrac{kx}{y^2}$. What happens to z when x and y are both doubled?

5. **a.** What happens to the circumference of a circle when you double the radius? Explain this in terms of direct variation.

    b. What happens to the area of a circle when you double the radius? Explain this in terms of direct variation.

REMEMBER THIS

1. Name the point where these two lines cross.

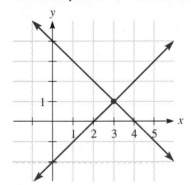

2. True or false? The ordered pair (1,4) is a solution to both $5x - y = 1$ and $x + y = 5$.
3. If $f(x) = 2x - 3$, find $f(\frac{1}{2})$.

4. Express the distance (y) that a car going 50 mi/hour will travel in t hours.
5. Graph $x + y < 3$ subject to the restriction that $x \geq 0$ and $y \geq 0$.

6. Write an equation in general form for the line through (0,0) and (1,5).

7. Evaluate $10^{-1} + 10^{-2} + 10^{-3}$.

8. Solve $a = \dfrac{b + c}{c}$ for c.

9. Solve $2x^2 - x - 6 = 0$.

10. Factor completely $6x^3 + 3x^2 - 3x$.

Chapter 6   SUMMARY

OBJECTIVES CHECKLIST   Specific chapter objectives are summarized below along with numbered example problems from the text that should clarify the objectives. If you do not understand any objectives or do not know how to do the selected problems, then restudy the material.

6.1 **Can you:**
1. **Determine if an ordered pair is a solution of an equation?**
Determine if the ordered pair $(1, -2)$ is a solution of $2x - y = 4$.   [Example 1a]

2. **Complete ordered pairs so that they are solutions of an equation?**
Fill in the missing component in the following ordered pair so that the pair is a solution of the equation $2x - y = 4$: (,6).   [Example 2a]

3. **Graph ordered pairs and determine coordinates of points?**
Graph the ordered pair and indicate its quadrant location: $(-2,3)$.   [Example 5a]

6.2 **Can you:**
1. **Graph linear equations by plotting ordered-pair solutions?**
Graph the equation $y = 2x + 1$.   [Example 1]

2. Graph linear equations by using intercepts?

Graph $3x + 2y = -6$ by using the x- and y-intercepts.   [Example 5]

3. Graph $x = a$ or $y = b$?

Graph $y = 2$.   [Example 6]

6.3 **Can you:**

1. Find and interpret the slope of a line?

Find the slope of the line through $(1,2)$ and $(3,3)$.   [Example 1]

2. Determine if lines are parallel, perpendicular, or neither by using slope?

A quadrilateral has vertices at $A(0,1)$, $B(6,4)$, $C(7,2)$, and $D(1,-1)$. Show that parallelogram $ABCD$ is a rectangle.   [Example 7b]

3. Interpret line graphs?

Refer to the figure and determine between which two hourly readings the temperature decrease was the greatest.   [Example 8c]

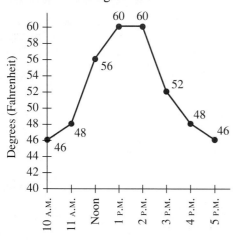

6.4 **Can you:**

1. Write an equation for a line given its slope and a point on the line?

Find an equation of the line through $(2,-3)$ with slope -4. Express the answer in the general form $Ax + By = C$.   [Example 1]

2. Write an equation for a line given two points on the line?

Find an equation for the line through the points $(-1,6)$ and $(2,5)$. Express the answer in the general form $Ax + By = C$.   [Example 2]

3. Find the slope and y-intercept given an equation for the line?

Find the slope and the y-intercept of the line defined by the equation $2x + 3y = -7$.   [Example 4b]

4. Graph and write an equation for a line given the slope and y-intercept?

The slope of a line is $-\frac{1}{2}$ and the y-intercept is $(0,3)$. (a) Graph the line, and (b) find its equation in slope-intercept form.   [Example 5]

5. Determine if the graphs of two linear equations are distinct parallel lines?

Determine if the graphs of $3x + 2y = 1$ and $6x + 4y = 9$ are distinct parallel lines.   [Example 6]

6.5 **Can you:**

1. Determine if an ordered pair is a solution of an inequality?

Determine if $(4,0)$ is a solution of $y \le x$.   [Example 1]

2. Graph linear inequalities?

Graph $2x - y > 4$.   [Example 5]

3. **Graph linear inequalities restricted to $x \geq 0$, $y \geq 0$?**
While on a diet, a teacher has yogurt and milk for lunch. If low-fat yogurt contains about 15 calories an ounce, skim milk contains about 10 calories an ounce, and lunch is to result in at most 180 calories, then the inequality

$$15x + 10y \leq 180$$

expresses the calorie restriction for lunch, where x and y represent the number of ounces eaten of yogurt and milk, respectively. Graph this inequality for $x \geq 0$ and $y \geq 0$. [Example 6]

6.6 **Can you:**
1. **Find the domain and range given a set of ordered pairs?**
In a certain class the correspondence between students' final averages and their final grades is given by $\{(80,B),(92,A),(74,C),(86,B)\}$. Find the domain and range of this relation. [Example 1]

2. **Determine if a set of ordered pairs is a function?**
Determine if the relation $\{(5,1),(6,1),(7,1)\}$ is a function. [Example 3a]

3. **Determine if an equation defines y as a function of x?**
Does the rule $y = x^2$ determine y as a function of x? [Example 4a]

4. **Find a formula that defines the functional relationship between two variables?**
Express the perimeter y of a square as a function of its side length x. [Example 5a]

5. **Use functional notation?**
If $y = f(x) = 5x - 1$, find $f(3)$, $f(-2)$, and $f(\frac{1}{2})$. [Example 6]

6.7 **Can you:**
Solve problems involving the following?
1. **Direct variation**
y varies directly as x, and $y = 4$ when $x = 3$. Find y when $x = 12$. [Example 2]

2. **Inverse variation**
If a car travels at a constant rate r, then the time t required to drive a fixed distance varies inversely as r. If it takes 13 hours to drive a certain distance at 55 mi/hour, how long will it take to drive this distance at 65 mi/hour? [Example 5]

3. **Variation of powers of variables**
y varies inversely as the cube of x, and $y = 32$ when $x = \frac{1}{2}$. Find y when $x = 2$. [Example 6]

KEY TERMS

Cartesian (or rectangular) coordinate system (6.1)	Line graph (6.3)	Rise (6.3)
Coordinates (6.1)	Ordered pair (6.1)	Run (6.3)
Dependent variable (6.6)	Origin (6.1)	Slope (6.3)
Domain (6.6)	Parallel lines (6.3)	Subscript (6.3)
Function (6.6)	Parallelogram (6.3)	Variation constant (6.7)
General form (6.2)	Perpendicular lines (6.3)	Vertical line (6.2)
Graph (6.1)	Quadrant (6.1)	x-axis (6.1)
Horizontal line (6.2)	Range (6.6)	x-intercept (6.2)
Hyperbola (6.7)	Rate of change (6.3)	y-axis (6.1)
Independent variable (6.6)	Rectangle (6.3)	y-intercept (6.2)
Linear inequality (6.5)	Relation (6.6)	

KEY CONCEPTS AND PROCEDURES

Section	Key Concepts or Procedures to Review
6.1	■ To find ordered-pair solutions to an equation in two variables, substitute specific values for one variable and then find the corresponding values of the other variable. ■ Any ordered pair can be represented as a point in the rectangular coordinate system. The first component indicates the distance of the point to the right or left of the vertical axis. The second component indicates the distance of the point above or below the horizontal axis.

6.2	■ Graph: The graph of an equation in two variables is the set of all points in a coordinate system that correspond to ordered-pair solutions of the equation.
	
	■ Linear equations in two variables: A linear equation in two variables is an equation that can be written in the general form
	
	$$Ax + By = C$$
	
	where A, B, and C are real numbers, with A and B not both zero. The graph of a linear equation in two variables is a straight line.
	
	■ To find intercepts: To find the x-intercept $(a,0)$, let $y = 0$ and solve for x. To find the y-intercept $(0,b)$, let $x = 0$ and solve for y.
	
	■ To graph $x = a$ and $y = b$
	
	  1. The graph of the linear equation $x = a$ is a vertical line that contains the point $(a,0)$. If $a = 0$, then the equation is $x = 0$, and the line is the y-axis.
	
	  2. The graph of the linear equation $y = b$ is a horizontal line that contains the point $(0,b)$. If $b = 0$, then the equation is $y = 0$, and the line is the x-axis.

6.3	■ Slope of a line: If (x_1,y_1) and (x_2,y_2) are any two distinct points on a line, with $x_1 \neq x_2$, then the slope m of the line is $$m = \frac{\text{rise}}{\text{run}} = \frac{\text{change in } y}{\text{change in } x} = \frac{y_2 - y_1}{x_2 - x_1}$$ ■ The slope of every horizontal line is 0. ■ The slope of every vertical line is undefined. ■ If the relationship between two variables x and y graphs as a line, then the slope of the line measures the rate of change of y with respect to x. ■ Two nonvertical lines are parallel if the slopes are equal. ■ Two nonvertical lines are perpendicular if the product of their slopes is -1.
6.4	■ Point-slope equation: An equation of the line with slope m passing through (x_1,y_1) is $$y - y_1 = m(x - x_1)$$ This equation is called the point-slope form of the equation of a line. ■ Slope-intercept equation: The graph of the equation $$y = mx + b$$ is a line with slope m and y-intercept $(0,b)$. ■ Two distinct parallel lines must have the same slope and different y-intercepts.
6.5	■ To graph a linear inequality **1.** Graph the linear equation that results when the inequality symbol is replaced by $=$. Draw this boundary line as a solid line if the inequality symbol is $\leq$ or $\geq$, and draw a dashed line for $<$ or $>$.

Section	Key Concepts or Procedures to Review

2. Shade the half plane on one side of the boundary line as follows.

Case 1: The inequality is solved for y or x.

Inequality Begins	Shade the Half Plane
$y <$	Below the line
$y >$	Above the line
$x <$	Left of the line
$x >$	Right of the line

Case 2: The inequality is not solved for y or x. Choose a test point that is not on the boundary line, and substitute the coordinates of this point into the inequality.

Resulting Inequality Is	Shade the Half Plane
True	Containing the test point
False	Not containing the test point

6.6
■ Relation definition: A relation is a set of ordered pairs. The set of all first components of the ordered pairs is called the domain of the relation. The set of all second components is called the range of the relation.

■ Function: A function is a relation in which no two different ordered pairs have the same first component.

■ The notation $f(x)$ is read "f of x" or "f at x" and means the value of the function (the y value) corresponding to the value of x.

6.7
■ y varies directly as x: The statement "y varies directly as x" means that
$$y = kx$$
where k is a constant called the variation constant.

■ y varies inversely as x: The statement "y varies inversely as x" means $xy = k$ or
$$y = \frac{k}{x}$$
where k is a constant called the variation constant.

CHAPTER 6 REVIEW EXERCISES

6.1

1. Is the ordered pair $(-1,2)$ a solution to $6x + 5y = 4$?

2. Fill in the missing component so that $(3, \quad)$ is a solution to $6x + 5y = 4$. $18 - 4 = 14$

3. Graph the point $(-3,2)$ and state its quadrant.

6.2

4. Graph the equation $y = -x + 4$ by plotting ordered-pair solutions.

5. Graph the equation $x + y = 10$ by using the intercepts.

6. Graph $y = 1$ and indicate whether it is horizontal, vertical, or neither.

7. The cost (y) of a van rental is $100 per week plus 50¢ per mile (x). Find and graph an equation for the total cost for one week.

6.3

8. Find the slope of the line which passes through the points (1,3) and (5,11).
9. A quadrilateral has vertices at $A(-3,2)$, $B(2,4)$, $C(4,-1)$, and $D(-1,-3)$. Show that it is a rectangle.

10. The line graph shows the progress of the price of a share of stock over five trading days. Between which two days did the stock price change the most?

Price ($)

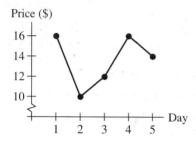

6.4

11. Find an equation for a line through the origin with slope 3.

$0,0$

12. Find an equation for the line through $(-1,1)$ and $(5,0)$.

13. Find the slope and y-intercept of the line defined by the equation $3x - y = 1$.
14. Determine if the graphs of these two lines are distinct parallel lines. $x + y = 2$
$$x - y = 2$$
15. The charge (y) for a certain long-distance phone call is $3 for the initial connection plus $0.25 per minute ($x$). This relationship graphs as a straight line. Find the equation for the line in slope-intercept form and draw the graph. (Assume that all calls are measured to the nearest minute.)

6.5

16. Is the point (2,3) a solution to $2x - 3y \le -1$?
17. Draw the graph of $x + y < 6$, restricting x and y to be nonnegative.

6.6

18. Does this set of ordered pairs determine a function? $\{(3,2),(4,3),(5,3)\}$.
19. Does the rule $x = y^2$ determine y as a function of x?
20. The length of a rectangle is twice the width (x). Express the area (A) of the rectangle as a function of its width.
21. If $y = f(x) = x^2 + 4$, find $f(-2)$.
22. The charge (C) in dollars for a printing job is given by the function $C = f(n) = 20 + 0.10n$, where n is the number of copies printed. Find $f(1,000)$ and interpret what it represents.

6.7

23. Write a variation statement for this relationship: pressure (p) varies directly as temperature (t).
24. Assume that y varies directly as x and that $y = 6$ when $x = 4$. Find y when $x = 6$.
25. The price of some cloth varies directly with its length. A piece 6 ft long costs $20. To the nearest cent, what is the price for 10 ft?
26. Assume that y varies inversely with x and that $y = 20$ when $x = -1.2$. Find y when $x = 6$.
27. Assume that y varies inversely with the square of x. If $y = 2$ when $x = \frac{1}{2}$, find y when $x = 1$.

CHAPTER 6 TEST

1. Which one of these ordered pairs is a solution to $2x - 4y = 1$?
   a. (4,2)   b. $(\frac{1}{2},\frac{1}{4})$ c. $(\frac{1}{2},0)$ d. $(0,\frac{1}{2})$
2. Which of these points is the highest, and which quadrant is it in?
   a. $(-3,5)$ b. $(-5,3)$ c. $(3,-5)$ d. $(5,-3)$

3. Graph this line and give its intercepts: $y = 2x - 3$.

4. Graph the line, $x = 4$.

5. The shipping cost for a catalog item charged by a mail-order company is $5 plus $2 per pound. Find an equation for the shipping cost (C) in terms of the weight of the item (w). Graph the equation.

6. Find the slope of the line through (3,5) and (3,10).
7. The equation of the line shown in the figure is $y = 5 + 2x$. Find the slope and interpret its meaning.

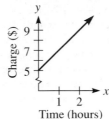

Time (hours)

8. Two lines are perpendicular. If the slope of one line is -3, what is the slope of the other line?
9. Find an equation in general form for the line through the point (2,5) which has slope 3.
10. Find an equation in general form for the line through $(2,-2)$ and $(-5,5)$.
11. Find the slope and the y-intercept of the line defined by $4x - 2y = 10$.

12. A line has slope $\frac{1}{2}$ and y-intercept (0,2). Find its equation and graph it.

13. Is the point (5,1) in the solution set of $x + 5y \le 10$?
14. For some sliced turkey and roast beef you cannot spend more than $12. The turkey is $3.50 per quarter pound and the roast beef is $4 per quarter pound. Write an inequality that expresses these restrictions. Let x and y stand for the amounts in pounds of turkey and roast beef, respectively.
15. Does the set of ordered pairs $\{(1,1),(2,2),(3,3),(4,4)\}$ determine a function?
16. Express the diameter (d) of a circle as a function of the circumference (C).
17. If $g(x) = 3x^2 - 2x + 1$, find $g(2)$.
18. The distance (d) in feet traveled by an object which falls for t seconds from a height is given by $d = f(t) = 16t^2$. Find $f(2)$ and explain what it means.
19. Write a variation equation for this statement: The speed (s) at which a wheel turns varies inversely with its weight (w).
20. Assume that y varies directly as the cube of x and that $y = 1$ when $x = \frac{1}{2}$. Find y when $x = 2$.

CUMULATIVE TEST 6

1. Evaluate $2 + 3[15 - 3(2 - 5)]$.
2. What property of real numbers is illustrated by this equation? $3m(m - 4) = 3m^2 - 12m$.
3. Given $x = -3$ and $y = 4$, evaluate $2x - y$.
4. How many terms are in this expression? $3x^2y$.
5. Solve and check: $2(x + 1) = -5$.
6. A shopper spent 20 percent of his cash on a shirt and 35 percent of his cash on a pair of pants, making a total of $66. How much cash was left after this?
7. Simplify $\dfrac{3x^{-2}y^4}{3x^3y^{-1}}$. Use only positive exponents in the answer.
8. Simplify $(3x - 4) - (2x - x^2) + (x + 5)$.
9. Find the product: $(2x + 5y)(2x - 5y)$.
10. What is the GCF for $12x^2$, $18xy$, and $15y$?
11. Factor completely: $a^3 + 7a^2 + 12a$.
12. Solve $x^2 - 2x - 3 = 0$.
13. Add $\dfrac{x}{2x - 1} + \dfrac{2x}{1 - 2x}$.

14. Simplify this complex fraction: $\dfrac{\dfrac{ab}{a + b}}{\dfrac{1}{a} + \dfrac{1}{b}}$.

15. A 4-ft pole casts a 6-ft shadow at the same time that a flagpole casts a 42-ft shadow. How tall is the flagpole?
16. Solve $\dfrac{3}{x} + \dfrac{1}{2} = \dfrac{1}{3}$.
17. Which of these equations have graphs that are straight lines?
    a. $2x - 4y = 7$ b. $x = 4$
    c. $y + 1 = 0$ d. $x = y$
18. Find the y-intercept of the graph of this line: $x + 2y + 1 = 0$.
19. Find an equation in general form for the line through the points (2,1) and (0,5).
20. If y varies directly as x^2 and $y = 2$ when $x = 2$, find y when $x = 3$. $y =$

Systems of Linear Equations and Inequalities

A company is trying to decide between two machines for packaging its new product. Machine A will cost $6,000 per year plus $2 to package each unit, so the total cost (*y*) of packaging *x* units annually is given by $y = 2x + 6{,}000$. Machine B will cost $4,000 per year plus $3 to package each unit, so $y = 3x + 4{,}000$ is the cost equation associated with this machine. How many units must be produced annually for the cost of the two machines to be the same? If the company plans to package more units than this, which machine should it purchase? (See Example 1 of Section 7.1.)

■■■■■■
THE COMPANY described in the chapter-opening problem must take into account two linear equations in two variables to make its decision. The pair of equations in this problem,

$$y = 2x + 6{,}000 \quad \text{and} \quad y = 3x + 4{,}000,$$

is an example of a **linear system of equations.** Here are some other examples of linear systems.

$$
\begin{array}{lll}
x + y = 25 & 2a + b = 8 & y = x - 1 \\
6x - y = 3 & b = a + 2 & x = 3
\end{array}
$$

Note that the pair of equations in the system must contain the same variables and that an equation like $x = 3$ may be written as $x + 0y = 3$ when expressed as a linear equation in two variables. In this chapter we will consider three elementary methods for solving such systems and apply these methods to analyze a wide variety of problems.

7.1   Solving Linear Systems Graphically

1 Solve a system of linear equations graphically and check to see whether an ordered pair is a solution of a system.

2 Recognize and solve linear systems that are inconsistent or dependent.

1 To understand the meaning of the solution of a system of linear equations, we begin by analyzing the chapter-opening problem, using a graphing approach.

EXAMPLE 1   Solve the problem in the chapter introduction on page 320.

Solution   A cost comparison between the two machines can be made by constructing the following tables and then graphing the two lines on the same coordinate system, as shown in Figure 7.1.

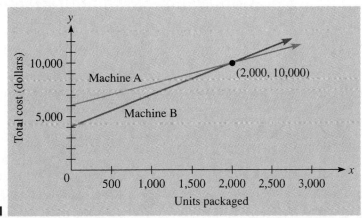

Figure 7.1

Machine A: $y = 2x + 6,000$

Units Packaged x	Cost (Dollars) y
500	7,000
1,000	8,000
1,500	9,000
2,000	10,000
2,500	11,000

Machine B: $y = 3x + 4,000$

Units Packaged x	Cost (Dollars) y
500	5,500
1,000	7,000
1,500	8,500
2,000	10,000
2,500	11,500

From the graph and the table we see that if 2,000 units are packaged annually, then both machines A and B have the same cost, which is $10,000. If the company plans to package more than 2,000 units, the graph shows that it should purchase machine A, since the y values are lower on the line associated with A beyond the intersection point.

PROGRESS CHECK 1   Redo the problem in Example 1, but assume that machine B will cost $7,000 per year plus $1 to package each unit, so $y = x + 7,000$ is the cost associated with this machine.   ⌐

Note in Example 1 that there are many ordered pairs that satisfy $y = 2x + 6,000$ and many ordered pairs that satisfy $y = 3x + 4,000$, but there is only one ordered pair that satisfies both equations. In general, the **solution of a system of linear equations** consists of all ordered pairs that satisfy both equations at the same time (simultaneously).

The next example shows how to determine whether a given ordered pair is a solution of a system.

EXAMPLE 2   Determine if the given ordered pairs are solutions of the system

$$x + y = 25$$
$$6x - y = 3.$$

a. $(4,21)$ **b.** $(10,15)$

Solution

a. $(4,21)$ means $x = 4$, $y = 21$. Replace x by 4 and y by 21 in each equation.

$$x + y \overset{?}{=} 25 \qquad\qquad 6x - y \overset{?}{=} 3$$
$$4 + 21 \overset{?}{=} 25 \qquad\qquad 6(4) - 21 \overset{?}{=} 3$$
$$25 = 25 \quad \text{True} \qquad\qquad 3 = 3 \quad \text{True}$$

$(4,21)$ is a solution of the system because it satisfies both equations.

b. To check $(10,15)$, replace x by 10 and y by 15 in each equation.

$$x + y \overset{?}{=} 25 \qquad\qquad 6x - y \overset{?}{=} 3$$
$$10 + 15 \overset{?}{=} 25 \qquad\qquad 6(10) - 15 \overset{?}{=} 3$$
$$25 = 25 \quad \text{True} \qquad\qquad 45 = 3 \quad \text{False}$$

Because $(10,15)$ is not a solution of $6x - y = 3$, it is not a solution of the system.

PROGRESS CHECK 2   Determine if the given ordered pairs are solutions of the system

$$5x + 2y = 11$$
$$x + 2y = -1.$$

a. $(5,-7)$ **b.** $(3,-2)$ ⌐

Because the graph of a linear equation provides a picture of its solutions, we can find all common solutions to a pair of linear equations by drawing their graphs on the same coordinate system. The solution is given by all points where the lines intersect, and we specify the solution by giving the coordinates of such points. Usually, the two lines will intersect at exactly one point, as in our next example.

EXAMPLE 3   Solve by graphing:

$$x + 2y = 6$$
$$y = x + 6.$$

Solution $x + 2y = 6$ is a linear equation in general form that is easily graphed by finding intercepts. Letting $x = 0$ gives a y-intercept of $(0,3)$, and letting $y = 0$ gives an x-intercept of $(6,0)$. The equation $y = x + 6$ is in slope-intercept form and may be graphed by recognizing and using the fact that the y-intercept is $(0,6)$ and the slope is 1. In Figure 7.2 we graph both of these equations on the same coordinate system,

Progress Check Answers

1. At 1,000 units both machines have a cost of $8,000; machine B
2. (a) No   (b) Yes

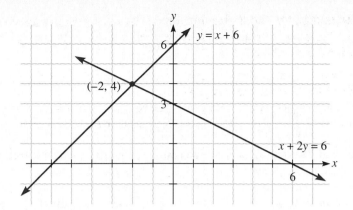

Figure 7.2

and it appears that the lines meet at $(-2,4)$. We check this apparent solution.

$$x + 2y = 6 \qquad\qquad y = x + 6$$
$$-2 + 2(4) \overset{?}{=} 6 \qquad\qquad 4 \overset{?}{=} -2 + 6$$
$$6 = 6 \quad \text{True} \qquad 4 = 4 \quad \text{True}$$

Thus, the solution is $(-2,4)$.

Note To obtain an accurate solution using the graphing method, draw the graphs carefully on graph paper. Even then, you may not be able to read the exact coordinates of a solution like $(\frac{5}{8}, -\frac{2}{7})$. Algebraic methods for finding exact solutions will be considered in the next two sections.

PROGRESS CHECK 3 Solve by graphing:

$$x + 2y = 4$$
$$y = -2x - 1.$$

2 It is possible for a linear system to have either no solution or infinitely many solutions. Examples 4 and 5 illustrate these cases.

EXAMPLE 4 Solve by graphing:

$$x - 2y = 5$$
$$x - 2y = -2.$$

Solution We find the x- and y-intercepts for each graph, as in the table below, and then graph the lines as in Figure 7.3.

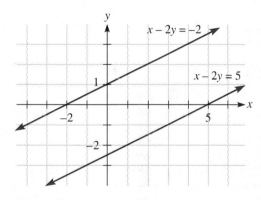

Figure 7.3

Equation	x-intercept	y-intercept
$x - 2y = 5$	$(5,0)$	$(0,-\frac{5}{2})$
$x - 2y = -2$	$(-2,0)$	$(0,1)$

The two lines appear to be parallel. Because parallel lines do not meet, there is no ordered pair that satisfies both equations, so the system has no solution.

We can verify that the graphs are distinct parallel lines by writing each equation in slope-intercept form.

$$x - 2y = 5 \qquad\qquad x - 2y = -2$$
$$-2y = -x + 5 \qquad\qquad -2y = -x - 2$$
$$y = \tfrac{1}{2}x - \tfrac{5}{2} \qquad\qquad y = \tfrac{1}{2}x + 1$$

Each line has a slope of $\frac{1}{2}$, so the lines are parallel. Because the y-intercepts are different, the lines are also distinct.

Progress Check Answer
3. $(-2,3)$

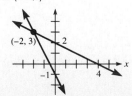

PROGRESS CHECK 4 Solve by graphing:

$$x - y = 2$$
$$y = x.$$

EXAMPLE 5 Solve by graphing:

$$2x + y = 6$$
$$x + \tfrac{1}{2}y = 3.$$

Solution For both equations the x-intercept is $(3,0)$ and the y-intercept is $(0,6)$, which means that the same line (see Figure 7.4) is the graph of both equations. Therefore, the graphical solution consists of all the points on that line, and there are an infinite number of ordered-pair solutions. This case occurred because multiplying both sides of the bottom equation by 2 produces the top equation, and so the two equations are equivalent.

An alternative method for establishing that the equations are equivalent is to compare the equations after both have been written in slope-intercept form. The result is $y = -2x + 6$ for both equations in this system.

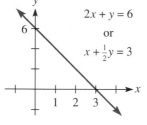

Figure 7.4

PROGRESS CHECK 5 Solve by graphing:

$$3x - y = 6$$
$$-x + \tfrac{1}{3}y = -2.$$

We have seen that it is possible for a system of linear equations to have exactly one solution, no solution, or infinitely many solutions. No other cases are possible. When a linear system has at least one solution (as in Examples 3 and 5), the system is called **consistent,** and a consistent system with an infinite number of solutions (as in Example 5) is called **dependent.** A system with no solution (as in Example 4) is called **inconsistent.** Consider carefully Figure 7.5, which summarizes the cases we have discussed.

Progress Check Answers

4. No solution; these are parallel lines.

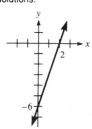

5. These are the same line; there are infinitely many solutions.

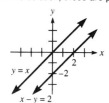

Possible Graph	Geometric Interpretation	Number of Solutions	Type of System
y graph	Graphs cross at exactly one point	Exactly one solution	Consistent system
y graph	Graphs are the same line	Infinitely many solutions	Consistent system that is called dependent
y graph	Graphs are parallel lines	No solution	Inconsistent system

Figure 7.5

EXERCISES 7.1

In Exercises 1–4, (a) find an (x,y) pair that satisfies *both* equations by completing the table of values; (b) draw graphs for both lines to confirm that they intersect at the point found in part **a**; and (c) answer the question.

1. The cost (y) to rent recording studio A is $100 plus $50 per hour for x hours. The cost to rent studio B is $50 plus $75 per hour for x hours. Equations giving respective total costs are, therefore, $y = 100 + 50x$ and $y = 50 + 75x$. For how many hours use do the studios have the same cost? After that amount of time, which is less expensive?

Studio A: $y = 100 + 50x$

Hours x	Cost (Dollars) y
1	
2	
3	
4	

Studio B: $y = 50 + 75x$

Hours x	Cost (Dollars) y
1	
2	
3	
4	

2. Redo Exercise 1, but for this problem assume that studio A charges $250 plus $100 per hour, while studio B charges $310 plus $80 an hour.

Studio A: $y = 250 + 100x$

Hours x	Cost (Dollars) y
1	
2	
3	
4	

Studio B: $y = 310 + 80x$

Hours x	Cost (Dollars) y
1	
2	
3	
4	

3. At mile marker 200 on a long interstate highway a car going 60 mi/hour is 15 mi behind a car going 55 mi/hour. If they travel at constant speed, at what mile marker will the faster car catch up to the slower one? Use the given equations.

Faster car: $y = 200 + 60x$

Time x	Mile Marker y
1	
2	
3	

Slower car: $y = 215 + 55x$

Time x	Mile Marker y
1	
2	
3	

4. Redo Exercise 3, but for this problem assume that the slower car is going 57 mi/hour.

Faster car: $y = 200 + 60x$

Time x	Mile Marker y
1	
2	
3	
4	
5	

Slower car: $y = 215 + 57x$

Time x	Mile Marker y
1	
2	
3	
4	
5	

In Exercises 5–12, determine if the given ordered pairs are solutions of the system.

5. $3x + 2y = 7$
   $3x - 2y = -1$
   a. $(1,2)$
   b. $(3,-1)$

6. $x - y = -4$
   $2x + 3y = 7$
   a. $(-1,3)$
   b. $(-5,-1)$

7. $x + 2y = -6$
   $2x - y = 3$
   a. $(-6,0)$
   b. $(0,-3)$

8. $5x - 4y = 4$
$4x - 5y = 5$
   a. $(4,4)$ **b.** $(0,-1)$

9. $2x + 3y = 2$
$6x - 6y = 1$
   a. $(\frac{1}{2},\frac{1}{3})$ **b.** $(\frac{1}{6},\frac{2}{3})$

10. $2x - 5y = 4$
$4x - 15y = 9$
   a. $(\frac{3}{2},-\frac{1}{5})$ **b.** $(\frac{1}{2},-\frac{3}{5})$

11. $x = 7$
$y = x - 1$
   a. $(7,0)$ **b.** $(7,6)$

12. $y = 3$
$x + y = 2$
   a. $(-1,3)$ **b.** $(0,2)$

In Exercises 13–34, solve the systems of equations by graphing. Check that the apparent solution satisfies both equations.

13. $3x + 2y = 5$
$3x - 2y = 1$

14. $3x + 4y = 14$
$4x - 3y = 2$

15. $y = 3x + 1$
$y = -3x + 1$

16. $y = 2x - 1$
$y = -2x - 1$

17. $x + 2y = 7$
$y = 3x$

18. $x - 2y = -8$
$y = x + 5$

19. $x - y = 0$
$y = 3x + 4$

20. $2y = 3x - 15$
$2y = -3x + 3$

21. $y = 2$
$x = 1$

22. $y = x$
$y = -x$

23. $x - 3y = 3$
$2x + 5y = 6$

24. $3y = x - 1$
$5y = -2x - 9$

25. $\frac{1}{2}x + y = -3$
$x + \frac{1}{3}y = -1$

26. $\frac{1}{2}x + \frac{1}{3}y = 2$
$\frac{1}{4}x - \frac{1}{6}y = 0$

27. $y = 2x + 1$
$y = 2x + 3$

28. $y = -x - 1$
$y = -x + 1$

29. $y - x = 1$
$2y - 2x = 1$

30. $2y = 4x + 1$
$y = 2x - 6$

31. $3x + y = 9$
$x = 3 - \frac{1}{3}y$

32. $2x - y = 3$
$3y = 6x - 9$

33. $4y = 8 + x$
$\frac{1}{4}x - y = -2$

34. $x + 2y = 3$
$-x - 2y = -3$

THINK ABOUT IT

1. One sales job pays $150 per week plus a 20 percent commission, while another pays $250 per week with no commission.
   a. Write an equation for each of these which expresses weekly pay (y) in terms of amount of sales (x).
   b. Solve the system of equations by graphing.
   c. At what amount of sales do the jobs have the same pay?
   d. If you think you will usually sell about $700 worth of merchandise in a week, which job is preferable?
2. a. If two lines have different slopes, what is the maximum number of points of intersection possible?
   b. If three lines all have different slopes, what is the maximum number of points of intersection possible?
   c. If four lines all have different slopes, what is the maximum number of points of intersection possible?
   d. Can you see what the pattern is? What is the answer for five lines?

3. It is possible to extend the study of systems of linear equations to include more variables. Here is a system of three equations in three variables.

$$x + y + z = 6$$
$$x - y + z = 2$$
$$x + y - z = 0$$

An ordered *triple* is a solution of the system if it is a solution to all three equations. In an ordered triple, the third component is the value for z. Show that $(1,2,3)$ is a solution of the given system.
4. Two lines through the origin have slope 1,000 and 5,000, respectively. What is their point of intersection?
5. Two lines can intersect anywhere in the coordinate system. It is not necessary that the point of intersection have integer or rational coordinates.
   a. Graph $\sqrt{2}x + \sqrt{3}y = 5$ and $\sqrt{2}x - \sqrt{3}y = -1$. (*Hint:* Use your calculator to estimate intercepts.)
   b. Show that the lines intersect at $(\sqrt{2}, \sqrt{3})$.

REMEMBER THIS

1. What expression results when you substitute $5 - y$ for x in $2x - y$? Give the answer in simplest form.
2. Solve the equation $5.5x = 4x + 24$.
3. Does the ordered pair $(2, -1)$ satisfy *both* of these equations? $2x + y = 3$ and $3x - 4y = 10$.
4. What equation results when you multiply both sides of $-x + 3y = 7$ by -1?
5. Solve $x + \frac{1}{2}(6 - 2x) = 3$.
6. What is the slope of a line which is perpendicular to $y = 2x - 3$?

7. Find an equation in general form for the line through $(3, -4)$ with slope 2.
8. Multiply $\dfrac{2x - 3y}{x + 1} \cdot \dfrac{x - 1}{3y - 2x}$.
9. Simplify $\dfrac{\dfrac{a^2}{3} + \dfrac{3}{a}}{\dfrac{a}{3} + \dfrac{3}{a^2}}$.
10. Simplify $8 - 4(2 - x)$.

7.2 Solving Linear Systems by Substitution

A cable company offers a "pay-per-view" club that entitles members to watch current hit movies at a price of $4 per movie. The annual membership fee is $24, so the total cost ($y$) for members to watch x movies annually is given by $y = 4x + 24$. Without the club, the charge is $5.50 for each viewing, so $y = 5.5x$ is the cost equation for nonmembers. How many movies must be viewed annually for the cost to be the same for members and nonmembers? (See Example 2.)

OBJECTIVES

1. Solve linear systems by substitution when at least one equation is solved for one of the variables.

2. Solve linear systems by substitution when neither equation is solved for one of the variables.

3. Solve, by substitution, linear systems that are inconsistent or dependent.

1 The graphing method for solving linear systems is good for illustrating the principle involved and for estimating the solution. However, algebraic methods are usually faster and provide an exact solution. When at least one equation in a linear system is solved for one of the variables, then the system may be solved efficiently by substitution, as shown in the next two examples.

EXAMPLE 1 Solve by substitution:

$$2x - y = 2$$
$$x = 5 - y.$$

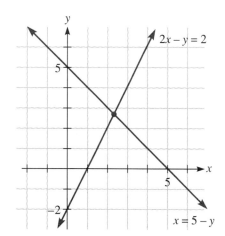

Figure 7.6

Solution First, note that if we attempt to solve the system by graphing, as shown in Figure 7.6, then it is difficult to determine the exact solution. We can overcome this difficulty by noting that x and $5 - y$ are different names for the same number, according to the bottom equation. Therefore, we can obtain one equation with one variable by substituting $5 - y$ for x in the top equation.

$$2x - y = 2 \quad \text{Given equation.}$$
$$2(5 - y) - y = 2 \quad \text{Substitute } 5 - y \text{ for } x.$$

Now, solve for y.

$$2(5 - y) - y = 2$$
$$10 - 2y - y = 2 \quad \text{Remove parentheses.}$$
$$10 - 3y = 2 \quad \text{Combine like terms.}$$
$$-3y = -8 \quad \text{Subtract 10 from both sides.}$$
$$y = \tfrac{8}{3} \quad \text{Divide both sides by } -3.$$

Thus, the y-coordinate of the solution is $\tfrac{8}{3}$.

To find the x-coordinate, substitute $\tfrac{8}{3}$ for y in the equation that is solved for x and simplify.

$$x = 5 - y \quad \text{Given equation.}$$
$$x = 5 - \tfrac{8}{3} \quad \text{Replace } y \text{ by } \tfrac{8}{3} .$$
$$x = \tfrac{7}{3} \quad \text{Simplify.}$$

Finally, we check that $(\tfrac{7}{3}, \tfrac{8}{3})$ is the solution.

$$2x - y = 2 \qquad\qquad x = 5 - y$$
$$2(\tfrac{7}{3}) - \tfrac{8}{3} \overset{?}{=} 2 \qquad\qquad \tfrac{7}{3} \overset{?}{=} 5 - \tfrac{8}{3}$$
$$\tfrac{14}{3} - \tfrac{8}{3} \overset{?}{=} 2 \qquad\qquad \tfrac{7}{3} = \tfrac{7}{3} \quad \text{True}$$
$$2 = 2 \quad \text{True}$$

Thus, the solution is $(\tfrac{7}{3}, \tfrac{8}{3})$.

PROGRESS CHECK 1 Solve by substitution:

$$5x + 3y = 5$$
$$x = 2 - y.$$

EXAMPLE 2 Solve the problem in the section introduction on page 327.

Solution The system of equations given in the problem is

$$y = 4x + 24$$
$$y = 5.5x.$$

We choose to substitute $5.5x$ for y in the top equation and then solve for x.

$$5.5x = 4x + 24 \quad \text{Replace } y \text{ by } 5.5x \text{ in the top equation.}$$
$$1.5x = 24 \quad \text{Subtract } 4x \text{ from both sides.}$$
$$x = 16 \quad \text{Divide both sides by 1.5.}$$

Progress Check Answer

1. $(-\tfrac{1}{2}, \tfrac{5}{2})$

Thus, both members and nonmembers are charged the same when they view 16 movies annually.

To check that the cost is the same, find the cost in each case.

$$y = 4x + 24 \qquad y = 5.5x$$
$$y = 4(16) + 24 \qquad y = 5.5(16)$$
$$= 88 \qquad\qquad = 88$$

The solution checks.

PROGRESS CHECK 2 Redo the problem in Example 2, but assume that nonmembers are charged $5.20 per movie, so that the cost equation for nonmembers is $y = 5.2x$.   ⌐

2   If neither equation is solved for one of the variables, then the substitution method requires us to solve for a variable in one of the equations as the initial step in the solution. In such cases we first look to solve for a variable with a numerical coefficient of 1.

EXAMPLE 3   Solve by substitution:

$$2x + y = 3$$
$$3x - 4y = 10.$$

Solution   We may solve for either variable in either equation. However, we choose to solve for y in the top equation, because this choice leads to an equivalent equation that does not contain fractions.

$$2x + y = 3 \qquad \text{Given equation.}$$
$$y = 3 - 2x \qquad \text{Subtract } 2x \text{ from both sides.}$$

Now, substitute $3 - 2x$ for y in the bottom equation and solve for x.

$$3x - 4y = 10 \qquad \text{Given equation.}$$
$$3x - 4(3 - 2x) = 10 \qquad \text{Substitute } 3 - 2x \text{ for } y.$$
$$3x - 12 + 8x = 10 \qquad \text{Remove parentheses.}$$
$$11x - 12 = 10 \qquad \text{Combine like terms.}$$
$$11x = 22 \qquad \text{Add 12 to both sides.}$$
$$x = 2 \qquad \text{Divide both sides by 11.}$$

Next, replace x by 2 in the equation that is solved for y.

$$y = 3 - 2x \qquad \text{Derived equation.}$$
$$y = 3 - 2(2) \qquad \text{Replace } x \text{ by 2.}$$
$$= -1 \qquad \text{Simplify.}$$

Finally, check the solution $(2, -1)$ in both of the *original* equations.

$$2x + y = 3 \qquad\qquad 3x - 4y = 10$$
$$2(2) + (-1) \overset{?}{=} 3 \qquad 3(2) - 4(-1) \overset{?}{=} 10$$
$$3 = 3 \quad \text{True} \qquad\qquad 10 = 10 \quad \text{True}$$

Thus, the solution is $(2, -1)$.

Caution   When students solve a linear system of equations, a common error is to solve for only one variable and write answers like $x = 2$. Remember that you must find *both* coordinates of a solution.

PROGRESS CHECK 3   Solve by substitution:

$$2x - 3y = 8$$
$$x + 2y = -3.$$

⌐

Progress Check Answers
2. 20
3. $(1, -2)$

23. $-2x + 7y = 13$
    $8x - 3y = -2$

24. $-3x + 5y = 14$
    $9x - 2y = -3$

25. $-5x - 3y = 6$
    $10x - 3y = 0$

26. $-4x - 3y = 5$
    $8x + 3y = -8$

27. $2x + 3y + 1 = 3x + 4y - 2$
    $x - y = 2x - 3y$

28. $3x + 4y + 5 = 4x + 5y$
    $2x - y = 3x - 2y - 1$

29. $\frac{2}{3}x + \frac{3}{4}y = 2$
    $\frac{4}{3}x + \frac{3}{8}y = \frac{5}{2}$

30. $\frac{1}{2}x + \frac{2}{3}y = 1$
    $\frac{1}{4}x + \frac{1}{6}y = \frac{1}{3}$

In Exercises 31–38, determine whether the system is inconsistent or dependent.

31. $2x - y = 1$
    $-2x + y = 1$

32. $3x + 2y = 6$
    $3x + 6 = -2y$

33. $4x - y = 3$
    $\frac{4}{3}x = \frac{1}{3}y + 1$

34. $x - 3y = 1$
    $6y + 2 = 2x$

35. $3x + 2y = 12$
    $x + \frac{2}{3}y = 4$

36. $y = 4x - 6$
    $\frac{1}{2}y = 2x - 3$

37. $y = 2x + 4$
    $\frac{1}{2}y = x + 1$

38. $2y = 3x - 5$
    $y = \frac{3}{2}x - 5$

In Exercises 39–42 a word problem is given. Use the substitution method to solve the appropriate linear system.

39. One video store rents tapes for an annual membership of $10 plus $2.50 per rental. The total annual cost (y) is then given by $y = 10 + 2.50x$ if x tapes are rented during the year. For nonmembers, the cost is $3.00 per tape, so annual cost is given by $y = 3x$. How many tapes must be rented annually for the cost to be the same for members and nonmembers? Which plan should you use if you think you will rent two tapes per month?

40. Redo Exercise 39, but assume that the membership now costs $13. This means that for members the total cost is given by $y = 13 + 2.5x$.

41. Economists have made mathematical models which project production of various resources over time. For country A, the production index is given by $y = 100 + 6x$, where x is the number of years elapsed from the present. For country B, the index is given by $y = 110 + 5x$. According to these indexes, how many years will it take for country A to catch up to country B?

42. Refer to Exercise 41. Country C has a production index given by $y = 90 + 5.5x$. How long will it take country C to catch up to country B?

THINK ABOUT IT

1. In a research laboratory a piece of metal is being heated. It is at 0 degrees Celsius at the beginning of the experiment and its temperature increases 3 degrees per minute. Starting at the same time, another piece of metal is being cooled down from 100 degrees at the rate of 5 degrees per minute.
   a. Express each temperature as a function of elapsed time to get a system of linear equations.
   b. How long will it take before the two pieces are the same temperature?
   c. At how many degrees will they be the same temperature?

2. Refer to Problem 1. At how many instants will the difference between the temperatures of the two pieces be 10 degrees? Find all such times and the corresponding temperatures.

3. a. Use the method of substitution to find the point of intersection of the lines $y = ax + b$ and $y = cx + d$.
   b. For what values of a, b, c, and d will the lines not intersect at all?

4. The lines $2x + 3y = 4$ and $3x + 4y = 5$ intersect at $(-1,2)$; so do the lines $3x + 4y = 5$ and $4x + 5y = 6$. Is it true that $ax + (a + 1)y = a + 2$ and $(a + 1)x + (a + 2)y = a + 3$ will always intersect at $(-1,2)$, no matter what value a is? Solve the system to find out.

5. Consider the system
$$y = x^2$$
$$y - x = 2.$$
   a. Why is this *not* a *linear* system?
   b. The two equations represent a parabola (see Section 9.5) and a line which intersect at two points. Solve the system by substitution. You will get a quadratic equation with two solutions. Solve it to find *both* points of intersection.

REMEMBER THIS

1. Add $x + 2y$ and $x - 2y$.
2. Multiply both sides of $2x - 7y = 1$ by -1.
3. Solve $2x + 3y = -9$ for x if $y = -1$.
4. Solve $0.15x + 9,000 = 30,000$.
5. In the expression $3x - y + 7$, what is the coefficient of y?
6. Why isn't the pair $(2,1)$ a solution to the system
$$3x + 2y = 8$$
$$3x - 2y = 7?$$

7. What is the equation for the line shown in the figure?

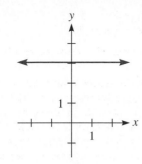

8. If $f(x) = 2x^2 - 1$, find $f(0)$.

9. Add $\dfrac{1}{x + 1} + \dfrac{2}{x + 2}$.

10. Solve $A = \dfrac{b + c}{d}$ for b.

7.3 Solving Linear Systems by the Addition-Elimination Method

Y ou win $30,000 in a state lottery and are advised to put some of the money in a risky investment that earns 15 percent annual interest and the rest in a safe investment that earns 7 percent interest. If you want the total interest to be $3,000 annually, then solving the system

$$x + y = 30{,}000$$
$$0.15x + 0.07y = 3{,}000$$

gives the amount to invest at each rate. Find x and y, which represent the amount invested in the risky and safe investments, respectively. (See Example 4.)

OBJECTIVES

1 Solve linear systems by addition-elimination for the case where one variable can be eliminated immediately.

2 Solve linear systems by addition-elimination when the multiplication property of equality is needed before one variable can be eliminated.

3 Solve linear systems that are inconsistent or dependent by the addition-elimination method.

4 Solve linear systems after choosing an efficient method.

1 Another algebraic method for solving linear systems of equations is based on the property that

if $A = B$ and $C = D$, then $A + C = B + D$.

In words, adding equal quantities to equal quantities results in equal sums. For this method to result in the elimination of a variable, the coefficients of either x or y must be opposites, as in the system in Example 1.

EXAMPLE 1   Solve by addition-elimination:

$$x + y = 9$$
$$x - y = 3.$$

Solution   In this system the coefficients of y are 1 and -1. If we add the left-hand sides and the right-hand sides of the two equations, the result is an equation that contains only one unknown.

$$
\begin{array}{r}
x + y = 9 \\
\underline{x - y = 3} \\
2x \quad\;\; = 12
\end{array}
$$

Now, we can solve $2x = 12$ for x.

$$2x = 12$$
$$x = 6$$

To find the y-coordinate, substitute 6 for x in either of the given equations.

$x + y = 9$	or	$x - y = 3$
$6 + y = 9$	or	$6 - y = 3$
$y = 3$	or	$-y = -3$
		$y = 3$

Thus, the solution is (6,3). Check the solution in the usual way.

PROGRESS CHECK 1   Solve by addition-elimination:

$$x + 2y = 11$$
$$x - 2y = -1.$$
⌐

2   We often first need to use the multiplication property to form equivalent equations in order to eliminate a variable when adding both sides of the equations.

EXAMPLE 2   Solve by addition-elimination:

$$2x + 3y = -9$$
$$2x - 7y = 1.$$

Solution   In both equations the coefficient of x is 2. If we multiply on both sides of either equation by -1, then we will obtain x terms with opposite coefficients. We choose to do this in the bottom equation, which gives

$$-1(2x - 7y) = -1(1) \rightarrow \quad \begin{matrix} 2x + 3y = -9 \\ -2x + 7y = -1. \end{matrix}$$

Now, we can proceed as in Example 1.

$$\begin{matrix} 2x + 3y = -9 \\ \underline{-2x + 7y = -1} \\ 10y = -10 \quad \text{Add left and right sides.} \\ y = -1 \quad \text{Divide both sides by 10.} \end{matrix}$$

To find x, replace y by -1 in one of the original equations.

$2x + 3y = -9$	
$2x + 3(-1) = -9$	Substitute -1 for y.
$2x - 3 = -9$	Simplify.
$2x = -6$	Add 3 to both sides.
$x = -3$	Divide both sides by 2.

The solution (which you should check in both *original* equations) is $(-3,-1)$.

PROGRESS CHECK 2   Solve by addition-elimination:

$$4x + 3y = 1$$
$$7x + 3y = 13.$$
⌐

The system in the next example requires us to rewrite both equations so that adding will eliminate a variable. It will simplify the discussion to label equations with numbers.

EXAMPLE 3   Solve by addition-elimination:

$$6x + 6y = 7 \qquad (1)$$
$$8x - 9y = -2. \qquad (2)$$

Solution   We choose to eliminate the x variable, so the initial goal is to obtain two equations with opposite x terms. Multiplying both sides of equation (1) by 4 and both sides of equation (2) by -3 gives

$$4(6x + 6y) = 4(7) \rightarrow \quad 24x + 24y = 28 \qquad (3)$$
$$-3(8x - 9y) = -3(-2) \rightarrow \quad -24x + 27y = 6. \qquad (4)$$

Now, add equations (3) and (4).

$$24x + 24y = 28$$
$$\underline{-24x + 27y = 6}$$
$$51y = 34 \quad \text{Add.}$$
$$y = \tfrac{34}{51} \quad \text{Divide both sides by 51.}$$
$$= \tfrac{2}{3} \quad \text{Simplify.}$$

To find x, we can replace y by $\tfrac{2}{3}$ in equation (1).

$$6x + 6y = 7$$
$$6x + 6(\tfrac{2}{3}) = 7 \quad \text{Replace } y \text{ by } \tfrac{2}{3}.$$
$$6x + 4 = 7 \quad \text{Simplify.}$$
$$6x = 3 \quad \text{Subtract 4 from both sides}$$
$$x = \tfrac{1}{2} \quad \text{Divide both sides by 6.}$$

The solution is $(\tfrac{1}{2}, \tfrac{2}{3})$. Check the solution in equations (1) and (2).

Note   Sometimes the determined value for one of the solution coordinates is a fraction or a decimal that is awkward to use for finding the other coordinate. In such cases, it may be easier to find the remaining coordinate value by returning to the original system of equations. For instance, in this example, after finding y, we may find x as follows.

$$3(6x + 6y) = 3(7) \rightarrow \quad 18x + 18y = 21$$
$$2(8x - 9y) = 2(-2) \rightarrow \quad \underline{16x - 18y = -4}$$
$$34x = 17$$
$$x = \tfrac{17}{34}, \text{ or } \tfrac{1}{2}$$

PROGRESS CHECK 3   Solve by addition-elimination:

$$6x + 10y = 7$$
$$15x - 4y = 3. \qquad \lrcorner$$

Fractions or decimals that appear in a linear system are eliminated first by multiplying both equation sides by some number. In the case of fractions, multiply using the least common denominator. For decimals, multiply using a multiple of 10 that converts all decimals to integers.

EXAMPLE 4   Solve the problem in the section introduction on page 333.

Solution x and y represent the amounts invested in the risky and safe investments, respectively, and the system of equations given in the problem is

$$x + y = 30{,}000 \qquad (1)$$
$$0.15x + 0.07y = 3{,}000. \qquad (2)$$

First, we multiply both sides of equation (2) by 100 to clear decimals.

$$100(0.15x + 0.07y) = 100(3{,}000) \rightarrow \quad 15x + 7y = 300{,}000 \qquad (3)$$

Progress Check Answer

3. $(\tfrac{1}{3}, \tfrac{1}{2})$

If we now multiply both sides of equation (1) by -7 and add the result to equation (3), we can eliminate y and determine x.

$$-7(x + y) = -7(30,000) \rightarrow \quad \begin{array}{rl} -7x - 7y = & -210,000 \\ 15x + 7y = & 300,000 \end{array} \quad \begin{array}{l} (4) \\ (3) \end{array}$$

$$\begin{array}{rl} 8x = & 90,000 \\ x = & 11,250 \end{array}$$

To find y, we can substitute 11,250 for x in equation (1).

$$\begin{array}{r} x + y = 30,000 \\ 11,250 + y = 30,000 \\ y = 18,750 \end{array}$$

Thus, you should invest \$11,250 in the risky investment and \$18,750 in the safe investment.

Note Analyzing a problem using two variables and a system of equations is often more natural than restricting the solution to just one unknown. Compare this example to Example 3 in Section 2.7.

PROGRESS CHECK 4 Redo the problem in Example 4, but assume that the risky investment earns 16 percent annual interest and the safe investment earns 6 percent annual interest. The associated system is

$$\begin{array}{r} x + y = 30,000 \\ 0.16x + 0.06y = 3,000. \end{array}$$ ⌐

3 As in the substitution method, if we eliminate a variable and obtain an equation with no solution, then the system is inconsistent. If the resulting equation is always true, the system is dependent.

EXAMPLE 5 Solve by addition-elimination:

$$\begin{array}{rl} 2x - 3y = 1 & \quad (1) \\ 6x - 9y = 4. & \quad (2) \end{array}$$

Solution If we multiply both sides of equation (1) by -3 and add the result to equation (2), both x and y are eliminated.

$$-3(2x - 3y) = -3(1) \rightarrow \quad \begin{array}{rl} -6x + 9y = & -3 \\ 6x - 9y = & 4 \end{array} \quad \begin{array}{l} (3) \\ (2) \end{array}$$

$$0 = 1 \quad \text{Add.}$$

The false equation $0 = 1$ indicates that there is no solution and the system is inconsistent.

PROGRESS CHECK 5 Solve by addition-elimination:

$$\begin{array}{r} -4x + 2y = 3 \\ 2x - y = 0. \end{array}$$ ⌐

EXAMPLE 6 Solve by addition-elimination:

$$\begin{array}{rl} -\tfrac{3}{2}x + y = -\tfrac{7}{2} & \quad (1) \\ x - \tfrac{2}{3}y = \tfrac{7}{3}. & \quad (2) \end{array}$$

Solution First, clear the equations of fractions by multiplying both sides of equation (1) by 2 and both sides of equation (2) by 3.

$$2(-\tfrac{3}{2}x + y) = 2(-\tfrac{7}{2}) \rightarrow \quad -3x + 2y = -7 \quad (3)$$
$$3(x - \tfrac{2}{3}y) = 3(\tfrac{7}{3}) \rightarrow \quad 3x - 2y = 7 \quad (4)$$

Now if we add the resulting equations, both variables are eliminated.

$$-3x + 2y = -7 \qquad (3)$$
$$3x - 2y = 7 \qquad (4)$$
$$0 = 0 \quad \text{Add.}$$

The equation $0 = 0$, which is always true, indicates that there are an infinite number of solutions and the system is dependent.

PROGRESS CHECK 6   Solve by addition-elimination:

$$\tfrac{3}{4}x - y = \tfrac{9}{4}$$
$$-x + \tfrac{4}{3}y = -3.$$

⌐

4   We have seen that a linear system of equations can be solved by graphing, by substitution, and by addition-elimination. Keep in mind the following guidelines concerning the efficient use of each method.

Graphing   The graphing method is useful for *estimating* the solution and for *comparing* visually the linear equations in a system. When solving systems, we usually use it in conjunction with one of the algebraic methods.

Substitution   The substitution method is most efficient when at least one equation is solved for one of the variables. It is also a good choice when the system contains a variable with a coefficient of 1 or -1. We avoid this method when it leads to significant work with fractions.

Addition-Elimination   The addition-elimination method is usually the easiest when neither equation is solved for one of the variables. If you have trouble choosing, select this method.

EXAMPLE 7   Solve using addition-elimination or substitution:

$$5x - 6y = 8 \qquad (1)$$
$$3x = 4y + 6. \qquad (2)$$

Solution   Neither equation is solved for one of the variables and neither equation has a variable with a coefficient of 1 or -1. Therefore, we choose the addition-elimination method. To write the equations in the same form, usually $ax + by = c$, subtract $4y$ from each side in equation (2) to obtain the system

$$5x - 6y = 8 \qquad (1)$$
$$3x - 4y = 6. \qquad (3)$$

Multiplying equation (1) by 3 and equation (3) by -5 enables us to eliminate x.

$$3(5x - 6y) = 3(8) \rightarrow \qquad 15x - 18y = 24 \qquad (4)$$
$$-5(3x - 4y) = -5(6) \rightarrow \qquad -15x + 20y = -30 \qquad (5)$$
$$2y = -6 \quad \text{Add.}$$
$$y = -3$$

We choose to find x by replacing y by -3 in equation (2).

$$3x = 4y + 6$$
$$3x = 4(-3) + 6$$
$$3x = -6$$
$$x = -2$$

Thus, the solution is $(-2, -3)$. Check this solution in equations (1) and (2).

Progress Check Answer
6. Same line; infinitely many solutions; dependent system

PROGRESS CHECK 7   Solve using addition-elimination or substitution:

$$2x + 3y = -21$$
$$y = -3x.$$

⌐

The addition-elimination method of this section is summarized next.

Addition-Elimination Method Summary

To solve a linear system by addition-elimination:

1. If necessary, write both equations in the form $ax + by = c$.
2. If necessary, multiply one or both equations by numbers that make the coefficients of either x or y opposites of each other. Add the two equations to eliminate a variable.
3. Solve the equation from step 2. A unique solution gives the value of one variable. If the equation is $0 = n$, where $n \neq 0$, the system is inconsistent. If the equation is $0 = 0$, the system is dependent.
4. Use the known coordinate value to find the other coordinate value through substitution in either of the original equations.
5. Check the solution in each of the original equations.

Progress Check Answer

7. $(3, -9)$

EXERCISES 7.3

In Exercises 1–32, solve the system using the addition-elimination technique.

1. $x + y = 8$
$\quad x - y = 4$

2. $x + y = 10$
$\quad x - y = 8$

3. $-x + y = 3$
$\quad x + y = 1$

4. $-x + 2y = 4$
$\quad x + y = -1$

5. $3x + 2y = -12$
$\quad x - 2y = 4$

6. $\quad 3x - 2y = -10$
$\quad -3x - y = 13$

7. $\quad x - 3y = 14$
$\quad 2x + 3y = 1$

8. $\quad 2x - y = 11$
$\quad -2x - 3y = 9$

9. $2x + 3y = 5$
$\quad 2x - 5y = -3$

10. $3x + 2y = 10$
$\quad 3x - 4y = -2$

11. $2x + 6y = 3$
$\quad 4x + 6y = 4$

12. $3x + 5y = 3$
$\quad 6x + 5y = 4$

13. $2x + 3y = 7$
$\quad 4x - 5y = -8$

14. $3x + 4y = 13$
$\quad 6x - 5y = -13$

15. $3x + 2y = 1$
$\quad 9x + 4y = 0$

16. $\quad 4x - 9y = -15$
$\quad 12x + 3y = -5$

17. $2x + 5y = 1$
$\quad 3x - 2y = 11$

18. $3x + 5y = -4$
$\quad 4x - 3y = -15$

19. $\quad 2x + 7y = -10$
$\quad -3x + 5y = 15$

20. $-5x - 2y = 4$
$\quad 4x + 3y = -6$

21. $\frac{1}{2}x + \frac{2}{3}y = 5$
$\quad \frac{2}{3}x + \frac{1}{9}y = 2$

22. $\frac{1}{3}x - \frac{3}{2}y = 1$
$\quad \frac{4}{3}x - \frac{1}{2}y = -7$

23. $0.25x + 0.1y = 2$
$\quad 3x - 2y = -8$

24. $1.3x - 2.4y = 38$
$\quad 1.2x - 2.2y = 35$

25. $6x + 8y = 123x$
$\quad 3x + 4y = 9$

26. $\quad 5x + 2y = 3$
$\quad -10x - 4y = 6$

27. $\frac{1}{2}x - y = \frac{3}{2}$
$\quad x - 2y = 1$

28. $\frac{1}{3}x - \frac{2}{3}y = -\frac{1}{3}$
$\quad x - 2y = 2$

29. $2x + 3y = -3$
$\quad 4x + 6y = -6$

30. $3x + 5y = -1$
$\quad 9x + 15y = -3$

31. $8x - 10y = -2$
$\quad 4x - 5y = -1$

32. $12x - 9y = 6$
$\quad 4x - 3y = 2$

In Exercises 33–38, solve the given system and answer the questions. These are similar to problems solved using one variable in Section 2.7.

33. A total investment of $50,000 is to be split into two bank accounts, one earning 3.5 percent annual interest and the other 6 percent. In order that the combined interest be $2,500, how much should be invested in each account? The system of equations to be solved is

$$x + y = 50{,}000$$
$$0.035x + 0.06y = 2{,}500.$$

34. Redo Exercise 33, but assume that the total investment is $60,000. Therefore, solve

$$x + y = 60{,}000$$
$$0.035x + 0.06y = 2{,}500.$$

35. Find two numbers whose sum is 10 and whose difference is 17. The associated system is

$$x + y = 10$$
$$x - y = 17.$$

36. Find two numbers whose sum is $\frac{5}{8}$ and whose difference is $\frac{13}{8}$. Solve:

$$x + y = \frac{5}{8}$$
$$x - y = \frac{13}{8}.$$

37. To find out how many ounces of a 10 percent acid solution and a 7 percent solution should be mixed to achieve 24 oz of an 8 percent solution, solve the system

$$x + y = 24$$
$$0.07x + 0.10y = 0.08(24).$$

38. One solution is 40 percent alcohol and another is 50 percent alcohol. How much of each should be used to get 100 milliliters (ml) of a solution which is 48 percent alcohol? Solve:

$$x + y = 100$$
$$0.40x + 0.50y = 0.48(100)$$

THINK ABOUT IT

1. A prize of $44,000 is split and invested in two funds. One earns 2.5 percent, the other 6 percent.
   a. How should the investment be split so that the interest from the 6 percent account is twice the interest from the other account? Set up the system of equations and solve it.
   b. What will the total earned interest be?
2. In Problem 1, how should the money be split so that the interest from the 2.5 percent account is twice the interest from the 6 percent account? Round to the nearest dollar.
3. a. Show that $x = \dfrac{ce - bf}{ae - bd}$ and $y = \dfrac{af - cd}{ae - bd}$ gives the solution to

$$ax + by = c$$
$$dx + ey = f.$$

To do this, substitute these expressions for x and y in the two equations and show that an identity results.
   b. Apply the formula to solve

$$2x + 3y = 4$$
$$3x + 4y = 5.$$

4. Solve: $\sqrt{2}x + 3y = 11$
$$\sqrt{2}x - y = -1.$$
5. For what values of a does the following system have exactly one solution? What is this solution in terms of a?

$$2x + 3y = 5$$
$$ax + y = 1$$

REMEMBER THIS

1. What is the sum of the angle measures in a triangle?
2. What is the formula for the perimeter of a rectangle?

3. If x represents total sales, and a salesperson's commission is 10 percent of total sales, what expression represents the dollar amount of the commission?
4. Multiply both sides of $0.03x + 0.06y = 0.6$ by 100.

5. Simplify $2(x + y) + 2(x - y)$.
6. Solve by substitution:

$$5x + 3y = 5$$
$$x = 1 - y.$$

7. Is this system consistent or inconsistent?

$$y = 2x + 4$$
$$y = 2x + 5$$

8. Graph $15x + 5y \le 45$, assuming that $x \ge 0$ and $y \ge 0$.

9. Factor $c^2 + 4c + 4$.
10. Solve $x(x - 5) = 0$.

7.4 Additional Applications of Linear Systems

A machine shop has two large containers that are each filled with a mixture of oil and gasoline. Container A contains 3 percent oil (and 97 percent gasoline). Container B contains 6 percent oil (and 94 percent gasoline). How much of each should be used to obtain 12 qt of a new mixture that contains 5 percent oil? (See Example 5.)

OBJECTIVES

Set up and solve linear systems to solve problems involving the following:

1. **Geometric figures**
2. **Comparisons of linear equations**
3. **Annual interest**
4. **Liquid mixtures**
5. **Uniform motion**

In this chapter the applied problems up to this point have given the linear system needed to analyze the problem. However, in this section we consider the broader objective of analyzing a variety of word problems, where it is necessary to both set up and solve a linear system. To solve a word problem using a system of equations, use the following steps, which are a modification of the word problem procedures first given in Section 2.6.

To Solve Word Problems Using Linear Systems

1. Read the problem several times. If possible, display the given information in a sketch or chart.
2. Let two variables represent two unknown quantities. Write down precisely what each variable represents.
3. Set up a system of linear equations that expresses two distinct relationships between the two variables in the problem.
4. Solve the system of equations.
5. Answer the question.
6. Check the answers by interpreting the solution in the context of the word problem.

---

1 Perimeter, area, or angle measure formulas are often needed to solve problems involving geometric figures. You can consult Section 1.3 (as needed) for a review of many such formulas.

EXAMPLE 1 Two angles are complementary. If the difference between the angle measures is 26°, find the angle measures.

Solution First, represent the unknown angle measures. We let

$$x = \text{larger angle measure}$$
$$y = \text{smaller angle measure}.$$

Angle measures of complementary angles add up to 90°, so

$$x + y = 90.$$

Because the difference between the angle measures is 26°,

$$x - y = 26.$$

These two equations give the system

$$x + y = 90 \qquad\qquad (1)$$
$$x - y = 26. \qquad\qquad (2)$$

To find both angle measures, we need both system coordinates. Use the addition-elimination method.

$$
\begin{aligned}
x + y &= 90 \\
x - y &= 26 \\
\hline
2x \phantom{{}+y} &= 116 \qquad \text{Add.}\\
x \phantom{{}+y} &= 58 \qquad \text{Divide both sides by 2.}
\end{aligned}
$$

To find y, we can replace x by 58 in equation (1).

$$
\begin{aligned}
x + y &= 90 \\
58 + y &= 90 \qquad \text{Replace } x \text{ by 58.}\\
y &= 32 \qquad \text{Subtract 58 from each side.}
\end{aligned}
$$

Thus, the angle measures are 58° and 32°.

Check   Because 58° + 32° = 90°, the angles are complementary. Since their difference, 58° − 32°, is 26°, the solution checks.

PROGRESS CHECK 1   If the difference between two complementary angles is 14°, find their angle measures.   ⌐

Many problems that were solved in Chapter 2 using one variable are also readily solved using two variables and a system of equations. Compare the next example to the solution in Example 1 of Section 2.7.

EXAMPLE 2   A piece of molding is 75 in. long. You wish to use it to make a rectangular frame that is twice as long as it is wide. What are the dimensions of the frame?

Solution   To find the dimensions of the frame, let

$$x = \text{width}$$
$$y = \text{length}.$$

Now, illustrate the given information, as in Figure 7.7.

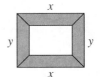

Figure 7.7

Because the molding is 75 in. long, the perimeter of the rectangle is 75 in., and equation (1) results from using $P = 2\ell + 2w$, the formula for the perimeter of a rectangle.

$$2x + 2y = 75 \qquad\qquad (1)$$

Relating the length and the width gives equation (2).

The length   is   twice the width.

$$y \qquad = \qquad 2x \qquad\qquad (2)$$

The resulting system,

$$2x + 2y = 75 \qquad\qquad (1)$$
$$y = 2x, \qquad\qquad (2)$$

is solved efficiently by the substitution method. Replace y by $2x$ in equation (1).

Progress Check Answer
1. 38°, 52°

$$2x + 2y = 75$$
$$2x + 2(2x) = 75 \quad \text{Substitute } 2x \text{ for } y.$$
$$6x = 75 \quad \text{Combine like terms.}$$
$$x = \tfrac{75}{6} \quad \text{Divide both sides by 6.}$$
$$= 12.5 \quad \text{Simplify.}$$

Then, replace x by 12.5 in equation (2).

$$y = 2(12.5) = 25$$

Thus, the width of the frame is 12.5 in. and the length is 25 in.

Check   The perimeter is $2(25) + 2(12.5) = 75$ in., while 25 in. is twice as long as 12.5 in. The solution checks.

PROGRESS CHECK 2   If the perimeter of a rectangle is 40 m and the length is 3 m greater than the width, find the dimensions of the rectangle.   ⌐

2   When the analysis of a problem involves the comparison of two linear equations, then it is useful to solve the linear system using both algebraic and graphing methods, as shown next.

EXAMPLE 3   You are trying to decide between two positions as a salesperson. The first offer pays $150 per week plus a 10 percent commission on sales, while the second offer pays $120 per week plus a 12 percent commission on sales.

  a. How much must you sell each week for the two jobs to pay the same?
  b. What is the weekly salary in this case?
  c. When sales exceed the number found in part **a**, which offer is better?

Solution   Let x represent the weekly amount of sales and y represent the weekly income. Then, translate each offer to a linear equation in two variables.

First offer: Income equals $150 plus 10 percent of sales.

$$y = 150 + 0.10x \tag{1}$$

Second offer: Income equals $120 plus 12 percent of sales.

$$y = 120 + 0.12x \tag{2}$$

The resulting system is

$$y = 150 + 0.10x \tag{1}$$
$$y = 120 + 0.12x. \tag{2}$$

a. To find x, replace y by $120 + 0.12x$ in equation (1).

$$120 + 0.12x = 150 + 0.10x$$
$$0.12x = 30 + 0.10x \quad \text{Subtract 120 from both sides.}$$
$$0.02x = 30 \quad \text{Subtract 0.10x from both sides.}$$
$$x = 1{,}500 \quad \text{Divide both sides by 0.02.}$$

The two jobs pay the same when weekly sales are $1,500.

b. To find y, replace x by 1,500 in either equation of the system.

$$y = 150 + 0.10x \qquad \text{or} \qquad y = 120 + 0.12x$$
$$y = 150 + 0.10(1{,}500) \qquad\qquad y = 120 + 0.12(1{,}500)$$
$$= 300 \qquad\qquad = 300$$

The weekly salary is $300 when the two jobs pay the same.

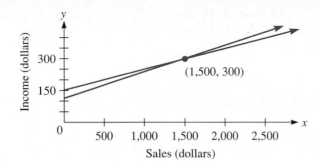

Figure 7.8

c. To compare the two offers, solve the system graphically, as in Figure 7.8. Because the *y* values are higher on the line associated with the second offer beyond the intersection point, the second offer is better when sales exceed $1,500.

Check The solutions shown in parts **b** and **c** confirm that at $1,500 in weekly sales both jobs pay $300. After this point, a higher commission rate results in a higher income. The solution checks.

PROGRESS CHECK 3   Answer the questions in Example 3, but assume that the first offer is a straight 15 percent commission on sales, while the second offer pays $140 per week plus an 8 percent commission on sales.   ⌐

3   To solve problems involving annual interest, use $I = Pr,$ where *I* represents annual interest, *P* represents principal, and *r* represents the annual interest rate. In our remaining examples it is helpful to analyze each problem in a chart format.

EXAMPLE 4   The board members of an employees' pension plan have $8 million to invest. They will split it into two accounts, one earning 9 percent annual interest and the other earning 7 percent. How much should they put into each account so that the total earnings are $650,000?

Solution   To find the amount to invest at each rate, we let

$$x = \text{amount invested at 9 percent}$$
$$y = \text{amount invested at 7 percent.}$$

Now we use $I = Pr$ and analyze each investment in a chart format.

Investment	Principal	·	Interest Rate	=	Interest
1st account	*x*		0.09		0.09*x*
2nd account	*y*		0.07		0.07*y*

The two principals add up to $8 million, and the sum of the interests from the two accounts is to be $650,000. So the required linear system is

$$x + y = 8,000,000 \tag{1}$$
$$0.09x + 0.07y = 650,000. \tag{2}$$

To solve this system, we first multiply both sides of equation (2) by 100 to clear decimals.

$$x + y = 8,000,000 \tag{1}$$
$$9x + 7y = 65,000,000 \tag{3}$$

If we now multiply both sides of equation (1) by -7 and add the result to equation (3), we can eliminate y and determine x.

$$-7x - 7y = -56,000,000$$
$$9x + 7y = 65,000,000$$

$2x$	$= 9,000,000$	Add.
x	$= 4,500,000$	Divide both sides by 2.

To find y, we can substitute 4,500,000 for x in equation (1).

$x + y = 8,000,000$	
$4,500,000 + y = 8,000,000$	Replace x by 4,500,000.
$y = 3,500,000$	Subtract 4,500,000 from each side.

Thus, the board members should invest $4,500,000 (or $4.5 million) in the 9 percent account and $3,500,000 (or $3.5 million) in the 7 percent account.

Check   The first account earns $405,000 (or 9 percent of $4.5 million), and the second account earns $245,000 (or 7 percent of $3.5 million). Thus, the total interest is $405,000 + $245,000 = $650,000, so the solution checks.

PROGRESS CHECK 4   A $16,000 retirement investment is split into two investments, one earning 11 percent annual interest and the other earning 6 percent. If the total annual interest was $1,400, how much was invested at each rate?   ⌐

4   As in Section 2.7, liquid mixture problems are analyzed using

$$\begin{pmatrix} \text{percent of} \\ \text{an ingredient} \end{pmatrix} \cdot \begin{pmatrix} \text{amount of} \\ \text{solution} \end{pmatrix} = \begin{pmatrix} \text{amount of} \\ \text{the ingredient} \end{pmatrix}.$$

Compare the solution of the section-opening problem that follows with the solution in Example 6 of Section 2.7.

EXAMPLE 5   Solve the problem in the section introduction on page 340.

Solution   To find the correct mixture, let

$$x = \text{amount used from container A}$$
$$y = \text{amount used from container B}.$$

Once again, we analyze the problem using a chart.

Solution	Percent Oil $\cdot$	Amount of Solution (qt)	= Amount of Oil (qt)
Container A	3	x	$0.03x$
Container B	6	y	$0.06y$
New Solution	5	12	0.05(12), or 0.6

To set up a system of equations, we first note that x qt from container A combine with y qt from container B to form 12 qt of the new solution, so

$$x + y = 12. \tag{1}$$

To obtain a second equation, we reason that the amount of oil in the new solution is the sum of the amounts contributed by the solutions from containers A and B.

Amount of oil from container A	plus	amount of oil from container B	equals	amount of oil in new solution.	
$\underbrace{0.03x}$	$+$ ↓	$\underbrace{0.06y}$	$=$ ↓	$\underbrace{0.6}$	(2)

To solve the resulting system, we first multiply both sides of equation (2) by 100 to clear decimals.

$$100(0.03x + 0.06y) = 100(0.6) \rightarrow \quad \begin{array}{ll} x + y = 12 & (1) \\ 3x + 6y = 60 & (3) \end{array}$$

If we now multiply both sides of equation (1) by -3 and add the result to equation (3), we can eliminate x and determine y.

$$\begin{array}{rl} -3x - 3y = -36 & \\ \underline{3x + 6y = 60} & \\ 3y = 24 & \text{Add.} \\ y = 8 & \text{Divide both sides by 3.} \end{array}$$

To find x, replace y by 8 in equation (1).

$$\begin{array}{rl} x + y = 12 & \\ x + 8 = 12 & \text{Add.} \\ x = 4 & \text{Subtract 8 from each side.} \end{array}$$

Thus, 4 qt from container A should be mixed with 8 qt from container B.

Check The new solution in total contains $4 + 8 = 12$ qt. Also, 4 qt from container A contain $0.03(4) = 0.12$ qt of oil, while 8 qt from container B contain $0.06(8) = 0.48$ qt of oil. Thus, the new mixture contains 0.60 qt of oil in 12 qt of mixture. Because $0.60/12 = 0.05 = 5$ percent, the new mixture does contain 5 percent oil and the solution checks.

PROGRESS CHECK 5 A chemist has two acid solutions, one 30 percent acid and the other 70 percent acid. How much of each must be used to obtain 10 liters of a solution that is 41 percent acid? ⌐

5 Our current methods may also be used to solve any of the uniform motion problems first considered in Section 2.7. In addition, the next example shows a uniform motion problem that requires two variables and a system of equations. You need to recall the formula $d = rt$, where d represents the distance traveled in time t by an object moving at a constant rate r.

EXAMPLE 6 It takes a boat 1 hour to go 12 mi downstream (with the current); the return trip upstream (against the current) takes 2 hours. What is the speed of the current? What would be the speed of the boat if there were no current?

Solution Let

$$x = \text{speed of the boat in still water}$$
$$y = \text{speed of the current.}$$

The downstream rate is then $x + y$ and the upstream rate is $x - y$. We can summarize the key facts about each trip as follows.

Direction	Distance	Rate	Time
Downstream	12	$x + y$	1
Upstream	12	$x - y$	2

Progress Check Answer

5. 7.25 liters of 30 percent solution and 2.75 liters of 70 percent solution

THINK ABOUT IT

1. In a certain store the average value of a sale is $50. In this store job A pays $200 per week plus $10 per sale. Job B pays $300 per week plus a 10 percent commission on each sale.
   a. For what number of sales would the two jobs be expected to pay the same?
   b. What is the pay for the number of sales in part **a?**
2. An author is offered two contracts. For one, her royalty rate is 12 percent of the retail price of the book. For the other it is 15 percent of the wholesale price.
   a. Which contract is better when the wholesale price is 40 percent less than the retail price?
   b. What relationship must there be between the retail and wholesale prices for these two contracts to be equivalent?

3. Two solutions are being mixed. One is 10 percent acid; the other is 15 percent acid. In what ratio should they be mixed to create a 12 percent solution?
4. It is not possible to mix a 5 percent acid solution with a 10 percent acid solution to get a 15 percent solution. Show this by setting up a problem where you try to determine how much of each you should use to get 60 oz of 15 percent solution.
5. Job A starts at $150/week while job B starts at $350/week. The commission on job B is 80 percent of the commission on job A. What must the two commission rates be so that the two jobs pay the same for $5,000 in sales?

REMEMBER THIS

1. Graph the solution set for $y > 2x$.

2. Does the point $(0,0)$ satisfy *both* of these inequalities?
$$4x + y \le 8$$
$$y \ge 2x$$
3. What inequality results when you multiply both sides of the inequality $2x - y < 6$ by -1?
4. Which of these inequalities corresponds to the words "x is at least 10"?
   a. $x \le 10$ **b.** $x \ge 10$ **c.** $x < 10$ **d.** $x > 10$

5. Which of these inequalities corresponds to the words "x is no more than 15"?
   a. $x \le 15$ **b.** $x \ge 15$ **c.** $x < 15$ **d.** $x > 15$
6. Solve.
$$2x - 5y = 8$$
$$3x + 5y = -13$$
7. If two lines are parallel, then they have the same
   a. slope   **b.** x-intercept   **c.** y-intercept   **d.** graph

8. What is the GCF for these expressions: x^2y^2, xy, $2y$?
9. What is the zero product principle?

10. What property is illustrated by $n(a + b) = (a + b)n$?

7.5   Solving Systems of Linear Inequalities

For a party you need to buy some ham at $8 per pound and some cheese at $4 per pound. You will need *at least* 1 lb of ham and *at least* 2 lb of cheese; and you wish to spend *no more than* $32 on these two items.
   a. Translate these requirements into a system of linear inequalities.
   b. Solve the system in part **a** graphically.
   c. Specify and interpret one solution that is in the solution set. (See Example 3.)

OBJECTIVES

1 Solve systems of linear inequalities graphically.

2 Solve word problems that translate to systems of linear inequalities.

1   When we translate the situation described in the section-opening problem, the phrases "at least" and "no more than" lead to a system of inequalities. By analogy to

systems of equations, the **solution set of a system of linear inequalities** is the intersection of the solution sets of all the individual inequalities in the system. The best way to specify such solution sets is in a graph. On the same coordinate system, we graph the solutions to each inequality in the system and then shade the overlap (intersection) of these half planes. This procedure is illustrated in Example 1, which includes a detailed review of how to graph a linear inequality (as first described in Section 6.5).

EXAMPLE 1   Graph the solution set of the system

$$4x + y \leq 8$$
$$y > 2x.$$

Solution   First, graph $4x + y \leq 8$ as in Figure 7.9(a). Use a solid line, since the

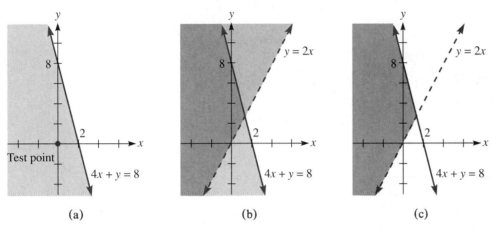

(a)   (b)   (c)

Figure 7.9

inequality symbol is $\leq$; so the points on the line are solutions. Shade the half plane that contains the origin, because choosing $(0,0)$ as a test point leads to a true inequality, as shown below.

$$4x + y \leq 8$$
$$4(0) + 0 \leq 8 \quad \text{Replace } x \text{ by 0 and } y \text{ by 0.}$$
$$0 \leq 8 \quad \text{Simplify.}$$

Then, on the same coordinate system graph $y > 2x$ as in Figure 7.9(b). Use a dashed line, since the inequality symbol is $>$; so the points on $y = 2x$ are not solutions. Shade the half plane above the line, because the inequality begins $y >$. The region with darkest shading gives the intersection of the two half planes. Thus, Figure 7.9(c) is the graph of the solution set of the system.

Note   With respect to the boundary lines, note that the solution set of this system includes a portion of the line $4x + y = 8$ and excludes entirely the line $y = 2x$.

PROGRESS CHECK 1   Graph the solution set of the system

$$x + 4y < 8$$
$$y \leq 2x.$$

EXAMPLE 2   Graph the solution set of the system

$$x + y > 3$$
$$2x - y > -4.$$

Progress Check Answer

1.

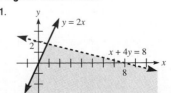

Solution   By choosing the origin as a test point in both cases, we determine that $x + y > 3$ graphs as the set of points above the line $x + y = 3$, while $2x - y > -4$ graphs as the set of points below the line $2x - y = 4$. The graph of the solution set of this system is the intersection of these two half planes, which is shown in Figure 7.10.

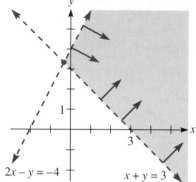

Figure 7.10 $2x - y = -4$ $x + y = 3$

PROGRESS CHECK 2   Graph the solution set of the system

$$x + y \geq 2$$
$$x - 2y \leq -4.$$

2   To solve the section-opening problem, we must both set up and solve a system of linear inequalities.

EXAMPLE 3   Solve the problem in the section introduction on page 348.

Solution

a. First, let

$$x = \text{amount of ham in pounds}$$
$$y = \text{amount of cheese in pounds.}$$

At \$8 per pound for ham and \$4 per pound for cheese, the cost limitation is given by

cost of ham	plus	cost of cheese	is no more than	32.
8x	+	4y	≤	32

The requirement of at least 1 lb of ham translates to $x \geq 1$. Requiring at least 2 lb of cheese gives $y \geq 2$. Thus, the requested system is

$$8x + 4y \leq 32$$
$$x \geq 1$$
$$y \geq 2.$$

b. With $(0,0)$ as a test point, the graph of $8x + 4y \leq 32$ is the half plane on or below the line $8x + 4y = 32$. Furthermore, the graph of $x \geq 1$ is the half plane on or to the right of the vertical line $x = 1$, while $y \geq 2$ graphs as the half plane on or above the horizontal line $y = 2$. The graphical solution to this system is the intersection that is shown in Figure 7.11 of these three half planes.

c. From the graph we can read that one solution is $(2,3)$. Thus, buying 2 lb of ham and 3 lb of cheese (at a total cost of \$28) satisfies the requirements of spending no more than \$32 to purchase at least 1 lb of ham and at least 2 lb of cheese.

PROGRESS CHECK 3   Answer the questions in Example 3, but assume that you wish to spend no more than \$45 to buy at least 2 lb of roast beef at \$9 per pound and at least 4 lb of potato salad at \$3 per pound.

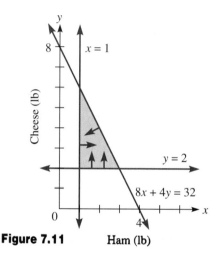

Figure 7.11   Ham (lb)

Progress Check Answers

2.

$x - 2y = -4$

$x + y = 2$

3. (a) $9x + 3y \leq 45$, $x \geq 2$, $y \geq 4$ (b)

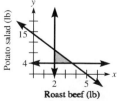

Roast beef (lb)

(c) One solution is 3 lb of roast beef and 5 lb of potato salad, for a total cost of \$42.

EXERCISES 7.5

In Exercises 1–20, graph the solution sets of the system.

1. $y \geq x + 1$
$y \leq 2 - x$

2. $y \leq -x$
$y \geq x - 3$

3. $2x - y > -1$
$2x + y > -4$

4. $y < x + 4$
$x - 2y < -6$

5. $x + 2y \leq 3$
$4x + y > 8$

6. $2x + y < -2$
$x - 3y \geq -9$

7. $y - \frac{1}{2}x > 2$
$y + \frac{1}{2}x \geq 3$

8. $y - \frac{1}{3}x \leq -3$
$y + \frac{1}{3}x < -1$

9. $4x + y > 4$
$x + 4y < 8$

10. $x - 2y > 4$
$2x + 3y > 3$

11. $y - 4 \leq 0$
$x + y > 2$

12. $y - 2 > 1$
$x + 3 < 2$

13. $x \geq 2y - 3$
$y \geq 0$

14. $y \geq -3x - 2$
$x \leq 0$

15. $y \leq x$
$y \geq x + 1$ no solution

16. $y \geq x$
$y \leq x + 1$

17. $y \leq x + 2$
$y < x + 1$

18. $x + y \geq 1$
$x + y > 0$

19. $2x + 3y < 6$
$x \geq 1$
$y \leq 1$

20. $x \geq 2$
$y > 2$
$x + y < 6$

For Exercises 21–24, (a) translate the problem into a system of linear inequalities; (b) solve the system graphically; and (c) specify and interpret one solution from the solution set.

21. For a costume design you need to buy silver ribbon at $2 per foot and gold ribbon at $3 per foot. You need at least 20 ft of each, and you wish to spend no more than $120 all together.

22. For slipcover material you will buy some at $9 per yard and some at $12 per yard. You can spend up to $108, but you need at least 3 yd of the $9 material and at least 4 yd of the $12 material.

23. You need to buy both large and small screws. The small ones each cost 3 cents and the large ones each cost 4 cents. You have $5 to spend, which must cover the material plus a 5 percent sales tax. You need at least 75 small and 50 large screws.

24. The maximum weight a certain shipper will handle for the contents of one package is 50 lb. One package is to contain some metal parts, which each weigh 1 lb, and some plastic parts, which each weigh $\frac{1}{2}$ lb. The package must contain at least 20 plastic and 10 metal parts.

THINK ABOUT IT

1. These three inequalities determine a triangular region.

$$y \leq x + 2$$
$$y \geq 5x - 14$$
$$y \geq -x + 4$$

   a. Find the three vertices.
   b. Show that this is a right triangle.
   c. Find the area of the triangle.

2. The area of mathematics called **linear programming** deals with optimizing functions given constraints. An example based on the chapter-opening problem follows. Given the conditions of the problem, what is the *most food* (in pounds) that you can buy? That is, what values of x and y which satisfy the conditions of the problem will maximize $x + y$? Find the answer by examining Figure 7.11. (*Hint:* The solution is on the boundary.)

3. For Exercise 23, what is the maximum number of screws you can buy?

4. Systems of inequalities can be extended to include equations that are not linear. The sketch shows the graphs of $y = x^2 + 1$ and $y = x + 2$. Shade the region which is the solution set to

$$y \geq x^2 + 1$$
$$y \leq x + 2$$
$$x \geq 0.$$

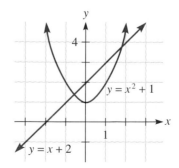

5. The graph of $y = |x|$ is shown in the figure. Use it to graph the solution set of the system

$$y \geq |x|$$
$$x \leq 0.$$

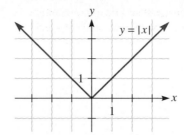

REMEMBER THIS

1. True or false? Both 5 and -5 are solutions to $x^2 = 25$.
2. True or false? Both 2 and -2 are solutions to $x^3 = 8$.
3. Compute $\sqrt{49} \cdot \sqrt{49}$.
4. Compute 5^5.
5. True or false? If x is a real number, then $x^2 + 1$ must be positive.
6. If two angles are complementary, then the sum of their measures is _____ .

7. What expression represents 2.5 percent of an amount x?
8. Solve:
$$x + y = 8$$
$$x - y = 3.$$
9. Write an equation for "M varies directly as n^2."
10. Simplify $\dfrac{3x^2y}{6xy^2}$.

Chapter 7 SUMMARY

OBJECTIVES CHECKLIST Specific chapter objectives are summarized below along with numbered example problems from the text that should clarify the objectives. If you do not understand any objectives or do not know how to do the selected problems, then restudy the material.

7.1 Can you:

1. **Solve a system of linear equations graphically and check to see whether an ordered pair is a solution of a system?**
Solve by graphing:

$$x + 2y = 6$$
$$y = x + 6.$$

[Example 3]

2. **Recognize and solve linear systems that are inconsistent or dependent?**
Solve by graphing:

$$x - 2y = 5$$
$$x - 2y = -2.$$

[Example 4]

7.2 Can you:

1. **Solve linear systems by substitution when at least one equation is solved for one of the variables?**
Solve by substitution:

$$2x - y = 2$$
$$x = 5 - y.$$

[Example 1]

2. **Solve linear systems by substitution when neither equation is solved for one of the variables?**
Solve by substitution:

$$2x + y = 3$$
$$3x - 4y = 10.$$

[Example 3]

3. **Solve, by substitution, linear systems that are inconsistent or dependent?**
Solve by substitution:

$$x - 2y = 5$$
$$y = \tfrac{1}{2}x.$$

[Example 5]

7.3 Can you:

1. **Solve linear systems by addition-elimination for the case where one variable can be eliminated immediately?**
Solve by addition-elimination:

$$x + y = 9$$
$$x - y = 3.$$

[Example 1]

2. **Solve linear systems by addition-elimination when the multiplication property of equality is needed before one variable can be eliminated?**
You win $30,000 in a state lottery and are advised to put some of the money in a risky investment that earns 15 percent annual interest and the rest in a safe investment that earns 7 percent interest. If you want the total interest to be $3,000 annually, then solving the system

$$x + y = 30,000$$
$$0.15x + 0.07y = 3,000$$

gives the amount to invest at each rate. Find x and y, which represent the amount invested in the risky and safe investments, respectively.

[Example 4]

3. **Solve linear systems that are inconsistent or dependent by the addition-elimination method?**
Solve by addition-elimination:

$$2x - 3y = 1$$
$$6x - 9y = 4.$$

[Example 5]

4. **Solve linear systems after choosing an efficient method?**
Solve by addition-elimination or substitution:

$$5x - 6y = 8$$
$$3x = 4y + 6.$$

[Example 7]

7.4 Can you:

Set up and solve linear systems to solve problems involving the following?

1. **Geometric figures**
Two angles are complementary. If the difference between the angle measures is 26°, find the angle measures.

[Example 1]

2. **Comparisons of linear equations**
You are trying to decide between two positions as a salesperson. The first offer pays $150 per week plus a 10 percent commission on sales, while the second offer pays $120 per week plus a 12 percent commission on sales.
a. How much must you sell each week for the two jobs to pay the same?
b. What is the weekly salary in this case?
c. When sales exceed the number found in part **a,** which offer is better?

[Example 3]

3. **Annual interest**
The board members of an employees' pension plan have $8 million to invest. They will split it into two accounts, one earning 9 percent annual interest and the other earning 7 percent. How much should they put into each account so that the total earnings are $650,000?

[Example 4]

4. **Liquid mixtures**
A machine shop has two large containers that are each filled with a mixture of oil and gasoline. Container A contains 3 percent oil (and 97 percent gasoline). Container B contains 6 percent oil (and 94 percent gasoline). How much of each should be used to obtain 12 qt of a new mixture that contains 5 percent oil?

[Example 5]

5. Uniform motion

It takes a boat 1 hour to go 12 mi downstream (with the current); the return trip upstream (against the current) takes 2 hours. What is the speed of the current? What would be the speed of the boat if there were no current? [Example 6]

7.5 **Can you:**

1. Solve systems of linear inequalities graphically?

Graph the solution set of the system

$$4x + y \leq 8$$
$$y > 2x.$$

[Example 1]

2. Solve word problems that translate to systems of linear inequalities?

For a party you need to buy some ham at $8 per pound and some cheese at $4 per pound. You will need *at least* 1 lb of ham and *at least* 2 lb of cheese; and you wish to spend *no more than* $32 on these two items.

a. Translate these requirements into a system of linear inequalities.
b. Solve the system in part **a** graphically.
c. Specify and interpret one solution that is in the solution set. [Example 3]

KEY TERMS

Consistent system (7.1)

Dependent system (7.1)

Inconsistent system (7.1)

Solution of a system of linear equations (7.1)

Solution of a system of linear inequalities (7.5)

System of linear equations (7.1)

System of linear inequalities (7.5)

KEY CONCEPTS AND PROCEDURES

Section	Key Concepts or Procedures to Review
7.1	■ Methods to solve a system of linear equations graphically
	■ Summary of types of systems of linear equations

Possible Graph	Geometric Interpretation	Number of Solutions	Type of System
	Graphs cross at exactly one point	Exactly one solution	Consistent system
	Graphs are the same line	Infinitely many solutions	Consistent system that is called dependent
	Graphs are parallel lines	No solution	Inconsistent system

Section	Key Concepts or Procedures to Review
7.2	■ Substitution method summary: To solve a linear system by substitution 1. If necessary, solve for a variable in one of the equations. Avoid fractions if possible by solving for a variable with coefficient 1 or -1. 2. Use the result in step 1 to make a substitution in the *other* equation. The result is an equation with one unknown. 3. Solve the equation from step 2. A unique solution gives the value of one variable. If the equation is never true, the system is inconsistent. If the equation is always true, the system is dependent. 4. Use the known coordinate value to find the other coordinate value through substitution in the equation in step 1. 5. Check the solution in each of the *original* equations.
7.3	■ Guidelines for choosing a method to solve a system of linear equations *Graphing:* The graphing method is useful for *estimating* the solution and for *comparing* visually the linear equations in a system. When solving systems, we usually use it in conjunction with one of the algebraic methods. *Substitution:* The substitution method is most efficient when at least one equation is solved for one of the variables. *Addition-elimination:* The addition-elimination method is usually the easiest when neither equation is solved for one of the variables. If you have trouble choosing, select this method. ■ Addition-elimination method summary: To solve a linear system by addition-elimination 1. If necessary, write both equations in the form $ax + by = c$. 2. If necessary, multiply one or both equations by numbers that make the coefficients of either x or y opposites of each other. Add the two equations to eliminate a variable. 3. Solve the equation from step 2. A unique solution gives the value of one variable. If the equation is $0 = n$, where $n \neq 0$, the system is inconsistent. If the equation is $0 = 0$, the system is dependent. 4. Use the known coordinate value to find the other coordinate value through substitution in either of the original equations. 5. Check the solution in each of the original equations.
7.4	■ To solve word problems using linear systems 1. Read the problem several times. If possible, display the given information in a sketch or chart. 2. Let two variables represent two unknown quantities. Write down precisely what each variable represents. 3. Set up a system of linear equations that expresses two distinct relationships between the two variables in the problem. 4. Solve the system of equations. 5. Answer the question. 6. Check the answers by interpreting the solution in the context of the word problem.
7.5	■ The solution set of a system of linear inequalities is the intersection of the solution sets of all the individual inequalities in the system. The best way to specify such solution sets is in a graph. On the same coordinate system, graph the solutions to each inequality in the system and then shade the intersection of these half planes.

CHAPTER 7 REVIEW EXERCISES

7.1

1. Determine which of these pairs is a solution to the given system of equations.

$$2x + 4y = 6$$
$$5x - y = 4$$

a. (1,1)   **b.** (3,0)   **c.** (2,6)

2. Solve this system of equations by graphing.

$$3x + y = 4$$
$$x - 2y = 6$$

7.2

3. Solve this system of equations by substitution.

$$y = 3x - 2$$
$$y = 3x + 2$$

4. Solve this system of equations by substitution.

$$3x - 5y = 2$$
$$3x + 10y = 11$$

7.3

5. Solve this system of equations by the addition-elimination method.

$$x + y = 1$$
$$x - y = -5$$

6. Solve this system of equations by the addition-elimination method.

$$3x + 5y = 7$$
$$2x + 3y = 4$$

7. Determine if this system is inconsistent or dependent.

$$x - y = 5$$
$$2x - 2y = 10$$

8. Find two numbers whose sum is 7 and whose difference is 13.

7.4

9. Two angles are complementary. The difference between the angle measures is 12°. Find the angle measures.

10. For what amount of sales do these two jobs pay the same? The first job pays $250 per week with a 10 percent commission. The second pays $300 per week with an 8 percent commission.

11. At a clothing sale red-tag items are discounted by 20 percent and green-tag items are discounted by 15 percent. A shopper bought some of each and paid a total of $181 after a combined discount of $39. After the discount, how much did the shopper spend on red-tag items?

12. One container holds a solution of 10 percent ammonia and 90 percent water, while another has 15 percent ammonia and 86 percent water. How much of each should be mixed to get 120 oz of a 12 percent mixture?

13. A boat goes downstream 35 mi in 2 hours with a steady current. Against the current it takes 2 hours and 48 minutes (or 2.8 hours). Therefore, if s equals the speed of the boat in still water, and c equals the speed of the current, then $s + c = \frac{35}{2}$ and $s - c = 35/2.8$. What is the speed of the current?

7.5

14. Graph the solution set of the system:

$$y \geq x - 5$$
$$y \leq x + 1.$$

15. You need to buy two sizes of nails at the hardware store. The small ones cost $1 per pound, and the large ones cost $2 per pound. There is no tax on the purchase, but you cannot spend more than $5. You must buy at least 1 lb of each. Translate this problem into a system of linear inequalities, and solve the problem graphically.

CHAPTER 7 TEST

1. A company is trying to decide between two machines for packaging its new product. Machine A will cost $6,000 per year plus $3 to package each unit, so the total cost (y) of packaging x units annually is given by $y = 3x + 6,000$. Machine B will cost $4,000 per year plus $4 to package each unit. Therefore, $y = 4x + 4,000$ is the cost equation associated with this machine. How many units must be produced annually for the cost of the two machines to be the same? If the company plans to package more units than this, which machine should it purchase?

2. Is the ordered pair (0,0) a solution of this system?

$$y = x$$
$$x - y = 1$$

3. By graphing, determine the solution of this system of linear equations.

$$x + y = 5$$
$$x + y = 6$$

4. Describe the solution set for this system.

$$2x - 3y = 12$$
$$6x - 9y = 36$$

5. Solve by substitution.
$$3x - y = 5$$
$$x = 1 - y$$

6. Solve by substitution.
$$4x + 5y = 3$$
$$-2x + 10y = 1$$

7. Solve by addition-elimination.
$$3x - 3y = 3$$
$$2x + 3y = 2$$

8. Solve by addition-elimination.
$$x - 2y = 5$$
$$x - 3y = 7$$

9. You win $60,000 in a state lottery, and you are advised to put some of the money in a risky investment that earns 15 percent annual interest and the rest in a safe investment that earns 5 percent interest. If you want the total interest to be $4,000 annually, then solving the system
$$x + y = 60,000$$
$$0.15x + 0.05y = 4,000$$
gives the amount to invest at each rate. Find x and y, which represent the amount invested in the risky and safe investments, respectively.

10. Solve by addition-elimination.
$$3x - 5y = 10$$
$$6x - 10y = 5$$

11. You are solving a system of two linear equations by the addition-elimination approach. After addition the resulting equation is $0 = 0$. What does this mean?

12. Two angles are complementary and the difference is 4 times the smaller angle. Find the measure of the larger angle.

13. A rectangle has perimeter equal to 96 cm and the width is one-third the length. Find the area.

14. A membership in one video rental store costs $10 per year, after which the videos rent for $2 each. In another store membership is free, but it charges $2.50 per video. For what number of rentals does the first deal become a better deal?

15. How should a $12,000 investment be split so that the total annual earnings are $750 if one portion is invested at 5 percent and the rest at 10 percent?

16. One solution is 10 percent acid, while a second is 18 percent acid. How should they be mixed to get 100 ml of a 12 percent solution?

17. It takes a motorboat $2\frac{1}{2}$ hours to go 40 mi upstream against a steady current, but only 1 hour 40 minutes to return downstream. Find the speed of the current, and give the time it would take the boat to make the trip in still water.

18. Graph the solution set of the system
$$3x + y \le 6$$
$$y > 3x.$$

19. For a party you need to buy some ham at $12 per pound and some cheese at $6 per pound. You will need *at least* 1 lb of ham and *at least* 2 lb of cheese; and you wish to spend *no more than* $36 on these two items. Translate these requirements into a system of linear inequalities and solve the system graphically.

20. Is the point $(1, -1)$ in the solution set of this system?
$$3x - y \ge 1$$
$$y < 3x - 4$$

CUMULATIVE TEST 7

1. True or false? $5 - 2 \cdot 3^2 \ge (5 \cdot 2) \cdot 3^2$.
2. If x represents the price of an item, what algebraic expression represents the total charge when there is a 6 percent sales tax?

3. Explain why 3 is a rational number.

4. Simplify $-1 - (-1 - 1)$.
5. What is the numerical coefficient of x in the expression $3x^2 - x + 5$?
6. Solve $2(x - 3) = 4x + 5$.
7. Using the formula $A = \frac{1}{2}h(a + b)$, find the value of b when $h = 10$ cm, $a = 10$ cm, and $A = 150$ cm².

8. Solve $-3x + 2 < 5$ and graph the solution set.

9. Multiply $(x^3 + 2)(x^3 - 2)$.
10. Simplify $(3x^2y)(2xy)^2$.
11. Factor completely $8x^2 - 18$.
12. Solve $5(2x - 1)(x + 4)(3x - 5) = 0$.
13. The perimeter of a square is 3 more than its area. Find the side length of the square. Assume that the units are meters.

14. For what value(s) of x is this expression undefined? $\dfrac{x - 2}{x + 3}$

15. Simplify $\dfrac{\dfrac{1}{2} - \dfrac{3}{y}}{\dfrac{2}{3} + \dfrac{1}{y}}$.

16. Solve $\dfrac{4}{x} + 1 = \dfrac{9}{2x}$.

17. One machine can do a job in 24 minutes, and another can do it in 36 minutes. How long will it take the two machines to do the job if they work together? (Round to the nearest minute.)

18. Graph the line through $(-2,5)$ and $(4,-1)$, and find an equation for the line.

19. y varies inversely as the square of x.

    a. Write this as a variation equation.

    b. What is the constant of variation if $y = 4$ when $x = 2$?

20. Solve this system of equations.

$$2x + 5y = 1$$
$$3x + 2y = 1$$

Roots and Radicals

90 ft

90 ft

A baseball diamond is a square that is 90 ft long on each side. To the nearest tenth of a foot, how far is it from home plate to second base? (See Example 6 of Section 8.1.)

IN SECTION 1.3 we saw that algebraic expressions are expressions that combine variables and constants using the operations of arithmetic. In this chapter we focus on the operation of extracting roots by introducing radical notation and then developing some important properties of radicals that enable us to multiply, divide, add, and subtract radicals. This chapter concludes by showing how to solve radical equations and by using our work with radicals to develop a consistent definition for exponents that are rational numbers.

8.1 Evaluating Radicals

O B J E C T I V E S

1 Find square roots and principal square roots.

2 Identify square roots as rational numbers, irrational numbers, or nonreal numbers.

3 Use square roots to solve problems involving the Pythagorean relation.

4 Find nth roots.

1 We already know how to find the second power of a number. For instance, consider the table below, which shows some examples of applying the squaring rule.

x	3	-3	5	-5	$\frac{2}{7}$	$-\frac{2}{7}$	0.4	-0.4
x^2	9	9	25	25	$\frac{4}{49}$	$\frac{4}{49}$	0.16	0.16

In many applications we need to reverse these assignments and ask

$$\text{if } x^2 = a, \text{ then what is } x?$$

To answer this question, it is useful to have a definition of a square root.

Definition of a Square Root

The number b is a square root of a if $b^2 = a$.

In other words, if squaring b results in a, then b is a square root of a. As suggested in the table above and as illustrated next, every positive number has two square roots.

EXAMPLE 1 Find the square roots of 25.

Solution Because $5^2 = 25$ and $(-5)^2 = 25$, the square roots of 25 are 5 and -5.

PROGRESS CHECK 1 Find the square roots of 9. ⌐

To avoid the ambiguity of having two square roots for a positive number, we define the **principal square root** of a positive number to be its *positive* square root. Thus, the principal square root of 25 is 5. The radical sign $\sqrt{}$ is used to symbolize the principal square root, so

$$\sqrt{25} = 5.$$

To symbolize the negative square root of 25 (which is -5), we write $-\sqrt{25}$. Because $0^2 = 0$, also note that $\sqrt{0} = 0$. Thus 0 has exactly one square root that is neither positive nor negative.

EXAMPLE 2 Find each square root.

a. $\sqrt{121}$ **b.** $-\sqrt{64}$ **c.** $\sqrt{\frac{4}{49}}$

Solution

a. $\sqrt{121}$ denotes the positive square root of 121. Because $11^2 = 121$, $\sqrt{121} = 11$.
b. $-\sqrt{64}$ denotes the negative square root of 64, so $-\sqrt{64} = -8$.
c. Because $(\frac{2}{7})^2 = \frac{4}{49}$, $\sqrt{\frac{4}{49}} = \frac{2}{7}$.

Progress Check Answer

1. 3, -3

PROGRESS CHECK 2   Find each square root.

a. $\sqrt{81}$ b. $-\sqrt{49}$ c. $\sqrt{\frac{9}{25}}$ ⌐

2   The key $\boxed{\sqrt{}}$ on a calculator gives the principal square root of a number. For instance, the keystroke for computing $\sqrt{121}$ is

$$121 \boxed{\sqrt{}} \boxed{\qquad 11}.$$

To compute $-\sqrt{64}$, we first compute $\sqrt{64}$ to be 8; then the answer is -8, because $-\sqrt{64}$ is the negative square root of 64. When $\sqrt{a}$ is a rational number, then a is called a **perfect square.** Recall from Section 1.4 that real numbers that are not rational are called irrational. When a is not a perfect square, then $\sqrt{a}$ is an irrational number that may only be approximated on a calculator to some desired level of accuracy.

EXAMPLE 3   Find each square root. Identify each number as rational or irrational, and approximate irrational numbers to the nearest hundredth.

a. $\sqrt{256}$ b. $\sqrt{10}$ c. $\sqrt{3.61}$

Solution

a. By calculator, $\sqrt{256} = 16$. Thus 256 (which is 16^2) is a perfect square, and $\sqrt{256}$ is a rational number. With respect to integers, note that the perfect square integers are 0, 1, 4, 9, 16, 25, 36, 49, 64, 81, 100, 121, 144, 169, 196, 225, 256, and so on.

b. 10 is not a perfect square integer, so $\sqrt{10}$ is irrational. By calculator,

$$10 \boxed{\sqrt{}} \boxed{3.1622777}.$$

The calculator display approximates $\sqrt{10}$ to seven decimal places. Rounding off to the nearest hundredth, we write $\sqrt{10} \approx 3.16$, where $\approx$ means "is approximately equal to."

c. By calculator, $\sqrt{3.61} = 1.9$. Thus, 3.61 is a perfect square (namely 1.9^2), and 3.61 is a rational number.

PROGRESS CHECK 3   Find each square root. Identify each number as rational or irrational, and approximate irrational numbers to the nearest hundredth.

a. $\sqrt{44}$ b. $\sqrt{196}$ c. $\sqrt{5.29}$ ⌐

We have discussed how to find square roots for positive numbers or zero. No real number is the square root of a negative number, such as -4, because the product of two equal real numbers is never negative. On your calculator, try computing $\sqrt{-4}$ as follows.

$$4 \boxed{+/-} \boxed{\sqrt{}} \boxed{\text{E} \qquad}$$

The display shows an error message because $\sqrt{-4}$ is not a real number. (In Section 9.4 we will extend the number system beyond real numbers to complex numbers, which will enable us to work with square roots of negative numbers.)

EXAMPLE 4   Find each square root that is a real number. If the number is not a real number, state this.

a. $-\sqrt{9}$ b. $\sqrt{-9}$ c. $\sqrt{0}$

Solution

a. $-\sqrt{9}$ denotes the negative square root of 9, so $-\sqrt{9}$ is -3.

b. $\sqrt{-9}$ denotes the square root of -9. Square roots of negative numbers are never real numbers, so $\sqrt{-9}$ is not a real number.

c. $\sqrt{0} = 0$, because $0^2 = 0$.

Progress Check Answers

2. (a) 9   (b) -7 (c) $\frac{3}{5}$

3. (a) Irrational; 6.63   (b) Rational; 14
(c) Rational; 2.3

PROGRESS CHECK 4   Find each square root that is a real number. If the number is not a real number, state this.

a. $\sqrt{-16}$ b. $-\sqrt{16}$ c. $-\sqrt{0.16}$ ⌐

3   Analyzing problems that involve right triangles and the Pythagorean theorem is an application of finding square roots. For reference, we restate the Pythagorean theorem that was first considered in Section 4.7.

Pythagorean Theorem

In a right triangle with legs of length a and b and hypotenuse of length c,

$$c^2 = a^2 + b^2.$$

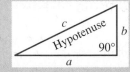

EXAMPLE 5   In a right triangle the hypotenuse measures 17 cm and one leg measures 15 cm. Find the length of the other leg.

Solution   First, sketch Figure 8.1, arbitrarily letting b represent the unknown side length. Now, use the Pythagorean relation and find b^2.

$$c^2 = a^2 + b^2$$
$$17^2 = 15^2 + b^2$$
$$289 = 225 + b^2$$
$$64 = b^2$$

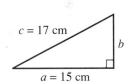

Figure 8.1

Because the unknown length must be positive, b is the positive square root of 64.

$$b = \sqrt{64} = 8$$

Thus, the length of the other leg is 8 cm.

PROGRESS CHECK 5   In a right triangle if the hypotenuse measures 13 ft and one leg measures 5 ft, find the length of the other leg.   ⌐

EXAMPLE 6   Solve the problem in the chapter introduction on page 360.

Solution   Consider the sketch of the problem in Figure 8.2, where x represents the distance from home plate to second base. Note that x is the length of the hypotenuse, so

$$c^2 = a^2 + b^2$$
$$x^2 = 90^2 + 90^2$$
$$= 8{,}100 + 8{,}100$$
$$= 16{,}200.$$

By calculator, $x = \sqrt{16{,}200} \approx 127.3$. Thus, to the nearest tenth the requested distance is 127.3 ft.

PROGRESS CHECK 6   A little league baseball diamond is a square that is 60 ft long on each side. To the nearest tenth, how far is it from home plate to second base?   ⌐

4   Our ideas about square roots may be extended to a more general question. For any positive integer n,

if $x^n = a$, then what is x?

To answer this question, we need a more general definition of root.

Figure 8.2

Progress Check Answers
4. (a) Not real   (b) -4 (c) -0.4
5. 12 ft
6. 84.9 ft

> **Definition of *n*th Root**
>
> For any positive integer n, the number b is an nth root of a if $b^n = a$.

In other words, if raising b to the nth power results in a, then b is an nth root of a.

The **principal *n*th root** of a is denoted by $\sqrt[n]{a}$; and when n is even and $a \geq 0$, then $\sqrt[n]{a}$ means the nonnegative nth root of a. In the expression $\sqrt[n]{a}$, which is called a **radical,** we say $\sqrt{}$ is the **radical sign,** a is the **radicand,** and n is the **index** of the radical. The index is usually omitted from the square root radical, and $\sqrt[3]{a}$ is called the **cube root of *a*.** Figure 8.3 provides powers of numbers that are useful for the examples and exercises in this chapter.

n	n^3	n^4	n^5
2	8	16	32
3	27	81	243
4	64	256	1,024
5	125	625	3,125
10	1,000	10,000	100,000

Figure 8.3

EXAMPLE 7 Find each root that is a real number.

a. $\sqrt[3]{27}$ **b.** $\sqrt[3]{-27}$ **c.** $\sqrt[4]{16}$ **d.** $\sqrt[4]{-16}$

Solution

a. $\sqrt[3]{27} = 3$, because $3^3 = 27$. Read $\sqrt[3]{27} = 3$ as "the cube root of 27 is 3."
b. $\sqrt[3]{-27} = -3$, because $(-3)^3 = -27$. Note that cube roots of negative numbers are negative numbers; and in general, odd roots of negative numbers are negative.
c. $\sqrt[4]{16}$ denotes the positive fourth root of 16. Since $2^4 = 16$, $\sqrt[4]{16} = 2$.
d. $\sqrt[4]{-16}$ is not a real number, because no real number raised to the fourth power gives -16. In general, when n is even, the nth root of a negative number does not exist in the set of real numbers.

PROGRESS CHECK 7 Find each root that is a real number.

a. $\sqrt[3]{125}$ **b.** $\sqrt[3]{-125}$ **c.** $\sqrt[4]{81}$ **d.** $\sqrt[4]{-81}$ ⌐

The root key $\boxed{\sqrt[x]{y}}$ on a calculator is used to take the xth root of the number y. For instance, a keystroke to compute $\sqrt[4]{16}$ is

$$16 \;\boxed{\sqrt[x]{y}}\; \boxed{4} \;\boxed{=}\; \boxed{2}.$$

On some calculators the root key looks like $\boxed{y^{1/x}}$, and the reason for this labeling will be seen in Section 8.7.

You should also be aware of two other special cases. Some calculators have a cube root key $\boxed{\sqrt[3]{}}$ that is used just like the square root key $\boxed{\sqrt{}}$. And some calculators require that y be positive to use $\boxed{\sqrt[x]{y}}$.

EXERCISES 8.1

In Exercises 1–6, find both square roots of the given number.

1. 1
2. 4
3. $\frac{9}{16}$
4. $\frac{25}{36}$
5. 1.44
6. 7.29

In Exercises 7–12, find the given square root.

7. $\sqrt{169}$
8. $\sqrt{441}$
9. $-\sqrt{196}$
10. $-\sqrt{225}$
11. $-\sqrt{\frac{1}{9}}$
12. $\sqrt{\frac{4}{25}}$

In Exercises 13–30, find each square root. Identify the number as rational or irrational, and approximate irrational numbers to the nearest hundredth. If the number is not real, state this.

13. $\sqrt{2}$
14. $\sqrt{3}$
15. $\sqrt{1.9881}$
16. $\sqrt{2.9929}$
17. $-\sqrt{40}$
18. $-\sqrt{90}$
19. $\sqrt{3.6}$
20. $\sqrt{36}$
21. $\sqrt{-1}$
22. $\sqrt{-9}$
23. $-\sqrt{\frac{-1}{-9}}$
24. $-\sqrt{\frac{-16}{-9}}$
25. $\sqrt{0.01}$
26. $\sqrt{0.9}$ 0.95;
27. $\sqrt{-1.21}$
28. $-\sqrt{-1.44}$
29. $\sqrt{-0}$
30. $-\sqrt{-0}$

In Exercises 31–36, give any irrational numbers to the nearest tenth.

31. Find the length of the diagonal.

1 in.

1 in.

32. Find the length of the diagonal.

2 cm

2 cm

33. Find the length of the leg marked x.

26 ft   24 ft

x

34. Find the length of the leg marked x.

19.5 yd

x

7.5 yd

35. The distance between bases in a softball diamond is 65 ft. Find the distance from first base to third base.

36. The dimensions of a rectangular park are 1 mi by $\frac{1}{2}$ mi, as shown in the figure. If you walk directly from A to C, how much do you save compared with going from A to B to C?

B $\frac{1}{2}$ mi

1 mi C

A

In Exercises 37–48, find each root that is a real number.

37. $\sqrt[3]{1}$
38. $\sqrt[3]{64}$
39. $\sqrt[4]{625}$
40. $\sqrt[4]{1296}$
41. $\sqrt[3]{-1}$
42. $\sqrt[3]{-8}$
43. $\sqrt[4]{-256}$
44. $\sqrt[4]{-1}$
45. $\sqrt[4]{0}$
46. $\sqrt[3]{0}$
47. $\sqrt[6]{-64}$
48. $\sqrt[6]{-729}$

THINK ABOUT IT

1. The number 1 is both a perfect square and a perfect cube.
   a. What is the next positive integer that is both a perfect square and a perfect cube?
   b. After 1, what is the smallest positive integer that is simultaneously a perfect square, cube, and fourth power?
2. Given that a is a real number, explain why you can't tell if $\sqrt{a}$ is real.
3. Is it correct to reduce a fraction by taking the square root of the numerator and denominator? Give an example to support your answer.

4. True or false? The principal square root of a positive number is always smaller than the number. Explain your conclusion.
5. Which of these qualifies as being *exactly* correct? The diagonal of a square with side 1 foot is
   a. $\sqrt{2}$ ft   **b.** 1.41 ft
   Explain your answer.

REMEMBER THIS

1. Compute $\sqrt{9} \cdot \sqrt{9}$.
2. True or false? $\sqrt{4} + \sqrt{9} = \sqrt{13}$.
3. True or false? $\sqrt{4} \cdot \sqrt{9} = \sqrt{36}$.
4. Solve:
$$x + y = 10$$
$$2x - 3y = -30.$$
5. Two angles are complementary and their difference is 8°. Find both angle measures.
6. Does the pictured line have positive or negative slope?

7. Prove that these two lines are perpendicular:
$$y = 3x - 1$$
$$y = -\tfrac{1}{3}x + 2.$$
8. Multiply $(3x + 2)(x^2 - 4)$.
9. Solve $3(x - 1) + 4(5 - x) = 6$.
10. Solve $\dfrac{4}{x} = \dfrac{x}{1}$.

8.2 Multiplication Property of Radicals

For maximum safety it is recommended that the distance between the base of a ladder and a building be one-fourth the length of the ladder. How far up a building does a 20-ft ladder reach (in simplified radical form) when it is set up following this guideline? (See Example 3.)

OBJECTIVES

1   **Multiply square roots.**

2   **Simplify square roots.**

3   **Multiply square roots and simplify where possible.**

4   **Multiply and simplify higher roots.**

1   To perform computations with radicals, we need to develop some properties of radicals. An important property that follows from the definition of $\sqrt{a}$ is

$$(\sqrt{a})^2 = a$$

for $a \geq 0$. For instance, $(\sqrt{9})^2 = 9$, and the result checks, since $\sqrt{9} = 3$ and $3^2 = 9$.

To see a second property, consider the products

$$\sqrt{4} \cdot \sqrt{25} = 2 \cdot 5 = 10 \qquad \text{and} \qquad \sqrt{4 \cdot 25} = \sqrt{100} = 10.$$

It follows that $\sqrt{4} \cdot \sqrt{25} = \sqrt{4 \cdot 25}$, which illustrates the multiplication property of square roots.

Multiplication Property of Square Roots

For nonnegative real numbers a and b,

$$\sqrt{a} \cdot \sqrt{b} = \sqrt{a \cdot b}.$$

In words, the product of square roots that are real numbers is the square root of the product of the radicands. To prove this property, first note that $(\sqrt{ab})^2 = ab$ for $a \geq 0$, $b \geq 0$. Also,

$$(\sqrt{a} \cdot \sqrt{b})^2 = (\sqrt{a})^2(\sqrt{b})^2 = ab.$$

Therefore, $(\sqrt{a} \cdot \sqrt{b})^2$ and $(\sqrt{ab})^2$ are equal. Since $\sqrt{a} \cdot \sqrt{b}$ and $\sqrt{ab}$ are both non-negative, this equality implies $\sqrt{a} \cdot \sqrt{b} = \sqrt{ab}$.

EXAMPLE 1   Find each product.

a. $\sqrt{10} \cdot \sqrt{7}$ **b.** $\sqrt{6} \cdot \sqrt{6}$ **c.** $8\sqrt{2} \cdot 7\sqrt{11}$ **d.** $(3\sqrt{5})^2$

Solution

a. $\sqrt{10} \cdot \sqrt{7} = \sqrt{10 \cdot 7} = \sqrt{70}$
b. $\sqrt{6} \cdot \sqrt{6} = \sqrt{6 \cdot 6} = \sqrt{36} = 6$. Alternatively, $\sqrt{6} \cdot \sqrt{6} = (\sqrt{6})^2 = 6$.
c. Reorder and regroup as shown. Then multiply.

$$8\sqrt{2} \cdot 7\sqrt{11} = (8 \cdot 7)(\sqrt{2} \cdot \sqrt{11}) = 56\sqrt{22}$$

d. $(3\sqrt{5})^2 = 3^2 \cdot (\sqrt{5})^2 = 9 \cdot 5 = 45$

PROGRESS CHECK 1   Find each product.

a. $\sqrt{2} \cdot \sqrt{3}$ **b.** $\sqrt{12} \cdot \sqrt{12}$ **c.** $4\sqrt{3} \cdot 5\sqrt{7}$ **d.** $(5\sqrt{3})^2$ ⌐

2   The multiplication property of square roots is used in the form

$$\sqrt{a \cdot b} = \sqrt{a} \cdot \sqrt{b}$$

to simplify square roots. The objective is to remove any factor of the radicand that is a perfect square integer, as shown in Example 2. Note that we will refer to perfect square integers simply as *perfect squares* from this point on.

EXAMPLE 2   Simplify each square root.

a. $\sqrt{50}$ **b.** $\sqrt{48}$

Solution

a. Rewrite 50 as the product of a perfect square and another factor and simplify.

$$\begin{aligned} \sqrt{50} &= \sqrt{25 \cdot 2} && \text{25 is a perfect square and } 50 = 25 \cdot 2. \\ &= \sqrt{25} \cdot \sqrt{2} && \text{Multiplication property of square roots.} \\ &= 5\sqrt{2} && \text{Simplify.} \end{aligned}$$

b. Both 4 and 16 are perfect square factors of 48. It is easier to choose the *larger* perfect square and proceed as before.

$$\sqrt{48} = \sqrt{16 \cdot 3} = \sqrt{16} \cdot \sqrt{3} = 4\sqrt{3}$$

Note   In part **b,** factoring 48 using the smaller perfect square gives

$$\sqrt{48} = \sqrt{4 \cdot 12} = \sqrt{4} \cdot \sqrt{12} = 2\sqrt{12},$$

which leads to

$$2\sqrt{12} = 2\sqrt{4 \cdot 3} = 2\sqrt{4} \cdot \sqrt{3} = 2 \cdot 2 \cdot \sqrt{3} = 4\sqrt{3}.$$

Progress Check Answers
1. (a) $\sqrt{6}$ (b) 12   (c) $20\sqrt{21}$ (d) 75

Both methods give the same result, but choosing the larger perfect square factor leads to an easier simplification.

PROGRESS CHECK 2   Simplify each square root.

a. $\sqrt{27}$ b. $\sqrt{72}$

EXAMPLE 3   Solve the problem in the section introduction on page 366.

Solution   First, sketch the situation, as in Figure 8.4. We let x represent how high the ladder reaches, and the known leg measures 5 ft, because one-fourth of 20 is 5. Using the Pythagorean relation, we obtain the following.

$$20^2 = 5^2 + x^2$$
$$400 = 25 + x^2$$
$$375 = x^2$$
$$\sqrt{375} = x \quad \text{(since } x \text{ must be positive)}$$

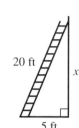

Figure 8.4   20 ft x 5 ft

Now, express this answer as a simplified radical.

$$\sqrt{375} = \sqrt{25 \cdot 15} = \sqrt{25} \cdot \sqrt{15} = 5\sqrt{15}$$

The exact distance up the building is $5\sqrt{15}$ ft. Check by calculator that an approximate answer is 19.4 ft.

PROGRESS CHECK 3   How high up a building does a 16-ft ladder reach (in simplified radical form) when it is set up according to the safety guideline described in the section-opening problem?

To simplify square roots whose radicand contains powers of variables, consider that

$$(a^1)^2 = a^2, \quad (a^2)^2 = a^4, \quad (a^3)^2 = a^6, \quad (a^4)^2 = a^8, \ldots$$

So for nonnegative a,

$$\sqrt{a^2} = a, \quad \sqrt{a^4} = a^2, \quad \sqrt{a^6} = a^3, \quad \sqrt{a^8} = a^4, \quad \text{and so on.}$$

In general, for $a \geq 0$ the square root of an even power of a is a raised to one-half that power.

EXAMPLE 4   Simplify each square root. Assume $x \geq 0, y \geq 0$.

a. $\sqrt{x^6y^2}$ b. $\sqrt{80x^5}$ c. $\sqrt{300x^{11}y^6}$

Solution

a. Even powers of variables are perfect squares, so

$$\sqrt{x^6y^2} = \sqrt{x^6} \cdot \sqrt{y^2} \quad \text{Multiplication property of square roots.}$$
$$= x^3y \quad \text{Simplify.}$$

b. Rewrite $80x^5$ as the product of its largest perfect square factor and another factor. The greatest perfect square that divides 80 is 16, and that divides x^5 is x^4. So factor $80x^5$ using $16x^4$ as one factor.

$$\sqrt{80x^5} = \sqrt{16x^4 \cdot 5x} \quad \text{Factor } 80x^5 \text{ using } 16x^4 \text{ as one factor.}$$
$$= \sqrt{16x^4} \cdot \sqrt{5x} \quad \text{Multiplication property of square roots.}$$
$$= 4x^2\sqrt{5x} \quad \text{Simplify.}$$

c. The largest perfect square factor of $300x^{11}y^6$ is $100x^{10}y^6$.

$$\sqrt{300x^{11}y^6} = \sqrt{100x^{10}y^6 \cdot 3x} \quad \text{Factor } 300x^{11}y^6 \text{ using } 100x^{10}y^6 \text{ as one factor.}$$
$$= \sqrt{100x^{10}y^6} \cdot \sqrt{3x} \quad \text{Multiplication property of square roots.}$$
$$= 10x^5y^3\sqrt{3x} \quad \text{Simplify.}$$

Progress Check Answers

2. (a) $3\sqrt{3}$ (b) $6\sqrt{2}$
3. $4\sqrt{15}$ ft, or about 15.5 ft

Note   In this example we assumed $x \geq 0$ and $y \geq 0$, so expressions like $\sqrt{x^6}$ and $\sqrt{y^2}$ simplify to x^3 and y, respectively. Without this assumption it is necessary to use absolute value and write

$$\sqrt{x^6} = |x^3| \qquad \text{and} \qquad \sqrt{y^2} = |y|.$$

For instance, $\sqrt{(-3)^2} \neq -3$. Instead, $\sqrt{(-3)^2} = \sqrt{9} = 3$, and this result may be obtained using $\sqrt{(-3)^2} = |-3| = 3$. Throughout this chapter we restrict radicands to nonnegative real numbers, so simplifications involving absolute value will not be necessary.

PROGRESS CHECK 4   Simplify each square root. Assume $x \geq 0$, $y \geq 0$.

a. $\sqrt{x^2y^8}$ **b.** $\sqrt{98y^7}$ **c.** $\sqrt{200x^9y^4}$ ⌐

3   Multiplying square roots sometimes results in an expression that can be simplified. Final answers should be simplified where possible.

EXAMPLE 5   Multiply and simplify where possible. Assume $x \geq 0$.

a. $\sqrt{3} \cdot \sqrt{6}$ **b.** $\sqrt{3} \cdot \sqrt{7}$ **c.** $\sqrt{2x} \cdot \sqrt{10x^5}$

Solution

a. First, multiply to get

$$\sqrt{3} \cdot \sqrt{6} = \sqrt{3 \cdot 6} = \sqrt{18}.$$

Now, simplify.

$$\sqrt{18} = \sqrt{9 \cdot 2} = \sqrt{9} \cdot \sqrt{2} = 3\sqrt{2}$$

Thus, $\sqrt{3} \cdot \sqrt{6} = 3\sqrt{2}$.

b. $\sqrt{3} \cdot \sqrt{7} = \sqrt{3 \cdot 7} = \sqrt{21}$. No factor of 21 is a perfect square, so we cannot simplify this result.

c. Multiply first.

$$\sqrt{2x} \cdot \sqrt{10x^5} = \sqrt{2x \cdot 10x^5} = \sqrt{20x^6}$$

Then, simplify.

$$\sqrt{20x^6} = \sqrt{4x^6 \cdot 5} = \sqrt{4x^6} \cdot \sqrt{5} = 2x^3\sqrt{5}$$

Thus, $\sqrt{2x} \cdot \sqrt{10x^5} = 2x^3\sqrt{5}$.

PROGRESS CHECK 5   Multiply and simplify where possible. Assume $x \geq 0$.

a. $\sqrt{6} \cdot \sqrt{2}$ **b.** $\sqrt{10} \cdot \sqrt{7}$ **c.** $\sqrt{5x^3} \cdot \sqrt{10x}$ ⌐

4   Square root properties are specific cases of nth root properties in the case when $n = 2$. A more general statement of the properties we have considered is given next.

---

*n*th Root Properties

For real numbers a, b, $\sqrt[n]{a}$, and $\sqrt[n]{b}$

   1. $(\sqrt[n]{a})^n = a$
   2. $\sqrt[n]{a} \cdot \sqrt[n]{b} = \sqrt[n]{a \cdot b}$

Examples 6 and 7 discuss some adjustments to the methods used for square roots that are needed for working with cube roots and higher roots.

Progress Check Answers

4. (a) xy^4 (b) $7y^3\sqrt{2y}$ (c) $10x^4y^2\sqrt{2x}$
5. (a) $2\sqrt{3}$ (b) $\sqrt{70}$ (c) $5x^2\sqrt{2}$

EXAMPLE 6 Simplify each radical. Assume $x \geq 0$.

a. $\sqrt[3]{81}$ b. $\sqrt[4]{48}$ c. $\sqrt[3]{x^5}$

Solution

a. To simplify cube roots, look for factors of the radicand from the perfect cubes

$$8, 27, 64, 125, \ldots$$

Seeing $81 = 27 \cdot 3$ and using the multiplication property of radicals gives

$$\sqrt[3]{81} = \sqrt[3]{27 \cdot 3} = \sqrt[3]{27} \cdot \sqrt[3]{3} = 3\sqrt[3]{3}.$$

b. Consider fourth powers of 2, 3, 4, and so on, to simplify fourth roots.

$$2^4 = 16, \qquad 3^4 = 81, \qquad 4^4 = 256, \ldots$$

A factor of 48 that is a perfect fourth power is 16.

$$\sqrt[4]{48} = \sqrt[4]{16 \cdot 3} = \sqrt[4]{16} \cdot \sqrt[4]{3} = 2\sqrt[4]{3}$$

c. Exponents that are multiples of 3 lead to perfect cubes.

$$\sqrt[3]{x^5} = \sqrt[3]{x^3 \cdot x^2} = \sqrt[3]{x^3} \cdot \sqrt[3]{x^2} = x\sqrt[3]{x^2}$$

PROGRESS CHECK 6 Simplify each radical. Assume $x \geq 0$.

a. $\sqrt[3]{48}$ b. $\sqrt[4]{162}$ c. $\sqrt[3]{x^4}$

EXAMPLE 7 Multiply and simplify where possible. Assume $y \geq 0$.

a. $\sqrt[3]{4} \cdot \sqrt[3]{32}$ b. $\sqrt[4]{y^5} \cdot \sqrt[4]{y^2}$

Solution

a. The product of the radicals is

$$\sqrt[3]{4} \cdot \sqrt[3]{32} = \sqrt[3]{4 \cdot 32} = \sqrt[3]{128}.$$

Then, 64 is the largest perfect cube factor of 128, so

$$\sqrt[3]{128} = \sqrt[3]{64 \cdot 2} = \sqrt[3]{64} \cdot \sqrt[3]{2} = 4\sqrt[3]{2}.$$

b. Multiplying gives

$$\sqrt[4]{y^5} \cdot \sqrt[4]{y^2} = \sqrt[4]{y^5 \cdot y^2} = \sqrt[4]{y^7}.$$

Perfect fourth powers involve exponents that are multiples of 4.

$$\sqrt[4]{y^7} = \sqrt[4]{y^4 \cdot y^3} = \sqrt[4]{y^4} \cdot \sqrt[4]{y^3} = y\sqrt[4]{y^3}$$

Thus, $\sqrt[4]{y^5} \cdot \sqrt[4]{y^2} = y\sqrt[4]{y^3}$.

Progress Check Answers

6. (a) $2\sqrt[3]{6}$ (b) $3\sqrt[4]{2}$ (c) $x\sqrt[3]{x}$
7. (a) $5\sqrt[3]{2}$ (b) $y\sqrt[4]{y}$

PROGRESS CHECK 7 Multiply and simplify where possible. Assume $y \geq 0$.

a. $\sqrt[3]{25} \cdot \sqrt[3]{10}$ b. $\sqrt[4]{y^3} \cdot \sqrt[4]{y^2}$

EXERCISES 8.2

In Exercises 1–16, find each product and simplify as much as possible.

1. $\sqrt{3} \cdot \sqrt{5}$
2. $\sqrt{3} \cdot \sqrt{7}$
3. $\sqrt{6} \cdot \sqrt{5}$
4. $\sqrt{6} \cdot \sqrt{7}$
5. $\sqrt{8} \cdot \sqrt{8}$
6. $\sqrt{11} \cdot \sqrt{11}$
7. $\sqrt{7} \cdot \sqrt{7}$
8. $\sqrt{13} \cdot \sqrt{13}$
9. $2\sqrt{3} \cdot 3\sqrt{5}$
10. $3\sqrt{2} \cdot 4\sqrt{3}$
11. $5\sqrt{10} \cdot 2\sqrt{3}$
12. $4\sqrt{3} \cdot 5\sqrt{5}$
13. $2\sqrt{3} \cdot 5\sqrt{3}$
14. $5\sqrt{7} \cdot 2\sqrt{7}$
15. $(3\sqrt{3})^2$
16. $(5\sqrt{5})^2$

In Exercises 17–22, simplify each square root.

17. $\sqrt{75}$
18. $\sqrt{45}$
19. $\sqrt{128}$
20. $\sqrt{80}$
21. $\sqrt{162}$
22. $\sqrt{243}$

In Exercises 23–30, express the answer (a) in simplified radical form and (b) in decimal form to the nearest tenth.

23. The two equal sides of an isosceles right triangle have length 2 in. Find the hypotenuse.

24. The two equal sides of an isosceles right triangle have length 3 in. Find the hypotenuse.

25. An isosceles right triangle has hypotenuse of length 2 in. Find the length of the two equal legs.

26. An isosceles right triangle has hypotenuse of length 4 in. Find the length of the two equal legs.

27. Find the hypotenuse of this triangle.

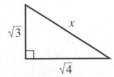

28. Find the hypotenuse of this triangle.

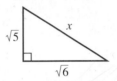

29. Carpenters need to cut plywood for a porch roof as shown. What are the dimensions of the rectangular shaded piece?

30. You have a 20-ft ladder, but the closest to the wall that you can put its base is 5 ft. If you can reach the wall 3 ft above the point where the top of the ladder rests, how high up on the wall can you reach?

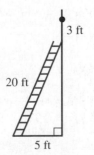

In Exercises 31–42, simplify each square root. Assume all variables are not negative.

31. $\sqrt{x^4 y^2}$

32. $\sqrt{x^4 y^6}$

33. $\sqrt{25 x^2 y^6}$

34. $\sqrt{16 x^8 y^4}$

35. $\sqrt{45 x^3}$

36. $\sqrt{27 y^5}$

37. $\sqrt{8x}$

38. $\sqrt{12y}$

39. $\sqrt{75 x^5 y^7}$

40. $\sqrt{63 x^3 y^5}$

41. $\sqrt{500 x^{20} y^{11}}$

42. $\sqrt{360 x^9 y^{12}}$

In Exercises 43–56, multiply and then simplify where possible. Assume all variables are nonnegative.

43. $\sqrt{2} \cdot \sqrt{6}$

44. $\sqrt{3} \cdot \sqrt{15}$

45. $\sqrt{6} \cdot \sqrt{10}$

46. $\sqrt{15} \cdot \sqrt{10}$

47. $\sqrt{3} \cdot \sqrt{5}$

48. $\sqrt{5} \cdot \sqrt{7}$

49. $\sqrt{6} \cdot \sqrt{35}$

50. $\sqrt{14} \cdot \sqrt{15}$

51. $\sqrt{3x} \cdot \sqrt{3x}$

52. $\sqrt{7y} \cdot \sqrt{7y}$

53. $\sqrt{6x} \cdot \sqrt{3x^3}$

54. $\sqrt{2x^3} \cdot \sqrt{10x^5}$

55. $\sqrt{20x^3 y} \cdot \sqrt{5xy^3}$

56. $\sqrt{24xy^5} \cdot \sqrt{6xy}$

In Exercises 57–70, simplify each radical. Assume variables are nonnegative.

57. $\sqrt[3]{40}$

58. $\sqrt[3]{56}$

59. $\sqrt[3]{54}$

60. $\sqrt[3]{128}$

61. $\sqrt[4]{32}$

62. $\sqrt[4]{64}$

63. $\sqrt[4]{324}$

64. $\sqrt[4]{243}$

65. $\sqrt[3]{x^6}$

66. $\sqrt[3]{y^9}$

67. $\sqrt[3]{x^{10}}$

68. $\sqrt[3]{y^7}$

69. $\sqrt[4]{y^6}$

70. $\sqrt[4]{x^7}$

In Exercises 71–82, multiply and simplify where possible. Assume variables are nonnegative.

71. $\sqrt[3]{4} \cdot \sqrt[3]{2}$

72. $\sqrt[3]{3} \cdot \sqrt[3]{9}$

73. $\sqrt[3]{25} \cdot \sqrt[3]{25}$

74. $\sqrt[3]{36} \cdot \sqrt[3]{3}$

75. $\sqrt[4]{9} \cdot \sqrt[4]{81}$

76. $\sqrt[4]{9} \cdot \sqrt[4]{18}$

77. $\sqrt[3]{x^4} \cdot \sqrt[3]{x^2}$

78. $\sqrt[3]{y^5} \cdot \sqrt[3]{y^4}$

79. $\sqrt[4]{xy} \cdot \sqrt[4]{x^3 y}$

80. $\sqrt[4]{8x^2 y^2} \cdot \sqrt[4]{2x^2 y^3}$

81. $\sqrt[3]{2xy^2} \cdot \sqrt[3]{4x^2 y^5}$

82. $\sqrt[3]{16x^4 y^5} \cdot \sqrt[3]{4x^5 y^5}$

THINK ABOUT IT

1. It is true for $a \geq 0$ and $b \geq 0$ that $\sqrt{a} \cdot \sqrt{b} = \sqrt{a \cdot b}$. Is it also true that $\sqrt{a} + \sqrt{b} = \sqrt{a + b}$? Try several values of a and b to find out.
2. True or false? The product of two irrational numbers is always an irrational number. Give an example.
3. For positive numbers a and b, $\sqrt{ab}$ is called the *geometric mean* of a and b.
   a. Find the geometric mean of 4 and 9.
   b. Show that a square whose side is the geometric mean of a and b has the same area as a rectangle with length a and width b.
4. Multiply $(\sqrt{a} - \sqrt{b})(\sqrt{a} + \sqrt{b})$ using FOIL. Assume $a, b > 0$.
5. a. By what number k would you have to multiply the side length of a square in order to get a square with double the area? See the figure.

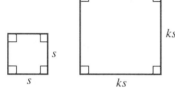

   b. By what number k would you have to multiply the side of a cube in order to get a cube with double the volume? See the figure.

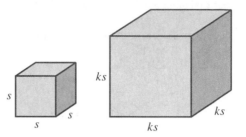

REMEMBER THIS

1. True or false? $\dfrac{\sqrt{100}}{\sqrt{4}} = \sqrt{25}$.
2. What is the result of multiplying both the numerator and the denominator of $\dfrac{1}{\sqrt{3}}$ by $\sqrt{3}$?
3. Solve $a = bc$ for b.
4. Find the exact length of the diagonal of a square whose side measures 1 in.
5. In a right triangle, the length of the hypotenuse is 2 cm and the length of one leg is 1 cm. Find the length of the other leg (in radical form).
6. Graph the inequality $y > x + 1$. (Shade the solution set.)
7. Solve:
$$2x + 3y = 1$$
$$2x - 5y = 0.$$
8. Which is the largest number in this selection?
   a. $|-3|^{-1}$ b. $(-3)^{-1}$ c. $(-1)^{-3}$ d. $|-1|^{-3}$
9. Simplify $(2x^2y)(2xy)^2$.
10. A tip of 15 percent on a meal amounted to $5.25. How much was the meal before the tip?

8.3 Quotient Property of Radicals

The picture tube of a popular model television set is square in shape and measures 20 in. diagonally, as shown in the accompanying diagram. Find the length of the side (s) of this picture tube given that

$$d = s\sqrt{2}$$

expresses the length of the diagonal (d) in terms of side length. (See Example 5.)

O B J E C T I V E S

1   Divide square roots.

2   Rationalize denominators.

3   Divide and simplify higher roots.

1   To develop a property of radicals with respect to division, consider that

$$\frac{\sqrt{100}}{\sqrt{4}} = \frac{10}{2} = 5 \quad \text{and} \quad \sqrt{\frac{100}{4}} = \sqrt{25} = 5.$$

We see that $\dfrac{\sqrt{100}}{\sqrt{4}} = \sqrt{\dfrac{100}{4}}$, which illustrates the quotient property of square roots.

Quotient Property of Square Roots

For nonnegative real numbers a and b with $b \neq 0$,

$$\frac{\sqrt{a}}{\sqrt{b}} = \sqrt{\frac{a}{b}}.$$

In words, the quotient of square roots that are real numbers is the square root of the quotient of the radicands. Consideration of the proof of the multiplication property of square roots given in the previous section suggests a method to prove this quotient property, and such a proof is requested in "Think About It" Exercise 5.

EXAMPLE 1   Divide and simplify where possible. Assume $x > 0$.

a. $\dfrac{\sqrt{50}}{\sqrt{2}}$
    b. $\sqrt{54} \div \sqrt{3}$
    c. $\dfrac{\sqrt{2x}}{\sqrt{x}}$

Solution

a. $\dfrac{\sqrt{50}}{\sqrt{2}} = \sqrt{\dfrac{50}{2}}$ Quotient property of square roots.

$\quad\quad = \sqrt{25}$ Divide.

$\quad\quad = 5$ Simplify.

b. Express the division in fraction form and divide.

$$\frac{\sqrt{54}}{\sqrt{3}} = \sqrt{\frac{54}{3}} = \sqrt{18}$$

Now, simplify.

$$\sqrt{18} = \sqrt{9 \cdot 2} = \sqrt{9} \cdot \sqrt{2} = 3\sqrt{2}$$

Thus, $\sqrt{54} \div \sqrt{3} = 3\sqrt{2}$.

c. $\dfrac{\sqrt{2x}}{\sqrt{x}} = \sqrt{\dfrac{2x}{x}}$ Quotient property of square roots.

$\quad\quad = \sqrt{2}$ Divide.

PROGRESS CHECK 1   Divide and simplify where possible. Assume $x > 0$.

a. $\dfrac{\sqrt{27}}{\sqrt{3}}$
    b. $\dfrac{\sqrt{35x^3}}{\sqrt{5x^2}}$
    c. $\sqrt{56} \div \sqrt{7}$

Progress Check Answers

1. (a) 3   (b) $\sqrt{7x}$ (c) $2\sqrt{2}$

2 It is often necessary to use the quotient property of square roots in the form

$$\sqrt{\frac{a}{b}} = \frac{\sqrt{a}}{\sqrt{b}}$$

to simplify square roots. For instance, the division

$$\frac{\sqrt{3}}{\sqrt{2}} \qquad \text{leads to} \qquad \sqrt{\frac{3}{2}}.$$

To eliminate the fraction in the radicand (or equivalently, to eliminate the square root denominator), we rewrite $\frac{3}{2}$ as an equivalent fraction whose denominator is a perfect square.

$$\sqrt{\frac{3}{2}} = \sqrt{\frac{3 \cdot 2}{2 \cdot 2}} = \sqrt{\frac{6}{4}} = \frac{\sqrt{6}}{\sqrt{4}} = \frac{\sqrt{6}}{2}$$

Thus, $\sqrt{3}/\sqrt{2} = \sqrt{3/2} = \sqrt{6}/2$. In general, we adopt the conditions that a simplified square root cannot have either a fraction in the radicand or a square root in the denominator. The process of eliminating square roots in the denominator is called **rationalizing the denominator.**

EXAMPLE 2 Simplify each expression by rationalizing the denominator. Assume $x > 0$.

a. $\sqrt{\dfrac{3}{7}}$ **b.** $\sqrt{\dfrac{5}{8}}$ **c.** $\sqrt{\dfrac{4}{x}}$

Solution

a. Rewrite $\frac{3}{7}$ as an equivalent fraction whose denominator is a perfect square and simplify.

$$\sqrt{\frac{3}{7}} = \sqrt{\frac{3 \cdot 7}{7 \cdot 7}} = \sqrt{\frac{21}{49}} = \frac{\sqrt{21}}{\sqrt{49}} = \frac{\sqrt{21}}{7}$$

b. Multiplying 8 by 2 gives the perfect square 16, so

$$\sqrt{\frac{5}{8}} = \sqrt{\frac{5 \cdot 2}{8 \cdot 2}} = \sqrt{\frac{10}{16}} = \frac{\sqrt{10}}{\sqrt{16}} = \frac{\sqrt{10}}{4}.$$

c. Since $\sqrt{x^2} = x$ for $x > 0$,

$$\sqrt{\frac{4}{x}} = \sqrt{\frac{4 \cdot x}{x \cdot x}} = \sqrt{\frac{4x}{x^2}} = \frac{\sqrt{4}\sqrt{x}}{\sqrt{x^2}} = \frac{2\sqrt{x}}{x}.$$

Note We attain the same result in part **b** if we simplify as follows:

$$\sqrt{\frac{5}{8}} = \sqrt{\frac{5 \cdot 8}{8 \cdot 8}} = \sqrt{\frac{40}{64}} = \frac{\sqrt{40}}{\sqrt{64}} = \frac{\sqrt{40}}{8},$$

which leads to

$$\frac{\sqrt{40}}{8} = \frac{\sqrt{4 \cdot 10}}{8} = \frac{\sqrt{4} \cdot \sqrt{10}}{8} = \frac{2\sqrt{10}}{8} = \frac{\sqrt{10}}{4}.$$

Notice that it is more convenient to change the denominator to the smallest possible perfect square.

PROGRESS CHECK 2 Simplify each expression by rationalizing the denominator. Assume $y > 0$.

a. $\sqrt{\dfrac{3}{5}}$ **b.** $\sqrt{\dfrac{11}{12}}$ **c.** $\sqrt{\dfrac{25}{y}}$

Progress Check Answers

2. (a) $\dfrac{\sqrt{15}}{5}$ (b) $\dfrac{\sqrt{33}}{6}$ (c) $\dfrac{5\sqrt{y}}{y}$

EXAMPLE 3   Simplify each expression. Assume $x > 0$ and $y > 0$.

a. $\dfrac{\sqrt{3}}{\sqrt{12}}$

b. $\dfrac{\sqrt{7}}{\sqrt{5}}$

c. $\dfrac{\sqrt{27xy^4}}{\sqrt{3x^5y}}$

Solution

a. $\dfrac{\sqrt{3}}{\sqrt{12}} = \sqrt{\dfrac{3}{12}} = \sqrt{\dfrac{1}{4}} = \dfrac{\sqrt{1}}{\sqrt{4}} = \dfrac{1}{2}$

b. $\dfrac{\sqrt{7}}{\sqrt{5}} = \sqrt{\dfrac{7}{5}} = \sqrt{\dfrac{7 \cdot 5}{5 \cdot 5}} = \sqrt{\dfrac{35}{25}} = \dfrac{\sqrt{35}}{\sqrt{25}} = \dfrac{\sqrt{35}}{5}$

c. $\dfrac{\sqrt{27xy^4}}{\sqrt{3x^5y}} = \sqrt{\dfrac{27xy^4}{3x^5y}} = \sqrt{\dfrac{9y^3}{x^4}} = \dfrac{\sqrt{9y^2}\sqrt{y}}{\sqrt{x^4}} = \dfrac{3y\sqrt{y}}{x^2}$

PROGRESS CHECK 3   Simplify each expression. Assume $x > 0$ and $y > 0$.

a. $\dfrac{\sqrt{8}}{\sqrt{18}}$

b. $\dfrac{\sqrt{11}}{\sqrt{3}}$

c. $\dfrac{\sqrt{80x^4y^3}}{\sqrt{5xy^5}}$ ⌐

In Example 3b the quotient $\sqrt{7}/\sqrt{5}$ led to a division that did not reduce. In this case it is easier to rationalize the denominator as follows.

$$\dfrac{\sqrt{7}}{\sqrt{5}} = \dfrac{\sqrt{7}}{\sqrt{5}} \cdot \dfrac{\sqrt{5}}{\sqrt{5}} = \dfrac{\sqrt{35}}{5}$$

Note that we multiplied by 1 in the form $\sqrt{5}/\sqrt{5}$ and that we chose $\sqrt{5}$ in the numerator and denominator because $\sqrt{5} \cdot \sqrt{5} = 5$. This method is also used to rationalize a square root denominator when the numerator does not contain a square root.

EXAMPLE 4   Rationalize each denominator. Assume $y > 0$.

a. $\dfrac{\sqrt{3}}{\sqrt{10}}$

b. $\dfrac{2}{\sqrt{3}}$

c. $\dfrac{x}{\sqrt{y}}$

d. $\dfrac{5}{\sqrt{32}}$

Solution

a. $\dfrac{\sqrt{3}}{\sqrt{10}} = \dfrac{\sqrt{3}}{\sqrt{10}} \cdot \dfrac{\sqrt{10}}{\sqrt{10}} = \dfrac{\sqrt{30}}{10}$

b. $\dfrac{2}{\sqrt{3}} = \dfrac{2}{\sqrt{3}} \cdot \dfrac{\sqrt{3}}{\sqrt{3}} = \dfrac{2\sqrt{3}}{3}$

c. $\dfrac{x}{\sqrt{y}} = \dfrac{x}{\sqrt{y}} \cdot \dfrac{\sqrt{y}}{\sqrt{y}} = \dfrac{x\sqrt{y}}{y}$

d. $\dfrac{5}{\sqrt{32}} = \dfrac{5}{\sqrt{32}} \cdot \dfrac{\sqrt{2}}{\sqrt{2}} = \dfrac{5\sqrt{2}}{\sqrt{64}} = \dfrac{5\sqrt{2}}{8}$

(Once again note that it is more convenient to change the denominator to the smallest possible perfect square.)

PROGRESS CHECK 4   Rationalize each denominator. Assume $x > 0$ and $y > 0$.

a. $\dfrac{\sqrt{3}}{\sqrt{5}}$

b. $\dfrac{1}{\sqrt{2}}$

c. $\dfrac{\sqrt{x}}{\sqrt{y}}$

d. $\dfrac{7}{\sqrt{8}}$ ⌐

EXAMPLE 5   Solve the problem in the section introduction on page 372.

Solution   Replace d by 20 in the given formula and solve for s.

$$d = s\sqrt{2} \qquad \text{Given formula.}$$
$$20 = s\sqrt{2} \qquad \text{Replace } d \text{ by 20.}$$
$$\dfrac{20}{\sqrt{2}} = s \qquad \text{Divide both sides by } \sqrt{2}.$$

Progress Check Answers

3. (a) $\dfrac{2}{3}$ (b) $\dfrac{\sqrt{33}}{3}$ (c) $\dfrac{4x\sqrt{x}}{y}$

4. (a) $\dfrac{\sqrt{15}}{5}$ (b) $\dfrac{\sqrt{2}}{2}$ (c) $\dfrac{\sqrt{xy}}{y}$ (d) $\dfrac{7\sqrt{2}}{4}$

Now, simplify the result by rationalizing the denominator.

$$\frac{20}{\sqrt{2}} = \frac{20}{\sqrt{2}} \cdot \frac{\sqrt{2}}{\sqrt{2}} = \frac{20\sqrt{2}}{2} = 10\sqrt{2}$$

Thus, the side length of the picture tube is $10\sqrt{2}$ in. Check by calculator that an approximate answer is 14.14 in.

PROGRESS CHECK 5   Find the length of the side of a square-shaped picture tube that measures 30 in. diagonally.

In Example 5 rationalizing the denominator gave a more convenient statement of the side length. Keep in mind that we have given three conditions for writing a simplified square root.

Simplified Square Root

To write a square root in simplest form
1. Remove all factors of the radicand that are perfect squares.
2. Eliminate all fractions in the radicand.
3. Eliminate square roots in the denominator.

3   The quotient property of square roots extends to other roots to give the quotient property of radicals.

Quotient Property of Radicals

For real numbers a, b, $\sqrt[n]{a}$, and $\sqrt[n]{b}$,

$$\frac{\sqrt[n]{a}}{\sqrt[n]{b}} = \sqrt[n]{\frac{a}{b}}.$$

As with square roots, a simplified radical cannot have either a fraction in the radicand or a radical in the denominator. Therefore, we may need to rationalize denominators when simplifying, as discussed in Example 6.

EXAMPLE 6   Simplify each expression.

a. $\sqrt[4]{\frac{81}{16}}$ b. $\sqrt[3]{\frac{4}{9}}$ c. $\frac{\sqrt[4]{96}}{\sqrt[4]{2}}$ d. $\frac{\sqrt[3]{15}}{\sqrt[3]{12}}$

Solution

a. Both 81 and 16 are perfect fourth powers, so

$$\sqrt[4]{\frac{81}{16}} = \frac{\sqrt[4]{81}}{\sqrt[4]{16}} = \frac{3}{2}.$$

b. Rewrite $\frac{4}{9}$ as an equivalent fraction whose denominator is a perfect cube, and simplify.

$$\sqrt[3]{\frac{4}{9}} = \sqrt[3]{\frac{4 \cdot 3}{9 \cdot 3}} = \sqrt[3]{\frac{12}{27}} = \frac{\sqrt[3]{12}}{\sqrt[3]{27}} = \frac{\sqrt[3]{12}}{3}$$

c. First, divide to get

$$\frac{\sqrt[4]{96}}{\sqrt[4]{2}} = \sqrt[4]{\frac{96}{2}} = \sqrt[4]{48}.$$

Now, simplify.

$$\sqrt[4]{48} = \sqrt[4]{16 \cdot 3} = \sqrt[4]{16} \cdot \sqrt[4]{3} = 2\sqrt[4]{3}$$

Thus, $\sqrt[4]{96}/\sqrt[4]{2} = 2\sqrt[4]{3}$.

d. Divide first.

$$\frac{\sqrt[3]{15}}{\sqrt[3]{12}} = \sqrt[3]{\frac{15}{12}} = \sqrt[3]{\frac{5}{4}}$$

To simplify, we see that $4 \cdot 2$ gives the perfect cube 8.

$$\sqrt[3]{\frac{5}{4}} = \sqrt[3]{\frac{5 \cdot 2}{4 \cdot 2}} = \sqrt[3]{\frac{10}{8}} = \frac{\sqrt[3]{10}}{\sqrt[3]{8}} = \frac{\sqrt[3]{10}}{2}$$

Thus, $\sqrt[3]{15}/\sqrt[3]{12} = \sqrt[3]{10}/2$.

PROGRESS CHECK 6 Simplify each expression.

a. $\sqrt[4]{\dfrac{1}{81}}$
b. $\sqrt[3]{\dfrac{3}{4}}$
c. $\dfrac{\sqrt[4]{128}}{\sqrt[4]{2}}$
d. $\dfrac{\sqrt[3]{25}}{\sqrt[3]{45}}$

Progress Check Answers

6. (a) $\dfrac{1}{3}$ (b) $\dfrac{\sqrt[3]{6}}{2}$ (c) $2\sqrt[4]{4}$ (d) $\dfrac{\sqrt[3]{15}}{3}$

EXERCISES 8.3

In Exercises 1–18, divide and simplify where possible. Assume variables are positive.

1. $\dfrac{\sqrt{72}}{\sqrt{2}}$
2. $\dfrac{\sqrt{75}}{\sqrt{3}}$

3. $\dfrac{\sqrt{3}}{\sqrt{48}}$
4. $\dfrac{\sqrt{2}}{\sqrt{32}}$ $\frac{1}{4}$

5. $\dfrac{\sqrt{54}}{\sqrt{9}}$
6. $\dfrac{\sqrt{35}}{\sqrt{7}}$

7. $\dfrac{\sqrt{12.3}}{\sqrt{4.1}}$
8. $\dfrac{\sqrt{7.5}}{\sqrt{2.5}}$

9. $\dfrac{\sqrt{36}}{\sqrt{2}}$
10. $\dfrac{\sqrt{36}}{\sqrt{3}}$ $2\sqrt{3}$

11. $\dfrac{\sqrt{30x}}{\sqrt{5x}}$
12. $\dfrac{\sqrt{24y^2}}{\sqrt{12y^2}}$

13. $\dfrac{\sqrt{21x^3}}{\sqrt{7x^2}}$
14. $\dfrac{\sqrt{30y^2}}{\sqrt{5y}}$

15. $\dfrac{\sqrt{60x^3y^3}}{\sqrt{15xy^2}}$
16. $\dfrac{\sqrt{45x^3y^5}}{\sqrt{5x^2y^3}}$

17. $\dfrac{\sqrt{2xy}}{\sqrt{8x^3y}}$
18. $\dfrac{\sqrt{2x^5y^3}}{\sqrt{18xy^3}}$

In Exercises 19–30, simplify each expression by rationalizing the denominator. Assume all variables are positive.

19. $\sqrt{\dfrac{2}{7}}$
20. $\sqrt{\dfrac{3}{5}}$ $\dfrac{5}{5}\dfrac{\sqrt{15}}{\sqrt{25}}$ $\dfrac{\sqrt{15}}{5}$

21. $\sqrt{\dfrac{1}{2}}$
22. $\sqrt{\dfrac{1}{3}}$

23. $\sqrt{\dfrac{5}{18}}$
24. $\sqrt{\dfrac{7}{12}}$

25. $\sqrt{\dfrac{15}{32}}$
26. $\sqrt{\dfrac{47}{50}}$

27. $\sqrt{\dfrac{1}{x}}$
28. $\sqrt{\dfrac{5}{y}}$

29. $\sqrt{\dfrac{3}{y^3}}$
30. $\sqrt{\dfrac{1}{x^3}}$

In Exercises 31–52, simplify each expression. Assume all variables are positive, and rationalize denominators.

31. $\dfrac{\sqrt{12}}{\sqrt{27}}$
32. $\dfrac{\sqrt{32}}{\sqrt{18}}$

33. $\dfrac{\sqrt{32}}{\sqrt{50}}$
34. $\dfrac{\sqrt{45}}{\sqrt{20}}$

35. $\dfrac{\sqrt{7}}{\sqrt{3}}$
36. $\dfrac{\sqrt{2}}{\sqrt{7}}$

37. $\dfrac{\sqrt{2}}{\sqrt{5}}$
38. $\dfrac{\sqrt{10}}{\sqrt{3}}$

39. $\dfrac{\sqrt{x}}{\sqrt{y}}$
40. $\dfrac{\sqrt{x}}{\sqrt{3}}$

41. $\dfrac{\sqrt{54xy^6}}{\sqrt{6x^7y}}$
42. $\dfrac{\sqrt{24x^6y}}{\sqrt{6x^3y^3}}$

43. $\dfrac{\sqrt{50x^2}}{\sqrt{8xy}}$
44. $\dfrac{\sqrt{8x^4y^2}}{\sqrt{18xy^3}}$

45. $\dfrac{3}{\sqrt{2}}$
46. $\dfrac{1}{\sqrt{2}}$

47. $\dfrac{x}{\sqrt{5}}$

48. $\dfrac{y}{\sqrt{6}}$

49. $\dfrac{3}{\sqrt{8}}$

50. $\dfrac{5}{\sqrt{18}}$

51. $\dfrac{2}{\sqrt{8}}$

52. $\dfrac{6}{\sqrt{18}}$

In Exercises 53–54, express the answer (a) exactly, in simplified form, and (b) approximately, in decimal form to the nearest hundredth.

53. Use the formula $d = s\sqrt{2}$ to find the side s of a square that measures 1 ft diagonally.

54. Repeat Exercise 53, but assume that the diagonal measures 10 in.

For Exercises 55–56, use the trigonometry fact that the sine of an acute angle in a right triangle is given by the ratio of the length of the side opposite the angle to the length of the hypotenuse.

55. Refer to the figure.

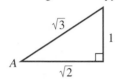

a. Use the Pythagorean theorem to confirm that the triangle is a right triangle.

b. Find the sine of angle A with rationalized denominator.

c. Find the sine of angle A in decimal form to four decimal places.

56. Repeat Exercise 55, but use this figure.

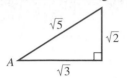

In Exercises 57–74, simplify each expression. Assume all variables are positive.

57. $\sqrt[4]{\dfrac{1}{16}}$

58. $\sqrt[4]{\dfrac{16}{625}}$

59. $\sqrt[3]{\dfrac{8}{27}}$

60. $\sqrt[3]{\dfrac{27}{64}}$

61. $\sqrt[3]{\dfrac{1}{4}}$

62. $\sqrt[3]{\dfrac{1}{25}}$

63. $\sqrt[4]{\dfrac{3}{8}}$

64. $\sqrt[4]{\dfrac{5}{27}}$

65. $\dfrac{\sqrt[4]{96}}{\sqrt[4]{3}}$

66. $\dfrac{\sqrt[4]{192}}{\sqrt[4]{3}}$

67. $\dfrac{\sqrt[3]{16}}{\sqrt[3]{12}}$

68. $\dfrac{\sqrt[3]{15}}{\sqrt[3]{50}}$

69. $\dfrac{\sqrt[4]{30}}{\sqrt[4]{12}}$

70. $\dfrac{\sqrt[4]{24}}{\sqrt[4]{120}}$

71. $\dfrac{\sqrt[3]{x}}{\sqrt[3]{y}}$

72. $\dfrac{\sqrt[3]{xy}}{\sqrt[3]{x^2}}$

73. $\dfrac{\sqrt[4]{a}}{\sqrt[4]{b}}$

74. $\dfrac{\sqrt[4]{a^2b^3}}{\sqrt[4]{ab^5}}$

THINK ABOUT IT

1. An important mathematical finding credited to the Pythagoreans of the sixth century B.C. is that the diagonal and the side of a square cannot both be whole numbers.

a. Show this is true for a square with side length equal to 1.

b. Show this is true for a square with diagonal length equal to 2.

2. a. Is it true that $\sqrt{10}$ is twice as large as $\sqrt{5}$? If not, what square root *is* twice as large as $\sqrt{5}$?

b. What cube root is twice as large as $\sqrt[3]{5}$?

3. The diagonal of the square is 2 in. Find the area of the inscribed circle. Express the answer exactly in terms of π and also as a decimal accurate to the nearest hundredth.

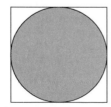

4. In right triangle ABC as shown, the run divided by the hypotenuse is called the cosine of angle A. On a set of coordinate axes, put point A at the origin and point B at (1,2), and draw triangle ABC with a right angle at C, as in the figure. Then find the cosine of angle A to four decimal places.

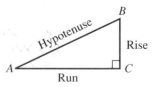

5. Prove the quotient property of square roots. Apply the method that was used in Section 8.2 to establish the multiplication property of square roots.

REMEMBER THIS

1. Simplify by combining like terms: $4x + 7x$.

2. Add $n + \dfrac{1}{n}$.

3. Simplify $\dfrac{3x^2y - 4y^2x}{xy}$.

4. Compute: $\sqrt[3]{7} \cdot \sqrt[3]{7} \cdot \sqrt[3]{7}$.

5. Compute $(3\sqrt{5})^2$.

6. Solve:
$$x - 2y - 9 = 0$$
$$2x - y + 3 = 0.$$

7. When a system of two linear equations is consistent but not dependent, then the graphs of the two lines intersect at
   a. 0 points   **b.** 1 point   **c.** more than 1 point

8. If the slope of a line is -4, then when x increases by 1,
   a. y increases by 4   **b.** y decreases by 4

9. Simplify $7 - 2(5 - 3)$.

10. The sum of a number and its reciprocal is equal to 3. Translate this statement to an equation. Do not solve it.

8.4 Addition and Subtraction of Radicals

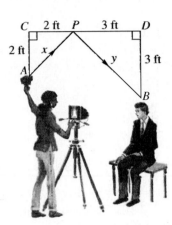

For a better lighting effect, a photographer bounces the light from a flash at A off a reflecting surface to take a picture of a client at B, as shown in the accompanying diagram. Determine the distance given by $x + y$ so that the photographer can set the camera at the correct f-stop. (See Example 3.)

OBJECTIVES

1 Add and subtract square roots.

2 Add and subtract square roots that involve rationalizing denominators.

3 Add and subtract higher roots.

1 In Section 2.1 we added algebraic expressions by combining *like terms* using the distributive property. For instance,

$$4x + 7x = (4 + 7)x \qquad \text{Distributive property.}$$
$$= 11x. \qquad \text{Simplify.}$$

Similarily, if x is replaced by $\sqrt{5}$,

$$4\sqrt{5} + 7\sqrt{5} = (4 + 7)\sqrt{5} \qquad \text{Distributive property.}$$
$$= 11\sqrt{5}. \qquad \text{Simplify.}$$

Thus, the distributive property indicates that we combine like square roots by combining their numerical coefficients. By definition, **like square roots** are square roots that have the same radicand, and note that we may only combine like square roots.

EXAMPLE 1   Simplify where possible. Assume $y > 0$.

a. $9\sqrt{2} + 6\sqrt{2}$ **b.** $\sqrt{3} - 14\sqrt{3}$ **c.** $4\sqrt{7y} + 5\sqrt{7y}$ **d.** $\sqrt{3} + \sqrt{2}$

Solution

a. $9\sqrt{2} + 6\sqrt{2} = (9 + 6)\sqrt{2}$ Distributive property.
$\qquad\qquad\quad = 15\sqrt{2}$ Simplify.

b. $\sqrt{3} - 14\sqrt{3} = 1\sqrt{3} - 14\sqrt{3}$ $\sqrt{3} = 1 \cdot \sqrt{3}$.
$\qquad\qquad\quad = (1 - 14)\sqrt{3}$ Distributive property.
$\qquad\qquad\quad = -13\sqrt{3}$ Simplify.

c. $4\sqrt{7y} + 5\sqrt{7y} = (4 + 5)\sqrt{7y}$ Distributive property.
$\qquad\qquad\quad = 9\sqrt{7y}$ Simplify.

d. $\sqrt{3}$ and $\sqrt{2}$ are not like square roots, and $\sqrt{3} + \sqrt{2}$ does not simplify.

PROGRESS CHECK 1   Simplify where possible. Assume $y > 0$.

a. $4\sqrt{7} + 11\sqrt{7}$ **b.** $\sqrt{5} + \sqrt{5}$ **c.** $\sqrt{5} - \sqrt{2}$ **d.** $2\sqrt{6y} - 9\sqrt{6y}$ ⌐

Simplifying square roots in a sum or difference sometimes results in like square roots, which can then be combined.

EXAMPLE 2   Simplify where possible. Assume $x > 0$.

a. $\sqrt{50} + \sqrt{18}$ **b.** $3\sqrt{5} + \sqrt{50}$ **c.** $4\sqrt{25x} - 6\sqrt{9x}$

Solution

a. First, simplify each square root.

$$\sqrt{50} = \sqrt{25 \cdot 2} = \sqrt{25} \cdot \sqrt{2} = 5\sqrt{2}$$
$$\sqrt{18} = \sqrt{9 \cdot 2} = \sqrt{9} \cdot \sqrt{2} = 3\sqrt{2}$$

The results are like square roots, so we combine as follows.

$$\sqrt{50} + \sqrt{18} = 5\sqrt{2} + 3\sqrt{2} \quad \text{Replace } \sqrt{50} \text{ by } 5\sqrt{2} \text{ and } \sqrt{18} \text{ by } 3\sqrt{2}.$$
$$= 8\sqrt{2} \quad \text{Combine like square roots.}$$

b. $3\sqrt{5}$ does not simplify, while $\sqrt{50} = 5\sqrt{2}$, as shown in part **a.** Thus,

$$3\sqrt{5} + \sqrt{50} = 3\sqrt{5} + 5\sqrt{2}.$$

We cannot simplify further because the sum does not contain like square roots.

c. Simplify each square root and then combine like square roots.

$$4\sqrt{25x} - 6\sqrt{9x} = 4 \cdot \sqrt{25} \cdot \sqrt{x} - 6 \cdot \sqrt{9} \cdot \sqrt{x} \quad \text{Multiplication property of square roots.}$$
$$= 4 \cdot 5\sqrt{x} - 6 \cdot 3\sqrt{x} \quad \sqrt{25} = 5 \text{ and } \sqrt{9} = 3.$$
$$= 20\sqrt{x} - 18\sqrt{x} \quad \text{Simplify.}$$
$$= 2\sqrt{x} \quad \text{Combine like square roots.}$$

Caution   Although $\sqrt{a}\sqrt{b} = \sqrt{ab}$ and $\sqrt{a}/\sqrt{b} = \sqrt{a/b}$ are square root properties, note that

$$\sqrt{a} + \sqrt{b} \quad \text{does not equal} \quad \sqrt{a + b}$$

for $a > 0$ and $b > 0$. For instance,

$$\sqrt{9} + \sqrt{16} = 3 + 4 = 7, \quad \text{while} \quad \sqrt{9 + 16} = \sqrt{25} = 5,$$

so $\sqrt{9} + \sqrt{16}$ does *not* equal $\sqrt{9 + 16}$.

PROGRESS CHECK 2   Simplify where possible. Assume $x > 0$.

a. $\sqrt{75} - \sqrt{12}$ **b.** $\sqrt{32} + \sqrt{28}$ **c.** $2\sqrt{36x} + 7\sqrt{4x}$ ⌐

EXAMPLE 3   Solve the problem in the section introduction on page 379.

Solution   Consider the sketch of the problem in Figure 8.5. In right triangles APC and BPD, x and y measure each hypotenuse, respectively, so the Pythagorean relation gives

$$x^2 = 2^2 + 2^2 \quad \text{and} \quad y^2 = 3^2 + 3^2$$
$$= 8 \qquad\qquad = 18$$
$$x = \sqrt{8} \qquad\qquad y = \sqrt{18}.$$

The distance $x + y$ is then

$$\sqrt{8} + \sqrt{18} = \sqrt{4 \cdot 2} + \sqrt{9 \cdot 2} \quad \text{Factor.}$$
$$= \sqrt{4} \cdot \sqrt{2} + \sqrt{9} \cdot \sqrt{2} \quad \text{Multiplication property of square roots.}$$
$$= 2\sqrt{2} + 3\sqrt{2} \quad \text{Simplify.}$$
$$= 5\sqrt{2}. \quad \text{Combine like square roots.}$$

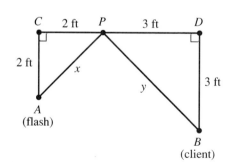

C   2 ft   P   3 ft   D
2 ft
x
y
3 ft
A
(flash)
B
(client)

Figure 8.5

Progress Check Answers

1. (a) $15\sqrt{7}$ (b) $2\sqrt{5}$ (c) Does not simplify
(d) $-7\sqrt{6y}$
2. (a) $3\sqrt{3}$ (b) $4\sqrt{2} + 2\sqrt{7}$ (c) $26\sqrt{x}$

The light from the flash travels $5\sqrt{2}$ ft to reach the client. Check by calculator that this distance is about 7.1 ft.

PROGRESS CHECK 3   Redo the problem in Example 3, but assume $AC = 1$ ft, $CP = 2$ ft, $DP = 4$ ft, and $BD = 2$ ft.   ⌐

2   Rationalizing denominators may also simplify radical expressions to a form that is more useful in addition and subtraction problems, as shown next.

EXAMPLE 4   Simplify each expression.

a. $\sqrt{125} - 10\sqrt{\dfrac{1}{5}}$

b. $\sqrt{2} + \dfrac{1}{\sqrt{2}}$

Solution

a. First, simplify.

$$\sqrt{125} = \sqrt{25 \cdot 5} = \sqrt{25} \cdot \sqrt{5} = 5\sqrt{5}$$

$$10\sqrt{\frac{1}{5}} = 10\sqrt{\frac{1 \cdot 5}{5 \cdot 5}} = 10\sqrt{\frac{5}{25}} = \frac{10\sqrt{5}}{\sqrt{25}} = \frac{10\sqrt{5}}{5} = 2\sqrt{5}$$

Then,

$$\sqrt{125} - 10\sqrt{\frac{1}{5}} = 5\sqrt{5} - 2\sqrt{5} = 3\sqrt{5}.$$

b. Rationalizing the denominator in $1/\sqrt{2}$ gives

$$\frac{1}{\sqrt{2}} = \frac{1}{\sqrt{2}} \cdot \frac{\sqrt{2}}{\sqrt{2}} = \frac{\sqrt{2}}{2}, \text{ or } \frac{1}{2}\sqrt{2}.$$

Then,

$$\sqrt{2} + \frac{1}{\sqrt{2}} = 1\sqrt{2} + \frac{1}{2}\sqrt{2} = \left(1 + \frac{1}{2}\right)\sqrt{2} = \frac{3}{2}\sqrt{2}.$$

PROGRESS CHECK 4   Simplify each expression.

a. $\sqrt{96} - 4\sqrt{\dfrac{3}{2}}$

b. $\sqrt{5} + \dfrac{1}{\sqrt{5}}$ ⌐

3   The distributive property is also the basis for adding and subtracting cube roots and higher roots. For example,

$$4\sqrt[3]{2} + 7\sqrt[3]{2} = (4 + 7)\sqrt[3]{2} \quad \text{Distributive property.}$$
$$= 11\sqrt[3]{2}. \quad \text{Simplify.}$$

Likewise,

$$9\sqrt[4]{5} - 3\sqrt[4]{5} = (9 - 3)\sqrt[4]{5} \quad \text{Distributive property.}$$
$$= 6\sqrt[4]{5}. \quad \text{Simplify.}$$

In general, **like radicals** are radicals that have the same radicand and the same index. The distributive property indicates that we combine like radicals by combining the numerical coefficients, and that we may only combine like radicals.

EXAMPLE 5   Simplify $\sqrt[3]{-24} + \sqrt[3]{81}$.

Solution   Use the methods in Section 8.2 to simplify each cube root.

$$\sqrt[3]{-24} = \sqrt[3]{-8 \cdot 3} = \sqrt[3]{-8} \cdot \sqrt[3]{3} = -2\sqrt[3]{3}$$
$$\sqrt[3]{81} = \sqrt[3]{27 \cdot 3} = \sqrt[3]{27} \cdot \sqrt[3]{3} = 3\sqrt[3]{3}$$

Progress Check Answers

3. $3\sqrt{5}$ ft, or about 6.7 ft

4. (a) $2\sqrt{6}$ (b) $\dfrac{6}{5}\sqrt{5}$

The results are like radicals, so we combine as follows.

$$\sqrt[3]{-24} + \sqrt[3]{81} = -2\sqrt[3]{3} + 3\sqrt[3]{3} \quad \text{Replace } \sqrt[3]{-24} \text{ by } -2\sqrt[3]{3} \text{ and } \sqrt[3]{81} \text{ by } 3\sqrt[3]{3}.$$
$$= (-2 + 3)\sqrt[3]{3} \quad \text{Distributive property.}$$
$$= 1\sqrt[3]{3}, \text{ or } \sqrt[3]{3} \quad \text{Simplify.}$$

Progress Check Answer

5. $-\sqrt[3]{2}$

PROGRESS CHECK 5 Simplify $\sqrt[3]{16} + \sqrt[3]{-54}$.

EXERCISES 8.4

In Exercises 1–30, simplify where possible. Assume all variables are positive.

1. $3\sqrt{3} + 5\sqrt{3}$
2. $6\sqrt{5} + 2\sqrt{5}$
3. $\sqrt{5} + 4\sqrt{5}$
4. $3\sqrt{7} + \sqrt{7}$
5. $\sqrt{2} - \sqrt{2}$
6. $\sqrt{3} - \sqrt{3}$
7. $2\sqrt{5} - 3\sqrt{5}$
8. $10\sqrt{3} - 11\sqrt{3}$
9. $6\sqrt{2} - 9\sqrt{2}$
10. $-5\sqrt{7} + 3\sqrt{7}$
11. $3\sqrt{2x} + 3\sqrt{2x}$
12. $2\sqrt{3y} + 2\sqrt{3y}$
13. $4\sqrt{5xy} - 3\sqrt{5xy}$
14. $-7\sqrt{10xy} + 3\sqrt{10xy}$
15. $2\sqrt{3} + 3\sqrt{2}$
16. $\sqrt{2} + \sqrt{11}$
17. $3\sqrt{7} - 2\sqrt{5}$
18. $2\sqrt{5} - 5\sqrt{2}$
19. $\sqrt{32} + \sqrt{18}$
20. $\sqrt{48} + \sqrt{27}$
21. $\sqrt{20} - \sqrt{45}$
22. $\sqrt{32} - \sqrt{50}$
23. $5\sqrt{3} + \sqrt{12}$
24. $2\sqrt{5} + \sqrt{20}$
25. $\sqrt{2} - \sqrt{200}$
26. $\sqrt{3} - \sqrt{300}$
27. $2\sqrt{16x} + 3\sqrt{4x}$
28. $3\sqrt{9y} + 2\sqrt{4y}$
29. $5\sqrt{36x^2y} - 2x\sqrt{9y}$
30. $\sqrt{xy^2} - 3y\sqrt{25x}$

In Exercises 31–34, give exact answers using radicals in simplest form.

31. Square $ABCD$ is inscribed in a larger square, as shown.
    a. Find the perimeter of $ABCD$.
    b. Find the area of $ABCD$.

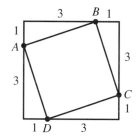

32. Square $ABCD$ is inscribed in a larger square, as shown.
    a. Find the perimeter of $ABCD$.
    b. Find the area of $ABCD$.

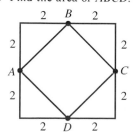

33. For the triangle pictured, find the hypotenuse of the following.
    a. The small triangle
    b. The large triangle
    c. Find the length of the piece marked x.

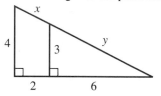

34. For the triangle pictured, find the hypotenuse of the following.
    a. The small triangle
    b. The large triangle
    c. Find the length of the piece marked x.

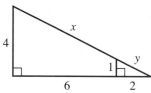

In Exercises 35–52, simplify each expression, rationalizing denominators when helpful. Assume all expressions represent real numbers.

35. $\sqrt{3} + \dfrac{1}{\sqrt{3}}$
36. $\sqrt{6} + \dfrac{1}{\sqrt{6}}$
37. $2\sqrt{3} - \dfrac{5}{\sqrt{3}}$
38. $3\sqrt{2} - \dfrac{5}{\sqrt{2}}$
39. $\sqrt{27} - \sqrt{\dfrac{1}{3}}$
40. $\sqrt{8} + 3\sqrt{\dfrac{1}{2}}$
41. $\sqrt{20} + 3\sqrt{\dfrac{1}{5}}$
42. $3\sqrt{18} - \sqrt{\dfrac{1}{2}}$
43. $5\sqrt[3]{7} - 2\sqrt[3]{7}$
44. $\sqrt[4]{5} + 3\sqrt[4]{5}$
45. $8\sqrt[4]{xy^2} + 9\sqrt[4]{xy^2}$
46. $5\sqrt[3]{x^2y^2} - 2\sqrt[3]{x^2y^2}$
47. $3\sqrt[4]{2} - \sqrt[4]{2}$
48. $6\sqrt[3]{5} + \sqrt[3]{5}$
49. $\sqrt[3]{-81} + \sqrt[3]{-24}$
50. $\sqrt[3]{256} + \sqrt[3]{-32}$
51. $\sqrt[4]{32} + 5\sqrt[4]{2}$
52. $\sqrt[4]{162} - \sqrt[4]{1,250}$

THINK ABOUT IT

1. This problem demonstrates that visual patterns are not always helpful when working with square roots.
   a. Is it true that $1 + 49 = 25 + 25$? Will the equality still be correct if you just write a square root symbol in front of each term? In other words, is it true that
   $$\sqrt{1} + \sqrt{49} = \sqrt{25} + \sqrt{25}?$$
   b. Is it true that $75 + 12 = 48 + 27$? Is it true that
   $$\sqrt{75} + \sqrt{12} = \sqrt{48} + \sqrt{27}?$$
   c. Given that $a + b = c + d$, then which of these statements is correct?

   i. $\sqrt{a + b} = \sqrt{c + d}$ **ii.** $\sqrt{a} + \sqrt{b} = \sqrt{c} + \sqrt{d}$

2. a. Using the figure, find the perimeter and area of the inscribed square.

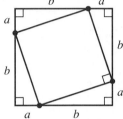

b. Find the area of the larger square and then explain, using the result from part **a**, why $(a + b)^2$ does not equal $a^2 + b^2$.

3. Simplify $\sqrt{a} + \dfrac{1}{\sqrt{a}}$, where $a > 0$.

4. Only one of these is true. Which is it? Explain.
   a. $\sqrt{-5} = -\sqrt{5}$
   b. $\sqrt[3]{-5} = -\sqrt[3]{5}$

5. If you allow factoring "over the irrationals" instead of just "over the integers," then you can factor $\sqrt{2}$ from both terms of $\sqrt{2} + \sqrt{6}$.
   a. What will the factored expression look like?
   b. Factor $\sqrt{3}$ from both terms of $2\sqrt{3} + \sqrt{6}$.

REMEMBER THIS

1. Multiply $(a - b)(a + b)$.
2. Multiply $(a - b)^2$.
3. Simplify $\dfrac{4 - 2x}{12}$.
4. Simplify $(\sqrt{3})^2 - (\sqrt{7})^2$.
5. Combine like terms: $3x + 7 - 4x - 8$.
6. Divide and simplify: $\dfrac{\sqrt{8}}{2}$.

7. One job pays $200 per week plus a 15 percent commission on sales, while another pays a straight 20 percent commission. For what amount of sales do the jobs have the same pay?
8. Find an equation for the line through $(-2,2)$ and $(4,0)$. Give the answer in slope-intercept form.
9. Reduce to lowest terms: $\dfrac{x^2 - 2x - 15}{x^3 - 9x}$.
10. Solve: $-x + 2 = 3 - x$.

8.5 Further Radical Simplifications

If a resistor of R_1 ohms is connected in parallel to a resistor of R_2 ohms, as shown in the diagram, then the total resistance of R is given by

$$R = \frac{R_1 R_2}{R_1 + R_2}.$$

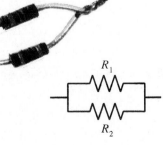

When R_1 and R_2 are related such that $R_1 = c$ ohms and $R_2 = \sqrt{c}$ ohms, find an expression for R in simplest radical form. (See Example 6.)

OBJECTIVES

1 Multiply radical expressions involving more than one term.

2 Use a special product formula to multiply certain radical expressions.

3 Divide radical expressions involving more than one term in the numerator and a single term in the denominator.

4 Use a conjugate to rationalize a denominator.

1 Multiplications involving radicals with more than one term are like multiplications involving polynomials with more than one term. That is, using the distributive property, we multiply each term of one radical expression by each term of the other radical expression; and then we combine like radicals.

EXAMPLE 1 Simplify each expression. Assume $x \geq 0$.

a. $\sqrt{3}(\sqrt{2} + \sqrt{5})$
b. $\sqrt{2}(\sqrt{6} - \sqrt{2})$
c. $(4 + \sqrt{5})(4 - \sqrt{5})$
d. $(2 + \sqrt{x})(3 + \sqrt{x})$

Solution

a. $\sqrt{3}(\sqrt{2} + \sqrt{5}) = \sqrt{3} \cdot \sqrt{2} \cdot + \sqrt{3} \cdot \sqrt{5}$ Distributive property.
$= \sqrt{6} + \sqrt{15}$ Multiplication property.

b. First, multiply.

$\sqrt{2}(\sqrt{6} - \sqrt{2}) = \sqrt{2} \cdot \sqrt{6} - \sqrt{2} \cdot \sqrt{2}$ Distributive property.
$= \sqrt{12} - \sqrt{4}$ Multiplication property.

Now, simplify. Because $\sqrt{4} = 2$ and $\sqrt{12} = \sqrt{4} \cdot \sqrt{3} = 2\sqrt{3}$,

$$\sqrt{2}(\sqrt{6} - \sqrt{2}) = 2\sqrt{3} - 2.$$

c. Multiply each term of $4 + \sqrt{5}$ by each term of $4 - \sqrt{5}$ and combine like radicals. The FOIL method (see Section 3.6) may be used here, as shown below.

$$
\begin{aligned}
(4 + \sqrt{5})(4 - \sqrt{5}) &= 4 \cdot 4 + 4(-\sqrt{5}) + (\sqrt{5})(4) + (\sqrt{5})(-\sqrt{5}) \\
&\qquad F \quad + \quad O \quad + \quad I \quad + \quad L \\
&= 16 \quad - 4\sqrt{5} \quad + 4\sqrt{5} \quad - \sqrt{25} \\
&= 16 \quad - 4\sqrt{5} \quad + 4\sqrt{5} \quad - 5 \\
&= 11
\end{aligned}
$$

d. Multiply each term of $2 + \sqrt{x}$ by each term of $3 + \sqrt{x}$ and combine like radicals.

$$
\begin{aligned}
(2 + \sqrt{x})(3 + \sqrt{x}) & \\
&= 2 \cdot 3 + 2(\sqrt{x}) + (\sqrt{x})(3) + (\sqrt{x})(\sqrt{x}) \quad \text{Multiply using FOIL.} \\
&= 6 + 2\sqrt{x} + 3\sqrt{x} + x \quad \text{Multiply and simplify.} \\
&= 6 + 5\sqrt{x} + x \quad \text{Combine like square roots.}
\end{aligned}
$$

PROGRESS CHECK 1 Simplify each expression. Assume $x \geq 0$.

a. $\sqrt{5}(\sqrt{7} + \sqrt{2})$
b. $\sqrt{3}(\sqrt{3} - \sqrt{6})$
c. $(2 - \sqrt{3})(2 + \sqrt{3})$
d. $(4 - \sqrt{x})(5 - \sqrt{x})$

2 The special product formulas from Section 3.6 are often used to multiply certain square root expressions. For instance, the expression in Example 1c is the product of the sum and the difference of two terms, so

$$
\begin{aligned}
(a + \quad b)(a - \quad b) &= a^2 - \quad b^2 \\
(4 + \sqrt{5})(4 - \sqrt{5}) &= (4)^2 - (\sqrt{5})^2 \\
&= 16 - 5 = 11.
\end{aligned}
$$

Note that this product is determined more easily by using

$$(a + b)(a - b) = a^2 - b^2.$$

Other useful special product formulas from Section 3.6 are

$$
\begin{aligned}
(a + b)^2 &= a^2 + 2ab + b^2 \\
(a - b)^2 &= a^2 - 2ab + b^2.
\end{aligned}
$$

Progress Check Answers

1. (a) $\sqrt{35} + \sqrt{10}$ (b) $3 - 3\sqrt{2}$ (c) 1
(d) $20 - 9\sqrt{x} + x$

EXAMPLE 2   Simplify each expression. Assume $y \geq 0$.

a. $(5 + \sqrt{2})^2$

b. $(\sqrt{y} + 3)(\sqrt{y} - 3)$

Solution

a. Use the formula for $(a + b)^2$ with $a = 5$ and $b = \sqrt{2}$.

$$
\begin{aligned}
(a + b)^2 &= a^2 + 2ab + b^2 \\
(5 + \sqrt{2})^2 &= (5)^2 + 2(5)(\sqrt{2}) + (\sqrt{2})^2 \\
&= 25 + 10\sqrt{2} + 2 \\
&= 27 + 10\sqrt{2}
\end{aligned}
$$

b. The product of the sum and difference of two terms is the square of the first term minus the square of the second term.

$$
\begin{aligned}
(a + b)\ (a - b) &= a^2 - b^2 \\
(\sqrt{y} + 3)(\sqrt{y} - 3) &= (\sqrt{y})^2 - (3)^2 \\
&= y - 9
\end{aligned}
$$

PROGRESS CHECK 2   Simplify each expression. Assume $y \geq 0$.

a. $(4 - \sqrt{7})^2$

b. $(6 - \sqrt{y})(6 + \sqrt{y})$

3   The next example shows methods for simplifying quotients involving radical expressions that have more than one term in the numerator and a single term in the denominator.

EXAMPLE 3   Simplify each expression.

a. $\dfrac{4 - 2\sqrt{3}}{12}$

b. $\dfrac{\sqrt{8} + \sqrt{28}}{\sqrt{2}}$

c. $\dfrac{5 + \sqrt{3}}{\sqrt{2}}$

Solution

a. Note that 2 is a common factor in the numerator and the denominator. Therefore, we factor and divide out the common factor.

$$
\dfrac{4 - 2\sqrt{3}}{12} = \dfrac{2(2 - \sqrt{3})}{12} \quad \text{Factor the numerator.}
$$

$$
= \dfrac{2 - \sqrt{3}}{6} \quad \text{Express in lowest terms.}
$$

b. Because $\sqrt{8}/\sqrt{2}$ and $\sqrt{28}/\sqrt{2}$ simplify easily, we choose the following method.

$$
\dfrac{\sqrt{8} + \sqrt{28}}{\sqrt{2}} = \dfrac{\sqrt{8}}{\sqrt{2}} + \dfrac{\sqrt{28}}{\sqrt{2}} \quad \dfrac{a + b}{c} = \dfrac{a}{c} + \dfrac{b}{c}.
$$

$$
= \sqrt{\dfrac{8}{2}} + \sqrt{\dfrac{28}{2}} \quad \text{Quotient property.}
$$

$$
= \sqrt{4} + \sqrt{14} \quad \text{Divide.}
$$

$$
= 2 + \sqrt{14} \quad \text{Simplify.}
$$

c. Multiply by 1 in the form $\sqrt{2}/\sqrt{2}$ to rationalize the denominator.

$$
\dfrac{5 + \sqrt{3}}{\sqrt{2}} = \dfrac{5 + \sqrt{3}}{\sqrt{2}} \cdot \dfrac{\sqrt{2}}{\sqrt{2}} \quad \text{Multiply by } \dfrac{\sqrt{2}}{\sqrt{2}}.
$$

$$
= \dfrac{5\sqrt{2} + \sqrt{6}}{2} \quad \text{Multiply the fractions.}
$$

Progress Check Answers

2. (a) $23 - 8\sqrt{7}$ (b) $36 - y$

PROGRESS CHECK 3   Simplify each expression.

a. $\dfrac{3 + 9\sqrt{5}}{6}$ b. $\dfrac{\sqrt{80} - \sqrt{20}}{\sqrt{5}}$ c. $\dfrac{\sqrt{3} + \sqrt{2}}{\sqrt{5}}$

4   To understand the procedure for rationalizing denominators that contain square roots and two terms, first recall from Example 1c that

$$(4 + \sqrt{5})(4 - \sqrt{5}) = 4^2 - (\sqrt{5})^2 = \overbrace{16 - 5}^{\text{a rational number}} = 11.$$

Then, note that for any nonnegative rational numbers a and b,

$$(\sqrt{a} + \sqrt{b})(\sqrt{a} - \sqrt{b}) = (\sqrt{a})^2 - (\sqrt{b})^2 = \overbrace{a - b}^{\text{a rational number}}.$$

In general, the expressions $\sqrt{a} + \sqrt{b}$ and $\sqrt{a} - \sqrt{b}$ are called **conjugates** of each other. Binomial denominators involving square roots are rationalized by multiplying the numerator and the denominator by the conjugate of the *denominator*.

EXAMPLE 4   Rationalize the denominator $\dfrac{6}{3 + \sqrt{7}}$.

Solution   The conjugate of $3 + \sqrt{7}$ is $3 - \sqrt{7}$, so in this fraction rationalize the denominator by multiplying the numerator and the denominator by $3 - \sqrt{7}$.

$$\dfrac{6}{3 + \sqrt{7}} = \dfrac{6}{3 + \sqrt{7}} \cdot \dfrac{3 - \sqrt{7}}{3 - \sqrt{7}} \qquad \text{Multiply using the conjugate of the denominator.}$$

$$= \dfrac{6(3 - \sqrt{7})}{(3)^2 - (\sqrt{7})^2} \qquad \text{Multiply fractions and use } (a + b)(a - b) = a^2 - b^2.$$

$$= \dfrac{6(3 - \sqrt{7})}{9 - 7} \qquad \text{Square in the denominator.}$$

$$= \dfrac{6(3 - \sqrt{7})}{2} \qquad \text{Simplify the denominator.}$$

$$= 3(3 - \sqrt{7}) \qquad \text{Express in lowest terms.}$$

$$= 9 - 3\sqrt{7} \qquad \text{Remove parentheses.}$$

PROGRESS CHECK 4   Rationalize the denominator $\dfrac{14}{3 + \sqrt{2}}$.

EXAMPLE 5   Rationalize the denominator $\dfrac{\sqrt{2}}{\sqrt{5} - \sqrt{3}}$.

Solution   The conjugate of $\sqrt{5} - \sqrt{3}$ is $\sqrt{5} + \sqrt{3}$.

$$\dfrac{\sqrt{2}}{\sqrt{5} - \sqrt{3}} = \dfrac{\sqrt{2}}{\sqrt{5} - \sqrt{3}} \cdot \dfrac{\sqrt{5} + \sqrt{3}}{\sqrt{5} + \sqrt{3}} \qquad \text{Multiply using the conjugate of the denominator.}$$

$$= \dfrac{\sqrt{2}(\sqrt{5} + \sqrt{3})}{5 - 3} \qquad \text{Multiply fractions and use } (a - b)(a + b) = a^2 - b^2.$$

$$= \dfrac{\sqrt{10} + \sqrt{6}}{2} \qquad \text{Subtract in the denominator and remove parentheses in the numerator.}$$

PROGRESS CHECK 5   Rationalize the denominator $\dfrac{\sqrt{5}}{\sqrt{6} - \sqrt{2}}$.

EXAMPLE 6   Solve the problem in the section introduction on page 383.

Solution   Replacing R_1 by c and R_2 by $\sqrt{c}$ in

$$R = \frac{R_1 R_2}{R_1 + R_2} \quad \text{gives} \quad R = \frac{c\sqrt{c}}{c + \sqrt{c}}.$$

To rationalize the denominator, multiply both the numerator and the denominator by $c - \sqrt{c}$, which is the conjugate of the denominator.

$$\frac{c\sqrt{c}}{c + \sqrt{c}} \cdot \frac{c - \sqrt{c}}{c - \sqrt{c}} = \frac{c\sqrt{c}(c - \sqrt{c})}{c^2 - c} \quad \begin{array}{l}\text{Multiply fractions and use} \\ (a + b)(a - b) = a^2 - b^2.\end{array}$$

$$= \frac{c\sqrt{c}(c - \sqrt{c})}{c(c - 1)} \quad \text{Factor the denominator.}$$

$$= \frac{\sqrt{c}(c - \sqrt{c})}{c - 1} \quad \text{Divide out the common factor } c.$$

$$= \frac{c\sqrt{c} - c}{c - 1} \quad \text{Remove parentheses.}$$

Thus, a formula for R for the given conditions is $R = \dfrac{c\sqrt{c} - c}{c - 1}$.

Note   In Examples 4–6 notice that we do not remove parentheses in the numerator until the fraction is expressed in lowest terms.

PROGRESS CHECK 6   Redo the problem in Example 6, but assume that R_1 and R_2 are related by $R_1 = a$ ohms and $R_2 = 2\sqrt{a}$ ohms.   ⌐

Progress Check Answer

6. $\dfrac{2a\sqrt{a} - 4a}{a - 4}$

EXERCISES 8.5

In Exercises 1–44, simplify each expression. Assume variables are not negative.

1. $\sqrt{2}(\sqrt{3} - \sqrt{7})$
2. $\sqrt{5}(\sqrt{6} - \sqrt{2})$
3. $\sqrt{3}(\sqrt{5} + \sqrt{10})$
4. $\sqrt{6}(\sqrt{5} + \sqrt{7})$
5. $\sqrt{x}(\sqrt{n} + \sqrt{a})$
6. $\sqrt{y}(\sqrt{a} - \sqrt{b})$
7. $\sqrt{6}(\sqrt{2} + \sqrt{3})$
8. $\sqrt{10}(\sqrt{5} + \sqrt{2})$
9. $\sqrt{3}(\sqrt{6} - \sqrt{3})$
10. $\sqrt{6}(\sqrt{12} - \sqrt{6})$
11. $(1 + \sqrt{2})(1 - \sqrt{2})$
12. $(1 + \sqrt{3})(1 - \sqrt{3})$
13. $(3 - \sqrt{5})(3 + \sqrt{5})$
14. $(4 - \sqrt{15})(4 + \sqrt{15})$
15. $(3 + 2\sqrt{3})(3 - 2\sqrt{3})$
16. $(5 + 3\sqrt{2})(5 - 3\sqrt{2})$
17. $(-2 + 3\sqrt{5})(-2 - 3\sqrt{5})$
18. $(-4 + 2\sqrt{2})(-4 - 2\sqrt{2})$
19. $(3 + \sqrt{x})(3 - \sqrt{x})$
20. $(8 - \sqrt{y})(8 + \sqrt{y})$
21. $(\sqrt{x} - y)(\sqrt{x} + y)$
22. $(\sqrt{x} - 3y)(\sqrt{x} + 3y)$

23. $(3 + \sqrt{2})(2 + \sqrt{2})$
24. $(2 + \sqrt{3})(3 + \sqrt{3})$
25. $(1 + \sqrt{2})(2 - \sqrt{2})$
26. $(1 + \sqrt{3})(3 - \sqrt{3})$
27. $(5 - \sqrt{2})(2 - \sqrt{2})$
28. $(4 - \sqrt{3})(3 - \sqrt{3})$
29. $(1 + \sqrt{x})(2 + \sqrt{x})$
30. $(3 - \sqrt{y})(4 - \sqrt{y})$
31. $(1 - \sqrt{xy})(2 - \sqrt{xy})$
32. $(4 + \sqrt{xy})(2 + \sqrt{xy})$
33. $(2 + \sqrt{x})(3 - \sqrt{x})$
34. $(3 + \sqrt{x})(2 - \sqrt{x})$
35. $(2 + \sqrt{2x})(3 + \sqrt{2x})$
36. $(1 + \sqrt{3x})(2 + \sqrt{3x})$
37. $(1 + \sqrt{2})^2$
38. $(1 + \sqrt{3})^2$
39. $(2 - \sqrt{3})^2$
40. $(3 - \sqrt{2})^2$
41. $(2 + \sqrt{x})^2$
42. $(2 - \sqrt{x})^2$
43. $(\sqrt{x} + \sqrt{y})(\sqrt{x} - \sqrt{y})$
44. $(\sqrt{x} + \sqrt{y})(\sqrt{y} - \sqrt{x})$

In Exercises 45–58, simplify the given quotient. Rationalize denominators in final answers, if necessary.

45. $\dfrac{3 - 3\sqrt{2}}{6}$

46. $\dfrac{2 + 4\sqrt{5}}{6}$

47. $\dfrac{10 + 5\sqrt{3}}{10}$

48. $\dfrac{6 + 18\sqrt{3}}{18}$

49. $\dfrac{\sqrt{6} + \sqrt{27}}{\sqrt{3}}$

50. $\dfrac{\sqrt{12} + \sqrt{15}}{\sqrt{3}}$

51. $\dfrac{3\sqrt{6} - \sqrt{12}}{\sqrt{6}}$

52. $\dfrac{5\sqrt{6} - \sqrt{18}}{\sqrt{6}}$

53. $\dfrac{1 + \sqrt{2}}{\sqrt{3}}$

54. $\dfrac{2 + \sqrt{3}}{\sqrt{5}}$

55. $\dfrac{1 + \sqrt{2}}{\sqrt{2}}$

56. $\dfrac{3 + \sqrt{3}}{\sqrt{3}}$

57. $\dfrac{1 - \sqrt{5}}{\sqrt{10}}$

58. $\dfrac{5 - \sqrt{6}}{\sqrt{3}}$

In Exercises 59–66, rationalize the denominator and simplify.

59. $\dfrac{1}{1 - \sqrt{2}}$

60. $\dfrac{1}{1 + \sqrt{2}}$

61. $\dfrac{\sqrt{2}}{\sqrt{3} - \sqrt{2}}$

62. $\dfrac{\sqrt{3}}{\sqrt{6} - \sqrt{5}}$

63. $\dfrac{\sqrt{5}}{\sqrt{7} - \sqrt{2}}$

64. $\dfrac{\sqrt{7}}{\sqrt{8} - \sqrt{5}}$

65. $\dfrac{\sqrt{7}}{\sqrt{2} - \sqrt{5}}$

66. $\dfrac{\sqrt{3}}{\sqrt{5} - \sqrt{8}}$

In Exercises 67–72, solve the problem and give the answer in simplest radical form.

67. An important expression in statistics is $\sqrt{\dfrac{p(1 - p)}{n}}$, which is used in computing the margin of error for a survey of n people, where p reflects the proportion who agree on some opinion.

Assume $p = 0.5$ and simplify the formula.

68. Repeat Exercise 67, but use $p = 0.1$.

69. The harmonic mean of two numbers a and b is given by $\dfrac{2ab}{a + b}$.

   a. What is the harmonic mean of n and $\sqrt{n}$?

   b. Compute the harmonic mean of 3 and 9 by both formulas.

70. The contraharmonic mean of two numbers a and b is given by $\dfrac{a^2 + b^2}{a + b}$.

   a. What is the contraharmonic mean of n and $\sqrt{n}$?

   b. Compute the contraharmonic mean of 3 and 9 by both formulas.

71. According to Ohm's law, if three resistors of R_1, R_2, and R_3 ohms are connected in parallel, then the total resistance of the circuit is given by

$$R = \frac{R_1 R_2 R_3}{R_1 R_2 + R_1 R_3 + R_2 R_3}$$

Suppose $R_1 = c$ ohms, while $R_2 = R_3 = \sqrt{c}$ ohms. Find an expression for R in simplest radical form.

72. Repeat Exercise 71, but assume that $R_1 = R_2 = c$ ohms, while $R_3 = \sqrt{c}$ ohms.

THINK ABOUT IT

1. Use the rectangle shown.
   a. Find the area.
   b. Find the length of the diagonal.

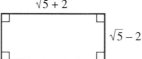

2. The diamond has vertices at the midpoints of the sides of the rectangle. The length of the rectangle is $\sqrt{5} + 2$, and the width is $\sqrt{5} - 2$.

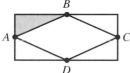

   a. Find the area of the shaded triangle.
   b. Use the answer from part **a** to find the area of the diamond.
   c. Draw segments AC and BD, and then explain why it is "obvious" that the area of the diamond is half the area of the rectangle.

3. **a.** Show that $(\sqrt{5} + 2)(\sqrt{5} - 2)$ equals 1.
   b. Show that $(\sqrt{10} + 3)(\sqrt{10} - 3)$ equals 1.
   c. Show that $(\sqrt{x} + y)(\sqrt{x} - y)$ equals 1 whenever x is one more than y^2.

4. Multiply $(1 + \sqrt[3]{2})(1 - \sqrt[3]{2} + \sqrt[3]{4})$.

5. Multiply $(1 + \sqrt[4]{2})(1 - \sqrt[4]{2} + \sqrt[4]{4} - \sqrt[4]{8})$.

REMEMBER THIS

1. Simplify $(\sqrt{3x})^2$. Assume $x \geq 0$.
2. Simplify $(\sqrt{x-3})^2$. Assume $x \geq 3$.
3. Is it true that if $x^2 = y^2$, then it must follow that $x = y$? Explain.
4. It is true that $1 + 2$ equals 3. Is it also true that $1^2 + 2^2$ equals 3^2?
5. Approximate $8/\pi^2$ to the nearest hundredth using a calculator.

6. Simplify $5\sqrt{3} + 4\sqrt{3}$.

7. Rationalize the denominator $\dfrac{1}{\sqrt{3}}$.

8. Explain why the point $(1,2)$ is a solution to the system
$$4y - 3x = 5$$
$$y = x + 1.$$

9. The price of a car plus 5 percent sales tax came to $7,140. What was the price of the car before the tax?
10. Factor $9x^2 - 12xy + 4y^2$.

8.6 Radical Equations

The period T of a pendulum is the time required for the pendulum to complete one round-trip of motion—that is, one complete cycle. When the period is measured in seconds and the pendulum length ℓ is measured in feet, then a formula for the period is

$$T = 2\pi \sqrt{\dfrac{\ell}{32}}.$$

To the nearest hundredth of a foot, what is the length of a pendulum whose period is 1 second? (See Example 7.)

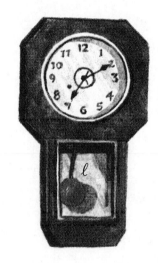

OBJECTIVES

1 Solve radical equations containing square roots.

2 Solve radical equations involving extraneous solutions.

3 Solve applied problems involving radical equations.

1 An equation in which the unknown appears in a radical is called a **radical equation.** In this section we focus only on radical equations that contain square roots, like

$$\sqrt{x-3} = 4, \qquad \sqrt{8x-10} = \sqrt{3x}, \qquad 1 + \sqrt{x+1} = x.$$

Such equations may be solved using these steps.

To Solve Radical Equations Containing Square Roots

1. If necessary, isolate a square root term on one side of the equation.
2. Square both sides of the equation.
3. Solve the resulting equation, and check all solutions in the *original* equation.

This procedure is illustrated in Examples 1–3.

EXAMPLE 1 Solve the equation $\sqrt{x-3}=4$.

Solution The square root term is already isolated on one side of the equation, so square both sides of the equation and solve for x.

$$\sqrt{x-3}=4$$
$$(\sqrt{x-3})^2=4^2 \quad \text{Square both sides.}$$
$$x-3=16 \quad \text{Simplify.}$$
$$x=19 \quad \text{Add 3 to both sides.}$$

To check, replace x by 19 in the original equation.

$$\sqrt{x-3}=4$$
$$\sqrt{19-3}\overset{?}{=}4 \quad \text{Replace }x\text{ by 19.}$$
$$\sqrt{16}\overset{?}{=}4$$
$$4\overset{\checkmark}{=}4$$

The proposed solution checks, and the solution set is $\{19\}$.

PROGRESS CHECK 1 Solve $\sqrt{x+4}=5$. ⌐

EXAMPLE 2 Solve $\sqrt{8x-10}=\sqrt{3x}$.

Solution A square root term is isolated on each side of the equation, and squaring both sides eliminates both square roots.

$$\sqrt{8x-10}=\sqrt{3x}$$
$$(\sqrt{8x-10})^2=(\sqrt{3x})^2 \quad \text{Square both sides.}$$
$$8x-10=3x \quad \text{Simplify.}$$

Now, solve for x and check.

$$8x-10=3x$$
$$-10=-5x \quad \text{Subtract 8x from both sides.}$$
$$2=x \quad \text{Divide both sides by }-5.$$

Check
$$\sqrt{8x-10}=\sqrt{3x}$$
$$\sqrt{8(2)-10}\overset{?}{=}\sqrt{3(2)} \quad \text{Replace }x\text{ by 2.}$$
$$\sqrt{6}\overset{\checkmark}{=}\sqrt{6}$$

Thus, the solution set is $\{2\}$.

PROGRESS CHECK 2 Solve $\sqrt{2x-3}=\sqrt{5x-18}$. ⌐

EXAMPLE 3 Solve the equation $2\sqrt{x}-1=8$.

Solution First, isolate the square root term on one side of the equation.

$$2\sqrt{x}-1=8$$
$$2\sqrt{x}=9 \quad \text{Add 1 to both sides.}$$

Now, square both sides of the equation and solve for x.

$$(2\sqrt{x})^2=9^2 \quad \text{Square both sides.}$$
$$2^2(\sqrt{x})^2=9^2 \quad \text{Product-to-a-power property.}$$
$$4x=81 \quad \text{Simplify.}$$
$$x=\tfrac{81}{4} \quad \text{Divide both sides by 4.}$$

Replacing x by $\frac{81}{4}$ in the left member of the original equation gives

$$2\sqrt{\tfrac{81}{4}} - 1 = 2(\tfrac{9}{2}) - 1 = 9 - 1 = 8.$$

So the proposed solution checks, and the solution set is $\{\frac{81}{4}\}$.

PROGRESS CHECK 3   Solve $3\sqrt{x} - 2 = 6$. ⌐

2   The check in step 3 in the procedure for solving radical equations containing square roots is *not optional,* because this procedure is based on the following property.

Squaring Property of Equality

If A and B are algebraic expressions, then the solutions of the equation $A = B$ are among the solutions of the equation $A^2 = B^2$.

   To illustrate that squaring both sides of an equation can produce additional solutions, consider the equation

$$x = 2, \text{ which has one solution, namely 2.}$$

Squaring both sides of $x = 2$ gives

$$x^2 = 4, \text{ which has two solutions, 2 and } -2.$$

The squaring property of equality guarantees that every solution of the equation $A = B$ is a solution of $A^2 = B^2$; but solutions of $A^2 = B^2$ may or may not be solutions of $A = B$, so checking is necessary. Solutions of $A^2 = B^2$ that do not satisfy the original equation are called **extraneous solutions.** Examples 4 and 5 illustrate this possibility.

EXAMPLE 4   Solve $\sqrt{x + 2} = x$.

Solution   A square root term is already isolated on one side of the equation, so square both equation members. Then solve the resulting quadratic equation by the methods of Section 4.6.

$\sqrt{x + 2} = x$	Given equation.
$(\sqrt{x + 2})^2 = x^2$	Square both sides.
$x + 2 = x^2$	Simplify.
$0 = x^2 - x - 2$	Rewrite the equation so one side is 0.
$0 = (x - 2)(x + 1)$	Factor the nonzero side.
$x - 2 = 0$ or $x + 1 = 0$	Set each factor equal to 0.
$x = 2$ $x = -1$	Solve each linear equation.

Now, check.

$$\sqrt{x + 2} = x \qquad\qquad \sqrt{x + 2} = x$$
$$\sqrt{2 + 2} \overset{?}{=} 2 \qquad\quad \sqrt{-1 + 2} \overset{?}{=} -1$$
$$\sqrt{4} \overset{?}{=} 2 \qquad\qquad\quad \sqrt{1} \overset{?}{=} -1$$
$$2 \overset{\checkmark}{=} 2 \qquad\qquad 1 \neq -1 \quad \text{Extraneous solution}$$

The check shows that 2 is a solution while -1 is extraneous. Thus, the solution set is $\{2\}$.

PROGRESS CHECK 4   Solve $\sqrt{12 - x} = x$. ⌐

   In the next example squaring both equation members leads to an extraneous solution in the case when the original equation has no solution.

EXAMPLE 5   Solve $5 + \sqrt{x - 2} = 1$.

Solution   First, isolate the square root term on one side of the equation. Then, square both sides and solve for x.

$$
\begin{array}{ll}
5 + \sqrt{x - 2} = 1 & \\
\sqrt{x - 2} = -4 & \text{Subtract 5 from both sides.} \\
(\sqrt{x - 2})^2 = (-4)^2 & \text{Square both sides.} \\
x - 2 = 16 & \text{Simplify.} \\
x = 18 & \text{Add 2 to both sides.}
\end{array}
$$

Now, check the proposed solution in the original equation.

$$
\begin{array}{ll}
5 + \sqrt{x - 2} = 1 & \\
5 + \sqrt{18 - 2} \overset{?}{=} 1 & \text{Replace } x \text{ by 18.} \\
5 + \sqrt{16} \overset{?}{=} 1 & \\
5 + 4 \overset{?}{=} 1 & \\
9 \neq 1 &
\end{array}
$$

Since $9 = 1$ is a false statement, 18 is not a solution of the original equation. Thus, $5 + \sqrt{x - 2} = 1$ has no solution, and the solution set is $\emptyset$.

Note   Isolating the square root term in this example led to the equation

$$\sqrt{x - 2} = -4.$$

At this point, if you recognize that the radical sign $\sqrt{}$ symbolizes the *nonnegative* square root, you know that the two sides can never be equal and that there is no solution.

PROGRESS CHECK 5   Solve $\sqrt{x + 5} + 2 = 0$.   ⌐

In the next example isolating the square root term leads to an equation with two terms on one side of the equation. To square a binomial, use

$$(a + b)^2 = a^2 + 2ab + b^2 \quad \text{or} \quad (a - b)^2 = a^2 - 2ab + b^2.$$

For instance, in Example 6 we expand $(x - 1)^2$ as follows.

$$
\begin{array}{ccccccc}
(a & - & b)^2 = & a^2 & - & 2ab & + & b^2 \\
\downarrow & & \downarrow & \downarrow & & \downarrow & & \downarrow \\
(x & - & 1)^2 = & (x)^2 & - & 2(x)(1) & + & (1)^2 \\
& & = & x^2 & - & 2x & + & 1
\end{array}
$$

EXAMPLE 6   Solve $1 + \sqrt{x + 1} = x$.

Solution   First, isolate the radical on one side of the equation.

$$
\begin{array}{ll}
1 + \sqrt{x + 1} = x & \\
\sqrt{x + 1} = x - 1 & \text{Subtract 1 from both sides.}
\end{array}
$$

Then, square both sides of the equation and solve the resulting quadratic equation.

$$
\begin{array}{ll}
(\sqrt{x + 1})^2 = (x - 1)^2 & \text{Square both sides.} \\
x + 1 = x^2 - 2x + 1 & \text{Remove parentheses.} \\
0 = x^2 - 3x & \text{Rewrite the equation so one side is 0.} \\
0 = x(x - 3) & \text{Factor the nonzero side.} \\
x = 0 \quad \text{or} \quad x - 3 = 0 & \text{Set each factor equal to 0.} \\
x = 3 &
\end{array}
$$

Now, check.

$$1 + \sqrt{x+1} = x$$
$$1 + \sqrt{0+1} \stackrel{?}{=} 0$$
$$1 + \sqrt{1} \stackrel{?}{=} 0$$
$$1 + 1 \stackrel{?}{=} 0$$
$$2 \neq 0 \quad \text{Extraneous solution}$$

$$1 + \sqrt{x+1} = x$$
$$1 + \sqrt{3+1} \stackrel{?}{=} 3$$
$$1 + \sqrt{4} \stackrel{?}{=} 3$$
$$1 + 2 \stackrel{?}{=} 3$$
$$3 \stackrel{\vee}{=} 3$$

Only 3 is a solution of the original equation, so the solution set is {3}.

PROGRESS CHECK 6   Solve $7 + \sqrt{x-1} = x$.

3   Formulas that contain square roots can lead to radical equations, as shown next.

EXAMPLE 7   Solve the problem in the section introduction on page 389.

Solution   Replacing T by 1 in

$$T = 2\pi\sqrt{\frac{\ell}{32}} \quad \text{gives} \quad 1 = 2\pi\sqrt{\frac{\ell}{32}}.$$

The square root term is already isolated on one side of the equation, so square both sides of the equation and solve for ℓ.

$$1^2 = \left(2\pi\sqrt{\frac{\ell}{32}}\right)^2 \qquad \text{Square both sides.}$$

$$1^2 = 2^2\pi^2\left(\sqrt{\frac{\ell}{32}}\right)^2 \qquad \text{Product-to-a-power property.}$$

$$1 = 4\pi^2 \cdot \frac{\ell}{32} \qquad \text{Simplify.}$$

$$1 = \frac{\pi^2}{8}(\ell) \qquad \text{Simplify.}$$

$$\frac{8}{\pi^2} \cdot 1 = \frac{8}{\pi^2} \cdot \frac{\pi^2}{8} \cdot \ell \qquad \text{Multiply both sides by } \frac{8}{\pi^2}.$$

$$\frac{8}{\pi^2} = \ell \qquad \text{Simplify.}$$

Now, approximate $8/\pi^2$ using a calculator.

$$8 \boxed{\div} \pi \boxed{x^2} \boxed{=} \boxed{0.8105694}$$

To the nearest hundredth, the length of the pendulum is 0.81 ft.

   To check this answer, replace ℓ by 0.81 in the given formula and compute $2\pi\sqrt{0.81/32}$.

$$0.81 \boxed{\div} 32 \boxed{=} \boxed{\sqrt{}} \boxed{\times} 2 \boxed{\times} \pi \boxed{=} \boxed{0.9996486}$$

Since ℓ was approximated to the nearest hundredth, round off the result to the same decimal place, to obtain 1.00. When $\ell = 0.81$ ft, $T = 1.00$ second, as given in the problem.

PROGRESS CHECK 7   To the nearest hundredth, what is the length of a pendulum whose period is 2 seconds?

Progress Check Answers
6. {10}
7. 3.24 ft

EXERCISES 8.6

In Exercises 1–28, solve the equation and check the solution.

1. $\sqrt{x-2} = 5$
2. $\sqrt{x+2} = 5$
3. $\sqrt{2x-3} = 1$
4. $\sqrt{3x-2} = 1$
5. $\sqrt{5-x} = 3$
6. $\sqrt{3-x} = 5$
7. $\sqrt{x+1} = -2$
8. $\sqrt{x-2} = -1$
9. $\sqrt{2x+3} = \sqrt{4x}$
10. $\sqrt{x+2} = \sqrt{5x}$
11. $\sqrt{2x-1} = \sqrt{3x-2}$
12. $\sqrt{3x+4} = \sqrt{4x+2}$
13. $\sqrt{7x+2} = \sqrt{8x+2}$
14. $\sqrt{5-x} = \sqrt{6-x}$
15. $3\sqrt{x} = \sqrt{x+16}$
16. $2\sqrt{x} = \sqrt{x+6}$
17. $\sqrt{x+5} = \sqrt{3x+9}$
18. $\sqrt{-2x} = \sqrt{x+9}$
19. $2\sqrt{x} - 4 = -1$
20. $3\sqrt{x} + 3 = 5$
21. $\sqrt{2x+3} = 0$
22. $\sqrt{x+1} - 2 = -2$
23. $5\sqrt{x+1} - 1 = 4$
24. $6\sqrt{5-x} + 2 = 8$
25. $2\sqrt{x-1} - 1 = 2$
26. $4\sqrt{x+3} + 7 = 17$
27. $3\sqrt{5x-1} - 3 = -2$
28. $6\sqrt{4x+3} - 5 = -3$

In Exercises 29–50, squaring may introduce extraneous solutions. Be sure to check all apparent solutions in the original equation.

29. $\sqrt{x+6} = x$
30. $\sqrt{x+20} = x$
31. $\sqrt{15-2x} = x$
32. $\sqrt{3-2x} = x$
33. $\sqrt{8x+12} = 2x$
34. $\sqrt{7-6x} = 4x$
35. $x = \sqrt{5x-6}$
36. $x = \sqrt{7x-12}$

37. $2x = \sqrt{9x-2}$
38. $3x = \sqrt{11x-2}$
39. $2 + \sqrt{x-1} = 1$
40. $4 + \sqrt{2x-4} = 2$
41. $1 - \sqrt{x+3} = 5$
42. $6 - \sqrt{5-x} = 7$
43. $4 + \sqrt{x+2} = x$
44. $5 + \sqrt{x-5} = x$
45. $\sqrt{x+4} - 2 = x$
46. $\sqrt{x+9} - 3 = x$
47. $2 + \sqrt{x-3} = x - 1$
48. $3 + \sqrt{x-2} = x + 1$
49. $\sqrt{x+1} + 5 = x - 6$
50. $\sqrt{x+5} + 4 = x - 3$

In Exercises 51 and 52, use the formula given in the section-opening problem.

51. To the nearest hundredth of a foot, what is the length of a pendulum whose period is 3 seconds?
52. To the nearest hundredth of a foot, what is the length of a pendulum whose period is $\frac{1}{2}$ second?
53. The physicist Robert Hooke (1635–1703) found that the formula $T = 2\pi\sqrt{\dfrac{m}{k}}$ can describe the period T of a mass m bobbing up and down on a spring. The period is the time it takes the mass to bounce up and down once after it is stretched and released. The constant k depends on the elasticity of the particular spring, and it is different for different springs. For a spring with $k = 10$, what mass will have a period of 1 second? Assume the mass is given in kilograms. Round to the nearest thousandth of a kilogram.
54. Repeat Exercise 53, but assume that the period is now 2 seconds.
55. There are many kinds of averages. One of them is based on the square root of the average of squares and is called the root-mean-square. The formula $A = \sqrt{\dfrac{m^2 + n^2}{2}}$ gives the root-mean-square for two positive numbers m and n. Find n if $m = 1$ and $A = \sqrt{5}$.
56. Repeat Exercise 55, but use $m = 1$ and $A = \sqrt{13}$.

THINK ABOUT IT

1. The formula $T = 2\pi r\sqrt{\dfrac{r}{GM}}$ is used in the study of the motion of planets. T is the time for one orbit, M is the mass of the sun, r is the distance from the planet to the sun, and G is a constant called the constant of universal gravitation. When distance is measured in meters and time in seconds, then the value of G is 6.67×10^{-11}. Use this formula to find the mass (in kilograms) of the sun. We know that the distance from the earth to the sun is 93 million mi (or 1.5×10^{11} m), and that it takes 1 year (or 3.15×10^7 seconds) for earth to orbit the sun.

2. The formula $m = \dfrac{m_0}{\sqrt{1 - \dfrac{v^2}{c^2}}}$ is used in the theory of relativity
to show how the mass of an object changes as it moves. The mass is m_0 when the object is still, v is the velocity at which it moves, and c is the speed of light.
   a. Find (to the nearest pound) the mass of a 100-lb object that moves at half the speed of light.
   b. What is the mass if it moves at 99 percent of the speed of light?

3. The formula for the escape velocity of a rocket (the speed it must reach to continue into outer space without falling back to earth) is $v = \sqrt{\dfrac{2GM_e}{R_e}}$, where G is the constant of universal gravitation and M_e and R_e are the mass and the radius of the earth. Solve this formula for R_e.

4. In Example 7 the formula $T = 2\pi\sqrt{\dfrac{\ell}{32}}$ gives the period of a pendulum. Is it true that doubling the length will double the period? Explain.

5. Given the equation $\sqrt{x + 5} + \sqrt{x} = \sqrt{6x + 1}$.
   a. Show by substitution that 4 is a solution to the equation.
   b. See if you can algebraically solve the equation and arrive at the solution set. You must go *twice* through the process of squaring both sides and simplifying.

REMEMBER THIS

1. Simplify $(3x^2)^3$.
2. Multiply $x^3(x^2 + y)$.
3. Multiply $3 \cdot \frac{1}{2} \cdot \frac{3}{2}$
4. True or false? $(2^3)^4 = (2^4)^3$.
5. Subtract $\frac{1}{3}$ from $-\frac{4}{3}$.

6. Rationalize the denominator $\dfrac{-2}{1 + \sqrt{3}}$.

7. Which one of these statements is the true one? Assume that m and n are positive numbers.
   a. $\sqrt{m} \cdot \sqrt{n} = \sqrt{mn}$
   b. $\sqrt{m} + \sqrt{n} = \sqrt{m + n}$

8. Graph the line given by $y = 2x - 4$.

9. Solve $3x(x - 3) = 0$.
10. According to the father's will, the estate is split among three children such that the middle child gets twice as much money as the youngest but only half as much as the oldest. If the total bequest is $91,000, how much does the oldest child inherit?

8.7   Rational Exponents

The compounded annual interest rate r that is required for an investment to double in n years is given by

$$r = 2^{1/n} - 1.$$

To the nearest hundredth of a percent, what compounded annual interest rate is needed for an investment to double in 10 years? (See Example 2.)

OBJECTIVES

1 Evaluate expressions containing rational exponents.

2 Use exponent properties to simplify expressions with rational exponents.

3 Use exponent properties to simplify radicals.

It is a goal of algebra to be able to use any real number as an exponent, and in this section we will give meaning to expressions with rational number exponents, such as

$$9^{1/2}, \qquad 8^{2/3}, \qquad \text{and} \qquad 32^{-4/5}.$$

The determining principle is that properties that apply for integer exponents should continue to work for rational number exponents.

1   For integral exponents, recall that to raise an exponential expression to a power, we use the same base and multiply the exponents. That is,

$$(a^m)^n = a^{mn}.$$

If this law is to hold for rational exponents, then consider

$$(9^{1/2})^2 = 9^{(1/2)2} = 9^1 = 9.$$

We see that squaring $9^{1/2}$ results in 9, so $9^{1/2}$ is a square root of 9. We choose to define $9^{1/2}$ as the principal square root of 9, so

$$9^{1/2} = \sqrt{9} = 3.$$

Similarly,

$$(8^{1/3})^3 = 8^{(1/3)3} = 8^1 = 8.$$

Raising $8^{1/3}$ to the third power results in 8, so we define

$$8^{1/3} = \sqrt[3]{8}, \qquad \text{so that} \qquad 8^{1/3} = 2.$$

In general, our previous laws of exponents may be extended by making the following definition.

Definition of $a^{1/n}$

If n is a positive integer and $\sqrt[n]{a}$ is a real number, then

$$a^{1/n} = \sqrt[n]{a}.$$

EXAMPLE 1   Evaluate each expression.

a. $36^{1/2}$ b. $64^{1/3}$ c. $32^{1/5}$ d. $(-8)^{1/3}$

Solution   Convert to radical form and then simplify.

a. $36^{1/2} = \sqrt{36} = 6$
b. $64^{1/3} = \sqrt[3]{64} = 4$
c. $32^{1/5} = \sqrt[5]{32} = 2$
d. $(-8)^{1/3} = \sqrt[3]{-8} = -2$

PROGRESS CHECK 1   Evaluate each expression.

a. $64^{1/2}$ b. $125^{1/3}$ c. $81^{1/4}$ d. $(-64)^{1/3}$ ⌐

When the indicated root is not a rational number, then $a^{1/n}$ may be approximated on a calculator, as shown next.

EXAMPLE 2   Solve the problem in the section introduction on page 395.

Solution   Replacing n by 10 in

$$r = 2^{1/n} - 1 \qquad \text{gives} \qquad r = 2^{1/10} - 1.$$

To approximate $2^{1/10} - 1$, recognize that $2^{1/10} - 1$ equals $\sqrt[10]{2} - 1$ and use the root key $\boxed{\sqrt[x]{y}}$ on a calculator.

$$2 \boxed{\sqrt[x]{y}} \; 10 \boxed{-} \; 1 \boxed{=} \; \boxed{0.0717734}$$

Because $0.0717734 = 7.17734$ percent, the required interest rate, to the nearest hundredth of a percent, is 7.18 percent.

Note   Based on the definition of $a^{1/n}$, some calculators use expressions like $\boxed{y^{1/x}}$ to label the root key.

PROGRESS CHECK 2   To the nearest hundredth of a percent, what compounded annual interest rate is needed for an investment to double in six years?   ⌐

To extend the definition of rational exponent to the case when the numerator is not 1, we extend the power-to-a-power property in the form

$$a^{mn} = (a^m)^n.$$

For instance, to evaluate $9^{3/2}$, we reason that

$$9^{3/2} = (9^{1/2})^3 = (\sqrt{9})^3 = 3^3 = 27,$$

or

$$9^{3/2} = (9^3)^{1/2} = 729^{1/2} = \sqrt{729} = 27.$$

By both methods, $9^{3/2} = 27$. Although either method may be used, note that if the root is a rational number, then it is easier to find the root first. This example suggests how to define any exponent that is a rational number.

Rational Exponents

If m and n are integers with $n > 0$, and if m/n represents a reduced fraction such that $a^{1/n}$ is a real number, then

$$a^{m/n} = (\sqrt[n]{a})^m = \sqrt[n]{a^m}.$$

EXAMPLE 3   Evaluate each expression.

a. $25^{3/2}$ **b.** $27^{2/3}$ **c.** $(-8)^{2/3}$ **d.** $16^{-1/2}$

Solution   Use $a^{m/n} = (\sqrt[n]{a})^m$, since each root is a rational number.

a. $25^{3/2} = (\sqrt{25})^3 = 5^3 = 125$
b. $27^{2/3} = (\sqrt[3]{27})^2 = 3^2 = 9$
c. $(-8)^{2/3} = (\sqrt[3]{-8})^2 = (-2)^2 = 4$
d. $16^{-1/2} = (\sqrt{16})^{-1} = 4^{-1} = \frac{1}{4}$

Note   The most straightforward, general way to find rational powers on a calculator is to use the power key $\boxed{y^x}$ and enclose the exponent within parentheses. By this method, $25^{3/2}$ is computed as

$$25 \boxed{y^x} \boxed{(} 3 \boxed{\div} 2 \boxed{)} \boxed{=} \boxed{\quad 125}.$$

PROGRESS CHECK 3   Evaluate each expression.

a. $100^{3/2}$ **b.** $8^{5/3}$ **c.** $(-32)^{2/5}$ **d.** $81^{-1/2}$ ⌐

In Example 3d an alternative method for evaluating $16^{-1/2}$ is to recognize that the negative exponent definition $a^{-n} = 1/a^n$ (where $a \neq 0$ and n is an integer) extends to give

$$a^{-r} = \frac{1}{a^r}, \quad \text{where} \quad a \neq 0 \text{ and } r \text{ is a rational number.}$$

Progress Check Answers
2. 12.25 percent
3. (a) 1,000   (b) 32   (c) 4   (d) $\frac{1}{9}$

Thus,

$$16^{-1/2} = \frac{1}{16^{1/2}} = \frac{1}{\sqrt{16}} = \frac{1}{4}.$$

The simplifications in Example 4 also use this definition.

EXAMPLE 4   Evaluate each expression.

a. $4^{-5/2}$ **b.** $(-27)^{-5/3}$

Solution

a. $4^{-5/2} = \dfrac{1}{4^{5/2}} = \dfrac{1}{(\sqrt{4})^5} = \dfrac{1}{2^5} = \dfrac{1}{32}$

b. $(-27)^{-5/3} = \dfrac{1}{(-27)^{5/3}} = \dfrac{1}{(\sqrt[3]{-27})^5} = \dfrac{1}{(-3)^5} = -\dfrac{1}{243}$

PROGRESS CHECK 4   Evaluate each expression.

a. $16^{-3/2}$ **b.** $(-64)^{-2/3}$

2 Because all previous laws of exponents hold for rational number exponents, we may use these extended properties to simplify expressions with rational exponents.

EXAMPLE 5   Perform the indicated operations and write the result with only positive exponents.

a. $2^{1/5} \cdot 2^{3/5}$ **b.** $(7^{1/3})^{3/2}$ **c.** $\dfrac{5^{-1/2}}{5^{3/2}}$

Solution

a. $2^{1/5} \cdot 2^{3/5} = 2^{1/5 + 3/5} = 2^{4/5}$
b. $(7^{1/3})^{3/2} = 7^{(1/3)(3/2)} = 7^{1/2}$
c. $\dfrac{5^{-1/2}}{5^{3/2}} = 5^{-1/2 - 3/2} = 5^{-2} = \dfrac{1}{5^2}$

PROGRESS CHECK 5   Simplify as directed in Example 5.

a. $7^{1/4} \cdot 7^{3/4}$ **b.** $(5^{2/5})^{1/2}$ **c.** $\dfrac{2^{-2/3}}{2^{7/3}}$

EXAMPLE 6   Perform the indicated operations and write the result with only positive exponents. Assume all expressions represent real numbers.

a. $x^{3/2} \cdot x^{1/2}$ **b.** $(x^5 y^3)^{1/3}$ **c.** $\dfrac{x^{-3/4} \cdot x^{7/4}}{x^{1/4}}$

Solution

a. $x^{3/2} \cdot x^{1/2} = x^{3/2 + 1/2} = x^2$
b. $(x^5 y^3)^{1/3} = x^{5(1/3)} y^{3(1/3)} = x^{5/3} y$
c. $\dfrac{x^{-3/4} \cdot x^{7/4}}{x^{1/4}} = x^{-3/4 + 7/4 - 1/4} = x^{3/4}$

PROGRESS CHECK 6   Simplify as directed in Example 6.

a. $y^{-3/2} \cdot y^{1/2}$ **b.** $(a^7 x^{10})^{1/5}$ **c.** $\dfrac{x^{-4/3} \cdot x^{7/3}}{x^{1/3}}$

3 To express a radical in simplest form, the index of the radical must be as small as possible. The next example discusses how to simplify $\sqrt[n]{a^m}$ when m and n have common factors (other than 1).

Progress Check Answers
4. (a) $\frac{1}{64}$ (b) $\frac{1}{16}$

5. (a) 7   (b) $5^{1/5}$ (c) $\dfrac{1}{2^3}$

6. (a) $\dfrac{1}{y}$ (b) $a^{7/5}x^2$ (c) $x^{2/3}$

EXAMPLE 7   Express each radical in simplest form.

a. $\sqrt[4]{3^2}$ **b.** $\sqrt[6]{4^3}$ **c.** $\sqrt[9]{x^6}$

Solution   Express the radical in exponential form, reduce the rational exponent, and then convert back to radical form.

a. $\sqrt[4]{3^2} = (3^2)^{1/4} = 3^{2/4} = 3^{1/2} = \sqrt{3}$
b. $\sqrt[6]{4^3} = (4^3)^{1/6} = 4^{3/6} = 4^{1/2} = \sqrt{4} = 2$
c. $\sqrt[9]{x^6} = (x^6)^{1/9} = x^{6/9} = x^{2/3} = \sqrt[3]{x^2}$

PROGRESS CHECK 7   Express each radical in simplest form.

a. $\sqrt[8]{5^6}$ **b.** $\sqrt[4]{9^2}$ **c.** $\sqrt[4]{x^2},\ x \geq 0$

Progress Check Answers
7. (a) $\sqrt[4]{5^3}$ (b) 3   (c) $\sqrt{x}$

EXERCISES 8.7

In Exercises 1–34, evaluate the given expression.

1. $25^{1/2}$ **2.** $49^{1/2}$
3. $100^{1/2}$ **4.** $144^{1/2}$
5. $-16^{1/2}$ **6.** $-1^{1/2}$
7. $1^{1/3}$ **8.** $0^{1/3}$
9. $27^{1/3}$ **10.** $216^{1/3}$
11. $(-1,000)^{1/3}$ **12.** $(-8,000)^{1/3}$
13. $16^{1/4}$ **14.** $256^{1/4}$
15. $(-32)^{1/5}$ **16.** $(-243)^{1/5}$
17. $4^{1/2} \cdot 4^{1/2}$ **18.** $9^{1/2} \cdot 9^{1/2}$
19. $8^{1/3} \cdot 8^{1/3} \cdot 8^{1/3}$ **20.** $27^{1/3} \cdot 27^{1/3} \cdot 27^{1/3}$
21. $25^{5/2}$ **22.** $9^{3/2}$
23. $(-8)^{4/3}$ **24.** $(-1)^{5/3}$
25. $25^{-3/2}$ **26.** $16^{-3/2}$

27. $4^{-7/2}$ **28.** $36^{-5/2}$

29. $(-27)^{-2/3}$ **30.** $(-8)^{-4/3}$
31. $(-2)^{1/2}$ **32.** $(-3)^{1/2}$
33. 0^{-1} **34.** 0^{-2}

In Exercises 35–76, perform the indicated operations and write the result with only positive exponents. Assume all expressions represent real numbers.

35. $2^{1/3} \cdot 2^{4/3}$ **36.** $3^{1/2} \cdot 3^{3/2}$
37. $5^{1/2} \cdot 5^{-1/2}$ **38.** $6^{2/3} \cdot 6^{-2/3}$
39. $7^{1/2} \cdot 7^{1/3}$ **40.** $11^{1/3} \cdot 11^{1/4}$
41. $8^{1/2} \cdot 8^{-3/2}$ **42.** $10^{1/3} \cdot 10^{-4/3}$
43. $(3^{1/2})^{2/3}$ **44.** $(2^{1/4})^{8/3}$
45. $(4^{2/3})^{3/2}$ **46.** $(5^{3/4})^{2/5}$
47. $(2^{1/4})^3$ **48.** $(3^{2/5})^3$
49. $(4^{1/3})^6$ **50.** $(5^{2/3})^3$
51. $\dfrac{3^{-1/2}}{3^{1/2}}$ **52.** $\dfrac{5^{-1/3}}{5^{2/3}}$

53. $\dfrac{6^{1/4}}{6^{-7/4}}$ **54.** $\dfrac{2^{2/3}}{2^{-7/3}}$

55. $\dfrac{4^{-2/3}}{4^{4/3}}$ **56.** $\dfrac{7^{-5/4}}{7^{3/4}}$

57. $x^{1/2}x^{5/2}$ **58.** $x^{2/3}x^{4/3}$
59. $x^3x^{1/2}$ **60.** $x^2x^{1/3}$

61. $(3x^{1/3})^2$ **62.** $(2x^{2/3})^3$
63. $(4x^3y^2)^{1/2}$ **64.** $(27x^2y^6)^{1/3}$
65. $(a^{-1/2}b^{3/2})^2$ **66.** $(a^{-2/3}b^{1/3})^3$
67. $\dfrac{x^{-1/3}x^{5/3}}{x^{1/3}}$ **68.** $\dfrac{y^{-1/2}y^{3/2}}{y^{1/2}}$

69. $\dfrac{y^{1/3}y^{-2/3}}{y^{5/3}}$ **70.** $\dfrac{x^{-3/4}x^{1/4}}{x^{-1/4}}$

71. $\dfrac{2^{1/6}x^{1/2}y^{1/3}}{2^{1/6}x^{1/3}y^{1/2}}$ **72.** $\dfrac{3^{1/5}xy^{2/3}}{3^{1/5}x^{1/3}y}$

73. $\dfrac{x^{-1/2}}{y^{2/3}} \cdot \dfrac{x^2}{y^3}$ **74.** $\dfrac{x^{-1/4}}{y^{1/2}} \cdot \dfrac{x}{y}$

75. $\dfrac{x^{-1/2}}{y^{-1/3}} \div \dfrac{x^{-1/3}}{y^{-1/2}}$ **76.** $\dfrac{x^{-1/4}}{y^{1/4}} \div \dfrac{y^{1/4}}{x^{-1/4}}$

In Exercises 77–86, use a calculator to evaluate the given expression. Round answers to the nearest thousandth.

77. $2^{1/10}$ **78.** $2^{1/3}$
79. $10^{0.2}$ **80.** $10^{0.3}$
81. $5^{1.4}$ **82.** $5^{2.8}$
83. $2^{-0.5}$ **84.** $2^{-0.4}$
85. $1.4^{2.3}$ **86.** $2.3^{1.4}$

87. The compound annual interest rate r that is required for an investment to triple in n years is given by $r = 3^{1/n} - 1$. To the nearest hundredth of a percent, what should this rate be for an investment to triple in 10 years?

88. Repeat Exercise 87, but assume that the investment should triple in five years.

89. If you start with u_0 grams of radioactive radium, the formula $u = u_0 \cdot 2.718^{-0.000447t}$ approximates how much is left after t years. How much of 100 original grams will be left after 1,550 years?

90. Repeat Exercise 89, but find out how much will be left after 3,100 years.

In Exercises 91–98, express each radical in simplest form. Assume all variables are positive.

91. $\sqrt[4]{5^2}$ **92.** $\sqrt[6]{5^3}$ **93.** $\sqrt[6]{x^2}$
94. $\sqrt[12]{y^4}$ **95.** $\sqrt[4]{4^2}$ **96.** $\sqrt[9]{8^3}$
97. $\sqrt[3]{a^6}$ **98.** $\sqrt[4]{b^{10}}$

THINK ABOUT IT

1. In this section of the text we extended the definition of exponents to include rational numbers. This definition can be further extended to include expressions like 2^π, where the exponent is an irrational number. Such expressions may be evaluated or approximated on a scientific calculator. Use a calculator to evaluate these to the nearest hundredth.
   a. 2^π (Enter π as accurately as you can. Explain why you would expect 2^π to be a little larger than 8.)
   b. $2^{\sqrt{2}}$
   c. π^π
   d. $\sqrt{2}^{\sqrt{2}}$

2. a. Explain why $(\sqrt{2}^{\sqrt{2}})^{\sqrt{2}}$ must be equal to 2 by using the power-to-a-power property of exponents. This shows that raising an irrational number to an irrational power *may* yield a rational result.
   b. Evaluate $(\sqrt{5}^{\sqrt{2}})^{\sqrt{2}}$.
   c. Evaluate $\left[\left(\sqrt[3]{3}^{\sqrt[3]{3}}\right)^{\sqrt[3]{3}}\right]^{\sqrt[3]{3}}$.

3. On a scientific calculator there is a key marked LN, which computes a value called the natural logarithm.
   a. Show by calculator that the natural logarithm of 2 is 0.6931, accurate to four decimal places.

b. Mathematicians have shown that the formula $(\sqrt[n]{x} - 1)n$ gives better and better approximations to the natural logarithm of x as you use larger values for n. Evaluate the formula when $x = 2$ and $n = 10$ to see how close the result is to 0.6931. Repeat for $n = 100$, 1,000, and 10,000 to show how much the result improves.

4. Use the formula in Exercise 3 to approximate the natural logarithm of 3. Use $n = 10$, 100, 1,000, and 10,000 to see how the result improves as n increases. The correct value to four decimal places is 1.0986.

5. A dangerous amount of radioactive strontium 90 was released in the Chernobyl accident in the Ukraine in 1986. The formula $N_0 \cdot 2.718^{-0.02424t}$ approximates how much of an original quantity N_0 of this strontium remains after t years. The number 2.718 is an approximation to a constant called e, which often appears in formulas describing growth or decay. (On a scientific calculator, use the e^x key with $x = 1$ to see a more accurate evaluation of e.) The value -0.02424 is called the decay constant for strontium. Each radioactive element has its own decay constant. Show that about half of the original amount of strontium 90 is still there in 28.6 years. For this reason the value 28.6 is called the **half-life** of strontium 90.

REMEMBER THIS

1. Which of these equations are called quadratic equations?
   a. $x^2 + 2x - 2 = 0$
   b. $2x - 3 = 0$
   c. $x^2 = 5$
2. Compute $-\sqrt{25} + \sqrt{16}$.
3. Is -3 a solution to $x^2 + x - 2 = 4$?
4. Is -3 a solution to $x^2 + 9 = 0$?
5. Factor $x^2 - 3x - 18$.
6. Solve $\sqrt{28 - 3x} = x$.
7. Which of these lines is parallel to $y = 3x - 5$?
   a. $y = 3x + 1$ b. $y = x - 5$

8. Simplify $\dfrac{3 + \dfrac{1}{n}}{n + \dfrac{1}{3}}$.

9. What is the GCF of $3x^2$, 3, and x^2?

10. Simplify $\dfrac{3.1 \times 10^{12}}{3.1 \times 10^{11}}$.

Chapter 8   SUMMARY

OBJECTIVES CHECKLIST   Specific chapter objectives are summarized below along with numbered example problems from the text that should clarify the objectives. If you do not understand any objectives or do not know how to do the selected problems, then restudy the material.

8.1 Can you:

1. **Find square roots and principal square roots?**
   Find the square roots of 25.   [Example 1]

2. **Identify square roots as rational numbers, irrational numbers, or nonreal numbers?**
   Find each square root. Identify each number as rational or irrational, and approximate irrational numbers to the nearest hundredth.
   a. $\sqrt{256}$ b. $\sqrt{110}$ [Example 3a and 3b]

3. **Use square roots to solve problems involving the Pythagorean relation?**
   In a right triangle the hypotenuse measures 17 cm and one leg measures 15 cm. Find the
   length of the other leg.   [Example 5]

4. **Find *n*th roots?**
   Find $\sqrt[3]{-27}$.   [Example 7b]

8.2 **Can you:**
1. **Multiply square roots?**
   Find $8\sqrt{2} \cdot 7\sqrt{11}$.   [Example 1c]

2. **Simplify square roots?**
   Simplify $\sqrt{50}$.   [Example 2a]

3. **Multiply square roots and simplify where possible?**
   Multiply and simplify the product. Assume $x \geq 0$. $\sqrt{2x} \cdot \sqrt{10x^5}$.   [Example 5c]

4. **Multiply and simplify higher roots?**
   Multiply and simplify the product. Assume $y \geq 0$. $\sqrt[4]{y^5} \cdot \sqrt[4]{y^2}$.   [Example 7b]

8.3 **Can you:**
1. **Divide square roots?**
   Divide and simplify the quotient. Assume $x > 0$. $\dfrac{\sqrt{2x}}{\sqrt{x}}$.   [Example 1c]

2. **Rationalize denominators?**
   The picture tube of a popular model television set is square in shape and measures 20 in.
   diagonally. Find the length of the side (s) of this picture tube given that $d = s\sqrt{2}$ expresses
   the length of the diagonal (d) in terms of side length. Rationalize the denominator for the
   answer.   [Example 5]

3. **Divide and simplify higher roots?**
   Simplify the expression and rationalize the denominator: $\sqrt[3]{\frac{4}{9}}$.   [Example 6b]

8.4 **Can you:**
1. **Add and subtract square roots?**
   Simplify $\sqrt{50} + \sqrt{18}$.   [Example 2a]

2. **Add and subtract square roots that involve rationalizing denominators?**
   Simplify $\sqrt{125} - 10\sqrt{\frac{1}{5}}$.   [Example 4a]

3. **Add and subtract higher roots?**
   Simplify $\sqrt[3]{-24} + \sqrt[3]{81}$.   [Example 5]

8.5 **Can you:**
1. **Multiply radical expressions involving more than one term?**
   Simplify $\sqrt{2}(\sqrt{6} - \sqrt{2})$.   [Example 1b]

2. **Use a special product formula to multiply certain radical expressions?**
   Simplify $(\sqrt{y} + 3)(\sqrt{y} - 3)$. Assume $y \geq 0$.   [Example 2b]

3. **Divide radical expressions involving more than one term in the numerator and a single term in
   the denominator?**
   Simplify $\dfrac{\sqrt{8} + \sqrt{28}}{\sqrt{2}}$.   [Example 3b]

4. **Use a conjugate to rationalize a denominator?**
   Rationalize the denominator $\dfrac{6}{3 + \sqrt{7}}$.   [Example 4]

8.6 **Can you:**

1. **Solve radical equations containing square roots?**
   Solve $2\sqrt{x} - 1 = 8$. [Example 3]

2. **Solve radical equations involving extraneous solutions?**
   Solve $\sqrt{x + 2} = x$. [Example 4]

3. **Solve applied problems involving radical equations?**
   The period T of a pendulum is the time required for the pendulum to complete one round-trip of motion—that is, one complete cycle. When the period is measured in seconds and the pendulum length ℓ is measured in feet, then a formula for the period is

   $$T = 2\pi \sqrt{\frac{\ell}{32}}.$$

   To the nearest hundredth of a foot, what is the length of a pendulum whose period is 1 second? [Example 7]

8.7 **Can you:**

1. **Evaluate expressions containing rational exponents?**
   Evaluate $25^{3/2}$. [Example 3a]

2. **Use exponent properties to simplify expressions with rational exponents?**
   Perform the indicated operation and write the result with only positive exponents: $2^{1/5} \cdot 2^{3/5}$. [Example 5a]

3. **Use exponent properties to simplify radicals?**
   Express the radical in simplest form: $\sqrt[4]{3^2}$. [Example 7a]

KEY TERMS

Conjugates (8.5)

Cube root (8.1)

Extraneous solutions (8.6)

Index of radical (8.1)

Like radicals (8.4)

Like square roots (8.4)

nth root (8.1)

Perfect square (8.1)

Principal nth root (8.1)

Principal square root (8.1)

Radical (8.1)

Radical equation (8.6)

Radical sign (8.1)

Radicand (8.1)

Rationalize a denominator (8.3)

Square root (8.1)

KEY CONCEPTS AND PROCEDURES

Section	Key Concepts or Procedures to Review
8.1	■ **Definition of a square root:** The number b is a square root of a if $b^2 = a$.
	■ **Pythagorean theorem:** In a right triangle with legs of length a and b and hypotenuse of length c, $$c^2 = a^2 + b^2$$
	■ **Definition of nth root:** For any positive integer n, the number b is an nth root of a if $b^n = a$.
8.2	■ **Multiplication property of square roots:** For nonnegative real numbers a and b, $\sqrt{a} \cdot \sqrt{b} = \sqrt{a \cdot b}$.
	■ **nth root properties:** For real numbers a, b, $\sqrt[n]{a}$, and $\sqrt[n]{b}$ 1. $(\sqrt[n]{a})^n = a$ 2. $\sqrt[n]{a} \cdot \sqrt[n]{b} = \sqrt[n]{a \cdot b}$

Section	Key Concepts or Procedures to Review
8.3	■ Quotient property of square roots: For nonnegative real numbers a and b with $b \neq 0$, $$\frac{\sqrt{a}}{\sqrt{b}} = \sqrt{\frac{a}{b}}$$ ■ **Simplified square root:** To write a square root in simplest form 1. Remove all factors of the radicand that are perfect squares. 2. Eliminate all fractions in the radicand. 3. Eliminate square roots in the denominator. ■ Quotient property of radicals: For real numbers a, b, $\sqrt[n]{a}$, and $\sqrt[n]{b}$ $$\frac{\sqrt[n]{a}}{\sqrt[n]{b}} = \sqrt[n]{\frac{a}{b}}$$
8.4	■ Like square roots are square roots that have the same radicand. They are combined by using the distributive property. This principle extends to cube roots and higher roots.
8.5	■ To rationalize a binomial denominator, multiply numerator and denominator by the conjugate of the denominator.
8.6	■ To solve radical equations containing square roots: 1. If necessary, isolate a square root term on one side of the equation. 2. Square both sides of the equation. 3. Solve the resulting equation, and check all solutions in the *original* equation. ■ **Squaring property of equality:** If A and B are algebraic expressions, then the solutions of the equation $A = B$ are among the solutions of the equation $A^2 = B^2$.
8.7	■ Definition of $a^{1/n}$: If n is a positive integer and $\sqrt[n]{a}$ is a real number, then $$a^{1/n} = \sqrt[n]{a}$$ ■ Rational exponents: If m and n are integers with $n > 0$, and if m/n represents a reduced fraction such that $a^{1/n}$ is a real number, then $$a^{m/n} = (\sqrt[n]{a})^m = \sqrt[n]{a^m}$$

CHAPTER 8 REVIEW EXERCISES

8.1
1. Find the square roots of 144.
2. Which of these is an irrational number?   **a.** $\sqrt{50}$ **b.** $\sqrt{1.44}$
3. To the nearest hundredth, find the length of the remaining leg of a right triangle if one leg is 1 m and the hypotenuse is 2 m.

4. Which is larger, $\sqrt[5]{32}$ or $\sqrt[3]{27}$?

8.2
5. Multiply $2\sqrt{2} \cdot 3\sqrt{3}$.
6. Simplify $\sqrt{48}$.
7. Multiply and simplify the product. Assume $x \geq 0$. $\sqrt{3x} \cdot \sqrt{6x^3}$.
8. Multiply and simplify the product. Assume $y \geq 0$.

8.3

9. Divide and simplify. Assume $x > 0$.

10. Solve for s and rationalize the denominator: $24 = s\sqrt{2}$.
11. Simplify and rationalize the denominator $\sqrt[3]{\dfrac{9}{5}}$.

8.4
12. Combine and simplify $\sqrt{50} - \sqrt{18}$.
13. Combine and simplify $\sqrt{27} - \dfrac{6}{\sqrt{3}}$.
14. Combine and simplify $\sqrt[3]{-128} + \sqrt[3]{16}$.

8.5
15. Multiply $\sqrt{3}(\sqrt{5} + \sqrt{6})$.
16. Multiply $(\sqrt{a} + \sqrt{b})(\sqrt{a} - \sqrt{b})$. Assume $a, b \geq 0$.
17. Simplify $\dfrac{\sqrt{20} + \sqrt{75}}{\sqrt{5}}$.

18. Rationalize the denominator $\dfrac{8}{4 - \sqrt{2}}$.

8.6

19. Solve $3\sqrt{x} + 2 = 3$.
20. Solve $\sqrt{x + 6} = x$.
21. Use the formula $T = 2\pi\sqrt{\dfrac{\ell}{32}}$ to determine the length ℓ in feet of a pendulum whose period is $T = 2$ seconds. Give the length to the nearest hundredth of a foot.

8.7

22. Evaluate $100^{3/2}$.
23. Simplify and write the result with only positive exponents:
    $3^{2/5} \cdot 3^{-4/5}$.
24. Express the radical in simplest form. Assume $x \geq 0$. $\sqrt[6]{x^2}$.

CHAPTER 8 TEST

1. Find $-\sqrt{81}$.
2. Why do we say that $\sqrt{-9}$ is not a real number?

3. If the two legs of a right triangle are 1 ft and 2 ft in length, find the length of the hypotenuse. Give the answer in exact radical notation and also as a decimal rounded to the nearest tenth.

4. Simplify $(3\sqrt{7})^2$.
5. Simplify $\sqrt{32}$.
6. Simplify $\sqrt{40x^3}$. Assume $x \geq 0$.
7. Multiply $\sqrt[3]{2a} \cdot \sqrt[3]{2a} \cdot \sqrt[3]{2a}$.
8. Simplify $\dfrac{\sqrt{48}}{\sqrt{2}}$.
9. Simplify by rationalizing the denominator. Assume $y > 0$.
   $\sqrt{\dfrac{5}{y}}$.

10. Simplify $\dfrac{\sqrt{75xy}}{\sqrt{5x^7y}}$. Assume $x, y > 0$.

11. Which of these are true?
    a. $\sqrt[5]{\dfrac{20}{31}} = \dfrac{\sqrt[5]{20}}{\sqrt[5]{31}}$
    b. $\sqrt[5]{20} + \sqrt[5]{31} = \sqrt[5]{20 + 31}$
    c. $\sqrt[5]{20} - \sqrt[5]{31} = \sqrt[5]{20 - 31}$
    d. $\sqrt[5]{20} \cdot \sqrt[5]{31} = \sqrt[5]{20 \cdot 31}$

12. Simplify $\dfrac{3x\sqrt{5} - \sqrt{45x^2}}{3x\sqrt{5} + \sqrt{45x^2}}$. Assume $x > 0$.

13. Combine and simplify. Rationalize the denominator in the answer. $\sqrt{3} - \dfrac{1}{\sqrt{3}}$.

14. Simplify $\sqrt[4]{32} - \sqrt[4]{162}$.
15. Multiply $(1 + \sqrt{x})^2$. Assume $x \geq 0$.
16. Solve $3\sqrt{2x - 1} = 7$.
17. Solve $\sqrt{6 - x} = x$.
18. Multiply $7^{1/2} \cdot 7^{3/2}$.
19. Evaluate $(-125)^{2/3}$.
20. The compounded annual interest rate r required for an investment to double in n years is given by $r = 2^{1/n} - 1$. What rate is needed for an investment to double in 14 years? (Round to the nearest hundredth of a percent.)

CUMULATIVE TEST 8

1. True or false? A number is always larger than its reciprocal.

2. If n represents a negative number, identify *all* of the following expressions that must also represent a negative number.
   a. n^3 b. $-n$ c. $-n^2$ d. $(-n)^2$ e. $1/n$

3. The equation $3x + 4 = 7$ is called
   a. a conditional equation
   b. an identity
   c. a false equation

4. A sales clerk has $1,800 in his monthly budget. One-quarter is for rent, one-third for food, and $90 for a car payment. There is some amount left over.
   a. Represent this situation by an equation, where x represents the amount left over.

   b. Solve the equation to determine the amount left over.

5. Find the product $(2x - 1)(x + 4)$.

6. Divide $x^2 + 2x + 3$ by $x - 1$.

7. Solve $2x^2 + 5x = 3$.
8. What is a prime polynomial?

9. Solve the equation for n: $\dfrac{a}{b} = \dfrac{m}{n}$.

10. Express this fraction in lowest terms: $\dfrac{x^2 + x - 6}{x^2 + 2x - 8}$.

11. Solve $\dfrac{1}{x} + \dfrac{1}{3} = \dfrac{5}{2x}$.

12. Find the slope of the line which passes through $(5,2)$ and $(9,2)$.

13. Examine the graph to determine the equation of the line shown. Give the equation in slope-intercept form.

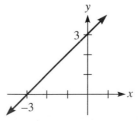

14. If $y = f(x) = x^2 + 9$, find $f(-1)$.

15. At what point do the graphs of these equations intersect?
$$2x + 3y = 1$$
$$3x - 2y = 1$$

16. When a system of two linear equations is inconsistent, which of these statements must be true?
   a. The two lines are distinct and parallel.
   b. There is really only one line.
   c. One of the lines is vertical.

17. Graph the solution set of the system
$$y < x - 5$$
$$y > 2x - 4.$$

18. Simplify $\dfrac{\sqrt{50} + \sqrt{18}}{2}$.

19. Solve $\sqrt{x} + 2 = x$.

20. In an isosceles right triangle, the length of each leg is s. Use the Pythagorean theorem to find and simplify a formula for the hypotenuse (h).

CHAPTER 9

Quadratic Equations

On a sailboat the wind causes a wind pressure gauge to register 4.8 lb/ft². If the pressure p (in pounds per square foot) of a wind blowing at v mi/hour is given by

$$p = 0.003v^2,$$

find the wind speed at that moment. (See Example 5 in Section 9.1.)

IN SECTION 4.6 we discussed how to solve quadratic equations by the factoring method. In this chapter we first review this method and then develop other techniques, because the factoring method is limited to only certain equations. When finished, we will be able to find *any* solution of *any* quadratic equation. This ability is important in the analysis of quadratic functions, as shown in the concluding section of this chapter.

9.1 Solving Quadratic Equations by the Square Root Property

1 Solve quadratic equations using the factoring method (of Section 4.6).

2 Solve quadratic equations using the square root property.

1 Recall that an equation that can be written in the form $ax^2 + bx + c = 0$, where a, b, and c are real numbers with $a \neq 0$, is called a **second-degree** or **quadratic equation.** Example 1 reviews the factoring method for solving such equations, which applies provided we can factor the polynomial described in step 2 of the following procedure.

Factoring Method for Solving Quadratic Equations

1. If necessary, change the form of the equation so that one side is 0.
2. Factor the nonzero side.
3. Set each factor equal to 0 and obtain the solution(s) by solving the resulting equations.
4. Check each solution by substituting it in the original equation.

Remember that the method is based on the zero product principle, which states that for any numbers a and b, $ab = 0$ implies $a = 0$ or $b = 0$.

EXAMPLE 1 Solve $x^2 - 3x = 18$.

Solution Follow the steps given.

$$x^2 - 3x = 18$$
$$x^2 - 3x - 18 = 0 \qquad \text{Rewrite the equation so one side is 0.}$$
$$(x - 6)(x + 3) = 0 \qquad \text{Factor the nonzero side.}$$
$$x - 6 = 0 \quad \text{or} \quad x + 3 = 0 \qquad \text{Set each factor equal to 0.}$$
$$x = 6 \qquad\qquad x = -3 \qquad \text{Solve each linear equation.}$$

To catch any mistakes, we check in the original equation.

$$x^2 - 3x = 18 \qquad\qquad\qquad x^2 - 3x = 18$$
$$6^2 - 3(6) \overset{?}{=} 18 \qquad\qquad (-3)^2 - 3(-3) \overset{?}{=} 18$$
$$36 - 18 \overset{?}{=} 18 \qquad\qquad\qquad 9 + 9 \overset{?}{=} 18$$
$$18 \overset{\checkmark}{=} 18 \qquad\qquad\qquad\qquad 18 \overset{\checkmark}{=} 18$$

Thus, the solution set is $\{6, -3\}$.

PROGRESS CHECK 1 Solve $x^2 + 3x = 10$.

2 Our work with square roots in Chapter 8 is the basis for solving quadratic equations if the factoring method does not apply. In that chapter, recall that the definition of square roots was motivated by the question

$$\text{if } x^2 = a, \text{ then what is } x?$$

Progress Check Answer
1. $\{-5, 2\}$

We saw that if squaring x results in a, then x is a square root of a. We also saw that every positive number a has two square roots, denoted by $\sqrt{a}$ and $-\sqrt{a}$. Thus, if $a > 0$, then the solutions of the quadratic equation $x^2 = a$ are $\sqrt{a}$ and $-\sqrt{a}$. This property is called the **square root property.**

Square Root Property

If a is a positive real number, then

$$x^2 = a \qquad \text{implies} \qquad x = \sqrt{a} \quad \text{or} \quad x = -\sqrt{a}.$$

We often abbreviate "$x = \sqrt{a}$ or $x = -\sqrt{a}$" by writing $x = \pm\sqrt{a}$. Read $\pm\sqrt{a}$ as "plus or minus $\sqrt{a}$."

EXAMPLE 2   Solve using the square root property.

a. $x^2 = 25$ **b.** $y^2 = 3$ **c.** $x^2 = 18$

Solution

a. $x^2 = 25$ implies

$$x = \sqrt{25} = 5 \qquad \text{or} \qquad x = -\sqrt{25} = -5.$$

Because $5^2 = 25$ and $(-5)^2 = 25$, the solutions check. The solution set $\{5, -5\}$ may be abbreviated $\{\pm 5\}$.

b. $y^2 = 3$ implies

$$y = \sqrt{3} \qquad \text{or} \qquad y = -\sqrt{3}.$$

Thus, the solution set is $\{\sqrt{3}, -\sqrt{3}\}$, or $\{\pm\sqrt{3}\}$. Use $(\sqrt{a})^2 = a$ to check these solutions.

c. $x^2 = 18$ implies $x = \sqrt{18}$ or $x = -\sqrt{18}$. Since

$$\sqrt{18} = \sqrt{9}\sqrt{2} = 3\sqrt{2},$$

the solution set is $\{3\sqrt{2}, -3\sqrt{2}\}$, or $\{\pm 3\sqrt{2}\}$. Check this answer.

PROGRESS CHECK 2   Solve using the square root property.

a. $x^2 = 9$ **b.** $x^2 = 26$ **c.** $y^2 = 32$ ⌐

Example 3 discusses the solution of quadratic equations of the type $x^2 = a$, when a is not a positive number.

EXAMPLE 3   Solve each equation.

a. $x^2 = 0$ **b.** $x^2 = -1$

Solution

a. $0^2 = 0$, and the zero product principle indicates 0 is the only number whose square is 0. Thus, the solution set is $\{0\}$. Note that the square root property may be extended to include this case, because it is true that

$$x^2 = 0 \qquad \text{implies} \qquad x = \sqrt{0} \quad \text{or} \quad x = -\sqrt{0}.$$

b. $x^2 = -1$ has no real number solutions, because no real number is the square root of a negative number. (*Note:* For equations like this, your answer should read "no real number solution." In Section 9.4 we will extend the number system beyond real numbers to complex numbers and solve equations like $x^2 = -1$.)

PROGRESS CHECK 3   Solve each equation.

a. $x^2 = -9$ **b.** $y^2 = 0$

Isolating x^2 and using the square root property is an efficient method for solving quadratic equations in which there is no first-degree term, as shown next. The general form for such equations is $ax^2 + c = 0$, with $a \neq 0$, because we are taking $b = 0$ for this special case.

EXAMPLE 4   Solve $5x^2 - 2 = 0$.

Solution   First, transform the equation so that x^2 is on one side by itself.

$$5x^2 - 2 = 0$$
$$5x^2 = 2 \quad \text{Add 2 to both sides.}$$
$$x^2 = \tfrac{2}{5} \quad \text{Divide both sides by 5.}$$

Then, using the square root property gives

$$x = \sqrt{\tfrac{2}{5}} \quad \text{or} \quad x = -\sqrt{\tfrac{2}{5}}.$$

In simplest radical form

$$\sqrt{\frac{2}{5}} = \sqrt{\frac{2 \cdot 5}{5 \cdot 5}} = \sqrt{\frac{10}{25}} = \frac{\sqrt{10}}{5}$$

and the solution set is $\{\sqrt{10}/5, -\sqrt{10}/5\}$. Check these solutions.

PROGRESS CHECK 4   Solve $3x^2 + 5 = 12$.

EXAMPLE 5   Solve the problem in the section introduction on page 406.

Solution   The pressure p registers 4.8 lb/ft², and replacing p by 4.8 in

$$p = 0.003v^2 \quad \text{gives} \quad 4.8 = 0.003v^2.$$

To find the wind speed v, first solve for v^2.

$$4.8 = 0.003v^2$$
$$\frac{4.8}{0.003} = v^2 \quad \text{Divide both sides by 0.003.}$$
$$1,600 = v^2 \quad \text{Simplify.}$$

Then, using the square root property gives

$$v = \sqrt{1,600} = 40 \quad \text{or} \quad v = -\sqrt{1,600} = -40.$$

Because the wind speed cannot be negative, we reject $v = -40$. So the wind speed at that moment is 40 mi/hour.

PROGRESS CHECK 5   During a hurricane a wind pressure gauge registers a pressure of 43.2 lb/ft². Find the wind speed at that moment.

Examples 6–8 illustrate that the square root property is also an efficient method for solving quadratic equations expressed in the form $X^2 = a$, where X is the binomial $px + q$. Note, in general, that if

$$(px + q)^2 = a,$$

where $a \geq 0$, then by the square root property

$$px + q = \sqrt{a} \quad \text{or} \quad px + q = -\sqrt{a}.$$

EXAMPLE 6   Solve $(x + 4)^2 = 49$.

Solution   By the square root property, $(x + 4)^2 = 49$ implies

$$x + 4 = \sqrt{49} \quad \text{or} \quad x + 4 = -\sqrt{49}.$$

Now, simplify $\sqrt{49}$ to 7 and solve each equation.

$$
\begin{array}{lll}
x + 4 = 7 & \text{or} & x + 4 = -7 \\
x = 3 & \text{or} & x = -11
\end{array}
$$

Since $(3 + 4)^2 = 7^2 = 49$, and $(-11 + 4)^2 = (-7)^2 = 49$, both solutions check, and the solution set is $\{3, -11\}$.

PROGRESS CHECK 6   Solve $(x - 6)^2 = 100$.   ⌐

EXAMPLE 7   Solve $(2y - 1)^2 - 7 = 0$.

Solution   First, transform the equation so that $(2y - 1)^2$ is on one side of the equation by itself.

$$
\begin{aligned}
(2y - 1)^2 - 7 &= 0 \\
(2y - 1)^2 &= 7
\end{aligned}
$$

Now, apply the square root property and solve each equation.

$$
\begin{array}{lll}
2y - 1 = \sqrt{7} & \text{or} & 2y - 1 = -\sqrt{7} \\
2y = 1 + \sqrt{7} & \text{or} & 2y = 1 - \sqrt{7} \\
y = \dfrac{1 + \sqrt{7}}{2} & \text{or} & y = \dfrac{1 - \sqrt{7}}{2}
\end{array}
$$

Next, we confirm that $(1 + \sqrt{7})/2$ is a solution.

$$\left[2\left(\frac{1 + \sqrt{7}}{2}\right) - 1\right]^2 - 7 = [(1 + \sqrt{7}) - 1]^2 - 7 = (\sqrt{7})^2 - 7 = 0$$

In a similar way, $(1 - \sqrt{7})/2$ checks, and the solution set is

$$\left\{\frac{1 + \sqrt{7}}{2}, \frac{1 - \sqrt{7}}{2}\right\}.$$

PROGRESS CHECK 7   Solve $(3t - 5)^2 - 2 = 0$.   ⌐

EXAMPLE 8   When P dollars is invested at a compound interest rate r for two years, the compounded amount A is given by

$$A = P(1 + r)^2.$$

To the nearest hundredth of a percent, what compound interest rate is needed for an investment to grow from \$1,000 to \$1,150 in two years?

Solution   Replace A by 1,150 and P by 1,000 in the given formula and solve for $(1 + r)^2$.

$$
\begin{aligned}
1{,}150 &= 1{,}000(1 + r)^2 \\
\frac{1{,}150}{1{,}000} &= (1 + r)^2 \\
1.15 &= (1 + r)^2
\end{aligned}
$$

Now, apply the square root property and solve each equation.

$$
\begin{array}{lll}
1 + r = \sqrt{1.15} & \text{or} & 1 + r = -\sqrt{1.15} \\
r = -1 + \sqrt{1.15} & \text{or} & r = -1 - \sqrt{1.15}
\end{array}
$$

Progress Check Answers

6. $\{16, -4\}$

7. $\left\{\dfrac{5 + \sqrt{2}}{3}, \dfrac{5 - \sqrt{2}}{3}\right\}$

Since r cannot be negative, we reject $r = -1 - \sqrt{1.15}$ as a solution. Therefore, $r = -1 + \sqrt{1.15}$, and we approximate this solution by calculator.

$$1 \boxed{+/-} \boxed{+} 1.15 \boxed{\sqrt{\ }} \boxed{=} \boxed{0.0723805}$$

Because $0.0723805 = 7.23805$ percent, the required interest rate, to the nearest hundredth of a percent, is 7.24 percent.

PROGRESS CHECK 8   To the nearest hundredth of a percent, what compound annual interest rate is needed for an investment to grow from \$5,000 to \$5,900 in two years?

EXERCISES 9.1

In Exercises 1–6, solve by factoring.

1. $x^2 - 5x = -6$
2. $x^2 - 7x = -12$
3. $x^2 + 3x = 70$
4. $x^2 + 4x = 45$
5. $x^2 - 2x = -1$
6. $x^2 + 6x = -9$

In Exercises 7–34, solve using the square root property. Give exact solutions in radical form, if needed.

7. $x^2 = 100$
8. $x^2 = 121$
9. $x^2 = 144$
10. $x^2 = 169$
11. $x^2 = 1.21$
12. $x^2 = 0.04$
13. $x^2 = \frac{1}{64}$
14. $x^2 = \frac{1}{81}$
15. $x^2 = 2$
16. $x^2 = 5$
17. $x^2 = 7$
18. $x^2 = 10$
19. $x^2 = 8$
20. $x^2 = 12$
21. $x^2 = 27$
22. $x^2 = 45$
23. $x^2 = -4$
24. $x^2 = -9$
25. $x^2 + 8 = 0$
26. $x^2 + 12 = 0$
27. $x^2 + 5 = 5$
28. $x^2 + 1 = 1$
29. $3x^2 - 4 = 0$
30. $5x^2 - 9 = 0$
31. $2x^2 - 7 = 0$
32. $3x^2 - 8 = 0$
33. $2x^2 + 1 = 0$
34. $3x^2 + 2 = 0$

In Exercises 35–38, give answers to the nearest tenth.

35. One of the most famous equations in the history of science comes from the work of the Italian astronomer and physicist Galileo (1564–1642), who investigated the motion of falling and rolling objects. This equation is $d = 16t^2$, where d is the distance covered by a freely falling object near the surface of the earth. Such an object falls d ft in t seconds. Using this equation, determine about how long it takes an object to fall from the top of a 1,000-ft-high skyscraper to the ground.

36. When a ball player spots a pop fly 100 ft high directly overhead, it is just beginning to fall. Use the equation in Exercise 35 to calculate how long it will take to hit the player's glove, which is 6 ft above the ground.

37. Galileo discovered, by rolling balls down inclined planes (see figure), that the equation $v^2 = 64h$ relates the velocity of this ball and the *vertical* distance it has covered. Starting from rest, a ball which has dropped h ft has a velocity of v ft/second. It is remarkable that this velocity has nothing to do with the steepness of the plane. Use the given equation to find the velocity of a rolling ball when its vertical height is 2 ft below its starting height.

38. Refer to Exercise 37. Find the velocity of a ball when its vertical height is 3 ft below its starting height.

In Exercises 39–60, use the square root property to solve the equations. Give the solution in simplest radical form.

39. $(x + 2)^2 = 25$
40. $(x + 3)^2 = 36$
41. $(x - 1)^2 = 1$
42. $(x - 4)^2 = 4$
43. $(3x + 5)^2 = 0$
44. $(2x + 1)^2 = 0$
45. $(2x - 3)^2 = 4$
46. $(3x - 2)^2 = 1$
47. $(5x + 1)^2 = 9$

48. $(4x + 3)^2 = 16$
49. $(x + 1)^2 = -4$
50. $(x - 2)^2 = -9$
51. $(x - 2)^2 - 5 = 0$
52. $(x - 4)^2 - 3 = 0$
53. $(2x - 3)^2 - 6 = 0$
54. $(4x - 1)^2 - 2 = 0$
55. $(3x + 5)^2 = 7$
56. $(2x + 3)^2 = 10$
57. $(5x - 2)^2 = 8$
58. $(4x - 1)^2 = 18$
59. $(2x - 4)^2 = 32$
60. $(3x - 6)^2 = 18$

In Exercises 61–64, give the answer to the nearest hundredth of a percent.

61. Refer to Example 8 in this section for a formula for compound interest. What rate r is needed for an investment to grow from $1,000 to $1,200 in two years?
62. Repeat Exercise 61, but assume that the investment should grow from $500 to $700 in two years.
63. If money is invested at interest rate r but it is compounded twice a year instead of just once, then the effective rate E is somewhat higher, since interest is earned on interest for the second half of the year. The formula for the effective rate is

$$E = \left(1 + \frac{r}{2}\right)^2 - 1.$$ What rate r is needed so that the

effective interest rate E will be 6 percent?
(*Hint:* Let $E = 0.06$.)
64. Repeat Exercise 63, but find the rate r that is needed so that the effective interest E will be 8 percent.

THINK ABOUT IT

1. Why is the equation $(x + 2)^2 = 3$ called a *second-degree* equation?
2. Every positive number a has two square roots. Find the sum and product of the two square roots.
3. If you extend the square root property to the cube root property, you can solve equations of the form $x^3 = A$ by writing $x = \sqrt[3]{A}$. Solve these equations.
  a. $x^3 = 27$
  b. $x^3 = -8$
  c. $(x + 5)^3 = 64$
  d. $(x + 2)^3 = 5$

4. Solve $(x^2 - 2)^3 = 8$.
5. When P dollars is compounded annually at an annual interest rate r for n years, the compounded amount A is given by $A = P(1 + r)^n$. To the nearest hundredth of a percent, what compound interest rate is needed for an investment to grow from $1,000 to $2,000 in five years? (*Hint:* You will need to take the fifth root of both sides of the equation.)

REMEMBER THIS

1. Factor $x^2 + 6x + 9$.
2. Factor $x^2 - x + \frac{1}{4}$.
3. Simplify $\sqrt{13^2}$.
4. Simplify $\sqrt{n^2}$ if n is a positive number.
5. In a rectangle the length is 2 m more than the width x. Write an algebraic expression for the area.

6. Simplify $(3x^{1/2})^2$.
7. Solve $\sqrt{x - 1} = x - 1$.
8. Solve

$$2x - y = 11$$
$$x + y = 1.$$

9. Is the graph of $x = 5$ a horizontal or a vertical line?

10. Solve for t: $A = p + prt$.

9.2 Solving Quadratic Equations by Completing the Square

The design for a rectangular solar panel specifies that the area is to be 6 m² and the length is to be 2 m greater than the width. To the nearest hundredth of a meter, find the length and the width of this panel. (See Example 4.)

OBJECTIVES

1 Complete the square for $x^2 + bx$ and express the resulting trinomial in factored form.

2 Solve $ax^2 + bx + c = 0$ by completing the square when $a = 1$.

3 Solve $ax^2 + bx + c = 0$ by completing the square when $a \neq 1$.

1   In the previous section we considered how to solve quadratic equations of the form

$$(x + \text{constant})^2 = \text{constant}$$

by using the square root property. Actually, any quadratic equation may be expressed in this form if we develop a technique called **completing the square.** Consider an expression like

$$x^2 + 6x.$$

What constant needs to be added to make this expression a perfect square trinomial? Since the factoring model for a perfect square trinomial may be written as

$$x^2 + 2kx + k^2 = (x + k)^2,$$

we set $2k = 6$, so that $k = 3$ and $k^2 = 9$. Thus, adding 9 gives

$$x^2 + 6x + 9 = (x + 3)^2.$$

More generally, to complete the square for

$$x^2 + bx$$

we set $2k = b$, so that $k = b/2$ and $k^2 = (b/2)^2$. Therefore, adding $(b/2)^2$ completes the square.

Completing the Square

To complete the square for $x^2 + bx$ with $b \neq 0$, add $\left(\dfrac{b}{2}\right)^2$, which is the square of one-half of the coefficient of x.

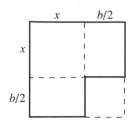

Figure 9.1

The diagram in Figure 9.1 may be used to obtain a geometric interpretation of the result, which shows why this procedure is called completing the square. This concept is discussed further in "Think About It" Exercise 4.

EXAMPLE 1   Determine the number that should be added to make the expression a perfect square. Then, add this number and factor the resulting trinomial.

a. $x^2 + 10x$ **b.** $x^2 - x$

Solution

a. The coefficient of x is 10, so $b = 10$. To complete the square, we add $\left(\dfrac{b}{2}\right)^2$, so we add

$$\left(\tfrac{10}{2}\right)^2 = 5^2 = 25.$$

Adding 25 to the expression and factoring gives

$$x^2 + 10x + 25 = (x + 5)^2.$$

b. The coefficient of x is -1. Thus $b = -1$, and $\left(\dfrac{b}{2}\right)^2 = \left(-\dfrac{1}{2}\right)^2 = \dfrac{1}{4}$. To complete the square, add $\dfrac{1}{4}$. Doing this and factoring yields

$$x^2 - x + \frac{1}{4} = \left(x - \frac{1}{2}\right)^2.$$

PROGRESS CHECK 1   Determine the number that should be added to make the expression a perfect square. Then, add this number and factor the resulting trinomial.

a. $x^2 + x$ **b.** $x^2 - 8x$ ⌐

2   The next example shows how completing the square can help solve a quadratic equation.

EXAMPLE 2   Solve $x^2 + 8x - 3 = 0$.

Solution   First, note that the factoring method does not apply to this equation. To solve by completing the square, begin by rearranging the equation so that the x terms are on one side of the equation and the constant term is on the other.

$$x^2 + 8x - 3 = 0$$
$$x^2 + 8x = 3 \qquad \text{Add 3 to both sides.}$$

Next, complete the square on the left. Half of 8 is 4 and $(4)^2 = 16$. By adding 16 to both sides of the equation, we then have

$$x^2 + 8x + 16 = 3 + 16 \qquad \text{Add 16 to both sides.}$$
$$(x + 4)^2 = 19. \qquad \text{Factor and simplify.}$$

Now, apply the square root property and solve each equation.

$$x + 4 = \sqrt{19} \qquad \text{or} \qquad x + 4 = -\sqrt{19}$$
$$x = -4 + \sqrt{19} \qquad \text{or} \qquad x = -4 - \sqrt{19}$$

Thus, the solution set is $\{-4 + \sqrt{19}, -4 - \sqrt{19}\}$. Check this answer.

PROGRESS CHECK 2   Solve $x^2 + 6x - 2 = 0$.   ⌐

EXAMPLE 3   Solve $x^2 - x = 2$.

Progress Check Answers

1. (a) $\frac{1}{4}$; $x^2 + x + \frac{1}{4} = (x + \frac{1}{2})^2$

(b) 16; $x^2 - 8x + 16 = (x - 4)^2$

2. $\{-3 + \sqrt{11}, -3 - \sqrt{11}\}$

Solution   The x terms and the constant term are already on different sides of the equation, so begin by completing the square. Half of -1 is $-\frac{1}{2}$ and $(-\frac{1}{2})^2 = \frac{1}{4}$. Add $\frac{1}{4}$ to both sides of the equation and simplify.

$$x^2 - x = 2$$
$$x^2 - x + \tfrac{1}{4} = 2 + \tfrac{1}{4} \qquad \text{Add } \tfrac{1}{4} \text{ to both sides.}$$
$$(x - \tfrac{1}{2})^2 = \tfrac{9}{4} \qquad \text{Factor and simplify.}$$

Now by the square root property

$$x - \tfrac{1}{2} = \sqrt{\tfrac{9}{4}} \qquad \text{or} \qquad x - \tfrac{1}{2} = -\sqrt{\tfrac{9}{4}} \ .$$

Since $\sqrt{\frac{9}{4}} = \frac{3}{2}$, we then have the following solutions.

$$x - \tfrac{1}{2} = \tfrac{3}{2} \qquad \text{or} \qquad x - \tfrac{1}{2} = -\tfrac{3}{2}$$
$$x = \tfrac{1}{2} + \tfrac{3}{2} \qquad \text{or} \qquad x = \tfrac{1}{2} - \tfrac{3}{2}$$
$$= 2 \qquad \text{or} \qquad = -1$$

Both solutions check, and the solution set is $\{2, -1\}$.

Note   The solutions in this example may also be obtained by the factoring method. (Confirm this.) Although the factoring method is easy to use here, keep in mind that the key feature of our current method is that it applies to *all* quadratic equations.

PROGRESS CHECK 3   Solve $x^2 + x = 6$.   ⌐

EXAMPLE 4   Solve the problem in the section introduction on page 413.

Solution   The length is to be 2 m greater than the width, so if we let

$$x = \text{width}$$
$$\text{then} \qquad x + 2 = \text{length}.$$

Now, illustrate the given information as in Figure 9.2. The panel is rectangular, and using $A = \ell w$ leads to the equation

$$x(x + 2) = 6.$$

Then by the methods of this section we solve the equation.

$$x^2 + 2x = 6 \qquad\qquad \text{Remove parentheses.}$$
$$x^2 + 2x + 1 = 6 + 1 \qquad\qquad \text{Add 1 to both sides.}$$
$$(x + 1)^2 = 7 \qquad\qquad \text{Factor and simplify.}$$
$$x + 1 = \sqrt{7} \qquad \text{or} \quad x + 1 = -\sqrt{7} \qquad \text{Square root property.}$$
$$x = -1 + \sqrt{7} \quad \text{or} \qquad x = -1 - \sqrt{7} \qquad \text{Subtract 1 from both sides.}$$

Because the width cannot be negative, reject $x = -1 - \sqrt{7}$. Thus, the width of the panel is $-1 + \sqrt{7}$ m, and the length is $(-1 + \sqrt{7}) + 2$ m, or $1 + \sqrt{7}$ m.
    To approximate this answer, use a calculator.

Width: 1 $\boxed{+/-}$ $\boxed{+}$ 7 $\boxed{\sqrt{}}$ $\boxed{=}$ $\boxed{1.6457513}$
Length: 1 $\boxed{+}$ 7 $\boxed{\sqrt{}}$ $\boxed{=}$ $\boxed{3.6457513}$

To the nearest hundredth, the panel is 1.65 m wide and 3.65 m long. This answer is reasonable since 3.65 is 2 more than 1.65, and $(3.65)(1.65) = 6.0225 \approx 6$.

PROGRESS CHECK 4   The design for a rectangular computer chip specifies that the area is to be 18 mm² and the length is to be 4 mm greater than the width. To the nearest hundredth of a millimeter, find the length and the width of this chip.   ⌐

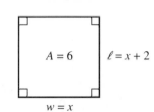

Figure 9.2

$A = 6$ $\ell = x + 2$

$w = x$

Progress Check Answers
3. $\{-3, 2\}$
4. Width = 2.69 mm, length = 6.69 mm

3 The procedure for completing the square requires that the coefficient of x^2 be 1. When the coefficient is not 1, then as a first step we divide both sides of the equation by the coefficient of x^2, as shown next.

EXAMPLE 5   Solve $2x^2 - 5x + 1 = 0$.

Solution   First, divide both sides of the equation by 2, which gives us a coefficient for x^2 of 1. Then write the x terms to the left of the equal sign and the constant term to the right.

$$2x^2 - 5x + 1 = 0$$
$$x^2 - \tfrac{5}{2}x + \tfrac{1}{2} = 0$$
$$x^2 - \tfrac{5}{2}x = -\tfrac{1}{2}$$

Now, complete the square. Half of $-\tfrac{5}{2}$ is $-\tfrac{5}{4}$, and $(-\tfrac{5}{4})^2 = \tfrac{25}{16}$. Add $\tfrac{25}{16}$ to both sides of the equation and simplify.

$$x^2 - \tfrac{5}{2}x + \tfrac{25}{16} = -\tfrac{1}{2} + \tfrac{25}{16}$$
$$(x - \tfrac{5}{4})^2 = \tfrac{17}{16}$$

By the square root property,

$$x - \frac{5}{4} = \sqrt{\frac{17}{16}} \quad \text{or} \quad x - \frac{5}{4} = -\sqrt{\frac{17}{16}}.$$

Finally, simplify $\sqrt{17/16}$ to $\sqrt{17}/4$ and solve both equations.

$$x - \frac{5}{4} = \frac{\sqrt{17}}{4} \quad \text{or} \quad x - \frac{5}{4} = -\frac{\sqrt{17}}{4}$$
$$x = \frac{5}{4} + \frac{\sqrt{17}}{4} \quad \text{or} \quad x = \frac{5}{4} - \frac{\sqrt{17}}{4}$$
$$= \frac{5 + \sqrt{17}}{4} \quad \text{or} \quad = \frac{5 - \sqrt{17}}{4}$$

Thus, the solution set is $\left\{\dfrac{5 + \sqrt{17}}{4}, \dfrac{5 - \sqrt{17}}{4}\right\}$. Check both solutions.

Note   It is not difficult to check the solutions in this example by calculator. For instance, to check to see if $(5 + \sqrt{17})/4$ is a solution, first store the computed value of this number.

$$5 \boxed{+} 17 \boxed{\sqrt{}} \boxed{=} \boxed{\div} 4 \boxed{=} \boxed{2.2807764} \boxed{\text{STO}}$$

Now, evaluate $2x^2 - 5x + 1$ where the stored value is x.

$$2 \boxed{\times} \boxed{\text{RCL}} \boxed{x^2} \boxed{-} 5 \boxed{\times} \boxed{\text{RCL}} \boxed{+} 1 \boxed{=} \boxed{\quad\quad 0}$$

Using this sequence, the end result in the display for some calculators may look like

$$\boxed{-1 \quad\quad -10}$$

which is scientific notation for -0.0000000001. This discrepancy from 0 is due to round-off errors and should be ignored.

PROGRESS CHECK 5   Solve $3x^2 - x - 5 = 0$.

Recall that we may abbreviate expressions of the form "$x = \sqrt{a}$ or $x = -\sqrt{a}$" by writing $x = \pm\sqrt{a}$, where the $\pm$ symbol is read "plus or minus." In the next example we use this abbreviation because it allows for a more compact method of solution.

Progress Check Answer

5. $\left\{\dfrac{1 + \sqrt{61}}{6}, \dfrac{1 - \sqrt{61}}{6}\right\}$

EXAMPLE 6   Solve $3x^2 - 4x - 2 = 0$.

Solution   In the following solution consider carefully each step, together with the associated side comment.

$$3x^2 - 4x - 2 = 0$$

$$x^2 - \frac{4}{3}x - \frac{2}{3} = 0 \qquad \text{Divide both sides by 3.}$$

$$x^2 - \frac{4}{3}x = \frac{2}{3} \qquad \text{Add } \frac{2}{3} \text{ to both sides.}$$

$$x^2 - \frac{4}{3}x + \frac{4}{9} = \frac{2}{3} + \frac{4}{9} \qquad \text{Add } \left[\frac{1}{2}\left(-\frac{4}{3}\right)\right]^2 \text{ to both sides.}$$

$$\left(x - \frac{2}{3}\right)^2 = \frac{10}{9} \qquad \text{Factor and simplify.}$$

$$x - \frac{2}{3} = \pm\sqrt{\frac{10}{9}} \qquad \text{Square root property.}$$

$$x - \frac{2}{3} = \frac{\pm\sqrt{10}}{3} \qquad \text{Simplify the radical.}$$

$$x = \frac{2}{3} + \frac{\pm\sqrt{10}}{3} \qquad \text{Add } \frac{2}{3} \text{ to both sides.}$$

$$= \frac{2 \pm \sqrt{10}}{3} \qquad \text{Combine the fractions.}$$

Thus, the solution set is $\left\{\dfrac{2 + \sqrt{10}}{3}, \dfrac{2 - \sqrt{10}}{3}\right\}$. Check them.

PROGRESS CHECK 6   Solve $3x^2 - 2x - 4 = 0$.

Progress Check Answer

6. $\left\{\dfrac{1 + \sqrt{13}}{3}, \dfrac{1 - \sqrt{13}}{3}\right\}$

EXERCISES 9.2

In Exercises 1–8, determine the number that should be added to make the expression a perfect square. Then, add this number and factor the resulting trinomial.

1. $x^2 + 12x$
2. $x^2 + 8x$
3. $x^2 - 2x$
4. $x^2 - 4x$
5. $x^2 + 3x$
6. $x^2 + 5x$
7. $x^2 - 7x$
8. $x^2 - 9x$

In Exercises 9–20, solve the equation by completing the square.

9. $x^2 + 4x - 2 = 0$
10. $x^2 + 6x - 4 = 0$
11. $x^2 - 2x - 1 = 0$
12. $x^2 - 2x - 2 = 0$
13. $x^2 - 6x + 7 = 0$
14. $x^2 - 8x + 13 = 0$
15. $x^2 - 6x + 9 = 0$
16. $x^2 + 8x + 16 = 0$
17. $x^2 - x - 72 = 0$
18. $x^2 - 2x - 48 = 0$
19. $x^2 - x + 1 = 0$
20. $x^2 - x + 2 = 0$

In Exercises 21–26, give the solution in (a) exact radical form and (b) decimal form to the nearest hundredth.

21. Find two numbers whose sum is 6 and whose product is 7.

22. Find two numbers whose sum is 8 and whose product is 11.

23. Find the dimensions of a rectangle whose length is 2 more than its width and whose area is 7 cm².

24. Find the dimensions of a rectangle whose length is 4 more than its width and whose area is 4 in.².

25. Find the lengths of all sides of a right triangle whose area is 7 in.². The sum of the two perpendicular sides is 8 in.

26. Find the lengths of all sides of a right triangle whose area is 9 cm². The sum of the two perpendicular sides is 10 cm.

In Exercises 27–38, solve the equation by completing the square. Remember to rewrite equations so that the coefficient of x^2 is 1.

27. $2x^2 - x - 2 = 0$

28. $2x^2 + 3x - 1 = 0$

29. $3x^2 + 5x + 1 = 0$

30. $3x^2 + 7x + 1 = 0$

31. $3x^2 + 6x + 2 = 0$

32. $2x^2 + 4x + 1 = 0$

33. $3x^2 + 4x - 5 = 0$

34. $5x^2 + 6x - 7 = 0$

35. $3x^2 + 2x - 1 = 0$

36. $4x^2 - 7x + 3 = 0$

37. $2x^2 + x + 1 = 0$

38. $2x^2 + 2x + 1 = 0$

THINK ABOUT IT

1. In quadratic equations of the form $x^2 + bx + c = 0$ (where the coefficient of x^2 is 1), it is always true that the sum of the two roots is equal to $-b$. Check that this is true for each of these equations.
   a. $x^2 + 2x - 15 = 0$
   b. $x^2 + 2x - 1 = 0$

2. In quadratic equations of the form $x^2 + bx + c = 0$, it is always true that the product of the roots is c. Check that this is true for each of these equations.
   a. $x^2 + 2x - 15 = 0$
   b. $x^2 + 2x - 1 = 0$

3. In the Italian Renaissance much progress was made in the algebraic solution of equations. A famous problem of that time was given by Cardano in his *Ars Magna* (1545). He asked whether one could split the number 10 into two parts, the product of them being 40. He said the problem was "manifestly impossible." Set up a quadratic equation to solve the problem and see what the difficulty is. Cardano was the first to publish a technique for solving third-degree equations.

4. The geometry of completing the square is instructive. We have seen that to make $x^2 + bx$ into a perfect square, we must add $(b/2)^2$. What does this have to do with completing the picture of a square? Consider the figure.
   a. Find expressions for areas A_1, A_2, and A_3 and their sum.
   b. What size is the missing corner piece that would be needed to complete the large square?
   c. What is the area of the large square after it is completed?

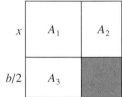

5. What would it mean to "complete the cube"?
   a. What number must be added to $x^3 + 6x^2 + 12x$ to make it a perfect cube? [*Hint:* The model is $(x + b)^3 = x^3 + 3x^2b + 3xb^2 + b^3$.]
   b. Add this number and show the expression in factored form.

REMEMBER THIS

1. In the expression $x^2 - 3x + 4$, what is the coefficient of x^2?
2. Evaluate $\sqrt{b^2 - 4ac}$ if $a = 1$, $b = -1$, and $c = -6$.
3. Simplify $\dfrac{2 - 10}{6}$.
4. True or false? $\sqrt{-12}$ is not a real number.
5. True or false? $\dfrac{2 + \sqrt{3}}{2}$ simplifies to $1 + \sqrt{3}$.

6. Solve $(x - 3)^2 = 25$.
7. Rationalize the denominator $\dfrac{1}{3 - \sqrt{2}}$.
8. What is the x-intercept of the line $3x + 4y = 15$?
9. Solve $\dfrac{3 - x}{5} = \dfrac{x}{2}$.
10. True or false? The square of $\frac{3}{5}$ is less than $\frac{3}{5}$.

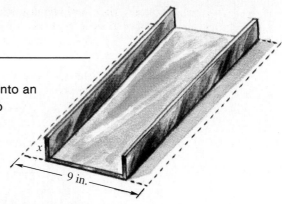

9.3 Solving Quadratic Equations by the Quadratic Formula

A long strip of galvanized sheet metal 9 in. wide is to be shaped into an open gutter by bending up the edges, as shown in the figure, to form a gutter with a rectangular cross-sectional area of 9 in.². Find the two possible heights (x) of this gutter by using the quadratic formula. (See Example 2.)

OBJECTIVES

1 Solve quadratic equations using the quadratic formula.

2 Solve quadratic equations after choosing an efficient method.

1 The key feature of solving quadratic equations by completing the square is that this method applies to *all* quadratic equations. Therefore, if we apply this method to the general quadratic equation $ax^2 + bx + c = 0$ with $a \neq 0$, then we obtain a formula for solving quadratic equations that always works. In the following derivation of this formula, we also display the solution of the particular quadratic equation $2x^2 + 7x + 1 = 0$ to illustrate in specific terms what is happening.

General Equation	Particular Equation	Comment
$ax^2 + bx + c = 0, a \neq 0$	$2x^2 + 7x + 1 = 0$	Given equation.
$x^2 + \dfrac{bx}{a} + \dfrac{c}{a} = 0$	$x^2 + \dfrac{7}{2}x + \dfrac{1}{2} = 0$	Divide on both sides by the coefficient of x^2.
$x^2 + \dfrac{bx}{a} = -\dfrac{c}{a}$	$x^2 + \dfrac{7}{2}x = -\dfrac{1}{2}$	Subtract the constant term from both sides.
$x^2 + \dfrac{b}{a}x + \dfrac{b^2}{4a^2} = -\dfrac{c}{a} + \dfrac{b^2}{4a^2}$	$x^2 + \dfrac{7}{2}x + \dfrac{49}{16}$ $= -\dfrac{1}{2} + \dfrac{49}{16}$	Add the square of one-half of the coefficient of x to both sides.
$\left(x + \dfrac{b}{2a}\right)^2 = \dfrac{b^2 - 4ac}{4a^2}$	$\left(x + \dfrac{7}{4}\right)^2 = \dfrac{41}{16}$	Factor on the left and add fractions on the right.
$x + \dfrac{b}{2a} = \pm\sqrt{\dfrac{b^2 - 4ac}{4a^2}}$	$x + \dfrac{7}{4} = \pm\sqrt{\dfrac{41}{16}}$	Apply the square root property.
$x + \dfrac{b}{2a} = \pm\dfrac{\sqrt{b^2 - 4ac}}{2a}$	$x + \dfrac{7}{4} = \dfrac{\pm\sqrt{41}}{4}$	Simplify the radical.
$x = \dfrac{-b}{2a} + \dfrac{\pm\sqrt{b^2 - 4ac}}{2a}$	$x = \dfrac{-7}{4} + \dfrac{\pm\sqrt{41}}{4}$	Isolate x on the left.
$x = \dfrac{-b \pm \sqrt{b^2 - 4ac}}{2a}$	$x = \dfrac{-7 \pm \sqrt{41}}{4}$	Combine fractions on the right.

In the particular equation, the two solutions are

$$x = \frac{-7 + \sqrt{41}}{4} \quad \text{and} \quad x = \frac{-7 - \sqrt{41}}{4}.$$

In the general equation, the two solutions are

$$x = \frac{-b + \sqrt{b^2 - 4ac}}{2a} \quad \text{and} \quad x = \frac{-b - \sqrt{b^2 - 4ac}}{2a}.$$

Our work with the general equation results in the quadratic formula.

Quadratic Formula

If $ax^2 + bx + c = 0$, and $a \neq 0$, then

$$x = \frac{-b \pm \sqrt{b^2 - 4ac}}{2a}.$$

Any quadratic equation may be solved with this formula. The idea is to substitute appropriate values for a, b, and c in the formula and then simplify.

EXAMPLE 1   Solve $3x^2 - 2x - 8 = 0$ using the quadratic formula.

Solution   By comparing the given equation with the general equation,

$$3 \quad - 2x - 8 = 0$$
$$\updownarrow \quad \updownarrow \quad \updownarrow$$
$$ax^2 + bx + c = 0,$$

we see that $a = 3$, $b = -2$, and $c = -8$. Substituting these values into the quadratic formula gives

$$x = \frac{-b \pm \sqrt{b^2 - 4ac}}{2a} = \frac{-(-2) \pm \sqrt{(-2)^2 - 4(3)(-8)}}{2(3)}$$

$$= \frac{2 \pm \sqrt{4 + 96}}{6}$$

$$= \frac{2 \pm \sqrt{100}}{6}$$

$$= \frac{2 \pm 10}{6}.$$

Now, convert from abbreviated form and solve each equation separately.

$$x = \frac{2 + 10}{6} = \frac{12}{6} = 2 \quad \text{or} \quad x = \frac{2 - 10}{6} = -\frac{8}{6} = -\frac{4}{3}$$

Thus, the solution set is $\{2, -\frac{4}{3}\}$. Check both solutions in the original equation.

PROGRESS CHECK 1   Solve $5x^2 + 3x - 2 = 0$ using the quadratic formula. ⌐

EXAMPLE 2   Solve the problem in the section introduction on page 419.

Solution   The sheet metal is 9 in. wide, so bending up edges of length x on each side converts the sheet metal to a gutter with dimensions as shown in Figure 9.3. The rectangular cross-sectional area is 9 in.^{2}, and using $A = \ell w$ leads to the equation

$$9 = (9 - 2x)x.$$

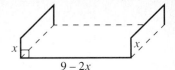

Figure 9.3

To solve this equation using the quadratic formula, first express the equation in the form $ax^2 + bx + c = 0$.

$$9 = 9x - 2x^2 \qquad \text{Remove parentheses.}$$
$$2x^2 - 9x + 9 = 0 \qquad \text{Add } 2x^2 \text{ and subtract } 9x \text{ on both sides.}$$

In this equation, $a = 2$, $b = -9$, and $c = 9$.

$$x = \frac{-b \pm \sqrt{b^2 - 4ac}}{2a} = \frac{-(-9) \pm \sqrt{(-9)^2 - 4(2)(9)}}{2(2)}$$
$$= \frac{9 \pm \sqrt{81 - 72}}{4}$$
$$= \frac{9 \pm \sqrt{9}}{4}$$
$$= \frac{9 \pm 3}{4}$$

Then,

$$x = \frac{9 + 3}{4} = \frac{12}{4} = 3 \qquad \text{or} \qquad x = \frac{9 - 3}{4} = \frac{6}{4} = \frac{3}{2}.$$

For applied problems we check to see if both answers are meaningful in the context of the problem.

If $x = 3$, then $9 - 2x = 3$, and $A = 3 \cdot 3 = 9$.

If $x = \frac{3}{2}$, then $9 - 2x = 6$, and $A = 6(\frac{3}{2}) = 9$.

Both solutions check, so the two possible heights for the gutter are 3 and $\frac{3}{2}$ in.

Note   To solve quadratic equations of the form $ax^2 + bx + c = 0$ using a calculator and the quadratic formula, it is usually best to start with the radical and store its value by substituting for a, b, and c in the following sequence.

$$b \boxed{x^2} \boxed{-} 4 \boxed{\times} a \boxed{\times} c \boxed{=} \boxed{\sqrt{}} \boxed{\text{STO}}$$

Then, one solution adds the stored number.

$$b \boxed{+/-} \boxed{+} \boxed{\text{RCL}} \boxed{=} \boxed{\div} 2 \boxed{\div} a \boxed{=}$$

The other solution subtracts the stored number.

$$b \boxed{+/-} \boxed{-} \boxed{\text{RCL}} \boxed{=} \boxed{\div} 2 \boxed{\div} a \boxed{=}$$

Check this method using the equation in this example.

PROGRESS CHECK 2   Redo the problem in Example 2, but assume that the cross-sectional area of the gutter is 7 in.^{2}.   ⌐

For the next example, note that the quadratic formula is easily adjusted when the variable in the quadratic equation is not symbolized by x.

EXAMPLE 3   Solve $y^2 - 4y - 3 = 0$ using the quadratic formula.

Solution   In this equation the variable is y, while $a = 1$, $b = -4$, and $c = -3$.

$$y = \frac{-b \pm \sqrt{b^2 - 4ac}}{2a} = \frac{-(-4) \pm \sqrt{(-4)^2 - 4(1)(-3)}}{2(1)}$$

$$= \frac{4 \pm \sqrt{16 + 12}}{4}$$

$$= \frac{4 \pm \sqrt{28}}{2}$$

Now $\sqrt{28} = \sqrt{4}\sqrt{7} = 2\sqrt{7}$, and we simplify.

$$y = \frac{4 \pm 2\sqrt{7}}{2} \qquad \text{Simplify the radical.}$$

$$= \frac{2(2 \pm \sqrt{7})}{2} \qquad \text{Factor the numerator.}$$

$$= 2 \pm \sqrt{7} \qquad \text{Divide out the common factor 2.}$$

Thus, the solution set is $\{2 + \sqrt{7}, 2 - \sqrt{7}\}$. Check this answer.

Caution   A common student error is to interpret the quadratic formula as

<table>
<tr><td align="center">**Wrong**</td><td></td><td align="center">**Wrong**</td></tr>
<tr><td align="center">$x = -b \pm \dfrac{\sqrt{b^2 - 4ac}}{2a}$</td><td align="center">or as</td><td align="center">$x = \dfrac{-b}{2a} \pm \sqrt{b^2 - 4ac}.$</td></tr>
</table>

Remember to divide the *entire* expression $-b \pm \sqrt{b^2 - 4ac}$ by $2a$.

PROGRESS CHECK 3   Solve $t^2 + 8t + 3 = 0$ using the quadratic formula.   ⌐

EXAMPLE 4   Solve $x^2 + 4 = 2x$.

Solution   To obtain the form $ax^2 + bx + c = 0$, first subtract $2x$ from both sides of the equation.

$$x^2 + 4 = 2x$$
$$x^2 - 2x + 4 = 0$$

From this equation, we see that $a = 1$, $b = -2$, and $c = 4$.

$$x = \frac{-b \pm \sqrt{b^2 - 4ac}}{2a} = \frac{-(-2) \pm \sqrt{(-2)^2 - 4(1)(4)}}{2(1)}$$

$$= \frac{2 \pm \sqrt{4 - 16}}{2}$$

$$= \frac{2 \pm \sqrt{-12}}{2}$$

Because $\sqrt{-12}$ is not a real number, this equation has no real number solution.

PROGRESS CHECK 4   Solve $2x^2 + 3 = x$.   ⌐

2   In this chapter we have shown several different methods for solving quadratic equations. Selecting an efficient solution method depends on the particular equation to be solved. The following guidelines will help in your choice of methods.

Progress Check Answers

3. $\{-4 \pm \sqrt{13}\}$

4. No real number solution

Guidelines to Solve a Quadratic Equation

Equation Type	Recommended Method
$ax^2 + c = 0$	Use the square root property. If $ax^2 + c$ is a difference of squares, consider the factoring method.
$ax^2 + bx = 0$	Use the factoring method.
$(px + q)^2 = k$	Use the square root property.
$ax^2 + bx + c = 0$	First, try the factoring method. If $ax^2 + bx + c$ does not factor or is hard to factor, use the quadratic formula.

Note that solving by completing the square is not recommended as an efficient method. Instead, this method is important because it is the basis for the quadratic formula and because in other problem-solving situations completing the square is used to convert an expression to a standard form that is easier to analyze.

EXAMPLE 5   Solve $16x^2 - 24x + 9 = 0$. Choose an efficient method.

Solution   The equation is of the form $ax^2 + bx + c = 0$, where $a = 16$, $b = -24$, and $c = 9$. First, we try the factoring method, and we find it applies.

$$16x^2 - 24x + 9 = 0$$
$$(4x - 3)(4x - 3) = 0$$
$$4x - 3 = 0 \quad \text{or} \quad 4x - 3 = 0$$
$$x = \tfrac{3}{4} \quad \text{or} \quad x = \tfrac{3}{4}$$

Thus, there is exactly one solution, and the solution set is $\{\tfrac{3}{4}\}$. If the factoring step in this solution seemed hard, then an alternative solution method is to use the quadratic formula.

$$x = \frac{-b \pm \sqrt{b^2 - 4ac}}{2a} = \frac{-(-24) \pm \sqrt{(-24)^2 - 4(16)(9)}}{2(16)}$$
$$= \frac{24 \pm \sqrt{576 - 576}}{32} = \frac{24 \pm \sqrt{0}}{32} = \frac{24 \pm 0}{32} = \frac{24}{32} = \frac{3}{4}$$

By either method, the solution set is $\{\tfrac{3}{4}\}$.

PROGRESS CHECK 5   Solve $9x^2 = 100$. Choose an efficient method.

Progress Check Answer
5. $\{\tfrac{10}{3}, -\tfrac{10}{3}\}$

EXERCISES 9.3

In Exercises 1–6, give the value for a, b, and c in the quadratic formula. Do not solve the equation.

1. $3x^2 - 4x + 5 = 0$
2. $2x^2 + x - 1 = 0$
3. $y^2 - 5 = 0$
4. $3y^2 + 4y = 0$
5. $y^2 = -3y + 4$
6. $5x^2 + 2x = 8$

In Exercises 7–12, give the value of $b^2 - 4ac$. Do not solve the equation.

7. $x^2 + 2x + 1 = 0$
8. $x^2 + 2x - 1 = 0$
9. $3y^2 - 2 = 0$
10. $2y^2 + 3 = 0$
11. $2x^2 - 5x + 3 = 0$
12. $3x^2 + 4x + 1 = 0$

In Exercises 13–18, the two solutions of a quadratic equation are given in condensed form. Show both solutions in simplest form.

13. $\dfrac{-2 \pm 4}{2}$

14. $\dfrac{-9 \pm 6}{3}$

15. $\dfrac{3 \pm 3}{4}$

16. $\dfrac{-3 \pm 3}{6}$

17. $\dfrac{4 \pm 2\sqrt{5}}{8}$

18. $\dfrac{6 \pm 12\sqrt{2}}{9}$

In Exercises 19–34, solve the equation by using the quadratic formula.

19. $3x^2 + 4x + 1 = 0$

20. $x^2 + 4x + 3 = 0$

21. $10x^2 + 3x = 1$

22. $12x^2 - x = 1$

23. $9x^2 = 6x - 1$

24. $9x^2 = 24x - 16$

25. $y^2 + 2y - 2 = 0$

26. $x^2 - 4x + 1 = 0$

27. $x^2 + 7 = 6x$

28. $x^2 + 13 = 8x$

29. $y^2 + y - 1 = 0$

30. $4y^2 + 4y - 5 = 0$

31. $4y^2 + 3 = 12y$

32. $9y^2 - 1 = 12y$

33. $x^2 + x + 1 = 0$

34. $2x^2 + 3x + 4 = 0$

In Exercises 35–50, choose whatever method is efficient for solving the equation.

35. $4x^2 - 12x + 9 = 0$

36. $9x^2 + 12x + 4 = 0$

37. $3x^2 - 12 = 0$

38. $4y^2 - 64 = 0$

39. $y^2 - \frac{9}{4} = 0$

40. $x^2 - \frac{4}{25} = 0$

41. $3x^2 = 15$

42. $5x^2 = 10$

43. $3x^2 + 4x = 0$

44. $5x^2 - 4x = 0$

45. $(2x + 1)^2 = 25$

46. $(3x - 1)^2 = 25$

47. $x^2 - x = 20$

48. $y^2 - y = 30$

49. $4y^2 - 16y + 1 = 0$

50. $4y^2 - 12y + 1 = 0$

In Exercises 51–52, give solutions to the nearest hundredth.

51. $1.2x^2 + 3.1x - 4.2 = 0$

52. $2.1x^2 + 3.2x - 4.3 = 0$

In Exercises 53–56, check that answers are sensible in the context of the problem.

53. A strip of metal which is 12 in. wide is to be shaped into an open gutter with cross-sectional area of 9 in.². But for the gutter to be useful, the height must be at least 1 in. (See Example 2 of this section for a sketch.) Find the height.

54. Repeat Exercise 53, but assume that the area must now be 8 in.².

55. Two adjacent building lots are each square, with a total area of 13,600 ft². The total street frontage for the two lots is 160 ft. (See the figure.) Find the dimensions of each square lot. (*Hint:* Write two equations and solve by substitution.)

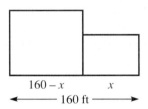

56. (Refer to the figure for Exercise 55.) Two adjacent building lots have a total area of 11,250 ft². The total street frontage is 150 ft. Find the dimensions of each square lot.

57. An ancient problem leading to a quadratic equation is given on an Egyptian papyrus from about 2000 B.C. The problem is as follows: Divide 100 square units into two squares such that the side of one of the squares is three-fourths the side of the other. You can solve the problem by the method of substitution.

58. Divide 10 square units into two squares such that the side of one of the squares is half the side of the other.

59. An object is thrown upward from a platform 10 ft above the ground with an initial velocity of 40 ft/second. The formula $y = 10 + 40t - 16t^2$ gives its height above the ground t seconds after it is released. After how many seconds will the object be 30 ft above the ground? Why are there two answers? (Round to the nearest hundredth of a second.)

60. Repeat Exercise 59, but answer these questions: After how many seconds will the object be 6 ft above the ground? Why is there only one answer? (Round to the nearest hundredth of a second.)

61. Babylonian mathematics texts from about 3000 B.C. contain problems asking for a number which added to its reciprocal gives a specified sum. They worked out a general procedure for such problems. Try this one. What number when added to its reciprocal equals 5? Give the answer in radical form and in decimal approximation to the nearest thousandth.

62. What number when added to its reciprocal equals 6? Give the answer in radical form and in decimal approximation to the nearest thousandth.

THINK ABOUT IT

1. The equation $x^2 + y^2 = 25$ determines a circle with radius 5 and with center at the origin. (See the figure.)

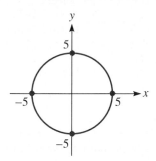

a. On the same set of axes, draw the line $y = x + 1$. How many points of intersection are there between the lines and the circle?

b. Solve the set of equations $x^2 + y^2 = 25$ and $y = x + 1$ by substituting $x + 1$ for y in the equation of the circle. This will lead to a quadratic equation. The two solutions for x will be the x-coordinates for the two points of intersection.

c. Find the two points of intersection.

2. If $x^2 + 6x + c = 0$, find all values for c that lead to solutions that are real numbers.

3. You can solve $x^2 + 3x - 1 = 0$ by using the quadratic formula with $a = 1$, $b = 3$, and $c = -1$. But if you first multiply both sides of the equation by -1, then you will get $-x^2 - 3x + 1 = 0$, for which $a = -1$, $b = -3$, and $c = 1$.

a. Show that you get the same solutions in both cases.

b. In the formula $x = \dfrac{-b \pm \sqrt{b^2 - 4ac}}{2a}$, replace each variable by its opposite and show that the value for x does not change.

4. Explain why the quadratic formula fails if $a = 0$. What kind of "quadratic equation" do you have if $a = 0$? For example, what is the equation if $a = 0$, $b = 2$, and $c = 5$?

5. **a.** Write both solutions for $ax^2 + bx + c = 0$ with $a \neq 0$ which are given by the quadratic formula.

b. Show that the sum of the two solutions is $-b/a$.

c. Show that the product of the two solutions is c/a.

d. Verify the results of parts **b** and **c** for the equation $2x^2 + 3x - 1 = 0$.

REMEMBER THIS

1. True or false? $\sqrt{7 \cdot 8} = \sqrt{7} \cdot \sqrt{8}$.

2. By which property is it true that $a\sqrt{8} = \sqrt{8}a$?

3. Which of these is *not* a real number?
   a. 0   **b.** $\sqrt{7}$ **c.** -3 **d.** $\sqrt{-16}$

4. Multiply $(3 - \sqrt{2})(3 + \sqrt{2})$.

5. Expand $(4 + 3x)^2$.

6. What number must be added to $x^2 - 10x$ to complete the square?

7. Solve $x^2 = 5$.

8. Which of these represents a line which is perpendicular to the graph of $y = 5x$?
   a. $y = -5x$ **b.** $y = \frac{1}{5}x$ **c.** $y = -\frac{1}{5}x$

9. What is the formula for the volume of a cylinder with radius r and height h?

10. Solve $d = rt$ for r if $d = 2.5 \times 10^{11}$ and $t = 5 \times 10^9$.

9.4 Complex Number Solutions to Quadratic Equations

Complex numbers are used extensively in electronics to designate voltage, current, and other electrical quantities, because such representations indicate both the strength and time (or phase) relationship of the quantities. An important formula in this discipline is Ohm's law for alternating current circuits, which states

$$V = IZ$$

where V is the voltage, I is the current, and Z is the impedance. If the current in a particular circuit is $3 - 4i$ amperes and the impedance is $5 + 2i$ ohms, find the complex number that measures the voltage. The unit of measure for voltage is volts. (See Example 4.)

OBJECTIVES

1 Express square roots of negative numbers in terms of i.

2 Add, subtract, and multiply complex numbers.

3 Determine if a complex number is a solution of a quadratic equation.

4 Find complex number solutions to quadratic equations.

1 To be able to find *any* solution of *any* quadratic equation we need to be able to solve equations like

$$x^2 = -9$$

by extending the number system beyond real numbers to include numbers like $\sqrt{-9}$. To do this, we first introduce a different type of number called an **imaginary number.** The base unit in imaginary numbers is $\sqrt{-1}$ and it is designated by i. Thus, by definition

$$i = \sqrt{-1} \quad \text{and} \quad i^2 = -1.$$

Square roots of negative numbers such as $\sqrt{-7}$ and $\sqrt{-9}$ may now be interpreted as

$$\sqrt{-7} = \sqrt{-1 \cdot 7} = \sqrt{-1} \cdot \sqrt{7} = i\sqrt{7},$$
$$\sqrt{-9} = \sqrt{-1 \cdot 9} = \sqrt{-1} \cdot \sqrt{9} = i \cdot 3 = 3i.$$

These results suggest the following definition.

Principal Square Root of a Negative Number

If a is a positive real number, then

$$\sqrt{-a} = i\sqrt{a}.$$

EXAMPLE 1 Express each number in terms of i.

a. $\sqrt{-10}$ b. $-\sqrt{-10}$ c. $\sqrt{-16}$ d. $\sqrt{-12}$

Solution

a. $\sqrt{-10} = i\sqrt{10}$
b. $-\sqrt{-10} = -i\sqrt{10}$
c. $\sqrt{-16} = i\sqrt{16} = 4i$
d. $\sqrt{-12} = i\sqrt{12} = i\sqrt{4}\sqrt{3} = 2i\sqrt{3}$

Note When we write expressions like those in Example 1, the position of the imaginary unit i may be chosen for convenience. In this example, we chose to write i in front of the radicals so that expressions like $\sqrt{10}\, i$ are not confused with $\sqrt{10i}$.

PROGRESS CHECK 1 Express each number in terms of i.

a. $\sqrt{-64}$ b. $\sqrt{-14}$ c. $-\sqrt{-14}$ d. $\sqrt{-18}$

Progress Check Answers

1. (a) $8i$ (b) $i\sqrt{14}$ (c) $-i\sqrt{14}$ (d) $3i\sqrt{2}$

2 Using real numbers and imaginary numbers, we may extend the number system to include a number like $1 + 3i$. Such a number is needed to solve equations like $(x - 1)^2 + 9 = 0$. (See Example 6.) This type of number is called a **complex number.**

> ### Definition of Complex Numbers
>
> A number of the form $a + bi$, where a and b are real numbers and $i = \sqrt{-1}$, is called a complex number.

The form $a + bi$ is called the **standard form** of a complex number. Because a and/or b may be zero, the complex numbers include the real numbers and the imaginary numbers. For instance, consider these examples.

Real number → $4 = 4 + 0i$
Complex number in standard form
Imaginary number → $3i = 0 + 3i$

The relationship among the various sets of numbers is illustrated in Figure 9.4.

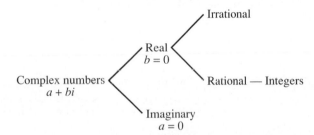

Figure 9.4

Operations with complex numbers are defined so that the properties of real numbers (like the distributive law) continue to apply. In many ways, computations with complex numbers are similar to computations with polynomials. The important difference is to always represent square roots of negative numbers in terms of i, remembering that $i^2 = -1$. The following examples illustrate how to add, subtract, and multiply with complex numbers.

EXAMPLE 2 Combine the complex numbers.

a. $\sqrt{-25} + \sqrt{-49}$ **b.** $(5 - 8i) + (3 + 2i)$ **c.** $(4 - i) - (7 - 10i)$

Solution

a. $\sqrt{-25} + \sqrt{-49} = i\sqrt{25} + i\sqrt{49}$
$= 5i + 7i$
$= 12i$

b. $(5 - 8i) + (3 + 2i) = (5 + 3) + (-8 + 2)i$
$= 8 - 6i$

c. $(4 - i) - (7 - 10i) = 4 - i - 7 + 10i$
$= (4 - 7) + (-1 + 10)i$
$= -3 + 9i$

PROGRESS CHECK 2 Combine the complex numbers.

a. $\sqrt{-100} + \sqrt{-4}$ **b.** $(7 - i) + (-8 + 5i)$ **c.** $(3 + 2i) - (1 - 9i)$ ⏌

Progress Check Answers

2. (a) $12i$ (b) $-1 + 4i$ (c) $2 + 11i$

EXAMPLE 3 Multiply the complex numbers.

a. $\sqrt{-4} \cdot \sqrt{-9}$ **b.** $(5 + 3i)(1 - 2i)$ **c.** $(4 + 3i)^2$

Solution

a. $\sqrt{-4} \cdot \sqrt{-9} = i\sqrt{4} \cdot i\sqrt{9}$
$$= 2i \cdot 3i$$
$$= 6i^2$$
$$= 6(-1) \quad \text{Replace } i^2 \text{ by } -1.$$
$$= -6$$

b. $(5 + 3i)(1 - 2i) = 5(1) + 5(-2i) + 3i(1) + 3i(-2i) \quad \text{Use FOIL.}$
$$= 5 - 10i + 3i - 6i^2$$
$$= 5 - 10i + 3i - 6(-1) \quad \text{Replace } i^2 \text{ by } -1.$$
$$= 5 - 10i + 3i + 6$$
$$= 11 - 7i$$

c. $(4 + 3i)^2 = (4)^2 + 2(4)(3i) + (3i)^2 \quad \text{Use } (a + b)^2 = a^2 + 2ab + b^2.$
$$= 16 + 24i + 9i^2$$
$$= 16 + 24i + 9(-1) \quad \text{Replace } i^2 \text{ by } - 1.$$
$$= 7 + 24i$$

Caution The property of radicals $\sqrt{a}\sqrt{b} = \sqrt{ab}$ does *not* hold when a and b are *both negative*. For instance, the solution in part **a** shows

$$\sqrt{-4} \cdot \sqrt{-9} = -6.$$

Therefore,

$$\sqrt{-4} \cdot \sqrt{-9} = \sqrt{(-4)(-9)} = \sqrt{36} = 6 \text{ is } \textbf{wrong.}$$

Always remember to express complex numbers in terms of i before performing computations.

PROGRESS CHECK 3 Multiply the complex numbers.

a. $\sqrt{-25} \cdot \sqrt{-1}$ **b.** $(3 - i)(6 + 4i)$ **c.** $(2 + 5i)^2$

EXAMPLE 4 Solve the problem in the section introduction on page 425.

Solution Replace I by $3 - 4i$ and Z by $5 + 2i$ in $V = IZ$ and multiply.

$$V = (3 - 4i)(5 + 2i)$$
$$= 15 + 6i - 20i - 8i^2$$
$$= 15 + 6i - 20i - 8(-1) \quad \text{Replace } i^2 \text{ by } -1.$$
$$= 15 + 6i - 20i + 8$$
$$= 23 - 14i$$

Thus, the measure of the voltage is $23 - 14i$ volts.

PROGRESS CHECK 4 Find the voltage in an alternating current circuit in which the current is $3 - 6i$ amperes and the impedance is $4 + i$ ohms.

3 By using the operations we have considered, we can check to see whether a complex number is a solution of a quadratic equation.

EXAMPLE 5 Is $2 + i$ a solution of the equation $x^2 - 4x + 5 = 0$?

Solution   Replace x by $2 + i$ and see if the resulting equation is true.

$$x^2 - 4x + 5 = 0$$
$$(2 + i)^2 - 4(2 + i) + 5 \overset{?}{=} 0 \quad \text{Replace } x \text{ by } 2 + i.$$
$$4 + 4i + i^2 - 8 - 4i + 5 \overset{?}{=} 0$$
$$i^2 + 1 \overset{?}{=} 0$$
$$-1 + 1 \overset{?}{=} 0 \quad \text{Replace } i^2 \text{ by } -1.$$
$$0 \overset{\checkmark}{=} 0$$

Thus, $2 + i$ is a solution of $x^2 - 4x + 5 = 0$.

PROGRESS CHECK 5   Is $2 - i$ a solution of $x^2 - 4x + 5 = 0$?   ⌐

4   In the remaining examples we solve quadratic equations whose solutions are not real numbers. In Example 6 we first consider quadratic equations that are solved efficiently using the square root property, which now extends to include the square roots of negative numbers.

EXAMPLE 6   Solve each equation.

a. $x^2 = -18$ **b.** $(x - 1)^2 + 9 = 0$

Solution

a. By the square root property, $x^2 = -18$ implies
$$x = \sqrt{-18} \quad \text{or} \quad x = -\sqrt{-18}.$$
Since $\sqrt{-18}$ is $i\sqrt{18} = i\sqrt{9}\sqrt{2} = 3i\sqrt{2}$, the solution set is $\{3i\sqrt{2}, -3i\sqrt{2}\}$. Check these solutions.

b. First, transform the equation so that $(x - 1)^2$ is on one side of the equation by itself.
$$(x - 1)^2 + 9 = 0$$
$$(x - 1)^2 = -9$$

Now apply the square root property.
$$x - 1 = \sqrt{-9} \quad \text{or} \quad x - 1 = -\sqrt{-9}$$
Since $\sqrt{-9}$ is $i\sqrt{9} = 3i$,
$$x - 1 = 3i \quad \text{or} \quad x - 1 = -3i$$
$$x = 1 + 3i \quad \text{or} \quad x = 1 - 3i.$$

We confirm that $1 + 3i$ is a solution next.
$$[(1 + 3i) - 1]^2 + 9 = (3i)^2 + 9 = 9i^2 + 9 = 9(-1) + 9 = 0$$
In a similar way, $1 - 3i$ checks, and the solution set is $\{1 + 3i, 1 - 3i\}$.

PROGRESS CHECK 6   Solve each equation.

a. $x^2 = -12$ **b.** $(x - 3)^2 + 4 = 0$ ⌐

In the next example we solve $x^2 + 4 = 2x$, as in Example 4 of the preceding section. However, we can now continue the solution and find the complex number answers, instead of stating only that the equation has no real number solution.

EXAMPLE 7   Solve $x^2 + 4 = 2x$.

Solution   To obtain the form $ax^2 + bx + c = 0$, first subtract $2x$ from both sides of the equation, to get

$$x^2 - 2x + 4 = 0.$$

From this equation, we see that $a = 1$, $b = -2$, and $c = 4$.

$$x = \frac{-b \pm \sqrt{b^2 - 4ac}}{2a} = \frac{-(-2) \pm \sqrt{(-2)^2 - 4(1)(4)}}{2(1)}$$

$$= \frac{2 \pm \sqrt{4 - 16}}{2} = \frac{2 \pm \sqrt{-12}}{2}$$

Now $\sqrt{-12} = i\sqrt{12} = i\sqrt{4}\sqrt{3} = 2i\sqrt{3}$, and we simplify.

$$x = \frac{2 \pm 2i\sqrt{3}}{2} \qquad \text{Simplify the radical.}$$

$$= \frac{2(1 \pm i\sqrt{3})}{2} \qquad \text{Factor the numerator.}$$

$$= 1 \pm i\sqrt{3} \qquad \text{Divide out the common factor 2.}$$

Thus, the solution set is $\{1 + i\sqrt{3}, 1 - i\sqrt{3}\}$. Check these solutions.

Progress Check Answer

7. $\{-1 + 2i\sqrt{2}, -1 - 2i\sqrt{2}\}$

PROGRESS CHECK 7   Solve $x^2 + 2x = -9$.   ⌐

EXERCISES 9.4

In Exercises 1–20, express each number in terms of i.

1. $\sqrt{-1}$
2. $\sqrt{-25}$
3. $\sqrt{-100}$
4. $\sqrt{-121}$
5. $-\sqrt{-4}$
6. $-\sqrt{-9}$
7. $\sqrt{-\frac{4}{9}}$
8. $\sqrt{-\frac{16}{25}}$
9. $\sqrt{-11}$
10. $\sqrt{-13}$
11. $\sqrt{-15}$
12. $\sqrt{-2}$
13. $-\sqrt{-5}$
14. $-\sqrt{-6}$
15. $\sqrt{-20}$
16. $\sqrt{-28}$
17. $\sqrt{-27}$
18. $\sqrt{-45}$
19. $-\sqrt{-63}$
20. $-\sqrt{-32}$

In Exercises 21–44, combine the complex numbers.

21. $4i + 5i$
22. $2i + 7i$
23. $3i - i$
24. $5i - i$
25. $2i - 5i$
26. $3i - 6i$
27. $\sqrt{-4} + \sqrt{-9}$
28. $\sqrt{-4} + \sqrt{-25}$
29. $\sqrt{-1} + \sqrt{-1}$
30. $\sqrt{-1} + \sqrt{-16}$
31. $\sqrt{-9} - \sqrt{-4}$
32. $\sqrt{-1} - \sqrt{-4}$
33. $(3 - 4i) + (2 + 5i)$
34. $(4 + 2i) + (1 - 3i)$
35. $(1 + i) - (2 - i)$
36. $(2 + i) - (1 - i)$
37. $(5 + 2i) - (5 - 2i)$
38. $(-2 + i) - (-2 - i)$
39. $(-3 - 2i) + (-2 + 2i)$
40. $(-2 + 6i) + (-1 - 6i)$
41. $(-1 + i) + (3 - 2i)$
42. $(-2 + 3i) + (3 - 2i)$
43. $(7 + i) - (2 - 3i)$
44. $(1 + 3i) - (4 - 4i)$

In Exercises 45–72, multiply the complex numbers.

45. $\sqrt{-1} \cdot \sqrt{-1}$
46. $\sqrt{-4} \cdot \sqrt{-4}$
47. $\sqrt{-9} \cdot \sqrt{-25}$
48. $\sqrt{-16} \cdot \sqrt{-9}$
49. $\sqrt{-2} \cdot \sqrt{-2}$
50. $\sqrt{-3} \cdot \sqrt{-3}$
51. $i \cdot i$
52. $2i \cdot 3i$
53. $-i \cdot i$
54. $-2i \cdot i$
55. $(3 + 2i)(5 + 4i)$
56. $(7 + 3i)(2 + 4i)$
57. $(2 + 3i)(3 + 4i)$
58. $(3 + 5i)(2 + 3i)$
59. $(3 + 4i)(3 - 2i)$
60. $(2 + 3i)(4 - 3i)$
61. $(1 + i)(2 - i)$

62. $(1 + i)(3 - i)$
63. $(2 + 5i)(2 - 5i)$
64. $(3 + 4i)(3 - 4i)$
65. $(3 + 2i)(2 + 3i)$
66. $(5 + i)(1 + 5i)$
67. $(1 + i)^2$
68. $(1 - i)^2$
69. $(2 + 3i)^2$
70. $(3 + 5i)^2$
71. $(4 - i)^2$
72. $(3 - i)^2$

Exercises 73–76 refer to alternating current circuits where the formula $V = IZ$ may be used to find the voltage V for the given current I and impedance Z. In each case, find the voltage.

73. Current is $2 - 3i$ amperes; impedance is $4 + i$ ohms.

74. Current is $2 - 3i$ amperes; impedance is $5 + 2i$ ohms.

75. Current is $1 + 2i$ amperes; impedance is $1 + 2i$ ohms.

76. Current is $3 - i$ amperes; impedance is $3 - 2i$ ohms.

In Exercises 77–82, determine whether the complex number is a solution to the given equation.
77. $x^2 - 2x + 2 = 0; 1 + i$
78. $x^2 - 2x + 2 = 0; 1 - i$
79. $x^2 - 6x + 10 = 0; 3 - i$
80. $x^2 - 6x + 10 = 0; 3 + i$
81. $x^2 - x + 1 = 0; 2 + i$
82. $x^2 - 3x + 3 = 0; 3 + i$

In Exercises 83–110, solve the quadratic equation by any efficient method. The solutions are complex numbers.
83. $x^2 = -25$
84. $x^2 = -36$
85. $x^2 = -20$

86. $x^2 = -24$
87. $x^2 + 7 = 0$
88. $x^2 + 5 = 0$
89. $2x^2 + 3 = 0$
90. $3x^2 + 4 = 0$
91. $(x - 5)^2 + 1 = 0$
92. $(x - 4)^2 + 4 = 0$
93. $(x + 1)^2 + 4 = 0$
94. $(x + 3)^2 + 16 = 0$
95. $(4x - 1)^2 + 9 = 0$
96. $(2x + 3)^2 + 4 = 0$
97. $(x - 2)^2 + 5 = 0$
98. $(x - 3)^2 + 6 = 0$
99. $x^2 + x + 1 = 0$
100. $x^2 + x + 2 = 0$
101. $2x^2 + x + 1 = 0$
102. $2x^2 + x + 2 = 0$
103. $x^2 + 2 = 2x$
104. $x^2 + 10 = 6x$
105. $x^2 + 6x = -17$
106. $x^2 + 10x = -33$
107. $4x^2 - 4x + 9 = 0$
108. $3x^2 - 4x + 4 = 0$
109. $2x^2 - 8x = -17$
110. $4x^2 + 4x = -13$

THINK ABOUT IT

1. Complex numbers are not ordered like real numbers. The relations "less than" and "greater than" do not apply to complex numbers. But they can be assigned a magnitude, similar to the concept of the absolute value of a real number. The absolute value of the complex number $a + bi$ is defined as $\sqrt{a^2 + b^2}$.
   a. Find the absolute value of $3 - 4i$.
   b. Find the absolute value of $1 + i$.
   c. Find the absolute value of $2 + i$.
2. Refer to the previous problem. The absolute value of the product of two complex numbers is always equal to the product of their absolute values. Show that this is true for $1 + i$ and $2 + i$.
3. Which of these numbers is a square root of $4i$? Check your result by squaring the answer to see that you get $4i$.
   a. 2   b. -2 c. $2i$ d. $-2i$ e. $\sqrt{2} + i\sqrt{2}$

4. Find a value for n that will make the product a pure imaginary number of the form $a + bi$, where $a = 0$: $(3 + 2i)(6 + ni)$.
5. The quadratic formula shows that two nonreal roots of a quadratic equation are a conjugate pair. That is, one has the form $m + ni$ and the other has the form $m - ni$. You can therefore make up a quadratic equation to have a desired pair of roots.
   a. Just as the equation $(x - 3)(x - 5) = 0$ has roots equal to 3 and 5, so the equation $[x - (1 + i)][x - (1 - i)] = 0$ has roots equal to $1 + i$ and $1 - i$. Multiply and simplify $[x - (1 + i)][x - (1 - i)] = 0$ to get a quadratic equation in standard form, and then use the quadratic formula to check that its roots are $1 + i$ and $1 - i$.
   b. Find a quadratic equation in standard form whose roots are $2 + i$ and $2 - i$.

REMEMBER THIS

1. Which of these is *not* a linear equation in two variables?
   a. $y = 7 - x$ **b.** $y = x^2$ **c.** $y = x$
2. True or false? The ordered pair $(2, -7)$ is a solution of $y = -2x^2 + 1$.
3. If $y = f(t) = 128t - 16t^2$, find $f(4)$.
4. True or false? The average of two real numbers is halfway between them.
5. Divide $\dfrac{b}{a}$ by 2.
6. Solve $x^2 - 2x - 1 = 0$ by the quadratic formula.

7. Solve $\dfrac{1}{x} = x + \dfrac{3}{2}$.
8. Write an equation for the line through the origin with slope 1.
9. Simplify $\dfrac{x^2 + 4x}{x^2 + 8x + 16}$.
10. Add $\dfrac{1}{b} + \dfrac{1}{c}$.

9.5 Graphing Quadratic Functions

The height (y) of a projectile shot vertically upward from the ground with an initial velocity of 128 ft/second is given by

$$y = f(t) = 128t - 16t^2,$$

where y is measured in feet and t is the elapsed time in seconds.
a. When does the projectile attain its maximum height?
b. What is the maximum height?
c. When does the projectile hit the ground?
d. Graph this function. (See Example 6.)

OBJECTIVES

|1| Graph quadratic functions by finding ordered-pair solutions.

|2| Graph quadratic functions using the vertex of the graph.

|3| Graph quadratic functions using the vertex and intercepts of the graph.

|4| Solve applied problems involving quadratic functions.

|1| In Section 6.2 we graphed linear equations in two variables. Recall that such an equation graphs as a straight line and that the line gives us a useful picture of all ordered-pair solutions of the equation. We now extend our coverage of graphing to obtain pictures of the solutions of quadratic equations in two variables of the form

$$y = ax^2 + bx + c, \quad \text{with} \quad a \neq 0.$$

Because each x value leads to exactly one y value in such equations, we may view the relations graphed in this section in the context of a function.

Definition of Quadratic Function

A **quadratic function** is a function defined by

$$y = f(x) = ax^2 + bx + c,$$

where a, b, and c are real numbers, with $a \neq 0$.

One method for graphing a quadratic function is based on obtaining some ordered-pair solutions of the equation. We will use this method in Examples 1–3.

EXAMPLE 1   Graph $y = x^2$.

Solution   Begin by substituting integer values for x from, say, 2 to -2 and making a list of ordered-pair solutions.

If $x =$	Then $y = x^2$	Thus, the Ordered-Pair Solutions Are
2	$y = 2^2 = 4$	(2,4)
1	$y = 1^2 = 1$	(1,1)
0	$y = 0^2 = 0$	(0,0)
-1	$y = (-1)^2 = 1$	$(-1,1)$
-2	$y = (-2)^2 = 4$	$(-2,4)$

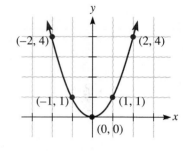

Figure 9.5

Now, graph these ordered pairs and draw a smooth curve through them. The resulting graph of $y = x^2$ is shown in Figure 9.5.

PROGRESS CHECK 1   Graph $y = x^2 - 1$.

EXAMPLE 2   Graph $y = -2x^2 + 1$.

Solution   Make a list of ordered-pair solutions, as shown below, and plot them. Then, graph $y = -2x^2 + 1$ by drawing a smooth curve through these points, as shown in Figure 9.6.

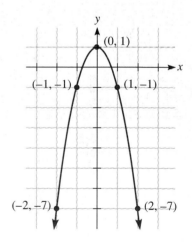

If $x =$	Then $y = -2x^2 + 1$	Thus, the Ordered-Pair Solutions Are
2	$y = -2(2)^2 + 1 = -7$	$(2,-7)$
1	$y = -2(1)^2 + 1 = -1$	$(1,-1)$
0	$y = -2(0)^2 + 1 = 1$	$(0,1)$
-1	$y = -2(-1)^2 + 1 = -1$	$(-1,-1)$
-2	$y = -2(-2)^2 + 1 = -7$	$(-2,-7)$

Figure 9.6

PROGRESS CHECK 2   Graph $y = 2 - x^2$.

Progress Check Answers

1.
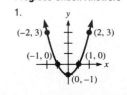

2.
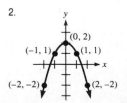

The graphs in Examples 1 and 2 are curves that are called **parabolas.** It can be shown that the graph of every quadratic function is a parabola, and we may use some features of parabolas to graph quadratic functions efficiently. First, note in the graph of $y = x^2$ in Figure 9.5 that the parabola has a minimum turning point and opens up like a cup. This occurs whenever the coefficient of x^2 is a positive number. If the coefficient of x^2 is a negative number, then the graph turns at the highest point on the graph, as shown in the graph of $y = -2x^2 + 1$ in Figure 9.6. We now know some important facts about the graph of a quadratic function.

Graph of a Quadratic Function

For the function defined by $y = ax^2 + bx + c$ with $a \neq 0$

1. The graph is a parabola.
2. If $a > 0$, the parabola opens upward and turns at the lowest point on the graph.
3. If $a < 0$, the parabola opens downward and turns at the highest point on the graph.

EXAMPLE 3   Graph $y = x^2 - 6x + 4$.

Solution   The equation defines a quadratic function, so the graph is a parabola. The coefficient of x^2 is 1, so $a > 0$ and the parabola opens upward. Now we determine some ordered-pair solutions using the same replacements for x as in Examples 1 and 2.

If $x =$	Then $y = x^2 - 6x + 4$	Thus, the Ordered-Pair Solutions Are
2	$y = (2)^2 - 6(2) + 4 = -4$	$(2, -4)$
1	$y = (1)^2 - 6(1) + 4 = -1$	$(1, -1)$
0	$y = (0)^2 - 6(0) + 4 = 4$	$(0, 4)$
-1	$y = (-1)^2 - 6(-1) + 4 = 11$	$(-1, 11)$
-2	$y = (-2)^2 - 6(-2) + 4 = 20$	$(-2, 20)$

When we graph these ordered pairs and draw a smooth curve through them, as in Figure 9.7(a), we find we must continue to substitute larger values for x to see the graph pass through its turning point.

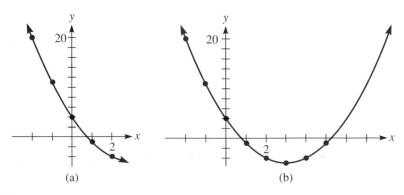

(a)   (b)

Figure 9.7

If $x = 3$, $y = (3)^2 - 6(3) + 4 = -5$, so $(3,-5)$ is a solution.

If $x = 4$, $y = (4)^2 - 6(4) + 4 = -4$, so $(4,-4)$ is a solution.

If $x = 5$, $y = (5)^2 - 6(5) + 4 = -1$, so $(5,-1)$ is a solution.

By using these additional points, we may now draw the desired graph, as shown in Figure 9.7(b).

PROGRESS CHECK 3   Graph $y = x^2 + 6x + 5$.   ⌐

2   Example 3 showed that we may have difficulty deciding which replacements for x will enable us to draw the parabola. To overcome this difficulty, we need some way to locate the turning point of the parabola, which is called the **vertex**. Consider Figure 9.8, which illustrates that the vertex in the graph of a quadratic function is located on a vertical line called the **axis of symmetry** of the parabola. This line divides the parabola into two segments such that if we make a fold on the line, the two halves will coincide.

To find the equation of the axis of symmetry, note that this line is halfway between any pair of points on the parabola that have the same y-coordinate. The easiest pair of points to analyze is the pair of points on the graph of $y = ax^2 + bx + c$ whose y-coordinate is c. Therefore, we replace y by c in this equation and solve for x.

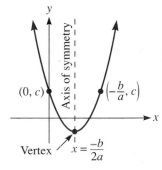

Figure 9.8

$$y = ax^2 + bx + c$$
$$c = ax^2 + bx + c \qquad \text{Replace } y \text{ by } c.$$
$$0 = ax^2 + bx \qquad \text{Subtract } c \text{ from both sides.}$$
$$0 = x(ax + b) \qquad \text{Factor the nonzero side.}$$
$$x = 0 \quad \text{or} \quad ax + b = 0 \qquad \text{Set each factor equal to 0.}$$
$$ax = -b$$
$$x = \frac{-b}{a}$$

The x-coordinate halfway between 0 and $-b/a$ is given by

$$x = \frac{0 + (-b/a)}{2},$$

so the equation of the axis of symmetry is

$$x = \frac{-b}{2a}.$$

Because the vertex lies on the axis of symmetry, the x-coordinate of the vertex is $-b/2a$. We may find the y-coordinate of the vertex by finding $f(-b/2a)$, which is the y value when $x = -b/2a$, and we have established the following formula.

Vertex Formula

The vertex of the graph of $y = f(x) = ax^2 + bx + c$, with $a \neq 0$, is located at

$$\left(\frac{-b}{2a}, f\left(\frac{-b}{2a} \right) \right).$$

Progress Check Answer

3.

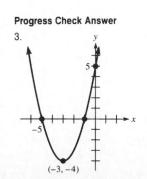

EXAMPLE 4   Consider the function defined by $y = f(x) = 2x^2 + 8x + 5$.

a. Find the equation of the axis of symmetry.
b. Find the coordinates of the vertex.
c. Graph the function.

Solution

a. Replacing a by 2 and b by 8 in the axis-of-symmetry formula gives

$$x = \frac{-b}{2a} = \frac{-8}{2(2)} = \frac{-8}{4} = -2.$$

Thus, $x = -2$ is the equation of the axis of symmetry.

b. The x-coordinate of the vertex is -2, since the vertex lies on the axis of symmetry. We find the y value of the vertex by finding $f(-2)$, the y value when $x = -2$.

$$f(x) = 2x^2 + 8x + 5$$
$$f(-2) = 2(-2)^2 + 8(-2) + 5$$
$$= 8 - 16 + 5$$
$$= -3$$

The vertex is located at $(-2, -3)$.

c. Since $a > 0$, the vertex is the lowest point on the graph. To obtain additional points, substitute integer values for x on both sides of $x = -2$ (which is the axis of symmetry).

If $x =$	Then $y = 2x^2 + 8x + 5$	Thus, the Ordered-Pair Solutions Are
-4	$y = 2(-4)^2 + 8(-4) + 5 = 5$	$(-4, 5)$
-3	$y = 2(-3)^2 + 8(-3) + 5 = -1$	$(-3, -1)$
-1	$y = 2(-1)^2 + 8(-1) + 5 = -1$	$(-1, -1)$
0	$y = 2(0)^2 + 8(0) + 5 = 5$	$(0, 5)$

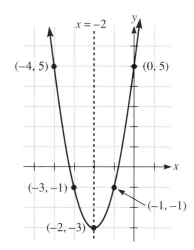

Figure 9.9

By drawing the parabola through these points and the vertex, we obtain the graph of the function shown in Figure 9.9.

PROGRESS CHECK 4   Redo the problems in Example 4 for the function defined by $y = f(x) = 2x^2 - 4x + 1$.

3   In the next example, after finding the vertex, we will find x- and y-intercepts for the graph instead of constructing a table of ordered-pair solutions. Recall that the points where the graph crosses the x- and y-axes are called the **x- and y-intercepts,** respectively. To find the y-intercept, we let $x = 0$ and solve for y. Doing this in the equation $y = ax^2 + bx + c$ gives

$$y = a(0)^2 + b(0) + c.$$

Thus, the y-intercept is always $(0, c)$.

To find any x-intercepts, replace y by 0 in the equation $y = ax^2 + bx + c$ and solve for x. Thus, the x-coordinates of any x-intercepts are found by solving the quadratic equation $ax^2 + bx + c = 0$.

Progress Check Answers

4. (a) $x = 1$ (b) $(1, -1)$ (c)

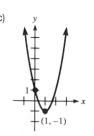

EXAMPLE 5   Consider the function defined by $y = f(x) = x^2 + 2x - 3$.

a. Find the equation of the axis of symmetry.
b. Find the coordinates of the vertex.
c. Find the coordinates of any x- and y-intercepts.
d. Graph the function.

Solution

a. In the given equation, $a = 1$ and $b = 2$. Therefore,

$$x = \frac{-b}{2a} = \frac{-2}{2(1)} = \frac{-2}{2} = -1.$$

The equation of the axis of symmetry is $x = -1$.

b. The x-coordinate of the vertex is given by $-b/2a$, so this coordinate is -1. We find the y-coordinate by finding $f(-1)$, the y value when $x = -1$.

$$f(x) = x^2 + 2x - 3$$
$$f(-1) = (-1)^2 + 2(-1) - 3$$
$$= 1 - 2 - 3$$
$$= -4$$

The vertex is at $(-1, -4)$.

c. The y-intercept is at $(0, -3)$, since $c = -3$ in the given equation. To find any x-intercepts, set $y = 0$ and use an efficient method to solve the resulting quadratic equation.

$$x^2 + 2x - 3 = 0$$
$$(x + 3)(x - 1) = 0$$
$$x + 3 = 0 \quad \text{or} \quad x - 1 = 0$$
$$x = -3 \qquad\qquad x = 1$$

The x-intercepts are at $(-3, 0)$ and $(1, 0)$.

d. By drawing a parabola through the vertex and the intercepts, we obtain the graph of the function shown in Figure 9.10.

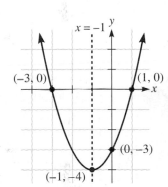

Figure 9.10

PROGRESS CHECK 5   Redo the problems in Example 5 for the function defined by $y = f(x) = x^2 - 6x + 5$.

⌐4⌐   The section-opening problem considers an application associated with quadratic functions.

EXAMPLE 6   Solve the problem in the section introduction on page 432.

Solution

a. In the formula $y = f(t) = 128t - 16t^2$, $a = -16$, $b = 128$, and t replaces x in the role of the independent variable. Since $a < 0$, the vertex is the highest point in the graph. Use the axis-of-symmetry formula to find the time when this highest point is reached.

$$t = \frac{-b}{2a} = \frac{-128}{2(-16)} = \frac{-128}{-32} = 4$$

The projectile attains its maximum height when $t = 4$ seconds.

b. We find the maximum height by finding the y value when $t = 4$ seconds.

$$y = f(t) = 128t - 16t^2$$
$$f(4) = 128(4) - 16(4)^2$$
$$= 512 - 256$$
$$= 256$$

Progress Check Answers

5. (a) $x = 3$ (b) $(3, -4)$ (c) $(0,5), (1,0), (5,0)$

(d)

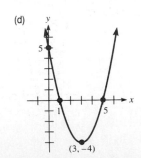

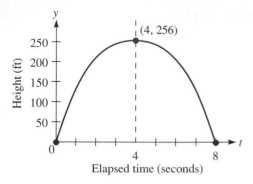

Figure 9.11

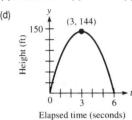

The projectile reaches a maximum height of 256 ft.

c. When the projectile hits the ground, the height (y) of the projectile is 0. Thus, we set $y = 0$, and solve the resulting quadratic equation.

$$128t - 16t^2 = 0$$
$$16t(8 - t) = 0$$
$$16t = 0 \quad \text{or} \quad 8 - t = 0$$
$$t = 0 \qquad\qquad 8 = t$$

The projectile is on the ground initially, and it hits the ground 8 seconds after launching.

d. Parts **a** and **b** indicate that the vertex is at (4,256), while part **c** indicates that the intercepts are (0,0) and (8,0). Drawing a parabola using these points gives the graph of the function shown in Figure 9.11. Note that we only consider values for t between 0 and 8, inclusive, because the formula is not meaningful outside this interval.

PROGRESS CHECK 6   The height (y) in feet of a projectile shot vertically up from the ground with an initial velocity of 96 ft/second is given by $y = f(t) = 96t - 16t^2$, where t is the elapsed time in seconds. Answer the questions in Example 6 (see the section introduction) with respect to this formula.   ⌐

EXERCISES 9.5

In Exercises 1–16, graph the given quadratic functions by making a list of ordered-pair solutions.

1. $y = x^2 + 1$

2. $y = x^2 + 2$

3. $y = -x^2 + 1$

4. $y = -x^2 + 2$

5. $y = 4x^2$

6. $y = 5x^2$

7. $y = -2x^2$

8. $y = -3x^2$

9. $y = 2x^2 - 3$

10. $y = 3x^2 - 2$

11. $y = x^2 + x + 1$

12. $y = x^2 - x + 1$

20. $y = f(x) = -x^2 - 2x + 1$

21. $y = f(x) = \frac{1}{4}x^2 - x + 2$

13. $y = -x^2 - x + 1$

14. $y = -x^2 + x + 1$

22. $y = f(x) = \frac{1}{2}x^2 + 2x + 4$

15. $y = x^2 - 2x + 3$

16. $y = x^2 + 4x - 2$

In Exercises 23–44 you are given a quadratic function.
  a. Find the equation of the axis of symmetry.
  b. Find the coordinates of the vertex.
  c. Find the coordinates of the x- and y-intercepts.
  d. Graph the function.
23. $y = f(x) = x^2 + 2x - 15$

In Exercises 17–22 you are given a quadratic function.
  a. Find the equation of the axis of symmetry.
  b. Find the coordinates of the vertex.
  c. Graph the function.
17. $y = f(x) = 3x^2 + 6x - 2$

24. $y = f(x) = x^2 + 4x - 12$

18. $y = f(x) = 2x^2 - 4x - 2$

19. $y = f(x) = -x^2 + 2x + 1$

25. $y = f(x) = -x^2 + 6x - 5$

30. $y = f(x) = x^2 - 8x + 12$

26. $y = f(x) = -x^2 - 2x + 3$

31. $y = f(x) = 9x^2 - 18x + 5$

27. $y = f(x) = x^2 - 9$

32. $y = f(x) = 4x^2 - 8x + 3$

28. $y = f(x) = 16 - x^2$

33. $y = f(x) = 4x^2 + 8x - 5$

29. $y = f(x) = x^2 - 6x + 8$

34. $y = 9x^2 + 18x - 16$

35. $y = f(x) = -x^2 - 6x$

36. $y = f(x) = -x^2 - 4x$

37. $y = f(x) = x^2 - x - 2$

38. $y = f(x) = x^2 - 3x - 4$

39. $y = f(x) = -x^2 - 3x - 2$

40. $y = f(x) = -x^2 - 5x - 6$

41. $y = f(x) = x^2 - 5$

42. $y = f(x) = x^2 - 6$

43. $y = f(x) = x^2 - 10x + 23$

44. $y = f(x) = x^2 - 6x + 7$

In Exercises 45–48, answer these questions.
  a. When does the projectile attain maximum height?
  b. What is the maximum height?
  c. When does the projectile hit the ground?
  d. Graph the function.

For each exercise the function $y = f(t)$ gives the height y (in feet) of a projectile shot vertically upward from the ground with the given initial velocity; t is the elapsed time (in seconds).

Initial Velocity	*Function*
45. 64 ft/second	$y = f(t) = 64t - 16t^2$

46. 32 ft/second $y = f(t) = 32t - 16t^2$

47. 48 ft/second $y = f(t) = 48t - 16t^2$

48. 80 ft/second $y = f(t) = 80t - 16t^2$

Exercises 49–52 concern projectiles. In the seventeenth century Galileo studied the motion of projectiles to help understand the path of a shell shot from a cannon. He found that the path was part of a parabola, as shown in the figure. The actual formula depends on both the angle at which the cannon is aimed and the initial velocity with which it is fired. The general formula is $y = K_1x - K_2x^2$, where K_1 and K_2 vary according to the angle and initial velocity; y gives the height (in feet) when the projectile is x ft from the gun. For Exercises 49–52, use the given values of K_1 and K_2 to find (a) the maximum height of the shell and (b) the horizontal distance traveled before it hits the ground.

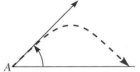

	Angle	*Initial Velocity*	K_1	K_2
49.	45°	40 ft/second	1	0.02
50.	45°	80 ft/second	1	0.005
51.	60°	40 ft/second	$\sqrt{3}$	0.04
52.	60°	80 ft/second	$\sqrt{3}$	0.01

THINK ABOUT IT

1. In the definition of quadratic function given in this section, why is it necessary to say $a \neq 0$?
2. The great Greek mathematician Archimedes (in 225 B.C.) showed that the area inside a parabola (see the figure) equals $\frac{4}{3}$ the area of the triangle. Use his discovery to find the area enclosed by the parabola $y = x^2$ and the line $y = 4$.

3. The x-intercepts of the parabola $y = ax^2 + bx + c$ are found by solving the quadratic equation $ax^2 + bx + c = 0$. The parabola $y = x^2 + x + 1$ has no x-intercepts. (Graph it!) What happens when you try to find them by solving the equation $x^2 + x + 1 = 0$?
4. For what values of c does the graph of $y = x^2 + x + c$ have two x-intercepts?

5. If you draw a line and a point not on the line, and then mark all other points that are equidistant from the original line and point, these new points will form a parabola. For example, if you draw the point $(0,1)$ and the line $y = -1$, then the collection of points equidistant from them forms the parabola whose equation is $y = \frac{1}{4}x^2$.
   a. Draw the point $(0,1)$ and the line $y = -1$ on the same set of axes.

   b. Now add the graph of $y = \frac{1}{4}x^2$ to the picture.
   c. The point $(2,1)$ is on the parabola. Show that its distance from $(0,1)$ is the same as its distance from the line $y = -1$.
   d. The point $(4,4)$ is on the parabola. Show that its distance from $(0,1)$ is the same as its distance from the line $y = -1$.

REMEMBER THIS

1. Solve $x^2 + 18 = 0$.
2. Evaluate $16^{3/4}$.
3. Solve

$$x + y = 0$$
$$x - y = 0.$$

4. What is the equation for this line?

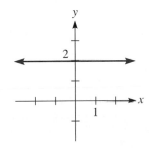

5. Simplify $\dfrac{a + \dfrac{1}{a}}{b}$.
6. Factor $x^2 - 36$.
7. Simplify $(3x^2y)(4x^{-1}y^{-2})$. Use only positive exponents in the answer.
8. Solve $3(x + 1) + 4 = -4$.
9. What property says that $3(x + y) = 3x + 3y$?
10. Is i a solution to $x^2 = -1$?

Chapter 9 SUMMARY

OBJECTIVES CHECKLIST   Specific chapter objectives are summarized below along with numbered example problems from the text that should clarify the objectives. If you do not understand any objectives or do not know how to do the selected problems, then restudy the material.

9.1 **Can you:**
1. **Solve quadratic equations using the factoring method?**
   Solve $x^2 - 3x = 18$.   [Example 1]

2. **Solve quadratic equations using the square root property?**
   Solve $(x + 4)^2 = 49$.   [Example 6]

9.2 **Can you:**
1. **Complete the square for $x^2 + bx$ and express the resulting trinomial in factored form?**
   Determine the number that should be added to make the expression a perfect square. Then, add this number and factor the resulting trinomial: $x^2 + 10x$.   [Example 1a]

2. **Solve $ax^2 + bx + c = 0$ by completing the square when $a = 1$?**
   Solve $x^2 + 8x - 3 = 0$.   [Example 2]

3. **Solve $ax^2 + bx + c = 0$ by completing the square when $a \neq 1$?**
   Solve $2x^2 - 5x + 1 = 0$.   [Example 5]

9.3 Can you:

1. **Solve quadratic equations using the quadratic formula?**
   Solve $3x^2 - 2x - 8 = 0$ using the quadratic formula.   [Example 1]

2. **Solve quadratic equations after choosing an efficient method?**
   Solve $16x^2 - 24x + 9 = 0$. Choose an efficient method.   [Example 5]

9.4 Can you:

1. **Express square roots of negative numbers in terms of *i*?**
   Express $\sqrt{-16}$ in terms of i.   [Example 1c]

2. **Add, subtract, and multiply complex numbers?**
   Multiply and simplify: $(5 + 3i)(1 - 2i)$.   [Example 3b]

3. **Determine if a complex number is a solution of a quadratic equation?**
   Is $2 + i$ a solution of the equation $x^2 - 4x + 5 = 0$?   [Example 5]

4. **Find complex number solutions to quadratic equations?**
   Solve $x^2 + 4 = 2x$.   [Example 7]

9.5 Can you:

1. **Graph quadratic functions by finding ordered-pair solutions?**
   Graph $y = x^2$.   [Example 1]

2. **Graph quadratic functions using the vertex of the graph?**
   Consider the function defined by $y = f(x) = 2x^2 + 8x + 5$.
   a. Find the equation of the axis of symmetry.
   b. Find the coordinates of the vertex.
   c. Graph the function.   [Example 4]

3. **Graph quadratic functions using the vertex and intercepts of the graph?**
   Consider the function defined by $y = f(x) = x^2 + 2x - 3$.
   a. Find the equation of the axis of symmetry.
   b. Find the coordinates of the vertex.
   c. Find the coordinates of any x- and y-intercepts.
   d. Graph the function.   [Example 5]

4. **Solve applied problems involving quadratic functions?**
   The height (y) of a projectile shot vertically upward from the ground with an initial velocity
   of 128 ft/second is given by $y = f(t) = 128t - 16t^2$, where y is measured in feet and t is
   the elapsed time in seconds.
   a. When does the projectile attain its maximum height?
   b. What is the maximum height?
   c. When does the projectile hit the ground?
   d. Graph this function.   [Example 6]

KEY TERMS

Axis of symmetry (9.5)

Complex number (9.4)

Imaginary numbers (9.4)

Parabola (9.5)

Perfect square trinomial (9.2)

Quadratic equation (9.1)

Quadratic function (9.5)

Second-degree equation (9.1)

Square root property (9.1)

Vertex of parabola (9.5)

x- and y-intercepts of a parabola (9.5)

KEY CONCEPTS AND PROCEDURES

Section	Key Concepts or Procedures to Review
9.1	■ Factoring method for solving quadratic equations

9.1

■ Factoring method for solving quadratic equations

    1. If necessary, change the form of the equation so that one side is 0.

    2. Factor the nonzero side.

    3. Set each factor equal to 0 and obtain the solution(s) by solving the resulting equations.

    4. Check each solution by substituting it in the original equation.

■ Square root property: If a is a positive real number, then

$$x^2 = a \quad \text{implies} \quad x = \sqrt{a} \quad \text{or} \quad x = -\sqrt{a}$$

9.2

■ Completing the square: To complete the square for $x^2 + bx$ with $b \neq 0$, add $(b/2)^2$, which is the square of one-half of the coefficient of x.

9.3

■ Quadratic formula: If $ax^2 + bx + c = 0$, and $a \neq 0$, then

$$x = \frac{-b \pm \sqrt{b^2 - 4ac}}{2a}$$

■ Guidelines to solve a quadratic equation

Equation Type	Recommended Method
$ax^2 + c = 0$	Use the square root property. If $ax^2 + c$ is a difference of squares, consider the factoring method.
$ax^2 + bx = 0$	Use the factoring method.
$(px + q)^2 = k$	Use the square root property.
$ax^2 + bx + c = 0$	First, try the factoring method. If $ax^2 + bx + c$ does not factor or is hard to factor, use the quadratic formula.

9.4

■ Principal square root of a negative number: If a is a positive real number, then

$$\sqrt{-a} = i\sqrt{a}$$

■ Definition of complex numbers: A number of the form $a + bi$, where a and b are real numbers and $i = \sqrt{-1}$, is called a complex number.

9.5

■ Definition of quadratic function: A quadratic function is a function defined by

$$y = f(x) = ax^2 + bx + c$$

where a, b, and c are real numbers, with $a \neq 0$.

■ Graph of a quadratic function: For the function defined by $y = ax^2 + bx + c$ with $a \neq 0$

    1. The graph is a parabola.

    2. If $a > 0$, the parabola opens upward and turns at the lowest point on the graph.

    3. If $a < 0$, the parabola opens downward and turns at the highest point on the graph.

■ Vertex formula: The vertex of the graph of $y = f(x) = ax^2 + bx + c$ with $a \neq 0$ is located at

$$\left(\frac{-b}{2a}, f\left(\frac{-b}{2a} \right) \right)$$

CHAPTER 9 REVIEW EXERCISES

9.1

1. Solve by factoring: $6x^2 - x - 12 = 0$.
2. Solve by using the square root property: $(x - 4)^2 = 49$.

9.2

3. Determine the number that should be added to $x^2 - 12x$ to complete the square. Then, add this number and factor the resulting trinomial.
4. Solve by completing the square: $x^2 - 6x + 7 = 0$.

5. Solve by completing the square: $4x^2 - 4x - 1 = 0$.

9.3

6. Solve by the quadratic formula: $x^2 - x - 1 = 0$.

7. A number plus its reciprocal equals 3. Express the number and

   its reciprocal in radical form.

9.4

8. Express $\sqrt{-20}$ in terms of i.
9. Multiply $(5 + 3i)(1 - 2i)$.
10. Show that $3 + i$ is a solution of $x^2 + 10 = 6x$.

11. Solve $(x - 3)^2 + 25 = 0$.
12. Solve $x^2 + 2x + 3 = 0$.

9.5

13. Graph $y = 3 - 2x^2$.

14. Consider $y = f(x) = 2x^2 + 4x + 1$.
    a. Find the equation of the axis of symmetry.
    b. Find the coordinates of the vertex.
    c. Find the y-intercept.
    d. Graph the function.

15. Graph $y = f(x) = x^2 + 4x - 5$. Label all intercepts and the vertex.

16. The height (y) in feet of a projectile shot vertically upward from the ground with an initial velocity of 100 ft/second is given by $y = f(t) = 100t - 16t^2$, where t is the elapsed time in seconds. Find its maximum height.

CHAPTER 9 TEST

1. Solve by factoring: $12x^2 + x - 6 = 0$.
2. Solve $x^2 - 10 = 0$.

3. Solve $(2x + 3)^2 + 1 = 6$.

4. Determine the number that should be added to $x^2 - 6x$ to complete the square. Then, add it and factor the resulting trinomial.
5. Solve by completing the square: $x^2 + 6x + 1 = 0$.

6. Solve by the quadratic formula: $3x^2 + x = 1$.

7. A right triangle with area equal to 3 in.^{2} has one leg twice the length of the other. Find the length of the hypotenuse.
8. Express $\sqrt{-4}$ in terms of i.
9. Simplify $3\sqrt{-4} + \sqrt{-9}$.
10. Identify all of these that are included in the set of complex numbers.
    a. 3   b. $5i$ c. $6 + 3i$ d. $\sqrt{7}$ e. $\sqrt{-11}$
11. Find the product of $3 - 2i$ and its conjugate.
12. True or false? The product of a complex number and its conjugate is always a real number.
13. Is i a solution to $x^2 + x + 1 = 0$?

14. Solve $1 - 3x^2 = 5$.

15. Solve $3x^2 - x = 2$ by using the quadratic formula. Identify the values for a, b, and c.

16. Solve $3x^2 + 2x + 1 = 0$ by using the quadratic formula.

17. True or false? The graph of $y = x^2 + 1$ has no x-intercept.
18. The graph of $x = y^2$ is shown. Why does this relation *not* define y as a function of x?

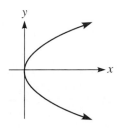

19. Graph $y = f(x) = 2x^2 - 6x + 4$. Label the vertex and all intercepts.

20. A projectile is shot vertically upward from the ground with an initial velocity of 200 ft/second. Its height (y) in feet is given by $y = f(t) = 200t - 16t^2$, where t is the elapsed time in seconds.
   a. Graph this function.

   b. How long will it take before the projectile hits the ground?

CUMULATIVE TEST 9 (FINAL EXAMINATION)

1. Which of these expressions is undefined?
   a. $\dfrac{0}{5}$ **b.** $\dfrac{5}{0}$

2. What is the sum of -2 and its reciprocal?
3. Simplify by removing parentheses and combining like terms: $4(x - 1) - 3(x - 2)$.
4. Solve $-3x + 5 = 6$.
5. Multiply $(x + 2)(x^2 - 2x + 3)$.
6. Simplify $(2x^{-1}y)^{-2}$ and write the result using only positive exponents.

7. Factor $9x^2 - 1$.
8. The area of a square is 45 more than the perimeter. If the dimensions are given in inches, find the side length of the square.
9. Write as a single fraction in lowest terms: $\dfrac{2x}{x - 4} + \dfrac{x - 1}{4 - x}$.

10. A pipe is feeding a swimming pool at the rate of 180 gal every half hour. At this rate, how long will it take to fill a pool which holds 30,000 gal? Represent this problem as a proportion; then solve and round the answer to the nearest hour.

11. Draw the graph of the line whose equation is $2x + 3y = 12$. Label the x- and y-intercepts.

12. The relationship between y and x is described by a line with slope equal to 3. What happens to y when x increases by 5 units?
13. At what point do the graphs of these equations intersect?
$$x - y = 5$$
$$3x - 5y = 18$$

14. Part of a pension fund of $50 million was invested at 9 percent annual interest and part at 6 percent annual interest. For one year it earned $4.2 million in interest. How much of the pension fund was invested at 9 percent?
15. Multiply and simplify: $(\sqrt{5} - \sqrt{3})(\sqrt{5} + \sqrt{3})$.
16. Solve and check your solutions: $\sqrt{x + 1} = x - 1$
17. Evaluate $(-27)^{5/3}$.

18. Solve $x^2 + x = 4$.

19. Find the product: $(5 + 3i)(5 - 3i)$.
20. Graph the parabola $y = f(x) = x^2 - 2x + 3$. Label the vertex and all intercepts.

Review of Decimals and Percents

For selling a house, a real estate agency receives a commission that is 6 percent of the selling price. What is the agency's commission when a house sells for $138,000? (See Example 9.)

OBJECTIVES

1 Write decimal numbers in words.

2 Add, subtract, multiply, and divide with decimals.

3 Convert between decimal numbers and percents.

4 Find a specified percent of a number.

1 Our practical need to make precise measurements of quantities such as weight, time, and length makes the idea of a decimal a familiar concept. To review this topic, first consider Figure A.1, which shows how we interpret the place value of each digit in the decimal number 72.815. Note in the decimal system that the place value of each position is one-tenth of the value of the place to its left. The position to the right of the decimal point is read like a whole number, followed by the name of the place of the digit furthest to the right. For instance, 72.815, which equals $72\frac{815}{1,000}$, is read "seventy-two and eight hundred fifteen–thousandths."

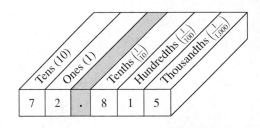

Figure A.1

EXAMPLE 1 Write each number in words.

  a. 0.7   **b.** 5.16   **c.** 12.004

Solution

  a. 0.7 is read "seven-tenths."
  b. 5.16 is read "five and sixteen-hundredths."
  c. 12.004 is read "twelve and four-thousandths."

Note For numbers that are less than 1, like 0.7, it is customary to place a zero to the left of the decimal point so that numbers like .7 are not misread as 7. However, the practice is arbitrary.

PROGRESS CHECK 1 Write each number in words.

  a. 0.07   **b.** 3.2   **c.** 16.105   ⌐

2 We review operations with decimals next.

To Add (or Subtract) with Decimals

  1. Place the numbers in a column so that the decimal points are beneath one another.
  2. Add (or subtract) as with whole numbers.
  3. Position the decimal point in the answer directly below the other decimal points.

EXAMPLE 2 Perform the indicated operations.

  a. 3.587 + 12.9 + 0.64   **b.** 4.9 − 1.075

Solution We follow the method given and attach final zeros where needed so that the numbers are the same length to the right of the decimal point.

  a.   3.587   **b.**   4.900
           12.900   − 1.075
         + 0.640   ───────
         ───────   3.825
          17.127

PROGRESS CHECK 2   Perform the indicated operations.

a. $0.18 + 4.025 + 1.9$ **b.** $8.04 - 3.509$

To Multiply with Decimals

1. Multiply the factors as with whole numbers.
2. Position the decimal point in the answer so that the number of digits to the right of the decimal in the product is equal to the *total* number of digits to the right of the decimals in the factors.

EXAMPLE 3   Multiply.

a. 3.15×2.7 **b.** 0.3×0.3

Solution

a.
$$
\begin{array}{r}
3.15 \\
\times\ 2.7 \\
\hline
2\ 205 \\
6\ 30 \\
\hline
8.505
\end{array}
$$

3.15 — 2 digits to the right of the decimal point
× 2.7 — 1 digit to the right of the decimal point
8.505 — 3 digits to the right of the decimal point

b. In the multiplication below, note in the answer that we must insert a 0 between the decimal point and the 9 to have the correct number of decimal places.

$$
\begin{array}{r}
0.3 \\
\times 0.3 \\
\hline
0.09
\end{array}
$$

0.3 — 1 digit to the right of the decimal point
×0.3 — 1 digit to the right of the decimal point
0.09 — 2 digits to the right of the decimal point

PROGRESS CHECK 3   Multiply.

a. 4.1×1.37 **b.** 1.6×0.02

To Divide a Decimal by a Whole Number

1. Divide as with whole numbers.
2. Position the decimal point in the answer directly above the decimal point in the dividend.

The next example displays the names of the key components in a division problem.

EXAMPLE 4   Divide $450.08 \div 16$.

Solution $450.08 \div 16 = 28.13$, as shown below.

$$
\begin{array}{r}
28.13 \quad \text{(quotient)} \\
16\overline{)450.08} \quad \text{(dividend)} \\
\end{array}
$$

(divisor) 16
$$
\begin{array}{r}
32 \\
\hline
130 \\
128 \\
\hline
20 \\
16 \\
\hline
48 \\
48 \\
\hline
0
\end{array}
$$

Progress Check Answers

2. (a) 6.105   (b) 4.531

3. (a) 5.617   (b) 0.032

PROGRESS CHECK 4   Divide $276.64 \div 14$.

To divide when the divisor is a decimal, we first move the decimal point in both the divisor and the dividend, as shown next.

EXAMPLE 5   Divide $3.175 \div 0.25$.

Solution   The divisor is 0.25, which is 25 hundredths, and $3.175 \div 0.25$ equals the fraction 3.175/0.25. If we multiply the numerator and the denominator in this problem by 100, then the decimal point moves two places to the right in both the dividend and the divisor, and we get

$$\frac{3.175}{0.25} = \frac{3.175 \times 100}{0.25 \times 100} = \frac{317.5}{25}.$$

Now we may use the procedure for dividing a decimal by a whole number.

$$\begin{array}{r} 12.7 \\ 25\overline{)317.5} \\ \underline{25} \\ 67 \\ \underline{50} \\ 175 \\ \underline{175} \\ 0 \end{array}$$

Thus, $3.175 \div 0.25 = 12.7$.

PROGRESS CHECK 5   Divide $8.505 \div 0.35$.

3   A common application of decimals is found in problems that involve percentages. The symbol % is sometimes used to represent **percent,** which means "per one hundred." Thus, 17 percent or 17% means "seventeen per hundred," 17/100, or 0.17. The relation 17 percent $= 0.17$ illustrates the following conversion rules.

> **Conversions Between Decimal Numbers and Percents**
>
> 1. To change a percent to a decimal number without the word *percent,* move the decimal point two places to the left and remove *percent*.
> 2. To change a decimal number to a percent, move the decimal point two places to the right and attach the word *percent*.

EXAMPLE 6   Write as decimal numbers.

a. 25 percent   **b.** 37.5 percent   **c.** 247 percent   **d.** 1 percent

Solution   Move the decimal point two places to the left and remove the word *percent*. Whole numbers are considered to end with a decimal point.

a. 25 percent $= 0.25$
b. 37.5 percent $= 0.375$
c. 247 percent $= 2.47$
d. 1 percent $= 0.01$

PROGRESS CHECK 6   Write as equivalent decimal numbers.

a. 32 percent   **b.** 175 percent   **c.** 2 percent   **d.** 0.2 percent

Progress Check Answers
4. 19.76
5. 24.3
6. (a) 0.32   (b) 1.75   (c) 0.02   (d) 0.002

EXAMPLE 7 Write as percents.

a. 0.48   b. 0.0717734   c. 0.005   d. 1

Solution Move the decimal point two places to the right and attach the word *percent*.

a. 0.48 = 48 percent
b. 0.0717734 = 7.17734 percent
c. 0.005 = 0.5 percent
d. 1 = 100 percent

PROGRESS CHECK 7 Write as percents.

a. 0.63   b. 2   c. 0.1225   d. 0.001   ⌐

4 Many application problems involving percentages take the following form:

What is *p* percent of *n?*

In this context the word *of* implies multiplication, as illustrated in Examples 8 and 9.

EXAMPLE 8 What is 15 percent of 80?

Solution Write the percent as a decimal, translate "of" to ×, and then multiply.

$$15 \text{ percent of } 80 = 0.15 \times 80 = 12$$

PROGRESS CHECK 8 What is 18 percent of 400?   ⌐

EXAMPLE 9 Solve the problem in the section introduction on page 448.

Solution The commission is 6 percent of $138,000. Thus,

$$\text{commission} = 0.06 \times 138,000$$
$$= 8,280$$

The agency's commission is $8,280.

PROGRESS CHECK 9 The sales tax in a particular city is 4.5 percent of an item's selling price. What is the sales tax on the purchase of a VCR that sells for $280? ⌐

Progress Check Answers

7. (a) 63 percent   (b) 200 percent
(c) 12.25 percent   (d) 0.1 percent

8. 72

9. $12.60

EXERCISES APPENDIX

In Exercises 1–12, write each number in words.

1. 0.8
2. 0.4
3. 0.01
4. 0.03
5. 0.002
6. 0.005
7. 5.04
8. 11.09
9. 31.145
10. 1.012
11. 10.06
12. 8.47

In Exercises 13–20, perform the indicated operations.

13. 1.234 + 54.32 + 0.08
14. 6.05 + 12.002 + 104.6

15. 8.62 − 2.26
16. 18.2 − 1.95
17. 0.4 + 0.05 + 0.006
18. 1.2 + 2.03 + 3.04
19. 104.1 − 2.1
20. 10.41 − 2.1

In Exercises 21–32, perform the indicated operations.

21. 1.23 × 1.4
22. 0.01 × 0.2
23. 3.05 × 4.2
24. 3.33 × 3.33
25. 1.2 × 2.3 × 3.4
26. 0.11 × 1.1 × 1.11
27. 14.4 ÷ 1.2
28. 20.8 ÷ 1.3
29. 31 ÷ 2.5
30. 36.96 ÷ 2.4
31. 15.2 ÷ 0.1
32. 12.87 ÷ 0.3

In Exercises 33–40, write the given percent in decimal notation.

33. 30 percent
34. 45 percent
35. 5 percent
36. 7 percent
37. 100 percent
38. 250 percent
39. 12.5 percent
40. 33.3 percent

In Exercises 41–50, write the given number as a percent.

41. 0.01
42. 0.08
43. 0.35
44. 0.68
45. 1.05
46. 2.15
47. 3
48. 5
49. 0.1234
50. 1.5678
51. What is 20 percent of 50?
52. What is 50 percent of 20?
53. What is 1 percent of 800?
54. What is 2 percent of 800?

55. What is 35 percent of 250?
56. What is 45 percent of 250?
57. What is 105 percent of 1,000?
58. What is 250 percent of 1,000?
59. The sales tax on restaurant meals is 8.25 percent in a certain city. What is the amount of tax on a $24 meal?
60. A patron calculates a 15 percent tip on a taxi ride which costs $12. How much is the tip?
61. A population grows by 2 percent over a period of time. It was originally 230,000 people.
   a. How many new people were added over this time?
   b. What was the population at the end of the period?
62. A bank account increases in value by 6.25 percent in one year. It started at $800.
   a. How much did it increase over the year?
   b. What is the total value of the account at the end of the year?
63. The profit in 1994 for a company is 4 percent lower than it was in 1993. The 1993 profit was $1.4 million. To the nearest thousand dollars, what is the 1994 profit?
64. In a certain region there were approximately 3 percent fewer motor vehicle accidents in 1994 than in 1993, when there were 1,590 accidents. To the nearest whole number, how many accidents were there in 1994?

Answers to Odd-Numbered Exercises and Review Questions

Chapter 1

Exercises 1.1

1.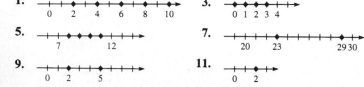

3.

5.

7.

9.

11.

13. $>$ 15. $<$ 17. $>$ 19. $<$ 21. $=$ 23. $>$ 25. True
27. True   29. False   31. True   33. True   35. True
37. $2 \cdot 3 \cdot 5$ 39. $2 \cdot 3 \cdot 3 \cdot 5$ 41. $2 \cdot 2 \cdot 2 \cdot 2$
43. $2 \cdot 2 \cdot 2 \cdot 2 \cdot 3 \cdot 3 \cdot 5$ 45. $5 \cdot 5 \cdot 5 \cdot 13$ 47. $2 \cdot 3 \cdot 5 \cdot 7 \cdot 11$
49. $\frac{2}{3}$ 51. $\frac{2}{7}$ 53. $\frac{4}{9}$ 55. $\frac{42}{125}$ 57. $\frac{16}{15}$ 59. $\frac{3}{4}$ 61. $\frac{8}{45}$
63. $\frac{3}{4}$ 65. 1   67. $\frac{5}{7}$ 69. $\frac{5}{2}$ 71. $\frac{13}{2}$ 73. $\frac{5}{3}$ 75. $\frac{1}{2}$ 77. 4
79. $\frac{5}{4}$ 81. 1   83. $\frac{1}{10}$ 85. 24   87. $\frac{10}{11}$ 89. 3   91. 2   93. 1
95. $\frac{5}{6}$ 97. $\frac{1}{6}$ 99. $\frac{25}{24}$ 101. $\frac{13}{6}$ 103. $\frac{5}{6}$ 105. $\frac{17}{6}$ 107. $\frac{5}{4}$
109. $\frac{3}{8}$ 111. $\frac{3}{8}$ 113. a. $\frac{25}{12}$ b. 1   115. 210   117. 525,000 lb
119. $612.50   121. $2,687\frac{1}{2}$ oz

Remember This (1.1)

1. 1.157625   2. 39,799.5   3. $\frac{27}{64}$ 4. 60; yes; no   5. 12; yes; no
6. 15   7. 36   8. 9   9. Product   10. Sum

Exercises 1.2

1. 16   3. 1   5. 0   7. 5,184   9. 729   11. $\frac{8}{27}$ 13. 27   15. 2
17. 54   19. 35   21. 98   23. 40   25. 46   27. 12   29. 10
31. 6   33. 4   35. 3   37. 12   39. $\frac{2}{3}$ 41. $\frac{19}{9}$ 43. 72
45. 60   47. 13.44   49. 0.24   51. 15   53. 1.78   55. 5
57. 11.17   59. 1,024   61. 0   63. 2,187   65. 438,976
67. 2.19   69. 14,758.49   71. 22,868.21   73. $28,864
75. $33   77. c   79. c   81. c   83. d

Remember This (1.2)

1. True   2. True   3.
4. a. 1   b. $3\frac{1}{24}$ 5. 10   6. $2 \cdot 3 \cdot 3$ 7. $2 \cdot 2 \cdot 2 \cdot 3$ 8. $\frac{1}{30}$
9. $6\frac{1}{2}$ 10. $10\frac{1}{2}$

Exercises 1.3

1. $x + 3$ 3. $s + $1,000$ 5. $a + 5$ 7. $c - 7$ 9. $p - 5
11. $A - 200,000$ acres   13. cd 15. $\frac{x}{y}$ 17. $3x$ 19. $\frac{1}{2}z$

21. $0.12r$ 23. $xy + 10$ 25. $x + 0.08x$ 27. $v - 0.10v$
29. $2(x + y)$ 31. $2a + 3b$ 33. $n(n + 2)$ 35. $\frac{x + 5}{10}$
37. The sum of twice x and y 39. a minus the product of 2
and b 41. Three times the sum of x and 1   43. The sum of 3
times x and 4   45. Divide the sum of a and b by 2.
47. The sum of a divided by 2 and b divided by 3
49. 9   51. 10   53. $\frac{3}{2}$ 55. 5   57. $\frac{3}{2}$ 59. 17   61. 38
63. 9   65. 0   67. 12 m; 9 m²   69. 20 mi; 25 mi²
71. 1.2 cm; 0.09 cm²   73. 18 in.; 20 in.²   75. 26 cm; 40 cm²
77. 12 ft; 8.75 ft²   79. 30 in.; 30 in.²   81. 9 ft; 3.9 ft²
83. 9.85 yd; 4 yd²   85. 25.12 in.; 50.24 in.²
87. 37.68 cm; 113.04 cm²   89. 65.31 ft; 339.62 ft²   91. 64 in.³
93. 864 ft³   95. 31.4 m³   97. 3.93 in.³   99. 4.19 ft³
101. 904.32 cm³   103. $A = 420$ ft²; no
105. $V = 1.125$ ft³; $W = 1,890$ lb
107. $V = 1,200$ ft³; $W = 144,000$ lb; 200,000 lb

Remember This (1.3)

1. 407.94   2.
3. $2,000   4. True   5. 23   6. $\frac{7}{12}$
7. 2^8 8. True   9. $(x + 3)(2y - 5)$ 10. $2 \cdot 2 \cdot 3 \cdot 5$

Exercises 1.4

1.   3.   5.

7.   9.

11.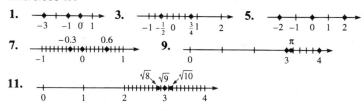

13. $<$ 15. $>$ 17. $>$ 19. $=$ 21. $>$ 23. $>$ 25. -3
27. 1.1   29. $-\frac{1}{3}$ 31. 3   33. $\frac{2}{3}$ 35. 1   37. 20 ft/second
39. -15 ft/second   41. 35 m/second; -35 m/second
43. a. $-1, 0, 1, 2$ b. $-1, 0, 1, 2$ c. None   d. $-1, 0, 1, 2$
45. a. $\frac{2}{2}$ b. $\frac{1}{2}, \frac{2}{2}, \frac{3}{2}$ c. None   d. $\frac{1}{2}, \frac{2}{2}, \frac{3}{2}$
47. a. 3.0   b. 1.2, 2.3, 3.0   c. None   d. 1.2, 2.3, 3.0
49. a. $\sqrt{4}$ b. $\sqrt{4}$ c. $\sqrt{2}, -\sqrt{3}$ d. $\sqrt{2}, -\sqrt{3}, \sqrt{4}$
51. a. 0   b. 0, -3.4 c. $-\sqrt{2}, \pi$ d. $-3.4, -\sqrt{2}, 0, \pi$
53. a. $-$200,000$ b. 1994   c. 1991

Remember This (1.4)

1. $p - 0.10p$ 2. One-fifth of the sum of 4 and 3 times a 3. 1
4. $P = 30$ in.; $A = 56$ in.²   5. 12.5 cm²   6. 0.785 m²   7. 0.956
8. $2,014   9. $<$ 10.

Exercises 1.5

1.
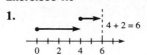
$4 + 2 = 6$

3. $5 + 0 = 5$

5. $1 + (-2) = -1$

7. $2 + (-2) = 0$

9. $-1 + 5 = 4$

11. $-3 + (-4) = -7$

13. 7   **15.** -7 **17.** -335 **19.** -2.8 **21.** -11.1 **23.** $\frac{4}{3}$

25. $-\frac{1}{6}$ **27.** 1   **29.** 1   **31.** 5   **33.** -1 **35.** 4   **37.** -2.2

39. -7 **41.** 0   **43. a.** -4 **b.** 7   **c.** $\frac{3}{5}$ **d.** -1.078 **e.** 0

45. 5   **47.** 13.2   **49.** Negative six   **51.** Four   **53.** Five

55. Negative four plus negative five equals negative nine.

57. Four plus negative five equals negative one.

59. The opposite of negative four plus negative five equals negative one.   **61.** True   **63.** False   **65.** False   **67.** $-\$65$

69. 0.4 degrees   **71.** -190 ft   **73.** -9 yd   **75.** \$474

77. Velocity $= 0$; no movement   **79.**

+	−1	−2	−3	−4
1	0	−1	−2	−3
2	1	0	−1	−2
3	2	1	0	−1

81. \$13,100 profit

Remember This (1.5)

1.
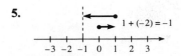

2. 6   **3.** $-\frac{1}{2}, \frac{1}{2}$

4. $A = \frac{1}{2}(5x + 2)x$ **5.** 7   **6.** False   **7.** $\frac{x}{2} + 10$

8. 0; the sum of zero plus any number is always that number.   **9.** 7

10. -7

Exercises 1.6

1. $s, 7 + (-4)$ **3.** $a, 7 - 2$ **5.** $a, -3 - 2$ **7.** $s, -8 + f$

9. $s, 1 + (-a)$ **11.** $a, d - (-2)$ **13.** 5   **15.** -8 **17.** -7

19. -8 **21.** -1.3 **23.** 1   **25.** $\frac{1}{5}$ **27.** $-\frac{4}{5}$ **29.** $\frac{7}{8}$

31. $-7 - 2 = -9$ **33.** $[8 + (-3)] - (-4) = 9$

35. $(-5 + 0) - (-6 + 7) = -6$

37. $-1 - [-1 + (-1)] = 1$ **39.** $-1 - [-9 + (-12)] = 20$

41. $-2 - [5 + (-6)] = -1$ **43.** Subtract negative 1 from 6.

45. Subtract b from the opposite of the negative of a.

47. From the sum of negative 2 and x, subtract 5.   **49.** -5

51. -5 **53.** 5   **55.** 6   **57.** 4   **59.** 0   **61.** -7.66 **63.** 8.68

65. 7   **67.** 65 degrees Fahrenheit   **69. a.** 131   **b.** 1,975

c. 211   **71.** 12 degrees Celsius   **73.** 35.68 degrees Celsius

75. 11,745 ft   **77.** 6.22   **79.**

a\b	−3	0	3
−2	1	−2	−5
0	3	0	−3
2	5	2	−1

81. Row 2: 2 4 6 8 10   Row 3: -2 -2 -2 -2 Row 4: 0 0 0

Remember This (1.6)

1. -9 **2.** 0   **3.**

4. $\frac{34}{10}$ **5.** $gh - 4$ **6.** 5   **7.** -0.27 **8.** 4^5 **9.** $2 \cdot 2 \cdot 2 \cdot 3 \cdot 5$

10. b

Exercises 1.7

1. -12 **3.** -96 **5.** 0   **7.** 20   **9.** 5   **11.** $-\frac{4}{7}$ **13.** 24

15. -8 **17.** 42   **19.** -1 **21.** -64 **23.** -81 **25.** -16

27. -1 **29.** 15   **31.** 1   **33.** -3 **35.** -34 **37.** 102

39. -30 **41.** 9   **43.** 936   **45.** 54   **47.** 5   **49.** 1   **51.** -3

53. 1   **55.** -1 **57.** -11 **59.** 5   **61.** 32   **63.** 7   **65.** 35

67. $-$ **69.** -2^4 **71.** True   **73.** 0   **75.** 4   **77.** -2

79. $5(-4 + 3) = -5$ **81.** $(5 + 5)(5 - 6) = -10$

83. $-[(-3)^2 + (-3)^3] = 18$ **85.**

·	0	1	2	3
0	0	0	0	0
−1	0	−1	−2	−3
−2	0	−2	−4	−6
−3	0	−3	−6	−9

87. -2; -1 **89.** 12; -7 **91.** 0; -5 **93.** 0

95. $P = 2(x + 4) + 2(x + 2)$, 8; $A = (x + 4)(x + 2)$, 3

97. $P = 2(x^2 - 2) + 2(x^2 - 3)$, 6; $A = (x^2 - 2)(x^2 - 3)$, 2

99. $P = 2(3x^2 - 2x) + 2(2x^2 - x)$, 16;

$A = (3x^2 - 2x)(2x^2 - x)$, 15   **101.** $V_{1989} = \$100,000$;

$V_{1997} = -\$60,000$ **103.** $V_0 = 300$ ft/second up;

$V_9 = 12$ ft/second up; $V_{10} = -20$ ft/second down   **105.** \$7,200

107. $V_{-7} = 182$; $V_0 = 0$; $V_7 = 112$ **109.** 2,125; 375

Remember This (1.7)

1. $x + (-a)$ **2.** $x - 4$ **3.** 5   **4.** -2 **5.** 21   **6.** $-5; \frac{1}{5}$

7. $\frac{3}{5}$ is a fraction with an integer in the numerator and a nonzero integer in the denominator   **8.** $|xy|$ **9.** All   **10.** $8; \frac{8}{3}$

Exercises 1.8

1. -4 **3.** -119 **5.** 4   **7.** 120   **9.** 6   **11.** 0

13. Undefined   **15.** $-\frac{9}{4}$ **17.** -1 **19.** -2 **21.** 5   **23.** $\frac{1}{2}$

25. 4   **27.** $\frac{1}{2}$ **29.** 1   **31.** -2 **33.** 0   **35.** -1

37. Undefined   **39.** -16 **41.** $\frac{8}{7}$ **43.** -1 **45.** -2 **47.** $-\frac{13}{5}$

49. $-\frac{1}{5}$ **51.** $-\frac{7}{2}$ **53.** 3   **55.** $-\frac{4}{5}$ **57.** 0.625   **59.** -0.8

61. -5 **63.** -4 **65.** $\frac{5}{2}$ **67.** $a \times \frac{1}{b}$ **69.** $x(\frac{1}{4f})$ **71.** $\frac{5}{x}$

73. -15 **75.** 7   **77.** 7   **79.** 2.21 percent; 12,521

81. -11.90 percent; about 8,800

Remember This (1.8)

1. 60   **2.** 5   **3.** 1   **4.** -7 **5.** 0   **6.** $d(-x + 2)$

7. 0.25 cm²; 2 cm   **8.** 8.47 square units

9.

10. $2 \cdot 3 \cdot 5$

Exercises 1.9

1. $(-a); m$ **3.** $ab; a$ **5.** $7; m$ **7.** $w + v; a$ **9.** $b + (-a); a$

11. $(-d)c; m$ **13.** $dg; m$ **15.** $y; a$ **17.** $-2s + 4f; a$

19. $(3x)y; m$ **21.** $(-2u + 7x) + s; a$ **23.** $(3a)(-5h); m$

25. a. $1 = 1$ **b.** $-2 = -2$ **27. a.** $-3.9 = -3.9$

b. $3.6 = 3.6$ **29. a.** $-\frac{1}{2} = -\frac{1}{2}$ **b.** $-\frac{1}{2} = -\frac{1}{2}$

31. a. $10 = 10$ **b.** $30 = 30$ **33. a.** $-3 = -3$ **b.** $12 = 12$

35. a. $\frac{3}{2} = \frac{3}{2}$ **b.** $-\frac{2}{3} = -\frac{2}{3}$ **37.** Commutative, addition
39. Associative, multiplication **41.** Associative, multiplication
43. -8; addition identity **45.** -2; multiplication identity
47. 1; multiplication inverse **49.** 0; addition inverse
51. 0; addition inverse **53.** b; multiplication inverse and identity
55. 3, 3 **57.** $-$ **59.** s, $3t$ **61.** $27 = 27$ **63.** $-5 = -5$
65. $0 = 0$ **67.** $5a + 5x$ **69.** $7x + 7y$ **71.** $6s + 2t$
73. $24x + 12y$ **75.** $-3x - 6y$ **77.** $-16x - 2y$ **79.** $2x - y$
81. $23x - 2$ **83.** $-3x - 4$ **85.** $3a + at + 5ad$
87. $-10 + 2x + 2xy$ **89.** $2x - 4 - 10y$ **91.** $3(9 + 11)$
93. $3(x - y)$ **95.** $(7 + 8)x$ **97.** $(2 - 6)x$ **99.** $(3 - 3)d$
101. $(2 + 5 - 1)x$ **103.** $-3x - 6$ **105.** $-x + 3$
107. $4 - 2x$ **109.** $2x - 3y + 4z$ **111.** $2x + 5y$
113. $3 - 2x + x^2$ **115. a.** Distributive **b.** 20
117. a. Associative, multiplication **b.** 62.8 **119. a.** Distributive
b. $16

Remember This (1.9)

1. -2 **2.** 0 **3.** 0 **4.** -1; -1; undefined; -1
5. -4 in each case **6.** -20 percent **7.** $-\frac{8}{5}$; $\frac{5}{8}$ **8.** $>$
9. $2 \cdot 2 \cdot 3$ **10.** $0.15(x + y)$

Chapter 1 Review Exercises

1.

3. True **5.** $\frac{2}{5}$ **7.** 36 **9.** 56 in. **11.** 13
13. c **15.** Three times x minus y
17. $P = 3$ ft $= 36$ in., $A = \frac{1}{2}$ ft$^2 = 72$ in.^{2} **19.** 41.9 lb **21.** $>$
23. -8 **25. a.** -4 degrees Fahrenheit
b. Mon., Tues., Thurs., Sat. **27.** -4 **29.** Plus **31.** False
33. $-4 + x$ **35.** $-1 - [3 + (-4)] = 0$ **37.** -5
39. $1{,}547 - (-431) - 1 = 1{,}977$ years **41.** 15 **43.** Negative
45. $x^2 + (-x)^2$ **47.** 16 units **49.** $-\frac{5}{13}$ **51.** $c\left(-\frac{1}{s}\right)$
53. -5.56 percent **55.** $5(xy)$ **57.** Distributive **59.** $-2x - 3$

Chapter 1 Test

1. False **2.** $\dfrac{2 \cdot 3 \cdot 3 \cdot 3}{2 \cdot 2 \cdot 2 \cdot 3 \cdot 5} = \dfrac{9}{20}$ **3.**

4. $\frac{3}{4} \cdot \frac{1}{2} = \frac{3}{8}$ **5.** 126 oz **6.** 3 **7.** $25,164 **8.** $m(m + 5)$
9. $\frac{11}{2}$ **10.** 84.78 cm^3 **11.** 4.5 **12.** 6
13. -3: integer, rational, real; $\frac{1}{2}$: rational, real; $\sqrt{3}$: irrational, real
14. -15 **15.** -2 **16.** 42 degrees Fahrenheit **17.** 37
18. $P = (x - \frac{1}{4}) + (x + \frac{1}{4}) + x$, 3; $A = \frac{1}{2}x(x - \frac{1}{4})$, $\frac{3}{8}$
19. -50 percent; yes **20.** $fh - fw$; both equal 16

Chapter 2
Exercises 2.1

1. $2x$, $4x$ **3.** x^2, $5x$, -2 **5.** $7xy$, $8x$, $-9y$ **7.** $-4xy$ **9.** $-\frac{4}{7}$
11. x^2y^2 **13.** 3 **15.** 1 **17.** -1 **19.** 1 **21.** $-\frac{2}{5}$ **23.** $\frac{1}{2}$
25. $8x$, $3x$ **27.** 5, 2 **29.** $\frac{x}{2}$, $-\frac{3x}{4}$, x **31.** None
33. $2x^2$, $3x^2$ **35.** None **37.** $18x$ **39.** $5x$ **41.** x **43.** x
45. $9x$ **47.** x **49.** $-13y$ **51.** 5 **53.** $-2x - 5$ **55.** 7
57. $-10x + y + 5$ **59.** $4x - 6y$ **61.** $8x$ **63.** $4x$
65. $6x + 3$ **67.** $3x + 1$ **69.** $31.4x$ **71.** $13.71x$

73. $9x + 42$ **75.** $8x - 4$ **77.** $3x + 14$ **79.** $3x - 3$
81. $x - 1$ **83.** $-0.01x + 11$ **85.** $-2x - 10$ **87.** -5
89. $-9x + 12$ **91.** $20x - 16$ **93.** $-2x - 3$ **95.** $2x$
97. $6x^2 + 3$ **99.** $-4x^2 + 2x$ **101.** $-2y^2 - 2$
103. $-2y^2 - 4y$ **105.** 0 **107.** $-y^2 - 10y + 11$
109. $P = 10x$

Remember This (2.1)

1. $-\frac{1}{4}$ **2.** $2 - 3$ **3.** $0.15x$ **4.** $5n - 2$ **5.** $A = \pi r^2$
6. -7 **7.** $-\frac{1}{5}$ **8.** xy^2 **9.** $4 + (-9)$ **10.** 0

Exercises 2.2

1. Conditional **3.** Conditional **5.** Conditional **7.** Identity
9. Identity **11.** False **13.** True **15.** False **17.** True
19. True **21.** False **23.** False **25.** $\{7\}$ **27.** $\{4.7\}$ **29.** $\{2\}$
31. $\{-2\}$ **33.** $\{0\}$ **35.** $\{\frac{5}{6}\}$ **37.** $\{5\frac{5}{8}\}$ **39.** $\{5.8\}$ **41.** $\{0\}$
43. $\{-6\}$ **45.** $\{\frac{8}{3}\}$ **47.** $\{\frac{37}{24}\}$ **49.** $\{-1.7\}$ **51.** $\{-2\}$
53. $\{-4\frac{1}{4}\}$ **55.** $9x = 9x$; identity **57.** $5x = 5x$; identity
59. $-3 = -3$; identity **61.** $-4 = -4$; identity
63. $4 \neq 0$; false **65.** $2 \neq -2$; false **67.** $18°$ **69.** 7 degrees
71. 11.8 psi **73.** 119.8 **75.** 951,000

Remember This (2.2)

1. x **2.** x **3.** x **4.** x **5.** x **6.** $6x$ **7.** $A = \frac{1}{2}bh$
8. $-x - 2$ **9.** $0.06x$ **10.** $x + 0.10x$

Exercises 2.3

1. $\{-7\}$ **3.** $\{3\}$ **5.** $\{0.4\}$ **7.** $\{-6\}$ **9.** $\{4.5\}$ **11.** $\{\frac{5}{2}\}$
13. $\{\frac{1}{5}\}$ **15.** $\{0\}$ **17.** $\{-0.9\}$ **19.** $\{1\}$ **21.** $\{12\}$ **23.** $\{4\}$
25. $\{\frac{2}{3}\}$ **27.** $\{-\frac{3}{4}\}$ **29.** $\{0\}$ **31.** $\{10\}$ **33.** $\{\frac{3}{8}\}$ **35.** $\{-1\}$
37. $\{2\}$ **39.** $\{0\}$ **41.** $\{1\}$ **43.** $\{3\}$ **45.** $\{-2\}$ **47.** $\{-5\}$
49. $\{0\}$ **51.** $\{-\frac{1}{8}\}$ **53.** $\{0.4\}$ **55.** $\{-\frac{7}{15}\}$ **57.** $\{-2.8\}$
59. $\{-\frac{17}{12}\}$ **61.** $\{7\}$ **63.** $\{-4\}$ **65.** $\{1\}$ **67.** $\{-2\}$ **69.** $\{0\}$
71. $\{0\}$ **73.** $\{-6\}$ **75.** $\{-2\}$ **77.** $\{-1\}$ **79.** $\{-6\}$
81. $\{-2\}$ **83.** $\{-5\}$

In Exercises 85–95 answers may vary.
85. Multiply by 5. **87.** Multiply by -1. **89.** Multiply by a.
91. Multiply by 2. **93.** Multiply by $\dfrac{b}{a}$. **95.** Add $\dfrac{a}{b}$.
97. $2,941 **99.** $22,815 **101.** 600 **103.** $10 million

Remember This (2.3)

1. $5x - 5$ **2.** $-6x + 8$ **3.** $4 + 6x$ **4.** $-5 + x$
5. $6x - 21$ **6.** $2x - 4$ **7.** $V = \ell \times w \times h$ **8.** $-\frac{1}{3}$
9. $\dfrac{n}{2} - 4$ **10.** $(x + 8)^2$

Exercises 2.4

1. $\{\frac{1}{2}\}$ **3.** $\{\frac{17}{5}\}$ **5.** $\{1\}$ **7.** $\{\frac{1}{2}\}$ **9.** $\{\frac{1}{3}\}$ **11.** $\{2\}$ **13.** $\{-2\}$
15. $\{0\}$ **17.** $\{-\frac{7}{8}\}$ **19.** $\{-20\}$ **21.** $\{-\frac{9}{2}\}$ **23.** $\{0\}$ **25.** $\{-6\}$
27. $\{0\}$ **29.** $\{1\}$ **31.** $\{0\}$ **33.** $\{\frac{10}{3}\}$ **35.** $\{24\}$ **37.** $\{1\}$ **39.** $\{3\}$
41. $\{10\}$ **43.** $\{2\}$ **45.** $\{\frac{20}{3}\}$ **47.** $\{1\}$ **49.** $\{-\frac{2}{5}\}$ **51.** $\{0\}$

53. $\{\frac{6}{7}\}$ **55.** $\{\frac{1}{3}\}$ **57.** $\{2\}$ **59.** $\{-\frac{1}{16}\}$ **61.** $\{-\frac{10}{9}\}$ **63.** $\{-\frac{97}{5}\}$
65. $\{-9\}$ **67.** $\{\frac{13}{3}\}$ **69.** $\{3\}$ **71.** $\{1\}$ **73.** False   **75.** Identity
77. Identity   **79.** False   **81.** False   **83.** Identity   **85.** $1,500
87. 2   **89.** False equation   **91.** -4 **93.** Jim's pay is $400;
John's pay is $500.

Remember This (2.4)

1. Add the additive inverse of ab.
2. Multiply by the multiplicative inverse of cd.   **3.** $\{\frac{3}{4}\}$
4. $\frac{1}{2}(x + 5)$ **5.** $bd + cd$ **6.** $C = 2\pi r = \pi d$ **7.** -8
8. -13 **9.** -100 **10.** $(3 + 4)^2$

Exercises 2.5

1. 2   **3.** 5   **5.** 625   **7.** $\frac{15}{4}$ **9.** $\frac{9}{2}$ **11.** $\frac{1}{2}$ **13.** 1   **15.** 10
17. 25   **19.** -10 **21.** 6   **23.** 100   **25.** 3   **27.** $\frac{1}{5}$ **29.** $\frac{1}{6}$
31. $\frac{2}{3}$ **33.** 0   **35.** 2.945   **37.** $r = \dfrac{d}{t}$ **39.** $\ell = \dfrac{P - 2w}{2}$
41. $\ell = \dfrac{V}{wh}$ **43.** $h = \dfrac{3V}{a^2}$ **45.** $z = \dfrac{X - \mu}{\sigma}$ **47.** $h = \dfrac{V}{\pi r^2}$
49. $a = 2A - b$ **51.** $a^2 = c^2 - b^2$ **53.** $b = \dfrac{2A - ha}{h}$
55. $y = \dfrac{11 - 3x}{2}$ **57.** $y = 2x$ **59.** $y = 3 + x$
61. $y = -2.5 - 1.25x$ **63.** $y = \dfrac{1 - x}{2}$ **65.** $y = 35 - x$
67. a. $k = \dfrac{m}{0.6214}$ **b.** 6,437 km   **69. a.** $a = \dfrac{A}{\pi b}$ **b.** 3.18 cm
71. a. $g_3 = \dfrac{A - 0.20g_1 - 0.30g_2}{0.50}$ **b.** 88

Remember This (2.5)

1. $3 + (-3) = 0$ **2.** Equation leads to $5 = 5$.   **3.** 3
4. $11y - x$ **5.** Distributive   **6. a.** $4x$ **b.** $x + 4$ **c.** $x - 4$
7. 89.5   **8.** $1, 0, -8$ **9. a.** True   **b.** True   **10.** False

Exercises 2.6

1. $125; $375   **3.** $80   **5.** 9.5 ft; 0.5 ft   **7.** 1 ft; 9 ft   **9.** 3.5 ft
11. a. 10,000 ft²; 10,000 ft²; 30,000 ft²   **b.** No   **13.** $8,000
15. 88   **17.** $37,500   **19.** -4 **21.** 54.6, 389.4   **23.** 12
25. 33, 34, 35   **27.** 45 more, or a total of 160 hits   **29.** 86 votes
31. Each expression equals $4x + 6$.   **33.** Each expression equals
$2x + 1$.   **35.** $3x + 3 = 100$ gives $x = \frac{97}{3}$, not an integer.

Remember This (2.6)

1. $P = 2\ell + 2w$ **2.** $90°$ **3.** $0.08x$ **4.** $\dfrac{x}{2} + 10,000$
5. $r = \dfrac{d}{t}$ **6.** $0.72 - 0.06x$ **7.** 4   **8.** 2   **9.** $\frac{11}{3}$ **10.** $5 + 2x$

Exercises 2.7

1. $1\frac{2}{3}$ in., $3\frac{1}{3}$ in.   **3.** 11 in., 8 in.   **5.** 29 in.   **7.** 2.5 cm
9. $b = 7.2$ in., $s = 14.4$ in.   **11.** 2.5 in.

13.

15.

17. $60°$

19. $67.5°, 22.5°, 90°$ **21.** $52.5°$ **23.** $50°, 60°, 70°$
25. $50°, 60°, 70°$ **27.** $20°, 60°, 100°$ **29.** $59°$
31. $x = 180°$; this leads to negative measures for angles.
33. $3,750 at 10 percent; $6,250 at 6 percent
35. $25,000 at 6 percent; $26,000 at 10 percent   **37.** $20,000
39. 4.5 percent   **41.** $2,195, $109.75   **43.** 5 million, 100,000
45. $16,300   **47.** $9.60   **49.** 4.95 hours   **51.** 6.25 hours
53. 8 hours   **55.** $33\frac{3}{4}$ mi/hour   **57. a.** 400 mi/hour,
600 mi/hour   **b.** 2,400 mi   **59.** 300 mi/hour   **61.** 3.75 lb 30
percent, 1.25 lb 50 percent   **63.** 3.33 liters of 25 percent, 1.67
liters of 40 percent   **65.** 1 gal, 2 gal   **67.** 0.4 liter   **69.** 2.77 oz
71. 2.9 percent

Remember This (2.7)

1.

2. True   **3.** 3, 4
4. average $= \dfrac{a + b + c}{3}$ **5.** $\{3\}$
6. Positive numbers are always greater than negative numbers.
7. $-2(3)$ **8.** Sum   **9.** $3x - 4 = 2x + 1$
10. a. Any positive number   **b.** 0   **c.** Any negative number

Exercises 2.8

1.

3.

5.

7.

9.

11.

13.

15.

17.

19. $\{x: x > 1\}$ **21.** $\{x: x < 3\}$ **23.** $\{x: 1 < x < 3\}$
25. $\{x: -5 \le x < -1\}$ **27.** $\{x: x < -\frac{13}{2}\}$ **29.** $\{x: x \ne 0\}$
31. $\{x: x > -5\}$ **33.** $\{x: x \le 8\}$ **35.** $\{x: x < 4\}$

37. $\{x: x \ge -1\}$ **39.** $\{x: x < 4\}$ **41.** $\{x: x > -7\}$

43. $\{x: x > -6\}$ **45.** $\{x: x > -2\}$ **47.** $\{x: x > 6\}$

49. $\{x: x > 2\}$ **51.** $\{x: x \le -6\}$ **53.** $\emptyset$

55. $\{x: -3 < x < -1\}$ **57.** $\{x: -\frac{9}{4} < x \le -\frac{5}{4}\}$

59. $\{x: 2 \le x \le 3\}$ **61.** $\{x: 2 < x < \frac{5}{2}\}$ **63.** $\{x: \frac{9}{4} < x < \frac{11}{4}\}$

65. $\emptyset$ **67.** $x > 14{,}286$ **69.** $27.21 \le k \le 35.83$
71. $5 \le F \le 23$ **73.** $68 \le x < 90.5$ **75.** From 30 to 35 people

Remember This (2.8)

1. $500; -256$ **2.** $-49; -49; -49$ **3.** $0; -25$ **4.** $\frac{1}{9}$ **5.** No
6. $5x^2 - 8$ **7.** $(m + n)^2$ **8.** $m^2 + n^2$ **9.** $r = \frac{C}{2\pi}$ **10.** 2.5

Chapter 2 Review Exercises

1. $4, -5y, -6y^2$ **3.** $3x^3, 6x^3$ **5.** $P = 14x$ **7.** $-12y^2 - 10y$
9. a. Conditional   **b.** Identity   **c.** False   **11.** $\{9\}$
13. $3x = 3x$ **15.** \$880   **17.** $\{14\}$ **19.** $\{7\}$ **21.** \$14,850
23. $\{-\frac{5}{3}\}$ **25.** $\{\frac{5}{7}\}$ **27.** $6 \ne 2$ **29.** $\frac{10}{3}$ **31.** 20
33. $w = \frac{P - 2\ell}{2}$ **35.** $\frac{1}{2}$ ft   **37.** $\frac{11}{2}$ **39.** 27 yd^2 **41.** \$3,000
43. 45 minutes   **45.** 0.25 qt   **47.**

49.

51. $n > 6{,}250$; no

Chapter 2 Test

1. $2x, -5x$; they have same variables with same exponents.
2. $-x + 6$ **3.** $P = 6x$

4. Yes; substitution yields $-2 = -2$.   **5.** $\{-6\}$ **6.** $\{10\}$
7. $\{6\}$ **8.** $\{4\}$ **9.** $\{-\frac{15}{8}\}$ **10.** $\{x: \frac{4}{5} < x < \frac{6}{5}\}$

11. $y = \frac{12 - 4x}{3}$
12. A false equation has no solution. $4 \ne -4$; $\emptyset$
13. $-\frac{9}{2}$ **14.** 8.33   **15.** 3.5 oz, 16.5 oz   **16.** -9
17. $A = 22°, B = 140°, C = 18°$
18. \$1,600 at 5 percent earned \$80; \$400 at 6.5 percent earned \$26.
19. 11.4 seconds   **20.** 10.5 g

Cumulative Test 2

1. $\frac{17}{20}$ **2.**

3. $\frac{10}{7}$ **4.** 2,954.91   **5.** 84.78 in.^{3}
6. 0   **7.** -15 **8.** $A = (x + 3)(x^2 + 4)$, 10 square units;
$P = 2(x^2 + x + 7)$, 14 units   **9.** -11.8 percent
10. $-3 + 3x - 4y$ **11.** $4 - x$ **12.** Yes; substitution
yields $-7 = -7$.   **13.** $\{\frac{7}{12}\}$ **14.** $\{\frac{35}{4}\}$ **15.** $\{x: 1 < x < 2\}$
16. $-\frac{16}{3}$ **17.** $c = ad - b$ **18.** $10°$ **19.** $\frac{1}{3}$ gal   **20.** 0.08

Chapter 3
Exercises 3.1

1. 4^5; 4; 5   **3.** $(\frac{3}{4})^3$; $\frac{3}{4}$; 3   **5.** $(-3)^4$; -3; 4   **7.** -3^4; 3; 4
9. $(a + b)^3$; $a + b$; 3   **11.** $\frac{1}{x^3}$; x; 3   **13.** $(1.4x)^2$; $1.4x$; 2

15. $(3 - x)^3$; $3 - x$; 3   **17.** $\frac{1}{(x + y)^2}$; $x + y$; 2   **19.** 17
21. -8 **23.** 2,000   **25.** -8 **27.** 1,000   **29.** 4
31. $3^9 = 19{,}683$ **33.** $(-3)^9 = -19{,}683$ **35.** $(-2)^4 = 16$
37. x^9 **39.** a^6 **41.** a^7b^7 **43.** $10x^2$ **45.** $12a^7$ **47.** $-24x^6$
49. 2^{n+3} **51.** 2^{6+2n} **53.** x^{2a} **55. a.** $12x^6$ **b.** $7x^3$
57. a. $15x^4$ **b.** $-8x^2$ **59. a.** $-x^4$ **b.** 0   **61.** $2^{12} = 4{,}096$
63. $(-2)^6 = 64$ **65.** $(\frac{1}{2})^6 = \frac{1}{64}$ **67.** a^6 **69.** a^8 **71.** -3^8
73. $9x^2$ **75.** $27x^3y^6$ **77.** $-x^3y^3$ **79.** $\frac{9a^4}{16x^2}$ **81.** $\frac{4x^4}{9y^2}$
83. $\frac{-a^6b^3}{8c^3}$ **85.** 5^{2n+2} **87.** 4^{6n^2} **89.** x^{18} **91.** $36x^4$
93. $108x^{12}$ **95.** $-245a^{15}$ **97.** $1{,}024a^{10}x^{11}$ **99.** $2^{2n}a^nx^{3n}$
101. 5   **103.** 9   **105.** 1   **107.** $f = \frac{1}{4}mnt^2$ **109.** $V = \frac{1}{4}\pi d^2h$
111. $A = 25x^2$ **113.** $A = 6x^2$ **115.** $A = 3n^4$ **117.** $V = 2x^3$

Remember This (3.1)

1. $75; 225$ **2.** $3n + 3$ **3.** $10; 100$
4. $\{x: x < -5\}$ **5.** $\{x: x < -1\}$

6. \$500; \$2,500   **7.** 0.4 hour or 24 min
8. $m = \frac{F}{a}$ **9.** 3.44 square units   **10.** 64

Exercises 3.2

1. 3   **3.** 1   **5.** 1   **7.** 4   **9.** 1   **11.** 1   **13.** 1   **15.** 3
17. -5 **19.** $\frac{1}{9}$ **21.** $\frac{4}{27}$ **23.** $-\frac{1}{3}$ **25.** 1   **27.** 16   **29.** $\frac{4}{3}$
31. $\frac{97}{36}$ **33.** $\frac{3}{7}$ **35.** x **37.** 2, 1   **39.** $\frac{25}{12}$, 1   **41.** 100.01, 1
43. $+$ **45.** $+$ **47.** $-$ **49.** $+$ **51.** $-$ **53.** $+$ **55.** x
57. $\frac{1}{x^2}$ **59.** $2^6 = 64$ **61.** 9   **63.** 1   **65.** $\frac{9}{49}$ **67.** $\frac{3y}{5x}$
69. $36xy^3$ **71.** $\frac{1}{64y}$ **73.** $\frac{100x^5}{y^8}$ **75.** $\frac{1}{x^7}$ **77.** $\frac{4}{9x^5}$ **79.** $\frac{x^6}{64}$
81. $\frac{9}{4x^2}$ **83.** $\frac{1}{x^9}$ **85.** x^{2n+4} **87.** 2^{1+n} **89.** x^{-2n-6}
91. 0.614   **93.** 1,588.455; 12, 8 percent   **95.** 0.310   **97.** 0.231
99. 10   **101.** $10^5 = 100{,}000$ **103.** $10^3 = 1{,}000$ **105.** 100
107. 10^{-9} second

Remember This (3.2)

1. 4^65^2 **2.** 118   **3.** -27 **4.** x^{10} **5.** $-15x^4$ **6.** $\frac{1}{64}$
7. $x^2 + y^2$ **8.** $77 + 18c$ **9.** 301.44 in.^{3} **10.** $\{\frac{2}{7}\}$

Exercises 3.3

1. 4.235×10^9 **3.** 3.45×10^5 **5.** 5.68×10^{-6}
7. 8.402×10^{-4} **9.** 1×10^3 **11.** 1×10^{-2} **13.** 1,350
15. 401   **17.** 0.001   **19.** 0.53028   **21.** 0.09807
23. 1,200,000,000   **25.** $\$4.01181 \times 10^{11}$ **27.** 4.04×10^{10}
29. 3.6×10^7 **31.** 602,200,000,000,000,000,000,000
33. 2,400,000,000,000,000,000,000   **35.** 4,500,000,000
37. 28,000   **39.** 9.2   **41.** 0.006171   **43.** 300
45. 300   **47.** 30,000   **49.** 5.865696×10^{12} mi
51. 6.21 eV; 6.21 eV > 4.5 eV   **53.** 1.28 seconds

Remember This (3.3)

1. 2^{-3} **2.** They are equal.   **3.** $\dfrac{1}{x}$ **4.** 1   **5.** 2, 3   **6.** $\left\{-\dfrac{3}{2}\right\}$

7. $a = \dfrac{b - c}{d}$ **8.** $\{x: -5 < x < 1\}$

9. 6   **10.** x^5

Exercises 3.4

1. Yes   **3.** Yes   **5.** Yes   **7.** No   **9.** No   **11.** No
13. 4, binomial   **15.** 3, trinomial   **17.** 2, trinomial
19. 0, monomial   **21.** 4, polynomial   **23.** 4, trinomial   **25.** -6
27. -7 **29.** -135 **31.** $\dfrac{3}{4}$ **33.** 1   **35.** 35.3
37. 16 ft, 64 ft, 144 ft   **39.** 0 ft, 4 ft, 0 ft   **41.** 1, 3, 6, 10, 190
43. $2x^2 + 3x - 2$ **45.** $-13x^3 - 4x - 4$
47. $-3x^3 + 3x^2 + 5x$ **49.** $3x + 3$ **51.** $4x^2 + 2x - 1$
53. $\dfrac{5x^2}{4} - \dfrac{5x}{6} + \dfrac{5}{8}$ **55.** $-16x + 8$ **57.** $4x^2 - 4$
59. $-3x^3 + 8x^2 - 2x + 6$ **61.** $3x^3 + 3x^2 - 5x - 2$
63. $-x^2 - x + 3$ **65.** $-\dfrac{x^2}{4} + \dfrac{7}{8}$ **67.** $x^2 + 5x - 2$
69. $13x^2 - 7x + 9$ **71.** $-2x^2 - 2x - 2$ **73.** $x^2 - 3x - 8$
75. $2x^3 + x$ **77.** $\dfrac{19x^2}{20} - \dfrac{8x}{15} + \dfrac{4}{3}$ **79.** $6x^2 + 2x + 8$
81. $2x^2 - 6$ **83.** $6x^2 + 3x + 1$ **85.** $\{2\}$ **87.** $\{0\}$ **89.** $\{1\}$

Remember This (3.4)

1. 6.92×10^{120} **2.** $2 \times 10^2 = 200$ **3.** 0.1   **4.** $3y^2$
5. a. True   **b.** False   **6.** a   **7.** 1.86×10^5 mi/second;
9.3×10^7 mi; 5×10^2 seconds   **8.** a, b, d   **9.** -2
10. 6 square units

Exercises 3.5

1. $6x^2$ **3.** $-20d^3$ **5.** $-6x^4$ **7.** $15x^5$ **9.** $6xy^2$ **11.** $-y^3$
13. $6x^2 + 3x$ **15.** $-3x^2 - 6x$ **17.** $-x^3 + 7x^2 + 9x$
19. $-6s^4 + 4s^3 + 2s^2$ **21.** $a^3b + a^2b^2 + ab^3$
23. $-6x^3y + 9x^3y^2 - 15x^2y^2$ **25.** $x^2 + 3x + 2$
27. $4a^2 + 8a + 3$ **29.** $x^2 - 9$ **31.** $7y^2 - 20y + 12$
33. $9x^2 - 24x + 16$ **35.** $18x^2 - 2$ **37.** $9x^2 + 30x + 25$
39. $ac + 3a + 2c + 6$ **41.** $3x^2 + 8xy - 3y^2$
43. $x^2 - 2xy + y^2$ **45.** $6x^2 + xy - y^2$ **47.** $x^4 + x^3 - x - 1$
49. $x^3 + 2x^2 + 2x + 1$ **51.** $3y^3 - 4y^2 + 2y - 1$
53. $8x^3 + 1$ **55.** $ac + ad + ae + bc + bd + be$
57. $x^3 + 2x^2y + 2xy^2 + y^3$ **59.** $2x^3 - 7x^2y - xy^2 + 21y^3$
61. $x^4 + 3x^3 + 6x^2 + 5x + 3$ **63.** $x^3 + 3x^2 + 2x$
65. $y^3 + 2y^2 - 5y - 6$ **67.** $a^3 + 6a^2 + 12a + 8$
69. $x^3 + x^2 - x - 1$ **71.** $x^4 + 4x^3 + 6x^2 + 4x + 1$
73. $3x^2 - 12x - 63$ **75.** $\dfrac{2x^2 + x}{2}$ **77.** $x^3 - x$
79. Revenue $= 40{,}000{,}000 + 1{,}000{,}000x - 50{,}000x^2$; price
$= \$3{,}900$; revenue $= \$40{,}950{,}000$ **81. a.** $p - p^2$ **b.** $\dfrac{1}{4} > \dfrac{9}{100}$

Remember This (3.5)

1. b, c   **2. a.** 3   **b.** 2   **c.** 1   **d.** 2   **3.** $-1, -1, 1$
4. $-2x^2 - x - 6$ **5.** $4x + 3$ **6.** 4.85×10^{-4}
7. $E = 2.07 \times 10^{22}$ **8.** $\dfrac{4z^2}{xy^6}$ **9.** 60   **10.** 17.78 ml

Exercises 3.6

1. $x^2 + 4x + 3$ **3.** $w^2 + 2w - 8$ **5.** $x^2 - 2x - 15$
7. $y^2 - 9y + 8$ **9.** $6x^2 + 11x + 3$ **11.** $12x^2 + 10x - 12$
13. $15 - 26c - 21c^2$ **15.** $\dfrac{1}{4}x^2 + 5x + 24$ **17.** $-8 + 2z^2$
19. $x^2 - 0.1x - 0.02$ **21.** $2x^2 - 11xy - 63y^2$
23. $\dfrac{1}{2}a^2 - \dfrac{3}{4}ab - \dfrac{1}{2}b^2$ **25.** $x^2 - 1$ **27.** $9w^2 - 25$
29. $4x^2 - 9y^2$ **31.** $16x^2 - 9u^2$ **33.** $\dfrac{9r^2}{16} - \dfrac{1}{16}$ **35.** $x^2 - 0.25$
37. $x^2 + 8x + 16$ **39.** $9y^2 - 30y + 25$ **41.** $4 + 24k + 36k^2$
43. $4x^2 + 0.8x + 0.04$ **45.** $9s^2 - 12sr + 4r^2$
47. $\dfrac{1}{16}j^2 + jk + 4k^2$ **49.** $(x + 5)$ **51.** $(m + n)$ **53.** $(x + y)$
55. $x^2 - (x^2 - 1) = 1$; if $x = 5$, then $25 - 24 = 1$; if $x \le 1$, then
the rectangle does not exist.   **57.** Square: 16×16; rectangle:
14×19 **59.** 4 square units

Remember This (3.6)

1. a. $2xy + 2xw$ **b.** $2y + 2w + xy + xw$ **2.** $4x^2 - \pi x^2$
3. $x^3 + 5x^2 + 11x + 10$ **4.** 3, 1, 1   **5.** b
6.   **7.** $x^2 - 11$ **8.** 3.45×10^{19}

9. 2 lb   **10.** $\left\{\dfrac{26}{7}\right\}$

Exercises 3.7

1. $8x$ **3.** 1   **5.** $-\dfrac{2}{y}$ **7.** $\dfrac{6y^2}{x}$ **9.** $-\dfrac{3}{4rs^2}$ **11.** $\dfrac{2}{w^3}$
13. $3x + 5$ **15.** $2h + 3$ **17.** $\dfrac{x}{2} - 2 + \dfrac{3}{2x}$
19. $-10x + 4 - \dfrac{2}{x}$ **21.** $-\dfrac{7y^2}{2} + \dfrac{3y}{2} + 2$
23. $4x^2 + 6xy + 9y^2$ **25.** $\dfrac{P}{2} - \dfrac{2W}{2} = \dfrac{P}{2} - W$
27. $\dfrac{9C}{5} + \dfrac{160}{5} = \dfrac{9C}{5} + 32$ **29.** $\dfrac{y}{5n} - \dfrac{5c}{5n} = \dfrac{y}{5n} - \dfrac{c}{n}$
31. $3x - 1 + \dfrac{9}{x + 2}$ **33.** $5y + 1 + \dfrac{4}{y - 1}$
35. $x + 4 - \dfrac{2}{3x + 2}$ **37.** $-4x - 2 + \dfrac{26}{4x - 2}$
39. $4n^2 - 8n + 14 - \dfrac{82}{3n + 6}$ **41.** $2x^2 + 1 + \dfrac{5}{x^2 + 2}$
43. $2y^3 + 5$ **45.** $x^2 + 2x + 4$
47. $x^4 - x^3 + x^2 - x + 1 - \dfrac{1}{5x + 1}$ **49.** False
51. Quotient $= 1$; remainder $= -1$
53. $2 = 2, 3 = 3, 4 = 4$; when $x = -1$, then the left-hand side is
undefined because divisor $= 0$.

Remember This (3.7)

1. $9x^2 - 30x + 25$ **2.** $16x^2 - 1$ **3.** $2x^3 - 5x^2 - 22x - 15$
4. $-x^2 - 4x + 12$ **5. a.** $\$5 \times 10^{11}$ **b.** $\$2 \times 10^3 = \$2{,}000$
c. 5×10^8 seconds ≈ 16 years   **6. a.** $\dfrac{1}{2}$ **b.** $\dfrac{1}{100}$ **c.** 27   **d.** 1
7. 15,625; yes   **8.**   **9.** $\$8.95$ **10.** $x = \dfrac{t}{r + s}$

Chapter 3 Review Exercises

1. a^4 **3.** $20a^5$ **5.** $6x^3$ **7.** $-\frac{1}{2}$ **9.** $\frac{1}{y^5}$ **11.** 0.41

13. 3.211×10^6 **15.** 1,500   **17.** **b** and **c**   **19.** 5

21. $2x^2 - 8x$ **23.** $3x^2 - 5x + 7$ **25.** $12y^2 - 16y - 3$

27. $2c^2 - 7c + 6$ **29.** $9v^2 - 12v + 4$ **31.** $\frac{y}{4x}$

33. $x = \dfrac{4y - 6}{3}$

Chapter 3 Test

1. $\dfrac{1}{(2x - 4)^2}$ **2.** -81 **3. a.** $7y^2$ **b.** $12y^4$ **4.** $\frac{1}{125}$ **5.** $\dfrac{4}{9x^2}$

6. $\dfrac{4y^6}{x}$ **7.** 1.35×10^{13} **8.** 552.25   **9.** 2.38×10^{33} **10.** 3

11. 190   **12.** $3x^2 - 4x + 9$ **13.** $6m^2 - 12m$

14. $3b^2 - 11b - 4$ **15.** $x^3 + 3x^2 + x - 2$ **16.** $25w^2 - 16$

17. $x^2 - 2x + 1$ **18.** Area of square is larger by 9 square units.

19. $3h^2 + 2h + 1$ **20.** $3x - 4$

Cumulative Test 3

1. -34 **2.** $0.08p$ **3.** Difference   **4.** $6 + 5$ **5.** $-\frac{1}{5}$

6. 6,300   **7.** $\frac{3}{20}$ **8.** $-3c + 5$ **9.** $5x - 1$ **10.** $\{\frac{7}{3}\}$

11. Because the solution set consists of all real numbers

12. $x + 2x + 4x = 64$ **13.** 6 minutes

14. $\{x : x > -1\}$ ┼┼┼◯──▶   **15.** a^9 **16.** $72x^8$
                        $-1\ \ 0$

17. $\frac{1}{2}$ **18.** $2x^2 - 3x$ **19.** $2x^3 - 5x^2 + 8x - 3$

20. a. $4n^2 + 12n + 9$; degree $= 2$ **b.** $8n + 12$; degree $= 1$

Chapter 4
Exercises 4.1

1. 6   **3.** 10   **5.** 1   **7.** 24   **9.** $2^2 \cdot 3$ **11.** $4x$ **13.** $5y$

15. $2xy$ **17.** 1   **19.** x **21.** $x + y$ **23.** $x + y$ **25.** $2x$

27. $-3x$ **29.** $x + 1$ **31.** $2(x - 3)$ **33.** $4(xz + 2uv)$

35. $3x^2(x^2 - 3x + 2)$ **37.** $y(6 + 5x - 30x^2)$

39. $12x^2y^2(3x + 2y)$ **41.** $y(2x + 1)$ **43.** $x(x^3 - 1)$

45. $4y(4y - 1)$ **47.** $(b + c)(3 + d)$ **49.** $(x + 7)(x - 2)$

51. $(x + 4)(2x + 1)$ **53.** $(3x - 5)(3x - 5)$ **55.** 1,300

57. 7,500   **59.** 0   **61.** $(x + 2)(3x + 5)$ **63.** $(y - 3)(y + 2)$

65. $(x + 2)(2x + 3)$ **67.** $(x + 2y)(4x + 5y)$

69. $(a - b)(a + b)$ **71.** $(2a + b)(4a - b)$

73. $A = 2ab(1 + a + b)$

75. $0.15(p_1 + p_2)$; cost is same as 15 percent discount on the total price.   **77.** $0.96x$; final price is 96 percent of original price.

Remember This (4.1)

1. $2x^2$ **2.** $24x^3$ **3.** $x^2 + 6x + 8$ **4.** False   **5.** $-3x + 12$

6. $2x + 5 + \dfrac{25}{x - 4}$ **7.** $\dfrac{5y^4}{x^2}$ **8.** $x + 1$ **9.** $\{-15\}$

10. 1.08×10^9

Exercises 4.2

1. 5   **3.** 6   **5.** $+$ **7.** $-, +$ **9.** $+, 10, -$ **11.** $+, -, 9$

13. $(x + 1)(x + 2)$ **15.** $(y + 4)(y + 3)$ **17.** $(t + 1)(t + 5)$

19. $(x - 5)(x - 4)$ **21.** $(y - 2)(y - 6)$ **23.** $(t - 3)(t - 6)$

25. $(x + 8)(x - 5)$ **27.** $(y + 4)(y - 3)$ **29.** $(t + 7)(t - 2)$

31. $(x + 1)(x - 8)$ **33.** $(y + 2)(y - 10)$

35. $(t + 6)(t - 12)$ **37.** $(x + 2b)(x + b)$

39. $(y - 5c)(y - 4c)$ **41.** $(z + 3d)(z - 5d)$

43. $2(x + 1)(x + 2)$ **45.** $4y(y + 3)(y + 4)$

47. $3x^2(x - 5)(x - 4)$ **49.** $2t^3(t + 2)(t - 7)$

51. $a(x - 2)(x - 1)$ **53.** $(x + 3)^3$ **55.** 6; $(x + 3)(x + 2)$

57. 24; $(y + 6)(y - 4)$ **59.** 2; $(a - 2)(a - 1)$

61. No two integers have a product of 1 and a sum of 1 with $1 \cdot 1$ and $(-1)(-1)$ as possible factor pairs.

63. No two integers have a product of 4 and a sum of 2 with $4 \cdot 1$, $2 \cdot 2$, $(-4)(-1)$, and $(-2)(-2)$ as possible factor pairs.

65. No two integers have a product of 7 and a sum of -2 with $7 \cdot 1$ and $(-7)(-1)$ as possible factor pairs.

67. No two integers have a product of -6 and a sum of 3 with $6(-1)$, $(-6)(1)$, $2(-3)$, and $(-2)(3)$ as possible factor pairs.

69. No two integers have a product of -8 and a sum of -4 with $8(-1)$, $(-8)(1)$, $4(-2)$, and $(-4)(2)$ as possible factor pairs.

71. No two integers have a product of -12 and a sum of -10 with $12(-1)$, $(-12)(1)$, $6(-2)$, $(-6)(2)$, $4(-3)$, and $(-4)(3)$ as possible factor pairs.   **73.** Prime   **75.** $(y - 3)(y + 2)$

77. Prime   **79.** $(x + 3)(x + 2)$ **81.** $(d + 8)(d + 7)$

x	$+$	3
x	x^2	$3x$
$+$		
2	$2x$	6

d	$+$	8
d	d^2	$8d$
$+$		
7	$7d$	56

83. $(x + 8a)(x + a)$ **85.** $(x - 12)(x - 13) = 0$

x	$+$	$8a$
x	x^2	$8ax$
$+$		
a	ax	$8a^2$

87. $16(t + 5)(t - 2) = 0$ **89.** $(m + 13)(m - 11) = 0$

Remember This (4.2)

1. $3x^2 + 11x + 6$ **2.** $18x^3 - 33x^2 - 6x$ **3.** $5x^2$

4. $3x^2y(9x + 5y)$ **5.** $(x + 2)(5x + 3)$ **6.** $\{x : \frac{4}{3} < x < \frac{8}{3}\}$

7. $\$3,500$ **8.** $\{-5\}$ **9.** 0   **10.** 1

Exercises 4.3

1. a   **3. c**   **5. b**   **7.** 3, 1   **9.** $-, +$ **11.** $-, 8, +, 3$

13. 1. **c**; 2. **b**; 3. **a** or **d**; 4. **a** or **d**   **15.** $(2x + 1)(x + 2)$

17. $(3x + 1)(x + 1)$ **19.** $(4x + 1)(x + 2)$

21. $(2x + 7)(x + 2)$ **23.** $(4x + 3)(x + 2)$

25. $(4x + 5)(5x + 6)$ **27.** $(5x - 1)(x - 1)$

29. $(5x - 6)(x - 2)$ **31.** $(3x - 1)(4x - 1)$

33. $(7x - 2)(2x - 3)$ **35.** $(2x - 1)(9x - 8)$

37. $(5x - 4)(2x - 5)$ **39.** $(3x + 5)(x - 1)$

41. $(5x - 1)(x + 11)$ **43.** $(5x - 3)(x + 2)$

45. $(4x + 13)(2x - 1)$ **47.** $(2x + 7)(2x - 5)$

49. $(2x + 3)(8x - 9)$ **51.** $(2x - 3)(x + 1)$

53. $(3x + 1)(x - 3)$ **55.** $(17x + 2)(x - 2)$

57. $(6x - 11)(2x + 1)$ **59.** $(6x + 3)(x - 7)$

61. $2(2x - 3)(9x + 5)$ **63.** $(2x + 7y)(x - 2y)$

65. $(3a - 5b)(4a - 5b)$ **67.** $(4c - 9d)(6c + d)$

69. $-(x - 2)(x + 1)$ **71.** $-(y + 4)(y + 6)$

73. $3(c - 1)(c + 2)$ **75.** $2x(2x + 3)(3x - 5)$

77. $-2(2x + 3)(7x - 4)$ **79.** $-6x^2(x + 1)^2$

81. $2y(x - 1)(8x + 3)$ **83.** $b(2a - b)(3a - b)$

85. $abc(2a + b)(3a - 4b)$ **87.** $2(x + 1)^3$ **89. a.** 11, 18

b. 9, 2   **91. a.** $-31, 240$ **b.** $-16, -15$ **93. a.** 7, -144
b. 16, -9 **95.** $(x + 1)(2x + 9)$ **97.** $(2x + 1)(5x - 2)$
99. $(2x + 5)(4x + 3)$ **101.** $(3x - 4)(3x - 2)$
103. $(5y + 6)(6y - 5)$ **105.** $6(x - 2)(2x + 1)$
107. $(2x - 5)(2x - 3) = 0$ **109.** $(x + 15)(4x - 15) = 0$
111. $(x - 1)(2x - 3) = 0$

Remember This (4.3)

1. $9x^2 - 12x + 4$ **2.** $x^3 + 3x^2 + 3x + 1$ **3.** $9x^2 - 25$

4. False   **5.** $xy(x^2y^2 + 1)$ **6.** $(3x - 4)(2x + 1)$ **7.** $\dfrac{-b^9}{27x^6}$

8. 2   **9.** $V = 10x^3 - 30x^2$ **10.** 1

Exercises 4.4

1. c   **3.** b   **5.** c   **7.** No   **9.** Yes   **11.** No   **13.** $a = x$, $b = 2$
15. $a = 4x$, $b = 5y$ **17.** $a = s$, $b = 6t$ **19.** $8x$ **21.** $4y$
23. $16xy$ **25.** $20ab$ **27.** $6a^2b^3$ **29.** $4xy^4z^2$ **31.** $(x + 2)^2$
33. $(y - 5)^2$ **35.** $(3a + 4)^2$ **37.** $(6x - 1)^2$ **39.** $(w + b)^2$
41. $(2r - 3s)^2$ **43.** $a = 3x$, $b = 4$ **45.** $a = 3f$, $b = 5g$
47. $a = 2x^3$, $b = 7y^4$ **49.** $(x + 1)(x - 1)$ **51.** $(9 + s)(9 - s)$
53. $(8a + b)(8a - b)$ **55.** $(7z + 2w)(7z - 2w)$
57. $(x^2 + y^3)(x^2 - y^3)$ **59.** $(6j^4 + k)(6j^4 - k)$
61. $a = x$, $b = 2$ **63.** $a = 3$, $b = 2x$ **65.** $a = 4x^5$, $b = 3y^4$
67. $(a + c)(a^2 - ac + c^2)$ **69.** $(d - 3)(d^2 + 3d + 9)$
71. $(2x - 1)(4x^2 + 2x + 1)$ **73.** $(2j + 3k)(4j^2 - 6jk + 9k^2)$
75. $(5g^3 + h^3)(25g^6 - 5g^3h^3 + h^6)$
77. $(6x^2 - 7y^3)(36x^4 + 42x^2y^3 + 49y^6)$
79. $(4x + 5y)(4x - 5y)$ **81.** Prime   **83.** $(h + 5)^2$
85. $(t + 2)(t + 6)$ **87.** $(x + 4)(x^2 - 4x + 16)$
89. $(z^2 + 9c)(z^2 - 9c)$ **91.** $(2x - 5y)^2$
93. $1{,}000s^2(s + 1)(s - 1)$ **95.** $(9x^2 + 4y^2)(3x + 2y)(3x - 2y)$
97. Prime   **99.** $a(x - a)^2$ **101.** $16c(x - 2c)(x^2 + 2cx + 4c^2)$
103. a. $(a + b)^2$ **b.**   **105.** $4(5 + 2t)(5 - 2t) = 0$

107. $V = \frac{4}{3}\pi(R - r)(R^2 + Rr + r^2)$

Remember This (4.4)

1. True   **2.** False   **3.** True   **4. a.** False   **b.** True

5. $x = \dfrac{y + 5c}{4n}$ **6.** $-\dfrac{2}{x} + \dfrac{1}{7} - 3x$ **7.** $x^3 - 11x^2 + 33x - 20$

8. $\dfrac{111}{1{,}000}$ **9.** $-\frac{1}{27}$ **10.** Less than

Exercises 4.5

1. a, b   **3.** e   **5.** None   **7.** $6(x - 2)(x - 6)$ **9.** $5(x - 2)^2$
11. $(7x + 2)(x - 8)$ **13.** $(2x - 7)(4x - 3)$
15. $(4 + y^2)(2 + y)(2 - y)$ **17.** $2cr(rt + s)$
19. $(x + 6)(x + a)$ **21.** $(1 - b)(x + 1)$
23. $(x + 2y)(x - 2y)(x^2 + 2xy + 4y^2)(x^2 - 2xy + 4y^2)$
25. $(3x + 11y)^2$ **27.** $(2x + 3)(4x^2 - 6x + 9)$
29. $(x^3 + 2)(x^3 - 2)$ **31.** Prime   **33.** Prime
35. $49(7u - 8)(u + 1)$ **37.** $A = (a + b)^2$
39. $A = 4h(s - h)$ **41.** $V = \pi h(R + r)(R - r)$

Remember This (4.5)

1. 0   **2.** True   **3.** False   **4.** $x(x + 1)(x + 2) = 0$

5. 0, 0, -12 **6.** 0, 0, 294   **7.** $\frac{73}{8}$ **8.** 3^{-4} **9.** $n = \dfrac{c}{a + b}$

10. \$2,000

Exercises 4.6

1. Yes   **3.** Yes   **5.** No; degree $= 1$ **7.** 0   **9.** 0, 5
11. 0, 1, -1 **13.** $\{2,3\}$ **15.** $\{3\}$ **17.** $\{\frac{3}{7}, -\frac{2}{5}\}$ **19.** $\{0,4\}$
21. $\{0,5\}$ **23.** $\{0,\frac{11}{3}\}$ **25.** $\{4\}$ **27.** $\{1,9\}$ **29.** $\{8,-3\}$
31. $\{2,4\}$ **33.** $\{6,-6\}$ **35.** $\{0,-1\}$ **37.** $\{-9,1\}$ **39.** $\{0,3\}$
41. $\{6,-6\}$ **43.** $\{8,-2\}$ **45.** $\{-7,2\}$ **47.** $\{0,5\}$ **49.** $\{3,-4,\frac{3}{2}\}$
51. $\{0,2,-1\}$ **53.** $\{0,-\frac{7}{4},\frac{7}{4}\}$ **55.** 5
57. a. 1 second   **b.** 2 seconds   **c.** No
59. a. When $n = 3$, $d = 0$; when $n = 4$, $d = 2$ **b.** 8
61. 30 units; side length is positive

Remember This (4.6)

1. $x + (x + 1) = -21$ **2.** 18 in.²   **3.** $A = \frac{1}{2}x(x - 3)$
4. True   **5.** False   **6.** $(x + 5)(x - 5)$; yes   **7.** Yes
8.   **9.** y **10.** $2x - 5$

Exercises 4.7

1. 0, 1; or -1, 0   **3.** 9, 10; or -10, -9 **5.** 2, 4.5; or -4.5, -2
7. 3, 4, 5; or -1, 0, 1   **9.** -2, 4; or -3, 3
11. True for all numbers   **13.** $s = 2$ m, $A = 4$ m², $P = 8$ m
15. $s = 7$ ft   **17.** 6 in.   **19.** 8 in. by 16 in.   **21.** 4 cm
23. $x = 6$ units   **25.** 15 in.   **27.** 36 units   **29.** 10 in.
31. a. $3^2 + 4^2 = 25$ and $5^2 = 25$ **b.** $4^2 + 5^2 = 41$, but $6^2 = 36$
33. Direct from A to B **35.** 50 m

Remember This (4.7)

1. True   **2.** True   **3.** Both equal 22.5; no   **4.** -1, 7, undefined
5. True   **6.** $\{1,-3\}$ **7.** $\dfrac{x + 2}{x + 3}$ **8.** $(2x + y)(4x^2 - 2xy + y^2)$
9. $2x^3 - 3x^2 - 8x - 3$ **10.** -7

Chapter 4 Review Exercises

1. $4x$ **3.** $8(x^2 - y)$ **5.** $A = \frac{1}{2}ah + \frac{1}{2}bh = \frac{1}{2}h(a + b)$
7. $(x - 12)(x + 6)$
9. No two integers have a product of 3 and a sum of 1 with $1 \cdot 3$
and $(-1)(-3)$ as the possible factor pairs.   **11.** $(3y + 2)(y + 1)$
13. $-3(3y + 5)(y + 1)$ **15.** $(3x + 2)(2x - 3)$ **17.** $6x$
19. $(2a + 5)^2$ **21.** $(x + 1)(x^2 - x + 1)$ **23.** Prime polynomial
25. $(a + 2b)(a^2 - 2ab + 4b^2)$ **27.** 0, 2, -3 **29.** $\{0,3\}$
31. $\{\frac{2}{3}, -\frac{1}{2}\}$ **33.** 4, 6.5; or -6.5, -4 **35.** $A = 108$ in.²

Chapter 4 Test

1. $4x^2$ **2.** $6a^2(a^2 - 3a + 2)$ **3.** $(x + 4)(x + 3)$
4. $(y - 6)(y - 2)$ **5.** $(x + 6)(x - 3)$ **6.** $(x - 8)(x + 2)$
7. Prime polynomial   **8.** $6x(x + 1)^2$ **9.** $(5x - 4)(2x + 1)$
10. $(a + 3)^2$ **11.** $(4c + 3)(4c - 3)$ **12.** $8(x + 1)(x - 1)$

13. $(a - 2)(a^2 + 2a + 4)$ **14.** $(x + 2)(3x + 2)$
15. $3x^2 + 18x + 2x + 12 = 3x(x + 6) + 2(x + 6)$
$= (x + 6)(3x + 2)$ **16.** $\{0,3\}$ **17.** $\{0, -\frac{1}{2}, \frac{4}{3}\}$ **18.** $\{3, -4\}$
19. 5 seconds   **20.** 5.5 ft

Cumulative Test 4

1.   **2.** 29   **3.** $\frac{5}{6}$ **4.** $10 - \pi$ m²

5. Commutative   **6.** $1.82, $9.10   **7.** $\{0\}$ **8.** 5.4 degrees
9. $250,000   **10.** $\{\frac{1}{3}\}$ **11.** $729a^{15}$ **12.** 1.24×10^2
13. $x^2 - x + 17$ **14.** $3x^3 - 20x^2 + 21x - 6$

15. $3x - 1 + \dfrac{9}{x + 2}$ **16.** $y(7 + 6x - 42xy)$

17. $(x + 9)(x - 2)$ **18.** $(3x + 1)(x - 5)$

19. $\{-5,3\}$ **20.** 12.5 in.

Chapter 5
Exercises 5.1

1. 2   **3.** 0   **5.** -2 **7.** None   **9.** 2   **11.** 3, -3 **13.** Yes
15. No   **17.** Yes   **19.** No   **21.** No   **23. a.** Yes   **b.** No

c. No   **d.** No   **e.** No   **25.** $\frac{3}{4}$ **27.** $\frac{1}{4}$ **29.** $3c^2$ **31.** $\dfrac{1}{x}$

33. $\dfrac{-1}{2a}$ **35.** -4 **37.** $\frac{3}{5}$ **39.** $\dfrac{3(x + 4)}{5}$ **41.** 3   **43.** $\dfrac{3}{x - 5}$

45. $\dfrac{1}{y + 1}$ **47.** $\dfrac{x + 2}{x + 3}$ **49.** $\dfrac{c - 3}{c - 4}$ **51.** $\dfrac{x}{x + 1}$

53. Lowest terms   **55.** $x - y$ **57.** $\dfrac{x + 2}{x + 3}$ **59.** -1

61. Lowest terms   **63.** $\dfrac{-1}{y + 2x}$ **65.** $-\dfrac{x}{b}$ **67.** $\dfrac{a^2 - ab + b^2}{a - b}$

69. $m^2 + mn + n^2$ **71.** $\dfrac{2 + c}{d}$ **73. a.** 4 percent   **b.** 0

c. Undefined

Remember This (5.1)

1. $\frac{8}{15}$ **2.** $-8x$ **3.** $\frac{14}{5}$ **4.** x **5.** False   **6.** 1   **7.** $\dfrac{5}{x - 1}$

8. $\{2, -1\}$ **9.** Distributive   **10.** True

Exercises 5.2

1. $\dfrac{18}{x^2}$ **3.** $\frac{3}{11}$ **5.** 1   **7.** $\dfrac{1}{16x}$ **9.** $2x^2$ **11.** $\dfrac{1}{x}$ **13.** 4

15. $\dfrac{1}{x + 3}$ **17.** $-\dfrac{1}{s}$ **19.** $-\dfrac{y}{x}$ **21.** 1   **23.** $\dfrac{(a + 2)(a + 5)}{(a + 7)(a + 3)}$

25. $\dfrac{5}{x}$ **27.** x **29.** $\dfrac{x + y}{x - y}$ **31.** 4   **33.** $\frac{9}{16}$ **35.** $\dfrac{2(x - 3)}{x + 3}$

37. $\dfrac{t + 5}{t - 5}$ **39.** $-\dfrac{x - 1}{x(y + 2)}$ **41.** $\dfrac{c + d}{a + b}$ **43.** $\dfrac{(x + 1)(x - 2)}{(x - 1)(x + 2)}$

45. $\dfrac{y^2}{2y + 3}$ **47.** $\dfrac{3d(y - b)}{c^2(y + b)}$ **49.** $4/\pi$

Remember This (5.2)

1. 2   **2.** True   **3.** 0   **4.** 2   **5.** $a = \dfrac{2c}{3b}$ **6.** True   **7.** 3

8. 17 cm   **9.**   **10.** $x^2 + (x + 1)^2$

Exercises 5.3

1. x **3.** $\dfrac{-2}{x + 4}$ **5.** 1   **7.** 1   **9.** $\dfrac{2n - m}{m + n}$ **11.** 0   **13.** $x - 1$

15. $\frac{1}{3}$ **17.** $\dfrac{1}{x + 2}$ **19.** $\dfrac{x}{y} + \dfrac{1}{y}$ **21.** $\dfrac{y^2}{x} + \dfrac{y}{x}$

23. $\dfrac{x}{x^2 + 1} + \dfrac{1}{x^2 + 1}$ **25. a**   **27. a**   **29. b**   **31.** $\dfrac{1}{y}$ **33.** $\dfrac{1}{x}$

35. 0   **37.** 1   **39.** $y + 3$ **41.** $x + 1$ **43.** $t + 2$ **45.** $\dfrac{1}{y + 1}$

47. $\dfrac{1}{x^2 - x - 1}$ **49.** $\dfrac{8}{x}$ **51.** $\dfrac{4}{x + 1}$ **53.** $\dfrac{5}{2 - x}$

55. $\dfrac{a - aw + bw}{c}$

Remember This (5.3)

1. 40   **2.** $-\frac{1}{120}$ **3.** $3y^2$ **4.** True   **5.** $\dfrac{y}{x^2}$ **6.** $-\dfrac{y - 1}{y(x + 1)}$

7. 4 gal   **8.** $\{0,1, -\frac{5}{2}\}$ **9.** $\dfrac{4}{xy}$ **10.** False

Exercises 5.4

1. 120   **3.** xy **5.** $6xy$ **7.** $30x^3y^3$ **9.** $20(x - 2)$

11. $3(y + 3)(y - 3)$ **13.** $\frac{47}{120}$ **15.** $\dfrac{5 + 2x}{5x}$ **17.** $\dfrac{3 + 2a}{3ab}$

19. $\dfrac{16 - 9x}{12x^2y}$ **21.** $\dfrac{2x + 3}{2x + 2}$ **23.** $\dfrac{4x - 3}{12(x + 1)}$ **25.** $\dfrac{5y + 2}{y(y + 1)}$

27. $\dfrac{x - 6}{x(x - 2)}$ **29.** $\dfrac{2x^2}{x^2 - 64}$ **31.** $\dfrac{-4}{y(y + 4)}$

33. $\dfrac{2d^2 + 6d + 5}{(d + 1)(d + 2)}$ **35.** $\dfrac{x^2 - 12}{(x + 3)(x + 4)}$ **37.** $\dfrac{2y - 2}{y^2 - 4}$

39. $\dfrac{x}{(x + 1)^2(x + 2)}$ **41.** $\dfrac{2x^2 - 4x - 4}{(x + 1)(x - 2)(x - 3)}$

43. $\dfrac{2x + 2y}{xy}$ **45.** $\dfrac{4x + 5y + 2xy}{x^2y^2}$ **47.** $\dfrac{2V}{hd} - \dfrac{bhd}{hd} = \dfrac{2V - bhd}{hd}$

Remember This (5.4)

1. $-\frac{1}{10}$ **2.** $4a + b$ **3.** $d - 3a$ **4.** $\dfrac{b}{a}$ **5.** $\frac{17}{24}$ **6.** $\dfrac{-x}{y - 4}$

7. $\dfrac{1}{(x - 1)(x + 2)(x + 3)}$ **8.** $c = \dfrac{ab}{b + 1}$

9.   **10.** 16 cm

Exercises 5.5

1. a. $\dfrac{11}{5} \div \dfrac{23}{10} = \dfrac{22}{23}$ **b.** $\dfrac{30 - 8}{3 + 20} = \dfrac{22}{23}$ **3. a.** $\dfrac{13}{2} \div \dfrac{5}{3} = \dfrac{39}{10}$

b. $\dfrac{30 + 9}{12 - 2} = \dfrac{39}{10}$ **5. a.** $\dfrac{9}{20} \div \dfrac{5}{6} = \dfrac{27}{50}$ **b.** $\dfrac{15 + 12}{30 + 20} = \dfrac{27}{50}$

7. a. $\dfrac{ab - 1}{b} \div \dfrac{ab + 1}{b} = \dfrac{ab - 1}{ab + 1}$ **b.** $\dfrac{b(a - 1/b)}{b(a + 1/b)} = \dfrac{ab - 1}{ab + 1}$

9. a. $\dfrac{ab - 1}{b} \div \dfrac{2ab - 1}{2b} = \dfrac{2ab - 2}{2ab - 1}$

b. $\dfrac{2b(a - 1/b)}{2b[a - 1/(2b)]} = \dfrac{2ab - 2}{2ab - 1}$

11. a. $\dfrac{a^2 + b}{a} \div \dfrac{b^2 + a}{b} = \dfrac{a^2b + b^2}{ab^2 + a^2}$

b. $\dfrac{ab(a + b/a)}{ab(b + a/b)} = \dfrac{a^2b + b^2}{ab^2 + a^2}$ **13.** $\dfrac{3ab}{2a + b}$ **15.** $\dfrac{14a}{11}$ **17.** $\dfrac{5a^2}{a - 3}$

19. $\dfrac{3xy}{(2x + y)(x + y)}$ **21.** $\dfrac{3x + 2y}{6}$ **23.** $\dfrac{(x + y)^2}{xy}$

25. $\dfrac{x^2 + xy + y}{2y}$ **27.** $\dfrac{8n^2 + 12n}{9n^2 - 12}$ **29.** $\dfrac{2n^2}{(n + 1)^2}$ **31.** $\dfrac{b}{a + 2b}$

33. $\dfrac{b}{a}$ **35.** -1 **37.** $M = \dfrac{2}{\frac{1}{6} + \frac{1}{12}} = \dfrac{24}{2 + 1} = 8$

Remember This (5.5)

1. $3x - 2$ **2.** $2x$ **3.** $30 + 3R$ **4.** Both equal 23,230.

5. $\dfrac{2A - ah}{h}$ **6.** 0   **7.** 4 ft   **8.** $\dfrac{x^4y^6}{z^2}$ **9.** -38

10. $(2a + 3)(a + 4)$

Exercises 5.6

1. $\{1\}$ **3.** $\{-25\}$ **5.** $\{-18\}$ **7.** $\{\frac{5}{6}\}$ **9.** $\{3\}$ **11.** $\{2\}$

13. $\{-\frac{28}{9}\}$ **15.** $\{5\}$ **17.** $\{-5\}$ **19.** $\{2\}$ **21.** $\{0\}$ **23.** $\emptyset$

25. $\{5\}$ **27.** $\{\frac{1}{2}\}$ **29.** $\{0\}$ **31.** $\{2\}$ **33.** $\{-4\}$ **35.** $\{\frac{1}{2}\}$

37. $\{-2,3\}$ **39.** $\{-\frac{1}{6},1\}$ **41.** $\{5\}$ **43.** $\{2,-5\}$ **45.** $\{-\frac{9}{2},6\}$

47. $\emptyset$ **49.** $\{20\}$ **51.** $\{-\frac{2}{3}\}$ **53.** $\emptyset$ **55.** $y = mx + 1$

57. $r = \dfrac{s\ell}{k}$ **59.** $a = 2A - b$ **61.** $r = \dfrac{S - a}{S}$ **63.** $x = \dfrac{1 - y}{y}$

65. $a = \dfrac{bc}{b + c}$ **67.** $x = -1$ **69.** 160 units

Remember This (5.6)

1. 10.5   **2. a.** False   **b.** False   **3.** $\frac{3}{8}$ teaspoon

4. $\dfrac{A_1}{A_2} = \dfrac{2xh}{xh} = 2$ **5.** $\dfrac{cd}{2d - 3c}$ **6.** $\dfrac{32y - 15x}{40x^2y^2}$ **7.** $\dfrac{1}{x + 1}$

8. $1\frac{1}{2}$ hours   **9.** $\{1,7\}$ **10.** $\emptyset$

Exercises 5.7

1. $\frac{1}{6}$ **3.** $\frac{12}{5}$ **5.** $\frac{10}{1}$ **7.** $\frac{1}{13}$ **9.** $15.9 \approx 16$ **11.** 32.5 minutes

13. $\frac{1}{2}$ cup   **15. a.** $x = 12(1.5) = 18$ **b.** 19.2 in.

17. a. 480 mi   **b.** Leave A before 7:24 A.M.   **19. a.** 160 minutes

b. $53\frac{1}{3}$ days $\approx$ 53 days

21.

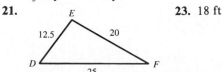

23. 18 ft

25. a. 40 men, 60 women   **b.** 20 in favor, 80 opposed

c. $x = 8$; fraction $= \frac{1}{5}$ **27. a.** 6 ft   **b.** 25 ft   **29.** 2.4 hours

31. Solve $\dfrac{1}{100} + \dfrac{1}{100} = \dfrac{1}{x}$; $x = 50$.   **33. a.** 40 minutes

b. 10 minutes   **c.** 8 minutes   **35. a.** 6 mi/hour   **b.** 4 hours

c. $3\frac{1}{4}$ hours   **37. a.** 6 mi/hour   **b.** 1 mi   **39.** 2 hours

Remember This (5.7)

1. $-3, -1, 1$ **2.** 0   **3.** No   **4.** $y = \dfrac{-3x - 5}{4}$ **5.** 40 m

6. $2(3x + 2y)(3x - 2y)$ **7.** $a^3 - 1$ **8.** $3\frac{1}{2}$ **9.** -2

10. $-\frac{5}{3}$ and $\frac{1}{3}$; $-\frac{1}{3}$ and $\frac{5}{3}$

Chapter 5 Review Exercises

1. -3 **3.** $\dfrac{x + 3}{x + 2}$ **5.** $-\dfrac{x - 6}{x(y + 2)}$ **7.** $\dfrac{x^2 - 2x + 3}{x + 3}$ **9.** $\dfrac{14}{x}$

11. $\dfrac{x^2 - 2x - 10}{(x + 2)(x + 3)}$ **13. a.** $\dfrac{19}{6} \div \dfrac{13}{3} = \dfrac{19}{26}$ **b.** $\dfrac{24 - 5}{2 + 24} = \dfrac{19}{26}$

15. $\{\frac{4}{5}\}$ **17.** $b = 3A - a - c$ **19.** $2\frac{5}{8}$ cups   **21. a.** 6 mi/hour

b. 3 hours

Chapter 5 Test

1. -1 **2.** $\frac{2}{3}$ **3.** $-x^2$ **4.** $\dfrac{n}{m(m - n)}$ **5.** $\dfrac{5}{x - 5}$ **6.** 1

7. $x + 5$ **8.** $\dfrac{x^2 - 4x - 18}{(x + 3)(x + 2)}$ **9.** $x + 2$ **10.** $\dfrac{2ab}{a + b}$

11. $\dfrac{b}{a - 3b}$ **12.** $\{\frac{21}{5}\}$ **13.** $\{2,-1\}$ **14.** $r = \dfrac{S - 5}{S}$

15. a. $x = 1$ **b.** 6 units by 3 units   **16.** 160 gal

17. Yes; in $\frac{6}{7}$ hour   **18. a** and **e**   **19.** Common factors can then be divided out by the fundamental principle of fractions.

20. $4/s$

Cumulative Test 5

1. 24   **2.** $-3x + 4$ **3.** True   **4.** $>$ **5.** $\frac{1}{2}$ **6.** $\{-1\}$

7. $2 \neq -15$ **8.** 7.1 oz   **9.** $\frac{13}{9}$ **10.** $8x^2 + 6x - 3$

11. $4x^2 - 9$ **12.** $x = 5y - 3$ **13.** $(b - 4)(b + 3)$

14. $(2y + 3)(2y - 3)$ **15.** $\{0,2,-\frac{3}{2}\}$ **16.** Either 2 or 14

17. None   **18.** $\dfrac{x + 2}{x + 3}$ **19.** $\{\frac{5}{18}\}$ **20.** $\frac{3}{4}$ hour

Chapter 6
Exercises 6.1

1. Yes   **3.** Yes   **5.** No   **7. a.** -18 **b.** -2 **9. a.** -2

b. 6   **11. a.** -3 **b.** -2 **13. a.** 6; 10 prints cost $6.

b. 26; for $10 you get 26 prints.

15. a. 43; the cost for 13 disks is $43.

b. 3; for $13 you get 3 disks.   **17. a.** Q_1 **b.** Q_1 **c.** Q_3

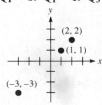

19. Each point is not located in a quadrant.

21. a. Q_4 **b.** Q_2 **c.** No quadrant   **23. a.** Q_1 **b.** Q_1 **c.** Q_2

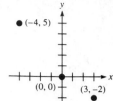

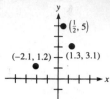

7.

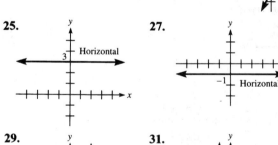

25.

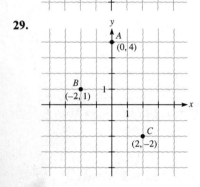

27.

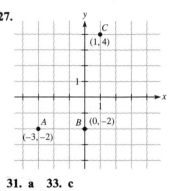

9.

11.

13.   **15.**   **17.**

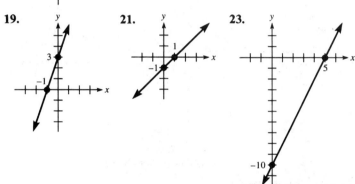

29.

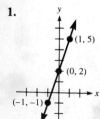

31. a   **33. c**

19.   **21.**   **23.**

Remember This (6.1)

1. $5, 1, -3$ **2.** True   **3.** $y = \dfrac{3x - 5}{4}$ **4.**

5. $x = -4$ **6.** 403,500 people killed   **7.** $\left\{\frac{1}{4}\right\}$ **8.** $\dfrac{b - ab^2}{a + a^2 b}$

9. $\left\{\frac{5}{2}, -1\right\}$ **10.** 4.58×10^{10}

25.

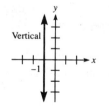

27.

29.

31.

Exercises 6.2

1.

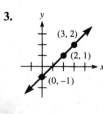

3.

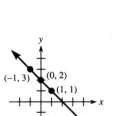

5.

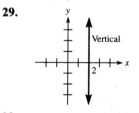

33.

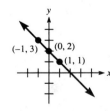

35.

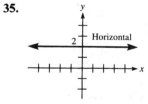

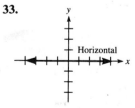

37.

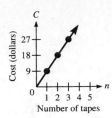

Number of tapes

39. a. $C = 3n + 4$

b.

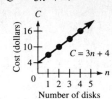

Number of disks

35.

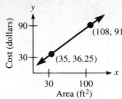

Area (ft²)

Slope = 0.75
Charge for cleaning is 75¢/ft².

37. a.

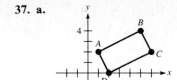

b. $m_{AB} = m_{DC} = \frac{1}{2}$,
$m_{BC} = m_{AD} = -2$

c. $m_{AB} \cdot m_{BC} = -1$

Remember This (6.2)

1. $-\frac{3}{4}$ **2.** Horizontal **3.** 4 **4.** -5 **5.** Yes **6.** $12x^2$
7. $V = \pi$ ft³ = 1,728 in.³ **8.** $3y^3 - y^2 + y + 2$ **9.** $3x$
10. $\{3\}$

Exercises 6.3

1.

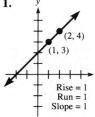

Rise = 1
Run = 1
Slope = 1

3.

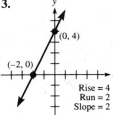

Rise = 4
Run = 2
Slope = 2

5.

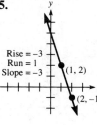

Rise = -3
Run = 1
Slope = -3

7.

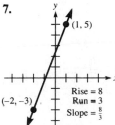

Rise = 8
Run = 3
Slope = $\frac{8}{3}$

9.

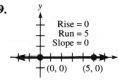

Rise = 0
Run = 5
Slope = 0

11.

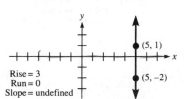

Rise = 3
Run = 0
Slope = undefined

13. 3 **15.** -3 **17.** 1

19. 0 **21.** 7 **23.** -6 **25.** $\frac{1}{6}$
27. Rise = 0, run = 4, slope = 0 **29.** -1

31.

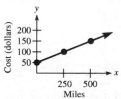

Miles

Slope = 0.20
Mileage charge = 20¢ per mile
y-intercept is (0, 50).
Minimum charge = $50

33.

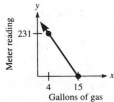

Gallons of gas

Slope = -21
Average miles per gallon = 21
y-intercept tells how far the car
can go without running out of gas.

39. a.

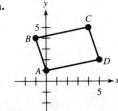

b. $m_{AB} = m_{DC} = -3$, $m_{BC} = m_{AD} = \frac{1}{5}$
c. $m_{AB} \cdot m_{BC} \neq -1$
41. a. -1 **b.** $\frac{1}{2}$ **c.** -4 **d.** Undefined **e.** -1.25
43. a. 22 degrees, 28 degrees **b.** 11 A.M. **c.** Noon and 1 P.M.
d. 0 degrees **45. a.** 6 to 7 **b.** Slope = 16.1; average rise in
temperature is 16.1 degrees per minute. **47. a.** 1960s
b. Slope = -5; population dropped an average of 5 people per year
between 1980 and 1990.

Remember This (6.3)

1. Add $4x - 3$ to both sides. **2.** Both become $x + 3y = 17$.
3. $y = \dfrac{c - ax}{b}$ **4.** $y = b$ **5.**

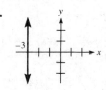

6. $(0, -6)$, $(8, 0)$ **7.** 137.5 cents per gallon **8.** 12.8 cm
9. True **10.** $\{2, -\frac{1}{2}\}$

Exercises 6.4

1. $y = x$ **3.** $3x + y = 5$ **5.** $2x - y = -7$ **7.** $x - y = 4$
9. $x - 2y = -1$ **11.** $x + 4y = 13$ **13.** $y = 1$ **15.** $x = 3$
17. $3x + 10y = 20$ **19.** $y = x$ **21.** $x - 3y = -5$
23. $x + 2y = 4$ **25.** $2x - 3y = 5$ **27.** $2x - y = 0$
29. $y = 3$ **31.** $x = 4$ **33.** $x - y = -1$
35. $8x - 15y = -1$ **37.** 3; (0,2) **39.** -1; (0,8)
41. $\frac{1}{3}$; (0,1) **43.** $-\frac{2}{3}$; (0,-1) **45.** 3; (0,-2)
47. $\frac{5}{3}$; (0,-2) **49.** $y = x$ **51.** $y = x + 2$

53. 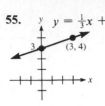 $y = -3x$ **55.** 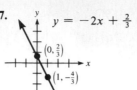 $y = \frac{1}{3}x + 3$ **57.** 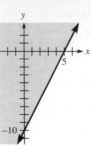 $y = -2x + \frac{2}{3}$ **29.**   **31. a.** 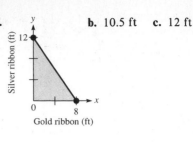   **b.** 10.5 ft   **c.** 12 ft

59. 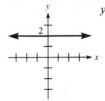 $y = 2$

61. Yes   **63.** Yes   **65.** No   **67.** No   **69.** Yes   **71.** Yes
73. a. $y = \frac{1}{4}x + 2$ **b.** \$2   **c.** 20
75. a. 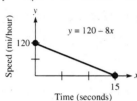   **b.** 120 mi/hour   **c.** 15 seconds

33. a. 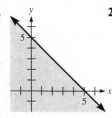   **b.** Yes   **c.** 1 lb

Remember This (6.5)

1. 1   **2.** 3   **3.** True   **4.** 5, −5   **5.** $3x - y = 1$
6. $y = x - 1$ **7.** 4   **8.** 5   **9.** $\{0, \frac{1}{2}, -1\}$
10. Either $-\frac{1}{2}$ and $\frac{3}{2}$ or $\frac{9}{2}$ and $\frac{13}{2}$

Remember This (6.4)

1. True   **2.**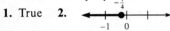

3. False   **4.** 1   **5.** $\frac{1}{3}$ **6.** All real numbers   **7.** $\dfrac{2x^2 + 5x - 2}{x^2 - 4}$

8. $3x + 1 + \dfrac{2}{x - 1}$ **9.** 3 oz   **10.** $-\dfrac{y^8}{x^9}$

Exercises 6.5

1. Yes   **3.** No   **5.** Yes   **7.** No   **9.** No   **11.** Yes

13.   **15.** 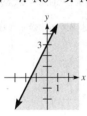   **17.**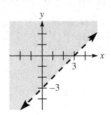

19. 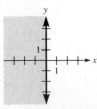   **21.** 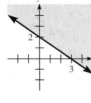   **23.**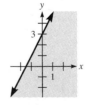

25. 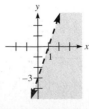   **27.**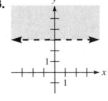

Exercises 6.6

1. Domain = {Mary, Ellen, Janice}, range = {1,2,3}
3. Domain = {Chicago, New York}, range = {Chicago, Los Angeles}   **5.** Domain = {1,2}, range = {0,1,−1}   **7.** Yes
9. No   **11.** Yes   **13.** Yes   **15.** No   **17.** Yes   **19.** Yes
21. Yes   **23.** No   **25.** No   **27.** Yes   **29.** No   **31.** $d = 2r$
33. $i = 12f$ **35.** $w = 65v + 5$ **37.** $A = \dfrac{\pi d^2}{8}$ **39.** $t = \dfrac{d}{55}$
41. $c = \frac{2}{3}w$ **43.** $f(1) = 6, f(0) = 2, f(-\frac{1}{2}) = 0, f(0.1) = 2.4$
45. $g(-1) = 0, g(1) = 0, g(0) = -1, g(0.5) = -0.75$
47. $h(1) = 0, h(-2) = 0, h(-1) = -2, h(0.3) = -1.61$
49. $f(3) = -36, f(-3) = -36, f(\frac{1}{2}) = -1, f(0.1) = -0.04$
51. $g(0) = 100, g(1) = 109, g(\frac{1}{2}) = 104\frac{3}{4}, g(0.2) = 101.96$
53. $h(1) = \pi + 1, h(0) = 0, h(-\pi) = \pi^3 - \pi$
55. $f(2) = 224$; the ball fell 224 ft in the first 2 seconds.
57. a. $f(10) = 154{,}745$; the house is worth \$154,745 after 10 years.
b. After 16 years

Remember This (6.6)

1. Yes   **2.** No   **3.** No   **4.**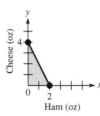

5. $y = -3x + 1$ **6.** $m = 0.10$; the cost is 10¢ per minute.
7. $(0, -9)$ **8.** $\dfrac{x^2 + 2x - 2}{x(x + 2)}$ **9.** {2}   **10.** $\dfrac{(a^2 - 1)b^2}{a^2(b^2 + 1)}$

Exercises 6.7

1. $p = kw$ **3.** $c = k\ell$ **5.** $d = kt, k = 400$ **7.** $\frac{15}{2}$ **9.** $\frac{5}{3}$

11. $-\frac{3}{2}$ **13.** \$13   **15.** \$4.86   **17.** 28.5 lb   **19.** $t = \dfrac{k}{s}$

21. $t = \dfrac{k}{n}$ **23.** $r = \dfrac{k}{d}$ **25.** 12   **27.** -4 **29.** $\frac{6}{5}$

31. 26 hours   **33.** $33\frac{1}{3}$ hours = 33 hours 20 minutes   **35.** $l = \dfrac{k}{d^2}$

37. $V = kr^3$ **39.** $E = kv^2$ **41.** 15   **43.** 320   **45.** -27
47. 100 ft   **49. a.** Weight increases 8 times.   **b.** 400 lb
51. Work increases 9 times.

Remember This (6.7)

1. $(3,1)$ **2.** True   **3.** -2 **4.** $y = 50t$ **5.**

6. $5x - y = 0$ **7.** $\dfrac{111}{1,000} = 0.111$ **8.** $c = \dfrac{b}{a - 1}$

9. $\{2, -\frac{3}{2}\}$ **10.** $3x(2x - 1)(x + 1)$

Chapter 6 Review Exercises

1. Yes   **3.**
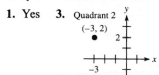

5.
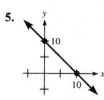
   7.
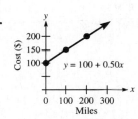

9. $m_{AB} = m_{CD} = \frac{2}{5}$; $m_{BC} = m_{AD} = -\frac{5}{2}$; $m_{AB} \cdot m_{BC} = -1$
11. $y = 3x$ **13.** Slope = 3, y-intercept = $(0, -1)$
15.
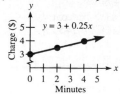
   17.

19. No   **21.** 8   **23.** $p = kt$ **25.** \$33.33   **27.** $\frac{1}{2}$

Chapter 6 Test

1. c   **2. a** $(-3,5)$; Q_2 **3.**

   4.
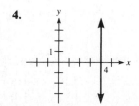

5. $C = 2w + 5$
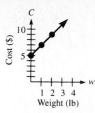
   6. Undefined

7. The slope is 2; the charge is \$2.00 per hour.   **8.** $\frac{1}{3}$
9. $3x - y = 1$ **10.** $x + y = 0$
11. Slope = 2; y-intercept = $(0, -5)$ **12.**

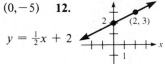

$y = \frac{1}{2}x + 2$

13. Yes   **14.** $14x + 16y \le 12$ **15.** Yes   **16.** $d = \dfrac{C}{\pi}$ **17.** 9

18. $f(2) = 64$; the object falls 64 ft in 2 seconds.   **19.** $s = \dfrac{k}{w}$

20. 64

Cumulative Test 6

1. 74   **2.** Distributive   **3.** -10 **4.** 1   **5.** $\{-\frac{7}{2}\}$ **6.** \$54

7. $\dfrac{y^5}{x^5}$ **8.** $x^2 + 2x + 1$ **9.** $4x^2 - 25y^2$ **10.** 3

11. $a(a + 3)(a + 4)$ **12.** $\{3, -1\}$ **13.** $\dfrac{-x}{2x - 1}$ **14.** $\dfrac{a^2b^2}{(a + b)^2}$

15. 28 ft   **16.** $\{-18\}$ **17.** All   **18.** $(0, -\frac{1}{2})$ **19.** $2x + y = 5$
20. $y = \frac{9}{2}$

Chapter 7
Exercises 7.1

1. a. $(2, 200)$ **b.**

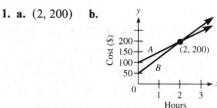

c. Same cost for 2 hours; after that, studio A is less expensive.
3. a. $(3, 380)$ **b.**
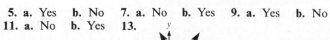
   c. At mile marker 380

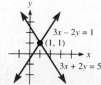

5. a. Yes   **b.** No   **7. a.** No   **b.** Yes   **9. a.** Yes   **b.** No
11. a. No   **b.** Yes   **13.**

15.

17.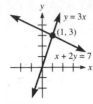

19.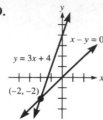

21.

23.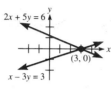

25. $x + \frac{1}{3}y = -1$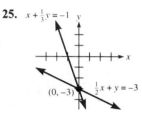

27.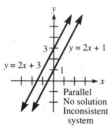

29.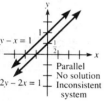

31.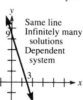

33.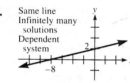

Remember This (7.1)

1. $10 - 3y$ **2.** $\{16\}$ **3.** Yes   **4.** $x - 3y = -7$
5. All real numbers   **6.** $-\frac{1}{2}$ **7.** $2x - y = 10$
8. $-\dfrac{x - 1}{x + 1}$ **9.** a **10.** $4x$

Exercises 7.2

1. $(\frac{1}{2}, \frac{1}{4})$ **3.** $(-2,7)$ **5.** $(0,-\frac{1}{2})$ **7.** $(\frac{3}{2}, 9)$ **9.** $(10,11)$
11. $(0,0)$ **13.** $(2,5)$ **15.** $(\frac{8}{7}, \frac{5}{7})$ **17.** $(-4,2)$ **19.** $(-1,2)$
21. $(3,4)$ **23.** $(\frac{1}{2}, 2)$ **25.** $(-\frac{2}{5}, -\frac{4}{3})$ **27.** $(2,1)$ **29.** $(\frac{3}{2}, \frac{4}{3})$
31. No solution; inconsistent system
33. Infinitely many solutions; dependent system
35. Infinitely many solutions; dependent system
37. No solution; inconsistent system
39. 20 tapes; use the membership plan.   **41.** 10 years

Remember This (7.2)

1. $2x$ **2.** $-2x + 7y = -1$ **3.** $x = -3$ **4.** $\{140,000\}$
5. -1 **6.** $(2,1)$ does not satisfy the second equation.   **7.** $y = 3$
8. -1 **9.** $\dfrac{3x + 4}{(x + 1)(x + 2)}$ **10.** $b = Ad - c$

Exercises 7.3

1. $(6,2)$ **3.** $(-1,2)$ **5.** $(-2,-3)$ **7.** $(5,-3)$ **9.** $(1,1)$
11. $(\frac{1}{2}, \frac{1}{3})$ **13.** $(\frac{1}{2}, 2)$ **15.** $(-\frac{2}{3}, \frac{3}{2})$ **17.** $(3,-1)$
19. $(-5,0)$ **21.** $(2,6)$ **23.** $(4,10)$
25. No solution; inconsistent system
27. No solution; inconsistent system
29. Dependent system; infinite number of solutions
31. Dependent system; infinite number of solutions
33. $20,000 at 3.5 percent and $30,000 at 6 percent
35. 13.5 and -3.5 **37.** 16 oz of 7 percent solution and
8 oz of 10 percent solution

Remember This (7.3)

1. $180°$ **2.** $P = 2\ell + 2w$ **3.** $0.10x$ **4.** $3x + 6y = 60$
5. $4x$ **6.** $(1,0)$ **7.** Inconsistent
8. 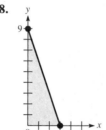   **9.** $(c + 2)^2$ **10.** $\{0,5\}$

Exercises 7.4

1. $45°$ and $135°$ **3.** $15°$ **5.** 6.75 cm²   **7.** 3 in. by 8 in.
9. a. $5,000   **b.** First job pays more.   **11. a.** $83\frac{1}{3}$ mi
b. One at $24.95 per day   **13. a.** 100 days   **b.** 200 liters total
15. a. 12.5 years   **b.** 25 years   **17.** $25,000 in each
19. $83.20   **21.** $94.50 on paperbacks, $119.76 on hardbacks
23. 96 oz of 5 percent solution, 32 oz of 9 percent solution
25. 9 oz of 1 percent solution, 15 oz of 1.8 percent solution
27. 4 containers of 80 percent alloy, 1 of 60 percent alloy
29. Plane speed $= 250$ mi/hour, wind speed $= 50$ mi/hour
31. 1 mi/hour   **33.** $\frac{1}{30}$ mi/minute, or 2 mi/hour

Remember This (7.4)

1.   **2.** Yes   **3.** $-2x + y > -6$ **4. b**   **5. a**

6. $(-1,-2)$ **7.** Slope   **8.** y
9. If the product of two numbers is 0, then at least one of those
numbers is 0.   **10.** Commutative property of multiplication

Exercises 7.5

1.

3.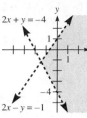

5.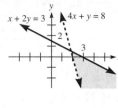

7.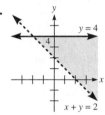

9.

11.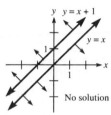

13.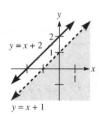

15. No solution

17.

19.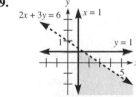

21. a. $2x + 3y \leq 120,\ x \geq 20,\ y \geq 20$

b.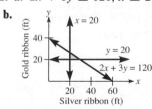

c. One solution is 25 ft silver ribbon and 22 ft gold ribbon; total cost $116.

23. a. $3x + 4y + 0.05(3x + 4y) \leq 500 \rightarrow 3.15x + 4.20y \leq 500,$ $x \geq 75,\ y \geq 50$ **b.**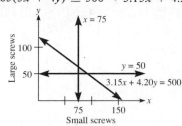

c. One solution is 80 small screws and 55 large screws; total cost $4.83.

Remember This (7.5)

1. True   **2.** False   **3.** 49   **4.** 3,125   **5.** True   **6.** 90°
7. $0.025x$ **8.** $\left(\frac{11}{2}, \frac{5}{2}\right)$ **9.** $M = kn^2$ **10.** $\dfrac{x}{2y}$

Chapter 7 Review Exercises

1. a   **3.** No solution; inconsistent system   **5.** $(-2, 3)$
7. Dependent; infinitely many solutions   **9.** 39°, 51°   **11.** $96
13. 2.5 mi/hour   **15.** $x + 2y \leq 5,\ x \geq 1,\ y \geq 1$

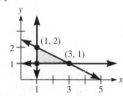

Chapter 7 Test

1. 2,000 units; after this, machine A costs less to operate.   **2.** No
3. No solution
4. These are the same line; there are infinitely many solutions.
5. $\left(\frac{3}{2}, -\frac{1}{2}\right)$ **6.** $\left(\frac{1}{2}, \frac{1}{5}\right)$ **7.** $(1, 0)$ **8.** $(1, -2)$
9. Invest $10,000 at 15 percent and $50,000 at 5 percent.
10. No solution; inconsistent system
11. System is dependent; there is only one line.
12. 75°   **13.** 432 cm²   **14.** 21 rentals
15. $9,000 at 5 percent and $3,000 at 10 percent
16. 75 ml of 10 percent solution and 25 ml of 18 percent solution
17. Current: 4 mi/hour; time in still water: 2 hours
18. 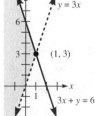   **19.** $12x + 6y \leq 36,\ x \geq 1,\ y \geq 2$

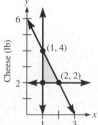

20. No

Cumulative Test 7

1. False   **2.** $1.06x$ **3.** Because 3 can be written as $3/1$, the ratio of two integers   **4.** 1   **5.** -1 **6.** $\{-\frac{11}{2}\}$
7. $b = 20$ cm   **8.** $\{x{:}x > -1\}$

9. $x^6 - 4$ **10.** $12x^4y^3$ **11.** $2(2x + 3)(2x - 3)$
12. $\{\frac{1}{2}, -4, \frac{5}{3}\}$ **13.** Either 1 m or 3 m   **14.** -3
15. $\dfrac{3y - 18}{4y + 6}$ **16.** $\{\frac{1}{2}\}$ **17.** 14 minutes
18.

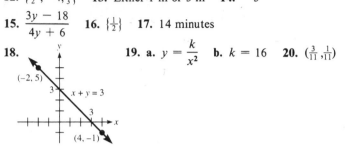

19. a. $y = \dfrac{k}{x^2}$ **b.** $k = 16$ **20.** $(\frac{3}{11}, \frac{1}{11})$

Chapter 8
Exercises 8.1

1. $1, -1$ **3.** $\frac{3}{4}, -\frac{3}{4}$ **5.** $1.2, -1.2$ **7.** 13   **9.** -14
11. $-\frac{1}{3}$ **13.** 1.41; irrational   **15.** 1.41; rational
17. -6.32; irrational   **19.** 1.90; irrational   **21.** Not real
23. $-\frac{1}{3}$; rational   **25.** 0.1; rational   **27.** Not real
29. 0; rational   **31.** 1.4 in.   **33.** 10 ft   **35.** 91.9 ft   **37.** 1
39. 5   **41.** -1 **43.** Not real   **45.** 0   **47.** Not real

Remember This (8.1)

1. 9   **2.** False   **3.** True   **4.** $(0, 10)$ **5.** $41°, 49°$
6. Negative   **7.** Since $3(-\frac{1}{3}) = -1$, the product of their slopes is -1.   **8.** $3x^3 + 2x^2 - 12x - 8$ **9.** $\{11\}$ **10.** $\{2, -2\}$

Exercises 8.2

1. $\sqrt{15}$ **3.** $\sqrt{30}$ **5.** 8   **7.** 7   **9.** $6\sqrt{15}$ **11.** $10\sqrt{30}$
13. 30   **15.** 27   **17.** $5\sqrt{3}$ **19.** $8\sqrt{2}$ **21.** $9\sqrt{2}$
23. a. $2\sqrt{2}$ in.   **b.** 2.8 in.   **25. a.** $\sqrt{2}$ in.   **b.** 1.4 in.
27. a. $\sqrt{7}$ **b.** 2.6   **29. a.** $2\sqrt{17}$ by 15 ft   **b.** 8.2 by 15 ft
31. x^2y **33.** $5xy^3$ **35.** $3x\sqrt{5x}$ **37.** $2\sqrt{2x}$ **39.** $5x^2y^3\sqrt{3xy}$
41. $10x^{10}y^5\sqrt{5y}$ **43.** $2\sqrt{3}$ **45.** $2\sqrt{15}$ **47.** $\sqrt{15}$ **49.** $\sqrt{210}$
51. $3x$ **53.** $3x^2\sqrt{2}$ **55.** $10x^2y^2$ **57.** $2\sqrt[3]{5}$ **59.** $3\sqrt[3]{2}$
61. $2\sqrt[4]{2}$ **63.** $3\sqrt[4]{4}$ **65.** x^2 **67.** $x^3\sqrt[3]{x}$ **69.** $y\sqrt[4]{y^2}$ **71.** 2
73. $5\sqrt[3]{5}$ **75.** $3\sqrt[4]{9}$ **77.** x^2 **79.** $x\sqrt[4]{y^2}$ **81.** $2xy^2\sqrt[3]{y}$

Remember This (8.2)

1. True   **2.** $\dfrac{\sqrt{3}}{3}$ **3.** $b = \dfrac{a}{c}$ **4.** $\sqrt{2}$ in.   **5.** $\sqrt{3}$ cm
6.   **7.** $(\frac{5}{16}, \frac{1}{8})$ **8.** d   **9.** $8x^4y^3$ **10.** \$35

Exercises 8.3

1. 6   **3.** $\frac{1}{4}$ **5.** $\sqrt{6}$ **7.** $\sqrt{3}$ **9.** $3\sqrt{2}$ **11.** $\sqrt{6}$ **13.** $\sqrt{3x}$
15. $2x\sqrt{y}$ **17.** $\dfrac{1}{2x}$ **19.** $\dfrac{\sqrt{14}}{7}$ **21.** $\dfrac{\sqrt{2}}{2}$ **23.** $\dfrac{\sqrt{10}}{6}$ **25.** $\dfrac{\sqrt{30}}{8}$
27. $\dfrac{\sqrt{x}}{x}$ **29.** $\dfrac{\sqrt{3y}}{y^2}$ **31.** $\frac{2}{3}$ **33.** $\frac{4}{5}$ **35.** $\dfrac{\sqrt{21}}{3}$ **37.** $\dfrac{\sqrt{10}}{5}$
39. $\dfrac{\sqrt{xy}}{y}$ **41.** $\dfrac{3y^2\sqrt{y}}{x^3}$ **43.** $\dfrac{5\sqrt{xy}}{2y}$ **45.** $\dfrac{3\sqrt{2}}{2}$ **47.** $\dfrac{x\sqrt{5}}{5}$
49. $\dfrac{3\sqrt{2}}{4}$ **51.** $\dfrac{\sqrt{2}}{2}$ **53. a.** $\dfrac{\sqrt{2}}{2}$ ft   **b.** 0.71 ft
55. a. $1^2 + (\sqrt{2})^2 = (\sqrt{3})^2$ **b.** $\sqrt{3}/3$ **c.** 0.5774   **57.** $\frac{1}{2}$
59. $\frac{2}{3}$ **61.** $\dfrac{\sqrt[3]{2}}{2}$ **63.** $\dfrac{\sqrt[4]{6}}{2}$ **65.** $2\sqrt[4]{2}$ **67.** $\dfrac{\sqrt[3]{36}}{3}$ **69.** $\dfrac{\sqrt[4]{40}}{2}$
71. $\dfrac{\sqrt[3]{xy^2}}{y}$ **73.** $\dfrac{\sqrt[4]{ab^3}}{b}$

Remember This (8.3)

1. $11x$ **2.** $\dfrac{n^2 + 1}{n}$ **3.** $3x - 4y$ **4.** 7   **5.** 45   **6.** $(-5, -7)$
7. b   **8.** b   **9.** 3   **10.** $x + \dfrac{1}{x} = 3$

Exercises 8.4

1. $8\sqrt{3}$ **3.** $5\sqrt{5}$ **5.** 0   **7.** $-\sqrt{5}$ **9.** $-3\sqrt{2}$ **11.** $6\sqrt{2x}$
13. $\sqrt{5xy}$ **15.** Does not simplify   **17.** Does not simplify
19. $7\sqrt{2}$ **21.** $-\sqrt{5}$ **23.** $7\sqrt{3}$ **25.** $-9\sqrt{2}$ **27.** $14\sqrt{x}$
29. $24x\sqrt{y}$ **31. a.** $4\sqrt{10}$ **b.** 10   **33. a.** $3\sqrt{5}$ **b.** $4\sqrt{5}$
c. $\sqrt{5}$ **35.** $\frac{4}{3}\sqrt{3}$ **37.** $\sqrt{3}/3$ **39.** $\frac{8}{3}\sqrt{3}$ **41.** $\frac{13}{5}\sqrt{5}$
43. $3\sqrt[3]{7}$ **45.** $17\sqrt[4]{xy^2}$ **47.** $2\sqrt[4]{2}$ **49.** $-5\sqrt[3]{3}$ **51.** $7\sqrt[4]{2}$

Remember This (8.4)

1. $a^2 - b^2$ **2.** $a^2 - 2ab + b^2$ **3.** $\dfrac{2 - x}{6}$ **4.** -4
5. $-x - 1$ **6.** $\sqrt{2}$ **7.** \$4,000   **8.** $y = -\frac{1}{3}x + \frac{4}{3}$
9. $\dfrac{x - 5}{x(x - 3)}$ **10.** $\emptyset$

Exercises 8.5

1. $\sqrt{6} - \sqrt{14}$ **3.** $\sqrt{15} + \sqrt{30}$ **5.** $\sqrt{nx} + \sqrt{ax}$
7. $2\sqrt{3} + 3\sqrt{2}$ **9.** $3\sqrt{2} - 3$ **11.** -1 **13.** 4   **15.** -3
17. -41 **19.** $9 - x$ **21.** $x - y^2$ **23.** $8 + 5\sqrt{2}$ **25.** $\sqrt{2}$
27. $12 - 7\sqrt{2}$ **29.** $2 + 3\sqrt{x} + x$ **31.** $2 - 3\sqrt{xy} + xy$
33. $6 + \sqrt{x} - x$ **35.** $6 + 5\sqrt{2x} + 2x$ **37.** $3 + 2\sqrt{2}$
39. $7 - 4\sqrt{3}$ **41.** $4 + 4\sqrt{x} + x$ **43.** $x - y$ **45.** $\dfrac{1 - \sqrt{2}}{2}$
47. $\dfrac{2 + \sqrt{3}}{2}$ **49.** $\sqrt{2} + 3$ **51.** $3 - \sqrt{2}$ **53.** $\dfrac{\sqrt{3} + \sqrt{6}}{3}$
55. $\dfrac{\sqrt{2} + 2}{2}$ **57.** $\dfrac{\sqrt{10} - 5\sqrt{2}}{10}$ **59.** $-1 - \sqrt{2}$ **61.** $\sqrt{6} + 2$
63. $\dfrac{\sqrt{35} + \sqrt{10}}{5}$ **65.** $-\dfrac{\sqrt{14} + \sqrt{35}}{3}$ **67.** $\dfrac{0.5\sqrt{n}}{n}$
69. a. $\dfrac{2n(\sqrt{n} - 1)}{n - 1}$ **b.** 4.5   **71.** $R = \dfrac{c(2\sqrt{c} - 1)}{4c - 1}$

Remember This (8.5)

1. $3x$ **2.** $x - 3$ **3.** No; x and y could be opposites.   **4.** No
5. 0.81   **6.** $9\sqrt{3}$ **7.** $\dfrac{\sqrt{3}}{3}$ **8.** Because it is a solution to
both equations.   **9.** \$6,800   **10.** $(3x - 2y)^2$

Exercises 8.6

1. $\{27\}$ **3.** $\{2\}$ **5.** $\{-4\}$ **7.** $\emptyset$ **9.** $\{\frac{3}{2}\}$ **11.** $\{1\}$ **13.** $\{0\}$
15. $\{2\}$ **17.** $\{-2\}$ **19.** $\{\frac{9}{4}\}$ **21.** $\{-\frac{3}{2}\}$ **23.** $\{0\}$ **25.** $\{\frac{13}{4}\}$
27. $\{\frac{2}{9}\}$ **29.** $\{3\}$ **31.** $\{3\}$ **33.** $\{3\}$ **35.** $\{2,3\}$ **37.** $\{\frac{1}{4},2\}$
39. $\emptyset$ **41.** $\emptyset$ **43.** $\{7\}$ **45.** $\{0\}$ **47.** $\{3,4\}$ **49.** $\{15\}$
51. 7.30 ft   **53.** 0.253 kg   **55.** $n = 3$

Remember This (8.6)

1. $27x^6$ **2.** $x^5 + x^3y$ **3.** $\frac{3}{2}$ **4.** True   **5.** $-\frac{5}{3}$ **6.** $1 - \sqrt{3}$
7. a   **8.**   **9.** $\{0,3\}$ **10.** \$52,000

Exercises 8.7

1. 5   **3.** 10   **5.** -4 **7.** 1   **9.** 3   **11.** -10 **13.** 2   **15.** -2
17. 4   **19.** 8   **21.** 3,125   **23.** 16   **25.** $\frac{1}{125}$ **27.** $\frac{1}{128}$ **29.** $\frac{1}{9}$
31. Not a real number   **33.** Not defined   **35.** $2^{5/3}$ **37.** 1
39. $7^{5/6}$ **41.** $\frac{1}{8}$ **43.** $3^{1/3}$ **45.** 4   **47.** $2^{3/4}$ **49.** 16   **51.** $\frac{1}{3}$
53. 36   **55.** $\frac{1}{16}$ **57.** x^3 **59.** $x^{7/2}$ **61.** $9x^{2/3}$ **63.** $2x^{3/2}y$
65. $\dfrac{b^3}{a}$ **67.** x **69.** $\dfrac{1}{y^2}$ **71.** $\dfrac{x^{1/6}}{y^{1/6}}$ **73.** $\dfrac{x^{3/2}}{y^{11/3}}$ **75.** $\dfrac{1}{x^{1/6}y^{1/6}}$
77. 1.072   **79.** 1.585   **81.** 9.518   **83.** 0.707   **85.** 2.168
87. 11.61 percent   **89.** About 50 g   **91.** $\sqrt{5}$ **93.** $\sqrt[3]{x}$ **95.** 2
97. a^2

Remember This (8.7)

1. a and c   **2.** -1 **3.** Yes   **4.** No   **5.** $(x - 6)(x + 3)$
6. $\{4\}$ **7.** a   **8.** $\dfrac{3}{n}$ **9.** 1   **10.** 10

Chapter 8 Review Exercises

1. $12, -12$ **3.** 1.73 m   **5.** $6\sqrt{6}$ **7.** $3x^2\sqrt{2}$ **9.** $x\sqrt{5}$
11. $\dfrac{\sqrt[3]{225}}{5}$ **13.** $\sqrt{3}$ **15.** $\sqrt{15} + 3\sqrt{2}$ **17.** $2 + \sqrt{15}$
19. $\{\frac{1}{9}\}$ **21.** 3.24 ft   **23.** $\dfrac{1}{32^{2/5}}$

Chapter 8 Test

1. -9 **2.** Because no real number when squared can equal
a negative number   **3.** $\sqrt{5}$ ft; 2.2 ft   **4.** 63   **5.** $4\sqrt{2}$
6. $2x\sqrt{10x}$ **7.** $2a$ **8.** $2\sqrt{6}$ **9.** $\dfrac{\sqrt{5y}}{y}$ **10.** $\dfrac{5}{x^3}$ **11.** a and d

12. 0   **13.** $2\sqrt{3}/3$ **14.** $-\sqrt[4]{2}$ **15.** $1 + 2\sqrt{x} + x$ **16.** $\{\frac{29}{9}\}$
17. $\{2\}$ **18.** 49   **19.** 25   **20.** 5.08 percent

Cumulative Test 8

1. False   **2.** a, c, e   **3.** a
4. a. $\dfrac{1,800}{4} + \dfrac{1,800}{3} + 90 + x = 1,800$
b. $x = 660$; there is \$660 left over.   **5.** $2x^2 + 7x - 4$
6. $x + 3 + \dfrac{6}{x - 1}$ **7.** $\{\frac{1}{2}, -3\}$
8. One that cannot be factored over the integers   **9.** $n = \dfrac{mb}{a}$
10. $\dfrac{x + 3}{x + 4}$ **11.** $\{\frac{9}{2}\}$ **12.** 0   **13.** $y = x + 3$ **14.** 10
15. $(\frac{5}{13}, \frac{1}{13})$ **16.** a
17.   **18.** $4\sqrt{2}$ **19.** $\{4\}$ **20.** $h = s\sqrt{2}$

Chapter 9
Exercises 9.1

1. $\{2,3\}$ **3.** $\{7, -10\}$ **5.** $\{1\}$ **7.** $\{\pm 10\}$ **9.** $\{\pm 12\}$
11. $\{\pm 1.1\}$ **13.** $\{\pm\frac{1}{8}\}$ **15.** $\{\pm\sqrt{2}\}$ **17.** $\{\pm\sqrt{7}\}$
19. $\{\pm 2\sqrt{2}\}$ **21.** $\{\pm 3\sqrt{3}\}$ **23.** No real number solution
25. No real number solution   **27.** $\{0\}$ **29.** $\left\{\pm\dfrac{2\sqrt{3}}{3}\right\}$
31. $\left\{\pm\dfrac{\sqrt{14}}{2}\right\}$ **33.** No real number solution   **35.** 7.9 seconds
37. 11.3 ft/second   **39.** $\{3, -7\}$ **41.** $\{0,2\}$ **43.** $\{-\frac{5}{3}\}$
45. $\{\frac{5}{2}, \frac{1}{2}\}$ **47.** $\{-\frac{4}{5}, \frac{2}{5}\}$ **49.** No real number solution
51. $\{2 + \sqrt{5}, 2 - \sqrt{5}\}$ **53.** $\left\{\dfrac{3 + \sqrt{6}}{2}, \dfrac{3 - \sqrt{6}}{2}\right\}$
55. $\left\{\dfrac{-5 + \sqrt{7}}{3}, \dfrac{-5 - \sqrt{7}}{3}\right\}$ **57.** $\left\{\dfrac{2 + 2\sqrt{2}}{5}, \dfrac{2 - 2\sqrt{2}}{5}\right\}$
59. $\{2 + 2\sqrt{2}, 2 - 2\sqrt{2}\}$ **61.** 9.54 percent   **63.** 5.91 percent

Remember This (9.1)

1. $(x + 3)^2$ **2.** $(x - \frac{1}{2})^2$ **3.** 13   **4.** n **5.** $A = x(x + 2)$
6. $9x$ **7.** $\{1,2\}$ **8.** $(4, -3)$ **9.** Vertical   **10.** $t = \dfrac{A - p}{pr}$

Exercises 9.2

1. 36; $(x + 6)^2$ **3.** 1; $(x - 1)^2$ **5.** $\frac{9}{4}$; $(x + \frac{3}{2})^2$
7. $\frac{49}{4}$; $(x - \frac{7}{2})^2$ **9.** $\{-2 + \sqrt{6}, -2 - \sqrt{6}\}$
11. $\{1 + \sqrt{2}, 1 - \sqrt{2}\}$ **13.** $\{3 + \sqrt{2}, 3 - \sqrt{2}\}$ **15.** $\{3\}$
17. $\{9, -8\}$ **19.** No real number solution   **21. a.** $3 \pm \sqrt{2}$
b. 4.41, 1.59   **23. a.** Width $= 2\sqrt{2} - 1$ cm; length $=$
$2\sqrt{2} + 1$ cm   **b.** Width $= 1.83$ cm; length $= 3.83$ cm

25. a. Shorter leg $= 4 - \sqrt{2}$ in.; longer leg $= 4 + \sqrt{2}$ in.; hypotenuse $= 6$ in. **b.** Shorter leg $= 2.59$ in.; longer leg $= 5.41$ in.; hypotenuse $= 6$ in. **27.** $\left\{\dfrac{1 + \sqrt{17}}{4}, \dfrac{1 - \sqrt{17}}{4}\right\}$

29. $\left\{\dfrac{-5 + \sqrt{13}}{6}, \dfrac{-5 - \sqrt{13}}{6}\right\}$ **31.** $\left\{\dfrac{-3 + \sqrt{3}}{3}, \dfrac{-3 - \sqrt{3}}{3}\right\}$

33. $\left\{\dfrac{-2 + \sqrt{19}}{3}, \dfrac{-2 - \sqrt{19}}{3}\right\}$ **35.** $\left\{\dfrac{1}{3}, -1\right\}$

37. No real number solution

Remember This (9.2)

1. 1 **2.** 5 **3.** $-\dfrac{4}{3}$ **4.** True **5.** False **6.** $\{8, -2\}$

7. $\dfrac{3 + \sqrt{2}}{7}$ **8.** $(5,0)$ **9.** $\left\{\dfrac{6}{7}\right\}$ **10.** True

Exercises 9.3

1. $a = 3, b = -4, c = 5$ **3.** $a = 1, b = 0, c = -5$
5. $a = 1, b = 3, c = -4$ **7.** 0 **9.** 24 **11.** 1 **13.** $-3, 1$

15. $0, \dfrac{3}{2}$ **17.** $\dfrac{2 + \sqrt{5}}{4}, \dfrac{2 - \sqrt{5}}{4}$ **19.** $\left\{-1, -\dfrac{1}{3}\right\}$ **21.** $\left\{-\dfrac{1}{2}, \dfrac{1}{5}\right\}$

23. $\left\{\dfrac{1}{3}\right\}$ **25.** $\{-1 \pm \sqrt{3}\}$ **27.** $\{3 \pm \sqrt{2}\}$ **29.** $\left\{\dfrac{-1 \pm \sqrt{5}}{2}\right\}$

31. $\left\{\dfrac{3 \pm \sqrt{6}}{2}\right\}$ **33.** No real number solution **35.** $\left\{\dfrac{3}{2}\right\}$

37. $\{\pm 2\}$ **39.** $\left\{\pm \dfrac{3}{2}\right\}$ **41.** $\{\pm \sqrt{5}\}$ **43.** $\left\{0, -\dfrac{4}{3}\right\}$ **45.** $\{2, -3\}$

47. $\{5, -4\}$ **49.** $\left\{\dfrac{4 \pm \sqrt{15}}{2}\right\}$ **51.** $\{0.98, -3.57\}$ **53.** 5.1 in.

55. Side of smaller lot $= 60$ ft; side of larger lot $= 100$ ft
57. Small square has side of 6; large square has side of 8.
59. 0.69 second; 1.81 second; one is for the way up and one
is for the way down. **61.** $\dfrac{5 + \sqrt{21}}{2}$; 4.791

Remember This (9.3)

1. True **2.** Commutative property of multiplication **3.** d **4.** 7
5. $16 + 24x + 9x^2$ **6.** 25 **7.** $\{\sqrt{5}, -\sqrt{5}\}$ **8.** c
9. $V = \pi r^2 h$ **10.** $r = 50$

Exercises 9.4

1. i **3.** $10i$ **5.** $-2i$ **7.** $\dfrac{2}{3}i$ **9.** $i\sqrt{11}$ **11.** $i\sqrt{15}$
13. $-i\sqrt{5}$ **15.** $2i\sqrt{5}$ **17.** $3i\sqrt{3}$ **19.** $-3i\sqrt{7}$ **21.** $9i$
23. $2i$ **25.** $-3i$ **27.** $5i$ **29.** $2i$ **31.** i **33.** $5 + i$
35. $-1 + 2i$ **37.** $4i$ **39.** -5 **41.** $2 - i$ **43.** $5 + 4i$
45. -1 **47.** -15 **49.** -2 **51.** -1 **53.** 1 **55.** $7 + 22i$
57. $-6 + 17i$ **59.** $17 + 6i$ **61.** $3 + i$ **63.** 29 **65.** $13i$
67. $2i$ **69.** $-5 + 12i$ **71.** $15 - 8i$ **73.** $11 - 10i$ volts
75. $-3 + 4i$ volts **77.** Yes **79.** Yes **81.** No **83.** $\{5i, -5i\}$
85. $\{2i\sqrt{5}, -2i\sqrt{5}\}$ **87.** $\{i\sqrt{7}, -i\sqrt{7}\}$ **89.** $\{i\sqrt{6}/2, -i\sqrt{6}/2\}$
91. $\{5 + i, 5 - i\}$ **93.** $\{-1 + 2i, -1 - 2i\}$

95. $\left\{\dfrac{1 + 3i}{4}, \dfrac{1 - 3i}{4}\right\}$ **97.** $\{2 + i\sqrt{5}, 2 - i\sqrt{5}\}$

99. $\left\{\dfrac{-1 + i\sqrt{3}}{2}, \dfrac{-1 - i\sqrt{3}}{2}\right\}$ **101.** $\left\{\dfrac{-1 + i\sqrt{7}}{4}, \dfrac{-1 - i\sqrt{7}}{4}\right\}$

103. $\{1 + i, 1 - i\}$ **105.** $\{-3 + 2i\sqrt{2}, -3 - 2i\sqrt{2}\}$

107. $\left\{\dfrac{1 + 2i\sqrt{2}}{2}, \dfrac{1 - 2i\sqrt{2}}{2}\right\}$ **109.** $\left\{\dfrac{4 + 3i\sqrt{2}}{2}, \dfrac{4 - 3i\sqrt{2}}{2}\right\}$

Remember This (9.4)

1. b **2.** True **3.** 256 **4.** True **5.** $\dfrac{b}{2a}$

6. $\{1 + \sqrt{2}, 1 - \sqrt{2}\}$ **7.** $\left\{\dfrac{1}{2}, -2\right\}$ **8.** $y = x$ **9.** $\dfrac{x}{x + 4}$

10. $\dfrac{b + c}{bc}$

Exercises 9.5

1. 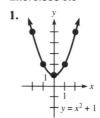 **3.** 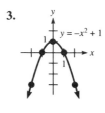 $y = -x^2 + 1$ **5.** 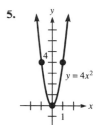 $y = 4x^2$

7. 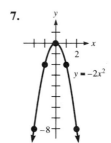 $y = -2x^2$ **9.** 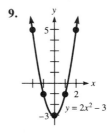 $y = 2x^2 - 3$ **11.** 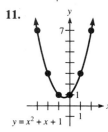 $y = x^2 + x + 1$

13. 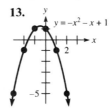 $y = -x^2 - x + 1$ **15.** 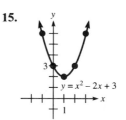 $y = x^2 - 2x + 3$

17. a. $x = -1$ **b.** $(-1, -5)$ **c.**
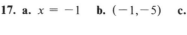
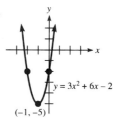
$y = 3x^2 + 6x - 2$
$(-1, -5)$

19. a. $x = 1$ **b.** $(1, 2)$ **c.**
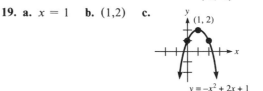
$(1, 2)$
$y = -x^2 + 2x + 1$

21. a. $x = 2$ **b.** $(2,1)$
c.

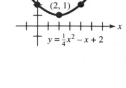

23. a. $x = -1$ **b.** $(-1,-16)$
c. $(0,-15), (-5,0), (3,0)$
d.

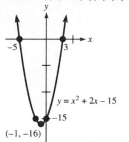

37. a. $x = \frac{1}{2}$ **b.** $(\frac{1}{2}, -\frac{9}{4})$
c. $(0,-2), (-1,0), (2,0)$
d.

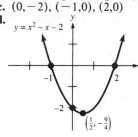

39. a. $x = -\frac{3}{2}$ **b.** $(-\frac{3}{2}, \frac{1}{4})$
c. $(0,-2), (-1,0), (-2,0)$
d.

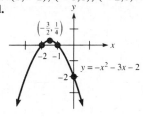

25. a. $x = 3$ **b.** $(3,4)$
c. $(0,-5), (1,0), (5,0)$
d.

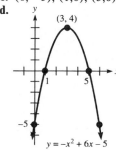

27. a. $x = 0$ **b.** $(0,-9)$
c. $(0,-9), (3,0), (-3,0)$
d.

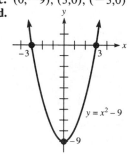

41. a. $x = 0$ **b.** $(0,-5)$
c. $(0,-5), (\sqrt{5},0), (-\sqrt{5},0)$
d.

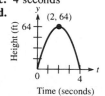

43. a. $x = 5$ **b.** $(5,-2)$
c. $(0,23), (5 + \sqrt{2},0), (5 - \sqrt{2},0)$
d.

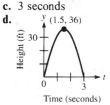

29. a. $x = 3$ **b.** $(3,-1)$
c. $(0,8), (2,0), (4,0)$
d.

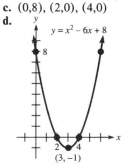

31. a. $x = 1$ **b.** $(1,-4)$
c. $(0,5), (\frac{1}{3},0), (\frac{5}{3},0)$
d.

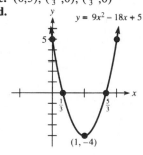

45. a. 2 seconds   **b.** 64 ft
c. 4 seconds
d.

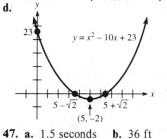

47. a. 1.5 seconds   **b.** 36 ft
c. 3 seconds
d.

49. a. 12.5 ft   **b.** 50 ft   **51. a.** 18.75 ft   **b.** 43.30 ft

Remember This (9.5)

1. $\{3i\sqrt{2}, -3i\sqrt{2}\}$ **2.** 8   **3.** $(0,0)$ **4.** $y = 2$ **5.** $\frac{a^2 + 1}{ab}$
6. $(x + 6)(x - 6)$ **7.** $\frac{12x}{y}$ **8.** $\{-\frac{11}{3}\}$ **9.** Distributive
10. Yes

Chapter 9 Review Exercises

1. $\{\frac{3}{2}, -\frac{4}{3}\}$ **3.** $36; (x - 6)^2$ **5.** $\left\{\frac{1 + \sqrt{2}}{2}, \frac{1 - \sqrt{2}}{2}\right\}$
7. $\frac{3 + \sqrt{5}}{2}$ and $\frac{3 - \sqrt{5}}{2}$ **9.** $11 - 7i$ **11.** $\{3 + 5i, 3 - 5i\}$

33. a. $x = -1$ **b.** $(-1,-9)$
c. $(0,-5), (\frac{1}{2},0), (-\frac{5}{2},0)$
d.

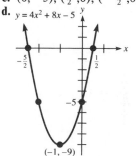

35. a. $x = -3$ **b.** $(-3,9)$
c. $(0,0), (-6,0)$
d.

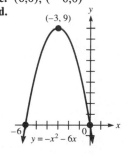

13.

15.

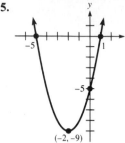

$(-2, -9)$

Chapter 9 Test

1. $\left\{\frac{2}{3}, -\frac{3}{4}\right\}$ **2.** $\left\{\sqrt{10}, -\sqrt{10}\right\}$ **3.** $\left\{\dfrac{-3 + \sqrt{5}}{2}, \dfrac{-3 - \sqrt{5}}{2}\right\}$

4. $9; (x - 3)^2$ **5.** $\left\{-3 + 2\sqrt{2}, -3 - 2\sqrt{2}\right\}$

6. $\left\{\dfrac{-1 + \sqrt{13}}{6}, \dfrac{-1 - \sqrt{13}}{6}\right\}$ **7.** $\sqrt{15}$ in.   **8.** $2i$ **9.** $9i$

10. All   **11.** 13   **12.** True   **13.** No   **14.** $\left\{\dfrac{2i\sqrt{3}}{3}, \dfrac{-2i\sqrt{3}}{3}\right\}$

15. $a = 3, b = -1, c = -2; \left\{1, -\frac{2}{3}\right\}$

16. $\left\{\dfrac{-1 + i\sqrt{2}}{3}, \dfrac{-1 - i\sqrt{2}}{3}\right\}$ **17.** True   **18.** Because some values of x correspond to more than one value of y

19.

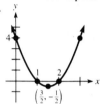

$\left(\frac{3}{2}, -\frac{1}{2}\right)$

20. a.

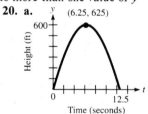

$(6.25, 625)$

Height (ft)

Time (seconds)

b. 12.5 seconds

Cumulative Test 9 (Final Examination)

1. b   **2.** -2.5 **3.** $x + 2$ **4.** $\left\{-\frac{1}{3}\right\}$ **5.** $x^3 - x + 6$

6. $\dfrac{x^2}{4y^2}$ **7.** $(3x + 1)(3x - 1)$ **8.** 9 in.   **9.** $\dfrac{x + 1}{x - 4}$

10. $\dfrac{180}{\frac{1}{2}} = \dfrac{30,000}{x}$; 83 hours   **11.**

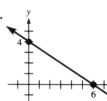

12. y increases by 15 units.   **13.** $\left(\frac{7}{2}, -\frac{3}{2}\right)$ **14.** \$40 million

15. 2   **16.** $\{4\}$ **17.** -243 **18.** $\left\{\dfrac{-1 + \sqrt{17}}{2}, \dfrac{-1 - \sqrt{17}}{2}\right\}$

19. 34   **20.**

$(1, 2)$

Exercises Appendix

1. Eight-tenths   **3.** One-hundredth   **5.** Two-thousandths
7. Five and four-hundredths
9. Thirty-one and one hundred forty-five–thousandths
11. Ten and six-hundredths   **13.** 55.634   **15.** 6.36   **17.** 0.456
19. 102   **21.** 1.722   **23.** 12.81   **25.** 9.384   **27.** 12   **29.** 12.4
31. 152   **33.** 0.30   **35.** 0.05   **37.** 1.00   **39.** 0.125
41. 1 percent   **43.** 35 percent   **45.** 105 percent
47. 300 percent   **49.** 12.34 percent   **51.** 10   **53.** 8   **55.** 87.5
57. 1,050   **59.** \$1.98   **61. a.** 4,600   **b.** 234,600
63. \$1.344 million, or \$1,344,000

Index

U

Uniform motion applications, 107, 345
Unlike terms, 74

V

Value of function, 305
Variable
    definition of, 18
    dependent, 304
    independent, 304
Variation
    constant of, 308
    direct, 308
    inverse, 309
Velocity, 25

Vertex of parabola, 435
Vertical line
    equation of, 279
    graph of, 279
Volume
    of cone, 202
    of cylinder, 22
    of rectangular solid, 22
    sphere, 22

W

Whole number, 3
Word problems, guidelines for solving, 100
Work problems, 258

X

x-axis, 271
x-intercept, 278

Y

y-axis, 271
y-intercept, 278

Z

Zero
    division by, 52
    exponent, 133
Zero product principle, 203

Exponents

1. $a^m \cdot a^n = a^{m+n}$
2. $(a^m)^n = a^{mn}$
3. $\dfrac{a^m}{a^n} = a^{m-n}$
4. $a^0 = 1$
5. $a^{-n} = \dfrac{1}{a^n} = \left(\dfrac{1}{a}\right)^n$

Radicals

1. $a^{1/n} = \sqrt[n]{a}$
2. $\left(\sqrt[n]{a}\right)^n = a$
3. $\sqrt[n]{a} \cdot \sqrt[n]{b} = \sqrt[n]{ab}$
4. $\dfrac{\sqrt[n]{a}}{\sqrt[n]{b}} = \sqrt[n]{\dfrac{a}{b}}$
5. $\sqrt[n]{a^m} = \left(\sqrt[n]{a}\right)^m = a^{m/n}$

Square Roots

1. $a^{1/2} = \sqrt{a}$
2. $\left(\sqrt{a}\right)^2 = a$
3. $\sqrt{a} \cdot \sqrt{b} = \sqrt{ab}$
4. $\dfrac{\sqrt{a}}{\sqrt{b}} = \sqrt{\dfrac{a}{b}}$
5. $\sqrt{a^2} = a$, if $a \geq 0$

Absolute Value

If a is a real number, then
$$|a| = \begin{cases} a & \text{if } a \geq 0 \\ -a & \text{if } a < 0. \end{cases}$$

Lines

1. Slope: $m = \dfrac{y_2 - y_1}{x_2 - x_1}$
2. Slope-intercept equation: $y = mx + b$
3. Point-slope equation: $y - y_1 = m(x - x_1)$
4. Parallel lines: $m_1 = m_2$; Perpendicular lines: $m_1 m_2 = -1$

Number Relations

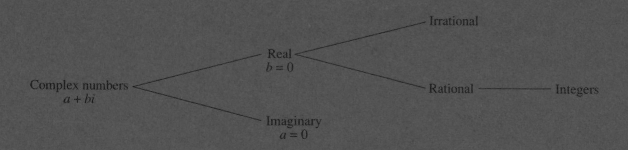

Quadratic Formula

If $ax^2 + bx + c = 0$ and $a \neq 0$, then
$$x = \dfrac{-b \pm \sqrt{b^2 - 4ac}}{2a}.$$

Square Root Property

If a is a real number, then
$$x^2 = a \text{ implies } x = \sqrt{a} \text{ or } x = -\sqrt{a}.$$